Vickery's Folk Flora

Vickery's Folk Flora

An A–Z of the Folklore and Uses of British and Irish Plants

ROY VICKERY

First published in Great Britain in 2019
by Weidenfeld & Nicolson

1 3 5 7 9 10 8 6 4 2

A CIP catalogue record for this book
is available from the British Library.

ISBN 9781474604628

Printed and bound by CPI Group (UK) Ltd,
Croydon, CR0 4YY

Weidenfeld & Nicolson

The Orion Publishing Group Ltd
Carmelite House
50 Victoria Embankment
London, EC4Y 0DZ
An Hachette UK Company

weidenfeldandnicolson.co.uk
www.orionbooks.co.uk

In memory of my grandparents:

Mary Blanche Vickery (born Eveleigh), 1884–1958

Ernest Albert Vickery, 1885–1964

Annie Fry Fry (born Nobbs), 1889–1980

Herbert Fry, 1890–1958

DISCLAIMER: Medicinal and other uses are given as they have been received from informants or found in publications. No attempt has been made to evaluate them; some may be beneficial, some may be harmless, and some may be harmful. You should undertake your own research before following any of the remedies and recipes included here.

Contents

Methods and Acknowledgements

Material used in this work has been gathered since the early 1970s, with focused collecting starting early in the 1980s. Since then some 7,620 items of information have been received from over 2,160 contributors. Each of these contributors, regardless of how, or how much, they contributed, are thanked.

Thus the work is a product to which many people have contributed in different ways. These contributions could have been utilised in a variety of ways; other people might have used the information to produce different results, especially when it comes to plants for which a large amount of information exists; what is used is only a small representative sample. Production of a work like this depends to a large extent on serendipity, picking up a newspaper, pulling down a volume, or overhearing a conversation – which leads to fresh discoveries. Other people would have similar experiences which would lead them along different paths. This work is by no means definitive, I hope other folk floras will emerge; this is simply one contribution to our knowledge.

Much of the early collecting was done via appeals in local newspapers. This had the advantage that it allowed all parts of the British Isles to be covered quickly and letters could be written whenever spare moments were available; the disadvantage was that only those who felt able to put pen to paper responded. Over the years such publications have decreased in number, so other methods have been employed. The most important of these has been giving talks and leading walks discussing plant folklore, hoping that participants will be stimulated to share their knowledge. Such activities have had mixed success, some audiences being far more willing to engage than others. Events held in the Natural History Museum's Wildlife Garden, in London, have proved particularly rewarding. They attract an international audience, so material contributed at them has helped develop an, admittedly vague, idea of how widespread British plant-lore is elsewhere. Thank you to Caroline Ware, the Museum's Wildlife Gardener, for her ongoing support.

Appeals have continued to be published wherever possible, but the results have been disappointing. All too often, it seems, people regard what they know as being known by everyone else, and therefore unworthy of recording. This is untrue. We are all unique individuals with our own memories and beliefs, and far more records are needed before we can adequately understand the distribution of beliefs and practices. Collecting continues and it is hoped that readers of this book will contribute to our greater understanding of British and Irish plant-lore. Please send information, comments and corrections via www.plant-lore.com.

It can be argued that the reliance on events such as talks and walks for information has resulted in a London bias. However, London is a great cosmopolitan city; people there have knowledge from all parts of the British Isles and beyond. Also, if one takes the view that areas which have the most 'folk' must, inevitably, have the most folklore, then perhaps London is under-represented. If one compares the population of London, estimated at 8.5 million, with that of Northern Ireland, estimated at 1.8 million, perhaps the latter is over-represented.

Two long-term attempts have been made to gather information. Eighty-five issues of the newsletter *Plant-lore Notes & News* were produced between 1988 and 2005, by which time it was receiving fewer contributions and attracting less interest. It was thought that a website would be more effective, and, eventually, in November 2010, the Plant-lore Archive website – www.plant-lore.com was created. Thanks to Trevor Maskery for his help in setting this up. Although the website is used a great deal and some useful information has been received from users, it appears that most users seek, but are reluctant to add, information.

Material collected since *c.*1975 forms the nucleus of the current work, so regardless of whatever shortcomings it might have, it contains a large amount of previously unrecorded information. This information has been supplemented by previously published material taken from a wide range of books, periodicals and ephemeral publications, and from unpublished archives, notably those held in the National Folklore Collection, University College Dublin. Thanks to all the librarians and archivists who care for this material, and so readily make it available.

The list of contributors to this work could be extended indefinitely. A full list of the helpers who are known would be prohibitively long and still omit those who are unknown, but whose contributions are equally valued and important. All who have contributed to the inception, development and completion of the work are profoundly thanked.

My contacts at Orion, Lucinda McNeile and Alan Samson, have been endlessly patient and have provided wise guidance. I am also grateful to Helen Ewing for her design input.

Finally, I thank *mi marido*, Carlos Bruzon, who, despite having little interest in plants or folklore, has faithfully supported me, particularly on travels in conjunction with this work.

To sum up, I quote from C.T. Prime's introduction to his *Lords and Ladies* (1960):

> Despite all the help so kindly given, I have ventured into so many fields that I fear I must have fallen into some errors or missed something important. Here I must beg the indulgence of the reader, and hope that [s]he will find some compensation in the text, and come to share an interest.

The Collection of Plant-lore in Britain and Ireland

The earliest information that we have on the folklore of plants is found in the writings of herbalists and antiquaries. In due course scientific botany diverged from herbalism and, much later, folklore emerged from what had become archaeology and local history. Occasionally a scholar ventured into both fields, the prime example of this being John Ray (1627–1705), who not only produced the first flora of the British Isles, and whose 1660 *Catalogus Plantarum circa Cantabrigiam nascentium* is considered England's first county flora, but also published a collection of English proverbs.

The writings of English herbalists were to a large extent derived from continental, often Mediterranean, sources, and contained few observations relevant to British plant-lore. Nonetheless, the 'father of English botany', William Turner (*c.*1508–68), mentioned the Glastonbury, or Holy, Thorn in his *New Herbal* of 1562. This tree was also mentioned, without enthusiasm, by John Gerard (1545–1612) in his *Herball* of 1597. In her classic study *Herbals* (1912, 2nd ed. 1938) Agnes Arber dismissed Gerard's work, suggesting that he used a translation of Rembert Dodoens's 1583 *Stirpium historiae pemptades sex* without acknowledging his debt. However, the *Herball* contains useful information on where various plants were found and some notes on local uses. For example, butterwort:

> groweth in a field called Cragge close, and at Crosbie, Ravenswaith in Wetmerland, upon Ingleborough fels, twelve miles from Lancaster . . . wives of Yorkshire do use [it] to anoint the dugs of their kine with the fat and oilous juice of the herbe butterwort, when they are bitten with venomous worm . . . and hurt by any other means.

In the seventeenth century antiquarians started travelling the country or delving into their home areas, seeking out antiquities and stories associated with archaeological sites. John Aubrey (1626–97) noted that dwarf elder – 'dane's blood' – grew abundantly at Slaughtonford in Wiltshire, formerly the site of a 'fight with the Danes', and John Taylor, in his *Wandering to See the Wonders of the West* (1649) described the destruction of the Holy Thorn by a Roundhead, who 'did cut it downe in pure devotion'.

The same century saw the publication of Ray's *Flora of Cambridgeshire* which contained much information on the uses of the plants listed, mainly derived from non-British sources. However, some of his notes appear to be original, such as the recommendation that water found in the leaf bases of teasel plants can be used to cure warts. He also mentioned that the Jew's-ear fungus (now known as jelly-ear fungus, *Auricularia auricula-judae*) is found on elder, and uses this as an opportunity to robustly denounce the Doctrine of Signatures, which suggested that the ear-shaped fungus was good for ears.

The accumulation of ethnobotanical information continued in later county floras, in which compilers usually repeated material from other works, without making much, if any, effort to collect local information. There were, however, a few admirable exceptions, such as John Lightfoot, who in his *Flora Scotica* (1777) included a mass of information which he had collected in the Highlands and Islands, and related this to material he had come across elsewhere.

In the mid-nineteenth century a number of semi-popular works on wild plants included valuable information on plant-lore. Such books included C.A. John's *Forest Trees of Britain* (2 volumes, 1847 and 1849) and Anne Pratt's *Wild Flowers* (1857) – both published by the Society for Promoting Christian Knowledge – and Lady Wilkinson's *Weeds and Wild Flowers: Uses, Legends, and Literature* (1858). This tradition continued with such works as A.R. Horwood's six-volume *New English Flora* (1921).

The 1870s and 1880s saw an unprecedented number of publications on plant folklore. In 1878, the year in which the Folklore Society was founded, the first part of the English Dialect Society's *Dictionary of English Plant-names*, compiled by James Britten and Robert Holland was produced. The *Dictionary*, completed in two further parts in 1880 and 1886, is an underrated work, which contains much information on plant folklore. Britten and Holland continued to accumulate plant-names until the end of their lives, but no supplement was ever produced.[1]

In September 1870 Britten appealed for information for inclusion in a 'small volume on folklore connected with plants'.[2] The volume never materialised, but Britten produced a large number of semi-popular articles on plant folklore, often in magazines where one would not expect to find such things. However, from October 1884, Britten, a devout – perhaps fanatical – convert to Roman Catholicism, threw his energy and enthusiasm into the revival and organisation of the Catholic Truth Society. This activity dominated most of his spare time throughout the remainder of his life, although he continued to produce occasional articles and book reviews about plant-lore subjects, mainly in the *Journal of Botany*, which he edited from 1879 until his death in 1924.[3]

Other works produced during the 1880s included Hilderic Friend's *Glossary of Devonshire Plant-names* in 1882, and *Flowers and Flower Lore* in 1884. Although the latter is typical of its period, being a lengthy accumulation of poorly authenticated material, Friend made some attempt to collect information from people he met as he strolled through the countryside in the course of his duties as a Methodist minister. These scraps of information ensure that Friend's work remains of some lasting value. He promised further volumes on the folklore of plants – notably on the plant-lore of the Far East, where he had lived as a missionary – but these too never materialised. Friend became increasingly interested in earthworms, becoming (at least in his own eyes) a leading authority on the group, so it is likely that this enthusiasm overwhelmed his earlier interest in plant-lore. Today Friend's work on earthworms is considered to be of little value, and is best forgotten or ignored; most of the supposedly new species he described are not considered to be valid.

Also published in 1884 was Richard Folkard's *Plant Lore, Legends and Lyrics* which uncritically threw together a wide range of material derived from a wide variety of sources. Many of the outlandish statements found in popular works on plant folklore can be traced back to this volume.

A little later, in 1889, T.F. Thiselton-Dyer published his *Folklore of Plants*, one of several books on various aspects of folklore that he produced. This volume attracted attention as careless reviews confused Thomas Firminger, a clergyman, with his better-known brother, William Turner Thiselton-Dyer, Director of the Royal Botanic Gardens, Kew. Thus the work erroneously became attributed to the British Empire's most eminent botanist.

The late nineteenth century also saw the publication of many dialect dictionaries, extensive compilations assembled by local enthusiasts, encouraged and often published by the English Dialect Society. These culminated in Joseph Wright's *English Dialect Dictionary*, published in six volumes between 1898 and 1905. Wright not only incorporated material from county dictionaries, but also utilised the enthusiasm and local expertise of a large number of 'voluntary readers', compilers of unprinted collections, and correspondents. James Britten acted as botanical advisor. The *Dictionary* contains a vast number of plant-names, usually carefully identified, and often giving explanations of why these names were used.

The Folklore Society was formed in 1878 and early volumes of its journal included a wealth of information sent in by collectors working in rural areas. The first volume of the journal contained an extensive article on the folklore of west Sussex, collected ten years earlier by Charlotte Latham, a local clergyman's wife. This was followed by an article in which Britten attempted to put Latham's plant-lore into context by citing similar beliefs and customs from other parts of the British Isles. Although much of the material included in the journal is valuable, its value is seriously diminished by the fact that little attempt was made to check the identities of the plants involved in various beliefs and remedies. What, for example, was 'red roger', which in 1897 was recorded as being 'used to stop bleeding at the nose' in County Down? Articles on the folklore of various parts of the British Isles continued to be published until about the end of the century. Then, until about the outbreak of the Second World War, came short notes contributed by correspondents in country villages. Later articles in the journal tended to be less concerned with recently collected material and concentrated on the interpretation of previously published works.

Between 1895 and 1914 the Society also produced a series of volumes on the folklore of various counties. These avoided newly collected material and reflected the Society's historical bias by restricting the contents to extracts from previously published works. Presumably the implication was that material collected and published in, say 1802, was more 'authentic' than anything gleaned a century later.

In the periods before both world wars Britain depended on Germany for pharmaceutical drugs.[4] When these drugs became unobtainable, especially during the 1939–45 war, the British were forced to re-evaluate their islands' heritage of medicinal and edible plants. This led to the publication of Mary Thorne Quelch's *Herbs for Daily Use* (1941) and

Florence Ranson's *British Herbs* (1949). This tradition continues with the production of Richard Mabey's *Food for Free* (1972 and frequently reprinted), and the large number of books on foraging wild foods which this in turn stimulated.

The growth of nationalism and interest in national identity resulted in conscious efforts being made to collect and preserve the folk culture, both oral and material, of Ireland, Scotland and Wales. In 1919 Michael Maloney published his *Irish Ethnobotany*, which 'endeavoured to indicate the wealth now lying hidden in the Gaelic nature creeds'. This publication appears to have introduced the term 'ethnobotany' – first used by J.W. Harshberger in a lecture to the Chicago University Archaeological Association in December 1895 – to the British Isles. The Irish Folklore Commission was established in 1935, the Welsh Folk Museum in 1948, the School of Scottish Studies in 1951, and the Ulster Folk Museum in 1961.[5] These organisations have devoted much effort to recording material in their countries, often concentrating their efforts on collecting in minority languages – Irish, Welsh and Gaelic. Pre-eminent among these organisations is the National Folklore Collection at University College Dublin, the successor to the Irish Folklore Commission, which holds some two million manuscript pages covering all parts of the country. There has been a tradition of professional collectors working in the field, and their collections were extensively supplemented by material contributed to the Commission's Schools' Scheme of 1937–8, when children attending Ireland's 5,000 primary schools were asked to collect local folklore. The material contributed to the scheme varies. Some head teachers thought they knew best and wrote essays; more satisfactory information comes from schools where the children contributed their knowledge. Among the contributed material is a wealth of herbal remedies and other aspects of plant-lore including a number of riddles, a subject which tends to be neglected in other collections.

As the historically dominant nation, the English have not considered it necessary to gather their folklore and use it to establish their identity. Thus there is no publicly funded national centre in England. While the Irish, Welsh, and Highland Scots are proud of their folklore and knowledge of it, the English tend to deny any acquaintance with such matters. However there has been intermittent activity in England. Late in the 1940s the English Dialect Survey was established at the University of Leeds, leading to the formation of the Leeds Folklife

Survey in 1960, and becoming the School of Dialect and Folk Life Studies in 1964. The school flourished for twenty years before closing in 1984. Similarly in Sheffield, the university's Survey of Language and Folklore, founded in 1964, became the Centre for English Cultural Tradition in 1975, and from 1997 the National Centre for English Cultural Tradition, before fading away in the first decade of the twenty-first century,[6] its archives being moved to the university library during the 2007–8 academic year.

In England, folklore studies have struggled along rather informally with folklore sometimes being included in the publications of local history and archaeology societies. Most of the well-established county societies have included some folklore in their journals. Particularly important is the Devonshire Association for the Advancement of Science, Literature and the Arts, which between 1876 and 2011 included 80 reports on folklore and 94 reports on dialect in its transactions. Since 2011 the reports have continued, amalgamated into one.

Though primarily a personal, rather literary, look at British plants, Geoffrey Grigson's *Englishman's Flora*, 1955, has provided bedside reading for people interested in plant-lore and plant-names ever since. Elegantly written and benefitting from wide reading and a thorough knowledge of plants and their habitats, the *Flora* exerted considerable influence, both positive and negative. Grigson alerted the reading public to the rich variety of local names which had been given to British plants (although he was perhaps ungenerous in acknowledging his great debt to nineteenth-century collectors of dialect words, especially Britten and Holland). On the negative side, his work unintentionally gave many the impression that all that could be said concerning British plant-lore had been safely garnered. There was no further work to be done, nothing more to be said. This spurious belief, together with the more recent, but widely held, impression that ethnobotany survives – or is only interesting – when 'primitive' isolated tribal peoples are concerned has seriously impeded collection and investigation.[7]

Four years later, in 1959, Iona and Peter Opie published their *Lore and Language of Schoolchildren*. This work, like Grigson's, was often considered to have brought together all that could be said on the subject and thus did not stimulate further work in its field. However, the Opies' work, together with the publications of Enid Porter of the Cambridge and County Folk Museum, demonstrated that the collection of British, and especially English, folklore could be a worthwhile pursuit.

There was still plenty of interesting material awaiting collection.

The 1960s revival of interest in traditional song, dance, music and customs, along with the availability of relatively cheap and portable tape-recorders, led to further collecting activities. Many of the collectors of this era, some of whom are still active, remained content merely to collect. They were unpaid with little spare time, and preferred the excitement of collecting, 'before it was too late', to organising and publishing their collections. What might be considered to be the last phase of this period was the publication of a series of 'county' folklore books, under the general editorship of Venetia Newall, then Honorary Secretary of the Folklore Society. The series ran to seventeen volumes published between 1973 and 1978, and involved a variety of contributors ranging from well-known folklorists to folk singers and journalists, all of whom tried to supplement information from earlier publications with newly collected material. Inevitably the results were mixed, depending on the contributor's knowledge, but the persistent weakness of such publications is that they tend to overstress what makes an area unique, concentrate on peculiar local customs, and ignore everyday practices and beliefs.

In March 1982 the Folklore Society initiated a survey of plants which were believed to produce misfortune if picked or taken indoors. The survey continued until October 1984, by which time information had been received on over seventy 'unlucky' plants, vividly demonstrating the widespread knowledge and distribution of such beliefs.[8] Many of those collected in the early 1980s – for example the belief that the picking of cow parsley led to the death of one's mother – had not been collected before. Had these beliefs come into being in comparatively recent times, or had they escaped collectors in the past? The survey also revealed how people wanted to record and pass on what they knew.

This stimulated a major attempt to collect plant-lore, through the newsletter *Plant-lore Notes & News*, eighty-six issues of which were produced between 1988 and 2007,[9] and led to the publication, in 1995, of Roy Vickery's *Dictionary of Plant-lore*, the precursor of the present *Flora*. At the same time the environmental group Common Ground promoted Richard Mabey's work, which led to the publication of his *Flora Britannica* in 1996.[10] This amazing bestseller is often considered to be the ultimate source of information on plant folklore, but it tends to emphasise how people as individuals react to plants, rather than how the 'folk', people as communities, do so. The *Dictionary* and the *Flora* being published approximately a century after the late nineteenth-century flurry of interest in the subject allow students of plant-lore to compare and contrast the methods used and the results obtained in the 1880s and the 1990s.

More recently a number of other initiatives have been launched. From 1993 to 2012 Ethnomedica, based at the Royal Botanic Gardens, Kew, collected remedies from around Britain. This led to the accumulation of over 6,000 records (of which the use of dock leaves to treat nettle stings was by far the most common); few publications have resulted from this collection and the records remain archived at Kew.[11] Between 1999 and 2004 William Milliken and Sam Bridgewater at the Royal Botanic Garden, Edinburgh, collected information on the uses and folklore of Scottish plants, leading to the publication of their magnificent *Flora Celtica*.

Two other noteworthy publications relied mainly on previously published works. David Allen and Gabrielle Hatfield's *Medical Plants in Folk Tradition*, also published in 2004, sought to analyse how native British plants were used in folk medicine. For this they assembled a huge amount of interesting material, though their emphasis on the distribution of remedies left little space to describe how different remedies were prepared and used. A decade later, in 2014, Peter Wyse Jackson, formerly of the National Botanic Garden, Glasnevin, published his monumental *Ireland's Generous Nature*, which sought to record the uses made of wild plants in Ireland, but, in fact, strays beyond his remit to include information on imported foods and crop plants.

We have come a long way since the 1960s when the folklore of plants was considered a rather frivolous matter – 'let's have a giggle about some of our ancestors' quaint beliefs' (even though many of the more bizarre ideas were those propounded by the literate elite, the scientists of their day, rather than the more pragmatic and practical 'folk'). We now have a shelf-full of serious, but, in their way absorbing and entertaining volumes, in which it's not considered necessary to finish every other sentence with an exclamation mark. Their contents are sufficiently interesting to render such prompts redundant.

Notes for readers

Lack of space has limited the scope of this volume to vascular plants: flowering plants, conifers, and ferns. Mosses, seaweeds, fungi and lichens are excluded.

Geographical scope

This work focuses on Great Britain and Ireland, the Channel Islands, and the Isle of Man, material from elsewhere has largely been relegated to endnotes.

Scientific names

Names for native and naturalised wild plants are those given in the 2010 edition of Clive Stace's *New Flora of the British Isles*; names for cultivated plants are taken from a variety of sources, primarily the 2008 edition of David Mabberley's *Plant-Book*.

A number of abbreviations have been used in connection with scientific names:

agg. – a group (aggregate) of closely related species, e.g. *Rubus fruticosus* agg., the common brambles or blackberries.
cv. – cultivar, a plant of cultivated origin, derived from a known species (e.g. both beetroot and mangold are cultivars of the wild beet, *Beta vulgaris*) or of uncertain parentage (e.g. banana, *Musa* cv.).
sp. – species (singular), used when the exact specific identification of the plant is unknown.
spp. – species (plural), used where a passage refers to two or more species.
syn. – synonym, used to indicate a scientific name which was formerly used but is now considered incorrect.

Local names

The 'local' plant-names included in this work are derived from a wide range of sources, the most important being:

'Verbal provincialisms' and later dialect reports, 1876 – present (since in the 2011 reports on dialect and folklore) in the *Report and Transactions of the Devonshire Association for the Advancement of Science.*

James Britten and Robert Holland, *A Dictionary of English Plant-names*, [1878–] 1886, and manuscript slips intended for a supplement to this.

A.S. Macmillan, *Popular Names of Flowers, Fruits*, etc., 1922.

Geoffrey Grigson, *The Englishman's Flora*, 1955 and frequently reprinted.

Obviously the collection of names has depended on local enthusiasm, consequently Somerset and its surrounding counties, on which Macmillan focused, are exceptionally well represented. Counties have come and gone, or their boundaries have been redrawn, since most of these publications were produced. It is regretted that it was impossible to update county names, so the lists of names include a number of counties which no longer exist. Where a name is listed in capitals this indicates that it has been used for more than one species.

Names are included in the present work without their sources being cited. If further information is required readers are advised to visit the 'Local Names' page on www.plant-lore.com, where sources and further information, can be obtained. Names given in the Plant-name Index (pp. 809–91) include only those which are used for plants in the main part of this book. Sometimes a name is more commonly used for a plant not included; for example, marshmallow, listed as a name for mallow (*Malva sylvestris*) and tree-mallow *(M. arborea*), is the 'official' name for *Althaea officinalis*, but as this species is not included else where in the book it is excluded from the index.

As a general rule only those names for which localities are recorded have been included. Thus most of the names recorded in Mabey's *Flora Britannica* are omitted, since he considered that 'the geographical mobility of contributors . . . and the mobility of the names themselves through mass media would have made this a misleading and potentially inaccurate qualification'. However, these names are included on the website, together with many names for plants which are not included in this volume.

Names recorded in minority languages are not included in either the website or this book. Their addition would have led to a massive extension to the lists provided, and, as the author lacks knowledge of these languages, would have undoubtedly led to the inclusion of numerous errors.

Description and distribution of plants

Only plants which significantly feature in British and Irish folklore are included in this volume. Entries begin with a very short description of the plant, an indication of its distribution within Britain and Ireland, and where appropriate when it was first introduced to our islands. Thus 'cultivated since late in the sixteenth century' means a plant has been grown in Britain and Ireland since that time; it may well have been cultivated elsewhere for much longer.

Illustrations

Black-and-white engravings included in the text are derived from Walter Hood Fitch and Worthington George Smith's *Illustrations of the British Flora*, first published in 1880 to provide images to supplement George Bentham's *Handbook of the British Flora*, first published in 1853 and frequently revised, notably by Joseph Dalton Hooker in 1886. Photographs in the plate sections are by the author.

Calendar customs

Customs, however 'traditional' they claim or appear to be, are in a state of constant change; some older customs rapidly die out when the families which have run them for decades lose enthusiasm, others change their dates often from weekdays to weekends, some are revived after many years, and others start and after a few years are regarded as traditional. People wanting to attend any of the customs mentioned in this book are advised to check carefully, using if possible several sources, before travelling.

THE FLORA

A

ABORTION – HOUSELEEK, JUNIPER, PENNYROYAL and TANSY produce.

ACHES – YARROW used to treat.

ACNE – LEMON used to treat.

ADDER (*Vipera berus*)

Britain's only venomous snake was associated with various plants, and especially fruits which are, or are believed to be, poisonous. Such fruits were often referred to as ADDER'S MEAT, or adder's food.

Adder's meat was a name for the fleshy fruits of STINKING IRIS,[1] and BLACK BRYONY[2] in Devon, and LORDS-AND-LADIES elsewhere.[3] Other plants which were given the name included cow parsley in Somerset[4] and dog's mercury in Hertfordshire.[5]

According to Macmillan, writing of Somerset, the name adder's food is:

> given to the red berries of a number of plants, which are poisonous or supposed to be so, particularly those of the wild arum [lords-and-ladies], the [stinking] iris, bryony, etc. The word adder in this and most [plant-names] . . . has nothing to do with snakes and reptiles at all. It is neither more nor less than the Anglo-Saxon word *attor*, which means poison. Attor-berries, meaning poison-berries (the very name still used in Sussex) was changed first to adder berries, then to adder's food or adder's meat, and finally in many cases to SNAKE'S FOOD.[6]

Other plants associated with adders include GREATER STITCHWORT; ASH was believed to be feared by adders; HAZEL was thought to be poisonous to adders.

ADDER'S TONGUE (*Ophioglossum vulgatum*)

Small, inconspicuous fern, widespread in dampish places.

Local names are all variants of the standard name: adder's spear in Surrey and Sussex, edder's tongue in Cumberland, SNAKE'S TONGUE and serpent's tongue.

> [Adder's-tongue fern] was as an ingredient in an ointment, which, under the name of adder's spear ointment, is still, or was until recently employed, among other purposes, as a healing application to the inflamed udders of cows.[1]

> Crushed and boiled in olive oil it is used as a dressing for open WOUNDS. Most gypsies . . . denied knowledge of it, but I have had it given to me by three old women in widely separated districts.[2]

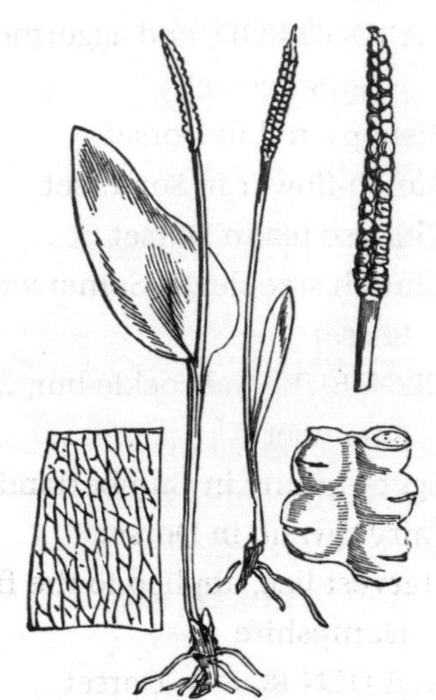

AFRICAN MARIGOLD (*Tagetes erecta*) and **FRENCH MARIGOLD** (*T. patula*)

Annual herbs with garish, usually orange, flowers, which despite their names are native to Mexico. The robust African marigold is believed to have been introduced via North Africa in 1535, while the smaller French marigold is said to have been brought to England by Huguenot refugees in 1572.[1] Both species are reputed to keep insect pests and BLIGHTS at bay.

> A tip for TOMATO-growers: put a French marigold among the tomato plants in the greenhouse to keep GREENFLY away.[2]

> African marigolds were planted along with CARROTS and ONIONS to keep away pests.[3]

> Plant small French marigolds in a greenhouse of tomatoes for they ward off blight.[4]

Similar practices are common in parts of France.[5]

AGRIMONY (*Agrimonia eupatoria* and *A. procera*)

Perennial, yellow-flowered herbs, widespread in rough grassy places such as roadsides, railway embankments and woodland margins; absent from much of Scotland.

Local names include:

AARON'S ROD, and aggermony, in west Somerset
Bishop's rod in Dorset
Bunch-flower in Somerset
Chinese tea in Dorset
Church steeples in Somerset and Sussex
CLOT-BUR, and cockle-bur, in Somerset
Eggremunny in Cumberland
Fairy's wand in Dorset
Harvest lice, applied to the fruits in Hampshire
GOLDEN ROD in Dorset
Lemon-flower, and lemonade, in Somerset; 'no doubt in consequence of the lemon-like perfume given off by the plant'
Lucky money in Berkshire
MONEY-IN-BOTH-POCKETS in Dorset
RATS' TAILS in Co. Durham and Wiltshire
Salt-and-pepper in Cornwall
SWEETHEARTS in Somerset; 'from the clinging receptacles of the fruit, cf. *Galium aparine*'
TEA-FLOWER in Berkshire.

> Tea plant . . . an old lady living at Mudford [Somerset] tells me she has always known the common agrimony . . . by this name and by no other. In her younger days all the TEA she drank was made from it . . . an old resident of Blackmore, West Buckland, would never take any other 'tea' and attributed her long life and great vigour to the use of this her favourite beverage.[1]

> Somewhere around 1914 . . . after cutting the hay, the carter went down the rows picking out . . . agrimony, and hanging a great sheaf of it on the tail-board of the cart. On asking him what it was for, he replied 'Agrimony tay, the best physic as there is come spring-time.'[2]

[Horseheath, Cambridgeshire] Agrimony tea was supposed to be good for LUMBAGO.[3]

When shepherd Tidmarsh [presumably of Ashton-under-Hill, Worcestershire] was ill he asked me to collect the little yellow flowers of agrimony for him. He used this as an infusion to make a kind of tea which he said stimulated the BLADDER.[4]

Aggermoney tea for chesty COUGHS.[5]

AGUE – Beaumont's Tree (an ELM) and GROUNDSEL cure.

ALDER (*Alnus glutinosa*)

Deciduous native tree, common beside streams and in other damp places. The fruits of alder are known as black knobs in the Peak District of Derbyshire, where they are much used in WELL-DRESSING.[1]

Local names include the widespread variants allar, aller and aller-tree, and:

Aer in western Scotland
Alls-bushes, in Devon
Arl in Gloucestershire
DOG-TREE in northern England
ELDER in Co. Donegal
ELLER in Sussex
Ellers in Cumberland
Gue in Yorkshire
Harrul in Gloucestershire
Irish mahogany in Co. Waterford
Orle in Gloucestershire
Orrel in Herefordshire
Owler in Craven, Yorkshire.

Known locally as aul, alder was watched by Herefordshire fishermen, according to whom:

> When the bud of the aul is as big as a trout's eye
> Then that fish is in season on the River Wye.[2]

There are scattered records of alder being used medicinally.

[Worcestershire] A superstition exists in some parts of the county that if pieces of the alder tree are carried in the waistcoat pocket, they will be a safeguard against RHEUMATISM.[3]

> In the late 1950s my father-in-law, Ernest, had a bad attack of GOUT. An elderly country farmer, well-known locally for herbal cures . . . came to see him. 'What's wrong wi you, Ern?' he asked. 'I don't know what's wrong, but I know tis bloody painful,' Ernest replied. 'All you want to cure that is some ripe alder cones, and boil in water till tis a rich brown colour, and then you drink a wine glass full of the juice very day for a wik, and thees'll be as right as rain, long before that.' And so, having faith (or fear of more pain), Ernest decided to do this. Our two sons went out and collected some cones which were boiled for a long time (two hours, I think), and when it was cold Ernest had his first glass. No result. The next day he had another glass, but still no result. On came the third day, and doubts about the cure began to creep in, but Ernest carried on. But after dinner poor Ernest began to feel giddy and could hardly stand. Funnily enough at that time the herbalist, Mr Isaiah Bowditch, arrived. He took one look at Ernest and said, 'You don't look no better, Ern.' By this time Ernest was in no mood to take that type of joke, and said, 'Tis all thy fault, it's that old muck you told I to take, I can't stand now.' Isaiah said, 'I think the best thing you can do is see the doctor quick.' Ernest did this, and was soon cured by conventional medicine.[4]

> Hall . . . grows on the banks of streams and rivers and resembles the hazel leaf. In spring time the leaves are very sticky, and if the tree is cut down the bark is very red. A handful of these leaves were boiled in water, left to cool and the liquid put in a jug. I well remember a local man telling my father his little girl had SORES on her scalp. In those pre-war days there were no antibiotics with the result the medics considered cutting off [her] hair which was a rich auburn colour. My father suggested he use[d] this herb, which consisted of putting the liquid in a saucer and dapping the liquid on to the scalp with cotton wool. It cured the little girl's scalp within a month, and hair was not cut.[5]

Alder wood was used in clog-making, as timber, and valued as a fuel.

> As a child, in Herefordshire [pre-Second World War], I saw, every few years, for a few weeks at a time, a small lorry passing thro' to

the station with loads of roughly shaped clog soles. The wood was a rich red colour. Gangs of men cut down the trees and cut lengths to shape in a temporary camp.[6]

Alders [are] constantly sought after for such buildings as lye constantly under water, where it will harden like a very stone; whereas being kept in any inconstant temper it rots immediately.[7]

Dorset woodmen apply the following adage to the alder poles, when peeled for rafters, viz:

Thatch me well, keep me dry,
Heart-of-oak I will defy.[8]

When we were living at Woolminstone, near Crewkerne [Somerset], we used a sleigh . . . to transport wood. We used to supply aller [alder] to Hinton House – there was a college evacuated to there during the War, and we supplied the wood, which was sawn up using a circular saw . . . Aller wood is very good for burning; it takes a year to dry, then it burns well.[9]

ALDER BUCKTHORN (*Frangula alnus*)

Deciduous shrub or small tree, scattered throughout England, scarce in west Wales, Scotland and Ireland.

Local names include the widespread black alder on the Isle of Wight, BLACK ALLER, DOGWOOD in Cornwall and STINKING ROGER in Cheshire.

The berries make a very powerful PURGATIVE. Once popular, this is very rarely used nowadays. A decoction of the bark is used [by gypsies] as a purgative, being mild in action.[1]

The wood yields by distillation in close vessels a very superior CHARCOAL for making gunpowder, for which purpose . . . it is planted in some parts of Kent and Sussex.[2]

ALE – made from HEATHER; GROUND IVY used when brewing.

ALEXANDERS (*Smyrnium olusatrum*)

Perennial herb, growing mostly by the sea in hedge-banks and on clifftops; native to southern Europe, said to have been introduced in

Roman times, and widely cultivated until replaced by CELERY in the fifteenth century.

Local names include:

Alick, 'men, women and children, sailors and country-folk all call it by [this] one name', around Dover, Kent
Allsanders in Cornwall
Hell-root, helroot, or helrut, in Dorset
HORSE-PARSLEY in Somerset
Megweed in Sussex
Melinroot in Dorset
Skeet and skit in Cornwall
Wild celery in the Isle of Wight.

> The flower buds of alexanders are delicious in salads, and the roots may be served instead of PARSNIPS. The leaves can be used as herbs or made into a white sauce, and the soft stems can be cooked like asparagus.[1]

> Alexanders is prolific by the sea. Some people locally use the very young shoots as a vegetable although I have no personal experience of this.[2]

> Wild plants as food . . . alexanders (said to be edible in all parts).[3]

ALEXANDRIAN LAUREL (*Danae racemosa*)

Small evergreen shrub, producing scarlet berries, native to south-west Asia, and occasionally cultivated, and sometimes used in churches on PALM SUNDAY,[1] possibly replacing the more traditional BOX.

ALLAH – AUBERGINE associated with.

ALLERGIES – NETTLE used to treat.

ALLEYN'S SCHOOL, Dulwich – CORNFLOWER associated with.

ALL FOOLS' DAY – see **APRIL FOOLS' DAY**.

ALL SOULS' DAY, 2 November – CHRYSANTHEMUM associated with.

ALPINE MEADOW-RUE (*Thalictrum alpinum*)

Inconspicuous perennial herb growing in damp places in upland areas, mainly in Scotland.

> Known as REDSHANK on the Shetland Islands, where it was used to produce gold and olive DYE.[1]

AMPHIBIOUS BISTORT (*Persicaria amphibia*)

Widespread perennial herb, growing in two forms, with floating leaves in lakes and slow-moving water, and erect on damp soil.

Local names include DOCK-FLOWER in Somerset and reed-willy, 'i.e. red willow', in Northumberland. Names given to the aquatic form include flatter-dock, 'flatter probably refers to the floating leaf', in Cheshire. Two names have been recorded for the land form in East Yorkshire: willow-grass, because it 'frequently forms a considerable part of the herbage of low, wet meadows in the north; the leaves resemble those of the willow in shape' and WILLOW-WEED.

> Formerly used [on the Shetland Islands] to produce a yellow DYE . . . as yallowin' girse or persicaria, and reported to us as still being used for this purpose in Fair Isle.[1]

ANGELICA (*Angelica* spp.)

Perennial herbs. Garden angelica (*A. archangelica*) cultivated since the sixteenth century, and wild angelica (*A. sylvestris*) which is widespread in damp places.

Names given to wild angelica include:

AIT-SKEITERS in Morayshire; 'children shoot oats through the hollow stems as peas are shot through a pea-shooter'
FIFES, and FLUTES, both used for the hollow stems, on Clare Island, Co. Mayo
GHOST KEX in Yorkshire
GROUND ASH in northern England and Berwickshire
GROUND ELDER in Cheshire
HOLY GHOST in Somerset
JACK-JUMP-ABOUT in Northamptonshire
Jellico in northern England
KEDLOCK in Cheshire and Lancashire
KEGLUS in Cheshire
KESK in Cumberland
KEWSE in Lancashire
KEWSIES in Lincolnshire
SCAB-FLOWER in Devon
Scoot, and scout, in Co. Donegal; 'its hollow tubes are used to make scoots or squirts by boys, with a ramrod'
SKYTES, 'from being used as skyters, i.e. shooters', in Scotland

SMOOTH KESH in Cumberland
Spoots and switiks in Shetland
Water-kesh in Cumberland.

> They used to hang angelica over their doors in the gypsy camp near Fremington [Devon] 'to ward off dark spirits'.[1]

> [Gypsies used to smoke HOGWEED] and wild angelica too, the latter sometimes filled with dried and crushed ELM leaves.[2]

> Wild angelica – swittik – makes black DYE. The hollow stems of hogweed (spootitrump) and wild angelica (swittik) were used as PEA-SHOOTERS.[3]

ANNE, QUEEN (1665–1714) – COW PARSLEY and WILD CARROT associated with.

ANNUAL KNAWEL (*Scleranthus annuus*)

Small annual or biennial herb, on dry open sandy soil, mainly in lowland areas; scarce in Ireland.

Other names include green-eyes in Dorset, and German knotgrass and KNOTGRASS, both of which are unlocalised:

> [Ireland] Given in URINARY complaints. Is a favourite herb with the herbalists of the present day. It is given by them in all diseases accompanied by a disordered urinary function.[1]

ANTIRRHINUM (*Antirrhinum majus*), also known as SNAPDRAGON

Short-lived perennial herb, native to south-west Europe, cultivated in British and Irish gardens as an ornamental since Elizabethan times, widely naturalised on old walls throughout lowland Britain.

Local names, most of which refer to the shape of antirrhinum's flowers, include:

Babies-in-the-cradle in Dorset
Bonny rabbit in Devon; 'i.e. bunny rabbit, a tautological children's name'
BOOTS-AND-SHOES in Dorset
BULLDOGS in Somerset
Bunny mouth in Kent and Surrey
BUNNY RABBITS in Devon
BUNNY-RABBITS' MOUTHS in Somerset
Catchfly in Lincolnshire
Chooky-pig in Somerset

CHUCKY-PIG in Devon and Somerset
Dog-mouth in Lincolnshire and Yorkshire
Dogs-d'-bark in Somerset
Dog's mouth in Wiltshire
Dog's nose in Somerset
Dog-snout in Norfolk
Dog's snout in Norfolk
Dragon's heads in Buckinghamshire and Somerset
DRAGON'S MOUTH in Lincolnshire
Frog's mouth in Dorset
GAP-MOUTH, 'a common name'
GRANNY'S BONNET in Somerset
GRANNY'S NIGHTCAP in Somerset
Horse's mouth in Somerset
JACOB'S LADDER in Somerset
LADY'S SLIPPER in Dorset
LION'S MOUTH in Sussex
Lion's snap in Somerset
Mappie's mou, i.e. rabbit's mouth, in Morayshire
Mappie's mouth in Scotland
MONKEY-CHOPS, monkey-faces, MONKEY-FLOWER, MONKEY-MOUTHS, and monkey-musk, in Somerset
Monkey-noses in Dorset
Monkey-sticks in Somerset
Old-man's face in Devon
Open jaws, open mouths, piggy-wiggy, and pig-o'-the-wall, in Somerset
Pig's chops in Somerset and Staffordshire
Pig's snouts in Somerset
RABBIT-FLOWER in Devon
RABBITS in Cheshire and Devon
Serpent's tooth in Cornwall
Snapdock in Devon
SNAP-JACKS in Somerset
TIGER'S MOUTH in Suffolk and Sussex
Yap-mouth in Somerset.

> Plants grown on or about the roof of the house to bring good LUCK and guard against FIRE were [BITING] STONECROP (*Sedum acre*) around Tranmore, County Waterford, and snapdragon (*Antirrhinum majus*) in county Westmeath.[1]

APHRODISIACS – BROAD BEAN, EARLY PURPLE ORCHID, LORDS-AND-LADIES, HAZEL nuts, and SPIGNEL thought to be so.

APPETITE – CENTAURY, DWARF CORNEL, ELECAMPANE, HOP, RUE and WORMWOOD stimulate.

APPLE (*Malus pumila*)

Long-cultivated fruit tree believed to have originated in the Tien-Shan mountains on the border of China and Kazakhstan; the most widely cultivated fruit in Britain and Ireland. Cultivated apples can be divided into three groups: dessert apples, which are eaten raw; cooking apples;

and cider apples, which are used in the production of the alcoholic drink. It is estimated that 2,000 varieties of dessert and cooking apples and hundreds of varieties of cider apple occur in the British Isles: 'you could eat a different apple every day of the year for more than six years and still not come to the end of the list.'[1] The National Fruit Collection at Brogdale, Kent, contains over 2,200 apple cultivars.

In common with other FRUIT TREES it was believed that if an apple flowered out of season misfortune or DEATH was foretold.

> Remarking an apple blossom a few days ago, month of November, on one of my trees I pointed it out as a curiosity to a Dorset labourer. 'Ah! Sir,' he said, 'tis lucky no women folk be here to see that'; and upon my asking the reason he replied, 'Because they's sure to think somebody were a-going to die.'[2]

> [South Moulton, Devon] we had apples and blossom on one branch of the tree one September, and were told it was a sign of death.[3]

> If an apple tree blossoms very much out of season it foretells a tragedy in the family before the year is out.[4]

> My mother has mentioned the superstition that apple blossom on the tree in August, with fruits, is a bad warning of death in the family. This was certainly then alive in the village of Crondall . . . where she spent her childhood in the 1920s.[5]

A belief recorded from Derbyshire[6] and elsewhere was that if the sun shone through the branches of apple trees on CHRISTMAS Day or OLD CHRISTMAS DAY an abundant crop of fruit was foretold. In Dorset:

> If wold Christmas Day be fair and bright Ye'd have apples to your heart's delight.[7]

At about the same time of year apple trees were wassailed:

> In certain parts of this country superstitious observances yet linger, such as drinking the health to the [apple] trees on Christmas and Epiphany eves, saluting them by throwing toasted crabs or toast round them, lighting fires &c. All these ceremonies are supposed to render the trees productive for the coming season.
>
> I once had the occasion to pass the preceding Twelfth day at a lone farm-house on the borders of Dartmoor, in Devonshire, and

> was somewhat alarmed at hearing, very late at night, the repeated discharge of fire-arms in the immediate vicinity of the house. On my inquiring in the morning as to what was the cause of the unseasonable noise, I was told that the farm-men were firing into the Apple-trees in the orchard, in order that the trees might bear a good crop.[8]

Although this custom seems to have been most prevalent in Devon and Somerset, it also occurred in other apple-growing areas, but appears to have been absent from the cider-making district around Hereford. According to Hutton[9] the earliest recorded date for the custom is 1585, at Fordwich in east Kent, outside its later core areas. In Sussex, where the custom was known as 'howling', the earliest reference dates back to 1670, when the rector of Horsted Keynes recorded in his diary on Boxing Day, 'Gave the howling boys sixpence.'[10] In the 1920s at Duncton, in West Sussex, 'Spratty' Knight was the chief wassailer. He was 'Captain' of a team which would assemble at the village pub and go around local farms asking each farmer 'Do you want your trees wassailed?'

> The gang, followed by numerous small children, then went to the orchard. Spratty blew through a cow's horn, which made a terrible sound. It was to frighten away any evil spirits that might be lurking around. Next one of the trees, generally the finest one, would be hit with sticks and sprinkled with ale. This was a gift to the god who looked after the fruit trees. Lastly all the company joined the wassailing song, the words of which were as follows:
>
> Stand fast, root, bear well, top,
> Pray, good God, send us a howling crop,
> Every twig, apples big; every bough, apples now;
> Hats full, caps full, five bushell sacks full,
> And a little heap under the stairs.
> Holloa, boys, holloa, and blow the horn!
>
> And hulloa they did, to the accompaniment of the horn. This completed the wassailing, and everyone trooped out of the orchard up to the farm-house door, where they were greeted by the farmer's wife with drinks and goodies. Sometimes money was given instead of good cheer . . . and then on around the village, till they arrived at the Cricketers' Inn, which was their last port of call.[11]

Similarly, in Devon in the 1940s:

> The sun on apple trees on Christmas Day would mean a good crop, and at Dunkerswell up the Culm Valley, on TWELFTH NIGHT they went out from the local pub at night to shoot into the branches and sing an old song.[12]

Wassailing went into 'fairly steady decline' from early in the nineteenth century,[13] but there have been a number of revivals, some of which have persisted for many years. Of these the best known is that at Carhampton in Somerset, where the custom continues to take place behind the Butcher's Arms Inn each January.[14] Elsewhere wassailing has been revived to provide publicity for cider companies, or by folk enthusiasts. In 1974 the Taunton Cider Company promoted a wassailing, complete with a Wassailing Queen, at Norton Fitzwarren in west Somerset.[15] For over fifty years Chanctonbury Ring Morris Men have been wassailing apple trees in West Sussex. In 1978 Dick Playl, a former Squire of the Men, explained:

> The practice is undoubtedly traditional to Sussex where it is known as 'Apple Howling' rather than 'Wassailing the Apple Trees', but the proceedings, as conducted by the Chanctonbury Ring Morris Men, are a compilation of traditions from various parts of the country rather than a purely Sussex tradition. We have used Christina Hole's *British Folk Customs* and records of the event at Carhampton, Somerset, to produce an hour long event which has proved a great success. As you will see in Geoffrey Palmer and Noel Lloyd's *A Year of Festivals* [1972] it is our team at Tandring, Hailsham, which is used to illustrate this traditional event! This was the first occasion that we did it, in 1967 (January 6th) . . . Chanctonbury Ring M.M. have now successfully done their apple howling for the past two years at Furner's Farm, Henfield, Sussex. We try to do it on the Eve of Epiphany, but since Friday clashes with a Folk Song Club commitment we have chosen a Thursday on both occasions. We had 300 spectators this year and in addition there were 25 morris men. Mr Whitstone, the owner of the orchard, is most co-operative! The morris men are enthusiastic!

This revival started at a time when there was a resurgence of interest in English folk music and dance. Later, in the 1990s, a time of 'green' and New Age ideas, further revivals reflected these concerns:

> There seems to be no record of apple tree wassailing in Yorkshire, and my attempts at reviving the custom will undoubtedly

> be frowned upon by some folklorists who are of the opinion that such a revival should not be contemplated in an area where the custom has not been previously performed. My reply is that apple trees have a right to be wassailed wherever they grow . . . There is a great affinity between trees and humans, but also, unfortunately at present, great isolation. Wassailing the apple trees is one way of helping to restore harmony and thus correcting this imbalance . . .
>
> This year I finally managed to find a suitable location in the Sheffield area, although only five people attended . . . Plans for 1994, apart from a repeat of the 1993 wassails, include a possible second wassailing location in the Sheffield area, and a possible one near Rotherham.[16]

Early in the twenty-first century community orchards became popular, leading to another spurt of wassailing activities. On 28 January 2017 such a wassailing, described as 'traditional, rural and family', was held at a community orchard in Mitcham, Surrey. A good number of people gathered to enjoy cider, foodstalls, music and displays of border-morris. The actual wassailing was performed by druids, who, after summoning of the spirits of the four points of the compass, poured lambswool (mulled cider) on the roots of a selected tree, and invited people to place toast soaked in lambswool in its branches 'for the robins'.[17]

Writing in 1884 Sabine Baring-Gould recorded a Devon belief that it was usual for there to be a late frost which severely damaged apple blossom on the nights of 19, 20 and 21 May. This period was known as ST FRANKIN'S DAYS.

The pre-Reformation church blessed each year's apple crop on ST JAMES'S DAY,[18] while in the seventeenth century it was recorded:

> In Herefordshire, and also in Somersetshire, on Midsommer's-eve, they make fires in the waies: *sc* – to Blesse the Apples. I have seen the same custome in Somerset, 1685, but they do it only for custome sake.[19]

In a letter written from Elton, Herefordshire, in 1880, it was recorded:

> Unless orchards are christened on ST PETER'S DAY [29 June] the crop will not be good; and there ought to be a shower of rain when the people go through the orchards, but no one seems to know for what purpose exactly.[20]

However, the most widespread and persistent date for the christening of apples was ST SWITHIN'S DAY (15 July).

> In the Huntingdonshire parish wherein I passed St Swithin's Day, 1865, we had not a drop of rain. A cottager said to me, 'It's a bad job for the apples that St Swithin hadn't rained upon 'em.' 'Why so?' 'Because unless St Swithin rains upon 'em they'll never keep through winter.'[21]

> When I was a lad we were told not to eat apples before St Swithin's Day or they would make us ill, as they had not been christened. This was in South Notts.[22]

> The Christening of Apples. This is a common expression for St Swithin's Day in the neighbourhood of Banbury in the middle of the last century. On that day apples were supposed to get big and mature quickly.[23]

> The Apple Christening Day is still a common folk-name given to St Swithin's Day in Surrey as well as in Berkshire and Oxfordshire, as I am told by many friends.[24]

> [Whitchurch, near Coalville, Leicestershire] Early apples are ready after St Swithin's Day.[25]

But the idea of apples being blessed on St James's Day also lingered:

> 'On St James's Day the Apples are Christened.' This saying is found among the people of Wiltshire and Somersetshire. Was St James considered to be patron of orchards? and was he invoked for a blessing on the infant fruit? as, at that season, May 1, the apple trees are in bloom.[26]

In 1913 a farmer living in Veryan, Cornwall, advised:

> Never pick apples before St James' Day, when they receive their final blessing.[27]

Clearly there is confusion over which St James's Day is intended: the 1856 passage refers to the feast of St James the Less (1 May) and the 1913 one seems to refer to the feast of St James the Great (25 July).

Since 1989 the environmental charity Common Ground has promoted 21 October as Apple Day, with the intention of stimulating a greater appreciation of local varieties of apple:

> We wanted to create a popular festival, a date in the calendar to alert people to our heritage, to broaden their knowledge and inspire action . . . We brought together forty stallholders with integrity and passion at Covent Garden.[28]

On 17 October 1992 *The Times* reported:

> More than 80 apple-promoting events have been planned around the country next week, from planting a new orchard of Cox's Orange Pippin, near Slough, home of its 19th-century founder . . . to cider-making demonstrations in Devon and Somerset, and children's apple activity day at Greenwich Borough Museum.

Apple Day has become a nationally recognised festival with events being held at National Trust properties, farmers' markets, and elsewhere each year, while increased interest in local varieties of apples has led to the planting of community orchards. On 26 October 2008 the Apple Day programme at London's trendy Borough Market invited people to:

> Come and enjoy a cornucopia of Autumn produce at Borough Market's celebration of Apple Day.
>
> Cheeses, meats, bread, delectable cakes and pies, cider, calvados, apple juice, mountains of apples of every imaginable variety and many more autumnal treats.
>
> Traders will be at Borough Market from 11 a.m. to 5 p.m. Music and festivities celebrating apples and autumn will be taking place throughout the day.

Across the road, in Southwark Cathedral:

> Apple Day will be marked . . . in the Choral Eucharist at 11.00 a.m. when the Cathedral will be decked with special displays. 'Fun with Apples' will be provided by the Ministry of Fun in the Churchyard after the service.
>
> The Cathedral's restaurant, The Refectory, will serve a bespoke apple-y menu.

On 15 October 2016 the Ely, Cambridgeshire, Apple Festival, promised an apple identification service, an apple market, morris dancing and folk music, and 'the ever-popular longest apple peeling competition', apple and spoon races, and apple shy.[29]

Particularly in Ireland, apples were widely used in HALLOWE'EN activities.

> [Co. Cork] Hallowe'en is celebrated on the night of 31st October . . . The children get a tub full of water and put it in the middle of the floor. They put an apple or a couple of apples into the tub. The children then kneel down around the tub and put their heads into the water, and try to catch an apple in their mouth. They enjoy themselves very much at this. This might be also known as 'Snap-Apple Night'.[30]

> I was born in 1946 and brought up in a working-class area of Manchester . . . [at Hallowe'en] occasionally during my childhood we bobbed for apples in a tin bath, or tried to eat them off strings with hands tied behind back. This happened once or twice in early childhood. Nowadays we have apple bobbing, apples on strings.[31]

In nineteenth-century Cornwall:

> The ancient custom of providing children with a large apple at Allhallow-eve is still observed to a great extent, at St Ives. 'Allan-day' as it is called is the day of days to hundreds of children, who would deem it a great misfortune were they to go to bed on 'Allan-night' without the time honoured Allan apple to hide beneath their pillows. A quantity of large apples are thus disposed of, the sale of which is signified by the term Allan Market.[32]

Similarly, according to a writer born in 1931:

> Allantide was still a popular occasion in my Newlyn childhood, and extra-large 'Allan Apples' very much in demand. The older girls put them under their pillows to dream of their sweethearts. While boys hung them on a string and took large bites.[33]

The peeling of an apple (or more rarely an ORANGE) so that the peel remained in one long strip, which was thrown over the shoulder to form the initial of a potential husband on the ground, was a widely recorded practice, particularly at Hallowe'en:

> 'At Midnight', says a 14-year-old in Aberdeen, 'all the girls line up in front of a mirror. One by one each girl brushes her hair three

> times. While doing this the man who is to be her husband is supposed to look over her shoulder. If this happens the girl will be married within a year.'
>
> 'After they have done this,' continues the young Aberdonian, 'each girl peels an apple, the peel must be in one piece, then she throws the peel over her left shoulder with her right hand. This is supposed to form the initial of her husband-to-be.'[34]

In Cornwall a more rough-and-ready method of LOVE DIVINATION was attempted:

> An apple pip flicked into the air indicated the lover's home, so long as this verse was used: North, south, east, west, Tell me where my love does rest.[35]

Similarly, in Lancashire (now Cumbria):

> In order to ascertain the abode of a lover, the anxious inquirer moves round in a circle, squeezing an apple pippin between the finger and thumb, which, on pressure being employed, flies from the rind in the supposed direction of the lover's residence. The following doggerel is repeated during the operation:
>
> Pippin, pippin, paradise,
> Tell me where my true love lies;
> East, west, north or south,
> Pilling brig or Cocker-mouth.
>
> That the reply may be corroborated, the inquirer afterwards shakes another pippin between the closed hands, and, on ascertaining the direction of the point of the pippin to the point of the compass, the assurance is supposed to be rendered doubly sure, if the charm works as desired, but not otherwise.[36]

Dorset girls could use an apple pip to test a lover's fidelity:

> If on putting it in the fire, it bursts with the heat she is assured of his affection; but if it is consumed in silence she may know that he is false. Whilst they anxiously await the effect the following couplet is usually announced:
>
> If you love me, pop and fly;
> If you hate me, lay and die.[37]

In 1882 it was recorded that on St Thomas's Day (21 December) Guernsey girls would use apples to obtain a glimpse of future lovers:

On that day you would take a golden pippin, and having walked backwards to your bed, and having spoken to no one, you must then place it underneath the pillow, and St Thomas will grant to you when asleep a vision of your future consort. On placing the pippin underneath the pillow, the following charm must be repeated:

Le jour de St Thomas,
Le plus court, et le plus bas,
Dieu, fais me voir en mon dormant,
Ce que j'aurai pour mon amant.
Montre moi et mon épousé
La maison ou j'habiterai.[38]

In Lancashire in the 1980s:

I remember twisting off apple stalks, each time the apple was turned a letter of the alphabet was said, when the stalk broke that letter was the initial of the Christian name of the one you were to marry. The broken stalk was then poked into the apple, counting letters again, when the skin broke that letter was his surname initial.[39]

The NURSERY BOGIES Awd Goggie and Lazy Laurence protected orchards and unripe fruit. In the East Riding of Yorkshire:

There is another wicked sprite, who comes in most usefully as a protector of fruit. His name is Awd Goggie, and he specially haunts woods and orchards. It is evident, therefore, that it is wise on the children's part to keep away from the orchard at improper times, because otherwise 'Awd Goggie might get them'. [40]

Further south, according to Katharine Briggs:

Lazy Laurence was a guardian spirit of the orchard, both in Hampshire and in Somerset . . . In Hampshire, he sometimes took the form of a colt and chased orchard thieves . . . In Somerset, Lazy Laurence seems rather to afflict the thieves with what is described in one of the night spells as 'Cramps and crookeing and fault in their footing'. The Somerset proverbial saying runs:

Lazy Lawrence let me goo,
Don't hold me Winter and Summer too.[41]

Here, as elsewhere in her publications, Briggs relied heavily on the

Somerset 'folklorist' Ruth Tongue (1898–1981), with whom she collaborated on several works. Tongue claimed to have collected a wealth of folk-tales and folklore in the West Country and elsewhere, but her work is not considered to be an authentic record of folk traditions. She was involved in the theatre, and as a sought-after teller of folk stories, she embellished genuine material with material which she had, consciously or unconsciously, taken from written sources. In her *Forgotten Folk-tales of the English Counties*,[42] she relates a story, supposedly set in Wareham, Dorset, in which a potential apple-thief is foiled by Lazy Laurence, a pixy-colt. J.B. Smith traces Lazy Laurence's pedigree back to an 1809 chapbook, *The History of Lawrence Lazy*, which was, perhaps, printed as early as 1670. This he sees as 'a social document, expressing as it does through the almost allegorical character of Lawrence the frustrations of down-trodden schoolboys and apprentices'. He questions whether Lazy Laurence ever grew into a 'spirit in the imaginations of the folk rather than folklorists', and concludes that although Tongue's story is an interesting artefact it is doubtful if it can be classified as a real folk-tale.[43]

Tongue also provides vague recollections of the Apple-Tree Man, which according to her was the oldest tree in the orchard.

> Pitminster was the place where in my childhood I was gravely and proudly conducted by a farm-child to a very old apple tree in their orchard and told mysteriously that it was 'the Apple-Tree Man'. In 1958 I heard of him again on the Devon-Somerset borders.[44]

According to Briggs, 'it seems that the fertility of the orchard was supposed to reside there'.[45]

The well-known proverb 'An apple a day keeps the doctor away' is said to have been coined by a Missouri fruit specialist for the 1904 St Louis World Fair.[46] However, an earlier saying, recorded in 1866, advised 'Eat an apple going to bed, and you'll keep the doctor from earning his bread.'[47] In some west Dorset farming families this injunction seems to have been taken literally, for apple dumplings formed a standard part of the daily evening meal, and:

> They used to keep apples from one year so that the last of them could be made into a pie eaten at the end of sheep shearing time [in May] the following year.[48]

In 1997 Food Chain, a charity which aimed to 'deliver nutritionally-balanced three-course meals across Greater London every Sunday to people who are housebound with HIV and AIDS-related illnesses', utilised an apple motif and 'An apple a day . . .' on its publicity material.

Apples were used in a variety of folk remedies.

> During my childhood in a Pennine village I was sent to the green-grocer for a rotten apple – a mouldy one – as a remedy for an obstinate STYE.[49]

> Thoroughly rotten apples were threaded onto CHILBLAINed toes to cool the burning and itching.[50]

> According to my 86-year-old aunt, an apple was placed in a room where there was SMALLPOX; as the apple went mouldy the small-pox was believed to be transferred from that patient to it.[51]

> Devon, 1990s. Local gypsies rubbed apples on WARTS and fed to pigs. This cured warts overnight.[52]

Burning apple wood 'will scent your room with an incense-like perfume,' [53] or 'fill your room with the gentlest of perfume.'[54]

APRIL FOOLS' DAY (All Fools' Day, 1 April) – game played with GOOSEGRASS.

ARBOR DAY (29 May) – a BLACK POPLAR at Aston-on-Clun decorated; see also OAK Apple Day.

ARTHRITIS – treated using BOGBEAN, CABBAGE, COMFREY, DANDELION, NETTLE and SAGE.

ARUM LILY (*Zantedeschia aethiopica*), also known as ALTAR LILY

Native to South Africa, introduced as an ornamental in 1731; now widely cultivated, becoming naturalised in many temperate areas, including the borders of former bulb-fields in Cornwall.

In many parts of the world arum lilies are associated with mourning and used to decorate graves. Consequently they were considered to be inappropriate for domestic flower arrangements or hospitals.

> In 1946, following demobilisation from the Royal Navy, I emi-

grated to New Zealand. I was lucky enough to obtain an apartment in St Heliers Bay, Auckland.

The rent included the garden, the maintenance of which I took on. At the bottom, in a sort of wild corner, was an enormous spread of arum lilies. They were breathtaking in their beauty, and finding a huge vase which stood on the floor, I cut some of the lilies and arranged them to make a pleasant display.

Next day my landlady arrived to check that I had everything I wanted . . .

On entering the living room, her hands flew to her mouth and she cried out: 'Oh, whatever have you done!' She turned to me, eyes blazing: 'You've brought arum lilies into the house. Don't you know you must never do that? They mean there will be a DEATH in the house.'

I confessed my ignorance of this 'old wives' tale' and said they looked so beautiful I could not resist bringing them in. She said that in New Zealand arum lilies grew like a weed and were treated as such. When she told a neighbour what I had done the other woman actually crossed herself, and, needless to say, I did not do that again.[1]

In hospitals they used to think lilies – you know, the big ones they used to have in wedding bouquets – arum lilies – UNLUCKY, and wouldn't allow them in the wards. When I worked in a hospital I ignored this twice, but each time, that night . . .[2]

Arum lilies . . . never bring them indoors or plant them in your garden.[3]

Arum lilies are frequently used to decorate churches on EASTER day, often bought with donations given in memory of the dead.

In 1947–8 the Rev. George Grove became Vicar of Thorncombe [west Dorset] and suggested to me that in order to achieve flowers in nice numbers cheaply for Easter parishioners might like to donate arums (the cost of them) in memory of someone known to them. This idea has continued ever since, though nowadays the cost of even one bloom can be too much for an individual, so we tend to ask people to give what they can towards flowers . . . We used to get about 50 flowers, and still achieve between 30 and 40.[4]

In recent years *Lilium longiflorum*, which has been given the name EASTER LILY, has tended to replace the arum lily for decorating churches at Easter.

However the Easter lilies associated with the 1916 Easter Rising in Ireland are arum lilies.

> The republican women . . . devised, made and sold the Easter lily emblems from 1926 onwards. Representing the arum lilies traditionally used to decorate churches at Easter, they were also partly inspired by the red POPPIES commemorating British victims of the First World War. Like them, and like the SHAMROCK, they are simple, personal, natural, religiously inspired emblems.[5]

In the late 1960s these Easter lilies became important emblems of the Nationalist groups in Northern Ireland. At their Easter parades:

> nearly all those attending display a single emblem. This is generally a paper Easter lily. The style in which it is worn indicates whether its owner supports the Officials or the Provisionals. When the republican movement split in 1969, the Officials chose to stick the lily to the coat lapel, while the Provisionals opted to use a pin . . . So important is this distinction that when, in the 1980s the Provisionals were driven by increasing costs to employ sticky badges themselves, they reputedly had a drawing of a pin superimposed in the design.[6]

According to a card provided at a stall selling paper lilies for wearing on lapels outside the General Post Office building in O'Connell Street, Dublin, on Easter Saturday 2017:

> Today, republicans continue to honour the heroic sacrifice made in 1916, when the IRA, hopelessly outnumbered and ill-equipped, took on the might of the British Army . . . Republicans wear the lily to honour all those who have given their lives in the cause of Irish freedom.

However, in 2017, comparatively few people were seen wearing lilies, although they continue to be frequently depicted on murals in Republican areas of Belfast.[7]

Despite their association with death and misfortune, arum lilies enjoy intermittent popularity in bridal bouquets.

ASCENSION DAY (fortieth day after EASTER) – ELM associated with.

ASH (*Fraxinus excelsior*)

Deciduous tree, widespread in hedgerows and woodland, readily colonising neglected cemeteries and similar habitats.

Local names include the widespread esh and:

Eshen tree in Norfolk
Hampshire weed
Skedge in Cornwall

Names given to ash fruits – 'keys' – include:

Ash candles in Dorset
BOOTS-AND-SHOES in Somerset
CATS-AND-KEYS in Somerset
Cat's keys in Teesdale
Cattikeyns in Wiltshire
Chat in Norfolk, Suffolk and Yorkshire
Eisch-keys in Lancashire
Esh-keys in Lancashire and Yorkshire
Pattikeys in Northamptonshire
Shacklers in Devon; shackle = to rattle
Widow-maker, 'because of its lethal habit of splitting as it is felled', in Essex.
WINGS in Somerset.

In the nineteenth century it was believed that if ash trees failed to produce keys disaster was foretold. In Yorkshire:

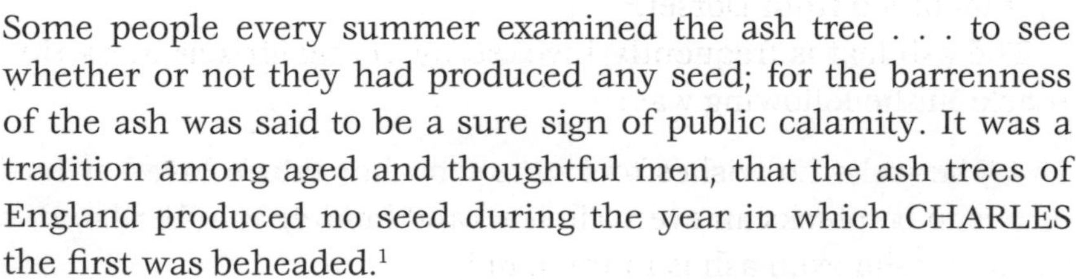

> Some people every summer examined the ash tree . . . to see whether or not they had produced any seed; for the barrenness of the ash was said to be a sure sign of public calamity. It was a tradition among aged and thoughtful men, that the ash trees of England produced no seed during the year in which CHARLES the first was beheaded.[1]

In East Anglia:

> The failure of the Crop of Ash-keys portends a death in the Royal Family . . . the failure in question is certainly, in some seasons, very remarkable; many an old woman believes that, if she were the fortunate finder of a bunch, and could get introduced to the king, he would give her a great deal of money for it.[2]

Mountain ash, or ROWAN, an unrelated tree with leaves similar to those of ash, was widely believed to provide PROTECTION, and occasionally ash itself was considered to be protective:

> Rowan and ash sticks were used to drive cattle . . . believed to be 'kindly' and both trees were believed to be endowed with properties that ensured no interference from harmful influences.[3]

More widely it was believed that carrying an ash stick would provide protection against ADDERS.

> The snakes were believed to avoid and dread them. Striking an adder with an ash branch was thought to kill it outright and a West Country cure for snake bite was not simply (and foolishly) to suck and spit, but also to intone a rhyme: 'Ashing tree, ashing tree, take this bite away from me.' This belief led to the rather dangerous notion that, if you walk through a snake-infested area carrying an ash branch, you are protected.[4]

In rural areas 'even' ash leaves leaves which lacked a terminal leaflet and therefore had an even number of leaflets – were used in LOVE DIVINATION. The earliest record of this is from Wales in 1813, when young people would search for an even-leaf and 'the first of either sex that finds one calls out Cyniver, and is answered by the first of the other that succeeds; and these two, if the omen fails not, are to be joined in wedlock.[5] This method of divination does not seem to have been recorded elsewhere, but in 1831 the usual form of divination was first recorded from Dorset.[6]

The ash leaf is frequently invoked by young girls as a matrimonial oracle in the following way:

> The girl who wishes to divine who her future lover or husband is to be plucks an even ash leaf, and holding it in her hand, says:
>
> The even ash is in my hand,
> The first I meet shall be my man.
>
> Then putting it into her glove, adds:

> The even ash leaf in my glove,
> The first I meet shall be my love.
>
> And lastly, into her bosom, saying:
>
> The even ash leaf in my bosom,
> The first I meet shall be my husband.
>
> Soon after which the future lover or husband will be sure to make his appearance.[7]

According to a 52-year-old Dorset woman who in 1976 described divination with even-ash leaves during her childhood:

> Start at the bottom leaflet on the left-hand side and say:
>
> An even ash is in my hand
> The first I meet will be my man
> If he don't speak and I don't speak,
> This even ash I will not keep.
>
> As each word is said, count a leaflet around the leaf until the rhyme is completed (this probably entails going round the leaf several times). When the rhyme is finished, continue reciting the alphabet until the bottom right-hand leaflet is reached. The letter given to this leaflet gives you the initial of your boyfriend. Two or three leaves may be used so that you get a wider range of letters.[8]

Although this practice seems to have been recorded most frequently from Dorset and adjacent counties, it was also known Northumberland in the 1840s, where after the rhyme was completed the leaf had to be placed in the enquirer's left shoe.[9]

In County Donegal:

> A girl carries a[n ash] leaf with an even number of leaflets. The first man she meets she asks the name of. His Christian name will be that of her future husband.[10]

In Lincolnshire, where ash was known as 'esh':

> there is a widespread opinion that if a man takes a newly-cut esh-plant not thicker than his thumb, he may lawfully beat his wife with it.[11]

Burning the ashen faggot – a faggot made from young ash stems – was a widespread Christmastide custom in Devon and Somerset during the nineteenth century. According to a writer late in the century, it was:

> an ancient ceremony transmitted to us from the Scandinavians who at their feast of Yuul were accustomed to kindle huge bonfires in honour of Thor. The faggot is composed of ashen sticks, hooped round with bands of the same tree, nine in number.[12]

However, Ronald Hutton gives the first two records of the custom as 1795 and 1806.[13]

In the first half of the nineteenth century:

> Some towns in Somerset held 'Ashen Faggot Balls'. The one in Taunton on January 2nd 1826 was 'most respectably attended by the principal families of the town and neighbourhood'. It was still held twenty years later, but by then the event was losing its appeal.[14]

Ashen faggots continue to be burnt in a few West Country pubs, and possibly miniature faggots are prepared for burning in private homes.

> On the evening of January 5th ('old' Christmas Eve) at Curry Rivel, a Somerset village situated on the southern edge of Kings Sedgemoor, the wassailers go 'visiting' around the parish with their wassail song and the ashen faggot is ceremoniously burned at the King William IV public house. The faggot is made from young ash saplings and bound with bonds ('fonds', 'fronds', 'thongs', or 'bonds') of withies (osiers); bramble has been used occasionally in the past. The number of bonds is variable but since the bursting of any one during the burning is a signal to 'drink up', decency and country logic demands a 'reasonable few'. Either five or six are normally used. At the appropriate moment the faggot is placed on the fire, traditionally by the oldest customer – one villager can recall the faggot being brought in a wheelbarrow as was 'right and proper' – and as each bond bursts there is much cheering and a general clamour for drink. The landlord, Mr John Cousins, prepares a bowl of hot punch for the occasion to augment the barrel of beer usually provided by the house Brewery. Until quite recently cider was consumed in large quantities; the 'brew' of cider and perry donated by (Langs) Hambridge Brewery in 1957 is particularly remembered.[15]

It is thought that the Curry Rivel's King William IV is the only place where the custom has been maintained without any breaks, but there are at least two well established revivals: at the Luttrell Arms in Dunster,

Somerset, since 1935, and at the Squirrel Inn, at Laymore on the Dorset/Somerset border since 1972:

> The Luttrell Arms occasion is the grandest, with a local carol sung as the flames rise and couples laying wagers upon which will be the first band to burst (a successful bet being said to bring luck in marriage).[16]

Although well-attended, the event at the Squirrel tends to be rather low key; people gather at the bar while the faggot burns almost unnoticed at the other end of the room. According to what appears to be a recent tradition, it is said that if a woman steps over the faggot she will become pregnant.[17] A local historian has suggested that this belief is a reminder of the Squirrel's possible former use as a venue for informal weddings, at which the couple, watched by witnesses, jumped over a besom.[18]

In parts of southern England ash twigs were carried by children on Ash Wednesday:

> In villages around Alton in Hampshire, and as far away as East Meon, near Petersfield, at Crowborough in Sussex, and doubtless in other places, children pick a black-budded twig of ash and put it in their pocket on this day. A child who does not remember to bring a piece of ash to school on Ash Wednesday can expect to have his feet trodden on by every child who possesses a twig, unless, that is, he or she is lucky enough to escape until mid-day.[19]

> [Heston, Middlesex, 1930s] On Ash Wednesday we all took a twig of ash tree to school and produced it when challenged or risked a kick – we had to get rid of it at 12 noon. We even risked the wrath of the teacher by rushing to the window to throw out our twigs as soon as the mid-day dinner bell rang.[20]

> When I was at school [*c*.1950], on the Hampshire, Sussex and Surrey borders, you had to have an ash twig tucked into your sock on ASH WEDNESDAY, until 12 o'clock; if you had one after 12 you were punished by having your feet stamped on. I assume the custom came to an end when schoolboys stopped wearing short trousers.[21]

According to Simpson and Roud, this custom 'must be related to the Catholic ritual of blacking one's brow with ashes, but whether as parody or as a misunderstanding of the word "ash" is impossible to say'.[22] The custom does not seem to have been recorded until the twentieth century and apparently died out last in Cranbrook, Kent, in the early 1970s.[23]

A widespread cure for HERNIA involved passing the patient through a split ash sapling, preferably one which had grown naturally from seed and had not previously been damaged by man. The tree was then tightly bound up and as it grew together the patient would be cured. This practice seems to have been first recorded by John Evelyn in the fourth (1706) edition of his *Sylva*:

> But (whether by the power of magick or nature, I determine not) I have heard it affirm'd with great confidence, and upon experience, that the rupture, to which many children are obnoxious, is healed, by passing the infant thro' a wide cleft made in the hole or stem of a growing ash-tree, thro' which the child is to be made pass; and then carried a second time round the ash, caused to repass the same aperture again, that the cleft of the tree suffer'd to close and coalesce, as it will, the rupture of the child, being carefully bound up, will not only abate, but be perfectly cur'd.[24]

A detailed description provided by the wife of a West Sussex clergyman in 1868 demonstrates how this cure, which required communal co-operation, was considered normal:

> A child so afflicted must be passed nine times every morning on nine successive days at sunrise through a cleft in a sapling ash tree, which has been so far given up by the owner of it to the parents of the child as that there is an understanding that it shall not be cut down during the life of the infant that is passed through it. The sapling must be sound of heart, and the cleft must be made with an axe. The child, on being carried to the tree, must be attended by nine persons, each of whom must pass it through the cleft from west to east. On the ninth morning the solemn ceremony is concluded by binding the tree tightly with a cord, and it is supposed that as the cleft closes the health of the child will improve. In the neighbourhood of Petworth some cleft ashes may be seen, through which children have very recently been passed. I may add that only a few weeks since, a person who lately purchased an ash-tree standing in this parish, intended to cut it down, was told by the father of the child who had some time before passed through it,

> that the infirmity would be sure to return upon his son if it were felled. Whereupon the good man said, he knew such would be the case; and therefore he would not fell it for the world.[25]

Similarly, according to the *Bath and Wells Diocesan Magazine* of 1886:

> A remarkable instance of the extraordinary superstition which still prevails in the rural districts of Somerset has lately come to light at Athelney. It appears that a child was recently born in the neighbourhood with a physical ailment, and the neighbours persuaded the parents to resort to a very novel method of charming away the complaint. A sapling ash was split down the centre, and wedges were inserted so as to afford an opening sufficient for the child's body to pass through without touching either side of the tree. This having been done, the child was undressed, and, with its face held heavenward, it was drawn through the sapling in strict accordance with the superstition. Afterwards the child was dressed and simultaneously the tree was bound up. The belief of those who took part in this strange ceremony is that if the tree grows the child will grow out of its bodily ills. The affair took place at the rising of the sun on a recent Sunday morning, in the presence of the child's parents, several of the neighbours, and the parish police-constable.

It appears that this practice died out in the 1940s,[26] but an example of an ash so used is preserved in the Somerset Rural Life Museum at Glastonbury.

A similar practice could be used to cure IMPOTENCE:

> In Wales the similar ritual was to split a young ash or HAZEL stem and hold it just fastened at the top. This made a symbolic vulva into which the impotent male introduced his recalcitrant organ. Binding up the tree again enabled it to heal, during which the impotence faded.[27]

In Cheshire a cure for WARTS was:

> to steal a piece of bacon and push it under a piece of ash-bark. Excrescences would then appear on the tree; as they grew, the warts would vanish.[28]

In Wiltshire NEURALGIA sufferers were advised to seek out a maiden (undamaged) ash and:

> cut off a piece of each finger and toe nail and a piece off your hair. Get up on the next Sunday morning before sunrise and with a gimlet bore a hole in the first maiden ash you come across and put the nails and hair in; then plug the hole up.[29]

In many areas 'shrew-ashes' were used to cure lameness in cattle and other illnesses. The earliest record of this cure, from Staffordshire in 1686, stated that OAK, ash or ELM trees could be utilised – 'it being indifferent which',[30] but later it seems that only ash was used. On 8 January 1776 Gilbert White noted:

> A shrew-ash is an ash whose twigs and branches, when gently applied to the limbs of cattle, will immediately relieve the pains which a beast suffers from the running of a shrew-mouse over the part affected . . . Against this accident, to which they were continually liable, our provident forefathers always kept a shrew-ash at hand, which, once medicated, would maintain its virtue for ever. A shrew-ash was made thus: Into the body of the tree a deep hole was bored with an auger, and a poor devoted shrew-mouse was thrust in alive, and plugged in, no doubt, with several quaint incantations long since forgotten.[31]

In the nineteenth century a particularly well-known shrew-ash grew in Richmond Park, Surrey. According to the park-keepers' tradition 'good Queen Bess [ELIZABETH I] had lurked under its shade to shoot deer as they were driven past'.[32] This tree was closely observed by Sir Richard Owen (1804–92), first director of the Natural History Museum, London, who from 1852 lived nearby at Sheen Lodge:

> Either the year he came to live in the park or the year after . . . he first encountered a young mother with a sick child accompanied by 'an old dame', 'a shrew-mother', or, as he generally called her a 'witch-mother'. They were going straight for the tree; but when they saw him, they turned off in quite another direction till they supposed he was out of sight. He, however, struck by their sudden avoidance of him, watched them from a distance, saw them return to the tree, where they remained some little time, as if busily engaged with it; then they went away. He was too far off to hear anything said, but heard the sounds of voices in unison on other occasions. He heard afterwards from the keeper of Sheen Gate . . . that mothers with 'bewitched' infants, or with young

> children afflicted with WHOOPING COUGH, decline, and other ailments, often came, sometimes from long distances, to this tree. It was necessary that they should arrive before sunrise . . . Many children were said to be cured at the tree. The greatest secrecy was always observed when visiting. This was respected by Sir Richard Owen, who, whenever he saw a group advancing towards it, moved away, and was always anxious that they should not be disturbed. He could not tell me in what year he last saw a group approach the tree to seek its aid. He could only say he had seen them often, and thought they continued to come for many years.[33]

> During a recent survey [of the Park] the site of the old shrew ash was identified. This proved to be . . . the spot where an ancient ash still stood in 1987. A sucker from its roots was still alive, although the tree itself was passé. The storm of autumn [? 1987] brought the trunk down. A railing has now been erected around the remains, which are to be left in the ground, and a young ash planted alongside the stump. Presumably it will eventually replace the old tree, but it means that the site at least will remain identifiable.[34]

It seems that beliefs concerning shrew-ashes died out late in the nineteenth century.

Other medicinal uses of ash depended less on magic and may have been of more practical benefit.

> The sap of a young ash sapling was used to cure EARACHE. A sapling was cut and put into a fire so that when the stick started to burn the sap came out the end and was caught on a spoon. This could be put on cotton wool and put into the ear.[35]

> RINGWORM was more common in my childhood . . . a remedy resorted to was to burn ash twigs in a tin box or similar container and allow the smoke from the smouldering twigs to envelop the affected part – usually arms, neck or face.[36]

> Ash leaves are used to combat viper [ADDER] bites. When an animal has been bitten farmers boil ash leaves and give the animal the resulting liquid and place the boiled leaves as a poultice on the bite. Works on people too![37]

Roy and Ursula Radford, writing of the West Country, which they define as stretching from Cornwall to the Cotswolds, record that 'feeding ash-tree leaves to cattle suffering from FOOT-AND-MOUTH DISEASE was a way of curing them'.[38]

In Ireland ash sticks were used as weapons:

> The Joyces are tinkers . . . they are wary and row among themselves. They do have some fierce fights in which the women join in. When they have each other's heads well cut with ash plants they settle down and are as friendly as ever.[39]
>
> Stories relating to Ireland's past tell of fair-day brawls where ash plants were used and blood flowed freely.[40]

In Desborough, Northamptonshire, in the 1920s, 'children made Rowell-Fair WALKING STICKS. These were made from straight unbranched young ash stems. Bark was removed all around the stem for about 2 inches, and this was repeated all down the stem at 2 inch intervals, giving quite an attractive effect'.[41]

Ash WOOD, which burns well even when damp, was a valued fuel. In east Devon, around Branscombe, ashes were pollarded 'primarily for the wood from the poles which provided high quality fuel for bread ovens of the local bakery. The old trees stand in old pastures and it seems likely that the cut branches were left for the local livestock to strip of leaves and bark as part of the seasoning of the wood prior to removal and burning.'[42]

According to Boulger:

> Few trees become useful so soon, it being fit for walking-sticks at four years' growth, for spade-handles at nine, and when three inches in diameter it is as valuable as the timber of the largest tree. In the Potteries it is largely used for crate-making, and in Kent for HOP-poles. Both the spokes and the felloes of wheels, carriage-poles and oars are made from it, and from its flexibility it is, in fact. 'the husbandman's tree' for every kind of agricultural implement.[43]

In many parts of Europe ash trees are pollarded and their leafy branches used as cattle fodder, but this practice, perhaps once widespread in the British Isles, seems to have died comparatively recently, as it is still possible to see trees which have obviously been pollarded in the Lake District, and in Worcestershire.[44] However, according to a report in the *Daily Telegraph* of 5 December 2012, each autumn free-ranging New Forest ponies veer off their normal routes to eat falling ash leaves, remembering locations year after year, and eating so many leaves that they ignore all other food.

A number of cultivated forms of ash are grown in parks and gardens. In 'Heterophylla' each leaf has only one leaflet, and 'Pendula', commonly known as weeping ash, has pendulous branches. In Leicestershire:

> Weeping ash is an important indicator of wealth/status when associated with an old farm or manor house. There is one in Dolesfoot Farm garden in Newborough, and in the Manor House at Clifton Camville, and in an old farmhouse on the Ashby Road on the outskirts of Ashby de la Zouch. If you could plant a weeping ash as opposed to the timber variety it was a sign you didn't need to concentrate on growing timber for construction. We do not have a weeping ash.[45]

For weather rhymes concerning oak and ash see OAK.

ASH WEDNESDAY – ASH twigs carried on; YEW burnt to provide ashes on.

ASPEN (*Populus tremula*)

Widespread deciduous tree, sometimes suckering to form thickets.

Local names include:
Ebble in Sussex
MOUNTAIN ASH in Inverness-shire
Old-wives' tongues, 'due to its constantly moving leaves', in Roxburghshire
OOMAN'S TONGUE, 'because motion is caused by the lightest breeze, and so they are always on the move', in Berkshire
Pipple on the Isle of Wight
Quaking ash in Inverness-shire
Quiggen espy, 'a corruption of quaking aspen', in Co. Antrim
Snapsa on the Isle of Wight.

> The Highlanders entertain a superstitious notion that our Saviour's CROSS was made of this tree, and for that reason suppose that the leaves of it can never rest.[1]

> The people of Uist [Hebrides] say '*gu bheill an crithionn crion air a chroiseadh tri turais*' – the hateful aspen is banned three times. The aspen is banned for the first time because it haughtily held up its head while all the other trees of the forest bowed their heads lowly down as the King of all created things was being led to Calvary.

And the aspen is banned the second time because it was chosen by the enemies of Christ for the cross upon which to crucify the Saviour of mankind. And the aspen is banned the third time – here the reciter's memory failed him. Hence the ever-tremulous, ever-quivering, ever-quaking notion of the guilty hateful aspen even in the stillest air.

Clods and stones and other missiles, as well as curses, are hurled at the aspen by the people. The reciter, a man of much natural intelligence, said that he always took off his bonnet and cursed the hateful aspen tree in all sincerity whenever he saw it. No crofter in Uist would use aspen about his plough or about his harrows, or about his farming implements of any kind. Nor would a fisherman use aspen about his boat or about his creels or about any fishing-gear whatsoever.[2]

ASTHMA – treated using BEETROOT, EYEBRIGHT, HEATHER, MULLEIN and NETTLE.

ASTROLOGICAL BOTANY

Popular books on plant folklore often contain statements which associate plants with different planets.

> Astrologically under the sign of Saturn, black nightshade [*Solanum nigrum*] is the birthday flower for 23 January, and in the LANGUAGE OF FLOWERS means 'Your thoughts are dark'.[1]

Such statements derive from systems expounded by seventeenth-century philosophers and herbalists, who believed that plants and illnesses were governed by a constellation or planet. A disease caused by one planet could be cured by the use of a herb belonging to an opposing planet. Alternatively, illnesses could be cured by sympathy – by the use of herbs belonging to the planet which was responsible for the disease.

> Every Planet cures his own diseases, as the Sun and Moon by their Herbs cure the Eys, Saturn the Spleen, Jupiter the liver, Mars the Gall and diseases of the Choller, and Venus disease in the Instruments of generation.

Conversely, the blessed thistle (*Cnicus benedictus*), a herb of Mars, cured the French pox (syphilis) 'by antipathy to Venus who governs it', while sanicle (*Sanicula europaea*), a herb of Venus, was recommended as a cure for wounds and 'other mischiefs Mars inflicteth upon the body of Man'.

The best remembered of these botanists is Nicholas Culpeper (1616–54). After having set himself up in practice as an astrologer physician in Spitalfields, Culpeper published his *Physicall Directory* in 1649. This work, he proclaimed, superseded all previous publications, which, in his opinion, were as 'full of non-sense and contradictions as an Egg is full of Meat'. The *Directory*, a pleasant little octavo volume, was followed in 1652 by his *English Physician*, a work which proved so popular that forty-five editions of it were produced before the end of the eighteenth century. A Welsh edition, translated by D.T. Jones of Llanllyfni, appeared in 1816/17,[2] and the book has continued to be in print in various forms for over 300 years. Although Culpeper's work is now invariably known as his 'Herbal', the word herbal did not appear in its title until over a century after its original publication, when Ebenezer Sibley published an illustrated edition entitled *Culpeper's English Physician, and Complete Herbal* in 1789.

Today astrological botany is derided, but those who do so fail to realise that the idea that plants are 'under the dominion' of planets was not isolated from other beliefs prevalent during the period in which it evolved. In the seventeenth century there was a widespread belief that astrology, avidly studied by many leading scholars, held the key to the understanding of the Universe.[3]

ATHEIST'S TOMBS – formerly marked by a FIG tree in Watford, and SYCAMORES at Aldenham and, still marked by ash and SYCAMORE at Tewin.

AUBERGINE (*Solanum melongena*)

Herb, native to tropical Asia, long cultivated in India, introduced to Spain by Arabs in the eighth century; now widely cultivated in tropical and warm temperate regions for its edible fruit.

Late in March 1990 it was reported that the word Allah, written in Arabic, had been found inside aubergine fruits in Muslim households in the English Midlands.

> The curious phenomenon of God's name appearing in Arabic inside aubergines has spread from Nottingham to Leicester, where three cases have been reported by devout Muslims in the past

week. As many as 5,000 pilgrims from all over the Midlands are reported to have visited the remarkable vegetables.

Tasleem Moulvi, of Kingnewton Street, Leicester, told the *Independent* yesterday that her mother found two significant aubergines on Friday night when she sliced them open after visiting another one, exposed in Bakewell Street.

One, sliced twice, shows the Arabic characters for Allah, repeated three times; the other, she said, appeared to contain a verse from the Koran, though this has not yet been deciphered.[1]

Thousands of Muslims are travelling to a terrace house in the backstreets of Leicester to see what is claimed to be a miraculous aubergine.

Farida Kassam asks visitors to take off their shoes as a mark of respect for the sliced vegetable, exhibited in a bowl of white vinegar in her front room. Beside it is a plate bearing the Arabic inscription Yah-Allah, meaning Allah is everywhere. Mrs Kassam, aged 30, proudly points out that the unusual seed pattern inside the aubergine appears to match the Arabic writing. The has fulfilled the faithful and confounded the curious, who have flocked to inspect the evidence. A magnifier has been thoughtfully provided . . . Mrs Kassam said: 'It is a miracle. This has happened to an ordinary family, that is why I am very proud of it. Allah never forgets anybody. We will preserve the aubergine as long as we can and then bury it in holy ground.'[2]

Elsewhere, on 14 April 1990, the *Nairobi Nation* reported:

A family in Nairobi got an Easter and Ramadan surprise when they cut open an aubergine and found the word Allah in Arabic. A week earlier, a member of the family studying in Britain had had a similar experience.[3]

Similar stories reappear from time to time. Thus on 4 June 1996 *The Times* carried a report that Muslims on a day trip to the Scottish Highlands had discovered a stone with the word Allah spelt out in white and brown markings. This was said to be the first time Allah's name had been discovered in stone in the British Isles, 'but it has been discovered several times in seed patterns within aubergines. The most recent case was in March when Salim and Ruksana Patel of Bolton, Greater Manchester, found the name inside an aubergine bought for 25p.'

Similarly, in 1999 a TOMATO displaying sacred Arabic scripts was discovered in Bradford.[4]

AUTUMN GENTIAN (*Gentianella amarella*)

Annual or biennial herb with dull purple flowers, on well-drained basic soils and dunes.

Autumn gentian was known on the Shetland Islands as dead-man's mittens, the 'half-open buds like livid finger-nails protruding through the green'.[1]

B

BABIES – found in CABBAGE and PARSLEY patches and under GOOSEBERRY bushes.

BACKS – 'achy' treated using YARROW.

BADGERS (*Meles Meles*)– CAPER SPURGE deters?

BALDNESS – treated using ONION.

BALM (*Melissa officinalis*), also known as lemon balm

Perennial, lemon-scented herb, native to the Mediterranean region and south-west Asia and cultivated as a culinary herb since late in the tenth century, becoming increasingly naturalised in southern England.

A few variations on the standard name have been recorded: bame and baume in Somerset, bawne in Cheshire, Cumberland and Lincolnshire, and bee-balm in west Cornwall, where it 'used to be planted by bee hives'.

> A brew of lemon balm could keep INSECTS away. My mother used it as a dog deterrent by bathing her bitch with it whilst it was in season.[1]
>
> In the villages around here lemon balm is taken as an infusion as a SEDATIVE, especially useful to reduce the itching in the later stages of CHICKENPOX.
>
> I use it regularly as a 'green tea' in the middle of a sleepless night. It soon induces drowsiness.[2]
>
> [Guernsey] Grown in gardens on account of its medicinal virtues. A tea made by pouring boiling water on the leaves is esteemed an excellent restorative.[3]

> Grown in [Forest of Dean] gardens, dried and made into tea – for STOMACH problems/COLIC.[4]

BANANA (*Musa* cv.)

Native to tropical Asia, long and widely cultivated in tropical regions for its edible fruit.

> Very popularly a banana is asked to decide whether a boy is being faithful. When the question has been asked, the lower tip of the fruit is cut off with a sharp knife, and the answer is found in the centre of the flesh, either Y meaning 'Yes' or a dark blob meaning 'No'.[1]

This means could also be used to predict the outcome of many other activities, or solve any problem, which required a straightforward yes or no answer.[2]

> Westernised Chinese people are described as bananas: 'yellow on the outside, white inside'.[3] Cf. COCONUT.

Banana skins are used to treat SKIN conditions.

> Banana skins heal ECZEMA miraculously. The inside of the banana skin is rubbed on the eczema. Although it does sting/itch the results are startling and achieved without scarring.[4]

> Put a piece of banana skin, pulp side, on the VERRUCA and put a plaster on it. Leave until it turns black (a couple of days). Repeat a few times until the verruca drops out.[5]

> Cyclists and marathon runners put bananas in their pants to stop them getting sore bums.[6]

Bananas are eaten to prevent CRAMP.

> My late husband cured his awful cramp in three weeks. Eat a good-sized banana a day.[7]

> For years I suffered from cramp . . . and often had to get out of bed and walk around to ease the terrible pain. The same thing used to happen in the swimming pool. Another swimmer approached me as I was trying to ease the cramp in my feet and said I should try eating three or four bananas a week. After a month the cramp

stops and now I eat the bananas has never returned.[8]

After years of leg and stomach cramp, I was told to eat two bananas a day to relax the muscles. They cured me.[9]

BARBERRY (*Berberis vulgaris*)

Spiny, yellow-flowered shrub, in hedgerows, coppices and waste ground, mainly in southern Britain, scarcely recorded in Ireland; as the alternate host of wheat-rust (*Puccinia graminis*) barberry was eradicated from many areas during the nineteenth century.

Local names include pipperidge in East Anglia, woodsore or WOOD-SOUR in Oxfordshire, and yellow-tree in Wales.

In many parts of Cornwall the barberry is called 'jaundice tree'. On the DOCTRINE OF SIGNATURES, an infusion of the yellow under-bark of this shrub is supposed to be a perfect cure for JAUNDICE, and I have known it to be frequently planted in gardens and shrubberies for this purpose.[1]

Mr T.C. (55+) of Meelick, Co. Clare, told me (2 July 1998) that an old man living up the road from him had a cure for jaundice. He would take a small sliver of the inner bark of barberry (*Berberis vulgaris*) and boil it in milk. This was given to the sufferer to drink and 'as soon as it hit the stomach, just as it was all drank' it caused a fairly violent emesis, but 'the jaundice would be gone'.[2]

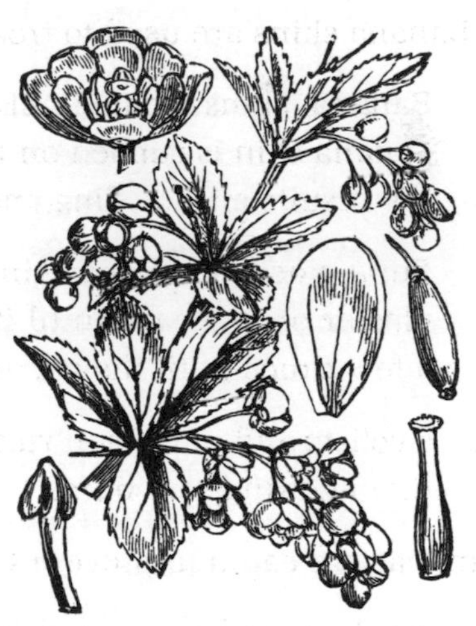

BARLEY (*Hordeum distichon*)

Annual grass, unknown as a wild plant; cultivated as a cereal for thousands of years.

Barley should be sown when BLACKTHORN or WHITLOW GRASS is in flower, or when ELM trees come into leaf.

> 'When the blackthorn is white, sow barley both day and night.' This came either from south Wiltshire or from mid-Hants. I learnt it working on a farm in 1942–44.[1]

> Alternatively, the 'temperature of the land for sowing barley in spring is right if the soil is warm on one's bare bottom'.[2]

Barley had the unusual distinction of having a traditional song devoted to it, or perhaps more precisely, to WHISKY distilled from it. An elaborate version of 'John Barleycorn' was printed early in the seventeenth century, and the song has remained in the oral tradition ever since.[3] The song starts with the sowing of barley seed and tells of its subsequent growth:

> He laid there till midsummer time of the year
> Till the weather was pleasant and warm
> And there Sir John how he grew a beard
> And he soon became a man.

Then came harvesting and the conversion of the grain into whisky:

> You can put Sir John in a nut brown jug
> And he'll make the merriest man.
> He'll make a maid dance around this room,
> Stark naked as she was born,
> He'll make a parson pawn his books
> With a little John Barleycorn.[4]

Barley water and wine had some medicinal uses:

> Barley wine was good for KIDNEY trouble.[5]

> I remember Nan used to make barley water by boiling pearl barley in a pan for several hours, topping up the water when necessary. This was flavoured with lemon and sugar and used for CYSTITIS. It was taken both hot and cold . . . Barley thickened broth was given to invalids or anyone with a COLD or flu.[6]

BARREN GROUND

A number of local legends tell how plants will not grow on patches of ground where evil deeds have taken place. A letter dated 17 July 1778 records the legend of 'the Brothers' Steps' in London:

They are situated in a field about half a mile from Montague House, in a North direction; and the prevailing tradition concerning them is, that two brothers quarrelled about a worthless woman, and . . . decided it by duel. The print of their feet is near three inches in depth, and remains totally barren . . . The number I did not reckon, but suppose they may be about ninety. A bank on which the first fell, who was mortally wounded and died on the spot, retains the form of his agonising posture by the curse of barrenness, while grass grows around it. A friend of mine showed me these steps in the year 1769, when he could trace them back by old people to the year 1686; but it was generally supposed to have happened in the early part of the reign of Charles II. There are people now living who well remember their being ploughed up, and barley sown, to deface them; but all was labour in vain; for the prints returned in a short time to their original form.[1]

In 1985:

[The patch on which the wounded brother fell] and the bank on which the woman sat are now gone, but it is said that the Brothers' Steps themselves can still be seen in the south-west corner of Tavistock Square Gardens, near the tree planted in 1953 in honour of Mahatma Gandhi. I am happy to report that there are bare patches there which could be taken for footprints.[2]

On Dragon's Hill, near the famous white horse at Uffington, Berkshire:

St George is said to have slain the Dragon . . . where the Dragon's blood was spilt, grass has never grown.[3]

Although shrubby vegetation thrives, trees will not grow in Ashdown Forest, East Sussex:

All attempts to restock it with trees have failed; this dates from the time when the first cannons were cast, because the trees were used for charcoal for iron smelting and whereas they were quite willing to help make tools or other objects of a peaceful nature, they refused to let themselves be used for making weapons. My mother was born in the house of Ralph Hogge, the founder of the first cannon in 1492.

It seems strange that trees have never grown extensively on the uplands again since the founding ceased in the seventeenth

> century, although they have on the lower reaches of the Weald. It also seems strange that there are quite extensive fires on (we never say 'in') the Forest around Easter each year. These fires are often deliberately started by gypsies who prefer the Forest to remain open land.[4]

In Ulster bare patches persist on the grave of a rioter shot in 1845, the bare earth in this instance presumably bearing testimony to innocence rather than a crime.

> A young man, John Boyle, was shot dead in a riot in Armagh on 12 July 1845. His grave is much visited and has always been an object of wonder. There are a number of bare round depressions on the grave, about four inches in diameter, on which grass never grows. Traditionally people say that these are the signs of the wounds he received (although he received one wound). The grave has been re-dug a number of times, as recently as last year when the whole graveyard was cleared and re-sown. Still these patches appeared.[5]

BARRENNESS in women – treated using DOCK seeds.

BASKETS – made from DOCK.

BATH ASPARAGUS (*Ornithogalum pyrenaicum*), also known as spiked star of Bethlehem.

Bulbous perennial, white-flowered herb, restricted to woodland and grassy banks in south central England.

Formerly sold in Gloucestershire and Somerset as Bath asparagus, French asparagus, French grass, French sparrow-grass, Prussian asparagus, sperage, wild asparagus or wild grass.

According to the *Gardeners' Chronicle* of 26 June 1873:

> Bath is famous for all sorts of things . . . Bath Asparagus, however, is not so well known. Passing through the streets at dusk one evening lately we saw what we took, at a glance, to be bunches of young Wheat-ears tied up. The morning's reflection convinced us that this could not be, and, moreover, revealed that they were the young flower-scapes of *Ornithogalum pyrenaicum*, and on visiting the market we saw a quantity of them, of which we purchased a sample, under the name 'Wild Asparagus' . . . all who partook

> declared it to be the best substitute for Asparagus yet tried . . . The abundance of the supply in Bath market was such that we can hardly imagine that it was all wild, some must surely have come from the cottage gardens.

Almost a century later:

> In the early 1970s Bath asparagus was sold in season in a greengrocer's at The Balustrade, London Road, Bath. The couple that ran it were at least in their 60s and lived in a Wiltshire village (I think Atworth) near a wood where Bath asparagus grew as profusely as bluebells. I actually bought a bunch once to try it out. They eventually sold the shop and I have not seen it elsewhere since. It is conceivable it sometimes makes an appearance in one of the west Wiltshire street markets, but I doubt it because stalls are usually held by commercial growers.[1]

BATTLE OF FLOWERS

In Jersey an annual Battle of Flowers is held on the second Thursday in August. First held in 1902 to celebrate the coronation of King Edward VII, in its current form the 'battle' consists of a procession of about thirty elaborately decorated floats. During the Battle's early years floats were decorated with flowers, mainly HYDRANGEAS, but in more recent years there has been an increasing tendency for paper to be used. Thus in 1987:

> Mr Graeme Rabet, Association Chairman, said that there is now almost an equal number of paper and fully floral entries in the parade. 'The standard of the paper floats is extremely high, and we regard the part they play as a vital one,' he said.

In earlier times it was customary for participants to throw flowers at each other, hence the name 'battle'. On 12 August 1909, immediately after the judging of floats had finished:

> On the 'commence fire' being sounded by buglers, the real fun commenced. We speak of the Battle of Flowers the whole year round, but the actual battle is only a ten-minute 'turn'. Still, people come from the north, the south, the east and the west to see it. They lose their individuality too!
>
> Staid old men become again as romping children, the stern

> official throws off his robe of dignity and austerity, and even the cleric forgets for a moment that he has a distinguishing collar that gives him away. If one threw a bunch of flowers at a fellow on an ordinary day we should most probably be summoned for assault, but with the Battle of Flowers everything is different.
>
> So it was yesterday; the aristocracy pelted the democracy, and vice versa. Alas for the erstwhile decorated cars! They were soon shorn of their grandeur and the roadway was ankle-deep in blooms. Alas for the poor flowers!

Violent hooliganism in 1960 brought such practices to an end, leaving a rather decorous, but extremely popular, parade of floats.[1]

For about forty years from 1971 there was a Battle of Flowers Museum at St Ouen, which, according to a leaflet produced in 1990 displayed:

> Only a fraction of the entries made by Miss F. Bechelet during the last 36 years in the Jersey Battle of Flowers. All her floats have the distinction of being made entirely of wild flowers grown in the west of the island . . . Miss Bechelet has won the highest awards in classes in which she has competed.

BAY (*Laurus nobilis*)

Evergreen shrub or small bushy tree, native to the Mediterranean region, cultivated for its aromatic leaves since at least late in the tenth century, and in recent years becoming naturalised, and a potential pest, in southern Britain.

> In the New Forest the bay tree was planted in gardens not only for the flavour of its leaves, but for the protection it afforded against all things evil. It had the power to fend off the DEVIL or WITCHES, as well as safeguard one from THUNDER, LIGHTNING and forest fires.[1]

> When my brother and sister-in-law moved to a house in south Devon . . . 30 years ago, they were told that the bay trees which grew at the entrance would protect the house, but my sister-in-law said they didn't, because she always said the house had a ghost.[2]

Cf. MYRTLE.

> Where the bay tree flourishes, there flourishes MONEY.[3]

[Hartland, Devon] On VALENTINE'S EVE take five bay leaves, and pin them to your pillow – one at each corner and one in the middle. Then lying down, say the following seven times, and count seven, seven times over at each interval:

Sweet guardian angels, let me have
What I most earnestly do crave –
A Valentine endued with love,
Who will both true and constant prove.

If you carefully observe this ceremonial, your future husband will appear to you in a dream.[4]

I went to school in Bournemouth in the 1950s–60s. A week before HALLOWE'EN we'd pick seven bay leaves each. On five we'd write boys' names – friends, fantasies, whatever – on one we'd write 'old maid' and in the remaining leaf there would be a question mark. The leaves would be put in an envelope, and kept under our pillows. Each morning we'd take out one leaf, without looking at it, and destroy it. On the last morning of course there would be only one leaf left – if it had a name on it, he would become your husband, the others are self-explanatory.[5]

BEDDING for farm animals – BRACKEN and EELGRASS used as.

BED-WETTING – DANDELIONS cause.

BEECH (*Fagus sylvatica*)

Large, deciduous tree, widespread on well-drained soils, also planted, often in garden hedges, where trimmed plants retain their dead golden-brown leaves throughout the winter.

Beech seems to have attracted very little folklore.

The disruptions on the bark of the smooth beech, where old growth has ceased, is thought in areas of Dorset to represent the 'evil eye', and the sinister beech grove at dusk is an unlucky place in which to be found alone.[1]

Children collected and ate young beech leaves, known as biscuit leaves in Somerset,[2] and beech nuts, commonly known as mast, and in Gloucestershire as beech-meats.[3]

We ate SORREL, beech leaves, briar tops.[4]

> When I was a child in Cornwall I used to eat young hawthorn leaves, which we called BREAD-AND-BUTTER, and young beech leaves.[5]

> [Wimborne St Giles, Dorset, 1930s] We ate beech nuts in small quantities, but as there was a grove of sweet CHESTNUTS in a field we preferred these.[6]

The nuts were also gathered to make necklaces:

> As children in Notts in the 1920s we gathered beech nuts and threaded them to make necklaces.[7]

> [Somerset, 1940s] Beechnuts – easy to thread using a needle and cotton. We made numerous necklaces until our families were sick of them. We then gave them to the pig.[8]

Copper beech (*Fagus sylvatica* 'Purpurea') is a cultivated form with purplish leaves.

> There used to be a Hanging Tree in Barnstaple, on the Castle Green. It was a massive copper beech. This was felled in recent years, but was a magnificent tree which fell foul of the Parks Dept.[9]

> A few years back the late Ivan Cresswell, who had a fine specimen [of copper beech] in his Rectory garden, told me that the reason they were so profuse in parsonage gardens is that they were offered free to incumbents who wished to plant them to commemorate the outcome of the Battle of Waterloo. He could not recall his source, but it interested me because for a period this house was a rectory and has a similar specimen, as does the other rectory in the village. I have seen others in like situations.[10]

BEER – BOGBEAN used in making; WORMWOOD improves.

BEES – ELDER used by bee-keepers when taking swarms.

BEETROOT (*Beta vulgaris* ssp. *vulgaris* cv.)

Cultivated form of wild beet, grown for its swollen red roots.

Although eaten elsewhere hot as a vegetable or in soups, in the British Isles beetroot is usually eaten cooked and cold as a salad vegetable; it also had a number of medicinal uses.

> Beetroot wine was very popular against ASTHMA, CHILBLAINS, EARACHE, and, some say, SNAKE BITES. I knew a gypsy named Penfold who told me his people kept beetroot by them for snake bites, and adders of course are common hereabouts. He said gypsies planted theirs at night. In fact, though travellers, I was surprised to learn they would have a variety of 'stewpot' vegetables growing just inside wood edges in their 'territory' . . . as well as helping themselves en route.[1]

BENT (*Agrostis capillaris*, syn. *A. tenuis*)

Perennial grass, widespread and abundant in grassland and waste places.

> As a child, evacuated to Elgin, Scotland, during World War II, I used to chew a grass which the local children called BREAD-AND-CHEESE. I have since identified this as the common bent, *Agrostis tenuis*. In fact a local mound was named after it, 'Bread-and-Cheese Hill'. Chewing it now, it doesn't taste a bit like bread and cheese.[1]

BERMUDA BUTTERCUP (*Oxalis pes-caprae*)

Bulbous, perennial, yellow-flowered herb, naturalised on arable and waste land, particularly bulb-fields, in the Isles of Scilly and the Channel Islands; native to South Africa (widely naturalised in the Mediterranean region), introduced in 1757, and first recorded in the wild in 1901.

In common with other species of *Oxalis* (such as WOOD SORREL), the stems of Bermuda buttercup are nibbled by children:

> The local children call them SOUR SAPS – chewing the sharp tasting stems of the flowers.[1]

> I was born in 1923 and lived in Scilly until I went away to school in 1935 . . . we used to eat sour-sabs – probably oxalis – it had yellow flowers – we ate the leaves.[2]

> [Isles of Scilly] Local children chew the flower stems as 'sorrel' for their lemony taste and leaves may be used in salads.[3]

BETONY (*Betonica officinalis*)

Purple-flowered perennial herb, widespread and common in hedge-banks, wood margins and unimproved grassland throughout most of England and Wales, scarce in Scotland and Ireland.

Local names include:

Bishop's wort in Gloucestershire and Somerset; Old English *bisceopwyrt*
Bitney in Devon
DEVIL'S PLAYTHING in Shropshire; 'children refuse to pick it'
Harry-nettle in Dorset; 'probably a corruption of hairy nettle, from the hairy or downy leaves'
Kill butcher in Somerset
Ragged nettle in Shropshire
Red-rod in Berkshire
Vitony in Cornwall
WILD HOP in Worcestershire
Wood-betony in Cheshire
WOODBINN in Cornwall.

Despite the *officinalis*, meaning 'used in medicine', in its Latin name, there is little evidence to suggest that betony was much used in British and Irish folk medicine. Allen and Hatfield list betony as being used to treat BURNS in Sussex, HEADACHE in Somerset, INDIGESTION in Cumbria, WOUNDS in Sussex and Wiltshire, and to purify the BLOOD in Shropshire.[1]

> [Derbyshire gypsies] Betony . . . infusion of leaves remedies STOMACH problems; ointment made from juice of fresh leaves and unsalted lard removes poison from STINGS and bites.[2]

BILBERRY (*Vaccinium myrtillus*), also known as whortleberry

Low shrub, common on acid moorland, heaths and in birch woodland.

Local names, most of which probably refer to the edible fruit, include:

Arts in Somerset and Wiltshire; 'the Semley end of Donhead Cliff grows arts in abundance, and is called Art Hill'
BLACKBERRY in Yorkshire
Blackhearts in the New Forest, Hampshire
Blaeberry, widespread; 'Old Scandinavian *blaa*, dark blue; in Norwegian *blaaber*
Blayberry, bleeaberry, both widespread
Bleeberry in Northumberland
Blueberry in Cumberland and Yorkshire
Brawlins in northern Scotland; 'sometimes applied; Gael. *braoilag* denotes a whortle [i.e. bil-]berry'

Brylocks in Scotland; 'Gaelic *braoilag*'
Coraseena in Co. Donegal
Cowberry in Somerset; 'cows are very fond of the fruit and bush, but it does not agree with them if eaten in any quantity'
CROWBERRY in Morayshire
Eartleberry in Devon
Erts in Devon
Fraughan in Ireland
Frocken in Devon; 'a slight variant of the Gaelic name'
Frughans in Co. Antrim and Co. Down; 'the old Irish name . . . poor women gather them in autumn and cry them about the streets of Dublin by the name'
Fruogs in Ireland
Hart-berry in Devon and Somerset
Hurtleberry, and hurts, in Devon
MOSS-BERRY in Co. Donegal
Urts in Cornwall
Whim-berry in Lancashire
Whin-berry in the South Wales valleys in the 1950s; 'in the summer whole families swarmed the hillsides picking whinberries which made the most delicious tarts and JAMS'
Whirtleberry in Kent
Whorts in southern England
Wimberry, 'i.e. wine-berry', widespread
Wim-wams in Cheshire and northern England
Wincopipe in northern England
Wurral in Devon.

Although its fruits are small, and the shrub's low stature makes their gathering a hard-on-the-back activity, bilberries were gathered on a commercial scale. On Exmoor early in the twentieth century, as soon as the berries were ripe:

> The headteacher would consult the rector, and school would be broken up then and there for five or six weeks.
>
> Children, sometimes accompanied by their mothers, would set off for the moor early in the morning with a basket and sandwiches to last all day . . . A large medicine bottle of cold tea was also carried, and this was secured in a handkerchief by a string . . .
>
> After picking had started it was a point of honour not to have our dinner until a certain quantity of whorts had been picked or as we said, 'we had to earn our dinner first'.

> A mark would sometimes be made with a squashed berry in the side of the basket, or sometimes a turned-up withy in the basket would be used to indicate when it was considered enough had been gathered . . .
>
> When several of the same family went, the younger members would pick into a cup or pint mug, and (when full) empty it into mother's or elder brother's or sister's basket . . .
>
> The first place where the worts ripened was on Grabbist. Not a favourite place as the berries grew between gorse, and fingers and legs were very sore at the end of the day, so we were always glad when they were ripe on Dunkery.
>
> Annicombe was also one of the first, and we always enjoyed getting there, where the whorts were much bigger and there was also the attraction of the stream in which we could paddle in the dinner break.
>
> Dunkery Hill Gate, Bincombe, and the slopes under Dunkery Beacon were all visited in turn.
>
> Sometimes when these places failed we used to walk through Cloutsham to Stoke Ridge, which took nearly two hours.
>
> The whorts were sold to Mr Tom Webber and his sister, who had a horse and open cart in which small barrels were carried . . . The price was 4d a quart at the beginning of the season, dropping to 3d or 2½d as they became more plentiful . . . Money earned was used to buy clothes for the winter . . . We were allowed to keep what money we earned on the last day of whorting, so we were usually very energetic on that day.[1]

Locally made bilberry jam is available from shops in the Exmoor area.[2]

In Co. Tipperary:

> In the woods that lie on the slopes of the Galtee mountain there are always a good supply of whorts.
>
> Every summer people go out into the wood picking these whorts, this has been done for about 200 years. They start picking on 29 June and continue to the end of August. Pickers, consisting sometimes of the whole family, leave early in the morning for the wood, and after picking for some hours one of their number will light a fire and prepare some tea, which they all sit down and enjoy about mid-day.
>
> Sometimes the fruit is plentiful and pickers have no difficulty in

filling their cans, but very often they are not so fortunate.

They often sing songs to lighten their work, which is not always pleasant as in warm days flies are a torment and the pickers have their heads covered with a handkerchief soaked in Jeyes Fluid to hunt them away.

Some people pick from three to seven gallons in the day and get from nine pence to one shilling per gallon.

The whorts are sent across to England where they are used to make DYE or mix with the cheaper kinds of jam.

Buyers meet the pickers every evening to buy their day's picking.[3]

In Ireland the gathering of bilberries was associated with the ancient festival of Lughnasa, which celebrated the beginning of the harvest, early in August each year:

> One of the customs connected with the festival is the picking of bilberries. There are many places in Ireland where all other features of the survival have disappeared but the festive outings to pick bilberries are still continued.[4]

While walking near Llanthony on the English/Welsh border in April, John Hillaby came across a shepherd who:

> had been up to the tops looking at the leaf . . . [which] is bilberry and sheep, he told me, love blueberry. As soon as it begins to sprout in the spring, the custom is to drive the flocks up into the hills from the valley folding, for the shepherds know the animals will be comfortable there and less likely to stray down again.[5]

In folk medicine:

> The berries have an astringent quality. In Arran and the Western Isles they are given in diarrhoeas and DYSENTRIES with good effect.[6]

> My late parents came from Poland after the Second World War and my mother occasionally used certain plants as herbal remedies . . . Fruits of the bilberry were preserved in sugar in glass jars and a few teaspoons of the concentrated juice diluted with water and warmed as a most effective cure for DIARRHOEA.[7]

> My dad, who is Polish, swore by bilberries in syrup for STOMACH upsets.[8]

BINDWEED (*Calystegia* spp.), also known as CONVOLVULUS; see also FIELD BINDWEED

Botanists failed to distinguish between the two vigorously climbing species of bindweed until 1948, and it is unlikely that non-botanists are able to do so.

Great bindweed (*Calystegia silvatica*) – perennial herb, with conspicuous white flowers, native to southern Europe, cultivated as an ornamental since 1815, being first found in the wild about fifty years later, now common in hedges, cultivated areas and waste ground throughout lowland areas.

Hedge bindweed (*Calystegia sepium*) – native perennial herb with white (or pink) flowers, common in hedges, scrub, and waste places in lowland areas.

Local names, some of which are undoubtedly shared with field bindweed, include the widespread BEAR-BIND, BETHWIND, HEDGE-BELLS, LADY'S SMOCK, SHIMMIES and WITHYWIND, and:

BEDWIND in Gloucestershire
BELL-BIND in Essex and Suffolk; 'i.e. the bindweed with bell flowers'
BELL-BINE in Suffolk
BELLWIND in Buckinghamshire and Surrey; 'cf. German *Windglocke*'
BELLWINE in Buckinghamshire
BETHWINE in south Buckinghamshire
BIND in East Anglia and Lincolnshire
BINE-LILY in Dorset
Bryony in Devon
Carn-lilies in Yorkshire; presumably a corruption of corn-lilies, see corn-lily below
CONQUERORS in Berwickshire
CORN-LILY in Yorkshire; if this name can be interpreted as a lily growing in a cornfield it seems probable that field bindweed is the species intended

Creeper in Nottinghamshire
CREEPING JENNY in Somerset
Cups in Somerset
Daddy white-shirt in Somerset
DEVIL'S GARTER in Co. Wexford

DEVIL'S GUTS in Norfolk
DEVIL'S NIGHTCAP in Dorset
Drill-drolls in Cornwall
FAIRY CAP in Ireland
Fairy paint-brushes in Cornwall
FAIRY TRUMPETS in Dorset
FLAGS in Hull, Yorkshire
Grandmother's nightcap in Devon, Somerset and Sussex
Granny-pop-out-of-bed in Dorset, and the Isle of Wight, where 'used until recently, referred to the manner in which the seeds could be squeezed out of the capsule';[1] probably a misunderstanding, see the children's pastime below
Granny's bonnets in Cornwall
Granny's night-bonnet in Somerset
GRANNY'S NIGHTCAP in Wiltshire
GROUND IVY in Devon
Harvest-lily in Surrey
Hedge-lily in Hampshire, Isle of Wight and Somerset
HONEYSUCKLE in Devon
Jack-run-in-the-hedge in Somerset
LADY'S NIGHTCAP in Wiltshire
Lady's shimmy, i.e. chemise, in Somerset
LADY'S UMBRELLAS in Somerset
Larger sunshade in Somerset
LILY in Wiltshire
Lily-bind in Yorkshire
Lily-flower in Hampshire
London bells in Devon
MILKMAID in Surrey and Sussex
MORNING-GLORY in Cheshire
Mother's nightcap in Devon
NIGHTCAPS in Lincolnshire, Somerset and Wiltshire
Nightshirts, and nit-clickers, in Somerset
Old-lady's smocks in Wiltshire
OLD-MAN'S NIGHTCAP in Somerset, Surrey and Sussex; 'i.e. devil's'
OLD-MAN'S SHIRTS in Dorset
OLD-WOMAN'S BONNET in Devon and Somerset
Our Lady's smock in Somerset
Piss-pots in the Weald of Kent and Sussex; 'cf. German *Pisspott*, *Pisspöttchen*, Dutch *piespotje*'
Poor-man's lilies in London
ROBIN-RUN-IN-THE-HEDGE in Ireland
SHIMMIES-AND-SHIRTS in Devon, Dorset, Somerset and Wiltshire
Shimmy-and-buttons in Devon, Dorset, Somerset and Wiltshire
SHIMMY-SHIRTS in Devon, Dorset, Somerset and Wiltshire
Smocks in Dorset and Somerset
SNAKE-FLOWER in Somerset
SNAKE-WEED in Somerset
TARE in Wiltshire
TRUMPET-FLOWERS in Somerset
Waywind in Oxfordshire
WHITE LILY in Cumberland
White shirts in Somerset
WHITE SMOCK in Devon
White weather-wind in Oxfordshire
Wild ivy in Dublin
WIRE-WEED in Surrey
WITH-WIND in Dorset, Gloucestershire, Hampshire and Wiltshire
WITHYWINE in Somerset in Devon, Dorset and Somerset
WOODBINE in Sussex.

> When you squeeze the green bit at the bottom of a convolvulus flower saying 'Granny pop out of bed' the white petals pop off. This came from my mother and aunt in London.[2]

> The children's game with convolvulus is known here [on the Isles of Scilly], but with the slight variation: 'Granny jump out of the rocking chair.'[3]

> [b. 1918] Great bindweed – we used to pick the flowers and hold the stem (receptacle) firmly close under the flower and then pinch hard saying 'Grandmother, grandmother, jump out of bed.' As you pinch hard the corolla is released as a separate piece and slowly falls to the ground, a bit like an old lady in a white night-gown jumping out of bed to the floor.[4]

> My mother, as a child during World War 2, used to say the rhyme 'Nanny Goat, nanny goat – pop out of bed!' whilst squeezing the base of large white bindweed flowers and making them jump off the stem. She lived in southeast London, I remember doing the same thing to them as a child.[5]

Around Inverness:

> The flower of the trailing plant convolvulus was called bee traps because as children we would watch a honey bee enter the flower when we would imprison it by holding the opening of the flower between finger and thumb, and listen to the insect wriggling and trying to get out and could feel the vibrations of its movements. May I say when it got too fierce we allowed the bee to escape, and stood well back. We never killed one.[6]

See also FIELD BINDWEED.

BIRCH (*Betula pendula*)

Native deciduous tree (Fig. 1), widespread, particularly on acidic soils, also widely planted in parkland.

Local names include:

Begh in Co. Donegal; Irish *beith*
Birk in Morayshire
Birk-tree in Cumberland
Bobbyn in Lothian
Burk, and burk-tree, in northern England
MAY in Scotland

Paper-beech, and paper-birch, in Wiltshire

Ribbon tree, 'because the bark of the young tree can be pulled off in long ribbon-like strips', in Lincolnshire

Silver-leaved tree in Somerset.

> A custom, dating from time immemorial, is that of using the branches and sprigs of the birch tree for decorating churches on WHIT SUNDAY.[1]

> Whitsuntide . . . is not with us [in Shropshire] the holiday time that it is in many counties. The only custom peculiar to the season . . . is that of decking the churches on Whit Sunday with birch branches stuck into holes in the tops of pews. Hordley Church was thus adorned up to the year 1857; at St Mary's, Shrewsbury, the custom was kept up until about the year 1865; at St Chad's it was continued up to 1855, and probably later.[2]

The custom is difficult to explain, but it seems as if birch has been associated with Whitsun since at least late in the sixteenth century, when, in 1583, Philip Stubbes in his *Anatomie of Abuses* reported:

> Against May, Whitsonday, or other time, all the yung men and maides, olde men and wives, run gadding over night to the woods, groves, hils, and mountains, where they spend all the night in pleasant pastimes; and in the morning they return, bringing with them birch and branches of trees to deck their assemblies withall.[3]

At present the custom survives in England only at the Church of St John the Baptist, in Frome, Somerset, where:

> The tradition of using birch branches at Pentecost [Whitsun] is still maintained . . . We are allowed the top of young silver birch trees in the woods of Longleat each year and we fix the resultant five or six foot branches to the pillars which separate the nave from the north and south aisles.
>
> We believe that the significance of the branches is twofold: being young growths they represent the renewal of life; and the stirring of the leaves resulting from the moving air currents in the church represent the sound of the 'rushing mighty wind' as the Holy Spirit descended on the Apostles.
>
> We have however nothing but tradition by word of mouth to support this interpretation nor can we say when the custom

> began, except to say we know that it was already in operation in 1836, since we have amongst our church records a contract by which the verger undertook to obtain HOLLY to decorate the church at EASTER and birch branches at Whitsun.[4]

Similarly, according to an information sheet displayed at St John's at Whitsun 2015, the branches 'symbolise the birth of the church. Also as the breeze runs through the leaves, it reminds us of the rushing wind of the Holy Spirit as it came upon the first apostles'.[5]

Birch sap can be fermented to make WINE.

> I have often drank a beautiful wine in this neighbourhood [East Ilsley, Berkshire], sparkling like champagne, which is made without sugar, water or spirit, being merely the sap of the birch-tree, boiled with honey and fermented with a little yeast. The birch grows abundantly about Bucklebury near here, so that on the advance of spring these trees are tapped, and a hollow tube inserted into the hole, through which the sparkling sap flows drop by drop into a vessel placed below. Fine weather is, of course, best for this operation.[6]

> I was born and brought up on a farm. At home we had a lot of birch trees. Each year my father used to tap the trees – that is cut a slot in the bark about three or four inches long, and insert a piece of ELDER wood, which was split and had its soft centre removed, at the lower end of the cut bark. The sap ran out and was collected in enamel buckets. When enough sap was collected it was made into wine, which was said to be of great help to RHEUMATISM sufferers.[7]

In recent decades there has been a revival of interest in birch wine (of which 42,700 tons were produced in Russia in 1986), with 7,000 bottles being produced at Moniack Castle, near Inverness, in 1991.[8]

It is said that while staying at Balmoral Queen VICTORIA drank large quantities of birch sap in an attempt to prevent the thinning of her HAIR.[9]

BIRD-CHERRY (*Prunus padus*)

Deciduous, white-flowered shrub or small tree, naturally occurring in Wales, East Anglia, central and northern England, Scotland and Northern Ireland; also widely planted as an ornamental.

Local names include:

Black dogwood in Surrey
Black merry in Hampshire, 'to distinguish it from a red variety which is known in Hampshire'
Eck-berry in Cumberland
Egg-berry in northern England
Hackberry in Cumberland, Westmorland (both now Cumbria) and Scotland
Hackers, and hacks, given to the fruit in northern England
Hagberry in Cumberland, Westmorland and Scotland
Heckberry in Cumberland
Heg-berry in Cumberland; 'children used to say "we call them heg-berries, because they heg oor teeth", i.e. set the teeth on edge'
HOG-BERRY, given to the fruit in Hampshire
Mazard in Co. Limerick.

Some people find the smell of bird-cherry flowers peculiarly offensive, which perhaps explains why it is sometimes considered inauspicious.

> [Northeast Scotland] The wood of hackberry or bird-cherry (*Prunus padus*) is not used as a staff or for any purpose, as it is looked on as the WITCH's tree.[1]

> We have a cottage in the Black Mountains – the Brecon Beacons National Park. We wanted to plant a hedge, so we ordered an equal number of *Prunus avium* and *P. padus* trees. They delivered the *P. avium*, but no *P. padus*. We were told later that the Welsh don't like *P. padus* – it's the tree the DEVIL hung his mother on.[2]

However, in Wester Ross bird-cherry shared some of the protective qualities of ROWAN:

> It was venerated for its ability to dispel EVIL. The wood was sometimes used to make the lunnaid (pin of the cow's fetter), and bird-cherry walking- sticks were believed to prevent people from getting lost in the mist.[3]

BIRD'S-FOOT TREFOIL (*Lotus corniculatus*)

Low-growing perennial herb (Fig. 2) with attractive yellow to orange flowers, widespread and common.

Although it has attracted very little folklore, bird's-foot trefoil has accumulated an extraordinarily large number of local names including the widespread EGGS-AND-BACON, FINGERS-AND-THUMBS, fingers-and-toes, LADY'S FINGERS, LADY'S SLIPPER, LAMB'S TOES, and tom thumbs, and:

Bellies-and-bums-fingers-and-thumbs in Essex
Bird's claws in Devon
BIRD'S EYE, and bird's foot, in Somerset
Bloom-fell in Scotland
BOOTS-AND-SHOES in Cornwall, Devon, Somerset and Sussex
Boxing gloves in Somerset
BREAD-AND-CHEESE in Somerset
BUNNY RABBITS, and bunny-rabbit's ears in Somerset
BUTTER-AND-EGGS in Gloucestershire and Somerset and Warwickshire
Buttered eggs in Cumberland
Butter-jags in northern England
Butter't eggs in Cumberland
CAMMOCK in Devon
Catcluke and catluke in Scotland; 'from some fanciful resemblance it has to a cat (cat's) or bird's foot. Danish = *Katte-cloe*, a cat's claw or clutch; Swedish *Katt-klor*, cat's claws'
Cat-in-clover in southern Scotland
Cat-poddish, or cat-puddish, in Cumberland
CAT'S CLAWS in Buckinghamshire
Cat's clover in Berwickshire
Catten-clover in southern Scotland
CHEESE-CAKE in Somerset, Warwickshire and Yorkshire
Cheese-cake grass in northern England
CLAVER in Co. Antrim
Cockies-and-hennies in Morayshire; 'from the varicoloured flowers'
COCKLES in Somerset
Craa's foot, and craa-taes, in Northumberland
CRAW-FOOT in Co. Donegal
Craw's toes in Scotland, 'from the spiky seed-pods'
CRAW-TAES [crow-toes] in Berwickshire and Edinburgh
Croobeen-cut in Co. Donegal
CROW-FEET in Gloucestershire, Somerset and Suffolk
CROW-FOOT in Gloucestershire
Crow's foot in Suffolk
CROW-TOES in Bristol and Scotland
CUCKOO'S BREAD-AND-CHEESE in Middlesex
CUCKOO'S STOCKINGS in Shropshire, Somerset and Sussex
Dead-man's finger in Hampshire
Devil's claws, and DEVIL'S FINGERS, in Bristol
Eggs-and-collop in Yorkshire
Fairies' footsteps in Co. Cork
Fell-bloom in Scotland
FIVE FINGERS in Cambridgeshire, Essex and Norfolk
God-Almighty's flowers in Devon
God-a'mighty's thumb-and-finger in Hampshire
GOD'S-FINGERS-AND-THUMBS, and golden midnights, in Somerset
Golden slippers in Devon and Somerset
Grandmother's slippers in Hampshire

Grandmother's toenails in Devon
Granny's toenails in Devon and Kent
Ground honeysuckle in Cheshire
HEN-AND-CHICKENS in Yorkshire
HONEYSUCKLE in Cheshire
HOP O' MY THUMB in Somerset and Wiltshire
JACK-JUMP-ABOUT in Northamptonshire
King's-finger grass in Middlesex
King's fingers, and kingfisher, in Buckinghamshire
Kitty-toe-shoes in Dorset
Knifes-an'-forks in Scotland; 'from the spiny seed-pods'
Lady Margaret's slipper in London
Lady's boots in Devon
LADY'S CUSHION in Devon, Dorset and Wiltshire
Lady's-finger grass in Hertfordshire
Lady's-fingers-and-thumbs in Somerset and Wiltshire
LADY'S GLOVES in Dorset and Northamptonshire
LADY'S PINCUSHION in Dorset
Lady's shoes-and-stockings in Buckinghamshire and Kent
Lady's-thumbs-and-fingers in Somerset
LAMB'S FOOT in Lancashire
LAMB-SUCKINGS in northern England
LOVE-ENTANGLED in Cornwall
Love-letters in Cheshire
MILKMAIDENS, and MILKMAIDS, in Sussex
No blame 'in the south of Ireland children gather this plant to take with them to school in the belief that the possession of it will save them from punishment, and it is said they will go miles out of their way to obtain it'
Old-woman's toenails in Devon
Pattens-and-clogs in Gloucestershire, Somerset and Sussex
Pea-thatch in Somerset; 'thatch' being probably a corruption of 'vetch'
Pig's foot in Suffolk
Pig's pettitoes in Sussex; according to the *English Dialect Dictionary*, in Cheshire 'pettitoes' means 'feet'
Pig-toes in Kent
PINCUSHION in Dorset
Rosy morn in Somerset
SHAMROCK, 'around Glasgow; possibly fitting any trefoil plant'
Sheep-foot in Cumberland
SHEPHERD'S PURSE in Somerset

SHOES-AND-STOCKINGS in southern England
Stockings-and-shoes in Devon
THIMBLES in Somerset
THUMBS-AND-FINGERS in Dorset and Somerset
Toes in Lancashire
Tommy tottles in Yorkshire
Tom Thumb's finger-and-thumbs in Dorset
Tom Thumb's honeysuckle in Wiltshire
Trefoy in Somerset
Wild calceolaria in Somerset and Co. Donegal; 'a gardener's term and a silly one'
Wild thyme in Somerset
YELLOW CLOVER in Yorkshire.

Allen and Hatfield provide two records of bird's-foot trefoil being used in folk medicine: as an EYEwash on South Uist in the Outer Hebrides, and to treat HORSES' cut legs in Somerset.[1]

BIRTH – SEA BEANS ease.

BISTORT (*Persicaria bistorta*; syn. *Polygonum bistorta*)

Perennial pink-flowered herb in damp pastures and on roadsides and riverbanks; also widely cultivated, particularly in old gardens; scarce in Ireland.

Local names include: dog-stinkers, 'it smelled of dog's urine during the summer months', in West Yorkshire; Easter-edges in Yorkshire, Easter magiants and eastermere giants in Cumbria; eastmore giants in West Yorkshire; poor-man's cabbage in Lancashire; snake-root in Ireland, and SNAKE-WEED in West Yorkshire.

> [Rochdale, Lancashire] My father told me him and his pals have known bistort as sweaty feet since he was a lad. He's nearly 68. I've found the smell to be only evident when the flower spikes are fully open. Thoroughly unpleasant.[1]

However, in north-west England bistort was widely gathered and eaten.

> When I used to work in Cumbria, everyone used to make Easter Ledges – there were various recipes, but they always contained *Polygonum bistorta* – and eat them during the spring. They were supposed to purify the BLOOD.[2]

> Easter ledges – or Herb Pudding or Yarby Pudding – depending on

which part of Cumbria one is in – seems to have been made early in the year when there are few vegetables, and eaten with 'Tatty Pot' which is mainly ONIONS, POTATOES and as much mutton as one had.

3 good handfuls of Easter Ledges
3 good handfuls of NETTLES
1 good handful of CABBAGE or broccoli or DANDELION leaves with a few RASPBERRY, BLACKCURRANT and/or GOOSEBERRY leaves
1 onion
1 large cup BARLEY – previously soaked overnight
Salt and pepper, or sometimes PARSLEY

Method: Wash greens, and chop onion. Put in a muslin bag and boil for one and a half hours. Empty bag into a bowl and beat in three eggs, bacon dripping, and about 2oz of butter. Reheat in oven.

In Cumberland they used OATS instead of barley and after it had been boiled put it on top of hard-boiled eggs.[3]

Here in the Halifax area there is a very active tradition of gathering the young leaves of *Polygonum bistorta* in spring to make 'dock' pudding, mixed with oatmeal, bacon fat, and sometimes nettles, all fried together.[4]

It is said that during the Second World War German radio propagandist, 'Lord Haw Haw', heard about people eating this dish and announced that the food situation was so critical in Yorkshire that people were reduced to eating grass.[5]

In parts of Yorkshire bistort was known as passion dock, from the custom of eating pudding made from it during Passiontide, the last two weeks in Lent.

Everyone has their own idea about what should go into dock pudding, but the basic method is to wash and chop the dock leaves, boil them with onions and then add oatmeal to thicken it up . . .

Full of iron and vitamins, it used to be eaten to 'clear the blood' in springtime like other spring greens, but tradition was adapted to the times. Nearly everyone said they made enough to freeze, and this was often then served as a special treat for breakfast on CHRISTMAS Day . . .

> The tradition of collecting and eating passion docks . . . has remained strong in Calder Valley for generations, but it was only in 1971 that the first World Championship Dock Pudding Contest was held as a prelude to the Calder Valley Festival of the Arts. This took place in Hebden Bridge and there were hundreds of entries. The *Halifax Courier*, the local newspaper, provided a trophy and there were other prizes for the winner and runners-up. Since that date the number of entrants decreased dramatically and it has not been held every year. It appears to be gaining popularity again; there were thirteen entrants in 1988, competing for prizes of £25, £15, £10 and, of course, the silver cup to be held for twelve months.[6]

In Warcop, Cumbria, bistort was used as a VERMICIDE: 'the moon must be full when the bistort is picked to cure worms'.[7]

BITING STONECROP (*Sedum acre*)

Small, perennial, succulent, yellow-flowered herb, in dry habitats, including sand dunes, rock faces, walls and roofs; widespread and common in England and Wales, scarce or absent from parts of Scotland and Ireland.

Local names include the widespread wall-pepper, and:

Bird's bread in Somerset, *cf.* the French *pain d'oiseau*
CANDLE-FLOWER in Staffordshire
Candles, and candlesticks, in Somerset
Creeping Charlie in Devon
Creeping Jack in Cheshire, Somerset and Wiltshire
CREEPING JENNY in Somerset
CREEPING SAILOR in Shropshire
Eating stonecrop in Suffolk
French (i.e. exotic) moss in Buckinghamshire
GINGER in Norfolk, Somerset and Suffolk, 'from its extreme pungency'
Gold dust in Suffolk
Golden carpet in Somerset
GOLDEN DUST in Cornwall and Suffolk
Golden moss in Oxfordshire, Warwickshire and Yorkshire
Golden stonecrop in Berkshire
Grapes, and HEN-AND-CHICKENS, in Somerset
Hen-chicken in Dorset
HUNDREDS-AND-THOUSANDS in Somerset
Irish moss in Yorkshire
Little house-leak in Cumberland and Northamptonshire, i.e. HOUSELEEK

LONDON PRIDE, and LOVE-ENTANGLED, in Somerset
Love-in-a-tangle in Cornwall
Moss in Warwickshire
Mousetail in Somerset
PICKPOCKET in Dorset
Pig's ears in Somerset
Plenty in Dorset
POOR-MAN'S PEPPER in Nottinghamshire and Sussex
Queen's cushion in Roxburghshire
Rock-crop in Cornwall
Rock-plant in Devon
Scath-weed in Lincolnshire
STAR-FLOWER in Somerset
Wall-ginger in Somerset and Yorkshire
Wall-grass in Devon
Wall-moss in Yorkshire
WALLWORT in Yorkshire
Welcome-home-husband-though-never-so-drunk in Dorset.

Stonecrop is seen on many thatched and other cottages and farmsteads in Wales. It was originally placed there as a protection against THUNDERbolts, LIGHTNING and WITCHES.[1]

In County Waterford biting stonecrop was 'grown on or about the roof of the house to bring good LUCK and guard against FIRE'.[2]

Crowdy-kit-o'-the-wall . . . the stems are by children stripped of their leaves and scraped across each other fiddle-fashion to produce a squeaking noise . . . The name is only known among old people now, as very few know what 'Crowdy-kit' means' but an old woman at Ipplepen [Devon], well-versed in herbs (eighty-eight years of age and still *yark*), gave me both the name and how it was to be explained. Her family used to be very musical, and could remember the fiddle being called crowdy.[3]

Cf. WATER FIGWORT.

BITTERSWEET (*Solanum dulcamara*), also known as woody nightshade, and, incorrectly, as deadly nightshade.

Scrambling shrub with purple flowers and red jewel-like fruits, growing in a wide range of dampish habitats in lowland areas.

Local names include:

Aw'food, 'i.e. half-wood' in Worcestershire
Bellydonya in Devon; possibly a mistake and deadly nightshade (*Atropa belladonna*) intended
Brook-brimble in Devon
CANDLE-FLOWER in Cornwall
DOGWOOD Lancashire

Fairy-bud in Somerset
Felon-wood, widespread, 'the berries were used against felons, or WHITLOWS', and, as fellon-wood, in Cumberland and Yorkshire
Felon-wort, 'from its use in curing whitlows, called in Latin *furnculi*'
FOOL'S CAP in Somerset
GRANNY'S NIGHTCAP in Devon and Somerset; 'probably owing to the kind of peak formed by the more or less conjoint stamens'
Guinea-goul in Co. Limerick
HALF-WOOD in Warwickshire and Worcestershire
Lady's necklace in Devon
LADY'S UMBRELLAS in Somerset
Mad-dog's berries in Morayshire
OLD-WOMAN'S NIGHTCAP in Somerset
POISON-BERRY, widespread, for the fruit
Poison-flower in Hertfordshire
POISONING-BERRIES in Yorkshire
Poisonous tea-plant in Oxfordshire, '? in distinction to *Lycium halmifolium* [= *L. barbarum*, Duke of Argyll's teaplant]'
Pushionberry in Berwickshire
Robin-run-i'-the-hedge in Lancashire 
Scaw-coo in Cornwall, 'Cornish *scaw cough*, scarlet elder'
Shady night in Lancashire
Skaw-coo in Cornwall
SNAKE-BERRY in Suffolk
SNAKE-FLOWER in Somerset
SNAKE'S FOOD, applied to the fruit, 'a variation of adder's food . . . which is a corruption of the Anglo-Saxon *attar* – poison'
Snake's poison food in Buckinghamshire
Terri-diddle, terry-diddle, terry-divle, and thether-devil, in Cheshire, 'where children chew the roots, and say they are like stick liquorice'
Witch-flower in Somerset.

Although the chewing of bittersweet roots 'like stick liquorice' seems improbable, Hodgson in his *Flora of Cumberland*, 1898, records school-boys keeping 'a stock of twigs in their pockets which they chew as their elders do tobacco'.[1]

> [Norfolk] TEETHING troubles: make a necklace of dried night-shade berries, let baby wear it, and it will prevent CONVULSIONS.[2]

> [Cotswolds] The berries of bittersweet . . . were used [to cure

> CHILBLAINS], well rubbed in; they were preserved in bottles for winter use.[3]

Late in the nineteenth century bittersweet could be seen in County Donegal 'trailed over cottages as an ornamental plant'.[4]

BITTER VETCH (*Lathyrus linifolius*)

Perennial herb, growing in wood margins, scrub and hedgerows, absent from areas of central and eastern England and central Ireland.

Local names, many of which are corruptions of the Gaelic *cairmeal*, include:

Caperoiles and carameile in Scotland
Carmeil, carmele, carmile, carmylie, cormeille and cormele, in the Scottish Highlands
Cornameliagh in Co. Donegal
Corra-meile in northern Scotland and the Hebrides
Fairies' corn in Co. Donegal
Gowk's (i.e. cuckoo's) gillyflower, 'referring to its attractive multi-coloured flowers', in Scotland
Heather-pease in Scotland
Knapperts in northern Scotland and the Hebrides
Napperty in Belfast
Napple in Galloway
Peasling in Yorkshire
Posy-pea in Shropshire.

> [Colonsay] The tuberous roots [of bitter vetch] were dug up and eaten raw, or tied in bundles and hung up to the kitchen roof to dry, and afterwards roasted. Used for flavouring whisky.[1]

In Somerset, roots of bitter vetch were known as nipper-nut, being 'nutritious and palatable' and often eaten by children.[2]. Similarly, in Berwickshire schoolboys called the roots liquory-knots, 'for, when dried, the taste of them is not unlike that of real liquorice',[3] while in Aberdeenshire and Morayshire, where they were known as gnapperts, 'the tubers are much sought after by children, who steep them in water overnight, which water they drink in the morning with great gusto, and then devour the tubers'.[4]

BLACK BRYONY (*Tamus communis*)

Perennial tuberous climbing herb, the only member of the predominantly tropical Yam Family (Dioscoreaceae) native to these islands, common in hedgerows in lowland areas of England and Wales.

Local names include the widespread POISON-BERRY, and:

ADDER'S MEAT, or ADDER'S MAIT, in Devon and Somerset
Adder's poison in Devon
Beadbine in Hampshire
Black bindweed in Yorkshire
Blacksmith's berries, and blacksmith's poison-berries, in Cornwall
Broyant in Montgomeryshire; 'used to rub on the joints of animals, especially pigs, that are lame from a disease which is there called broyant'
CHILBLAIN-berry in the Isle of Wight and Somerset
Devil's berries in Cornwall
Isle-of-Wight vine on the Isle of Wight
Ladyseal in Somerset
Little-girl's curls in Sussex
MANDRAKE, 'distinguishing it from the woman-drake of *Bryonia dioica'*, in Lincolnshire and Yorkshire
Murrain-berries, and murren-berries, on the Isle of Wight
Ox-berry in Herefordshire, Shropshire and Worcestershire; 'in west Cheshire the berries are collected by the farmers as a cure for barrenness in cattle'
POISONING-BERRIES in Yorkshire
RED BERRIES in Devon.

> The berries, when steeped in gin, [are a] popular remedy for CHILBLAINS in this island [of Wight] , where the power they possess, in common with the root, of removing superficial discolorations of the surface from BRUISES, SUNBURNS, &c., is equally well known and applied in practice.
>
> My friend Lady Erskine informs me that the black bryony is called in Wales 'Serpent's Meat', and the idea is there prevalent that those reptiles are always lurking near the spots where the plant grows.[1]

BLACKBERRY – see BRAMBLE.

BLACK CURRANT (*Ribes nigrum*)

Small shrub, believed to have been introduced early in the seventeenth century; widely cultivated in gardens and on a commercial scale for its fruit, and widely naturalised in damp open woodland and hedgerows.

Even before its Vitamin C content was discovered, black currant was used to prevent or cure COLDS.

> My [Cornish] great grandmother died at 86 years old in 1932, her knowledge of vitamins would be nil, obviously, but in winter she always kept a jug of hot water with big dollops of black currant jam in it on top of the 'slab' (Cornish name for kitchen range) and we always as children had to drink a cup of black currant tea to 'protect our LUNGS and keep away colds'.[1]

> As children (in south Scotland) my mother made black currant concentrate as prevention against colds.[2]

> Black currant tea was for HOARSENESS and sore THROATS.[3]

BLACK MEDICK (*Medicago lupulina*)

Annual, or short-lived perennial, herb, common in open grassy places throughout lowland areas.

Local names include:

Black grass in Buckinghamshire
Black nonsuch, or black nonesuch, in Norfolk
Black seed in Buckinghamshire
Black trefoil in Norfolk
Dog-clover in Somerset
FINGERS-AND-THUMBS in Dorset
Hop-medick in Northumberland
LAMB'S TOES in Staffordshire
Natural grass on the Isles of Scilly
Nonesuch in Hampshire and Sussex
Nonsuch in Wensleydale, Yorkshire
Nonsuch clover in Suffolk
Sanfoin in Buckinghamshire
Wild shamrock in Somerset
YELLOW CLOVER, 'a frequent name'

Black medick was occasionally considered to be SHAMROCK:

> The plant that I grew as shamrock, sent home on ST PATRICK'S DAY – sometime about 1915 – was *Medicago lupulina*.[1]

BLACK POPLAR (*Populus nigra*)

Native tree, growing beside water and on moist low-lying land, most frequent in east Wales, central England and East Anglia.

Local names include:

Cat-foot poplar, 'name . . . in use amongst cabinet-makers, and refers to the dark knots in the wood, which are said to resemble the marks of cat's feet', in Lancashire; cotton-tree, 'the seeds being enveloped in a beautiful white cotton', in Suffolk; pepillary, popilary, and poppilary, in Cheshire, and willow-poplar in Cambridgeshire.

> The fallen male catkins of black poplar . . . are known as DEVIL'S FINGERS and in many country areas it is thought to be unlucky to pick them up.[1]

However:

> When I was little [*c*.1940s] my older relatives urged me to collect poplar catkins and put them in a little saucer carefully mixed with water, for the FAIRIES to eat at night (while I was asleep!). Needless to say the catkins had gone by the morning . . . I believe it was a fairly common practice among children earlier this century.[2]

At Aston-on-Clun, Shropshire, a black poplar, known as the Arbor Tree, is decorated with flags attached to wooden flagpoles each year on 29 May, Arbor Day. These flags remain in position until they are replaced by new ones the following year.[3]

> The custom of dressing a tree with flags survives an era when it was done in worship of Bridget, goddess of fertility. Later sanctified as St Bridget or St Bride, the goddess had a tree for a shrine, on which tribal emblems and prayer flags were hung. A Bride's Tree survives at Aston-on-Clun, but it is unclear whether the name derives from Bridget or the following:
>
> On 29 May 1786 John Marston, the squire of Aston-on-Clun, married Mary Carter by special licence at Sibdon Carwood parish church. As the wedding couple were returning home in their carriage, exuberant villagers stopped it at the parish boundary, dragged it to the Bride's Tree, and the couple were presented with boughs as symbols of good luck.
>
> Mary was so enchanted with the dressing of the Bride's tree that she saw to it that the custom would continue.[4]

However, recent research shows that the first mention of the Arbor Tree was made in 1898, and has discovered no reference to it being decorated before 1912, and it appears that any association of the tree with Bride arose considerably later, in the 1950s.[5]

> Originally it appears that the Marston estate was responsible for the tree dressing. Following the sale of the estate in 1950 the Parish Council agreed to take responsibility, but in 1954 the Council agreed that the 'decorations be left in abeyance'. This led to a revival of interest in the custom. Formerly the tree had been dressed without much ceremony, but in 1955 an event which 'involved a prayer, hymns, the twenty-third psalm, and a new pageant' was instituted. The Pageant of the Tree included participants representing John Marston and his bride and other characters supposedly associated with the tree.[6]

Since 1977 an annual fête to raise money to pay for the flags has been held on a Sunday near to 29 May, and this is 'rapidly becoming a more substantial event than the one it is intended to finance'.[7]

As 29 May is also OAK Apple Day, it is possible that Arbor Day is in some way derived from celebrations to commemorate the restoration of the monarchy in 1660. In the 1950s pieces of the Bride's, or Arbor, Tree were given to couples on their wedding day. According to a local man interviewed for a television programme broadcast in November 1991:

> The parish used to present couples in the village with a cutting from this tree, as a sort of good luck thing, and they stopped doing it because all the women got pregnant within the year . . . They didn't necessarily want that to happen so quickly in a relationship so the vicar stopped doing it . . . People used to write to the parish asking for a cutting from this tree because they couldn't have children themselves, and in the 1950s or 60s there was a kind of outbreak of people wanting a cutting from this magic tree, because they heard that all the village maidens were having babies very easily.[8]

Thus on 12 April 1960 the *Birmingham Post* reported:

> There is unlikely to be a pageant this year at the annual dressing of the ancient Arbor Tree . . .
>
> This has been decided by the Parish Council, which is anxious to avoid what it feels has been 'garbled publicity' about the event in the past. The tree will continue to be decorated with flags on Arbor Day, 29 May.

> In recent years a pageant, in which schoolchildren have taken part, has been organised as part of the celebration. It is said that the custom goes back to ancient fertility rites. Matters came to a head after last year's ceremony when the Rector, the Revd T.S.D. Barrett, who is also chairman of the Parish Council, received letters from women in Italy and the United States who thought the tree might be able to help them have children.
>
> Mr Barrett said yesterday: 'We want to bring it down to its proper level. A small local affair for this dressing of flags would be nice, but we do not want all this nonsense about fertility rites.'

The ancient tree, reputed to be at least 300 years old, collapsed in 1995, and was replaced by a young tree grown from a cutting taken from the old tree and given to Marina Harding, who played the 'bride' in the 1986 pageant, which celebrated the two hundredth anniversary of the wedding of John Marston and Mary Carter.[9] The young tree is too immature to support flagpoles, so these are now attached at an angle to four posts set around it. The flags used in 2001 were those of the United Kingdom, the USA, Canada, Australia, New Zealand, St George, St Andrew, St Patrick and St David.[10]

BLACK SPLEENWORT (*Asplenium adiantum-nigrum*)

Small evergreen fern, common and widespread on walls and in stony places.

> [Guernsey] A strong decoction of this plant, to which is added plenty of brown sugar, or sometimes dried FIGS, is considered a sovereign remedy for COUGHS.[1]

BLACKTHORN (*Prunus spinosa*), also known as sloe

Deciduous, white-flowered shrub or small tree; widespread in hedges, scrub and woodland; also planted in hedges.

Local names include:

BLACK ALLER in Somerset
Black haw in Ireland
Blackthorn-may in Middlesex
Buckthorn in Lincolnshire
Bullister in Cumberland, Scotland and Ireland
Egg-peg-bush in Gloucestershire
GRIBBLE, 'a young blackthorn, or a knobbly walking-stick made of it', in Dorset

SCROG, or SCROGG, in Nottinghamshire and northern England
Sion in Northamptonshire
Slaa-thorn in Yorkshire
Slacen-bush in Northamptonshire
Slae in Scotland
Slaigh in Lancashire
Slaun-bush, and slaun-tree, in Leicestershire
Slaythorn in Lancashire
Slea-thorn, and slea-tree, in Cumberland
Slon-bush in Leicestershire and Northamptonshire
Slon-tree in Cornwall, Leicestershire and Somerset
Snag-blooth, and snag-blowth, applied to the blossom
Snag-bush in Dorset
Wayside beauty in Somerset
Wild damson in Wiltshire.

Names given to the astringent, bluish-black fruit, commonly known as sloes, include the widespread slan and snag, and:

Ballams in Somerset
Bullens in Shropshire
Bullies in Lincolnshire; 'it is probably more correctly assigned to the bullace or larger sloe'
Bullum in Cornwall and Devon; also applied in Dorset to 'big' fruit, bullace or larger sloe presumably bullace [*Prunus domestica* ssp. *insititia*]'
Castings in Devon
Hedge-peg, and hedge-pick, in Hampshire and Wiltshire
Hedge-pigs in Gloucestershire
Hedge-speaks, or HEDGE-SPECKS, in Wiltshire
HEG-PEGS in Gloucestershire
Hilp in Wiltshire
KEX in Hampshire
Pick in Wiltshire
Slac in Co. Donegal
Slags in Westmorland
Slane in Dorset
Slea, sleah, or sleea, in northern England
Sloan in Cornwall
Slone in Devon, where blossom known as slone-bloom
Sloos in Dorset
Sluies in Lincolnshire
Sneg, given to 'small' fruit in Dorset; cf. bullum
Winter-kecksies in Sussex.

Like HAWTHORN, with which it appears to be confused when in flower, blackthorn blossom is often banned from indoors.

> I remember that in April 1948 when I was in the maternity home at Bishop Auckland, Co. Durham, my husband brought a spray of blackthorn into the ward for my enjoyment, to the great consternation of the nursing staff. Although the word 'DEATH' was not mentioned, I think the term 'bad LUCK' was only used out of consideration for any patients who were within earshot. One nurse certainly mistook it for hawthorn, and could not be convinced that it was a different plant – it is possible the others also misidentified it, but my recollection is that both plants were highly suspect, and the blackthorn was thrown out.[1]

> [Hardwicke, near Gloucester] It was considered to be bad luck to pick blackthorn and on no account was it to be taken into the house. A scratch from the thorns could cause BLOOD POISONING. It was thought to be because Christ's CROWN OF THORNS was made from it.[2]

> I once put some blackthorn in the church as a decoration, but they were all mad with me and made me take it out. They thought it was hawthorn.[3]

> We were never allowed to take blackthorn into the house (Crown of Thorns). Hawthorn was okay.[4]

The flowering of blackthorn is said to coincide with a COLD SPELL of weather.

> [Isle of Wight] A period of weather, which happens commonly whilst the sloe is blossoming is called by the country people here the 'blackthorn winter'.[5]

> As a youth my late father worked on the land, as did his father. A keen gardener [he made sure] . . . nothing of a tender nature was planted out until after the blackthorn winter.[6]

> We have just been experiencing the blackthorn winter. The weather always turns cold when the blackthorn comes into bloom on the hedgerows.[7]

However, the period when blackthorn flowered was widely considered to be the ideal time for sowing BARLEY. In Berwickshire:

When the slae tree is white as a sheet
Sow your barley, wither it be dry or wet.[8]

In Gloucestershire:

When the blackthorn blossoms white
Sow your barley day and night.[9]

In Sandwich, Kent, each incoming mayor of the town is presented with a blackthorn stick. *The Customal of Sandwich*, written by Adam Champneys, town clerk in 1301, and now in the Kent County Archives in Maidstone, states:

> When the mayor of the preceding year and the jurats and commonalty are assembled in the church, and the sergeant has brought his horn, the mayor takes his stick and the horn from the sergeant, and the keys of the chest from the two jurats the keepers, and puts them near him.

Almost four centuries later:

> When the mayor goes in form to the court-house or to church, he is preceded by the common wardman and the two sergeants-at-mace, in liveries, each bearing on his shoulder a mace of silver gilt; and he carries himself in his hand a black knotty stick, as a badge of office.[10]

Another two centuries later:

> Tradition has it . . . that his blackthorn wand is carried to safeguard the holder against evil spells cast by WITCHES and by its very presence, repels any evil witch who may wish to cast a spell. Today (and certainly for the past hundred years) the Town Sergeant is responsible for selecting an appropriate wand from the local hedgerows and for drying the same and preparing it for presentation to each new mayor on the occasion of his taking office, in return for which duty he is presented by the mayor with a crown (five-shilling piece). Each mayor retains his wand after his year of office, and a new one is prepared for the new mayor.[11]

Three such sticks are in the collections of the Sandwich Guildhall Museum.

In nineteenth-century Ireland blackthorn sticks (shillelaghs) were favoured as weapons in faction fights.[12] In the 1990s miniature versions of such sticks were widely sold as souvenirs for tourists.[13]

> Sometimes little sloe bushes grew in a group in the corner of a field, and one of them might be cut down and used to clean the CHIMNEY by pulling the bush up and down. Long pieces in the hedgerow were cut and made into blackthorn WALKING STICKS; some people do them still, as they are bought by tourists; long ago old men used them as walking sticks.[14]

The fruits of blackthorn – sloes – are gathered to flavour sloe gin.

> The sloes are called bullums in these parts. Sloe gin is widely made here, at home, and commercially for 200 years. Most people roughly follow this method: Take a large spirit bottle. Fill it with sloes whose skins have been well pricked. Add half a pound of white sugar. Top up bottle with gin; Plymouth gin is best of course! Screw on the cap. Shake again once or twice a week for 2–3 months. Rack off and bottle, if you can resist drinking it. Sloe gin made in the autumn will be ready for Christmas.[15]

> Some people mention eating sloes, but it is probable that they were tried only once or twice, being far too astringent for most people's taste. In the 1930s, at Cawsand, Cornwall, sloes were known as bulloms, and the coarse red wine drunk by Brittany crabbers, who frequently anchored in the bay, was known as 'Johno Frenchman's Bollom Juice'.[16]

> I remember from my childhood in the 20s and 30s in Longbridge Deverill, Wilts . . . sloe gin – three years old at least – for COLDS and COUGHS. Strongly recommended.[17]

> [Somerset, 1940s] Sloe wine was a wonder drug for adults – as it cured everything. As it was home-made, it wasn't considered to be alcoholic, so even the strictest chapel-goer drank it.[18]

In County Carlow dried blackthorn leaves, known as Irish tea, were used as a substitute for TOBACCO,[19] while in wartime Cumbria, 'blackthorn leaves added to the pot made TEA go further and, some said, improved the taste'.[20]

BLADDER problems – treated using AGRIMONY, CHERRY, GOOD KING HENRY, PARSLEY PIERT and PELLITORY OF THE WALL.

BLADDER CAMPION (*Silene vulgaris*)

Perennial, white-flowered herb, widespread in open grassy places. Local names include:

Adder-and-snake-plant in Devon
Billy-bashers in Somerset
Birds' eggs in Shropshire
Bladder-bottle, and bladders-of-lard, in Somerset
Bladder-weed in Dorset
Blether-weed in Dorset; 'on account of the form of the calyx: blether = bladder'
BULL-RATTLE, on the Isle of Wight; 'the name rattle probably refers to the sound made by the dry inflated calyx'
Clapweed in Hertfordshire
COCKLE in Warwickshire
Corn-pop in Wiltshire
Cow-bell, and cow-cracker, in Scotland
Cowmack in northern Scotland; 'supposed to have great virtue in making the cows desire a male'
Cow-paps, 'evidently derived from the shape and size of the turgid seed-capsules', in Berwickshire
Cow-rattle in Buckinghamshire
Crackers in Sussex
Crow-cracker in Dumfriesshire
Fat-bellies in Somerset
KISS-ME-QUICK in Somerset
POP-GUNS in Dorset
POPPERS in Somerset and Wiltshire
Poppy in Wiltshire; 'the calyxes are popped by children'
RAGGED ROBIN in Somerset
RATTLE-BAGS in Devon
RATTLE-WEED in Wiltshire
Round campion by English gypsies
Shackle-backle, snags, and SNAP-JACKS, in Somerset
Snappers in Kent
Spathing poppy, and spattling poppy, in Cumberland
THUNDERBOLTS in Kent; 'children snap the calyxes, which explode with a slight report'
White bachelor's-buttons in Suffolk
White bottle in Cambridgeshire and Somerset
White cockle in Berwickshire
White cock-robin, and white hood, in Somerset
White-flower of-hell, and white-hell-flower, in Dorset; 'in consequence of the poison supposed to be contained in its leaves and bladders . . . a misconception. The leaves are frequently eaten by children'

White-hood in Somerset
White mint-drops in Northumberland
White riding-hood in Devon
White-Robin-Hood in Wiltshire.

> The young shoots have the odour and flavour of green peas, and are sometimes boiled as asparagus, though the writer found them too bitter to prove agreeable. They are, however, perfectly wholesome, and the bitterness might be removed by blanching.[1]

BLEEDING – KIDNEY VETCH and RIBWORT PLANTAIN used to check.

BLIGHTS – prevented using AFRICAN and FRENCH MARIGOLDS.

BLINDness – GERMANDER SPEEDWELL and POPPY produce.

BLISTERS – CELERY-LEAVED BUTTERCUP used to produce.

BLOOD – BETONY, BISTORT, DOCK, GROUND IVY, NETTLE and RUE purify; CALVARY CLOVER, EARLY PURPLE ORCHID and LORDS-AND-LADIES leaves, REDSHANK and TUTSAN fruit, stained by; DWARF ELDER, FIELD ERYNGO, FOXGLOVE, FRITILLARY, LILY OF THE VALLEY, PASQUE FLOWER, POPPY and WINTER ACONITE grow, or BARREN GROUND, where blood has been spilt.

BLOOD POISONING – BLACKTHORN causes; MALLOW used to treat.

BLOOD PRESSURE – NETTLE and SHEPHERD'S PURSE used to lower.

BLUEBELL (*Hyacinthoides non-scripta*)

Bulbous, blue-flowered plant characteristic of deciduous woodland.

Local names, many of which are also given to EARLY PURPLE-ORCHID, include the widespread BLUE BOTTLE, and:

ADDER'S FLOWER in Somerset
Bell-bottle in Buckinghamshire
BLOODY-MAN'S FINGERS in Gloucestershire; 'some error is to be suspected here'
BLUE BONNETS in Somerset
Blue goggles in Wiltshire
Blue googoo in Cornwall
Blue granfers-greygles in Dorset and Somerset, in contrast to early purple-orchid
Blue rocket in Co. Fermanagh
Blue trumpets in Somerset
Bummuck, and burmuck, in Co. Donegal
CRAKE-FEET in northern England
Crake's feet in Yorkshire
Crathies in Dumfriesshire
CRAW-FEET in Lancashire and Westmorland

CRAW-TAES, and craw-tees, in Scotland
Crocus in Cornwall
CROSS-FLOWER in Devon
CROW-BELLS in Hampshire and Wiltshire
CROW-FLOWER in Devon, Hampshire, Somerset and Wiltshire
CROW-FOOT in Cumberland, Lincolnshire and Radnorshire
Crow-picker in Co. Donegal
Crow's legs in Wiltshire
CROW-TOES in Co. Down
CUCKOO in Cornwall and Devon, and cuckoos on the Isles of Scilly
CUCKOO-FLOWER in Cornwall, Devon and Somerset
Cuckoo's boots in Dorset and Shropshire
CUCKOO'S STOCKINGS in Derbyshire, Northamptonshire and Somerset
CULVERKEYS in Kent, Northamptonshire and Somerset
Culvers in Essex and Oxfordshire
FAIRY-BELLS in Somerset
Goocoos in Devon
Goosey-gander in Devon, Dorset and Somerset
Gowk's hose in Dumfriesshire; 'gowk = cuckoo'
Gramfer-griggle in Dorset and Somerset
Grammer-greygle in Devon and Dorset
Granfer-gregors in Dorset
GRANFER-GRIGGLES in Dorset and Somerset
Granfer-grigglesticks in Somerset
Greggles in Dorset
Greygles in Dorset and Wiltshire; 'A.S. *graeg*, grey, graegl, or greygle, means what is greyish blue'
Griggles in Dorset
GUCKOO in Cornwall
HAREBELL in Devon and Co. Derry
HYACINTH in Scotland
LOCKS-AND-KEYS in Somerset
Pentecostal bells in East Anglia
Pride-of-the-wood in Somerset
Ring o'bells in Lancashire
Rook's flowers in Devon
Single gussies, and snake's flower, in Somerset
Snap-grass in Kent; 'probably in allusion to the brittle flower-stalks', or 'from the rubbing, clicking noise of stalks when gathered'
Wood-bells in Buckinghamshire.

Although designated by the charity Plantlife as the national flower for the United Kingdom in 2004, and since 1979 used as an emblem of the Botanical Society of Britain and Ireland (formerly the Botanical Society of the British Isles), though nor readily recognisable in the Society's current (2019) logo, the bluebell has attracted little folklore.

> [Hartland, Devon] It is UNLUCKY to bring bluebells into the house.[1]

> As a child [in Yorkshire] I used to pick masses of bluebells, but my mother would never let me bring them into the house . . . I tried to dismiss this kind of thing as superstitious nonsense, but found it most disturbing when my small son brought me a bunch of bluebells last year and insisted that we had them inside.[2]

> It is unlucky to take into your house . . . the bluebell.[3]

However, such beliefs appear to have been relatively rare. More usually:

> [Leamington Spa, Warwickshire, 1940s and 50s] Massive bunches of bluebells and PRIMROSES were taken home usually after a long cycle ride to the woods and a country picnic.[4]

Bluebell sap was used as glue:

> Our ancestors . . . produced a thick, pungent glue by grinding the bulbs into a pulp, then heating the mush to drive off excess water. When squeezed, the stems too produce a glue – so strong that fletchers used it to cement the feathered flights to the shafts of arrows. I still have a football scrapbook dated 1949 which has pictures stuck in with bluebell glue.[5]

BOG ASPHODEL (*Narthecium ossifragum*)

Perennial herb (Fig. 3) with attractive orange-yellow flowers, widespread on peat and damp moorland, mainly in western areas.

Local names include: bent in Co. Tyrone, clove-flower in Shetland, farfia and glashurlana in Co. Donegal, limerick in Shetland, MAIDEN'S HAIR in Lancashire where used as a blonde hair DYE, moor-golds in West Yorkshire, MOOR-GRASS in Hertfordshire, and yellow grass in Scotland.

Bog asphodel was also known as cruppany-grass in County Donegal, where it was believed that it gave SHEEP 'a stiffness of the bones, known as cruppan.[1]

Elsewhere:

> It was believed that if [bog asphodel] is eaten by sheep it would give them stiffness in their bones.[2]

> I lived for seven years in the Lake District. The fell farmers particularly feared bog asphodel in respect of their sheep, as it was a plant which 'softened their bones'. I must say I came across very many dead sheep on my fell walks, tangled in thorns, etc., but I have been told the belief that they could have been weakened by bog asphodel is sheer superstition.[3]

In the western Highlands of Scotland in *c.*1966, bog asphodel was known as break-bone.[4] Similarly the specific epithet *ossifragum*, given to bog asphodel by Linnaeus in his *Species Plantarum* (1753), is derived from the Latin *os* (bone) and *frangere* (to break).

The belief that bog asphodel weakens sheep's bones is rather more than superstition for it has been shown that it is poisonous to both sheep[5] and cattle.

> Fifteen cows among a herd of 50 suckler cows and calves rapidly lost body condition and became dull and anorexic after grazing pasture containing bog asphodel . . . The affected cows had evidence of kidney damage characterised by elevated plasma urea and creatinine concentrations. Eleven cows died and diffuse renal tubular necrosis was present in three cows which were examined post mortem. Similar renal lesions were reproduced experimentally by feeding bog asphodel to a healthy calf.[6]

Cf. BUTTERWORT and SUNDEW.

BOGBEAN (*Menyanthes trifoliata*), also known as buckbean

Rhizomatous perennial with pale pink flowers, growing in shallow still, or slow-moving, water, sometimes cultivated as an ornamental.

Names given to bogbean, most of which refer to its trifoliate leaves, include:

Bogbane in Ireland

Bog-hop, 'in the North, 'from its use

in making BEER'
Bog-nut in Scotland and Ireland
Bog-trefoil in Yorkshire
Craw-shoe in Orkney
Doudlar in Roxburghshire
Marsh trefoil in Ireland
Threefold in Yorkshire, Kirkcudbrightshire and Wigtownshire
Trefold in Shetland
Water-fluff in Norfolk, from the appearance of the pinkish white-fringed petals
Water-trefoil in Warwickshire
Water-triffle in Scotland.

Duncan Napier (1831–1921), an Edinburgh herbalist, recalled:

> During the months of April and May I used to get up very early in the morning to get my herbs, and I might have been standing in Duddingston Loch any time in the morning from four to seven o'clock, with my shoes and stockings off and trousers rolled up as far as they would go, pulling up buckbean. It was as much as I could do in those cold mornings to stand in the cold water, but I usually carried home a heavy bag full of buckbean, well up for a hundredweight, I got a good heat before I got home. It took me two days before I got each load tied into bunches and hung up to dry.[1]

Medicines prepared from bogbean were so notorious for their bitter taste that one suspects that their use was, in part, due to the idea that 'if it tastes bad, it must do you good'. For example, in the Hebrides, where bogbean was used to treat STOMACH problems:

> The roots of bogbean boiled for an hour in water and left to cool. The dose – one wine-glass full twice daily. When my neighbour John was given this remedy by his sister he took one gulp and said, 'Since I was young I have suffered from a bad stomach, even had a perforated ulcer and rushed to hospital but never have I been in such distress as I am at this moment,' and so saying rushed out of the room.[2]

> Bogbean seems to have been the most widely used plant medicinally in these parts [Shetland]. Its taste was foul, but it seems to have worked.[3]

Despite their taste, preparations of bogbean were used to treat an extraordinarily wide range of ailments.

> [Co. Cork] Bogbean (leaves boiled) juice is good for KIDNEYS.[4]

> [Co. Donegal] Bogbean was very valuable for SKIN diseases.[5]

> Granda told me bogbean was boiled as a medicine for CONSTIPATION.[6]

> I worked on a mixed farm . . . in the Swansea Valley . . . In 1949 . . . I had to go down to the wet boggy land below the farmhouse to collect bogbean. It was boiled to make a herbal tea. The farmer and his sister-in-law both drank the bogbean tea, which they swore gave relief to inflamed finger joints. They both suffered from ARTHRITIS or a similar condition.[7]

> Bogbean – *gulsa girse* – once used in the treatment of JAUNDICE. The Shetland name for jaundice is 'gulsa' from the Old Norse *gukustt*, meaning yellow sickness. The plant makes a green DYE.[8]

> I remember my mother (born 1916) talking about her mother and the root of the bogbean. This was boiled, I presume with water, and the resultant liquid used to ease congestion with bad COLDS. This was used by the rural people in the Glens of Antrim.[9]

In 1850 a traveller reported:

> Its leaves . . . are much used in the preparation of various bitters, which seem to be much relished on the Continent. Bitters are unusually fashionable, during my visit, on account of the prevalence of the cholera, which was then raging fiercely in Hamburg and Lübeck, and over a considerable part of Germany.[10]

BOG MYRTLE (*Myrica gale*), also known as gale

Short, erect shrub with aromatic leaves, common on damp moorland, especially in north-western areas.

Local names include:

Black sallow, bog sally, Dutch myrtle, dwarf sallow, and dwarf sally, in Ireland
GOLD in Hampshire
Golden osier on the Isle of Wight
Golden willow in Co. Donegal; 'I suppose from the gold of the pollen, conspicuous in early summer'
Golden withy on the Isle of Wight
Goule in Kent
GOWAN in Cumberland and Ulster
Moor myrtle in Yorkshire
Sweet gale, and sweet willow, in Sussex and Ireland
SWEET WITHY on the Isle of Wight
Wild sumac in Ireland.

An Irish name for bog myrtle was Our Saviour's rods;[1] the plant 'had dwindled to a low shrub . . . because Jesus was scourged with it by Pilate before he was delivered to crucifixion'.[2]

In impoverished moorland areas bog myrtle was a valued resource:

> [North Wales] The poor inhabitants are not inattentive to its virtues; they term it Bwrle, or the emetic plant, and use it for this purpose. An infusion of the leaves as tea, and an external application of them to the abdomen, is considered as a certain and efficacious VERMIFUGE . . . It furnishes a yellow DYE for woollen cloth; and by its powerful odour is fatal to MOTHS and bugs.[3]

> 'Gale' beer brewed from a plant growing on the moor above Ampleforth, in Yorkshire, is made and sold by Mrs Sigsworth of the 'Black Horse', the best public house in that long village. It bears a high local celebrity for its regenerative powers.[4]

> In Isla and Jura the inhabitants garnish their dishes with it, and lay it between their linen and other garments, to give a fine scent, and drive away moths.[5]

> [Islay] Bog myrtle gathered and hung up in kitchen to keep FLIES away.[6]

> [Donegal] It is gathered and put under beds to keep off vermin.[7]

> Sprigs of bog myrtle were frequently placed among bed-clothes by the Northumbrian housewife as a cure for FLEAS.[8]

Hence it was locally known as flea-wood.

With its widespread use in folk medicine and as an insect repellent it is unsurprising that bog myrtle is intermittently suggested as a source of income for Highlanders. Thus on 12 February 2007 the *Independent* excitedly reported that bog myrtle was 'very rare in its ability to drive away blood-sucking pests', and:

> Now scientists have discovered it is extremely useful in fighting acne and helping to delay ageing. In fact, Scotland could be on the verge of a new oil boom – this time in the shape of the first essential oil to be developed in the UK for commercial use for more than 40 years.

> CANDLE wax was once extracted [from bog myrtle] in Scotland. The shoots were washed in warm salted water, causing the wax to float to the surface like scum, and the candles made therefrom shed a rare perfume as they burned.[9]

However, recent writers consider that the 'one or two' references to candle wax being made from bog myrtle 'may be erroneous'; wax can be extracted from the plant's catkins by boiling them in water, 'but in such small quantities, that the process can hardly be realistic'.[10]

In County Donegal, where it was known as raidsogagh or rideokut, bog myrtle was 'held to be excellent for TANNING leather'.[11]

> [Hebrides] An infusion of the leafy tops [of bog myrtle] was given to children as a remedy for WORMS.[12]

> [Barra] Bog myrtle was hung, dried, infused among tea and given to children suffering from gastric or SKIN conditions.[13]

BOILS – treated using BRAMBLE, COMFREY, DAISY, DOCK, ELDER, HOUSELEEK and ORANGE LILY.

BONES – BOG ASPHODEL weakens, broken treated using COMFREY and DAISY.

BONFIRE (or GUY FAWKES) NIGHT (5 November) – MANGOLD associated with.

BOX (*Buxus sempervirens*)

Evergreen shrub or small tree probably native on chalk in southern England; planted in gardens, parks and woodlands, and naturalised, elsewhere; suffering by being eaten by caterpillars of the box-tree moth *(Cydalima perspectalis*), native to Eastern Asia, first recorded in Europe in 2006, reaching the United Kingdom in 2008, and now a serious pest in gardens in the Home Counties.

Britten and Holland in their *Dictionary of English Plant-names* list only one alternative name for box: bush-tree, recorded in Scotland late in the sixteenth century.[1] Writing of Somerset and surrounding counties in 1922 A.S. Macmillan records a number of names all apparently referring to the fruit, which were presumably used in furnishing dolls' houses at that time:[2]

Chairs-and-tables
Crocks-and-kettles, referring to the 'seeds' (probably fruits): 'I understand a game is played with these seeds . . . but do not know what form the game takes'
Furniture
Kettles-and-crocks
Milk-stools, referring to the 'flowers'
Pots-and-kettles
Tables-and-chairs.

It is sometimes considered UNLUCKY to take box twigs indoors:

> [Well into my sixties now, I was born in Mitcham, Surrey; my grandma] threw up her apron in alarm if box or HAWTHORN neared the front door.[3]

> Don't bring box into the house or someone will go out in a box.[4]

In 1868 a reporter sent to cover the aftermath of a colliery disaster at Hindley Green, near Wigan, wrote:

> I find an old Lancashire custom observed in the case of this FUNERAL. By the bedside of the dead man, the relatives, as they took their last look at the corpse, have formed a tray or plate, upon which lay a heap of sprigs of box. Each relative has taken one of these sprigs, and will carry it to the grave, many of them

dropping it upon the coffin. Ordinarily the tray contains sprigs of ROSEMARY or THYME, but these poor Hindley people not being able to obtain those more poetical plants, have, rather than give up the custom, contented themselves with stripping several trees of boxwood, hence it is that the mourners carry the bright green sprigs which I have seen.[5]

Although box does not seem to have been traditionally used on PALM SUNDAY in the British Isles, it is used in some London churches and, more rarely, elsewhere. In the late 1970s the French Catholic Church of Notre Dame de France, in Soho, blessed box as 'palm' on this day,[6] and it appears that this practice is standard in at least some parts of France:

> In Normandy . . . on Palm Sunday everybody, more or less, buys box in bunches, it is blessed in church, and then people visit the cemetery and leave a sprig of box on the graves of their dear departed. They also renew the sprig they tied to their crucifixes the previous year.[7]

Similarly in west and south-west Germany box 'was and still is much used' on Palm Sunday.[8]

> We use box on Palm Sunday, together with traditional palm-leaf crosses (for those who are keen on that). It is a tradition I started here. I brought this from the Netherlands. I like the idea that we get real little branches from the area. I use the blessed branches also for blessing the holy water.[9]

Early in the 1980s bunches of box, SALLOW and DAFFODILS were sold by young people in the forecourt of the Polish Catholic Church of Christ the King in Balham, London, on Palm Sunday. These were bought by members of the congregation, and at the conclusion of the service the priest walked down the aisle and back sprinkling holy water on the bunches of 'palm'.[10]

Box was used in veterinary medicine mainly as a VERMICIDE:

> Dried and powdered the leaves are still given to HORSES for the purpose of improving their coats. The powder is regarded by carters as highly poisonous, to be given with great care. In Devonshire, farriers still employ the old-fashioned remedy of powdered box leaves for bot-worm in horses.[11]

In the early 1930s my father, who was a ploughman, discovered that one of his horses had worms. His employer told him to treat it with the following remedy. Bake some box leaves in a tray in the oven until dry and crisp, rub to dust, then mix with the horse's feed of oats and chaff, and feed last thing at night. My father pointed out that the box was poisonous; the boss said that was the idea, the worms would feed off the box and die. Reluctantly my father carried the orders out. It is debatable whether the worms died or not, but one thing is for sure, the horse did. When my father went to the stables in the morning, there it was stretched out dead.[12]

BOXING DAY (26 December) – POTATOES planted on.

BRACKEN (*Pteridium aquilinum*)

Widespread (the most widely distributed fern in the world) deciduous, rhizomatous fern, frequently dominant on dry acid soils; often simply known as fern.

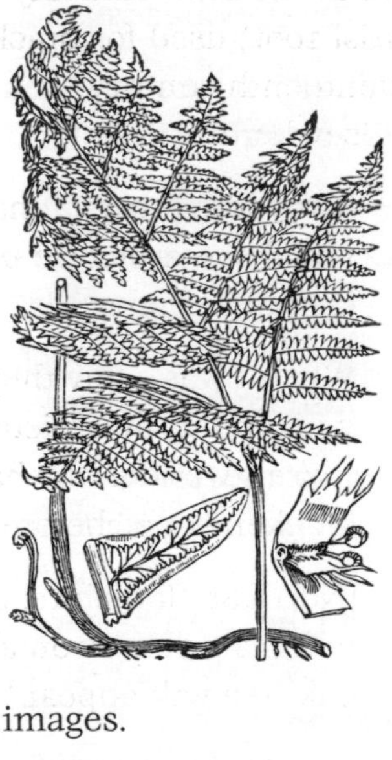

Local names include:
Adder's spit in Sussex
ADDER'S TONGUE in Berkshire
Brakes in Yorkshire
Emfern in Scotland
Farn in Gloucestershire
Great fern in Scotland
Mary's fern in Ireland
SNAKE-FERN in Dorset and Wiltshire
Thane, and vairdens in Devon
Vern in Hampshire.

Cutting through a stem or rhizome of bracken supposedly revealed a variety of images.

Bracken was originally described as *Pteris aquilina*, eagle fern, by Linnaeus in 1745, who noted: 'Cut across obliquely, the root contains a fair likeness of the Imperial Eagle'. This referred to the two-headed or double eagle which from the twelfth century formed the German Emperor's coat of arms, and in 1806, became the emblem of the Austrian Empire.[1]

> [West Sussex] Our custom of cutting the common brake or fern just above the root to ascertain the initials of a future wife's or husband's name.[2]

Similar practices have been recorded from East Anglia in about 1830, and Cornwall in 1887.[3]

In Scotland:

> My mind went back over some 70 years to a childish game we used to play called 'Holy Bracken'. Selecting a fat, juicy specimen, I used my pocket knife to sever it close to the ground and . . . there was: a perfect example of the most famous initials in the word – JC [Jesus Christ] . . . It was considered very lucky to find a good example.[4]

In Ireland, where a name for bracken was fern-of-god,[5] there was 'an old belief that if the stem is cut into three pieces there will be seen on the first slice the letter G, on the second O, and on the third D, and similar ideas presumably led to the name *Jesus-Christus-Wurzel* (Jesus Christ root) used for bracken in Germany from the second half of the seventeenth century'.[6]

Elsewhere:

> [Croydon, Surrey, and other places] Cut a fern-root slantwise and you'll see a picture of an oak tree: the more perfect, the luckier chance for you.[7]

> When I was a boy they used to say that if you split a bracken stem you would see a picture of King CHARLES hiding in his oak tree. I often wondered what would have been seen by those who split bracken stems before King Charles had hid in his oak tree.[8]

> [North Staffordshire, early 1940s] I remember being shown how if you make a horizontal cut through the stem (not root) of a fern an oak tree will appear.[9]

The idea that images of oak trees can be found in bracken stems presumably explains the name oak-fern which was used in west Somerset[10] and Norfolk.[11]

The fact that bracken, like other ferns, lacks flowers and seeds gave rise to a number of folk beliefs. In 1660 John Ray recorded:

> Many superstitious practices are associated with the gathering of

> the female fern [i.e. bracken] which some affirm ought to be the day of solstice, others on St John's Day.[12]

Presumably this referred to fern 'seed' which developed at midsummer and was believed to have the property of making those who gathered it invisible.

> A respectable countryman at Heston, in Middlesex, informed me in June 1793, that when he was a young man, he was often present at the ceremony of catching Fern-seed at mid night of the Eve of St John Baptist. The attempt, he said, was often unsuccessful, for the seed was to fall into a plate of its own accord, and that too without any shaking the plant . . . Our ancestors imagined that this plant produced seed that was invisible. Hence, from an extraordinary mode of reasoning, founded on the fantastic DOCTRINE OF SIGNATURES, they concluded that they who possessed the secret of wearing this seed about them would become invisible.[13]

In the seventeenth century it was believed that the burning of bracken – either for the preparation of potash or for its control – would lead to rain. On 1 August 1636 the Lord Chamberlain wrote to the High Sheriff of Staffordshire:

> His Majesty taking notice of an opinion entertained in Staffordshire that the burning of Ferne doth draw rain, and being desirous that the country and himself may enjoy fair weather as long as he remains in those parts, his Majesty hath commanded me to write unto you, to cause all burning of Ferne to bee forborne, until his Majesty be passed the country. Wherein not doubting but in consideration of their own interest, as well as that of his Majesty, will invite the country to a ready observance of his Majesty's command.[14]

The use of bracken in folk medicine appears to have been restricted to gypsies, who commonly used it as a cure for CONSTIPATION,[15] and used a 'decoction of sliced roots taken in wine' to expel WORMS.[16]

Although bracken is now considered to be a weed, in the past it had many uses, including the preparation of potash (which was used in the manufacture of glass and SOAP), as a fuel, as THATCH, as a packing material, as bedding for pigs and cattle, as compost and as a mulch,[17] and, in some parts of the world, as food.[18]

[18th-century Scotland] In several places in the north the inhabitants mow it green, and burning it to ashes, make those ashes up into balls, with a little water, which they dry in the sun, and make use of them to wash their linen with instead of soap.[19]

[Co. Leitrim] In this district, long, long ago the old people made their own soap. First of all they gathered faded ferns from the mountain side. Then they burned the ferns to ashes. Then they moistened the ashes with a drop of clean water and baked it into cakes. After a while they took this up and they shaped it into rings. When this stiffened they bored a hole and left it up to dry. When it was dry they took it down and used it as soap. This soap was made not later than half a century ago. This soap was used for all household purposes.[20]

In Glen Elg, in Inverness-shire, and in other places we observed that the people thatch'd their houses with the stalks of this fern . . . sometimes they used the whole plant for the same purpose, but it does not make so durable a covering.[21]

It was valued for packing purposes, growing conveniently handy in hilly slate-producing areas for the use of protecting the slates. It could be used around a variety of small goods carried in panniers by pack animals and farmers often sold it to pottery manufacturers to put round their finished wares – the potteries demanded huge quantities of packing materials, especially in the days before smooth canal transport and pneumatic tyres on land.[22]

In the Forest of Dean, Gloucestershire, I was surprised by some girls bringing a quantity of recently cut *Pteris aquilina*, or 'farn', which they retailed about two pence per bushel. On enquiring the use to which it was put, I was informed that it was extensively employed in the Forest for feeding pigs, which are very fond of it. For this purpose, however, it must be cut while the fronds are still uncurled, and a quantity of them boiled in a furnace. The slushy or mucilaginous mass thus produced is then consigned to the wash-tub, or any other receptacle, and in this state it will keep as pig-food for a considerable length of time. I was informed that it was found very serviceable, especially to cottagers, as coming in at an early period of the summer, when produce of the garden is but scanty.[23]

> When we were at Northleigh, near Colyton [Devon], when I was 13 or 14, we used to harvest bracken which grew on the hill. In those days, before the use of herbicides, there was lots of bracken around. We used to cut it with scythes – it was too steep for anything else. When it was dry we would take it back using a sleigh – two pieces of wood, turned up at the front, with iron on their undersides and slats between – pulled by a horse. The sleigh would be loaded with the bracken overlapping by about 2ft all round. If the sleigh moved too fast and banged against the horse's hind legs the overlapping bracken would act as a cushion and prevent the wood of the sleigh from hurting the horse. The bracken was used as bedding for pigs; I can't remember it ever being used for animals [cattle]. The pigs would root around and break it down into little pieces, like a mouse's nest.[24]

> In the late 1950s or early 1960s I remember visiting one of my father's cousins who farmed on the border of Dartmoor. His cows had bracken for bedding.[25]

Today bracken is usually considered to be a serious pest.

> [Co. Clare] Ferns grow on arable land and spread rapidly and impoverish the soil. This weed can be banished by being mown twice each year, first in June and later in September. A continual repetition of this for a few years banishes the fern.[26]

> [Llanthony, English–Welsh border, April] A young shepherd to whom I mentioned the rain looked up surprised, as if he hadn't noticed it before, although he had left his jacket folded up under a tree, to keep it dry. He had been up on the tops looking at the leaf and 'kicking that dreadful bracken to bits' . . . As for bracken, they kick off the snake-headed sprouts until, exhausted by constantly drawing on its roots, the vegetable pest curls up and dies, leaving room for more digestible fodder.[27]

However, the presence of bracken was believed to indicate fertile SOIL:

> Under GORSE – copper
> Under BRAMBLES – silver
> Under bracken – gold
> Roughly the value of the crop if the above three are ploughed under.[28]

BRAMBLE or **BLACKBERRY** (*Rubus fruticosus* agg.)

An aggregate of over 320 microspecies, more of which are frequently described; semi-deciduous shrubs with prickly stems, common throughout most of Britain and Ireland, but scarcer in northern Scotland.

Local names, some of which refer primarily to the fruit, include the widespread BRUMMEL and bumblekites, and:

Black begs in Yorkshire
Blackbern in Lancashire
Black bides in Kirkcudbrightshire and Wigtownshire
Black bowours and black bowwewers in Berwickshire
Black boyds in Scotland
Blackites in Cumbria
Black kites in Cumberland and Northumberland
Black spice in Yorkshire
Blaggs in Sheffield
Bleggs in Yorkshire
Boyds in western Scotland
Bramblin in Yorkshire
Brammelkite in Durham
Brammle in Yorkshire, Aberdeenshire and Dumfriesshire
BREAR, BREER and BRERE in northern England
Brier in Northumberland and Yorkshire
Brimble/s in Devon and Dorset
BRIMMLE in Devon
Broomles in Cumberland
Brumble in Norfolk
Brumleyberry bush in Berwickshire
Brumleys in Wensleydale, Yorkshire
Brummel kite in Cumberland
Bullkites in Durham
Bummekites, and bummelkites, in Cumberland, Yorkshire and Berwickshire
Bummell in Cumberland
Bummelty-kites, and bummely-kites, in Cumberland and Westmorland
Bummle-kittes in Cumberland, Yorkshire and Berwickshire
Bummull in Hampshire
CAT'S CLAWS in Somerset
COCK BRAMBLE in Norfolk
Cock brumble in Norfolk and Suffolk
Country lawyers in Leicestershire
DOCTOR'S MEDICINE in Somerset
Ewe-bramble in Somerset
Gaitberry on the Scottish Borders
Garten-berries in Scotland
Gatter-berry in Roxburghshire

GATTER-TREE in Roxburghshire
Hawk's-bill bramble in East Anglia
He-brimmel, applied to 'a bramble of more than one year's growth' in Somerset
Lady garten-berries in Roxburghshire
Lady's garters in Roxburghshire
Land-briars, applied to the 'long tangled shoots' in Shropshire
LAWYERS in Sussex and Warwickshire; 'when once they gets a holt of ye, ye doant easy get shut of 'em'
Moocher in Wiltshire
Mooches in Gloucestershire
Mouches in Hampshire
MULBERRY, 'the universal name among the lower orders' in Norfolk and Suffolk
Mushes in Devon
Scald-berry, 'from their supposed quality of giving scald [infected with a scrofulous disease, eczema or ringworm] heads to children'
Thief in Leicestershire
Yoe-brimble, or yoe-brimmel, in Somerset.

Although blackberries are probably the most widely gathered wild fruit, it seems that in some places they were formerly considered unfit for eating. In Herefordshire:

> An old gardener at Dadnor, near Ross, used to say that blackberries were not good to eat, 'the trail of the serpent is over them,' he said. My informant added that blackberries were not eaten in the district until comparatively recent times.[1]

> During the First World War blackberries were gathered by children in County Monaghan, presumably for export to the UK, but very few local people ate them; 'the one man who did we thought a bit touched on that account'.[2]

Elsewhere:

> Some years ago, before World War II, I was told by a friend living in Caen, France, that [the bramble is a symbol of the CROWN OF THORNS] and therefore the fruit was not gathered.[3]

> The French don't eat blackberries because Christ's crown of thorns was made of brambles. I don't know if that's true, but that's what I was told by a teacher when I was at prep school.[4]

Throughout the British Isles it was believed that blackberries should not be eaten after a certain date.

> [Lake District] The belief that when blackberries have been frosted

they become Devil's Fruit and are no longer fit for human consumption is still held locally.[5]

Pick blackberries before the end of August or WITCHES will poison the fruit.[6]

Blackberries should not be eaten after MICHAELMAS DAY (29 September) as they have the DEVIL in them after that. This has much truth in it in that a fungus attacks the plants about then, I believe. Personally I don't eat them after that day because I imagine they are probably unpalatable! They are usually wet and nasty anyway.[7]

According to my grandmother and mother ('rural Somerset'): never pick blackberries after Michaelmas because the devil peeps over the hedgerows and blasts them.[8]

According to my mother, born Buxton, Derbyshire, 1901, blackberries should not be picked after September 30th, because then 'the witch got into them'.[9]

[Somerset, 1940s] Blackberries – I was told never to pick blackberries after October 6th because after then the Devil had passed over them, or pissed over them, depending on who told you![10]

It was reckoned to be unlucky to pick blackberries after Michaelmas Day (old Michaelmas Day, October 1st), as, on that day, the Devil spits on them.[11]

[Midlands] Never pick blackberry after 31 October. The witches pee on them![12]

[Co. Offaly] The POOKA is out on HALLOWE'EN. It is supposed to crawl on blackberries and after that no one will eat a blackberry.[13]

It was said that blackberries shouldn't be eaten after Hallowe'en (31st October) as the Pooka (a kind of naughty fairy) spits upon them on that night, or does worse, depending on who is telling the story.[14]

Blackberries should not be picked after November 1st (some people say October 1st) so the Devil may have his share.[15]

In north-east England blackberries ripen, and can be picked, later than in the south:

> Blackberry week used to be common in the North East when children had a week off school to collect blackberries – usually coincided with half-term, about the second or third week in October – after Michaelmas.[16]

> R: '. . . we'll get it finished by autumn half-term.'
> M: 'Blackberry week, that's what we call it up north.'
> R: 'I've never understood that, because blackberries are always over by then, there's none left.'
> M: 'Not up north, you must remember we're about seven weeks later up there; there are still some left.'[17]
> My mother [aged 92] asserts that Blackberry Week was a school holiday in late September or early October when she was a youngster in Hylton (near Sunderland), circa 1930. Probably equivalent to half-term now. Children were expected to pick the 'free' fruit to be made into JAM, etc., 'so that it wasn't wasted'.[18]

In Hampshire a spell of fine weather 'generally experienced' at the end of September and early October when blackberries were ripening was known as the 'Blackberry Summer'.[19] In some areas it was thought that the period when blackberries are ripe was inauspicious.

> [Devon] Cats are never very well at blackberry time. Reported by Mrs M.C.S. Cruwys, as told by a man from Cruwys Morchard. HORSES, also unwell at this time; a widespread belief, reported by Miss K.E.F. Bate of Chudleigh Knighton. Chicken also included in this category in many parts. No doubt the observation refers to the slight physiological change many animals undergo in preparation for the winter.[20]

> Jack Hurley relates that, when attending the funeral of a young man who had committed suicide at Watchet [Somerset], a woman said to him: 'Ah, you know what they say, the blackberries be about'; inferring, he assumed, that the fall of the year symbolised the waning of man's powers, his life's autumn and depression, sometimes leading to suicide could be expected.[21]

> Kittens born in September are known as 'blackberry kittens' and are usually small and weak and difficult to rear. They are also very mischievous and naughty – more so than those born in other seasons.[22]

However, in west Dorset, 'blackberry chickens' were 'always the best':

> We couldn't afford to support a hundred and eighty fowls through the winter . . . so I got old Gappy from the village to come up one day . . . we put fifty of the oldest hens in wooden crates, and he took them off . . . A few days later, an ageing hen who had stolen a nest walked out of the bushes with eleven tiny chicks tottering behind her . . . 'They'm blackberry chickens,' cried old Daisy delightedly when she saw them, 'always the best, born late – but I never knowed 'em born so late as this in the year!'[23]

Occasionally brambles were placed, or planted, on graves.

> Brambles were planted on graves to keep the dead from walking.[24]

> The Old Man of Braughing [Hertfordshire] can rest in peace. For once more the villagers have completed the traditional ceremony of sweeping Fleece Lane and putting brambles on the old man's grave.
>
> The old man 'died' at a ripe old age and the bearers were carrying his coffin down Fleece Lane to the churchyard when they stumbled on some stones and the coffin fell and broke open. The Old Man sat up and soon afterwards was married again. When he died eventually he left cash for Fleece Lane to be swept clean of stones each year and for brambles to be placed on his grave to keep the sheep off.[25]

Brambles send out long arching shoots – stolons – which root when their tips reach the ground to form new plants. Passing under one of these shoots was believed to cure a variety of ailments.

Early in the seventeenth century:

> If your HORSE be shrew-runne [paralysed by a shrew], you shall looke for a briere which grows at both endes, and draw your horse thorow [sic] it and he will be well.[26]

> [Wales] Children troubled with RICKETS were put to crawl or creep under blackberry bushes three times a week, and the same remedy was used for infants slow to walk.[27]

> A [Herefordshire] woman cured her grandson of WHOOPING COUGH by holding him up to inhale the breath of a piebald horse. The boy's sister had the cough very badly, but when she

> got it there was no piebald horse . . . so they passed her under a bramble-bush rooted at both ends, for nine mornings. She got better . . . The bramble bush was supposed to be quite effectual in a recent case at Weobley, but the child was passed under nine times only, and an offering of bread and butter was placed beneath the bramble arch. 'She left her cough there with the bread and butter,' said my informant.[28]

> A widespread cure for whooping cough was to creep under a bramble or briar that had formed an arch by rooting itself at the tip . . . This cure was still in use in 1937 at Wolvesnewton in Monmouthshire. One suspects that a pun on 'whoop' and 'hoop' may have contributed to the popularity of the charm.[29]

> [A Staffordshire cure for whooping cough] was to find a briar on a bramble bush that was growing into the ground at both ends, and pass the child under and over it nine times on three mornings, before sunrise, while repeating:
>
> Under the briar, and over the briar,
> I wish to leave the chin cough here.[30]

> [Somerset] HERNIA could be cured by passing the patient under a blackberry bramble which re-rooted a distance from the original root.[31]

> [Dorset] To creep under a bramble three mornings following against the sun, just as it rises, is said to afford a complete cure for BOILS.[32]

> In Zennor [Cornwall] a certain cure for BLACKHEADS was to crawl nine times around a bramble bush.[33]

In parts of Ireland it was believed that crawling under a bramble arch would bring luck to card-players.

> [Co. Kerry] If a person go under a briar that has both ends growing in the ground and give themselves to the devil, it is supposed that person will have great luck in card-playing.[34]

> [Co. Kerry] If you go under a briar that has taken root at both ends it is said that you will be very lucky playing cards. Others say that going under such a briar gives one exceptional strength and vigorous health.[35]

> W: 'In Ireland in the 1940s they used to crawl under bramble arches to get luck playing cards. Before a big card-party they would do it.'
> R: 'But weren't they afraid of being carried away by the Devil?'
> W: 'They didn't mind about that; they just wanted the money.'[36]

Swellings caused on bramble stems by the gall wasp *Diastrophus rubi* were thought to ease painful joints.

> My mother spent her childhood in Dorset, mainly in Yetminster, and called the galls on brambles 'CRAMP thorns'. We collected them for an elderly lady who suffered dreadfully from painful joints and legs, and she swore by them. She used to wear those thick lisle stockings, so you can imagine they must have been pretty uncomfortable. My mother also used them.[37]

Blackberry vinegar was prepared for medicinal use:

> My mother used to keep a store of blackberry vinegar in the pantry that she made every autumn to ease sore THROATS and COUGHS. This was in the 1960s, and it was something I did when my own children were little in the 1980s, and is something I do now when I have the time because it is so delicious. We also used it with sugar on Yorkshire pudding, pancakes and steamed batter pudding.
>
> Blackberry vinegar: Soak one and a half pounds of blackberries in one pint malt vinegar for two days. Mash and strain. Add one pound sugar to every one pint liquid. Boil till thick and bottle.[38]

> My Grandma used to make blackberry vinegar. I had a bad CHEST so I was made to drink it – it was awful, made one feel really sick.[39]

In addition to eating blackberries, children also ate young shoots of brambles:

> On the way home from school [in the village of Crondall, Hampshire, in the 1920s] she and her friends . . . broke off the young growing shoots of the suckers [of brambles] peeled off the skin and ate the rest. It was sweet and juicy.[40]

> As children in South Wales – Glamorgan – we used to peel young stems of brambles and eat them. We called them pork.[41]

BREASTS – sore, CABBAGE provides relief; infected, CELERY cures; SAGE used to dry up breast milk.

BRASSICACEAE (formerly known as Cruciferae)

A family of mostly herbaceous plants, which includes important vegetables such as CABBAGE, RADISH and TURNIP, and ornamentals such as HONESTY, STOCK and wallflower, Brassicaceae are characterised by their flowers having four petals, which when seen from above form a cross.

> I was told by my father, born in 1872 on a farm in Wiltshire, that it is safe to eat all cruciferous plants – they have the sign of the Cross and so would bring you no harm.[1]

Conversely:

> I knew a lady who would never have any 'cross' flowers (Cruciferae, LILAC, etc.) in her house because they were associated with the CRUCIFIXION.[2]

BRIDGET, 'goddess of fertility' – tenuously associated with a BLACK POPLAR at Aston-on-Clun.

BROAD BEAN (*Vicia faba*)

Cultivated in the Middle East for over 8,000 years, in gardens as a vegetable since 1200, and in fields for animal feed.

> I have heard it said that every seventh year the bean turns in the pod. The normal attachment of the bean to the pod is reversed at the seventh year and then reverts to normal attachment the following season. My father believed this to be true and gives the following account of the learning of this fact. 'In the year 1919, my brother George and myself were carting stone to a road-building site at Westwoodside. Old Edmund Cooper came up to us and said, "Do you know that this year is the year when the beans hang t'other side of 't pod?" We had never heard of it so he said, "Come down this lane to a field of beans and I'll show you what I mean." We were shown that the beans were indeed hanging the wrong way in their pod.' My father claims he has checked this phenomenon in later years and has found it to be quite true.[1]

The scent of broad bean flowers was considered to be an APHRODISIAC.

[Oxfordshire, *c*.1920] There ent no lustier scent than a beanfield in bloom.[2]

[Suffolk, between the two World Wars] Peas and beans inflame lust . . . best of all traditional aphrodisiacs was the scent of the bean flower, for this not only stimulates passion in the man, but extreme willingness in the girl. The frustrated lover was always told 'take her into a bin field boy, and if there's a thorn bush or bit of barbed wire, back her up agin it and she'll keep a' comen farrad to ye'.[3]

I remember when working on a farm near Thetford beans were sown on the field beside our row of houses. It was a common chant to me for the weeks as they grew and flowered giving off their delicious scent that 'It wouldn't be long before the missus was calving'. It had no effect on me I must say. The chant came from the two young owners who were Suffolk bred people.[4]

Around Newcastle-on-Clun, Shropshire, it was thought that broad beans should be planted on the shortest day.[5] In Huntingdonshire:

On ST VALENTINE'S DAY
Beans should be in clay.[6]

In Wiltshire the first two days in March were preferred:

Sow PEAS and beans on David and Chade
Whether the weather be good or bad.[7]

It was widely recommended that four bean seeds should be planted for each seedling which would eventually reach maturity.

One for rook, one for crow;
One to rot, one to grow.[8]

The white fluffy inside skin of broad bean pods was frequently used to cure WARTS.

To cure warts – break open a broad bean pod – eat the beans and then rub the warts with the inside of the skin. Finally bury the skin, by the time it has rotted your warts will have disappeared. This is more effective when done by moonlight.[9]

A cure which I found very useful in my own family on several occasions is for warts – the ones that appear on hands. Rub the furry inside of a broad bean pod case on the wart for two or three nights running, and they simply shrivel and disappear.[10]

Less widely recorded cures included:

> Carry a few horse beans [a variety of broad bean] in waistcoat pocket – a sure cure for THIRST.[11]
>
> Remedy for WHOOPING COUGH: carry the babe through a field of beans in blossom, walking up and down the rows and let the child inhale the powerful scent.[12]
>
> One country cure that my grandmother used – she had a large family (10 children) and five were boys. As they grew up, from time to time they had barbers' RASH – a condition which one never hears about nowadays . . . However, my grandmother found the cure which was soothing and cooling, etc. – it was the inside layer of a broad bean pod – which as you know is soft and cool. Just rub all over the face and the rash disappeared quickly![13]

In west Dorset:

> John W came down last night. He had a COUGH, and if you caught a cough in March you never lose it until the broad beans are in flower.[14]

BRONCHITIS – treated using COLTSFOOT, MULLEIN, NETTLE and YARROW.

BROOM (*Cytisus scoparius*)

Short-lived, yellow-flowered shrub (Fig. 4), common, and, when in flower, conspicuous, mostly on sandy soils; often planted beside motorways.

Local names, many of which relate to the former use of broom for making brooms or besoms, include:

Banadle in Wales
Banathal, and bannal, in Cornwall; 'from the old Cornish and Welsh name, bannadle'
Basam, or basom, in Devon; 'as yellow as a basom' is a common south Devon expression
Bassam in Devon; 'from its use in making brooms or besoms'
Beem in Aberdeenshire
Beesom or beeson, in Devon
BESOM in Cornwall
Bisom in Devon

Bissom in Somerset
Bizzam, or bizzom, in Devon
BREEAM in Whitby, Yorkshire, where an infusion, known as breeam teea, was used as a DIURETIC
BRUM in Shropshire
Brume, 'of the ballads' in Scotland
BRUSHES in Dorset
Cat's peas in Cornwall
GOLDEN CHAIN in Dorset and Somerset
GOLDEN ROD, and green, besom in Somerset
Green broom in Hertfordshire
Green-wood in Somerset
LADY'S SLIPPER in Somerset
LING in the East Riding of Yorkshire
Little fair-one, and woodwax, in east Devon.

> [West Sussex] A good old gentleman of my acquaintance . . . strictly forbade green brooms being used during the month of May, and, as a reason for this prohibition, used to quote the adage:
>
> If you sweep the house with broom in May,
> You'll sweep the head of that house away.
>
> and this superstitious association between broom and DEATH in the month of May is extended to its blossom. A poor girl, who was lingering in the last stage of consumption, but whose countenance had always lighted up with pleasure at the sight of flowers, appeared one morning so exceedingly restless and unhappy, after a fresh nosegay of gay spring flowers had been laid on her she exclaimed, 'they are very nice indeed to smell; but yet I should be very glad if you could throw away that piece of yellow broom; for they do say that death comes with it if it is brought into the house in blossom during the month of May.[1]
>
> Hilda [b. Reading, Berkshire, *c.*1900] remembers her mother telling her:
>
> 'Don't bring any broom home and don't bring any may [HAWTHORN]. She said:
>
> If you bring broom into the house in May
> It will sweep the family away.'[2]

Broom was used as a remedy for JAUNDICE and RHEUMATISM.

> [Norfolk] An infusion of broom (flowers, stalks and root), boiled and strained, is given in small doses as a cure for jaundice.[3]
>
> A herbalist, practising in Tunstall, Staffordshire, about 50 years ago, used to collect sprigs of broom from the local countryside,

> which he then cut up into short pieces, placed in packets and sold as a cure for rheumatism, instructing the customer to make a tea from the sprigs by pouring boiling water over them.[4]

PLANTA GENISTA, the plant from which the Plantagenet dynasty derived their name is usually considered to be broom.

BRUISES – BLACK BRYONY, COMFREY, DEVIL'S-BIT SCABIOUS, FAIRY FLAX, GREATER PLANTAIN, MALLOW, POTATO, ROYAL FERN, SOAPWORT, SOLOMON'S SEAL, and YELLOW HORNED-POPPY used to treat.

BUCK'S-HORN PLANTAIN (*Plantago coronopus*)

Low-growing perennial herb, common in coastal areas, and spreading inland along salt-treated roads.

Alternative names include the unlocalised grace-of-god, and, at one time in East Anglia star-of-the-earth or earth-star:

> [*c*.1683] Star of ye Earth grows on Newmarket heath and is of extraordinary and admirable virtue in curing the bitings of mad dogs either in Beast or Man being infus'd in wine with treacle and one or two more simples.
>
> Near Elden they call it Earth Star and give ye whole plant bruised and rowled up in Butter and water to Sheep bitten by mad-dog. The cure performed two or three times.[1]

> [A well-remembered cure for HYDROPHOBIA] was buck's-horn plantain. This formerly grew in abundance . . . in South Glamorgan . . . the root and leaves made into a decoction sweetened with honey, and administered to the patient.[2]

BULRUSH (*Typha latifolia*), formerly known as reed-mace

Rhizomatous perennial herb, common and conspicuous beside ponds, rivers, ditches and canals in lowland areas.

Local names include:

Baccobolts on the Isle of Wight; 'from a resemblance in the spike to a roll of tobacco'
Blackamoor on the Isle of Wight
BLACK BOYS in Wiltshire
Black-headed-laddies in Berwickshire
BLACK HEADS on the Isle of Wight

and in Somerset
Blackie-toppers in Somerset
Black cap in Cumberland and Somerset
Black puddings on the Isle of Wight and in Somerset
Black sticks, and blacky-more, in Somerset
Bullsegg in Scotland
Candlewick in the Lake District
CAT-O'-NINE-TAILS in Lincolnshire and Warwickshire
CAT'S TAILS in Hampshire, Lincolnshire and Yorkshire
CHIMNEY-SWEEP in Somerset
Devil's poker in Somerset
Dod in northern England
FLAGS, applied to the leaves, in Hampshire
Flax-tail in Kent; 'the fruiting heads are downy like finely-combed flax'
Flue-brushes in Somerset
Gladden in Norfolk
HARD-HEAD in Lancashire
Holy pokers in Devon
Lewers, and lyver, in Somerset
POKERS in Cheshire and Kent
PUSSIES, and pussy's tails, in Somerset
Rush-cane in Hampshire
SEG in Worcestershire
Sweeps in Somerset
Whiteheads in Devon; 'when the downy matter [of the spikes] has ripened and lost the colour which gave them the name of blackheads'

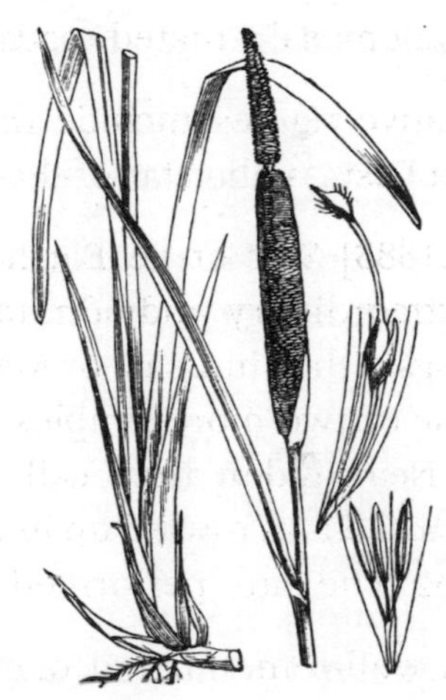

There are occasional records of bulrushes being considered unlucky and it is probable that this superstition was encouraged by adults who wanted to deter children from the potentially dangerous places in which the plant grew (cf. DUCKWEED). However, as they dry the seed heads of bulrush disintegrate releasing masses of tiny fluffy seeds which drift around and attach themselves to furnishings, making them unwelcome in the homes of even the moderately house-proud.

> Bulrush – near River Itchen, Winchester, Hants, in Irish household, about 1967/8. Child brought bulrushes into house and same day scalded herself with kettle of boiling water. An aunt who was visiting (southern Irish origin) immediately broke the bulrushes

into pieces and threw them out of the house saying they were the cause of the accident.[1]

We once had a vase of bulrushes in a corner of the room and on seeing them my mother was horrified and advised us to remove them.[2]

[Birmingham] I was told never to bring bulrushes into the house as they bring DEATH in the family.[3]

In Ulster:

[For HALLOWE'EN] we collected *Typha* (which we called bulrushes then), soaked the heads overnight (upside down in a suitable container of course) in paraffin oil and used them as torches. Our parents didn't like this, as they thought it was dangerous.[4]

BURDOCK (*Arctium* spp.)

Coarse perennial herbs (Fig. 5), growing on waysides and waste ground, with characteristic stiff hooks on the flower-buds and seed heads.

Many local names refer to the seed heads, others refer to the plant's resemblance to RHUBARB:

Bardog in Shetland
Bazzies in Kent
Beggar's buttons in Dorset
Billy button in Devon; 'boys are fond of sticking them down the front of their coats to give them the appearance of a "man of buttons", or a page'
BOBBY BUTTONS in Somerset
Buddy-weed in northern England
BUTTER-DOCK in Buckinghamshire
BUTTONS in Somerset
CACKEY MONKEYS in Flintshire
Castor-oil plant in Yorkshire
Clay-bur in Orkney
CLEAVERS in Somerset
Clingers in Dorset
CLITCH-BUTTON in Devon
Clite in Gloucestershire and Somerset
CLOGWEED in Wiltshire
CLOT-BUR in Cumberland and Ireland
Cloud-burr in Cumberland and Yorkshire
CLOUTS, and CLUTS, in northern England
COCKLE in Dorset
Cocklebells in Cornwall
Cockle-button in Cornwall and Dorset
Cocos in Dorset
Cradan, and credan, in Co. Donegal
Cuckle/s in Dorset, Somerset and Ireland

Cuckle-buttons in Devon
Cuckle-dock in Cornwall
Cuckold in Somerset
Cuckold buttons in Somerset
Cuckold dock in Cornwall, Devon and Somerset
Cuckoldy-burr-busses in Cumberland
CUCKOO-BUTTONS in Devon and Somerset
Cuckow in Dorset
DONKEYS in Somerset
Eddick in Cheshire
FLAPPER-BAGS in Scotland
Gookoo-buttons in Somerset
Gypsy comb in Berkshire and Nottinghamshire
GYPSY'S RHUBARB in Somerset
HARD-HEAD in Cornwall
HAYRIFF in Somerset
Kali-monkeys in Liverpool
Kisses, KISS-ME-QUICK, and loppy major, in Somerset
OLD-MAN'S BUTTONS in Devon
Pig's rhubarb, and RHUBARB, in Dorset
Roe-dock in Devon
SNAKE'S RHUBARB in Dorset
SOLDIER'S BUTTONS in Devon, Somerset and Wiltshire; 'c.f. the French *boutan de soldat*'
STICKLE-BACKS in Lancashire; 'because of their habit of attaching themselves to people's clothes'
Stick poppies in Cheshire
STICKY BOBS in Cheshire; 'applied more particularly to the infructescences which boys and girls are so fond of throwing at one another'
STICKY BUDS in Dorset
Sticky burrs in Cornwall
STICKY BUTTONS in Devon
Sticky jacks in Somerset
Sticky monkeys in Liverpool
Sticky willow in Argyllshire
SWEETHEARTS in Dorset
TOUCH-ME-NOT and TURKEY-RHUBARB in Somerset
TUZZY-MUZZY in Devon and Somerset
WILD RHUBARB in Somerset.

> [Cornwall] 'Piskies' or 'PIXIES' a race of fairies or 'small people' are said to amuse themselves at night by riding colts furiously around the fields and plaiting their manes, or tangling them with 'BILLY BUTTONS' [i.e. the burrs] of burdock.[1]

> As children in Essex we threw the burrs of burdock on to the backs of unsuspecting friends – if they stuck you had a sweetheart; if they fell off after a short while their affection would not be reciprocated. I lived in the then countryside of Chigwell/Hainault area, but my children played the same game 20 years later at Witham, Essex.[2]

Burdock burrs are an essential feature of the costume of the Burry

Man who appears at South Queensferry, Lothian, on the second Friday in August each year, when he 'perambulates the town visiting the houses and receiving cheerful greeting and gifts of money from householders'.[3]

> Burrs are collected in sacks the week before, dried and cleaned then assembled into square or rectangular patches each comprising around 500 burrs. On the Bury Man's Day, he starts dressing at 7 a.m., first putting on a set of long underwear and hood, then standing patiently while the patches are pressed on and individual burrs are carefully placed in sensitive areas, such as the oxters (armpits) and crotch. It takes two hours until all the burrs, plus flowers at the shoulders, hips and knees and a flowery hat are in place. Two men support the Burry Man, who walks slowly and stiffly around the town from 9 a.m. to 6 p.m., able to see a little through peepholes in his face mask and drink through a straw, neat whisky being his traditional tipple.[4]

Nothing is known about the origin of this enigmatic custom, which does not seem to have been recorded before the mid-nineteenth century. According to one popular book on folk customs:

> There are two theories suggesting an explanation of this strange custom. The first is that the Burry Man is a manifestation of the Spirit of Vegetation in another guise, similar to the 'Little Leaf Man' of Central Europe and 'Jack-in-the-Green' of England. The other theory is that it is a survival of the scapegoat of antiquity. Like the scapegoat, the Burry Man was probably believed to carry away all the evil influences from the community and would be driven from the village.[5]

Alternatively, 'it commemorates the landing at South Queensferry of Queen Margaret, the saintly wife of King Malcolm Canmore [*c*.1031–93], from whom the town derives its name'.[6]

Gypsies used burdock to prevent or cure RHEUMATISM:

> Infusion of leaves or flowers, or better still of crushed seeds, relieves and will cure rheumatism . . . Some gypsies carry the seeds in a little bag slung round the neck as a preventative of rheumatism.[7]

There are rare records of burdock being eaten as a vegetable:

I had a great-aunt who had done this in *c.*1895–1910 in the Chesham area of Buckinghamshire. I have recollections of reading that burdock 'greens', which were unpleasantly glutinous, were one of the disadvantages of life in the workhouse.[8]

BURNET ROSE (*Rosa spinosissima*)

Erect suckering shrub, with yellowish-white flowers and blackish-purple fruits, found mainly in coastal areas and also on inland heaths.

Local names include:

Barrow-rose in Pembrokeshire
Brid rose in Cheshire; 'i.e. bird rose'
Cant-robin in Fifeshire
Cat-hep or CAT-HIP in northern England and Berwickshire
Cat-rose in Yorkshire
CAT-WHIN in northern England
Fox-rose – in Warwickshire
SOLDIER'S BUTTONS in Kirkcudbrightshire.

On the small islands of Steep and Flat Holes, Sully and Barry, in the Bristol Channel, the blossoming of the burnet rose out of its proper season was regarded as an omen of shipwreck and disaster.[1]

BURNS – treated using BETONY, CELERY, ELM, GREATER PLANTAIN, HART'S TONGUE, HOUSELEEK, IVY, LAUREL, NAVELWORT, PRIMROSE and ST JOHN'S WORT.

BUTCHER'S BROOM (*Ruscus aculeatus*)

Short, evergreen shrub (Fig. 6) with scarlet berries, native to southern England, planted as a curiosity and sometimes naturalised elsewhere.

With its persistent 'leaves' (really cladodes, flattened stems), each of which has a sharp spine at its apex, and rather sparse red berries, butcher's broom was often compared with holly, giving rise to local names such as:

Box-holly on the Isle of Wight and in Somerset
Butcher-holly in Jersey
French holly on the Isle of Wight
HOLM in Wiltshire
Jerusalem thorn in Lancashire
Jew's myrtle in Essex; 'as the material of the CROWN OF THORNS
Kneed holly in Essex
Knee-holly in Essex and Kent; 'should this rather be ne-holly, i.e. not holly?' A more probable

explanation is that butcher's broom is a 'holly' that grows to no more than knee-height.
Knee-holm in Hampshire, the Isle of Wight and Sussex
Knee-hull in Essex
No holm, and no home, in Sussex; 'not holly'
Roman holly, 'in the neighbourhood of ROMAN roads', in Dorset
Shepherd's myrtle on the Isle of Wight.

> Butchers are said to make use of it in some parts of England for driving away, and perchance impaling on its sharp spines, the FLIES that settle on their meat and chopping-blocks. The more gentle of the craft with us [on the Isle of Wight] are content to deck their mighty CHRISTMAS sirloins with the berry-bearing twigs, and it contributes at that festive season, with other evergreens, to the decoration of our churches and dwellings.[1]

> [In 1966 I noted down from a great-aunt born in 1881] a cure for DROPSY was butcher's broom boiled and put on the affected part in poultice form.[2]

Butcher's broom growing in the grounds of the Carmelite Monastery in Notting Hill, London, is said to 'provide a link with vernacular uses as it was originally planted . . . to supply berries for use as hatpin ends'.[3]

BUTTERBUR (*Petasites hybridus*)

Perennial herb producing early spring flowers followed by large rhubarb-like leaves, widespread on moist fertile soils throughout lowland areas.

Local names include:
BATTER-DOCK in Cheshire
Bog-horns in Lincolnshire; 'children used the hollow stems as horns or trumpets'
Bog rhubarb in Lincolnshire and Somerset; 'from the likeness of its leaves to rhubarb, and its growth in swampy places'
Burbleck in Westmorland
Burn-blade in Kirkcudbrightshire and Wigtownshire; 'i.e. stream leaf'
Butcher's rhubarb
Butterburn in Buckinghamshire, Cambridgeshire and East Anglia
BUTTER-DOCK in Cheshire
Cap-dockin in Cleveland, Yorkshire
Cleat in Lincolnshire and Yorkshire
Clots in Cheshire and northern England

CLOUTS in Cheshire
CLUTS in northern England
Dunnies in Hampshire
Early mushroom in Dorset
Elden in Northumberland
Eldin-docken in northern England, Berwickshire and Roxburghshire
Eldins in Cumberland
Ell-docken in Northumberland
FLAPPER-BAGS in the Scottish Borders
Gallon in Belfast
Gaun in Lanarkshire
Gaund in Dumfriesshire
GYPSY'S RHUBARB in Hampshire
KETTLE-DOCK in Cheshire
Paddy's rhubarb in Scotland
Poison rhubarb in Yorkshire
SHALAKI in Orkney
Sheep-root in Scotland
SNAKE'S FOOD in Somerset
SNAKE'S RHUBARB in Dorset
SON-BEFORE-FATHER in Clackmannanshire; the 'flower appearing before the leaf'
TURKEY RHUBARB in Somerset
Umbrella-leaves in Yorkshire
Umbrella-plant in Somerset
UMBRELLAS in Flint and Somerset
Water docken in Cumberland
WILD RHUBARB in Wiltshire and Yorkshire.

A number of victims of the PLAGUE were buried in Veryan churchyard [Cornwall], on their graves nothing but plaguewort – butterbur – will grow.[1]

As children we considered butterbur and convolvulus (or FIELD BINDWEED) as absolute poison and no one would have dared pick them. When the leaves of the butterbur appeared we knew them as rat-leaves and could safely play amongst them.[2]

I remember as a child having . . . a cream being made from butterbur and applied to spots and SORES.[3]

BUTTERCUP (*Ranunculus* spp.)

Yellow-flowered herbs, common in pastures and other grassy areas.

Names which seem to have been applied indiscriminately to more than one species include, crae-taes and CRAW-FOOT in Scotland, eggs-and-butter in Cheshire, and sugar basins in Somerset.

Names given to **meadow buttercup** (*Ranunculus acris*), which is widespread in damp, rarely ploughed, fields, include:

BACHELOR'S BUTTONS, probably given to a double-flowered cultivated form
BLISTER-CUP in Lincolnshire
Blister-plant, 'used by herb-women for blisters', in Lincolnshire,

presumably used to produce, rather than treat, blisters.
Clovewort in Northamptonshire
COWSLIP in Devon
CROW-FLOWER in the Midlands, Rutland and Staffordshire
Goldy in Somerset
Gowan in Wigtownshire
KINGCUP in Yorkshire
LADY'S SLIPPER in Somerset
Pigeon's foot in west Somerset
Ram's glass in Somerset; 'no doubt a corruption of rams'-claws'
YELLOW BACHELOR'S-BUTTONS, applied to the double-flowered garden form, in Ayrshire.

Creeping buttercup (*Ranunculus repens*), an extremely common perennial herb with creeping stems which enable it to rapidly spread as a weed, has been given names which include:
BUTTER-AND-CHEESE in Yorkshire
Cat-clawks in Lancashire
Cat-claws in Lancashire
CRAW-FEET in Yorkshire and Scotland
Craw-tee in Donegal
Craw-tone, 'the Devon pronunciation of crow toes'
CRAZY, 'the smell . . . was formerly supposed to induce madness'
Crazy-more in Wiltshire
Creeping crazey in Gloucestershire
Crow-toe in Donegal
Delticups in Wiltshire
DEVIL'S GUTS in Northumberland; 'indicating its troublesomeness and its habit of throwing out runners'
Gold-knobs in Dorset
Granny-threads in Yorkshire
Kraatae in Shetland; 'crow's toes, from the shape of the leaves'
Lantern leaves in western England
Lawyer-weed in Somerset
Many-feet in Yorkshire
Meg-many-feet in Cumberland
Meg-wi'-many-teaz in the Lake District
Meg-wi'-mony-feet in Cumberland
Old-wife's threads in Yorkshire
RAM'S CLAWS in Dorset, Somerset and Wiltshire
Set-fast in Donegal
SIT-FAST in Scotland and northern Ireland; 'from the tenacity with which its roots cling to the ground rendering the plant difficult to eradicate'
Sit-sicker ('in allusion to its close adherence to the ground of its rooting stems'), and sit-sikker ('i.e. it sits sure – it is hard to get rid of'), in Scotland
Tangle-grass, and toad-tether in Yorkshire

Bulbous buttercup (*Ranunculus bulbosus*) which closely resembles creeping buttercup, but grows from a corm and lacks creeping stems, grows in similar habitats, mainly in lowland Britain. It is probable that few people other than botanists differentiate between the two species, but names which have been recorded as being given to bulbous buttercup include:

CROW-BELLS in Warwickshire
Crow-pickel in Northamptonshire
Crow-pigtel in Bedfordshire and Northamptonshire
Cuckoo-buds in Northamptonshire
EGGS-AND-BACON in Cheshire
FAIR-GRASS in Roxburghshire
GILDCUP, and giltcup, in Dorset
Maiden-in-the-meadow in Somerset

'Rustics in the Midland Counties' knew the 'common meadow buttercup' by the name of crazy:

> 'Throw those nasty flowers away' said a countrywoman to some children who had gathered their handfuls of buttercups, 'for the smell of them will make you crazy'.[1]

> [Gloucestershire] A gardener, who was with two generations of my family for 40 years, called *Ranunculus repens* 'CRAZIES'.[2]

A widespread childhood pastime is placing a buttercup under playmates' chins to see if they like butter:

> Buttercups held under the chin to see if one is as good as gold or likes butter.[3]

> If you hold a buttercup under your chin and you get a yellow colour that means you like butter. I did it to my grandmother the other day, and there was yellow, and she said she did like butter.[4]

> [North Hampshire, *c.*1940s] Hold a buttercup under the chin and say, 'Do you like butter?' If there is a yellow reflection on the skin, it is yes. 'Do you like cheese?', same again, I think. 'Do you like sitting on housemaid's knee?' Occasion for much giggling.[5]

Butter was a luxury which the poor rarely tasted, and thus it became be a symbol of wealth.

> [After a christening at Shottery, Warwickshire, in the 1930s] Else held a buttercup under his chin. ''E's goin to be rich!' she said.[6]

Sometimes it was thought that buttercups were responsible for giving butter its rich golden colour:

> If you see the buttercup weed plentiful on a pasture field it was supposed to be rich as it enriches the milk of cows and improves the colour of the butter made from such milk.[7]

> If someone had a grudge they would pull buttercups from a field, and that would make the milk go away. I'm not too sure about this, but it was done an awful lot in Ireland.[8]

Due to their acrid properties, buttercups were widely used in folk medicine.

> [In Cornwall creeping buttercup was] sometimes known as 'Kennel Herb' or 'KENNING HERB' from its use in making an ointment for the cure of 'Kennels' or 'Kennings', the local name for ulcers of the eye [STYES].[9]

> Crowfoot or buttercup (*Ranunculus repens*) if applied and held in position with a bandage would cause rheumatic joints to blister and was said to cure this complaint.[10]

> Put Vaseline into a pan with as many buttercup flowers (without stems) as can possibly be pressed into it. Allow to simmer, not boil, for ¾ hour. While hot strain through muslin into small pots. It is ready when cold and is very good for all SKIN trouble.[11]

> Wild buttercups were used in the treatment of PILES for many generations. The ground up roots were boiled with lard to make an ointment for external use, and two handfuls of leaves boiled in a pint of water made a decoction that was reputed to be of great benefit. A wine-glass full of the strained liquid was taken three times a day . . . Buttercup petals were used as a cure for JAUNDICE in the belief that since the complaint causes a yellowing appearance in the sufferer a yellow remedy will provide the treatment.[12]

> CHILBLAINS – rub with buttercup at night.[13]

It appears that the 'roots' of bulbous buttercup were occasionally eaten.

> Another piece of history is recalled in the name St Anthony's turnip for the bulbous buttercup, alluding to the swollen stem-base. These 'bulbs' were gathered as food in famine years but the toxins burnt the mouth, reminding people of St Anthony's fires of torment.[14]

> Some types of buttercup have a bulb like a spring onion (very hot! like a radish).[15]

'Are you interested in things like people digging up the roots of *Ranunculus bulbosus* and eating them like radishes?'
'Yes. I've never heard of that one. Have you done it?'
'No, I haven't, but I know people who have. I imagine they're very hot.'[16]

BUTTERWORT (*Pinguicula vulgaris*)

Perennial, rosette-forming, insectivorous herb, common in damp nutrient-poor habitats throughout much of Ireland, Wales, Scotland and north-west England.

Local names include:
Butterplant in Selkirkshire
Clowns in Roxburghshire
Eccle-grass in Orkney
Kerry violet in Co. Kerry
Mountain sanicle in Ireland
Steepweed in Ulster
Yirnin girse in Shetland; 'used for yarning (curdling) milk'

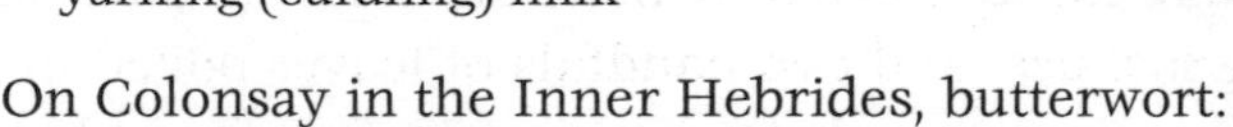

On Colonsay in the Inner Hebrides, butterwort:

> Together with whin [GORSE] and JUNIPER, was believed to act as a charm against WITCHCRAFT. COWS that ate it were said to be safe from elfish arrows and supernatural ailments that were supposed to make much havoc in olden times. It was believed that a healthy, nice-looking baby was sometimes coveted and, when the opportunity occurred, even carried off by the FAIRIES and a languishing, old-fashioned creature left in its place. Some women, as the story goes, who were watching a new-born infant in a house in Machrins to make sure that the child would not be changed heard two fairies coming to the window, and the following conversation take place. 'We will take it,' said one. 'We will not, we cannot,' said the other; 'its mother partook of the butter of the cow that ate the butterwort.'[1]

Elsewhere butterwort was associated with early Christian missionaries.

> I do recall one piece of the lore first mentioned to me by an

> aunt on the Island of Soay (Skye), but later traced to the island story-teller, one John MacRae, whose nephew, a retired windjammer sailor and fisherman passed on to me many of the local tales John had told him (*circa*) 1840s . . . this item referred to a little heath flower that grew among the peat heather of Soay – I'm ashamed to admit I do not know its name. The saying went that where the flower grew was where St Moalrudha touched the ground with his staff. I suppose that it was meant to signify the extent of St Moalrudha's travels, which were, indeed, very wide. Professor Charlesworth [1889–1972], the well-known Northern Irish Geologist, once told me that when he was young in Antrim the flower was known as St PATRICK's spit, or St Patrick's staff, attached to which was the story that while crossing a wide bog St Patrick lost his STAFF and unable to find a tree came upon one of these plants whose stem had grown so long and strong that he was able to use it as a staff. The flower on the head (which is shaped with a curve like a staff) never faded and afterwards whenever the staff touched the ground the flower sprang up.[2]

As butterwort grows in desolate, unproductive areas, perhaps these legends suggest not only the saints' wide travels, but also their concern for impoverished communities.

In common with other plants, such as BOG ASPHODEL and SUNDEW, which grow in similar habitats, butterwort was believed to cause ill health in cattle and sheep. This belief gave rise to such names as rot-grass and sheep-rot in northern Britain.[3]

Butterwort was widely used as a substitute for RENNET in cheese-making. In Lanarkshire it was known as earning grass:

> [Earning is a North Country word for cheese rennet] and to earn means to curdle milk. The plant is so called because of this property.[4]

C

CABBAGE (*Brassica oleracea*)

First recorded as a cultivated vegetable in the mid-sixteenth century; believed to be native on sea-cliffs around parts of Britain.

Throughout Ireland cabbages were used for LOVE DIVINATION at HALLOWE'EN.

> On Hallow Eve Night the boys and girls go to somebody's cabbage-garden, and each one pulls up the first head of cabbage he meets. By this he can tell the future. If there is a bit of clay on it he will get a rich wife. If it is a long straight stalk she will be tall. If it is a short stalk she will be small. If it is crooked then so will she be. The owners of cabbage-gardens have often to sit up all night, watching the garden or all the cabbages would be destroyed.[1]

> [Co. Down] The girls were blindfolded and went out in pairs, hand in hand, to the garden or field and told to pull the first cabbage they found. Its size and shape – whether it was big or small, straight or crooked – would indicate the shape and stature of their future spouse. If much earth adhered to the root they would have plenty of money; if there was only little they would be poor. The taste of the 'custoc', i.e. the heart, would tell them his temper and disposition, according to whether it was sweet or bitter. Finally the 'runts' or stems were hung above the door; each was given a number and the name of a boyfriend, for example Barney might be the name given to the third runt. If Barney was the third person to enter the house on the night, this was considered to be a good omen.[2]

In England the following verse was often written in Valentine cards and schoolgirls' autograph albums:

> My love is like a cabbage
> Often cut in two,

The leaves I give to others
The heart I give to you

On the Shetland Islands:

> Cabbage plants, complete with roots, were thrown through open doors and down chimneys at Hallowe'en. This was usually accepted in good spirits, although some house-proud women may have regarded it in a different light.[3]

Various traditional practices were followed to ensure that cabbage plants thrived.

> One always planted cabbages on the new MOON or the next day and we would plant SAGE or THYME in the rows to keep away pests.[4]

> A slice of RHUBARB placed in the bottoms of a 'dibbed' hole will prevent club root in brassicas. (It does!)[5]

Adults told inquisitive children that BABIES came from cabbage patches:

> [Co. Monaghan, *c*.1910] I knew the secret of birth before I was five years old . . . The small children at school and on the lanes debated the whereabouts of the finding the latest baby.
> 'Under a stalk of cabbage he was found.'
> 'Lord, and do ye know where I was found?'
> 'Where?'
> 'Under a stalk of cabbage.'
> Nearly all the children seemed to derive from the caterpillars. I didn't mind the lie but the lack of originality and variety rather annoyed me.
> When I was near six years I could stand it no longer.
> 'Head of cabbage,' I jeered at a fellow. 'That's not where ye come from.'
> He looked at me surprised that I should presume to doubt the authenticity of the legend . . .
> 'Was I not found in Brennan's cabbage-garden?'
> 'You were not,' I said. 'Your mother had you.'[6]

> [Belfast, *c*.1930] When my brother was born, he's four years younger than me, I asked my father where I'd come from. He told me from under a cabbage.[7]

Easily obtainable, cabbages were valued in folk medicine, and at present are used to provide relief from sore breasts.

> FEVER: heat cabbage leaves and place them on the soles of the feet.[8]

> [Gypsies] bind a cabbage leaf round the leg for swellings . . . an elderly woman I know had a swollen knee and tied a cabbage leaf round it and she said afterwards that it helped. That was only last year.[9]

> An orthopaedic surgeon at Bristol's Southmead Hospital was rather taken aback when a 72-year-old-woman with severe ARTHRITIS of the knee lifted her skirt to reveal a large cabbage leaf firmly strapped in place – which, she claimed, was the 'only thing' that relieved her pain.[10]

> [Ulster] Varicose ULCERS, and ulcers in general, sometimes treated with a plaster of fresh cow dung. Another, more usual, method was to treat the ulcer with fresh, green cabbage leaves.[11]

> If anyone in the family got a gathering [inflamed swelling] . . . mum would get a clean piece of cabbage leaf and bind it to the place required to bring it to a head. I did this recently – using a brussel sprout leaf as the gathering was small.[12]

> As a boy I wandered the countryside with a poacher just after World War II. My old friend always drank cabbage water after a HANGOVER, as a cure that is. If he had a HEADACHE, say from colds or flu, he would pick cabbage leaves and chew them, as quite a few locals did also.[13]

> My daughter has just been recommended to put cabbage leaves in her bra to prevent/ameliorate sore nipples after childbirth![14]

> When I was really ill [with cancer] in about 2000–1 they gave me cabbage leaves to put in my bra.[15]

A riddle which was frequently contributed to the Irish Folklore Commission's 1937–8 Schools' Folklore Scheme ran:

> 'Patch upon patch without any stitches; riddle me that and I'll buy you a pair of breeches.'
> 'A cabbage.'[16]

In the nineteenth century and at times of scarcity wild (or sea) cabbage was occasionally gathered for food.

> Though very bitter in their uncooked state, they [the leaves] may, by repeated washings, be rendered fit for food, and they are often boiled and eaten at sea-coast towns. At Dover they are gathered by boys from the cliffs, and carried about for sale.[17]

> I have had a call from a man whose father while on duty in Kent in the First World War stole the flour which was used during target practice and made dumplings with it to eat with 'cliff cabbage'.[18]

During the nineteenth century and the early part of the twentieth a giant cabbage, the stems of which were reputed to reach up to twenty feet in height, was widely cultivated on Jersey. The leaves of these plants were pulled off and fed to cows. When the growing points were about to run to flower they were cut off and either used for human consumption or fed to cattle, leaving the woody stalks, which were unsuitable for food, and used for making walking sticks. A leading manufacturer of these was Henry Charles Gee, who sold them at his shop in St Helier from the 1870s until 1928. During the late Victorian period Mr Gee sold some five or six hundred sticks a year, but by the late 1930s, when his daughter ran the business, only about one hundred and fifty were sold each year. In the sixth (*c.*1907) edition of Ward Lock's guide to the Channel Islands it was reported that 'the craving by visitors for these sticks is humorously known as the Jersey fever. Nearly all holiday-makers succumb.' Until at least as late as 1969 it was possible to see small clumps of giant cabbage growing on the island.[19] However, well into the 1980s L'Etacq Woodcrafts continued to make walking sticks, and also made from the stalks 'thimbles, fly swats, shoehorns, corkscrews, collector's eggs, keyrings and lighters'.[20]

CACTI

Succulent, usually leafless and often spiny perennials, mostly native to the New World, cultivated as houseplants of fluctuating popularity.

> Between 1964 and 1966 three instances have been recorded, two in Cambridge and one in Grantchester, of the belief that it is UNLUCKY to bring any species of cactus into the house. One of the informants said that not only would she never have one of

> these plants, but she would never give one to anybody else, 'they are so unlucky'.[1]

> Cacti are considered unlucky inside the house in Hungary. An aunt emigrated to England from Hungary and started a collection of cacti. Her sister came to visit her and was horrified to see the cacti. My aunt assured her that it was only a Hungarian superstition . . . A few weeks after her visit to England her sister died. My aunt threw out all her plants when she heard the bad news, and she never allows one in the house now.[2]

Having evolved in arid conditions, cultivated cacti often suffer from overwatering.

> You should stop watering cacti in September, when there is an R in the month. Don't water them again when there are Rs in the month; I did this to mine and they flowered really well this year. It's a useful rule of thumb.[3]

> On CHRISTMAS morning they always water the cactus as a sign of good luck.[4]

CAGE-BIRDS – fed on CHICKWEED, GREATER PLANTAIN and GROUNDSEL.

CALVARY CLOVER (*Medicago intertexta*)

Annual herb, with yellow flowers, spiny seedpods (hence the alternative name hedgehog medick), and, in cultivated forms, red markings on its leaves; native to the Mediterranean region.

> A certain garden clover has red spots on its leaves, which are said to have been caused by drops of Christ's blood when He hung on the cross on Mount Calvary. The dried seed-pod when bent into circular form resembles a miniature CROWN OF THORNS, and, of course, the trifoliated leaf is emblematic of the Trinity.[1]

> The restoration of St Bartholomew the Great Church, West Smithfield, London, has been promoted to the extent of £120 by the sale during Lent for the last few years of pods of a kind of trefoil called Calvary Clover, at the price of sixpence each pod. It is in many ways an interesting plant; the leaves have a blotch at the base of

each leaflet, bearing quite a resemblance to a spot of fresh blood, which gradually dies away as the plant grows. The pod is spirally wound into a ball, bearing numerous interlacing thorns on its margin, and, when unwound, which is easily done, is remarkably like a crown of plaited thorns. It seems to be the custom to sow the seeds on GOOD FRIDAY.[2]

CAMPION (*Silene* spp.)

Annual or perennial herbs.

> Red and white campions were always known to me as pudding bags, after their shape, similar to the boiled puddings in cloths eaten by the country families. The suet dumplings or puddings were the daily diet for farm labourers; also the pudding known as the Bedfordshire clanger, which consisted of a suet dough roll covering meat at one end and jam at the other, divided by a section of the dough.[1]

See also BLADDER CAMPION, RED CAMPION and SEA CAMPION.

CANCER – GOOSEGRASS, RAGWORT, SAGE, VIOLET and YEW used to treat; DAFFODIL associated with cancer charities.

CANDLES – BOG MYRTLE used to make; SCOTS PINE used as a substitute for.

CANNABIS – see HEMP.

CANUTE, King (*c*.995–1035) – reduced to eating CLOUDBERRY.

CAPER SPURGE (*Euphorbia lathyris*)

Biennial herb, believed to be an ancient introduction, frequently popping up in gardens and on disturbed ground elsewhere.

> Some people swear by it as a MOLE deterrent but it has never been effective in our garden . . . It was *herbe d'chorchi* to a Jerseyman because a plant in the garden provided protection against WITCHES.[1]

> Her father [from Somerset] always grew giant spurge to deter moles.[2]

This belief has given rise to such names as mole-plant[3] and, in Germany, *maulwurfvertreiker* (mole deterrent).[4] However, a note published in 2012 explains that caper spurge prefers dry soils where worms are scarce and therefore unlikely to attract moles.[5]

In Germany it was thought that caper spurge 'keeps little mice away',[6] and it is probable that it is used to deter badgers in Somerset:

> When I had badgers in my garden someone told me to get wild spurge. The white sap in it stings the badgers' noses and they go away. It's a spiky plant, I had some when I moved in three years ago, but had none last year, there are some plants this year. You have to keep breaking the stems to expose the sap.[7]
>
> An alternative explanation of the name mole-plant is 'because its sap was used to burn "moles" or beauty spots on Edwardian and Victorian courtiers'.[8]

In common with other spurges, caper spurge produces white sap which can be used for removing WARTS.[9]

CARNATION (*Dianthus caryophyllus*)

Said to have been introduced by the Normans,[1] the tough, easily available flowers of carnation have long been favoured by florists and popular as buttonholes and in bridal bouquets. Perpetual-flowering varieties, developed in France, were introduced to the British Isles early in the twentieth century. By the mid-1950s varieties had been bred which produced flowers throughout the year over several years and were at that time the only flower which florists had in their shops on every day of the year.[2]

Students at Oxford University wear carnations when sitting their final examinations.

> Students do not take their final exams in their own colleges, they are divided by gender and dress in subfusc – gowns and black and white for the girls, and the boys wear black suits with white bow ties and waistcoats, which are sometimes very colourful. On the first day of their finals they wear a white carnation, on the second a pink, and on the third a red one.[3]

In fact the exams last for several days, so white flowers are worn on the first day and pink on subsequent days until the final day, when red ones are worn.[4] It seems that this is a comparatively recent tradition, and it is said that it originally involved white carnations which were placed in red inkwells overnight so that as the exams progressed they became darker in colour.[5]

In the early 1890s, notably at the first night of Oscar Wilde's *Lady Windermere's Fan* in February 1892, green carnations were worn on the lapels of 'men who loved men'. Although Wilde claimed to have invented the green carnation (white flowers which had absorbed a green aniline dye), the craze had, in fact, started in Paris the previous summer.[6] Occasionally this tradition has been revived:

> Some friends of mine had everyone wearing green carnations when they celebrated their civil partnership.[7]

CARROT (*Daucus carota* ssp. *sativus*)

Root vegetable cultivated since prehistoric times, probably of Middle Eastern origin, the now standard orange form having been developed in the Netherlands in the seventeenth century.

During the Second World War carrots were promoted as being useful to improve night vision:

> There was a glut of carrots, so carrot jam, carrot flan, carrot this and carrot that were made. The airmen were fed carrots for they were supposed to help their sight [to] see in the dark.[1]

> We were told to eat raw carrots to enable us to see in the dark. I believe it was a tale in the last World War, but we did it a lot here to try and improve our night vision as most of the village had no electricity until after the War.[2]

> Carrots improve night vision – a belief now widely held – dates back to World War II as a story put about by the security services to hide the existence of radar used by night fighters.[3]

In common with other familiar vegetables, carrots were used in folk medicine:

> In the early [1939–45] war years in the north-east of Scotland we had a series of exceptionally cold and snowy winters, and we chil-

> dren were constantly suffering from CHILBLAINS, particularly on the toes. My grandmother (a native of the Aberdeen region born in 1878) used a carrot-based salve to treat these.[4]
>
> Carrot poultice for SORES and WOUNDS.[5]

AFRICAN MARIGOLD can be grown with carrots to keep pests away from them.

CATARRH – EYEBRIGHT and ST JOHN'S WORT used to treat.

CATHARTIC – GROUNDSEL used as.

CATTLE (see also COWS) – fed HOLLY; tempted with IVY when sick; *Caesalpinia* seeds (see SEA BEANS) and PRIMROSES protect; HOGWEED used to treat rheumatism in; IRISH SPURGE 'a grand physic for'; SOAPWORT used to treat 'wounds in', YARROW used to treat stomach problems.

CATS – unwell when BRAMBLES fruit; CREEPING CINQUEFOIL cures diarrhoea in; CUCUMBER and ORANGE deter; VALERIAN used to trap.

CEDAR (*Cedrus libani*)

Large, evergreen coniferous tree, native from Lebanon to south-west Turkey, introduced in the seventeenth century and a conspicuous ornamental tree in large gardens and parkland.

> The oldest, if not the oldest Cedars of Lebanon in England is that standing in Bretby Park, Derbyshire . . . planted in February, 1676 . . . This tall Cedar has lost many limbs, and is now scantily provided. The stump of every lost limb has been carefully sealed with lead, and each remaining branch is supported by chains.
>
> There is a legend that a limb of this tree falls at the death of a member of the family . . . The family history is a mournful one. The last Earl died a young man and childless on returning home after a visit. His sister was the wife of Lord Carnarvon, and her death will be remembered. The fall of these recent limbs has left the old tree a wreck – a lofty and noble trunk, almost naked, except for a few remaining branches on the top, supported by artificial means.[1]

According to J.H. Wilks in his *Trees of the British Isles in History and Legend* (1972):

> A cedar at Bretby, Derbyshire, has it main branches chained and braced to prevent the collapse of any part of the tree. At one time this tree belonged to the Caernarvon [*sic.*] family, and prophesied the death of a member of that family should a major branch fall. The last fall of a limb is said to have taken place on the death of Lord Caernarvon after his discovery of the tomb of Tutankhamun.

As Bretby Hall passed out of the family's ownership in 1926, and has subsequently been converted to luxury flats, it seems that Wilks' use of the present tense in 1972 is incorrect because the tree died in 1954.

CELANDINE (*Ficaria verna*, formerly known as *Ranunculus ficaria*) formerly known as pilewort

Perennial spring-flowering herb, with glossy dark green leaves and golden yellow flowers, widespread in damp, often shaded, soils.

Local names include the widespread BUTTERCUP, and:

Bright-eye in Devon
Butter, BUTTER-AND-CHEESE, and butter-chops, in Somerset
Cheese-and-butter, and cheese-cups, in Somerset
CRAZIES in Buckinghamshire
CRAZY in Wiltshire
CRAZY BETT in Wiltshire
Crazy-cup in Somerset
Cream-and-butter in Devon
Crowpightle in Bedfordshire
DILL-CUP in Dorset and Wiltshire
Farrabun in Co. Donegal
Foal-foot in Ayrshire
Fog-wort in Dorset and Somerset
Fox-wort in Somerset
Gentleman's cap-and-frills in Somerset
GILCUP in Dorset, Somerset and Wiltshire

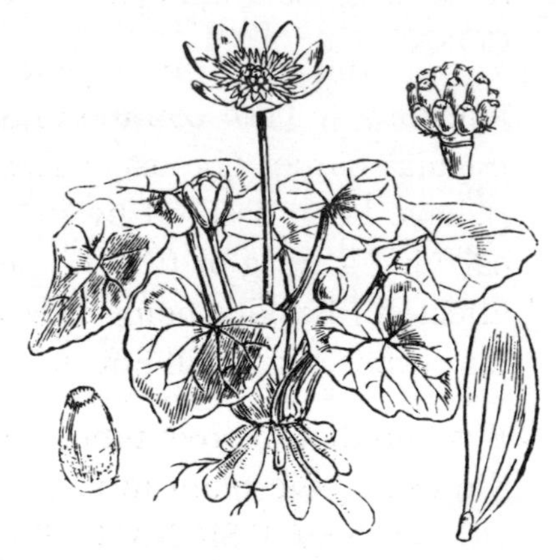

GILDCUP, and gilding cup, in Dorset
GILTY CUP, in Dorset, and 'the usual name' in Somerset
Golden butter in Dorset
Golden cap-and-frill in Somerset

GOLDEN CUP in Devon and Somerset
Golden drinking-cup in Somerset
Golden guineas in Northamptonshire
GOLDEN STARS in Somerset
Herb of St Ternan, in Aberdeenshire, 'a rarely used Deeside name . . . Ternan being a native of Banchory'
KINGCUP, and king's evil, in Devon
Legwort in Somerset
Marsh pilewort in Northamptonshire; it seems likely that this is an error and the name belongs to MARSH MARIGOLD
Power-wort, presumably a corruption of pilewort, in Somerset
Solandene in Scotland
Spring-messenger in Dorset
STAR-FLOWER in Somerset
STAR-LIGHT in Somerset
WART-GERSE, and WART-GRASS, in Cumberland.

In his *Names of Herbes* (1548), William Turner referred to celandine as fygwort – 'fig' being a former name for PILES (haemorrhoids). More recently celandine has been known as pilewort.

> [Guernsey] a remedy for piles is made by boiling fresh lard and straining through the flowers of this plant.[1]

> [Colonsay; celandine] roots are still used as a cure for piles, CORNS, etc.[2]

> Pilewort . . . the common country remedy for piles, hence the popular name. The usual gypsy remedy for the same complaint. (I have been assured by more than one gypsy that, by merely carrying a sprig or two in one's pocket, a complete cure may be effected.) Besides being used as an ointment, an infusion taken four times a day will effect a cure.[3]

> It is said that the root tubers of celandine resemble piles, and thus the use of the plant to treat this condition is an example of the DOCTRINE OF SIGNATURES. As late as the 1980s celandine was included in haemorrhoid creams.[4]

Like BUTTERCUPS, celandine has been used to treat SKIN ailments:

> Skin cleanser: drop a handful of celandine leaves into a pint of boiling water, strain and allow to cool. Apply the liquid to the face with lint, to tighten the skin, close pores, and remove wrinkles.[5]

CELERY (*Apium graveolens*)

Biennial herb, growing mainly in maritime situations in southern areas; also cultivated as a vegetable.

On 11 December 1996 *The Times* reported:

> Football fans at Second Division Gillingham [Kent] are being subjected to celery searches. Anyone caught in possession is threatened with a life ban because of a trend among fans to wave sticks of it while chanting an obscene anthem.

Variations of this 'anthem' include:

> Cel-er-y, cel-er-y,
> If she don't cum
> We'll stick it up her bum.[1]

In home medicine:

> If you have RHEUMATISM boil celery seed and drink it.[2]

> I recently met an old friend of my father, who, on hearing from me that father was having some URINARY trouble told me to tell him to eat plenty of celery, saying 'finest thing there is for watter [*sic.*] trouble'.[3]

> An ointment good for all SORES: the green leaves of celery cooked in home dried lard (to make sure there is no salt in it), strained, and put in pots for general use.[4]

> Wild celery has also been found effective in the healing of BURNS. An Exeter gentleman witnessed the following cure just after the end of the First World War: his sister, who was about 15 at the time, was very badly burned on the back of her head and shoulders, and the local doctor felt powerless to help. An old herbalist living nearby heard about the accident and claimed to be able to heal the child: he gathered wild celery and boiled it in water, and then used the liquor to bathe the burns. He subsequently applied the wild celery as a plaster. The treatment proved to be effective, and is said to be equally useful for healing infected BREASTS.[5]

CELERY-LEAVED BUTTERCUP (*Ranunculus scleratus*)

Annual herb, common on damp mud or in shallow water throughout lowland areas.

Local names include: BLISTER-CUP in Yorkshire, Germans-and-English in Somerset, and WATER-BLOB in Northamptonshire.

> [Scotland] The whole plant has a most acrimonious quality; if bruised and laid upon any part of the body it will in a few hours raise a BLISTER. Strolling beggars have been known sometimes purposely to make sores with it, in order the more readily to move compassion.[1]

CENTAURY (*Centaurium erythraea*)

Small, biennial, pink-flowered herb (Fig. 7), widespread in open grassy places throughout lowland areas.

Local names include the widespread sanctuary, and:

Blood-wort in Shropshire
Dramwe-na-murrough in Co. Donegal
FEVERFEW in Cheshire and Yorkshire
Feverfouillie in Scotland
Gentian in Sussex and Scotland
MOUNTAIN FLAX in Cumberland
Senna-pods in Dorset
Sentry in Kent
Spikenard in Somerset.

> [Known as gentian in Scotland] on the shores of the Moray Firth, where an infusion is drunk as a TONIC.[1]

> As a boy on holiday in Towyn, Merioneth [*c.*1915] I watched an old lady gathering a large bunch of centaury on the sand dunes, and was informed that this was to make 'tonic'.[2]

> Common centaury . . . was used as a 'NERVE calmer' by drinking an infusion of leaves and flowers.[3]

Vesey-FitzGerald in his survey of English gypsy medicine lists the names Christ's ladder, fellwort, feverwort and gall-of-the-earth, and states that an 'infusion of leaves good for jaundice . . . regarded as a first-rate tonic and very good for CONSUMPTION'.[4]

CHAMOMILE (*Chamaemelum nobile*)

Perennial, white-flowered herb, occasional in short grassland in southern England and south-west Ireland; introduced elsewhere, and cultivated in herb gardens.

> Chamomile root is a cure for TOOTHACHE. It should be put down on the aching.[1]

> Chamomile flowers made into an infusion [for] STOMACH and HEADACHES.[2]

> Dried chamomile flowers, stewed, added to the last rinse for blondes, made a nice shine; also drunk for NEURALGIA.[3]

CHANGELINGS – FOXGLOVE used to detect.

CHARCOAL – prepared from ALDER BUCKTHORN.

CHARLES II, King (1630–85) – associated with BRACKEN, FIELD MAPLE and OAK.

CHARLOCK (*Sinapis arvensis*), known in Ireland as *praiseach* or prushia

Yellow-flowered annual herb, common and widespread on disturbed soil.

Local names, many of which are variations of the standard name, include the widespread cadlock, carlock and KELK, and:

Brassics, brassock, brazocks (or brazzocks) in Yorkshire.
Bread-and-marmalade in Somerset
Cabbage-seed in Somerset
Callock in Northamptonshire
Calves' feet in Gloucestershire and Warwickshire
Charlick in Cornwall and Hampshire
Chedlock in Yorkshire
Churlick in Hampshire
Cradlick in Kent and east Sussex
Craps, applied to the seedpods, in Roxburghshire

Curlick in Buckinghamshire and Oxfordshire
Field kale in Cumberland
Gold-lock in Wiltshire
Hadridge in Pembrokeshire
Harlock in Essex
Headridge in Pembrokeshire
KEDLOCK in Northamptonshire
Ketlock in Lancashire
Kilk in Somerset
Kinkle in Kent
POPPLE in Cumberland
Prassha bwee in Co. Donegal
Prassia in Ireland
Presha, presha bhwee, and prushus, in Co. Antrim and Co. Down
Rape in Yorkshire
Runch in northern England
Runch-balls in Cumberland, Yorkshire and Roxburghshire
Runchie, and runchik, in Scotland, Orkney and Shetland
Rungy in Shetland
Scaldricks in Scotland
Scalies in Forfarshire and Stirlingshire
Scallock in Stirlingshire
Sheldock in Scotland
Shirts, applied to the seeds, in Scotland
Skedlock, skellock, and skellocks, in Scotland
Skillocks, and skillogs, in Co. Donegal
Turnip in Buckinghamshire
Warlock on the Isle of Wight
Wild kale in Scotland and Northern Ireland
Wild mustard in Somerset
WILD TULIP in Yorkshire
Willie-run-hedge in Lanarkshire
YELLOW in Northumberland
Yellow flower in Cheshire
Yellow top in Northumberland
YELLOW WEED in northern England and Scotland
Zeg in Somerset
Zenry in Cornwall; 'from *seneve*, a French derivative of the Graeco-Latin *sinapi*'
Zingreen in Somerset

Although now considered to be a weed, charlock was formerly used as a food at times of famine. In 1757 Philip Skelton wrote:

> Went out into the country [of County Donegal] to discover the real state of the poor . . . he was then a witness to many scenes of sorrow . . . in one cabin he found the people eating boiled prushia by itself for breakfast, and tasted this sorry food which seemed nauseous to him. Next morning he gave orders to have prushia gathered and boiled for his own breakfast, that he might live on the same sort of food with the poor. He ate this for one or two days; but at last his stomach turning against it, he set off immediately for Ballyshannon to buy oatmeal for them.[1]

Early in the nineteenth century it was noted that in springtime poor people in County Kilkenny were often reduced to eating charlock and 'a few other weeds',[2] while in the 1930s it was remembered that during times of famine:

> they would walk for miles gathering stuff called presha. This would bloom in the summer with a yellow blossom and people would gather burdens and boil it.[3]

A.T. Lucas suggests that the widespread use of charlock as a famine food in Ireland 'points back to a period when it had the status of a regular item of diet'.[4] Although it seems that in Ireland the foliage of charlock was eaten, in Scotland the seeds were eaten in times of scarcity. Thus, in 1884 it was recorded:

> for about three months of the year, when grain supplies had run out . . . any bread eaten was made from the seed of wild mustard or charlock, and was called 'reuthie' bread – 'REUTH' being the local Orkney name for such seeds.[5]

Perhaps it is significant that recent writers on wild plants which can be foraged for food make little mention of charlock. For example, Richard Mabey in his *Food for Free* (1972) has no entry for charlock, but mentions under CHICKWEED that it can be used to 'bulk' early spring salads.

CHERRY (*Prunus* cvs and spp.)

Cultivars of *Prunus avium* long-cultivated fruit trees, and more recently other species, mostly from eastern Asia, much planted as ornamentals.

A traditional English carol, *The Cherry Tree*, tells how Joseph, customarily much older than his wife Mary, finds her to be pregnant. Walking through a cherry orchard Mary asks Joseph to pluck a cherry for her, but he, 'with words most reviled', replied, 'Let he pluck thee a cherry that brought thee with child.' Speaking from within his mother's womb, Christ asks the cherry trees to lower their branches so that his mother can gather their fruit. The trees obey, Joseph repents,[1] and Christ has performed his first miracle.

The 11th Hussars, who in 1969 amalgamated with the 10th Royal Hussars to form the Royal Hussars (Prince of Wales's Own), were 'nicknamed the Cherry Pickers after an incident in a Spanish orchard in 1811. Some of the men dismounted to pick cherries in the orchard to take back to their camp. Unfortunately they were surprised by some French cavalry and had to fight them off on foot. Hence the nickname of Cherry Pickers'. Later nicknames given to the 11th Hussars included Cherubims, Cherry-Breeches and Cherrybums, all of which refer to the 'very tight pink trousers that were forced on the officers in 1840 by their regimental commander'.[2]

At Newington in Kent cherry orchards were formerly blessed on Rogation Sunday, but in October 1998 the incumbent of the parish reported that this custom had ceased many years ago. There were no longer any commercial cherry orchards near the church, the few remaining trees being privately owned. A Rogation Sunday service is still held, and the final blessing of the service takes place in the open air.[3]

Japanese cherry (*Prunus serrulata*) – the common showy flowering cherry much planted in suburban streets – is sometimes used in remembrance of the many thousands of deaths resulting from the dropping of atomic bombs on Hiroshima and Nagasaki in 1945. The anniversaries of the bombings (6 and 9 August) are occasionally marked by the planting of cherry trees,[4] or gathering around trees planted in previous years. For many years London Region CND have gathered near a tree planted in 1967 in Tavistock Square on 6 August, and 'participants are invited to bring flowers to lay beneath the tree'.[5]

Remedies which use cherries include:

> My mother always found that an infusion of cherry stalks was the perfect cure for inflammation of the BLADDER. Between the wars I remember being sent out to purchase some from a London chemist.[6]

And, GOUT can be prevented by eating cherries, 'six a day is enough and I have mine in cherry yogurt'.[7]

CHEST problems – treated using BRAMBLE.

CHESTNUT (*Castanea sativa*)

Large tree, native to the Mediterranean region, said to have been introduced in Roman times, now widespread throughout lowland England and Wales.

Local names given to the spiny fruits or edible nuts include:

French nut in Devon
Fuzz-beard, 'from the fuzzy and beard-like appearance of the prickles on the outside casing of the nut', also in Devon
Lady-nut in Somerset
Meat-nut in Devon
Polly-nut in Somerset
Stover-nut, 'on account of the abundance of chestnuts in Stover Park', in the Newton Abbot area of Devon
Vuzzy-veer in the Northam area of Devon, 'the only name by which this fruit was known in that part of Devon'.

> [Kent, 1940s] As children whenever we came to a sweet chestnut tree we would pick a leaf and strip out the leaf tissue from between the veins to make fish bones (skeletons). Useful if playing at 'keeping home' and cooking –very realistic.[1]

> The powdered nuts are good for PILES. Some gypsies wear the nuts in a little bag round their necks as preventative of piles. The bag must never be made of silk.[2]

> [Sussex during World War II] One lady asked them (them being my cousin and her charges, the schoolchildren) to collect sweet chestnut leaves, as she made a cough mixture from them.[3]

CHEWING GUM – GREATER PLANTAIN used as.

CHICKEN – unwell when BRAMBLE fruit; TANSY used to treat 'pip' in.

CHICKENPOX – BALM reduces itching.

CHICKWEED (*Stellaria media*)

Ubiquitous annual herb, with inconspicuous white flowers, preferring damp, nutrient-rich habitats.

Local names include the widespread CHICKENWEED, and:

Arva, and arvi, in Shetland
BIRD'S EYE in Somerset
Chickeney-weed, or chickeny-weed in Devon
Chicken's meat in East Anglia
Chickenwort in Scotland
Chick-wittles in Suffolk; 'i.e. victuals'
Chuckenwort in Aberdeenshire; 'i.e. chickenwort'
Clickenwort in Berwickshire
Clucken-weed, cluckweed, and cluckenwort in Northumberland
Cuckenwort, DUCK'S MEAT, and ervi in Scotland
Fox-tails in Dorset
Hen's inheritance in Nairnshire
Mischievous Jack in Somerset

MURREN in Yorkshire
Pokeweed, and RAM'S CLAWS, in Somerset
Schickenwir in Shetland
Tongue-grass in Ireland
White bird's-eye in Buckinghamshire.

In Shetland:

> Arvi (common chickweed) is said to grow wherever the SOIL is good quality.[1]

Traditional remedies which utilise chickweed include:

> [Berwickshire] This weed is a popular remedy, applied fresh, to allay the swelling caused by the STING of a bee.[2]

> [Orkney] The leaves when bruised are applied as poultices in cases of inflammation.[3]

> A painful arm or leg: chickenweed boiled and the juice rubbed on the place in the arm or leg where the pain is.[4]

> Chickweed can be made into an ointment for CHILBLAINS, RASHES, and stiff JOINTS. Wash ½lb chickweed and simmer in ½lb lard for two hours, strain through muslin and put in jars.[5]

> [Worcestershire, mid-1930s] A gypsy who came round about once a year recommended boiled chickweed for CONSTIPATION.[6]

In the 1840s chickweed and GROUNDSEL were hawked in London for feeding to CAGE-BIRDS;[7] more recently:

> Canaries like chickweed. My gran used to send us out to collect chickweed for her canaries.[8]

> Chickweed: I was told by my neighbour to collect this for my budgie, she said that birds like a bit of green stuff as well as seed.[9]

Although writers on wild foods often mention chickweed,[10] which is frequently at its best during the dank days of winter, it seems that it was not much appreciated in the past. Rather atypically:

> We used to collect chickweed from my father's allotment. We would bring it home, wash it in a sieve, then fry it with butter, one onion and some cheese. Very tasty.[11]

CHICORY (*Cichorium intybus*)

Perennial blue-flowered herb, believed to have been introduced in ancient times as a fodder-crop, widespread in rough grassland in southern and central England, scattered elsewhere.

Local names include bunks in East Anglia, strip-for-strip in Somerset, and swine-thistle in Yorkshire.

Gypsies used a decoction of chicory root as a cure for JAUNDICE.[1]

CHILBLAINS – treated using APPLE, BEETROOT, BITTERSWEET, BLACK BRYONY, BUTTERCUP, CARROT, CHICKWEED, HOLLY, MADONNA LILY, NAVELWORT, ONION and POTATO.

CHILEAN MYRTLE (*Luma apiculata*, syn. *Amomyrtus luma*)

Evergreen shrub or small tree, native to Argentina and Chile, cultivated for its scented flowers since the 1840s and naturalised in warmer areas since 1970.

On Guernsey, where common MYRTLE is rarely grown, Chilean myrtle, which thrives and 'seeds readily all over the place':

> was carried by . . . brides in their bouquets. As the threshold was crossed a sprig was planted by the front door. If it grew the MARRIAGE would be happy and fruitful.[1]

CHIMNEYS – BLACKTHORN, GORSE and HOLLY used to sweep.

CHINESE NEW YEAR – LETTUCE and TANGERINE associated with.

CHIVES (*Allium schoenoprasum*)

Perennial herb, rare as a native in parts of England and Wales, widely grown in gardens for its edible leaves.

Years ago when my son – then about four years of age – was

having a violent fit of WHOOPING COUGH an old lady stopped and asked if I had any chives in my garden. She told me to make some bread and butter sandwiches, filling them with chopped chives, and make him eat some after each fit of coughing, and he would be well in four days.

I did as bid and my son was well as she said.[1]

CHRISTMAS (25 December) – BUTCHER'S BROOM, HOLLY, IVY, LAUREL and POINSETTIA used for decoration; see APPLE; CACTI watered; dock (BISTORT) pudding eaten; ELDERberry wine drunk; if GRASS thrives death foretold, HOLY THORN flowers on.

CHRISTMAS EVE (24 December) – HOLLY not brought in before; HEMP seed used for LOVE DIVINATION on; HOLY THORN flowers on.

CHRISTMAS GREENERY

Evergreens, including HOLLY, IVY and MISTLETOE, have long played an important part in Christmas celebrations, and according to some writers the use of such greenery at this time of year is a survival of extreme antiquity.

> Long before the Christian era began, evergreens, which flourish when everything else in nature is withered and dead, were regarded as symbols of undying life, and used in magical rites to ensure the return of vegetation. The sacred buildings of Europe and Western Asia were decked with them for the Winter Solstice rituals. In ancient Rome, houses were adorned with laurels and BAY at the Kalends of January, and green garlands were worn and given as presents during the week-long celebrations of Saturnalia in December.[1]

However, regardless of what might or might not have happened in ancient times it is probable that bringing evergreens indoors at Christmas is a practice which developed simply because evergreens were readily available to bring extra colour to homes at a festive time. In theory, if not in practice, Christmas greenery should not be brought indoors before Christmas Eve, and should be removed by Twelfth Night (6 January).

> A: CHRISTMAS TREES should not be brought indoors before Christmas Eve.

> B: Holly must be taken down before Twelfth Night.
> A. Yes, but it's lucky to leave a bit up.[2]

> All greenery had to be removed from inside the house before Twelfth Night or bad LUCK would come.[3]

Alternatively:

> [Little Birch, Herefordshire, 1930s and 40s] don't take out greenery – holly or mistletoe – burn in house.[4]

> The holly and ivy that decorates the houses at Christmas is taken down after Twelfth Day and stored up until Pancake Night and burned under the pancakes when they are baking.[5]

> [West Yorkshire, 1961–2] Evergreens must be removed before New Year's Day, except for a piece of holly, which must be burned under the pancakes on Shrove Tuesday.[6]

CHRISTMAS TREE

Although Christmas trees form an important part of Christmas festivities, decorating homes, shopping centres and civic buildings, their history in Britain and Ireland is not lengthy. The idea of Christmas trees is attributed to Martin Luther (1483–1546):

> After wandering one Christmas Eve under the clear winter sky lit by a thousand stars, he set up for his children a tree with countless candles, an image of the starry heaven whence Christ came down.[1]

However, the first record of a Christmas tree such as we would recognise today is dated 1605, when:

> At Christmas they set up fir trees in the parlours at Strasburg and hang thereon roses cut out of many-coloured paper, apples, wafers, gold-foil, sweets, etc.[2]

The Christmas tree seems to have first appeared in Britain late in the eighteenth century; 'from 1789 to 1840 it was regularly mentioned as used in England by German settlers, guests, or governesses'.[3] In 1800 Queen Charlotte celebrated

> Christmas evening with a German fashion. A fir tree, about as

> high again as any of us, lighted all over with small tapers, several little wax dolls among the branches in different places, and strings of almonds and raisins alternately tied one to the other, with skipping ropes for the boys, and each bigger girl has muslin for a frock, a muslin handkerchief, and a pretty necklace and earrings besides. As soon as all the things were delivered out by the Queen and Princesses, the candles on the tree were put out, and the children set to work to help themselves, which they did very heartily.[4]

By the 1820s Christmas trees, formerly used only by the German community in Manchester, were starting to be used by local people.[5] However it was not until 1841, when Queen Victoria and Prince Albert's tree in Windsor Castle was widely reported and illustrated in the press, that Christmas trees started to become popular, although as late as 1912 they were described as being only 'a luxury for the well-to-do'.[6]

Trees were set up in the homes of the wealthy and cottagers and their children might be invited in to marvel at their splendour. Later, as electric lights became available, trees were also put up outside, so that today even the most tight-fisted local authority places lighted trees in its town centres. What is believed to be the first of these public trees was set up in Pasadena, California, in 1909.[7] In Britain the best-known of these trees is the one erected in Trafalgar Square, London, each year. Since 1947 this tree has been donated by the citizens of Oslo to the citizens of London, in appreciation of Britain's help to Norway during the Second World War.

Although HOLLY was (and perhaps still is) occasionally used as a Christmas tree, it is usual for a species of conifer to be used. In the 1990s Norway spruce (*Picea abies*) was most often selected. According to *The Times* of 12 December 1992:

> More than five million are expected to be sold in Britain over the Christmas season, with prices ranging from about £2 a foot for a Norway Spruce to about £4 a foot for a Nordmann fir [*Abies nordmanniana*].

Eighteen years later the Danish Christmas tree farmers' association reported that they expected 70 million trees to be sold in Europe in 2010. This, they claimed, was 20 million more trees than were sold fifteen years earlier, the increase being due to 'urbanisation and prosperity in Eastern Europe'.[8] At present Nordmann fir, which retains its foliage well in homes with central heating, is most often used.

Towards the end of the twentieth century many churches started holding Christmas tree festivals, at which local organisations are invited to dress a tree to raise money for charity. In 2008 Christ Church, Erith, in the London Borough of Bexley, held its third festival, at which seventy-five trees were displayed and money raised to support Greenwich and Bexley Cottage Hospice.[9] In 2014 at Ilminster, Somerset, fifty-four trees, 'as many as the Minster can hold', were displayed:

> We have everything from traditional trees to trees made from interesting bits of wood, including a rolling pin, and one made of just bits of twigs.
>
> The sheer variety is just unbelievable.[10]

For about a month from early December almost a hundred Christmas trees, decorated by local businesses, charities and campaign groups, fill the cloisters of Worcester Cathedral, attracting about 10,000 visitors each year. Admission is free, but visitors are encouraged to donate £2 to the Daisychain Benevolent Fund Trust, which organises the event.[11]

In recent years charities have also offered to recycle Christmas trees in exchange for a donation. St Richard's Hospice, Worcester, offered to collect and recycle Christmas trees on 13, 14 and 15 January 2017, and asked people wanting them to do so to register online, making a donation, ranging from £11 which 'could pay for a Hospice at Home Healthcare Assistant for an hour', to £25 which 'could pay for five patients to enjoy a nutritious meal in our Patient Unit'.[12] In East Sussex, Lewes Barbican Rotary Club offered to collect and recycle trees on Sunday 8 January 2017, in exchange for a donation of £5.[12]

CHRYSANTHEMUM (*Chrysanthemum* cvs)

Perennial herbs of unknown, probably Chinese, origin, widely grown for their showy, robust flowers, introduced to England late in the eighteenth century. Breeders around the world have produced thousands of varieties, and until the 1960s large-flowered forms were popular as winter cut-flowers: 'chrysanthemums had no competition, not even roses, during the winter'.[1] More recently spray chrysanthemums, which have a greater number of smaller flowers, have gained popularity and are available (and invariably included in inexpensive bouquets) throughout the year. Such flowers are commonly used by florists who supply funeral flowers. White-flowered forms are dyed or sprayed to produce flowers of any colour which can be used in FUNERAL floral tributes which represent the deceased's interests: pets, motor vehicles, pints of Guinness, shamrocks, and even McDonald's restaurants.[2]

In some parts of Europe the chrysanthemum is a favourite funeral flower and is associated with All Souls' Day. Consequently it is sometimes considered to be an UNLUCKY flower to have indoors.

> My hairdresser is an Italian, but she's married an Englishman and lived here for many years, so she's picked up the English ways. But her mother from Italy, who was staying here, was very upset when the hairdresser was given some chrysanthemums. 'They all wish you dead, they wish you dead,' she said. In Italy chrysanthemums are used at funerals and therefore associated with the dead, so if they are given to anyone, it's sort of saying: 'I wish you were dead.' The hairdresser said it was only an Italian superstition, and she didn't mind. She's been over here so long, and was so young when she married and came here that she doesn't mind.[3]

> In Germany chrysanthemums are considered unlucky; they're associated with funerals.[4]

Being relatively inexpensive, robust and colourful, chrysanthemums are much used in the Corpus Christi floral carpet prepared at Arundel Cathedral, West Sussex, each year,[5] and at church flower festivals elsewhere.

CIGARETTES – HOGWEED used as a substitute for.

CINERARIA (*Pericallis hybrida*)

Herb with gaudy flowers, a hybrid of horticultural origin, introduced in 1839 and popular as a conservatory plant and for planting in window-boxes.

> [Shetland] cineraria – sometimes called 'da devil's flooer' and UNLUCKY to keep in the house.[1]

CLAN BADGES

Although it has been claimed that the wearing of various plants as symbols of different Scottish clans is of ancient origin, it appears that the practice dates mainly from 1822, when a number of somewhat spurious traditions were revived or invented in celebration of King George IV's visit to Edinburgh.

> In the early days of our own history we find the rudest symbols were sufficient to answer the purpose of distinguishing one man, or band of men, from another. The Scottish clans were generally particularised by the pattern or colours of their tartan plaid.
>
> But this was found to be insufficient without the aid of floral emblems, and they therefore adopted a plan of ornamenting their bonnets or helmets with a sprig or branch of a plant as a symbolical badge of their various bodies. This ancient custom was again revived when his majesty visited his northern capital in the year 1822. His loyal Scottish subjects on that joyful event, paid their respects to their sovereign, at the palace of Holyrood House, each wearing the heraldic emblem of his clan.[1]

Although there is agreement about which plant was claimed by most clans, in some cases there is confusion. The following list is an edited version of that given in Henry Phillips' *Floral Emblems* (1825):

Buchanan BIRCH
Cameron OAK
Campbell BOG MYRTLE
Chisholm ALDER
Colquhoun HAZEL
Cumming sallow (GOAT WILLOW)
Drummond HOLLY
Farquharson FOXGLOVE
Ferguson POPLAR
Forbes BROOM
Fraser YEW
Gordon IVY
Graham SPURGE LAUREL
Grant CRANBERRY
Gunn ROSEROOT
Lamont CRAB APPLE
MacAllister HEATHER (*Calluna vulgaris* or *Erica cinerea*)
MacDonald HAREBELL
MacDonnell LING (*Calluna vulgaris* or *Erica cinerea*)
MacGregor SCOTS PINE
MacIntosh BOX
MacKay club rush
Mackenzie deergrass
MacKinnon ST JOHN'S WORT
MacLachlan ROWAN
MacLean CROWBERRY
MacLeod cowberry
MacNab 'ROSE bush berries'
MacNeill bladder wrack
MacPherson variegated BOX
MacQuarrie BLACKTHORN
MacRae fir clubmoss
Menzies ASH
Munroe eagle feathers
Murray JUNIPER
Ogilvie HAWTHORN
Oliphant SYCAMORE
Robertson BRACKEN

Rose DOG ROSE
Ross bearberry
Sinclair CLOVER
Stewart THISTLE
Sutherland smaller cat's-tail grass.

A more recent list complicates matters by providing additional plants for some clans; for example, the Fergusons claim rock rose and SUNDEW (neither of which would have been conspicuous when worn as a badge), as well as poplar, for their clan.[2]

See also WHITE HEATHER.

CLEAVERS – see GOOSEGRASS.

CLOUDBERRY (*Rubus chamaemorus*)

Low-growing perennial shrub growing on mountains in northern England, Wales, Scotland and Co. Tyrone.

Local names include:
Aivern in Morayshire
Aivrons in Scotland
Averin in Banffshire
Cnout-berry in Lancashire; 'there is a tradition that King Canute or Cnout, being reduced to great extremity, was preserved by eating this fruit'
Everocks in Scotland
Evron in Banffshire
Fintock in Perthshire
Knot-berry in Berwickshire
Knout-berry in northern England and Scotland
Noops in Berwickshire
Noutberry in Cumberland
Nowtberries in Yorkshire; 'because they taste of nowt'
Nub, and nub-berry, in Dumfriesshire
Nut-berry in Cumberland
Oot berry in Cumbria; 'found out bye as opposed to in bye'
Outberry in Co. Durham.

> The taste of the fruit is very peculiar, and I cannot readily suggest the slightly acrid flavour to those who have never tasted it. There is no suggestion of the sweetness of either strawberry or

> the BRAMBLE in it, rather is there a hint of the acidity characteristic of certain GOOSEBERRIES mixed with the tang of peat and HEATHER. As in the form of JAM the fruit is very agreeable, the preserve is much sought after.[1]

> Abundant on the peat-bogs and moory ground of Ben Lawers, both in flower and fruit. The latter, which are called avrons by the shepherds, are esteemed for their nutritious properties as well as their agreeable flavour.[2]

> A local farmer who farms the area of the Northern Pennines at Cross Fell . . . told me that the time to return the SHEEP to the fells was when the cloudberry (*Rubus chamaemorus*) was coming into leaf in May. Because of this the plant was called lamb's leaf.[3]

CLOVER (*Trifolium* spp.)

Perennial or annual herbs (Fig. 8) with leaves characteristically composed of three leaflets, common in grassy places. Although approximately thirty species of clover can be found wild in the British Isles, 'clover' usually refers to the two most common species: RED CLOVER and WHITE CLOVER.

Clovers (invariably white clover when found in the wild) which produce leaves with four instead of the usual three leaflets have long been considered to be 'LUCKY'.

> [1507] He that fyndeth the trayfle [trefoil] with foure leues, and kepe it in reuerence knowe for also true as the gospel yt he shall be ryche all his lyfe.[1]

> [1620] If a man walking in the fields, find any four-leaved grasse, he shall in a small while finde some good thing.[2]

> [Wales] It was considered LUCKY, and a token of marriage to find the four-leaved variety. Worn upon the person, or placed under the pillow, it induced cheerfulness of mind and made people light hearted. It is given and accepted as an emblem of good luck.[3]

> In my childhood Hilda [born Reading, Berkshire, about the turn of the century] always used to tell me 'Always wish when you see a four-leaved clover.[4]

A four-leaved clover should bring one luck if one should find one.[5]

[Horseheath, Cambridgeshire] A girl who put a four-leafed clover in her shoe did so in the knowledge that she would marry the first man she met.[6] The four leaves represent health, wealth, success and a lover.[7]

Judging by the hundreds of postcards produced throughout Europe and North America in the early part of the twentieth century, four-leaved clovers were widely recognised as good luck symbols. A rhyme recorded in 1965 from an 'ordinary factory-worker' in Kingston, Jamaica, probably originated on a postcard or a greetings card:

> Luck is a question of pluck, doing things over and over;
> Patience and skill, perseverance, and will, are the four leaves of a clover.[8]

Four-leaved clovers also enabled one to see FAIRIES and break the powers of enchantment. According to Michael Aislabie Denham, in an article published in the 1840s:

In South Northumberland a great deterrent as well as a revealer of fairies and a preventative of their influence, was the 'four-neuked clover' (a quadrifoil) although a 'five-neuked' specimen (a cinquefoil) is reckoned equally efficacious. This I learned from the people. Mr Chatto furnishes an instance.

'Many years ago, a girl who lived near Netherwitton, returning from milking with a pail on her head, saw many fairies gambolling in the fields, but which were invisible to her companions, though pointed out to them by her. On reaching home and telling what she had seen, the circumstance of her power of vision being greater than that of her companions was canvassed in the family, and the cause at length discovered in her weise [circular pad of grass placed under a pail when carried on the head], which was found to be of four-leaved clover – persons having about them a bunch, or even a single blade, of four-leaved clover being supposed to possess the power of seeing fairies, even though the elves should wish to be invisible; of perceiving in their proper character evil spirits which assumed the form of men, and detecting the arts of those who practised magic, necromancy and witchcraft.'[9]

In a collection of Cornish folklore first published in 1865, Robert Hunt gives a rather similar tale:

> A farmer lived in Bosfrancan in St Burrien, who had a very fine red-and-white cow called Daisey. The cow was always fat with her dewlaps and udder sweeping the grass. Daisey held her milk from calf to calf; had an udder like a bucket, yet she would never yield more than a gallon or so of milk, when one might plainly see that she still had at least two gallons more in her udder . . .
>
> One midsummer's day in the evening, the maid was later than usual milking . . . Daisey was the last cow milked, and the bucket was so full she could scarcely lift it to her head. Before rising from the milking-stool, the maid plucked up a handful of grass and clover to put in the head of her hat, that she might carry the bucket the steadier. She had no sooner placed the hat on her head, than she saw hundreds and thousands of Small People, swarming in all directions about the cow, and dipping their hands into the milk, taking it out on clover blossoms and sucking them . . .
>
> Her mistress came out into the garden between the field and the house, and called to know what was keeping the maid so long. When the maid told her what she had seen, her mistress said that she couldn't believe her unless she found a four-leaved grass. Then the maid thought of the handful of grass in her hat. In looking it over by candlelight, she found a bunch of three-leaved grass, and one stem with four leaves.[10]

Another tale in Hunt's collection tells how four-leaved clovers enabled one to see through enchantment:

> Many years since, there lived as housekeeper with a celebrated squire . . . one Nancy Tregier . . . Nancy left Pendeen one Saturday afternoon to walk to Penzance, for the purpose of buying a pair of shoes. There was an old woman, Jenny Trayer, living in Pendeen Cove – who had the reputation of being a witch . . . Nancy first called on the old woman to inquire if she wished anything brought home from Penzance. Tom, the husband of Nancy's friend, did no work . . . When Nancy went into Jenny's cottage, Tom was there, and right busy was she preparing some ointment, and touching her husband's eyes with it: this Jenny tried to hide in the mouth of the oven at the side of the chimney. Tom got up and said he must be off, and left the two women together. After a few idle com-

pliments, Jenny said that Nancy must have something to drink before she started for Penzance, and she went to the spence for the bottles. Nancy, ever curious, seized the moment, dipped her finger into the pot of green ointment, and, thinking it was good for eyes, she just touched her right eye before Jenny returned . . .

Penzance Market was in those days entirely in the street . . . Nancy walked about doing a little business and a great deal of gossiping; when amongst the standings in Market-Jew Street, whom should Nancy see but Tom Trayer, picking off the standings, shoes, stockings, hanks of yarn, and pewter spoons – indeed some of all the sorts of things which were for sale. Nancy walked up to him, and taking him by the arm, said, 'Tom, ar'then't ashamed to be carrying on such a game? However thee canst have the impudence, I can't think, to be picking the things from the standings and putting them in thy pocket in broad daylight, and the people all around thee.' Tom looked much surprised when Nancy spoke to him. At last he said, 'Is that you, Nancy? – which eye can you see me upon?' Nancy shut her left eye, this made no difference; then she shut her right eye, and there was Tom as before. She winked, and winked, and was surprised, you may be sure, to find that she could not see Tom with either eye. 'Now, Nancy,' said Tom, 'right or left.' 'Well,' said Nancy, ''tis strange; but there is something wrong with my left eye.'

'Oh, then, you see me with the right, do you?'

Then Tom put his finger on her right eye, and from that moment she was blind on that side . . .

Jenny's ointment had been made with four-leaved clover, gathered at a certain time of the moon. This rendered Fairyland visible, and made men invisible.[11]

In Ireland:

There was a great fair being held in Dingle one day long ago . . . there was a showman there, and the trick he had was a cock walking down the street ahead of him, drawing a big, heavy beam tied to his leg. At least all the people thought that it was a beam, and everyone was running after him . . . the crowd was getting bigger all the time . . . There came up the street a small old man carrying a load of rushes on his back. He wondered what all the people were looking at. All he could see was a wisp of straw being dragged along by a cock. He thought everyone had gone mad, and he asked them why they followed the cock like that.

Some of them answered him, 'Don't you see the great wonder?' they said.

'That great beam of wood being dragged after him by that cock . . .'

'All that he's pulling is a wisp of straw,' replied the old man.

The showman overheard him saying this. Over to him he went, and he asked him how much he wanted for the load of rushes . . . The old man named a figure . . . The showman gave it to him. He would have given him twice as much. As soon as the showman took the load of rushes off the old man's back, the man followed after the crowd, but all he could see was the cock pulling a heavy beam tied to his leg. He followed him all over Dingle.

What happened was that the old man had a four-leaved shamrock, unknown to himself, tied up in the load of rushes. That's what made what he saw different from what the other people saw, and that's why the showman paid him three times the value of the rushes.[12]

Versions of this story, of which fifty-seven have been recorded in Ireland,[13] are known throughout Europe from Scandinavia to Romania and have been traced back to the thirteenth century.[14]

> In Irish tradition four-leaved clover is said to be found only where a mare drops her first foal.[15]

> [Co. Wexford] If a mare foals her first foal in open air a four-leafed shamrock is supposed to grow on the spot if the mare herself was a first foal.[16]

According to a summary of a recording made on the Hebridean island of Barra in 1976:

> When a foal is born, it sneezes before trying to get to its feet so as to dislodge *dubhliath* [looks like a cormorant or rabbit's liver, about the size of a crown coin] from its nostril. [The informant] kept one to prove its existence to young people. If it is kept for seven years, a four-leaved clover will grow from it. A person searching for a four-leaved clover will not find one.[17]

Although finding a four-leaved clover is usually considered to be lucky:

> Another thing in Ireland that several people count unlucky is the four-leaf clover. I remember one year I couldn't put a foot down but I picked up a four-leaf clover. Then one Sunday in church I had the clovers in my prayer book and a lady sitting near me whispered 'Get rid of those quickly – they bring bad luck' and speaking for myself I couldn't think that I had any good luck while I had them.[18]

As the Irish consider species of clover to be SHAMROCK, the three leaflets of which represent the Holy Trinity, perhaps it is not surprising that some people should feel uneasy about four-leaved clovers.

Although it was once necessary to be either very lucky or to search very carefully to find a four-leaved clover, plants which produce four leaflets are now grown commercially.

> Looking over his many millions of four-leaf clovers is Charles T. Daniels, who grows nothing else on his Florida farm. His success obviously owes more to judgement than luck, although his stock probably originates from a single four-leaf plant, a chance mutation from the usual three-leaf variety. Mr Daniels encases his plants in plastic and sells them as a series of good luck novelties exported all over the world.[19]

> There are clover farms in the USA that specialise in producing four-leaved clovers. One of these clover farms covers 1.5 acres, with two large greenhouses and innumerable clover plants. A secret ingredient (biogenetically treated) is added to the feed to produce many four-leaved clovers on the plants.
>
> About 10,000 leaves are harvested daily and each is enclosed in plastic and sold as 'Good Luck' charms. Plants produced on the farm are not for sale, only the leaves, and the secret ingredient is jealously guarded.[20]

However, it is possible to obtain four-leaved clover plants, but these are usually species of wood-sorrel (*Oxalis*) rather than true clovers.

> I was given a four-leaved clover (a pretty little rock plant with variegated leaves) and told to plant it in my garden. Should it not flourish I would have bad luck, but luckily for me it thrived. I have now moved to another house and have left the plant behind. I am told that I have to be given a root myself! All this knowledge comes from the original donor who was a life-long farmer![21]

> Last April when I was visiting a rare-breeds farm in Kent, I bought a 'grow your own and keep good luck in your house' kit for growing 'the magic 4 leaf clover'. When I got home I followed the instructions given and waited for my four-leaved clover to grow. Nothing happened, so I wrote to the firm in Surrey that produced the kits and asked for a replacement. In reply I received a standard Xeroxed letter of regret, signed 'Tracy, Customer Services' – who presumably was very used to dealing with such complaints – and a replacement kit. This kit was somewhat more professionally produced than the original kit and was stated to have been 'made in Holland'. However, once again nothing grew. In both cases the bulb from which the clover was supposed to sprout appeared to be a species of *Oxalis* rather than anything to do with any clover, and the illustration on the Dutch kit appeared to represent an *Oxalis* rather than a clover.[22]

In February 2016 a 'Lucky Look four-leaved clover' kit purchased in Putney, London, depicted *Oxalis tetraphylla*, iron cross, on its packaging, and on following the instructions several of these plants were grown.

In recent years stylised four-leaved clovers have been used to promote lotteries.[23]

There are occasional records of clover leaves which have only two leaflets being considered magical:

> The following charm is used in the county of Cambridge by young men and women who are desirous of knowing the name of their future husbands or wives. The 'clover of two' means a piece of clover with only two leaves on it.
>
> A clover, a clover of two,
> Put it in your right shoe
> The first young man (woman) you meet
> In field, street, or lane,
> You'll have him (her) or one of his (her) name.[24]

CLUB ROOT (disease of CABBAGE and other BRASSICACEAE) – RHUBARB prevents.

COCKSPUR THORN (*Crataegus crus-galli* and *C. persimilis*)

Small trees, native to North America, much planted in parks for their abundant white flowers and persistent red fruit.

Here in Grange Park [Enfield, London], there are two houses with cockspur thorns at the drive entrance to tangle up evil.[1]

COCONUT (*Cocos nucifera*)

Palm, native to eastern Malesia, but now widely cultivated for a variety of uses, and naturalised throughout the tropics.

The actor Cedric Hardwicke, born 1893, remembered the fairs he attended during his childhood in the Black Country:

> If he won a coconut 'it was a rare and exciting, exotic prize' for 'you never saw a coconut at any other time'.[1]

Coconut-shies continue to be a feature of fairs, but coconuts are now widely sold in supermarkets and by greengrocers in ethnically diverse areas. For people of Indian heritage:

> The coconut is considered auspicious and the symbol of fertility in South India. They are given to married women visitors to your home along with betel leaves, areca nut and turmeric tubers 'tambulam' to represent a wish for health and progeny.
>
> It is a very versatile fruit and every part of it is used. The shell is used to fashion utensils and adornments, the husk as a dish scrubber, the pulp is a culinary delicacy and the water is very refreshing on a hot day.[2]

Thus coconuts occupy a prominent role in Hindu rituals.

> Several months ago when passing a small newsagent's shop in Mitcham Lane, Streatham, London, I saw a group of Indians cracking a coconut on the doorstep. When the coconut was opened they appeared to be delighted, and grinned widely while shaking hands. Yesterday I asked one of them about this, apparently throughout India the custom of cracking a coconut on the steps of a new business is considered to be very lucky.[3]

Coconuts are considered to be particularly suitable as offerings at Holi bonfires. At such a fire held on Streatham Common, South London, on the evening of 26 March 1994:

> Polythene bags containing foods, and many coconuts, were

thrown on the fire. On being questioned about the significance of the coconuts, participants said:

'Coconuts are considered to be a good offering at all [Hindu] festivals, even weddings because we think they are a sort of holy fruit, and the water inside them is pure.'

'Coconuts are given as *prasad* – that's an offering because they are so fruitful and so useful, because you can use all parts of them – they are a sort of fertility symbol as well – so we used them at all of our festivals, including marriages.'

Although the posters advertising the event stated that coconuts must not be removed from the fire, many, if not most, of the nuts were removed by men using long-handled scoops. The smouldering nuts were 'taken home and eaten' or 'taken home and placed beside the temple'. Some nuts were cracked open and eaten on the site.[4]

A writer on Indian plant-lore has suggested:

> As often happens with customs the world over, the meaning behind a ritual is lost but the symbol remained. So it is with the offering of the coconut fruit. Long, long ago, human sacrifice used to take place to propitiate the deity, particularly at the temple of Bhadra-Kali. But as time passed and people got enlightened, human sacrifice gave place to animal sacrifice and ultimately to the symbolic offering of a coconut which, with its round and fibrous outer covering, the epicarp, resembles a human head and the two dark spots on it represent the two human eyes. This is the closest resemblance of any member of the vegetable kingdom to a human head. For this reason it is offered as a symbolic human sacrifice.[5]

People of African descent who have adopted Western attitudes are said to be 'coconuts – brown on the outside, but white within'.[6] Cf. BANANA.

COFFEE substitutes – include YELLOW IRIS.

COLDS – treated using BARLEY, BLACK CURRANT, BOGBEAN, ELDER, ELM, HORSERADISH, LEMON, MADONNA LILY, MAIDENHAIR FERN, MOUSE-EAR HAWKWEED, MUGWORT, MULLEIN, NETTLE, ONION, RHUBARB, RUE, TURNIP and YARROW; GOOD KING HENRY and RASPBERRY create resistance to.

COLD SPELLS – coincide with BLACKTHORN, LABURNUM and DAMSON flowering; see also ST FRANKIN'S DAYS.

COLIC – treated using BALM.

COLTSFOOT (*Tussilago farfara*)

Perennial, yellow-flowered herb (Fig. 9), common and widespread on waste ground and roadsides, and an earlier coloniser of disturbed areas.

Local names include the widespread foal's foot, and:

Ass's foot in Somerset
Baccy-plant in Gloucestershire
Bull's foot in Buckinghamshire and Devon
Calf's foot in Yorkshire
CALVES' FOOT in Somerset
Cat-pee in Sheffield
Clatter-clogs in Cumberland
Clatterdock in Lancashire
Cleat/s in Cumberland, Lincolnshire and Yorkshire; 'because it grows among clods and clots'
Clicks in Lancashire
CLOTE in East Anglia
Cofleyblowse in Staffordshire
Cold-foot in Co. Donegal
Coughwort in Ireland
Could-foot in Co. Donegal
Cow-leaves in Selkirkshire
Dilly leaves on the Gower Peninsula, Glamorgan
Dishilago in Scotland
Dissilago in Orkney; 'probably a corruption of the Latin'
Dour docken, and dove-dock, in Caithness-shire
Dummy-leaf, and dummy-weed, in Hertfordshire
Dunny nettle in Berkshire
Fairies' pennies in Wiltshire
FLAPPER-DOCK in Lancashire
Fohanan in Co. Donegal
HOGWEED in Yorkshire
Hoofs in Gloucestershire
Horse-hoof in Northamptonshire and Somerset
ONE-O'CLOCK in Somerset, 'presumably from its seed heads, like those of DANDELION, being used by children 'to tell the time'
Pish-the-bed in Ireland, from confusion or equation with dandelion
SHALAKI in Orkney
Shellaggy, or shellogy, 'a curious example of the way its Latin name has been corrupted by common speech'
Son-afore-father in Cumberland and Scotland; 'due to the flowers appearing before leaves'
Sow-foot in Yorkshire
Spunk in Dublin
Sweep's brushes in Somerset
Tushalan in Northumberland

Tushie-luckie in Orkney; 'from the plant's traditional role in treating COUGHS (*tussis* in Latin)'
Tushy-lucky-gowan in Dumfriesshire
WILD RHUBARB, yellow stars, and Yellow trumpets, in Somerset

The leaves have been used medicinally as an infusion for coughs, and the practice of smoking them like TOBACCO is still very widespread.[1]

For BRONCHITIS – simmer a handful of coltsfoot leaves in a quart of water and allow to cool. Take a small dose every two hours until relief is obtained. The mixture can be sweetened with honey and also used as a gargle.[2]

Coltsfoot tea – very good for coughs.[3]

A cough medicine is still produced from coltsfoot in Anglesey. Indeed, my sister, who lives on the Lleyn Peninsula, has a bottle of the stuff, a thick brown liquid, in her kitchen. Sometime last year I saw a coltsfoot sweet for sale on a market stall at Ashbourne, but it didn't taste particularly pleasant.[4]

[On Colonsay, coltsfoot leaves] were smoked as a substitute for tobacco.[5]

During World War II coltsfoot was dried and used as a tobacco substitute.[6]

COLUMBINE (*Aquilegia vulgaris*)

Perennial herb, with violet-blue, or less usually pink, flowers, scattered throughout southern areas, also widely cultivated as an ornamental.

Approximately sixty vernacular names have been recorded, including the widespread GRANNY'S BONNET and GRANNY'S NIGHTCAP, the unlocalised acheley, akely, collobyn, dove's foot, doves-in-the-air, hawk's foot, ladies' bonnets, naked-woman's foot, old-woman's bonnet, RAGS-AND-TATTERS, and:

Baby's shoes in Somerset
BACHELOR'S BUTTONS in Wiltshire
Bonnets in Somerset
BOOTS-AND-SHOES in Cornwall and Somerset
Bubblyjocks in Dumfriesshire; 'the Scots word for turkey'
CAIN-AND-ABEL in Wiltshire
CULVERKEYS in Somerset

Dancing fairies, in Somerset and Yorkshire
Dolly-caps in Suffolk
Dolly's bonnets, dolly's shoes, and doves-at-the-fountain, in Somerset
Doves-in-a-dish in Hampshire; 'a local name . . . that I think may have originated with Vita Sackville-West'
Doves-in-the-ark, and doves-round-a-dish, in Somerset
Folly's flower in Dorset
FOOL'S CAP in Yorkshire; 'from the shape of the spurred petals suggestive of the jester's head-dress'
Grandmother's bonnet in Dorset and Somerset
Granny hoods in Lincolnshire and Yorkshire
GRANNY-JUMP-OUT-OF-BED in Wiltshire
Granny's mutch in Morayshire
HEN-AND-CHICKENS in Norfolk
LADY'S PETTICOAT in Dorset
Lady's shoes in Somerset
LADY'S SLIPPER in Cornwall
Mother Hubbard's bonnets in Lincolnshire
Mother Hubbard's nightcap in Yorkshire
Mother-in-a-hood in Lincolnshire
NIGHTCAPS in Somerset and Wiltshire
Old-woman's basket, and old lady's bonnet, in Somerset
OLD-WOMAN'S NIGHTCAP in Devon and Oxfordshire
Primrose-soldiers in Wiltshire; 'children play soldiers with it . . . fighting head against head'
Shoes-and-socks in Somerset
SHOES-AND-STOCKINGS in Cornwall and Somerset
SKULLCAP in Cornwall
SNAPDRAGON in Devon
SOLDIER'S BUTTONS in Somerset and Wiltshire
THIMBLES, and two-faces under-a-hat, in Somerset
Widow's weeds in Wiltshire.

The juice of columbine is used as a cure for swellings and the leaves are also made into poultices for this purpose.[1]

COMFREY (*Symphytum officinale*)

Coarse perennial herb, widespread on rough ground, roadsides and open damp places throughout lowland areas, also cultivated in gardens traditionally for medicinal use and more recently for use in composting.

Local names include:
Abraham-Isaac-and-Jacob in Lincolnshire
Alum, 'from its astringent properties'

Black-root in Co. Durham
Bruisewort in Sussex
Coffee-flowers in Somerset
GOOSEBERRY-PIE in Devon, Dorset, Suffolk and Wiltshire
JOSEPH-AND-MARY in Cornwall
LORDS-AND-LADIES in Northumberland; 'because of the pink and purple flowers'
Needle-cases in Dorset
PIGWEED in Wiltshire
Snake in Dorset
SNAKE'S FOOD, suckers, and sweet suckers, in Somerset.

Comfrey is remembered as having many medicinal uses.

The only local plant-name I knew in Lancashire was HEARTS-EASE; it's comfrey – the plant they grow in back gardens for herbal cures.[1]

[Derbyshire and South Lincolnshire] comfrey leaves had boiling water poured on them, then applied to BOILS, infected CUTS and GRAZES. It does actually heal large boils very rapidly. They can also be pounded using a pestle and mortar and combined with purified lard or lanolin and used to alleviate ECZEMA and skin RASHES. I've recently seen this work where modern drugs/remedies have not.

Comfrey was always abundant in old cottage gardens and used in a strong infusion to cure DIARRHOEA (SCOUR) both in animals and humans.

Comfrey leaves were used to line bean and root-crop trenches at planting time as comfrey was considered a good booster to plant growth.[2]

My husband . . . a Casualty Officer in the Royal Hospital, Sheffield, in the 1950s, it was not uncommon for patients suffering from suspected breakages and SPRAINS to arrive with the affected bones swathed in comfrey leaves, held in place by bandages.[3]

Comfrey was collected by a fellow student of mine as a boy in the 1940s and taken to the local white witch at Beer, Devon – I think she gave him six pence a large bag.

A friend of ours had her sprained ankle treated with a comfrey poultice in the Hebrides a few years ago – and it worked a treat![4]

Comfrey leaves spread on ARTHRITIS will soothe the pain and take any inflammation out.[5]

My cousin had an ULCER on her leg from her ankle to knee. Ointment from the doctor did her no good. Someone told her to bathe her leg with comfrey – boiling the leaves – bathe two or three times a day. In just one week it had gone down to [the size of] a postage stamp.[6]

I learned from a Lancashire miner that comfrey was widely used for BRUISES and sprains (boiling the leaves to get a solution).[7]

Knitbone or boneset – comfrey (bandaged over broken BONES).[8]

Comfrey – recommended as a cure for persistent nappy-rash by a lady living in Germany. I have not tried it myself, but she said it was the only thing that worked![9]

COMPLEXION – ELDERflower water good for; see also FRECKLES.

CONSTIPATION – treated using BOGBEAN, BRACKEN, CHICKWEED, GROUNDSEL and RHUBARB.

CONSUMPTION (tuberculosis) – CENTAURY, DANDELION, MUGWORT, MULLEIN and YARROW used to treat.

CONTRACEPTIVE – NETTLE used as.

CONVULSIONS – BITTERSWEET and PEONY prevent.

CORIANDER (*Coriandrum sativum*)

Annual herb, probably native to the eastern Mediterranean area, but widely cultivated for culinary use.

[From a Sikh gardener, 1991] Plant the seed directly on the ground, and break open with your bare feet, and stamp it into the ground. This works.[1]

CORK OAK (*Quercus suber*)

Evergreen tree native to south-west Europe and north-west Africa, where commercially cultivated for its bark which is used for wine-bottle stoppers and as an insulating material; occasionally cultivated as a curiosity in parks and gardens.

Fragments of bark from a 'wishing cork tree' growing in Coombe-in-Teignhead (now usually written as Coombeinteignhead), in Devon were reputed to bestow good fortune. An account given in a 1980s Christmas card which contained a fragment of the bark read:

> For the past 350 years a fine old cork oak has flourished in the village . . . and surrounded itself with a strange power to bring good LUCK to those observing certain rituals dating back to the time of the Great Plague of London in 1665. At that time, people came from all parts of the country to walk around the tree three times and as they walked to make a wish. Some came for better health, some for better fortune, and others for a wife or husband. It was said that few were disappointed . . . Even to this day, people from all over the world write for a piece [of its cork].

It is unknown when this tree first became commercially exploited, there are postcards of it which appear to date from the 1940s, and the Christmas card appears to have been first produced in the 1950s. No trace of the tree could be found in May 2012.

Corks are firmly believed to prevent CRAMP:

> A bag of bottle corks were put into the bed at night to prevent getting cramp – a family cure, always used.[1]

> Cure for leg cramp: Don't ask me how it works – but it really does – but putting a cork in your bedding (I tuck one into the bottom sheet) seems to do the trick. It lasts for months. I had excruciating cramp last week. Changed the cork and away it went. I don't really care that it might be just psychological as long as it does the trick – and it does.[2]

CORN CLEAVERS (*Galium tricornutum*)

Annual herb in cereal fields and on disturbed ground, once common, but now rare, in south central and south-east England.

In Bedfordshire the fruits of corn cleavers were formerly known as pin burs, and in the Podington area were collected by lace-makers to cover the heads of pins to protect their fingers.[1]

CORN COCKLE (*Agrostemma githago*)

Annual purple-flowered herb, probably native to the eastern Mediterranean region, introduced as a grain contaminant during the Iron Age; formerly a widespread weed of cornfields, but distribution drastically declined during the twentieth century due to improved seed-cleaning; now usually found as a result of wildflower seed being sown.

Local names include:
Cat's ears in Dorset
Cockerel in Suffolk
Cockleford in Gloucestershire
COCKLES in Somerset
Cokeweed in Scotland
Corn pink in Northamptonshire
Gye in Lancashire and Suffolk
HARD-HEADS in Northumberland
Little-and-pretty in Dorset
Pink in Scotland and Co. Down
Pook-needle in Sussex
Popille in Scotland
Puck needles in Sussex
ROBIN HOOD in Dorset.

In Herefordshire corn-showing was an EASTER Day activity, the purpose of which appears to have been to rid cornfields of corn cockle seedlings.

> At Easter the rustics have a custom of corn-showing. Parties are made to pick out cockle from the wheat. Before they set out they take with them cake and cider, and, says my informant, a yard of toasted cheese. The first person who picks the first cockle from the wheat had the first kiss of the maid and the first slice of cake.[1]

CORNFLOWER (*Centaurea cyanus*)

Annual herb with attractive bright blue flowers, an ancient, presumably accidental, introduction, and formerly a serious weed of arable ground; now extremely rare in such habitats, but cultivated as an ornamental in gardens and occasionally occurring on waste ground, or where wildflower seed mixes have been sown.

Local names include the widespread BLUE BONNETS and BLUE BOTTLE, and:

BACHELOR'S BUTTONS in Derbyshire and Yorkshire
BLAVER, or blavert, in Berwickshire and Roxburghshire
BLAWORT in Scotland
Blow-flower, in Somerset; 'a corruption of blue-flower'
Blue blaw in Northumberland
Blue blow in Dorset
BLUE BOBS in Somerset
BLUE BUTTONS in Somerset
BLUE CAPS in Kent and Northamptonshire
BLUE JACK, blue poppy, and BOBBY'S BUTTONS, in Somerset
Bottle-of-sorts in Yorkshire
BRUSHES in Stirlingshire
Corn-blinks in Devon
Corn blue-bottle in Ireland
CORN-BOTTLE in Devon and Northumberland
Cuckoo-hood in Scotland
HURT-SICKLE, '"because" says Culpepper, "with its hard wiry stem it turneth the edge of the sickle that reapeth corn"'
KNOBWEED and knotweed, in Nottinghamshire
Ladder-love in Somerset
LOGGERHEADS in Northamptonshire; 'from the resemblance of its knobbed involucres to a weapon so called, consisting of a ball of iron at the end of a stick'
Miller's delight in Dorset
Paint-brushes in Warwickshire
PINCUSHION in Suffolk
THUMBLE in Edinburgh
WITCH-BELLS, and witches' bells, in northern England and Edinburgh
Witch's needles in northern England.

Cornflower is associated with at least four long-established schools: Alleyn's School, Harrow School, James Allen's Girls' School and Winchester College.

At Alleyn's School, in Dulwich, London, it is believed that the tradition of wearing cornflowers on Founder's Day (the Saturday nearest 21 June) started in 1620 and has continued ever since. Similarly on the following Saturday James Allen's Girls' School celebrates its Founder's Day by wearing cornflowers.

> Thousands of the flowers are used by us every June. Three huge wreaths, each 2½ feet across, are laid by the three schools on the Founder's Grave and over 6000 people are wearing them in Dulwich over that weekend (i.e. staff, pupils, former pupils and parents). I understand that it is impossible to buy cornflowers elsewhere in London then.[1]

Cornflower is also considered to be the 'traditional flower' of Harrow School and:

> Is frequently worn by Old Harrovians at 'special events' such as Ascot, as well as part of everyday dress. Indeed it has almost become a badge of recognition and this ordinary little flower inspires much affection amongst Old Boys. The cornflower is much in evidence during the annual Eton versus Harrow cricket match . . . Many of my friends go to great length to obtain a flower for the match, some try to grow their own.[2]

It is thought that this tradition started during the nineteenth century,[3] and certainly it was well established in the 1930s, when:

> One of the high spots of our social year was the Eton and Harrow cricket match at Lord's each July . . . The men wore top hat, tail coats and suitable button-holes, Etonians sporting CARNATIONS dyed light blue, and Harrovians cornflowers.[4]

The first reference to cornflowers being associated with Winchester College dates from the 1890s, when they were worn for the 'Eton Match'.[5]

> Cornflowers are also worn at equestrian events, especially coach-driving.

> The cornflower is the emblem of the Coaching Club . . . The tradition started during the reign of George III when each year the Royal Mail Guards (who incidentally were the only employees of the Post Office on the equipage) were supplied with their new livery. On the nearest Sunday to, I believe, 1st July all the Royal Mail Coaches would assemble, possibly on the Embankment, drive up Whitehall and parade past the King at St James's Palace. Prior to the coaches moving to the assembly point, the Guards would go into the field nearby and pick wild cornflowers to wear as a button-hole (the blue of the cornflower against the scarlet livery looks wonderful). At the first meet of the Coaching Club on 27th June 1871 its President, the Duke of Beaufort, wore a button-hole of cornflowers in commemoration of the old coachmen, and this then became accepted as the emblem of the Coaching Club and is worn by its members to this day.[6]

In France cornflowers are worn in remembrance of servicemen who died during the World Wars, and in the United Kingdom it was an emblem of the Ypres League, which commemorated the 50,000 British troops who were slaughtered during the various battles of Ypres between October 1914 and October 1918. The League 'was started in the early 1920s and disbanded after the Second World War. Around the end of October each year they used to have a "smoking concert" at Caxton Hall, London and on the following Sunday placed a wreath on the Cenotaph'.[7]

CORN MARIGOLD (*Glebionis*, formerly *Chrysanthemum*, *segetum*)

Yellow-flowered annual herb, an ancient introduction, widespread on arable and disturbed ground; formerly a serious weed, now frequently introduced in wild-flower seed mixes.

Local names include the widespread boodle, GOLD ('perhaps more frequently used in the plural form, golds'; 'shortened for marigold'), and goldings, and:

Biggold in Somerset
Bollom in Dorset
Bossell in Wiltshire
Botham/s in Dorset, Hampshire and the Isles of Scilly
BOTHEM in Dorset and Hampshire
Botherem, or botherum, in Dorset
Bothul
Bottle in Dorset
Bozen in Hampshire
Bozzel in Hampshire and Wiltshire
BOZZOM, or BOZZUM, on the Isle of Wight
Buddle in East Anglia, Hertfordshire and Northamptonshire
Budland in Norfolk
Cargoulds in Scotland
Carkets in Orkney
Carr-gold in Lancashire
Corn daisy in Cornwall
Dunwich buddle in Suffolk
ESPIBAWN in Belfast
FAT HEN in Hampshire
Field marigold in Shropshire
Geal-gowan, and geal-seed, in Co. Donegal
Geelgowans, and gilgowans, in Co. Antrim
Geld, at Gilmorton, Leicestershire, where 'formerly such a pest that a fee of 3d was awarded by the farmers for its eradication'
Gill-gowan in the 'north of Ireland'
Golden daisy in Somerset
GOLDEN FLOWER in Somerset
Gold flower, 'a fairly general name'
Goles in Scotland
GOLLAND in Yorkshire
Gool in Scotland, where 'gool-riding . . . was the custom of riding through the parish to observe the growth of this plant and to fine the negligent farmer'; 'meaning gold'
Goold in Scotland
Gouland in Yorkshire

Gould in Lincolnshire
Gouldes in Dorset
Gouls in the Midlands
Gowans in Orkney
Gowlan in northern England
Gowland in Yorkshire
Guild and guile in Scotland
Guills in Somerset
GULD and gule, 'meaning gold', in Scotland; 'the Gule, the Gordon and the Hoodie Craw are the three warst things that Moray ever saw'
HARVEST FLOWER in Somerset
HORSE-DAISY in Cornwall
Manelet in Scotland
MARIGOLD in Cheshire
Marigold-goldings, or marigold-goldins, in Belfast
MARY-GOWLAN in Northumberland
Mogue Tobin in Co. Carlow; 'from a farmer who was driven out of his farm when unable to grow anything else'
Molly in Cornwall
MOONS in Northamptonshire
SUNFLOWER in Northamptonshire
TANSY in Gloucestershire
WILD CHRYSANTHEMUM in Ulster
Wild marigold in Northern Ireland
Yellow bottle in Kent
Yellow bozzum on the Isle of Wight
Yellow bussell in Berkshire
Yellow gold in Oxfordshire, as opposed to OXEYE DAISY, white gold
Yellow gould in Cumberland
YELLOW GOWAN in northern Scotland
YELLOW GULL in Cumberland
Yellow horse-daisy in Cornwall
Yellow moons in Dorset and Warwickshire
Yellow oxeye in Somerset, Yorkshire and Ireland.

On Colonsay, in the Hebrides, corn marigold was 'used to soothe throbbing PAINS'.[1]

CORNS – CELANDINE, GREATER PLANTAIN, HOUSELEEK, IVY and NAVELWORT used to treat.

CORN SPURREY (*Spergula arvensis*)

Common annual herb on light, often sandy, soils, in arable fields, on sea-shores and on waste ground; currently considered to be a weed, but formerly grown as a fodder plant and for its edible seeds

Local names include:
BEGGAR-WEED in Bedfordshire
Blore in Cumberland
Bottle-brush in Yorkshire
Cat's hair in Cornwall
COW-QUAKES in eastern England
Devil's beard in Cornwall and Devon
DEVIL'S FLOWER in Cornwall

DODDER in Cheshire and Cumberland
Dother/s in Cheshire, Lancashire and Berwickshire
Farmer's ruin in Yorkshire
Granyah in Co. Donegal
Guano-weed, 'many Cornish farmers believe that this plant was first introduced with guano'
MEADOW-FLAX in Dorset
MOUNTAIN FLAX in Dorset, Shropshire and Yorkshire
Mountain-vlix in Dorset
PICKPOCKET in Cornwall
PICK-PURSE in East Anglia and Lincolnshire
POVERTY in Berkshire
POVERTY-WEED in Yorkshire
REUTH, given to the seeds, in Orkney
Sandweed in Norfolk
TAILOR'S NEEDLE in Cornwall
TOADFLAX, toad's brass, and toad's gress, in Cheshire; 'i.e. toads' grass'
Wild broom in Dublin
YAR-NUT in Lancashire, Banffshire and Stirlingshire
YAWL in Cumberland and Berwickshire
Yor-nut in Lancashire
YOWIE-YORLIN in Cumberland.

> In Shetland in historic times they [corn spurrey seeds] were ground into meal – hence the Shetland name of meldi.[1]

> A profusion of meldi (corn spurrey) is a sign that more manure needs to be applied [to the SOIL].[2]

> To me hunger-weed is corn spurrey. My father used it [as an indicator] to which parts of our vegetable garden, on very sandy soil where Surrey, Sussex and Hampshire join, needed dunging. He said it was so called because on such soils it was used as a crop by the desperate.[3]

CORPUS CHRISTI

Celebrated by the Roman Catholic Church since the thirteenth century, the feast of Corpus Christi, observed on the Thursday, or a nearby Sunday, after Trinity Sunday, commemorates the institution of the

Eucharist. After solemn Mass, Christ's Body (Corpus Christi), in the form of bread and wine, is carried in procession as a public witness to the belief in the presence of Christ in the Eucharist.

> In the ancient world it was the custom to strew flowers in the path of important persons as a sign of respect. This custom was adopted by the Church in honour of the Blessed Sacrament, carried in procession in the festival of Corpus Christi. In some places in Europe this practice was extended so that whole streets were carpeted with flowers. Even today, in some towns in Italy, a carpet of flowers is laid the entire route of the procession, in intricate patterns and pictures depicting scenes from the gospels.[1]

In Britain the best known of these carpets is that prepared at the Roman Catholic cathedral in Arundel, West Sussex, where:

> The tradition of the carpet of flowers was taken from the village of Sutri outside Rome and introduced to Arundel by the 15th Duke of Norfolk in 1877. It seems not to have been carried on during the First World War, but was revived in 1919 and has continued ever since; the whole work being undertaken by parishioners.[2]

In recent decades the carpet has become extremely popular and many people visit the Cathedral on Tuesday and Wednesday to see the carpet being laid, and, particularly, on Thursday before a service of Mass late in the afternoon.

> A different design is used each year, but the plants employed are usually the same: mainly CHRYSANTHEMUMS to give large splashes of bright colours, CARNATIONS to provide more intricate highlights, and Lawson's CYPRESS which provides a sombre background.[3]

Since just before or just after the Second World War, a similar carpet is prepared at the Anglo-Catholic church of All Saints, Notting Hill, London.

> It doesn't last long. Usually it is made the day before and by the next evening you can usually see signs that it is beginning to go. The carpet's primary function is to be a carpet for the beginning of the Blessed Sacrament procession and it is not walked upon until that stage at the end of the Mass (after which it is destroyed by those who walk upon it in the procession).[4]

Another Anglo-Catholic church, St Mary's, Bourne Street, Belgravia, prepares a rather different carpet. St Mary's was built in 1874 and in 1987 its Corpus Christi procession was said to go 'back beyond living memory'. Here flowers are not used, and in the 1980s the carpet was made of leaves and herbs:

> Most of the herbs are straight kitchen herbs and then occasionally the odd spot of verbena or scented geranium gets included . . . the smell of the church is like something in an especially aromatic wood.

It is said that in those days the herbs were provided by a wealthy parishioner, who grew them at her country home. More recently the herbs have been replaced by leafy twigs of a variety of trees and shrubs, which are said to have been collected from the grounds of the Royal Hospital, Chelsea. The highly theatrical procession of the Blessed Sacrament is preceded by two children scattering ROSE petals on the carpet along its route.

COTTON GRASS (*Eriophorum* spp.)

Perennial herb with characteristic white, cottony fruiting-heads, common on open peaty ground.

Local names include:
Bog down, and bog silk, in Ireland
CATS' TAILS in Co. Tyrone
COTTON-FLOWER in Lincolnshire
Crushy-bracken in Ireland
Davy white-yeads in Shropshire
Flowans in Ireland
GHOST GRASS in Yorkshire
Hare's tail in Ireland
Moscrops in Cumberland
Moss-coach in Co. Donegal
Pisky wool in Cornwall
Pixies' flax in Devon
RAG in Yorkshire.

> Bog cotton, or cotton sedges – lukki-minnie's oo – very UNLUCKY to pick and bring indoors.[1]

> I am over 70 years old and was born and bred in Inverness-shire. I spent many summers in the company of my grandmother who was very knowledgeable about potions and cures from indigenous plants . . . In very early times children did not wear nappies during the day, but at night they had wads of bog-cotton heads to help keep them dry.[2]

Clothing made from cotton-grass fibres is mentioned in Scottish folk tales. In one a girl refuses a suitor unless he procures a gown of *canach* (cotton grass) down. In another, after a bewitched prince becomes a 'creature neither man nor beast', his father asks local girls to weave three shirts from *canach* down. Only one succeeds in doing this. When the prince receives the shirts he reverts to his 'handsome old self' and, inevitably marries the girl.[3]

However, it appears that the making of cloth from cotton-grass was never a practical proposition.

> The Great Exhibition of 1851 featured a number of items of clothing made from *canach* (bog cotton) down. These garments, woven by crafting women from Ross and Inverness-shire, were said to have been much admired for their beauty and fine texture.[4]

In the 1850s Berwickshire black-faced SHEEP were said to be especially fond of cotton grass:

> They will spend days in the middle of mosses, browsing on the favourite luxury, which is famous also for its renovating powers. Some [shep]herds maintain that if a weak sheep obtain but two or three mouthfuls of the fresh herbage, there will be no danger of its recovery. The farmers of an age scarce expired were accustomed to cast out their feeble sheep in the spring to the mosses, and leave them to range at will, till, by the aid principally of this grass, they had recovered.[5]

COTTON THISTLE (*Onopordum acanthium*)

Biennial herb, believed to be an ancient introduction, now widespread in waste places and around farm buildings, mainly in south-east England.

Local names include pig-leaves in northern England, rough dashle in Devon, and SCOTCH THISTLE in Berwickshire.

Considered by some writers to be the Scottish thistle, cotton thistle was also associated with Mary, Queen of Scots, who was executed at Fotheringhay, Northamptonshire, in 1587:

> [1875] on returning through the village an old dame enquired if we had been gathering Queen Mary's thistle, alluding to *Onopordum*, which tradition says was brought to Fotheringhay . . . by Mary's attendants.[1]

In 2012 it was reported that cotton thistle continues to grow on the earthworks of Fotheringhay Castle.[2]

COUCH (*Elytrigia repens*)

Rhizomatous perennial grass, common as a much-disliked weed on cultivated and disturbed ground.

Local names include the widespread cooch-grass, scutch, squitch, twitch, whicks, and wicks, and:

Blackawton clover in Devon
Brome grass in Ireland
Cassocks on the Somerset–Wiltshire border
Clitch grass, and clutch grass, in Somerset
Cooch in Dorset
Dog-grass in Cumberland; 'a dog will select with unerring instinct from a variety of grasses the leaves . . . as medicine'
Fanny grass in Northumberland
Felt in Scotland
Harl grass in Orkney
Hound's tooth in Somerset
Kwigga in Scotland; 'kvika means quick, vital'
Lonachies, and lonnachs, in northern Scotland
Quicken-grass in northern England
Quickens in Shetland
Quick-grass in Warwickshire; from the Anglo-Saxon '*cwic*, in allusion to the great vitality of its creeping underground stems'
Quicks in Norfolk
Quitch in Leicestershire
Quitch grass in Leicestershire and Ireland

Rack in Lothian, where also applied to 'weeds gathered from land, and generally piled up in heaps for being burnt'
Scutch-grass in Cheshire and Ireland
Shear-grass in Leicestershire
Sheep's cheese in Lothian and Roxburghshire
Skoil in Cornwall, where 'also applied to weeds generally'
Spear grass in East Anglia
Sproil on Dartmoor, Devon
Squatch in Cheshire
Squitch-grass in Warwickshire
Strap-grass, and strial, in Dorset
Stroil in Cornwall, and Devon; 'always used by my father (1912–92) . . . keen gardeners always bemoaned the presence of stroil'
Stroyl, 'because it do destroyle everything' in Cornwall, and 'the usual Devon word for this Jekyll and Hyde . . . valuable herb and vigorous weed'
Switch grass, and swutch grass, in Ireland
Twike in Lincolnshire
Twitch grass in Cornwall, Cumberland, East Anglia and Northamptonshire
Whickenins in Cumberland
Whickens in Durham and Yorkshire
Whigga, 'an alternative old name', in Scotland
Wick, and wicken-grass, in Yorkshire
WICKENS in Durham and Yorkshire
Windle-straw, and winnelstrea, in Cumberland
Wizzards in Morayshire
Wrack in Berwickshire and Roxburghshire; 'an alternative old name'
YAWL on the Isles of Scilly.

Richard Mabey records grandmother-grass as a name for couch in Guildford, Surrey, where:

> One plucks the head off the grass and sticks it in another head, still on its stem. A flip of the hand holding the stem and 'Grandmother, grandmother, jump out of bed' is recited as the first head springs out of its nest.[1]
>
> Gypsies considered a decoction of couch grass in cold water to be useful for reducing temperature and treating gall-stones.[2]

COUGHS – treated using AGRIMONY, BLACK SPLEENWORT, BRAMBLE, CHESTNUT, COLTSFOOT, ELDER, EYEBRIGHT, GROUND IVY, HEATHER, LEMON, MOUSE-EAR HAWKWEED, MUGWORT, MULLEIN, NETTLE, ONION, PEONY, RAMSONS, RUE, SWEDE, SWEET CICELY, TURNIP and YARROW; a March cough will not go until BROAD BEAN flowers.

COURTSHIP – HONEYSUCKLE STICK ensures successful.

COW PARSLEY (*Anthriscus sylvestris*)

Widespread white-flowered, perennial herb, conspicuous on roadsides in early summer.

Local names include the widespread KELK and RABBIT'S MEAT, and:

AIT-SKEITERS, 'children shoot oats through the hollow stems as peas are shot through a pea-shooter', in Morayshire
BAD-MAN'S OATMEAL, 'i.e. the devil's oatmeal', in Yorkshire
BILDERS in Cornwall
Bun in mid Yorkshire
Bunker in Norfolk
Cauliflower-flower, 'used by my father . . . though brought up around the Clent Hills [Worcestershire], he does not remember where he picked it up'
Cicely in northern England
Ciss in Lancashire
Cis-weed in Yorkshire
Coney-parsley in Sussex
Cow-mumble in Cambridgeshire, Essex and Norfolk
COW-WEED in Essex
Da-ho in Co. Antrim and Co. Down
Dead-man's flourish in Edinburgh
Dead-man's flower in Dublin
Deil's meal in Dumfriesshire; 'and other hedge Umbelliferae; diel [devil] from its supposed poisonous qualities . . . meal, in the sense of ground corn, from the light powdery appearance of its bunches of flowers'
Devil's bread in Yorkshire
DEVIL'S OATMEAL in Surrey
Devil's parsley in Cheshire
DOG-OAK in Hertfordshire
Dog's carvin in Shetland
Eldrot, and ELTROT, in Dorset
FIFES, and FLUTES, applied to the hollow stems on Clare Island, Co. Mayo
GHOST KEX in Yorkshire
GROUND ASH in north-east England and Berwickshire
GROUND ELDER in Cheshire
Gypsy curtains, and gypsy lace, in Somerset
Gypsy laces in Dorset and Somerset
GYPSY'S PARSLEY, and gypsy's umbrella, in Somerset
Ha-ho in Ireland
Hare's parsley in Somerset and Wiltshire

Helltrot in Dorset
Hemlock in Norfolk
Hi-how in Ireland
HILL-TROT in Wiltshire
HOLY GHOST in Yorkshire
HONITON LACE in Devon and Somerset
Humlock in Lancashire
Jack-jump-about in Northamptonshire
June-flower in Somerset
KADLE-DOCK, 'occasionally', in Cheshire
KECK in Ireland
KEDLOCK in Derbyshire
Keeshion in Co. Antrim and Co. Down
KEGGAS in Cornwall
Kellock in Lincolnshire
KESK in Cumberland
KETTLE-DOCK in Cheshire
KEWSIES in Lincolnshire
KEX in Co. Antrim; 'and/or other coarse umbelliferous weeds'
LACE-FLOWER in Somerset
LADY'S NEEDLEWORK in Gloucestershire
MAYWEED in Worcestershire
Moonlight in Wiltshire
Mother's dead in Lancashire and Yorkshire; 'might be because it resembles hemlock and might be picked by mistake; "never bring it into the house"'
Mother's die in Manchester; 'mother went absolutely mad when I brought cow parsley into the house . . . she feared it would predict her demise'
Naughty-man's oatmeal in Birmingham; 'and other hedge Umbelliferae'
OLD-ROT in Somerset
PIG'S PARSLEY in Somerset
Poor-man's oatmeal, 'in my childhood (80 years ago) we used cow parsley flowers as food at our out door "tea parties"', in Sussex
Queen Anne's lace-handkerchief in Dorset
Rabbit-keck in Nottinghamshire
RABBIT-MEAT in Lincolnshire, Sussex and Yorkshire
RABBIT'S FOOD in Buckinghamshire
Rat's bane, and SCABBY HANDS, in Somerset
SCAB-FLOWER in Cumberland
SCABS in Somerset
Sheep's parsley in Kent and Norfolk
SMOOTH KESH in Cumberland
Sweet ash in Gloucestershire
White meat, and WHITE WEED, in Yorkshire
Wild carraway in Banffshire
Wild parsley in Radnorshire.

In South Glamorgan it was believed that if cow parsley was brought indoors 'snakes will follow'.[1] However, the most frequently recorded superstition associated with the plant in recent years is the belief that to pick or bring indoors its flowers will lead to the death of one's mother.

> As a child in Yorkshire, we would never pick the tallish, very small white flowers which grew by the wayside. They were very pretty, tall, graceful plants, and many times I was tempted to pick them, but was told not to as it was called stepmother's blessing or MOTHER-DIE. Needless to say, as I was very fond of my mother, I did not wish to acquire a stepmother, which was what would happen, I was informed, if I picked them.[2]

> I myself was born in Wakefield, West Yorkshire, in 1927. I think I always knew cow parsley as mother-die – we children told each other one's mother would die if one picked it.[3]

> [Bradford, Yorkshire] Stepmother blossom – hedge parsley, cow parsley and other white umbellifers of a similar size. In the 1930s this was the only name we knew.[4]

> My childhood was spent in the Chigwell, Ongar and Upminster areas of Essex. My mother called cow parsley kill-your-mother-quick, and would never allow it in the house – or she would die. QUEEN ANNE'S LACE [a widespread name for the plant] is generally understood to refer to its lace-like appearance, but also to her (Queen Anne's) tragic child losses.[5]

> Wild cow parsley, which grows beside the road, we were not allowed to pick it. We called it BREAK-YOUR-MOTHER'S-HEART.[6]

These beliefs could have resulted from parents worrying about children confusing cow parsley, which is edible, with poisonous HEMLOCK, but that species appears to have attracted few negative attributes. It is also possible that house-proud women wanted to discourage their children from bringing cow parsley – the tiny petals of which rapidly fall making a mess – indoors. Whatever the reason the name mother-die does not appear to have been recorded for cow parsley in the nineteenth century. However, recently collected names which provide evidence for cow parsley being considered inauspicious include:

> As a child in the Ipswich district, I always called cow parsley dead-man's-flesh – I assume because so much of it grows in grave-yards. I didn't know of any other name for the plant until much later.[7]

> Cow parsley was known as bad-man's [i.e. the devil's] baccy.[8]

> An old man . . . who died a few years ago aged 80, called cow parsley shit parsley.[9]

Other 'negative' names include: ADDER'S MEAT in Cornwall,[10] blackman's oatmeal in Yorkshire,[11] dead-man's oatmeal in Northumberland,[12] devil's bread in north-east Ireland,[13] devil's porridge in Dublin,[14] and SNAKE-PLANT in Monmouthshire.[15]

In central west Scotland cow parsley was known as dog's flourish – 'as a local in her 70s said, "well it grows on the verges where the dogs have been!"'[16]

There appears to be a vague tradition which associates cow parsley with ST MARY THE VIRGIN. A Dorset name for the plant was My Lady's Lace,[17] while Our Lady's lace has been recorded from Norfolk, where it is said to be associated with the shrine of Our Lady of Walsingham,[18] and LADY'S LACE has been recorded from both Somerset[19] and Ireland, where:

> They used to call cow parsley Lady's lace, and I remember a maid saying 'Lady's lace looks lovely on the altar.' May was the month of Our Lady, and I remember that as children we used to make a MAY ALTAR in the house.[20]

A widespread name for cow parsley, especially its dry stems, was kex, and variations thereof, including keck and kecksy. Such stems were used by children as PEA-SHOOTERS:

> During and just after the War many things were unobtainable, and we used the hollow stems of cow parsley as pea-shooters.[21]

> We used cow parsley as pea-shooters, using HAWTHORN berries – agars – as peas.[22]

In the Forest of Dean:

> Pig weed was cow parsley. All Foresters kept at least one cottage pig and bunches of cow parsley were often fed to it, together with new BRACKEN tops, BLACKBERRY tips, etc., as a change of diet from kitchen scraps. But you mustn't feed this to a little pig or you would 'stitch 'un'. In other words it would grow too quickly for its skin.[23]

COWSLIP (*Primula veris*)

Perennial herb (Fig. 10), locally common on well-drained grasslands throughout much of southern Britain and central Ireland, producing honey-scented yellow flowers in early summer.

Local names include the widespread cove keys, and:

Boys-and-girls in Dorset
Bunch-of-keys in Somerset
Carslope in Yorkshire
Cooslip in Berwickshire
Cooslop in Lincolnshire
Cower-slop in Shropshire
COW-FLOP in Devon
Cow-paigle in Hertfordshire
Cowslap in Northamptonshire
COWSLOP in Cheshire, East Anglia and Northamptonshire; 'an old form of the word'
Cowslup in Warwickshire and Worcestershire
Cow's mouth in Scotland
Cow-stripling, and cow-stropple, in northern England
Cow-struplin in Yorkshire
Cow-strupple in northern England
Creivel, and crewel, in Dorset
Crows in Somerset
Cruel in Devon and Somerset
CUCKOO in Cornwall; '*coucou* is also a French name'
CULVERKEYS in Kent; 'wine made from it is called culverkey-wine'
Fairies' basins, fairies' flower, and FAIRY-BELLS, in Somerset
FAIRY-CUPS in Lincolnshire
Freckled face, golden bells, and GOLDEN DROPS, in Somerset
Herb Peter, in Cheshire and Somerset; 'said to be from its resemblance to a bunch of keys, which is the badge of St Peter'
Hodrod, and holrod, in Dorset
Horse-buckles in Kent and Wiltshire; 'probably a corruption of paigles, the East Anglian name for cowslips'
Keslop in Lancashire
Lady-keys in Kent
Lady's bunch-of-keys in Somerset
LADY'S FINGERS in Fifeshire
LADY'S KEYS in Somerset; 'also known by this name in Germany'
Long legs in Somerset
MILKMAIDENS and MILKMAIDS, in Lincolnshire
Oddrod in Dorset
PAIGLE in Essex
Peagles in Dorset and Somerset
Pingle in Dorset
Racconals in Cheshire
ST PETER'S HERB in Yorkshire.

There was a widespread belief that cowslips, and PRIMROSES, if planted upside down would produce red flowers.

[Cheshire] a curious belief about cowslips was that if they were planted upside down they would come up red.[1]

Also formerly widespread was a method of LOVE DIVINATION which used tissty-tossties, balls of cowslip flowers.

[Herefordshire] Make a ball of cowslip blossom, and toss it, using the same words [Rich man, poor man, beggar man, farmer; tinker, tailor, plough-boy, thief] over and over, till at the right one the ball falls to the ground. Or the ball is tossed to:

Tisty-tosty, tell me true,
Who shall I be married to?

Then the names of the actual or possible lovers are recited until the ball falls.[2]

You would go out, pick bunches of cowslips, and bring them back home. You'd have two chairs and tie a piece of string, about 12 inches long between them (like when you're making a cord). Then you would pick off the flower heads and hang them on the string with about half the flowers on either side – the more flower heads the better. Twist the string and tie the ends together to form a ball.

Toss the ball backwards and forwards continually saying:

Tissty-tossty tell me true,
Who am I going to be married to?
Tinker, tailor, soldier, sailor,
Rich man, poor man, beggar man, thief.

You would eventually marry a man with the same occupation as the one named when the last flower fell out.[3]

Cowslips are still used as a pretty test by children . . . who make the blossoms into flower-balls. These they toss up and catch with the right hand only, while repeating:

Pistey, postey, four-and-forty,
How many years shall I live?
One, two, three four

and so on, until the ball falls at the final number.[4]

In the late 1920s or early 1930s:

[On my grandmother's farm in Arlingham] the meadows were full of cowslips. My aunt took us children into the fields where we would sit and watch her make cowslip balls. I was probably 3–4 years old at the time and I cannot remember how she made them, but each ball was approximately 4 inches across and all flowers! We tossed them in the air and she would recite:

Tisty Tosty cowslip ball
Tell me where you're going to fall?
Dursley, Uley, Coaley, Cam,
Frampton, Fretherne, Arlingham?

. . . all the places mentioned are in the Severn Vale.[5]

Elsewhere cowslip balls were made simply as toys, or used as a welcoming gesture for guests.

A dodge ball is made from cowslips. A piece of twine about a foot long is got. The shanks are taken off the cowslips and the petals divided as evenly as possible and placed on the string head downwards. When the twine is almost full the ends are drawn together and thus forming a ball. Children have great sport dodging with their hands in the air seeing how long they are able to keep doing so.[6]

As a child [*c*.50 years ago] I made cowslip – paigle – balls . . . place the ball in a saucer of water.[7]

[In the early 1930s] we lived in Cradley, Herefordshire . . . In those days the visitor's room was devoid of running water, but had a beautiful marble-topped washstand on which stood a very large and pretty porcelain bowl, with an equally large matching water jug standing on it . . . The cowslip ball was then placed carefully in the visitor's water jug, and the strings tied round the handle to suspend it there. We were told that this would 'make the water smell nice'. I don't think it actually imparted any detectable odour, but at least it was a nice welcoming gesture.[8]

Cowslip Sunday was celebrated at Lambley, Nottinghamshire:

A few weeks ago I visited Lambley church . . . I was impressed by the beautifully embroidered altar-frontal, decorated with acorns, vine leaves and cowslips . . . The rector explained to me . . . the cowslips commemorate the village's Cowslip Sunday. Formerly the children gathered bunches on or for the first Sunday in May when the village street was lined with stalls. Coachloads of visitors came from Nottingham to buy the cowslips. Fortunately, prior to 1970, this custom was discontinued.[9]

Although this event was revived in 2009, and is now claimed to be 'an ancient and world-famous event', cowslips do not feature in it a great deal, the main attraction being a parade through the village during the afternoon, followed by the 'now legendary cowslip ceilidh' in the evening. It appears that even in Victorian times many people who visited Cowslip Sunday were not greatly interested in cowslips; in the 1860s there were reports of rowdy crowds and police being needed 'to sort things out'. And, although publicity stresses that the event is an ancient festival, it seems probable that it developed following the opening of the village railway station in 1852.[10]

On 1 April 1885, Lady Knightley of Fawsley Hall, Northamptonshire, noted:

> The Moreton Pinkney women and girls go into Buckinghamshire, and as far as Stowe, to gather cowslips and the next day take them to Daventry for sale, sometimes passing through Preston by 4 o'clock in the morning. They sell them picked at from 8d to 10d a gallon for cowslip wine. Now all the cottage windows are full of COLTSFOOT laid out to dry also for wine; later they will get DANDELIONS.[11]

Not surprisingly cowslips were often included in MAY GARLANDS. Gertrude Jekyll recorded making an 'immense cowslip ball two feet in diameter', when she attempted to improve the garlands which village children in the Thames Valley made 'in a scrappy sort of way'.[12]

Cowslip flowers were sucked by children and gathered by country wine-makers.

> I am now nearly 75 . . . [as children we] pulled the flowers out of cowslips to suck out the sweet drops of nectar.[13]

> Remedies which my mother used: cowslip wine (home-made) when we had MEASLES.[14]

> Cowslip wine was good for JAUNDICE.[15]

COWS (see also CATTLE) – BUTTERWORT protects, but also causes ill health.

CRABS – GORSE used to catch.

CRAB APPLE (*Malus sylvestris*)

Small tree, scattered in woods and hedgerows throughout England, Wales and Ireland, scarce in Scotland.

Local names include:

Grab in Cornwall and Devon
Grabstock, and GRIBBLE, in Dorset
Grindstone-apple in Wiltshire; 'used to sharpen reap-hooks, its acid biting into the steel'
Hang-downs in Somerset
Scrab in north-east Scotland
Screyb, and scribe, in Clydesdale
Scribe-tree in Ayrshire
SCROG, and scrog-apple in Berwickshire
SCROGG in north-east Scotland; 'i.e. a scrubby tree or bush'
Stub-apple in eastern England
Wharre in Cheshire
Wild-grass nettle, and wilding, in Shropshire
Wood-blades in Sheffield.

> At this season [MICHAELMAS] village maidens in the west of England go up and down the hedges gathering crab-apples, which they carry home, putting them into a loft, and forming with them the initials of their supposed suitors' names. The initials which are found on examination to be most perfect on Old Michaelmas Day, are considered to represent the strongest attachment, and the best for the choice of future husbands.[1]

A widespread saying, at least in southern England, concerning crab apples (and, perhaps, sour cultivated fruits) is applied to a young woman who having had many eligible boyfriends eventually chooses a husband who is considered to be inferior.

> As a child on the Dorset/Devon/Somerset borders during the early 1960s, I heard it said of one of my aunts that she 'had searched the orchard through and through, until she found the crab'.[2]

On the eastern edge of Dartmoor, Devon, a spinster who married late in life was said to have 'searched the orchard for the sweetest apple and picked the grab in the end'.[3] Similar expressions were used in Kent, where they were applied to philanderers of both genders.[4]

Although inedible when raw, crab apples are sometimes collected for culinary use. Typically:

> [*c*.1935] Crab apples were sour, but mum made a wonderful jelly with them.[5]

CRACK WILLOW (*Salix fragilis*)

Deciduous tree, widespread and common in damp areas; scarce in northern Scotland.

Local names include: CAT'S TAILS applied to the catkins in Somerset, snap-willow in southern England, and LAMB'S TAILS given to the catkins in Cornwall.

> Two Sussex men who were felling a crack willow tree on Bookham Common gave the tree the name widow's willow as it is liable to shed its branches without warning and injure anyone who attempts to cut it down.[1]

CRAMP – prevented using BANANA, CORK and POTATO; treated using BRAMBLE, CREEPING THISTLE and GARLIC MUSTARD.

CRACKNUT SUNDAY – celebrated in Surrey by cracking HAZEL nuts.

CRANBERRY (*Vaccinium oxycoccus*)

Trailing dwarf shrub, in moist, usually upland, moorland, often in sphagnum bogs.

Local names include:
Bogberry in Ireland
Cranes, and crones, in Cumberland
Kraanberry in Ireland
Mea-berry, and mea-wort, in Yorkshire
Moorberry, MOSS-BERRY, and red whorts, in Ireland.

> Bred and born here at the foot of the Cumbrian Fells (I am now retired) . . . We pick cranberries here soon [i.e. later in the summer] in the bogs near Cogra Moss, a small tarn near us, and at another 'secret' place, also in a bog on the Fells.[1]

Locally known as knupes in the Cheviot Hills of Northumberland early in the eighteenth century:

> The little hills the humble knupes produce,
> Which cure the SCURVY with their wholesome juice.[2]

CREEPING CINQUEFOIL (*Potentilla reptans*)

Yellow-flowered trailing herb, common on roadsides and other rough ground throughout most of lowland Britain and Ireland.

Local names include:
CREEPING JENNY in Somerset
FIVE-FINGER BLOSSOM in Suffolk
FIVE-FINGER GRASS on the Isle of Wight
FIVE FINGERS in Essex and Suffolk
FIVE-LEAVED GRASS in Buckinghamshire, Warwickshire and Worcestershire
Golden blossom in Devon.

A decoction of *Potentilla reptans* roots is given to Manx CATS with DIARRHOEA (they are very prone to this).[1]

I was brought up in East Ham, London, E6, by my grandparents, who came from Swanscombe in Kent. Every summer we used to take a trip to Leigh-on-Sea, Essex. There was a path from Leigh to Chalkwell by the sea with the railway on one side of the path and the sea on the other. We collected the runners of a plant called cinquefoil from beside this path in the summer and dried them. In the winter when people had a high temperature this was infused with boiling water and we had to drink it – it was awful! When I was five – I am now 68 – I had scarlet FEVER, and [after drinking this] there was no fever, which in those days was very remarkable.[2]

CREEPING SOFT-GRASS (*Holcus mollis*) – see YORKSHIRE FOG.

CREEPING THISTLE (*Cirsium arvense*)

Perennial, clump-forming herb with attractive but seldom appreciated pale purple flowers, common and widespread in grassy places.

Despite being a ubiquitous weed, creeping thistle has gathered few local names: DISLE or dysle in Devon, sheep's thistle in Dorset and Somerset, and sow dashle in Devon.

[A Warwickshire cure for] CRAMP: the swollen stems of the

cramp-thistle, i.e. the gall [caused by the fly *Urophora cardui*] not infrequent on the flowering shoots of SOLDIERS (the corn thistle).[1]

CREEPING WILLOW (*Salix repens*)

Low-growing shrub on acid heaths, moors and dunes.

Local names include bogall and cran-commer in Co. Donegal, sand-saugh in Yorkshire, and wild myrtle in Cumberland.

In Glen Alla, County Donegal, a decoction of creeping willow leaves was considered 'good for pains in the head, and much prized'.[1]

CRESTED DOG'S-TAIL (*Cynosurus cristatus*)

Perennial grass, widespread on dry grassland.

Local names include:

Dog's grass in East Anglia, Hampshire and Sussex
Gold seed in Somerset
Hendon Bent in Middlesex; 'the hay of Middlesex is often of good quality; Hendon, perhaps, produces the hay which has the best name in the market'
Pipe-stapple in Lothian; 'the stiff stalks are used for cleaning pipes'
SCRATCH-GRASS in Berkshire
Traleen grass, traneen, and traneen-grass, in Ireland
Wimble-straw in Northamptonshire
Windles in Scotland
Windle-straes in Cumberland
Windlestraw in Yorkshire.
WYCH-WOOD in Scotland; 'i.e. windlestraw'

[Yorkshire] During summer the children were fond of making 'trees' or 'dollies' out of dogtail grass. Gathering a handful, they twist other individual heads of grass round the stalks of the former, binding the bunches of more heads at intervals down the stems, so that they stick out from the sides in a fancied resemblance to branches of trees. In Lincolnshire, the men used to amuse themselves by making similar devices on Sunday afternoons. Beyond passing the time, it did not appear to be done for any definite purpose, though 'the tree' was often presented to the lady-love when completed.[1]

My mother, last summer, showed me how to make what she called 'Trees'. They are probably distant relatives of the corn dollies. They can only be made from one particular grass, the crested

> dog's-tail . . . One begins with three strands of the grass (including the inflorescence) and winds the inflorescence of another strand around the three stalks . . .
>
> After the inflorescence has been wound round the stalk is passed once round the others and knotted by passing the free end through the loop thus formed. The stalk is then bent down with the first three and another inflorescence is wound around them. By this means they increase in girth as they get longer. My mother then explained that they could be joined in groups to form trees. However she was vague as to the precise form of the trees . . .
>
> I discovered another local woman (local to Haslemere in Surrey where my mother lives) who also remembers making less complicated things she called Rats Tails which were made up in the same way but were unbranched.[2]

Writing of Ireland in 1825 James Townsend Mackay noted that crested dog's tail had been 'found of late to be one of the best grasses for making ladies bonnets'.[3] The collections of the National Museum of Ireland include a besom made of the grass,[4] and Anne O'Dowd in a list of toys made from straw, hay and rushes in Ireland includes 'peculiar playthings in the form of hens' feet made from the flowering stems of crested dog's tail'.[5]

CROSS, Christ's – made from ASPEN or ELDER.

CROWBERRY (*Empetrum nigrum*)

Low-growing evergreen shrub, on upland moorland and mountains.

Local names include:
Berry-heather in Orkney and Shetland
Crake-berry in Cumberland, Yorkshire and Ireland; 'crake = the common carrion crow'
Crow-ling in Yorkshire
Crow-peas in Morayshire
Deer's grass in Co. Donegal
Ling-berry in Cumberland
Monnocs-heather in Ulster
SHE-HEATHER in Co. Donegal.

Known as heather-berry in Northern Ireland:

> On [Heather-berry Day] the first Sunday in August the country people from Derry, Buncrana, and all around gather and go up Slieve Snacht in Inishowen. Hundreds of them go. They gather the berries of *Empetrum nigrum* (not BILBERRIES) and the whiskey is not forgot.[1]

In Scotland late in the eighteenth century:

> The highlanders frequently eat the berries, but they are not very desirable fruit. Boil'd in alum-water they will DYE yarn a black fuscous colour.[2]

CROWN IMPERIAL (*Fritillaria imperialis*)

Bulbous perennial, native to south-west Asia, introduced late in the sixteenth century and grown as an ornamental for its yellow, orange or red flowers.

Local names include crown-of-pearls in Buckinghamshire, roundabout-gentleman in Dorset and stink-lilies in Somerset.

> The crown imperial lily has water in it always, because it weeps for refusing to bow down when Our Lord passed by.[1]

> The way it appears to hang its head in shame is said to have derived from the bold way the flower originally stared at Christ on His way to the Crucifixion.[2]

Crown imperials were particularly sought after for inclusion in MAY GARLANDS.

> On May Day the children went round the village [Long Crendon, Buckinghamshire] with garlands of flowers, sometimes tied to sticks in the shape of a half loop, sometimes tied round a straight stick, with a crown imperial stuck at the top. In those days, most gardens had a group of those lilies. The little girls wore their Sunday best and made the rounds of the larger farm houses collecting pennies.[3]

CROWN OF THORNS – made from BLACKTHORN, BRAMBLE, BUTCHER'S BROOM and HAWTHORN; CALVARY CLOVER associated with.

CRUCIFIXION, Christ's – BRASSICACEAE, PASSION FLOWER and RED-SHANK associated with.

CUCKOO-FLOWER (*Cardamine pratensis*), also known as LADY'S SMOCK

Perennial herb with attractive pale pink flowers, common and wide-spread on damp soils.

Local names include the widespread MAY-FLOWER and MILKMAIDS, and:

Adder's meat on the Gower Peninsula, Glamorgan
APPLE-PIE in Yorkshire
Bird's eye in Cumberland, Shropshire and Yorkshire
Bog flower in Yorkshire
Bog spinks in Berwickshire
Bonny bird-eye in Cumberland
BREAD-AND-MILK in Glamorgan; 'c.f. the French *pain-au-lait*'
CARSONS in south-west Scotland; 'growing on carse, low, rich, damp land'
CUCKOO in Devon, Gloucestershire and Somerset
CUCKOO-BREAD in Devon and Somerset
CUCKOO-PINT in Leicestershire, Sussex and Wiltshire
CUCKOO-PINTLE in Leicestershire and Somerset
Cuckoo-spice in Yorkshire
CUCKOO-SPIT in Gloucestershire
Cuckoo's shoes-and-stockings in Somerset and South Wales; 'white flowers are the stockings, the more lilac ones shoes'
Gillie-flower in Nottinghamshire
GUCKOO in Cornwall
HEADACHE in Cambridgeshire; 'scent supposed to bring in a headache'
Lady-flock in Nottinghamshire
Lady's cloak in Somerset
Lady's glove in Northamptonshire
Lady-slip in Cheshire
LADY'S MANTLE in Somerset
LADY'S MILK-SILE, 'to sile milk is to strain it, and the tin sieve, in form like the wide part of a funnel through which it is strained is called a milk-sile . . . it is possible . . . the name is derived from a fancied resemblance in shape to the milk-strainer'
LAMB'S LAKENS in Cumberland
Laylocks 'i.e. lilacs', and lonesome lady, in Devon

Lucy locket in Derbyshire, 'when the children gather it they say "Lucy Locket lost her pocket in a shower of rain/Milner fun' it, Miller grum it, in a peck of grain"'
MAY in Cornwall
MAY-BLOB in Nottinghamshire
Meadow in Yorkshire
Meadow-flower in Cumberland
Meadow-kerses in Dumfriesshire, 'i.e. cresses'
Meadow-pink, milk-girl, and MILKIES, in Devon
Milking-maids in Somerset
Milkymaids in Devon
MOLL-BLOB in Northamptonshire
My lady's smock in Devon
NAKED LADIES in Somerset
Nightingale-flower in the New Forest, Hampshire
PAIGLE in Suffolk
PEE-THE-BED, 'because of its affinity for wet ground', in the Forest of Dean, Gloucestershire
Pick-folly in Northamptonshire
Pig's eye in Yorkshire
Pinks in Berwickshire
Potatoes-and-herrings on the Isle of Man; 'were the staple food of Manx crofters before the development of the visiting industry, but I cannot connect the name with the flower'
SHOES-AND-STOCKINGS, and SMELL-SMOCK, in Buckinghamshire
Smick-smock in Gloucestershire, Hampshire and Oxfordshire
Spinks in Berwickshire
Water-cuckoo in Wiltshire, 'distinguishing it from dry cuckoo, *Saxifraga granulata*'
WATER-LILY in Norfolk
Wild gillyflower in Shropshire.

My mother was not superstitious, and loved cuckoo-flowers, and we picked masses for her. Neighbours though would not have them in their house, as they brought 'bad LUCK'.[1]

Cuckoo-flowers . . . never taken indoors by my grandmother – 'They bring SICKNESS'[2]

If one picked a cuckoo-flower one would be attacked by a snake. This was a popular belief here, as cuckoo-flowers grew abundantly in The Close at Stradbrook and grass snakes were also found in this field.[3]

Although cuckoo-flowers are usually omitted from modern [MAY] GARLANDS with as much care as in the past, the rule is not observed everywhere. In Oxford the cross-shaped garlands carried around the streets on May morning by little bands of children almost always contain a profusion of these forbidden flowers, and no one seems aware that they are supposed to be UNLUCKY.[4]

CUCUMBER (*Cucumis sativus*)

Sprawling annual herb widely cultivated for its edible fruit; believed to have been first cultivated in India; known in Europe in classical times, and in the British Isles by AD 995.

A Guernsey rhyme advised:

> *Seume tes coucaömbres en mars, tu n'éras pas d'faire de pouque ni sac,*
> *Seume les en avril, tu-n-éras aen p'tit*
> *Mé, j'les seum'rai en mai, et j'en érai pus-s-que té*
> Sow your cucumbers in March, you'll need neither bag nor sack,
> Sow them in April, and you will have a few.
> But I will sow mine in May, and I will pick more than you.[1]

Cucumber has been recommended to treat a variety of EYE conditions. Typically:

> Cucumber slices for puffy eyes.[2]
> Cucumber slices on the eyes relieved eye strain.[3]

On 10 October 2017 a correspondent to the *i* newspaper from Rochdale, Lancashire, stated that she could vouch for the effectiveness of hanging a cucumber in the garden to deter CATS; 'for some reason, cats are frightened of them'.

CUTS – treated using COMFREY, DOCK, FOXGLOVE, GREATER PLANTAIN, GROUNDSEL, HORSERADISH, JERSEY LILY, LILY OF THE VALLEY, ONION, ORANGE LILY, ST JOHN'S WORT, SELFHEAL, VALERIAN and WATER FIGWORT.

CYCLAMEN (*Cyclamen* spp.), also known as SOWBREAD

Perennial herbs, mostly native to the Mediterranean region, cultivated as ornamentals with some species becoming widely naturalised in southern Britain.

Many herbalists recommended cyclamen to ease childbirth:

> The root hung about a Woman's Neck in Labour occasions speedy Delivery. It is very dangerous for Women with Child to make use of it, or step over it.[1]

CYPRESS (*Chamecyparis* spp., especially *C. lawsoniana*, and X *Cuprocyparis leylandii*)

Tall, fast-growing evergreen trees, native to North America, commonly planted as screens and windbreaks.

> [1920s] to plant two small cypress trees in one's garden would mean peace and prosperity to the household.[1]

> In County Mayo cypress was stuck into the ridge when POTATOES were planted. Its flat, robust, green twigs provide a suitable background for CORPUS CHRISTI carpets, and, in Ireland, particularly in urban areas, cypress is used as palm on PALM SUNDAY.[2]

CYSTITIS – treated using BARLEY.

D

DAFFODIL (*Narcissus* spp. and cvs; especially *N. pseudonarcissus*)

Bulbous, yellow-flowered herbs (Fig. 11). Wild daffodil (*N. pseudonarcissus*) native to England, found in woodlands, pastures and on riverbanks. Also hundreds of cultivars, many of which are planted beside country roads, or survive from old gardens, becoming naturalised.

Local names include the widespread lenticups and:

ADDER'S TONGUE in Dorset
BELL-FLOWER, and bell-rope, in Somerset
Bell-rose in Devon, Northamptonshire and Somerset
Bulrose in Lancashire
BUTTER-AND-EGGS in Devon
Churn in Somerset
Daffadoondilly in Cheshire
Daffidowndilly in Buckinghamshire and Cumberland
Daffy in Lancashire; 'contraction of daffadowndilly'
Daffydilly, 'very generally used by children' in Somerset
Daffydown in Derbyshire
DAFFY-DOWN-DILLY, dilly, and dilly-daffs, in Somerset
Dilly-dally in Buckinghamshire
Dong-bell, down-dilly, and Easter-lily, in Somerset
EASTER ROSE in Devon
Giggary in Ayrshire
Glacey petticoats in Wiltshire
Glens in Somerset
Gold bells in Devon
Golden trumpet on the Isle of Man
GOOSE-FLOPS, goose-leek, gracie-daisy, GRACY-DAISY, grassy-daisy, greasy-daisy, greysidaisy, and HEN-AND-CHICKENS, in Devon
Julians on the Isle of Wight
King's spear in Warwickshire
Lanthorn-lilies on the Isle of Wight
Lent-cocks, and Lenten lilies, in Devon
Lentils in Devon and Somerset
Lent-lily in Devon
Lent-pitcher in Somerset
LENT-ROSE in Cornwall
Lent-rosen in Somerset
Lents in Cornwall
Lenty-cups in Somerset
LILY in Scotland

Nebbits in Norfolk
Queen Anne's flowers in Wales
St Peter's bells in Co. Monaghan and Co. Tyrone
Scrambled eggs in Somerset
Sun bonnets in Dorset
Tags in Devon
Trumpets in Yorkshire
WHIT-SUNDAY, wild jonquil, and YELLOW BACHELOR'S-BUTTONS, in Somerset.

In common with PRIMROSES and SALLOW, daffodils were sometimes banned from the house by POULTRY-keepers.

> [Herefordshire] if daffodils be brought in when hens are sitting, they say there will be no chickens.[1]

> [Hartland, Devon] the number of goslings hatched and reared is governed by the number of wild daffodils in the first bunch of the season brought in the house.[2]

> [Isle of Man] unlucky to take daffodils into the house before the goslings are hatched.[3]

Wild daffodils are sometimes thought to indicate the former sites of religious foundations. At Frittlestoke, near Torrington, Devon, it was recorded in 1797 that:

> the people of the village call these plants Gregories, a name that struck us on account of its coinciding with the appellation of the order to which the neighbouring monastery belonged (the Canons of St Gregory).[4]

> In both Hampshire and the Isle of Wight, it was generally said that wild daffodils indicated the site of a monastery. St Urian's Copse, a short distance from Brading on the Island, is well known for its primroses and daffodils. There is a tradition that daffodils grow in profusion on only one side of a track running through the copse because a religious house once stood there.[5]

The only sizeable population of wild daffodils in the London area grows at Abbey Wood, a locality whose name commemorates Lesnes Abbey.[6]

Daffodils were associated with Australian soldiers:

> During the World War, 1914–1918, there was a large camp of Australian soldiers at Fovant, Wilts. Many died during the flu epidemic of 1918; they were buried in a nearby churchyard. Since that date

> to the present time local schoolchildren on a certain day each year lay a single daffodil on each grave.[7]

It appears that this practice died out many years ago. In June 1994 neither the rector of Fovant nor the head teacher of the local school could find any recollection of it. However, on 25 April 2015, the centenary of ANZAC Day, 'Fovant schoolchildren placed posies of spring flowers on the graves of the oldest and youngest Australian First World War soldiers buried in St George's Churchyard'.[8]

In the Second World War Australian soldiers were unflatteringly associated with daffodils:

> Australian war veterans have angrily rejected newly released War Office papers blaming the cowardice of Australian soldiers for the fall of Singapore in 1942. 'The Australians were known as daffodils, beautiful to look at but yellow all through,' says one of the documents.[9]

Both the daffodil and the LEEK are national symbols of Wales.

> The daffodil is associated with St David because it is traditionally said to bloom first on his day. It is an easier emblem to wear than the older leek, and every schoolchild in Wales sports one, real or artificial on March 1st.[10]

Since the 1980s daffodils have become associated with raising funds for cancer charities. In the UK the charity Marie Curie Cancer Care has promoted National Daffodil Day, selling fresh, and, more usually, artificial daffodils, to raise funds.

> Saturday is National Daffodil Day. Millions of spring blooms will be given away around the UK in exchange for a donation to Marie Curie Cancer Care. It is hoped to raise more than £20,000.
>
> The daffodil has been adopted by Marie Curie as a symbol of new hope and life, and as a reminder of the many positive developments in cancer care, treatment and research.[11]

The Irish Cancer Society 'borrowed' the idea of using the daffodil as a 'symbol of hope' from the Canadian Cancer Society and has 'run a very successful fund-raising campaign since the spring of 1988'.[12]

Elsewhere:

> Daffodil Days run from March 13–18. A bunch of daffodils is $6.

To order daffodils contact the American Cancer Society . . . All proceeds benefit the American Society's fight for a cure.[13]

The Cancer Council Australia is gearing up for another successful Daffodil Day. This year's fundraiser will take place on August 20, and the council hopes to raise $9.5 million by selling two million fresh daffodils and a range of merchandise.[14]

In bygone days when the picking of wildflowers was considered to be acceptable, large numbers of daffodils were gathered each year.

Fields known as 'the Channels' in Eardisley, Herefordshire, were famed for 'daffs' when I was a schoolgirl, pre World War II. They were picked in abundance and sold at Hay-on-Wye market just prior to PALM SUNDAY. Welsh dealers from the valleys bought them and they were used in graves on what the Welsh then called Flowering Sunday. Welsh families in Whitney-on-Wye, where I lived, made a great effort to wash tombstones, clip grass and generally tidy family graves and dress them with bunches of daffodils.[15]

Near Roewen in the Conwy Valley [Gwynedd] there is a place called Daffodil Mountain where tiny wild daffodils grow . . . When I was a child [1960s] the locals used to sweep out the tiny church (used as a sheep shelter in winter) and work together to decorate it with wild daffodils and pussy willow [SALLOW] (when available) for PALM SUNDAY. Every one of the graves in the churchyard was also decorated with PRIMROSES and daffodils.[16]

On the Isles of Scilly:

Prince Charles is paid one daffodil annually as RENT for the untenanted lands of Scilly – paid by the local Environmental Trust.[17]

In County Donegal daffodil roots were 'held to be useful as an EMETIC'.[18]

DAISY (*Bellis perennis*)

Perennial, white-flowered herb, common on mown and well-grazed grassland.

Approximately fifty local names have been recorded, including the widespread DOG-DAISY, and:

Baby's pet in Somerset
Bairnwort in Yorkshire; 'perhaps because children gather it; but an older name is banwort, of which it is probably a form'
Baiyan-flower in Lancashire
Banewort in Northumberland
Banwood in Yorkshire
Bennergowan in Dumfriesshire
Bennert, and benwood, in Cumberland
Bessy-bairnwort, and bessy-banwood, in Yorkshire
Boneflower in northern England
Cat-posy in Cumberland
Cockiloorie in Shetland
CURL-DODDY in south-west Scotland, 'i.e. curly head'
Day's eye in Somerset
Dazey in Cumberland
Dicky-daisy in Cheshire, 'a general name for wild flowers among children . . . applied more especially to *Bellis perennis*'
Ewe-gan in Yorkshire
Ewe-gollan, and ewe-gowan, in northern England and Scotland
Eye-of-the-day, and flower-of-spring, in Somerset
GOLLAND in Derbyshire, northern England and Caithness
GOWAN in Berwickshire, Inverness-shire and Morayshire
Gowerns in Somerset; 'evidently a corruption of gowan'
Gowlan in Derbyshire
GRACY-DAISY in Devon
INNOCENT in Somerset
Kokkeluri in Shetland
Little-open-star, LITTLE STAR, LOVE-ME-LOVE-ME-NOT, in Somerset
Marg in Berkshire; 'village children call field daisies margs, abbreviated without doubt from the French marguerite'
May-gowab in Berwickshire and Forfarshire
May-gowlan in Northumberland
Miss Modesty in Somerset
Nails in Wiltshire
Nooneen in Co. Donegal
Open-eye in Somerset
Shepherd's daisy in Northamptonshire
Silver penny, STAR, and TWELVE DISCIPLES, in Somerset
Wallie in Scotland
White frills in Somerset.

The wild daisy has been taken into cultivation and 'improved' leading to the production of monstrous red-petalled varieties. Names given to some of the older cultivated forms include:

BACHELOR'S BUTTONS and BILLY-BUTTONS in Shropshire
Gentleman's button, given to the 'double cultivated form' in Somerset
Grandma's daisy, given to the 'red double form' in Somerset
Mother Carey's chickens, given to a 'double form' in Somerset
SWEEPS, given to a 'dark red garden variety' in Cleveland, Yorkshire.

The 'hen-and-chickens' variety, in which a large flower is surrounded by a number of smaller flowers, seems to have been particularly popular, attracting names such as the children's daisy in Yorkshire; 'should perhaps be the childing daisy, i.e. the daisy producing young ones' and mother-of-a-thousand in Northamptonshire.

In many parts of the country daisies were thought to be harbingers of spring.

> 'It ain't spring,' said an old cottager to me 'until you can put your foot upon twelve daisies'.[1]

> Spring had arrived when you could put your foot on seven, or, in some places, nine daisies – the number could vary even between neighbours.[2]

Possibly such ideas stimulated a custom reported in the *Munster Express* of 8 January 1943:

> An old New Year's custom in the form of 'Penny for Daisy' was carried out as usual by the children on NEW YEAR'S DAY and large numbers of children collected the first flower of 1943 and received pennies for them.

Daisies, including OX-EYE DAISIES were often used in LOVE DIVINATION.

> In Wales the daisy is generally selected by the doubting maiden who is wishful to test the fidelity of her lover. Gathering a daisy, she commences plucking the petals off, saying with each one, 'Does he love me? – much – a little – devotedly – not at all,' and the last petal decides the question.[3]

> Pluck daisy petals – 'He loves me, he loves me not.'[4]

> I remember pulling each individual petal from a daisy whilst chanting 'He loves me, he loves me not', rather like 'tinker, tailor, soldier . . .' with cherry stones, or 'soldier brave, sailor blue, dashing airman . . .' (a more adolescent version).[5]

Similarly, in France:

> The French say as they pluck the ray florets from a daisy, either a *marguerite* (*Leucanthemum vulgare*) or a *pâquerette* (*Bellis perennis*):

> *'Il/Elle m'aime, un peu, beaucoup, tendrement, passionnément, à la folie, pas du tout.'*
>
> (He/she loves me, a little, a lot, tenderly, passionately, madly, not at all).[6]

The making of daisy chains, by slitting the stem of a daisy and inserting a second daisy through the slit, is a well-known and widespread childhood pastime.

> As children . . . we used to make endless daisy chains to wear round our necks and in our hair.[7]
>
> Daisy chains were made by children to hang round their necks. The end of each stalk was split a little way with the finger nail to make an opening big enough to poke the next daisy's head through and so on until the chain is long enough to go round one's neck.[8]

Rather unconvincingly:

> It is sometimes said that the habit of dressing children in daisy chains and coronals comes from a desire to protect them against being carried off by the FAIRIES. Daisies are a sun symbol and therefore protective magic.[9]

Between 1903 and 1958 24 May (Queen Victoria's birthday) was celebrated as Empire Day, when the 'non-party, non-sectarian, non-aggressive' Empire Movement encouraged people to 'fly the Union Jack and wear daisies, ox-eyed daisies, bachelor's buttons, marigolds, or marguerites'.[10]

> Another recollection [of schooldays in the early 1920s] was of Empire Day, 24 May – lessons were excused and all sang patriotic songs – with the Town Mayor and Mayoress and governors came; all the girls wore daisy chains.[11]
>
> We too held a special ceremony on Empire Day, and having been told that the lawn daisy grew in every country of the Empire, there was much striving to have a few to pin to your gym slip. Lawns were a bit scarce so my friends and I made our way to the nearby Olympia, at the site of which was a semi-private road which had a narrow strip of grass backed by iron railings, great was our joy on seeing it liberally sprinkled with daisies and great were our efforts

to get our arms far enough through the railings to pick some. We always managed it![12]

[South London, 1920] 'D'you remember the way we used to go round the big houses before Empire Day, asking if we could pick their daisies?'

'Remember?' . . . I saw the two of us stepping on the tiled paths towards great doors . . . I heard small voices asking, 'Please, can we pick your daisies?'

Daisy lawns at the top of Earlsfield were white as milk for Empire Day and many of the rich people who lived among them were glad to have them picked green. We took them to school to celebrate the Empire. The word had gone out: 'Gather daisies. The daisy is a symbol of our greatness.'

Indeed it was. The golden centre was us – Great Britain; the petals were the colonies, absolutely inseparable and dependent on us.[13]

I'm a 77-year-old widow . . . In the past when we celebrated Empire Day at school [we] were expected to wear some daisies pinned to our chests or buttonhole. My sister and I would rise early to collect our dew-fresh 'buttonhole daisies', feeling proud as we walked to school waving a Union Jack![14]

In the seventeenth century daisies were valued for the treatment of broken BONES:

The small Daisie is of greater Reputation than the other [ox-eye daisy], because it helpeth bones to knit again. It is therefore called by our people in the North of England Banwort, by which name I knew it forty years ago at Keibergh in the Parish of Kirk-oswald, and County of Cumbria, where I drew my first Breath, May the last 1676.[15]

Early in the twentieth century:

[On Colonsay daisies were] one of the principal ingredients used in the preparation of healing ointments.[16]

[My Dorset friend] gave me a cure for BOILS: 'Find a place where you can cover seven or nine daisies with your foot. Then pick and eat them.[17]

DAME'S VIOLET (*Hesperis matronalis*)

Biennial or perennial herb, native to southern Europe and western Asia, cultivated for its showy, white or purple, delightfully fragrant, flowers since late in the fourteenth century; widely naturalised.

Local names include: red rocket in Cheshire, summer lilac in Somerset, and Whitsun gillyflower in Cornwall.

> In Mitcham, Surrey (1928–39) we said that it was lucky to have some dame's violet in the garden – near the back gate and back door – we called it sweet rocket.[1]

DAMSON (*Prunus domestica* ssp. *insititia*)

A variety of PLUM with small blackish fruit, of domesticated origin; frequently naturalised.

> Damsons, sweeter and larger than sloes, grow all over Lancashire, North Wales and Cumbria. Collected and sweetened they make excellent JAMS, sweet wine, and, my favourite, damson gin.[1]

> 'Plum winter' – that time of the flowering of plum blossom (in this area invariably damson) that was held to always coincide with a COLD SPELL. Damsons grew everywhere locally, and although the sloe (BLACKTHORN) did also, we never called it 'blackthorn winter', which seems to be the norm elsewhere. Here it was usually the first half of April.[2]

> Shropshire, Staffordshire and Cheshire sent [damson] fruits for use in DYEing to Lancashire and Yorkshire mills (they gave their colour to khaki fatigues)[3]

During the 1930s Lancashire bus operators organised coach trips to see blossoming damson orchards in the Lyth Valley in Westmorland (now Cumbria).[4]

Due to deliberate selective breeding and unintentional hybridisation damsons vary a great deal. In Horseheath, Cambridgeshire:

> There was always plenty of good home-made jam in the village. For winter use 'Crickseys' (a sort of white damson) were baked in the 'stick' (brick oven) with a little sugar and water. Damsons were also made into cheese by putting the fruit into a pot which was stood in the oven all night; then the juice was poured off and the pulp put through a fine sieve; a pound of loaf sugar was added to each pint of the pulp which was boiled until it set, then it was put in shapes and tied down; kernels from the fruit stones were sometimes added.[5]

DANDELION (*Taraxacum officinale* agg.)

A complex of numerous microspecies; perennial golden-flowered herbs, common and conspicuously abundant in grassy places throughout the British Isles.

Local names include:
Bitter aks in Shetland
CANKER in Gloucester
Combs-and-hatpins in Somerset
Conquer-more in Dorset
Dandy-daddies, applied in Lincolnshire to the seed heads
Devil's milk-pail in Somerset
Devil's milk-plant in Kirkcudbrightshire
DINDLE in East Anglia
Dog-posy in Lancashire and Yorkshire
Golden sun in Somerset
HORSE-GOWAN in Scotland
Irish daisy in Yorkshire
Lay-a-bed in Somerset
Live-long in East Anglia
Male in Dorset
Milk-gowan in Scotland
Milky dashoes in Cornwall
Priest's crown in Lincolnshire
Stink-daisie in Clackmannanshire
Wild william in Scotland
Wishes in Wiltshire
Witch-gowan in Scotland.

The belief that picking or handling dandelion flowers leads to bed-wetting is widespread and well known.

> When I was a child in Brixton (1950s) we believed that if you picked dandelion flowers you would wet your bed.[1]

> Dandelions: not to be picked. Very unlucky. Children in Fife (1950s) called them PEE-THE-BEDS and anyone who picked them was mocked.[2]

> [1950s] as a child every other child I knew lived in horror of picking a dandelion – it was widely accepted as a fact this would lead to bed-wetting.[3]

> [I'm now in my 60s] mum wouldn't let us touch dandelions after 4 p.m. as she said it would make us go to the loo all night, but before 4 p.m. it was alright.[4]

This belief has given rise to a large number of local names, such as pee-beds in Cumbria, pee-in-bed in Lancashire, piss-i-beds and pissimire in Humberside, piss-in-the-beds in Co. Offaly, pisterbed in Co. Longford, pittly beds in Northumberland, and wet-the-bed, in Stockport, Greater Manchester. Elsewhere such names as the Dutch *pisse-bed* and the French *pissenlit* suggest the wider distribution of such beliefs.

Children use dandelion 'clocks' – ripe dandelion seed heads – to tell the time, or, less frequently, predict the future.

> When dandelions lost their yellow petals and grew that fluffy material, children used to pluck them and by blowing it they imagined they could tell the time. Each blow was counted as an hour, starting at one o'clock. When all the fluffy material had gone, that counted as the time of the day.[5]

> [1920s] The clocks or seedheads were childishly used to 'tell the time'. If it took three blows to blow the seedhead away it was three-o'clock.[6]

Local names which refer to this pastime include the widespread ONE-O'CLOCK and:

CLOCK in Wiltshire
Clocks-and-watches, and clock-flower, in Somerset
Doon-head in Scotland
Fairy clocks, and farmer's clocks, in Somerset
Four-o'clock in Devon and Somerset
Old-man's clock in Devon
One-o'clock-two-o'clock in Somerset
Schoolboy's clocks, and SHEPHERD'S CLOCKS, in Somerset
Time-flower, time-teller, watches-and-clocks, and what-o'clock in Somerset.
Time-tables in Hampshire

We used to blow dandelions seeds and count – 'This year, next year, sometime, never' (to get married).[7]

Dandelion seedheads when perfect were used to find out whether someone loved you or not by blowing short breaths at the plant and with each breath reciting 'He loves me, he loves me not', until all the seeds had blown away and the last blow decided the result![8]

Occasionally the seeds of dandelions, like autumn leaves, were considered to be lucky if caught.

Floating seeds of dandelions and similar plants are called FAIRIES by young children, and it was thought to be lucky to catch one.[9]

As a child, evacuated to Elgin, Scotland, during World War II . . . another 'taboo' plant was the common dandelion, it was supposed to make you wet the bed! . . . And yet, like all children, we didn't think anything of picking off the seed heads, which are supposed to tell you the time, according to how many times you blew on it to release all the little 'parachutes'. We believed the flying seeds were fairies, and blowing them released them from capture! If you could catch a passing fairy, you could make a wish before releasing it, then let it fly away on the wind.[10]

In Aberdeen these [dandelion] seeds were known as hairy witches. If you caught one it was considered lucky and you could make a wish.[11]

I (now aged 38) still make a wish if I ever catch a fluffy dandelion seed . . . but that's just for fun, because I think it's kinda cute, I don't expect my wish to come true.[12]

Swine's snout, a name given to dandelion in Somerset is said to refer 'to the form of its receptacle' after all the seeds have blown away. Elsewhere, in Somerset, Wiltshire and Ireland, the name monk's head likened the naked receptacle to a monk's tonsure.

Manuscript collections held in the National Folklore Collection, University College Dublin, indicate that in Ireland dandelions were used to treat a wide range of illnesses.

The juice that comes from a danelion [*sic.*] is a cure for every disease.[13]

Dandelion: Boil the leaves and the water in which they are boiled may be drunk. It is said to be a cure for anything. The leaves can also be eaten raw. Mr Sheehan has used it for his STOMACH.[14]

Roots of dandelion boiled, and strained, and drunk is good for CONSUMPTION.[15]

KIDNEY troubles: the leaf of dandelion is chewed in the mouth and the juice is swallowed. Mrs Griffin told me that this treatment cured Nora O'Callaghan . . . of the complaint about 30 years ago.[16]

People used to go out . . . and gather dandelion. They brought them in and boiled them, the juice of the dandelions were a good cure for weak HEARTS.[17]

Further evidence for this use is provided by the Ulster name of heart-fever grass.[18]

[Dandelion] was a great cure for JAUNDICE, to boil it along with buttermilk and when it is boiled take out the weed and drink the mixture.[19]

Dandelion Tea: First they put a knife under it and lifted it from the roots. They saved it in the sun until it got quite hard. Then they boiled the kettle and poured the boiling water on it, strained it, and put it into bottles. Then they drank it, and it was very good for the NERVES.[20]

To cure the sting of a NETTLE: look around and if there is a dandelion beside it you would rub the juice of it to it, it will cure the sting.[21]

Elsewhere in the British Isles dandelion appears to have been less valued in folk medicine, although its sap was widely used to cure WARTS.

For warts: squeeze the white milky juice of the dandelion onto the wart and allow to dry. Repeat the application as often as possible. The wart will blacken and eventually drop off.[22]

My son was troubled with a recurring wart; the doctors froze it out, and it came back; cut it out and it came back . . . within one week of rubbing dandelion stalks (although the roots are better) it went and never came back.[23]

Other uses included dandelion tea as a general tonic,[24] dandelion wine for INDIGESTION and kidney troubles,[25] and eating dandelion leaves 'to help ARTHRITIS sufferers'.[26]

> Dandelion was valued as food for TURKEYS, making them 'strong and healthy'.[27]

> We used dandelion leaves a lot in feeding young turkeys which were very delicate and hard to rear. The leaves were chopped up and mixed with scrambled or hard-boiled eggs.[28]

Rightly or wrongly dandelion leaves were considered to be good food for pet rabbits.

> Local plant names from an elderly friend in Porthnockie . . . dainties or denties – both flower and leaves of dandelion, the leaves picked to feed pet rabbits.[29]

> Dandelions were fed to the rabbits, but on the other hand we were told not to pick them because they would make you wet the bed (not true).[30]

Dandelion flowers are used to make a favourite homemade WINE.

> In wine-making circles, traditionally, dandelion wine is made on 23 April.[31]

> You should pick dandelion flowers for your dandelion wine on ST GEORGE'S DAY.[32]

Necessity being the mother of invention, during the Second World War:

> When stationed with the RAF at Tain in Scotland during the War, we could build a bicycle from a pile of bits and pieces kept in a large shed. One problem was a shortage of tyre valve tubing. As a stop gap some bright spark discovered that a three-quarters of an inch length of dandelion stem would keep us going for a few miles, and the rest of the stem was kept as spare.[33]

DANES – DWARF ELDER, FIELD ERYNGO, FRITILLARY, PASQUE FLOWER and TUTSAN associated with.

DEATH – ST JOHN'S WORT used to 'ward away'; ARUM LILY, BLACKTHORN, BULRUSH, COW PARSLEY, EARLY PURPLE ORCHID, FOXGLOVE, IVY, JUNIPER, PEONY, PEAR blossom, RED AND WHITE FLOWERS,

SNOWDROP, WHITE FLOWERS and wild THYME associated with; foretold when APPLE and FRUIT TREES flower out of season, GRASS thrives at Christmas; PARSLEY is transplanted, PRIMROSE blooms in winter.

In nineteenth-century Devon and Cornwall it was customary to deck house-plants with black crepe when death occurred in a household.

> [Cornwall] following death . . . plants would be put in mourning and swathed in black crepe, otherwise they too would drop their heads and die.[1]

In her *Walks about St Hilary* (1838) Charlotte Pascoe recorded:

> I saw with my own eyes a little black flag attached to our church-woman's bits of mignonette, which she assured me had begun to quail since her poor grandson was burnt to death, but had revived after she had put on it a piece of mourning.
>
> Not only had the woman's mignonette suffered, but plants belonging to her daughter, who lived in Penzance, had begun to droop after the accident, and had been revived only after a piece of black cloth had been tied to each one.
>
> Almost 90 years later a woman who admired a large geranium in a north Norfolk cottage was told that it had belonged to its owner's father, and when he died it had begun to sicken, but revived when a 'bit o'black' had been tied to it.[2]

DEPRESSION – HORSERADISH and ST JOHN'S WORT used to treat.

DEVIL – BAY and ST JOHN'S WORT protect against; contaminates BRAMBLE fruit; BINDWEED, DEVIL'S-BIT SCABIOUS, ELDER, FRUIT TREES flowering out of season, MONKEY PUZZLE, and SELFHEAL associated with; appears if HAZEL nuts are gathered on Sundays; dances on HOLLY left indoors after Twelfth Night; made shirts from NETTLE; PARSLEY and PARSNIP seed visit.

DEVIL'S-BIT SCABIOUS (*Succisa pratensis*; syn. *Scabiosa succisa*)

Perennial herb (Fig. 12) with attractive bluish-violet flowers, widespread in grassy places.

Local names include the widespread BLUE BOBS and BLUE BONNETS, and:

Angel's pincushions in Hampshire
BACHELOR'S BUTTONS in Co. Donegal
Bee-flower in Co, Derry
Bitin-billy in Sussex
Biting billy in Hampshire
Blue ball in Scotland
BLUE BUTTONS in Shropshire
BLUE CAPS in Sussex
Blue heads in Worcestershire
Blue kiss, and BLUE TOPS, in East Anglia
BUNDS in Co. Donegal
BUNDWEEDE in southern Scotland and Northern Ireland
CARL-DODDIE in Cumberland
Hog-a-back in Co. Donegal
LAMB'S EARS in Somerset
Pickerel-weed in Dorset
Pincushion in Cheshire
Sailor's buttons in Cornwall
Water-pine in Co. Donegal.

> The scabious . . . was once called forebitten more or bitten-off root. In order to account for this strange appearance in the root it was asserted that it had been bitten off by someone, and surely no one but the DEVIL could perform that, underground. So the tale started that he did it out of malice, for he saw that the herb was good for all manner of diseases, and he begrudges man the use of such a valuable medicine. The plant now bears the name of devil's bit in English; in German it is similarly known as *teufels abiss* . . . [Alternatively] with this root the devil practised such power that the Mother of God [ST MARY THE VIRGIN], out of compassion, took from him the means to do so with it any more. In the great vexation he felt at being thus deprived of his power, he bit off the root, which has never grown again.[1]

According to Joseph Wright in his *English Dialect Dictionary*, children in Northern Ireland knew devil's-bit scabious as CURL-DODDY, and would 'twist the stalk of this flower and, as it slowly untwists in the hand, say to it "Curl doddy in the midden, Turn around an' tak' my biddin"'.[2]

Children in Fife used the rhyme:

> Curl-doddy, do my bidden
> Soop [sweep] my house and [s]hool [shovel] my midden,

as if the plant gave them power to summon a brownie to sweep the house and shovel the dung and drudge for them in the brownie's manner.[3]

In the nineteenth century in the Rossendale Valley, Lancashire, scaly eruptions, known as devil's bites, were treated using devil's-bit scabious.[4] Uses elsewhere included the treatment of DOG BITES in west Somerset,[5] BRUISES and sore THROATS on the Isle of Man,[6] and 'all manner of ailments' in Cornwall.[7]

> In Gloucestershire the name [fire-leaves] is given to the leaves of PLANTAINS, more especially *Plantago media* [HOARY PLANTAIN]; and we have heard it in Herefordshire used for *Scabious succisa* (Devil's bit), which is very prevalent on the flats of the Wye. Both are named fire-leaves on the same principle, for we have seen a farmer in Gloucestershire with a Plantain leaf and he of Herefordshire with a Scabious leaf, select specimens, and violently twist them to ascertain if any water could be squeezed out of them. If so, this moisture is said to induce fermentation in newly carried hay sufficient to fire the rick. Both are mischievous in pasture, because such thick-leaved plants take longer to dry than the Grasses.[8]

DEVON WHITEBEAM (*Sorbus devoniensis*)

Tree with white flowers and brownish fruits, endemic to the British Isles – east Cornwall, Devon, South Somerset and south-east Ireland, naturalised in North Wales and north-east Ireland.

The botanist W.P. Hiern records the names French ales and French hailes on the labels of specimens collected in North Devon in the autumn of 1878, now in the herbarium of the Natural History Museum, London.

In their *Dictionary of English Plant-names* Britten and Holland record French hales as a name for a fruit sold in Barnstaple market 'for a halfpenny a bunch', which they identified as *Sorbus intermedia*.[1] At the time they were compiling their *Dictionary* whitebeams were poorly understood, and the fruits were undoubtedly those of Devon whitebeam. 'Hales' as a plant-name is most often used for the fruits of hawthorn, commonly known as haws, so French hales can be interpreted as 'exotic haws'.

On the Heligan estate, near Mevagissey in Cornwall, fruits collected from trees known as otmast, were used for 'stuffing cooked pheasants' early in the twentieth century. Early in the twenty-first century, one of these trees was refound and identified as Devon whitebeam.[2]

A number of other endemic whitebeams occur in the British Isles, but as these are small populations growing in inaccessible places (including the cliffs of Cheddar Gorge and the Avon Gorge), it's unlikely that they have any folk traditions associated with them, or acquired any local names.

DIABETES – PELLITORY OF THE WALL used to treat.

DIARRHOEA – treated using BILBERRY, COMFREY, HEATHER, HEDGE VERONICA, MULBERRY, NUTMEG, OAK, RASPBERRY, TORMENTIL and WOOD AVENS; in calves treated using MEADOWSWEET; in Manx cats treated using CREEPING CINQUEFOIL.

DIPHTHERIA – PRIVET associated with.

DISASTER – foretold when BURNET ROSE flowers out of season.

DISRAELI, Benjamin – PRIMROSE League founded to commemorate.

DIURETIC – BROOM and ORPINE used as.

DOCK (*Rumex obtusifolius* and other *Rumex* spp.)

Perennial herb, common on neglected cultivated land, roadsides and riverbanks.

Local names include the widespread bulmint, and:

BATTER-DOCK, and broad-dock, in Shetland
BULWAND in the Lake District
BUTTER-DOCK, and butter-docken, in Yorkshire
Celery seed, given to the plant 'when in seed', in Berwickshire
Cow-dock, cushycows, dockan, and docken, in northern England and Scotland
Dockens in Somerset
Docking in Devon
DOCTOR'S MEDICINE in Somerset
DONKEY'S OATS in Lancashire
Elgin in Shropshire
GOOD KING HENRY in Wiltshire

KETTLE-DOCK in Ayrshire
Land-robber, rantytanty, and REDSHANK, in Scotland
Smair-dock in Co. Donegal.

The use of dock leaves to treat nettle STINGS is the most widely known folk remedy in the British Isles.[1]

> Up to the age of 11 I attended the local primary school . . . on 29 May all the children had to wear a piece of OAK, preferably with an oak apple on it (thereby known as Oak Apple Day). Punishment for not wearing a piece of oak leaf was being stung by a nettle of which there were many to perform this task. Whoever was stung rubbed the place with a dock leaf, whilst saying the phrase 'Dock leaf, dock leaf, you go in; Sting nettle, sting nettle, you come out'.[2]

> Farmer in west Wicklow showed me the parts of dock to use for nettle stings, at the very centre of the growing bud there is a sticky gelatinous sap which when applied to nettle sting immediately stops the sting.[3]

> Between Mendham Hill . . . and All Saints Church [Suffolk] were several dried-up ditches and stanks (Fr. *étang*?) copiously cropped not only with common nettles but also with fewer docks. As children we were told that 'God planted the docks to soothe us if we got stung by the nettles'.[4]

> [Clare, Suffolk] for stinging nettle stings take a dock leaf, scrunch up a little and spit on it, and wipe the sting.[5]

Although it is sometimes claimed that a substance in the leaf is responsible for the reduction of pain, it is more probable that the cure works on a psychological level – searching for a dock leaf provides a distraction from the pain, while the rubbing action may help disperse the irritant. When talking to people in Britain and Ireland about herbal remedies, the usual response is 'oh, I don't know any'. But if the same people are asked 'what do you do if you're stung by a nettle?', they invariably reply, 'oh, rub on a dock leaf, of course'. Despite this the cure seems to be rarely known throughout the rest of Europe. Many people know of no cure for nettle stings. In Hungary, 'we just wait for them to go away'.[6] In Austria,[7] and Bulgaria,[8] urine was recommended, but 'I've never tried it'; in France,[9] Estonia[10] and the Netherlands,[11] GREATER PLANTAIN was used.

According to William Coles in his *Art of Simpling*, published in 1656:

> The seeds of docke tyed to the lefte arme of a woman do help Barrennesse.

More recently in County Donegal:

> No woman may fear to be barren who carries a bag filled with docken seeds under her left oxter.[12]

Although now regarded mainly as a troublesome weed, dock had many other medicinal uses.

> Men working in the ironstone quarries near Deddington [Oxfordshire] often get a peculiar SORE on their arms, and this they treat by cutting a dock-root across and rubbing the sore with the fresh cut surface . . .
>
> To purify BLOOD an infusion made by pouring boiling water on to young dock roots or COLTSFOOT leaves is sometimes recommended.[13]

> A cure for BOILS: Take as many dock roots as you can find and boil them in water until you get a thick gooey liquid, which the patient should drink.[14]

> My husband used young dock roots to clear his blood; we were lucky and had a garden with lots of nettles and where there are nettles there are usually dock roots.
>
> Back in 1943 I was expecting my daughter and the misery of STYES on my eyes, no sooner did one get better than another one started. It must have been the wrong time of year for young dock roots as the only ones he could find were old ones. I was pretty desperate, so scrubbed them and boiled them, and smelt them; the smell was enough. I thought what if it's poison.
>
> Oh well, here goes, John said one small wine-glass, but seeing they were old and smelling so strong I thought a table-spoon would be enough. Ugh! It was horrible, but by morning so much easier I took a second dose and a further that evening, and believe it or not the second morning they had gone and I have never had another stye since. But believe me it really tastes vile. Dock root tea is a wonderful cure for spots and PIMPLES.[15]

> For RHEUMATISM: Gather dock leaves, carefully dry them and bind them around the affected joint. (Said to be a complete cure).[16]

> When I was a child (I am now 91) we lived on the Isle of Man, we were quite hard up and couldn't afford doctors' bills. My father was mowing grass one day and the scythe slipped and cut him

> very badly on the leg, he didn't dare stay home from work, so with my mother's help, doctored himself; every day after school I collected large dock leaves, which mother crushed with a rolling pin, then applied the leaves straight on to the CUT which in time healed and caused no further trouble.[17]

> As children in the 1920s dock leaves were picked to ease nettle stings and SUNBURN . . . we would wrap these around arms and legs.[18]

> My father was a builder and in going to work in hot weather he always used to place a dock leaf in each boot (veins uppermost). These were taken out on returning home. My father used to say this was good for perspiring FEET . . . he never had bad feet, and he used to walk great distances.[19]

> Dock leaves . . . inside shoes for sore feet.[20]

Other uses included:

> Farm workers put a dock leaf in their TOBACCO pouches to keep contents moist.[21]

> I know someone who uses dock leaves to keep his tobacco moist – he's Jamaican.[22]

> If the old horse-men found a dock root when they were ploughing they would pick it up and feed it to their HORSES . . . to bring up the 'hammer marks' [dapple marks] on their coats.[23]

> Docken – the seeds stripped off and lightly boiled – were widely used as an addition to normal POULTRY feeding, and the stalks were used to make BASKETS. The stalks had to absorb salt to keep them pliable, and this was done either by immersion in the sea for some hours, or by sprinkling them with salt.[24]

DOCTRINE OF SIGNATURES

A theory elaborated in the sixteenth and seventeenth centuries claimed that plants had characters or 'signatures' which provided guidance on which diseases they were capable of curing. It is difficult to ascertain the extent to which the Doctrine was accepted by the educated classes from whom it might have been passed on to the mostly illiterate poor.

It is probable that the Doctrine has received more than its fair share of attention in popular publications on plant folklore.

A useful summary of the Doctrine's history is provided by Agnes Arber in her study *Herbals*, originally published in 1912. The first writer to propound the theory was Theophrastus Bombast von Hohenheim (1493–1541), better known by his Latinised name of Paracelsus. He had a meagre knowledge of plants and experienced a varied career which culminated in a short-lasting appointment as a professor at Basle. His ideas were enthusiastically taken up and expanded by Giambattista Porta in his *Phytognomonica*, published in Naples in 1588. This work suggested that eating long-lived plants would lengthen a man's life, whereas eating short-lived plants would shorten the eater's life.

In England the Doctrine was wholeheartedly commended in the publications of William Coles. In chapter 17 of his *Art of Simpling* of 1656 Coles declared that although sin and Satan had plunged mankind into an ocean of infirmities, God in his mercy had made herbs for the use of man, and had 'given them signatures whereby man may read even in legible Characters, the use of them'. However, even Coles was unable to find signatures on all plants. In chapter 18 he attempted to deal with this problem by explaining that although God had provided guidance by imprinting signatures on some herbs, as man had not been created to be 'like an idle loiterer or Truant' other herbs had been left unmarked so that man might discover the virtues by his own ingenuity. A year later Coles published his larger work, entitled *Adam in Eden*. Here he included such observations as:

> The milky juyce which Issueth from the wounded stalks and Leaves [of LETTUCE] is sufficient, that this Herb, if it be eaten boyled or raw, maketh plenty of milk in Nurses.[1]

> A Decoction of the long Mosse that hangs upon Trees, in a manner like hair is very profitable to be used in the falling off of hair, and this it doth by Signature.[2]

At the time Coles was writing it was not usual to distinguish between mosses and lichens, and it seems probable that his 'long moss' was in fact a 'beard' lichen of the genus *Usnea*. A shampoo containing usnic acid, derived from *Usnea* spp. and other lichens, is still occasionally available.[3]

The Doctrine was repudiated before the end of the sixteenth century by writers such as Rembert Doedens (1517–85), but continued to be mentioned in works published in the eighteenth century. More recently writers on popular books on folklore have mentioned the Doctrine, but it is difficult to ascertain if it ever strayed far from published works and became a genuine folk belief.

Although the idea of signatures did not necessarily conflict with the beliefs held by ASTROLOGICAL BOTANISTS, Coles thought he should expose the fallacies he found in the writings of the leading astrological-herbalist, Nicholas Culpeper. He did this by citing the opening verses of the bible. God created plants on the third day and the heavenly bodies on the fourth, so these later creations could not be expected to influence the earlier ones.[4]

Gabrielle Hatfield suggests that people discovered that a plant was effective against an illness, and then sought a character of the plant to act as a reminder of this. The cure came before the signature, not vice-versa:

> Thus, if the BARBERRY plant was useful for liver complaints, it would be natural to seek a feature of the plant by which to remember this. The bright yellow bark could be associated with JAUNDICE, so this would be a useful mnemonic.[5]

DODDER (*Cuscuta epithymum*)

Annual, leaf-less parasitic herb, with red stems and pale pink flowers, on gorse and other moorland shrubs, mainly in southern England.

Local names include:

Adder's cotton in Cornwall
Ail-weeds in Buckinghamshire; cf. hellweed
BEGGAR-WEED in Wiltshire
DEVIL'S GUTS in Ashdown Forest, East Sussex
Devil's net in Kent
Epiphany in Cornwall
Hairy bind in Hampshire
HELLWEED, 'called by the country people hell-weede, because they know not how to destroy it'
Mazzle-wort in Lincolnshire
Wicked tree in Dorset.

> *Herbe d'emeute* . . . so named from its powerful properties, which are utilised by Guernsey farmers on particular occasions in the treatment of horned cattle. A handful of the fresh plant is placed on a CABBAGE leaf, which is then rolled up and given to a cow to eat.[1]
>
> In Cornwall some folk with a liking for homemade saffron cake use dodder as saffron.[2]

Two names, devil's saffron, and SAFFRON, recorded from St Just-in-Penwith, in the far west of Cornwall,[3] presumably refer to this practice.

DOG BITES – DEVIL'S-BIT SCABIOUS used to treat.

DOG ROSE (*Rosa canina*, and other wild *Rosa* spp.)

Common thorny shrubs with attractive pink and white flowers and scarlet fruit (hips), in hedgerows and woodland edges and colonising neglected grassland.

Local names include the widespread CANKER, and:

Bird-brier in Cheshire, 'because the hips are eaten by birds'
Brandy-bottles in Somerset (presumably referring to the fruit)
BREAR, BREER and BRERE, in northern England
Briar-rose in northern England and southern Scotland
BRIMMLE in Shropshire
Buck-breer in Co. Antrim and Co. Down
Buckie-briar in Co. Antrim, Co. Down and Co. Fermanagh
Bucky in Belfast
CANKER-ROSE in Devon and Essex
Cat-choops in Cumbria
CAT-WHIN in northern England
Choop-rose in Cumberland
COCK-BRAMBLE in East Anglia
Dike-rose in Cumberland
Dog-brear in Yorkshire
Dog-brier in Dublin
Dog's briar in Hampshire
Ewe-brimmel, and ewe-mack, in Devon
Head-speaks in Gloucestershire
Hep-brier in Cheshire
Hip-briar in Shropshire
Hip-rose in Gloucestershire
Hip-tree in Gloucestershire and Northumberland
Hook-brimble in Devon
Horse-bramble in East Anglia
Klonger, and klunger, in Shetland
LAWYERS in Cheshire and Surrey
Neddy grinnel in Worcestershire
Pig-rose in Cornwall and Devon
Pig's noses, and pig's rose, in Devon
Sugar-candy, given to 'both the tender shoots and the fruits', in Somerset and Wiltshire

Titty-bottles in Somerset (presumably referring to the fruit)
Yew-bramble in Devon
Yew-brimble in Devon and Somerset
Yew-mat in Devon.

Names given to the fruit include:
Buckie-berries in Co. Antrim and Co. Down
Buckies in Aberdeenshire
Bumble-berry in Wiltshire
Canker-berry in Kent
Cankers in Dorset and East Anglia
CAT-HIP in Devon
Cattijugs in Yorkshire
Choops in northern England and Scotland
Choups in northern England
Conker-berries in Dorset and Somerset
Dog-berries in Hampshire and Yorkshire
Dog-choops, and dog-chowp, in Yorkshire
Dog-hip in Scotland
Dog-hippens in Aberdeenshire
Dog-job, and dog-jumps, in Yorkshire
Dog's hippens in Aberdeenshire
Hagisses in Hampshire
Hawps in northern Scotland
Hedgy-pedgies in Wiltshire
Hippans in Morayshire
Hipson in Oxfordshire
Huggans in Yorkshire
LOCKS-AND-KEYS in Somerset
NIPPER-NAILS, and nips, in Cheshire; 'cf. Norwegian *nypen*'
PIXY-PEARS in Devon
Puckies in Aberdeenshire and Ireland
RED BERRIES in Yorkshire
SOLDIERS in Kent.

> I am now 53. Rose itching powder was a great favourite at school – then known as Thetford Secondary Modern. Several boys would take the hips to school in their pockets in a complete state. This stopped the 'vital ingredients' affecting the collector. On spotting an unwary victim one or two hips would be broken open, most of the husk discarded, and the resulting seeds and irritating hairs thrust down his/her back via the collar. On certain days the vari-

> ous gangs dared not turn their backs on their rivals. I can assure you it is very uncomfortable, especially if you had a sports period afterwards and became hot and sweaty. We used the hips of dog rose (*Rosa canina*) as well as those of sweet briar (*R. rubiginosa*). They were both far more effective if the hips were left until they were very old and had become soft and brown.[1]

> A school prank in east Hertfordshire, 1945–50 . . . the most awful itching material was made from rose hips – the hairs on the seeds being put down someone's neck – a bath and a complete change of clothing being the only cure.[2]

This practice was extremely widespread, resulting in names such as buckie-lice in southern Scotland and Ireland, cow-itches in Cheshire, itching-berries in Lancashire, itchy pips on the Hampshire/Sussex border, and tickler and tickling tommy in Devon.

Dog rose hips and young shoots were eaten by children.

> [Inverness] rose hips are edible, once one removed the rather furry growth around the seeds, which was inclined to stick in the throat. The skins however were soft and very sweet. They were called muckies.[3]

> We ate SORREL, BEECH leaves and briar tops.[4]

> My cousin, now aged 78, remembers chewing new wood from the base of wild roses when they first began to shoot and called this bacon.[5]

> My garden helper – a Suffolk man of 83 . . . remembers . . . EGGS-AND-BACON – shoots of wild rose, eaten by children.[6]

Following the outbreak of the Second World War there was concern about inadequate supplies of vitamin C, due to a shortage of imported ORANGES. Initially it was hoped that home-grown BLACK CURRANTS could make up for this deficit, but in 1941 Dr Ronald Melville, a botanist at Kew, suggested 'It may be worthwhile to consider the possibility of using wild rose hips as they are richer in ascorbic acid [vitamin C] than black currants'.[7] Later that year, throughout the rest of the War, and for some years afterwards, children in the north-east of England and elsewhere collected rose hips for use in the manufacture of rose-hip syrup.

It is possible that Melville's suggestion was stimulated by reports from Germany:

> chemists have found that the hips of *Rosa canina* are a rich source of the anti-scorbutic vitamin C. The German State Railways are therefore using their tracks for growing this species; half a million plants are to be acquired for this alone, and other waste land is to be utilised.[8]

In the United Kingdom:

> Children in Northumberland and parts of north Durham gathered rose hips and sold them; for several years I and several friends did this each autumn. The man who started this practice was a friend of my family. He was Mr Norman Pattison and he worked for the firm of Scott and Turner, the makers of Andrews Liver Salts, at their factory in Gallowgate, Newcastle-upon-Tyne.
>
> In his early years with the firm he was a delivery driver, and when the company first began to make Delrosa, before the Second World War, he gathered wild rose hips for them. Then his wife joined in, and more of his friends and relatives. Mrs Pattison was a former teacher of girls' P.E. at Jarrow Grammar School, and was a close friend of my aunt, also a teacher. So when I started at that school as a pupil in 1941, Mrs Pattison had enlisted the help of at least two of her former colleagues on the staff, who organised groups of pupils (almost every one of us girls) to go out in late September and October, on Saturdays (I don't recall Sundays) and gather the hips from the hedgerows. Two teachers – the Geography and Botany mistresses – combined the hip gathering with field studies in their own subjects, and we enjoyed these outings, apart from the scratches! Our rose hips were weighed and Scott and Turner paid us 3d a pound for them; I know we received a letter telling us how much we were contributing to the war effort by enabling more rose hip syrup to be made – a valuable source of vitamin C, which would maintain our health during wartime.
>
> I went on these expeditions until 1948 when I left school for college.[9]

> I'm 48 and have lived within a mile of my present address all my life.
>
> For several years, up to the time I left school aged 15, myself

and other of my fellow scholars who felt so inclined, gathered hips in our spare time. We then took them to school where they were weighed by a particular teacher (Miss Temperley) and we were paid 3d (old money) for every pound collected. The accumulation of hips collected was made up in sacks, which were at intervals collected by a company, I think its name was Scott and Turner of Gallowgate in Newcastle-upon-Tyne, who made the hips into rose hip syrup.

I was quite a keen collector and having a bicycle I often ventured alone and further afield to places rich in hips and unknown to my school pals.

This resulted in my final year at school, in my gathering 23 stone of hips. I ended up being the highest collector that year, which earned me the princely sum of over £4, and as the Company gave me a bonus of a free bottle of syrup for every 50 pounds gathered, also six free bottles of syrup plus a badge (long gone) proclaiming me top collector.[10]

Further south, in Sussex, in 1946, it was reported:

The need for the collection of Rose Hips continues; with starvation in Europe the vitamin C which they contain is of the greatest value. In collecting them we can add to the great life saving effort that is afoot to preserve the life of many children.

The railway companies have allowed a special concession this year. EVERY passenger train station will accept filled sacks for transport with no charge to the senders. Prices offered by Messrs Paines and Bryne are as follows:

To the collector 3d per pound, to the sub-organisers 1d per pound, to County Committees making payments to the sub-organisers 1d per pound.

It is greatly hoped that the collection will be as good this year as in former years.[11]

Twenty years later *The Times* noted:

Since the last war, when a scheme for voluntary paid harvesting was devised the young collectors have gone on 'hipping' . . . Hundreds of tons are picked every autumn. The children are paid by the pound, and the school, which usually acts as the weighing and dispatching depot, received an additional payment a pound for its amenities fund.

> Some schools appoint a 'King' or 'Queen Hip' to supervise the depot operations; others have rose hip clubs with a merit badge for the child with the most poundage, and, of course, the most rose hip money in his pocket.[12]

However, in 1986, by which time Scott and Turner had been taken over by Winthrop Laboratories, it was reported that:

> The traditional method of obtaining rose hips via schoolchildren collection system ceased many years ago.
>
> We currently obtain our rose hips through a UK Agent, who obtains supplied on the international market, and these are supplied de-seeded, halved and dried. The most common source at the present time is Chile.[13]

The gall variously known as bedeguar, robin's pincushion and moss gall, is conspicuous and common on wild roses and was formerly used to prevent or cure a variety of illnesses.

> Hang round the patient's neck the excrescence often found upon the briar-rose, and called here in Sussex Robin Redbreast's Cushion; it is the finest thing known for WHOOPING COUGH.[14]

> Isle of Axholme . . . a green 'tossel' from a wild rose briar gathered and hung up in the house will prevent whooping cough.[15]

> [Shropshire] the wild-rose gall is . . . considered good for TOOTHACHE. 'If you light on a briar-rose *accidental* w'en yo' an' the toothache, an' wear it in your boasum, it'll cure it'.[16]

> [Wiltshire, canker rose] the mossy gall on the dog rose, formed by *Cynips rosae* [now known as *Diplolepis rosae*]; often carried in the pocket as a charm against RHEUMATISM.[17]

> In Wales they say if this [a moss gall] is placed under the pillow of a person who cannot SLEEP, it will perfectly restore him. But it was necessary to remove it at a given time, or, according to the old story, he would never awake.[18]

> Briar-ball: an excrescence from the briar, placed by [Northamptonshire] boys in their coat cuffs, as a charm to prevent flogging.[19]

DOG'S MERCURY (*Mercurialis perennis*)

Perennial herb, common in woodland shade throughout lowland areas in mainland Britain; scarce in Ireland.

Local names include: ADDER'S MEAT in Hertfordshire, boggart-flower and boggart-posy in Yorkshire, SNAKE-FLOWER in Dorset and Somerset, and SNAKE'S BIT and SNAKE'S VICTUALS in Sussex.

> It is called on the Isles of Sky *Lus-glen-Bracadale*, and I was informed that it was there sometimes taken by way of infusion to bring on salivation.[1]

> Dog's mercury – to make yellow (straw-colour) DYE.[2]

DOG VIOLET (*Viola riviniana*; also *V.canina* and *V. reichenbachiana*)

Low-growing, perennial herbs with bluish-violet flowers; common and widespread in hedgebanks and woodland.

Local names include:

BLUEBELL in Dublin
Blue mice in Somerset
Blue violet in Cheshire and Devon
Butter-pats, given to the seed capsule, 'from its resemblance to the half-pounds in which butter is made up for market in the locality' of Preston in Lancashire in 1882
Cuckoo's shoe in Shropshire
CUCKOO'S STOCKINGS in Caithness
Gypsy-violet in Somerset
HEADACHES in Dorset
Hedge-violet in Devon
Hypocrites in Somerset
PIGNUT, given to the seed capsules in Cheshire, 'where children are in the habit of eating them'
Pig-violet in Cheshire
SHOES-AND-STOCKINGS in Pembrokeshire
Snake violet in Dorset
Summer violet in Warwickshire.

> [Reepham, Norfolk] if a child swallowed one of its milk teeth, you had to make him eat a dog violet petal, or his adult tooth would be long like a dog's tooth.[1]

> Boil wild violets and drink the juice and it would cure a pain in the head.[2]

DRAGONS – GORSE associated with.

DRAKE, Sir Francis (*c*.1540–96) – associated with ORANGE Race at Totnes, Devon.

DREAMS, erotic – HONEYSUCKLE produces.

DROPSY – cured using BUTCHER'S BROOM and FOXGLOVE.

DROVERS – routes marked by SCOTS PINE.

DRUIDS – OAK and YEW associated with.

DUCKS – GOOSEGRASS fed to.

DUCKWEED (*Lemna minor*)

One of the world's smallest plants, rarely producing its almost invisible flowers in Britain and Ireland, where, apart from in northern Scotland, it is widespread, often forming dense floating mats on still water.

Local names include the widespread DUCK'S MEAT, and:

Boggart in Warwickshire
Creed in Wiltshire
Dig-meat in Cheshire
Duck-meat in Cheshire and Warwickshire
Duck-pond weed in Cumberland
Groves, grozen, and limpets-crimp in Somerset
Mardlens, and mardling in Suffolk
Toad-spit in Lincolnshire
Water-lentils in Berkshire.

In the Liverpool area the NURSERY BOGEY Jenny Greenteeth was associated with duckweed-covered pools, where until recently she was used to frighten children away from dangerous places. According to some people Jenny (or Jinny, as she is more usually known outside books) Greenteeth is simply a name for duckweed.

> At this day in all east Lancashire the older inhabitants call the green moss which covers the surface of stagnant ponds 'Jenny Greenteeth'. Further, I have often been told by my mother and nurse that if I didn't keep my teeth clean I should some day be dragged into one of these pools by Jenny Greenteeth, and I have met many elderly people who have had the same threat applied to them.[1]

> I was brought up in the Upton/Cronton area of the west side of Widnes in Lancashire (now Cheshire), about 12 miles inland from Liverpool. It was, and still is, largely a farming area, and many of

> the fields contain pits – never ponds – which, I believe, are old marl pits. Some of them have quite steep sides. Jinny was well known to me and my contemporaries and was simply the green weed, duckweed, which covered the surface of the stagnant water. Children who strayed too close to the edge of these pits would be warned to watch out for Jinny Greenteeth, but it was the weed itself which was believed to hold children under the water. There was never any suggestion that there was a witch of any kind there.[2]

> As a child in Cheshire I heard the name Jenny Greenteeth given to the bright green water plant that lies on the surface of stagnant ponds (the minute leaves are rather like tiny teeth) and imagined that if one fell into the pond, the green scum-like plant would close over one's head, thus Jenny (or Jinny) Greenteeth had 'got you'.[3]

> [Maghull] All us kids (I was born in 1951 . . .) knew about Jinny Greenteeth who inhabited the canal and, who, if you were daft enough to be tempted to the edge would pull you in and hold you under, she was personified . . . by common duckweed (*Lemna minor*), which treacherously disguised large parts of the water surface as smooth green lawn.[4]

According to others the growth of duckweed on a pond was a certain indication that Jenny lurked in its depths.

> As a child, about 50 years ago in the Liverpool area, I was frightened by Jenny Greenteeth, a sort of fairy, who would drag people down into deep pools. Jenny was particularly associated with pools covered with duckweed.[5]

> I remember [I am now 34], as a very small child, being told by my mother to stay away from ponds as Ginny Greenteeth lived in them. However, I only recall Ginny living in ponds which were covered in a green weed of the type which has tiny leaves, and covers the entire surface of the pond.
>
> The theory was that Ginny enticed little children into the ponds by making them look like grass and safe to walk on. As soon as the child stepped onto the green it, of course, parted, and the child fell through into Ginny's clutches and was drowned. The green weed then closed over, hiding all traces of the child ever being there. This last point was the one which really terrified me and kept me well away from ponds, and, indeed, my own children have also

> been told about Ginny, although ponds aren't as numerous these days.
>
> As far as I know Ginny had no known form, due to the fact that she never appeared above the surface of the pond.[6]

However, according to a 68-year-old woman, writing in 1980, the Jenny Greenteeth who was believed to inhabit two pools beside Moss Pitts Lane in Fazakerley, 'had pale green skin, green teeth, very long green locks of hair, long green fingers with long nails, and she was very thin with a pointed chin and very big eyes'.[7]

Occasionally Jenny could be found well away from watery places. In the 1930s Liverpool children would rush past the old St James's Cemetery, which was reputedly her home, while a decade later children in Cheshire were told that Jenny would seize them if they ventured too near railway lines.[8]

DWARF CORNEL (*Cornus suecica*)

Low-growing perennial herb on upland moors in northern England, and especially Highland Scotland.

> The berries have a sweet waterish taste, and are supposed by the Highlanders to create a great APPETITE, whence the Erse name [*Lus-a-chraois*, plant-of- gluttony].[1]

Alternatively, in 1996:

> Glutton-berry (sharp tasting, it made an aperitif before a Highland banquet), still occasionally in use as a name in the north-east.[2]

DWARF ELDER (*Sambucus ebulus*)

Robust perennial herb (Fig. 13), infrequent on waysides and waste ground throughout lowland areas.

Local names include: GROUND ELDER on the Isle of Wight and in southern England, WALLWORT in Shropshire, and water-elder, 'from Latin *sambucus palustris*', in Northamptonshire.

Other names such as DANE'S BLOOD in Cambridgeshire and Wiltshire, Dane-weed in Somerset and Suffolk, and Danewort in Berkshire, witness the widespread belief that dwarf elder grew where DANES had been slaughtered. In the seventeenth century John Aubrey wrote:

> Danes-blood (*ebulus*) about Slaughtonford [Wiltshire] in plenty. There was heretofore a great fight with the Danes, which made the inhabitants give it that name.[1]

This belief was also known in Scandinavia. In May 1741 Carl Linnaeus examined what was considered to be a mysterious plant growing in Småland:

> On a field grew almost nothing but HOUND'S TONGUE and the plant which is called *Mannablod* [man's blood].
>
> This *mannablod* or *manna-wort* [man's herb] is a plant which is much talked about in Sweden . . . for it was said that it grows in no other place in the world but here at Kalmar Castle, where it once grew up from the BLOOD of Swedes and Danes, killed in warfare on this field. We were much taken aback when we realised that the plant was nothing but common *Ebulus* or *Sambucus herbacea* . . . which grows wild in the greater part of Germany, around Vaxjo and in gardens.[2]

In Norfolk it was believed that dwarf elder, known in the local dialect as blood-hilder, had been brought to England by Danish invaders, 'and planted on the battle fields and graves of their country-men'.[3] At Ailsworth, Northamptonshire, it was recorded in 1953 that 'blood-elder' according to local tradition 'grew from the blood of Romans buried at the side of the road'.[4]

An alternative explanation for dwarf elder being given 'Danes' names is provided by John Parkinson in his *Theatrum Botanicum* of 1640:

> It is supposed it tooke the name Danewort from the strong purging quality it hath, many times bringing them that use it into a fluxe, which then we say they are troubled with the Danes.[5]

On the Isle of Wight:

> The plant is, I understand, sought by farriers and horse-doctors as a stimulant and to improve the coats of HORSES, which may account for its present scarcity in some localities, as between Chine cottage and Rose cliff, where a countryman informed me he had formerly seen it in abundance.[6]

In Ireland:

> About Williamstown and Mullincross in mid Louth . . . dwarf

elder (*Sambucus ebulus*), there called she-elder, was used to make a preparation for the treatment of ULCERS on cows' udders and teats.[7]

DYE – prepared from ALPINE MEADOW-RUE, ANGELICA, BILBERRY, BOG ASPHODEL, BOGBEAN, BOG MYRTLE, CROWBERRY, DAMSON, DOG'S, MERCURY, DYER'S GREENWEED, FUCHSIA, GORSE, GREEN ALKANET, HAZEL, HEATHER, LADY'S BEDSTRAW, NETTLE, PINEAPPLE-WEED, POPPY, WALNUT, WELD (see MULLEIN), WHITE WATERLILY and YELLOW IRIS.

DYER'S GREENWEED (*Genista tinctoria*)

Low-growing deciduous shrub (Fig. 14) with yellow flowers, in rough grassy places.

Many of dyer's greenweed's local names relate to its former use as a DYE plant:

ALLELUIA in Shropshire
Base-broom, 'referring not to its low growth, but to its being used as a base to prepare woollen clothes for the reception of scarlet and other dyes'[1]
BRUMMEL in Cornwall
Dyer's weed in Cumberland
Greening weed in East Anglia
Green-weed in East Anglia and Gloucestershire
Sarrat in Westmorland
She-broom in Cheshire and Yorkshire
Weld in Scotland
Woad in Yorkshire
Woadwax in Devon, Somerset and Wiltshire
Woadwaxen in Somerset
Woadwex in Dorset and Wiltshire
Woadwise in Kirkcudbrightshire and Wigtownshire
Woodas in Westmorland
WOODWAX in Dorset and Wiltshire
Woodwesh in north-east England
Woodwex in Dorset
Wudwise in southern Scotland
YELLOW in the Midlands; 'cf. German *Gilbe*'.

Apparently referring to Gloucestershire early in the nineteenth century, Knapp recorded:

> Our poorer people a few years ago, used to collect it by cart loads about month of July, and the season of woad-waxen was a little harvest for them, but it interfered with our hay-making. Women could gain about two shillings a day clear of expenses by gathering it.[2]

Further north:

> A plant which is known to have abounded in the neighbourhood of Kendal [Cumbria] many years ago, though it be now nearly uprooted . . . commonly called 'Dyer's Broom', was brought in large quantities to Kendal, from neighbouring commons and marshes, and sold to dyers. The plant, after being dried, was boiled for the colouring matter it contained, which was a beautiful yellow. The cloth was first boiled in alum water, for the mordant, and then immersed in the yellow dye. It was then dried, and submerged in a blue liquor extracted from woad [to produce the famous Kendal green].[3]

DYSENTERY – treated using BILBERRY and TORMENTIL; WOOD ANEMONE causes in cattle.

E

EARACHE – POPPY causes and used to treat; treated using ASH, FEVERFEW, FIG, HOUSELEEK, ONION and WILLOW.

EARLY PURPLE ORCHID (*Orchis mascula*)

Perennial, tuberous herb, widespread in hedgerows, deciduous woodland and permanent grassland.

Early purple orchid has accumulated an extraordinarily large number of local names, some of which it shares with green-winged orchid (*Orchis morio*) and marsh-orchids (*Dactylorhiza* spp.). Widespread names include ADAM-AND-EVE 'from the "male" and "female" tubers', adder's grass, CUCKOO, DEAD-MAN'S FINGERS, dead-man's hand, goosey-ganders, LADY'S FINGERS, and RED ROBIN. Names of more restricted distribution, many of which are shared with BLUEBELL, include:

Aaron's beard in Berwickshire
ADDER'S FLOWER in Cornwall, Hampshire and Somerset
Adder's mouths in Somerset
ADDER'S TONGUE in Cheshire, Devon and Dorset
Baldeeri in Shetland
Beldairy in Aberdeenshire
BLOODY FINGERS in Gloucestershire
BLOODY-MAN'S FINGERS in Cheshire, Gloucestershire and Worcestershire
Blue butcher in Somerset
Boldeeri in Shetland
Butcher in Herefordshire
Butcher-boys in Dorset
Butcher-flower in Somerset
CAIN-AND-ABEL in Berwickshire and Cumberland
CANDLESTICKS in Dorset
CLOTHES PEGS in Somerset
Cock-flowers in Hampshire
Cock's kame in Berwickshire
COWSLIP in Rutland
CRAKE-FEET in northern England
CRAW-FEET in Lincolnshire and Yorkshire
Craw-toes in Cumberland
CROW-FEET in Cumberland
CROWFOOT in Lincolnshire and northern England

Cuckoo-bud in Northamptonshire
CUCKOO-COCK in Essex
CUCKOO-FLOWER in southern England and East Anglia
Curlie-doddie in Shetland
Dandy-goslings in Wiltshire
Deil's foot in Berwickshire
Dog's dogger in Clackmannanshire; 'i.e. dog's dung'
Drake's feet in Lincolnshire
Ducks-and-drakes, and fried candlesticks, in Dorset
FROG'S MOUTH in Somerset
Gander-gause, and gandi-goslings, in Wiltshire
GEUKY-FLOWER in Devon
Gillyflower in Wiltshire
Gilly-gander in Dorest and Isle of Wight
Goo-goo in Cornwall
Goose-and-goslings in Somerset
GOSLINGS in Wiltshire
Gossips in Somerset
Grammer-griggles in Dorset
Grampha-griddle-goosey-gander, and granfer-goslings, in Wiltshire
Granfy-griggle, and gussets, in Dorset
HEADACHE-FLOWER in Berkshire
Hens, and hen's kames, in Berwickshire
Jessamine in Warwickshire
Johnny-cocks in Dorset and Somerset
Jolly soldiers in Devon
Keek-legs in Kent
Keet-legs in Somerset
Kettle-cap on the Isle of Wight
Kettle-case in southern England
Kettle-pad in Hampshire
King-finger in Buckinghamshire, Leicestershire, Northamptonshire, and Warwickshire
King-orchis in Northamptonshire
Kite's legs in Kent
Kite's pan in Wiltshire
Kittle-cases on the Isle of Wight
Locks-and-keys in Devon, Somerset and Sussex
LONG PURPLES in Devon, Somerset and Sussex
LORDS-AND-LADIES in Dorset
Mogolyeen-mire in Dublin
Naked nannies in Somerset
Paddock's spindle in Perthshire
Poison-more in Devon
Poor-man's blood in Kent
Priest's pintel in Cheshire, Cumberland and Warwickshire
Purple hyacinth in Somerset
RAM'S HORNS in Sussex
RED BUTCHER in Kent
Red googoos in Cornwall
Red granfer-gregors in Dorset
Regals in Dorset
Sammy-gussets in Somerset
Scabgowk in Durham
Single castles in Dorset
Single guss in Somerset and Wiltshire
SNAKE-FLOWER in Somerset
SOLDIERS in Dorset and Somerset
Soldier's cap in Somerset
Soldier's jacket in Dorset
Spotted dog in Somerset
Standergrass in Ireland
Standing gussets in Somerset

Stinkers in Sussex; 'from the tom-cat smell of the flowers after fertilisation, or at night after it has been picked'

Underground shepherd in Wiltshire

WAKE-ROBIN in Cheshire.

> One species of orchis, which in Cheshire is called GETHSEMANE, is said to have been growing at the foot of the cross, and to have received some drops of BLOOD on its leaves: hence the dark stains by which they have ever since been marked.[1]

> I was told by Mrs D (a devout Catholic) of Chillington, Ilminster [Somerset], in the 1950s that the red spots on the leaves of orchids are where the blood dropped from Christ when he was on the Cross.[2]

Presumably the Devon name, CROSS-FLOWER refers to this legend.

> [Hardwicke, near Gloucester] Tom Thumb, or early purple orchid, were a bit suspect in the house. They were connected with DEATH.[3]

The root system of early purple orchid typically consists of two swollen tubers which somewhat resemble testicles. These gave rise to a number of 'bag' and 'stones' names being given to the plant: bull-seg and bull's bag in Scotland, dog-stones in Somerset, and fox-stones in Dorset and Somerset. Presumably these tubers were also responsible for the plant's association with love-making and procreation.

> The decoction of the roots drank in Goats-milk mightly provokes Venery, helps conception, and strengthens the Genital parts.[4]

> [North-east Scotland] to gain LOVE there were various methods. The roots of the orchis were dug up. (The old root is exhausted, and when cast into water, floats – this is hatred. The new root is heavy and sinks when thrown into water – this is love, because nothing sinks deeper than love.) The root – love – was dried, ground, and secretly administered as a potion. Strong love was the result.[5]

> In Co. Wicklow the early purple orchis is called Mogra-myra, and is supposed to be most efficient as a love-potion.[6]

> The orchid vies with MANDRAKE as an APHRODISIAC – excessive ardour can be cooled with strawberry leaf tea.[7]

According to many commentators, the long purples in Shakespeare's *Hamlet* were early purple orchids, but see discussion under LORDS-AND-LADIES.

EASTER – ARUM LILY and EASTER LILY associated with; corn-showing to remove CORN COCKLE seedlings took place at; GOAT WILLOW should not be brought indoors before; graves decorated for (see FLOWERING SUNDAY); rain at means a good HOP harvest.

EASTER EGGS – GORSE and ONION used to dye.

EASTER LILY (*Lilium longiflorum*)

Native to Japan, the pure white flowers are popular for decorating churches at EASTER, largely replacing the ARUM LILIES which were formerly used.

> At Easter churches – or at least our church – are decorated with lilies – arum lilies, and mostly nowadays *Lilium longiflorum* – bought in memory of the dead.[1]

> Last Easter we filled Holy Trinity Church with the scent of over one hundred and fifty lilies donated by many members of our Church community in memory of their loved ones. We would like to repeat this during Easter 2003.[2]

The Easter lily associated with the 1916 Easter Rising in Ireland is arum lily.

EASTERN GLADIOLUS (*Gladiolus communis*)

Perennial, purple-flowered herb, introduced from southern Europe in the sixteenth century, and grown as an ornamental becoming naturalised mainly in south-west England and on the Isle of Man; formerly cultivated on the Isles of Scilly for sale as cut-flowers, now replaced by more showy varieties, but persisting as a troublesome weed.

> [Scilly] I have heard farmers refer to it as 'Jacks' and 'Rogues'. On St Martin's Geoffrey Grigson was told in 1940 that they call it 'Squeakers'. This no doubt has the same origin as 'Whistling Jacks' which Mr P.Z. MacKenzie tells me is the usual local name because the children use the leaves as reeds to WHISTLE.[1]

ECZEMA – treated using BANANA, COMFREY, GOOD KING HENRY, IVY and OAT.

EELGRASS (*Zostera marina*)

Maritime perennial herb, growing in the subtidal zone, sometimes cast up on the shore in great quantities.

Local names include:
Barnacle-grass in Co. Derry
Drew on the Shetland Islands
Grass-weed on the Isle of Wight
MALLOW in Orkney
Marlak, and marlie, in the Shetland Islands
Sleech in Co. Antrim and Co. Down
Sweet grass in Co. Donegal
Widgeon-grass in Dublin.

> Marlie used to be so abundant towards the head of Weisdale Voe [Shetland] that a channel sometimes needed to be cut to allow the passage of small boats . . . The autumn gales would drive masses of eelgrass on to the beaches from where it was harvested for bedding cattle and for stuffing MATTRESSES; for the latter purpose it had to be very carefully dried. In the Weisdale area these practices continued into the early 1920s. In Orkney it was formerly used as a MANURE for fields and as a THATCH.[1]

> Collected and dried as a stuffing for mattresses as it was believed to be proof against FLEAS.[2]

During a botanical expedition to Schleswig-Holstein in August 1850 W. Lauder Lindsay observed that eelgrass 'is thrown up in such abundance on the coasts that it is employed in the manufacture of mattresses.[3]

David Mabberley notes that dried eelgrass was used for stuffing pillows,[4] and it appears that the plant's use as bedding was widespread. For example, the herbarium of the Natural History Museum, London, holds a sample of labelled 'Seaweed from New Holland, used in the

Isle of Pines for making beds'. Although neither the date of collection nor the collector's name is recorded, it is probable that it was collected by James Everard Home (1798–1853) in New Caledonia, not New Holland (Australia).

In the early part of the twentieth century eelgrass was used to insulate buildings, although it seems that it was imported from North America, rather than collected from British shores, for this purpose. The commercial harvesting of eelgrass for insulating buildings started on the coasts of New England, and then from 1906 until the early 1960s (peaking late in the late 1920s) on the coasts of Nova Scotia, Canada. Leaves were gathered from tidal flats and sold for processing into quilts which had much extolled sound-deadening and heat-insulating properties. The Centre for Economic Botany, at the Royal Botanic Gardens, Kew, holds a sample presented to it by the Imperial Institute, and bearing the printed label:

British Empire product
'Riverbank'
Quilted Building Blanket
Sound deadening – heat and cold insulation
Joseph Stephenson & Co. (London) Ltd.

A hand-written label adds: 'Grass wrack or eelgrass – used for deadening sound in the new Bank of England, the London Library and elsewhere'.[5]

ELDER (*Sambucus nigra*)

Small deciduous tree (Fig. 15), producing showy creamy-white flowers and small, dark purple fruit, common, but in northern Scotland found probably only where formerly planted.

The island of Tresco, in the Isles of Scilly, was early in the fourteenth century called Trescau ('homestead of elder-trees'), due to trees said to have been brought to the island by monks in the eleventh century.[1]

Local names include the widespread boor-, bore-, or bour-tree, ELLER and eller-tree, and:

ALDER in Co. Donegal
Aldern in Wiltshire
Alderne in Devon
Arn-tree in Scotland
Baw-tree in Lincolnshire
Bertery in Yorkshire
Bitter flower in Devon
Bitter medicine in Somerset

Boon-tree, and borral, in Northumberland
Borral-tree in Scotland
Bothery tree in Holderness, Yorkshire, where 'toy POP-GUNS made from the branches called bothery-guns'
Bottery, or bottry, in Yorkshire; 'a corruption of bore-tree'
Bottery-tree in Northumberland
Boun-tree in Northumberland and Scotland; fruit known as bountree-berries
Bour in Somerset
Bull-tree in Cumberland and Dumfriesshire
Bur-tree in Cheshire and northern England; 'i.e. bore-tree; pop-guns made from it are called bur-tree guns or bur-tree puffers'
Buthery-tree, and buttery, in Yorkshire
Cauliflowers, given to the flowers, in Somerset
Devil's wood in Derbyshire
DOG-TREE in Yorkshire
Eldern in eastern England and Northamptonshire
Ellane in Herefordshire
Ellar in Kent, Lincolnshire and Sussex
Ellen in Radnorshire
Ellen-tree in Yorkshire
Ellern in Dorset
Ellet, and ellot, in Sussex
Elren in northern England
God's stinking-tree in Dorset; 'since it was used for the Cross'
Judas-tree in Kent; 'there is an old tradition that Judas hanged himself upon it'
Purple-berries, given to the fruit in Somerset
Scaw, scawen, scaw-tree, skaw, and skew, in Cornwall, where scawsy-bud has been recorded for the flowers
TEA-FLOWER, and tea-tree, in Somerset
Thrumman in Co. Donegal
Umbrellas, given by schoolchildren to the flowers in Somerset
Whit-aller, and whit-eller, in Somerset
Winlin-berries, given to the fruit in Berwickshire
Witches' tree in Cornwall.

Elder is one of the most enigmatic plants in the folklore of Britain and Ireland. On the one hand it is feared and associated with WITCHES, on the other it is valued for its protective qualities, as a fly repellent, and its many medicinal uses.

> The whole plant hath a narcotic smell; it is not well to sleep under its shade.[2]

> Elder – unlucky to bring either flowers or wood into a house: (a) because it is the witches' tree, (b) because it was believed that JUDAS ISCARIOT hanged himself from an elder tree, (c) because

LEFT 1. Birch branch decorating St John the Baptist Church, Frome, Somerset, Whitsunday, 25 March 2015

BELOW 2. Bird's-foot trefoil, Happy Valley, Croydon, Surrey

FAR BELOW 3. Bog asphodel, Easedale, Cumbria

ABOVE LEFT
4. Broom, Derwent, Cumbria

ABOVE RIGHT
5. Burdock: Burry Man, South Queensferry, West Lothian, 11 August 2017

LEFT 6. Butcher's broom, Therfield, Hertfordshire

OPPOSITE PAGE
ABOVE LEFT 7. Centaury, Stonegate, East Sussex

OPPOSITE PAGE
ABOVE RIGHT 8. Four-leaved clover, card posted in Newham, Gloucestershire, 1907

RIGHT 9. Coltsfoot, Tooting Common, Wandsworth, London

May Luck
be yours!

LEFT 10. Cowslip, Corfe Castle, Dorset

ABOVE 11. Welsh rugby supporter wearing (artificial) daffodils on her way to Wales v. South Africa, Twickenham, Middlesex, 17 October 2015

OPPOSITE PAGE LEFT
12. Devil's-bit scabious, Hilly Fields Park, Enfield, London

OPPOSITE PAGE RIGHT
13. Dwarf elder, Queen Elizabeth Olympic Park, London

TOP 14. Dyer's greenweed, Hartland Quay, Devon

ABOVE 15. Elder, The Shire Country Park, Birmingham

RIGHT 16. Everlasting pea, Rushton, Northamptonshire

1700
FIRST PRIZE YELLOW

OPPOSITE PAGE ABOVE LEFT 17. Foxgloves decorating St Nectan's Well, Welcombe, Devon, 22 June 2017

OPPOSITE PAGE ABOVE RIGHT 18. Gooseberry Show, Egton Bridge, North Yorkshire, 1 August 2017

OPPOSITE PAGE FAR LEFT 19. Goosegrass, Dartford, Kent

OPPOSITE PAGE LEFT 20. Gorse, Calton Hill, Edinburgh

ABOVE 21. Greater stitchwort, Birdsmoorgate, Dorset

RIGHT 22. Holy Thorn, Glastonbury, Somerset

ABOVE LEFT

23. 16th Hampstead Heath Conker Championships, London, 1 October 2017

ABOVE

24. Houseleek, Tenbury Wells, Worcestershire

LEFT

25. Hydrangea sepals used in well-dressing, Stoney Middleton, Derbyshire, 25 July 2015

if you fall asleep under elder flowers the scent will poison you or you will never wake up.[3]

[Leitrim, Waterford and the south of Ireland] The elder or 'bore' tree is believed to have been the tree from which Judas Iscariot hanged himself. The proof of which is the fact that its leaves have an 'ugly' smell, and, moreover, that its fruit has since degenerated from its original size and excellent flavour to become worthless both as to size and taste.[4]

Alternatively, elder provided the wood for Christ's CROSS:

Elder growing near the gate
Never crooked, never straight,
Never a bush and never a tree
Since our Lord was hung on thee.[5]

Another inauspicious aspect of elder is its association with witches, and, particularly on the Isle of Man, FAIRIES.

It was said at Beckley [Oxfordshire] that if you burn elder wood you will become bewitched. You never cut it down. In Wootton they say that elder is a witch tree. You should not mend a wattle hedge with it, as it will give the witches power. If you cut it, it will bleed.[6]

Elder – it is alright to pick the flowers for wine or culinary use, but the tree is friend of witches and the wood should never come into the house.[7]

Normally in the Isle of Man elder is the fairies' tree which is UNLUCKY to cut down, or burn when fallen. I was told in 1992 by a forestry worker of his pleasure that a large elder had blown over into a field adjoining his garden and thus relieved him of the need to find someone willing to remove it.[8]

More generally:

Collecting firewood from the hedges surrounding the cottage and returning happily laden, and being accused of bringing bits of elder into the house – it was unlucky to use these to light a fire.[9]

The only unlucky plant that I have heard of is the elder tree, which the old people looked upon as unlucky . . . it was unhealthy

> to have an elder tree growing near the house as it was often noted the inhabitants seemed more prone to TUBERCULOSIS or 'consumption' as it was known in Ireland in the old days.
>
> However, as TB was rampant all over the country at that time, I don't know if the belief would have any significance. My own people would not cut down an elder bush or burn it no matter how old or rotten it was. Nor allow an elder stick in the house, and it would be an unforgivable act to strike a child or even an animal with one.[10]

> The family name dies out on the property where the elder grows in the kitchen garden.[11]

> Elder wood – burning it on a domestic fire caused the DEVIL to sit on the chimney.[12]

> Do you know the Rollright Stones in Oxfordshire? You can't count them; you never get the same number twice. In the next field there is a big stone called King Stone, and there are various stones called after his Knights around. There are some elder bushes nearby. We used to go there as children on our bicycles and try to count the stones. We were told that if we picked a flower or a berry from these elderberry bushes we would be turned into stone. We used to dare each other to pick a berry or a flower, but no one ever did.[13]

In the early part of the nineteenth century:

> On MIDSUMMER EVE, when the 'eldern' tree was in blossom, it was a custom for people to come up to the King Stone and stand in a circle. Then the 'eldern' was cut, as it bled 'the King moved his head'.[14]

Sometimes it was thought that elder wood, flowers or berries could be safely gathered only if the tree's permission had first been sought.

> Hearing one day that a baby in a cottage close to my own was ill, I went across to see what was the matter. Baby appeared right enough, and I said so; but its mother promptly explained. 'It were all along of my maister's thick 'ed; it were in this how: t'rocker cummed off t' cradle, and he hedn't no more gumption than to mak' a new 'un out of illerwood without axing the Old Lady's leave, an' in coorse she didn't like that, and she came and

pinched t' wean that outrageous he were a'most black i' t' face; but I bashed 'un off, and putten an 'esh on, an' t' wean is as gallus as owt agin.'

This was something quite new to me, and the clue seemed worth following up. So going home I went straight down to my backyard, where old Johnny Holmes was cutting up firewood – 'chopping kindling' as he would have said. Watching the opportunity, I put a knot of elder-wood in the way and said, 'You're not feared of chopping that are you?' 'Nay,' he replied at once, 'I bain't feared of choppin' him, he bain't wick (alive), but if her were wick I dussn't, not without axin' the Old Gal's leave, not if it were ever so' . . . You just says 'Old Gal, give me of thy wood, and Oi will give some of moine, when I graws inter a tree.'[15]

If you chop an elder tree or fell it, you should bow three times and say:

Old Woman, Old Woman.
Give me some of your wood
And when I'm dead
I'll give you some of mine.[16]

[Staffordshire, 1930s] my mother said it was the thing if one wanted blossoms or fruit from an elder tree to say 'Please Mother may I have . . . [17]

In my part of the country – Pembrokeshire – you had to ask permission [from the tree] before you cut elder.[18]

Conversely there are many records of elder being a beneficial, protective tree.

[Northumberland] an old man told me that his aunt used to keep a piece of bour tree, or elder, constantly in her kist (chest) to prevent her clothes from malign influence.[19]

[Wales] In the past an elder planted before the door of a cowshed or stable protected the cows and horses from witchcraft and sorcery.[20]

[In Scotland elder was] often planted near old crofts and cottages as protection from witches.[21]

[In Guernsey, elder] was planted as near as possible to the back door, the most used entrance, since it was a sacred tree and a good protection against witchcraft.[22]

[Ireland] It was considered lucky to have an elderberry bush grow near your house, especially if it is 'self-set'.[23]

Elder: this was called boortree . . . the leaves were boiled and the water used to dose pigs. For this purpose, and because it was supposed to be a protection against LIGHTNING, there was a tree at every house. It can still be seen growing in places where there are no houses now, but where houses were years ago.[24]

Family folklore passed on to me . . . one should plant a ROWAN and elder tree and never cut them down, in order to keep witches away.[25]

Twenty-five years ago it was still common amongst local people here in S. Lincs to put a small posy of rowan, HAWTHORN and elder flowers by or over the door-frame to deter bad luck from entering.[26]

I can remember as a child elder growing around the wooden bottom-of-the-garden 'lavvy' at my uncle's farm near Brentwood, Essex, and many other similar loos with elder adjacent. I was told that elder would live 'almost for ever', and if one root died off another would spring up from a fallen branch or twig. They were treated with 'respect' as they kept away bad magic – no one used the word 'witches' – but the inference was there.[27]

However James Nicolson states that widdie (elder) was planted outside Shetland crofts simply to provide a windbreak, 'although it too required the shelter of a dry stone wall'.[28]

More usually elder was planted in the vicinity of toilets and other buildings to deter FLIES.

Elder bushes are invariably to be seen outside dairy windows on the north side of old-fashioned farmhouses in the Midlands. This is done because elder-trees are supposed to be very objectionable to flies, wasps and other insects, the tree thus provided both shade and protection. For the same reason a switch of elder with leaves is used when taking or driving a swarm of bees.[29]

Inspecting a slaughter house [in Cornwall] a summer or two ago, I

> commented on the absence of flies, and was told that this was due to a large elder bush growing some feet away and that branches of elder in any building would keep flies away.[30]

> According to some friends of mine elderberry bushes were planted by water butts and outside privies so that the smell would keep flies away.[31]

> As a youth my late father worked on the land . . . often handling horses it was common practice to tie branches of elder leaves to the harness to ward off flies.[32]

> My wife, who comes from Northumberland, tells me that her mother used to make a concoction with elder flower when she was a child. All the family washed their faces in it to keep virulent Northumbrian MIDGES at bay. She remembers it smelling not too pleasant, and tended to keep other children away as well, so she would take the first opportunity to wash it off![33]

> About 12 years ago, in Girton, Cambridge, a small swarm of bees (apparently known as a 'cast') settled on a plum tree in our garden, about six feet up. A neighbour, Mr C.G. Puck (now 84 years old) a retired shepherd and lifelong bee-keeper, came to collect the bees. He removed the queen bee from the swarm and placed her under a small open wooden box inverted on the ground under the tree. He then asked for a sprig of elder and laid this about nine inches above the swarm, saying that the smell of it was disliked by bees, and by early evening all the bees had moved into the box . . . He learned the use of elder in this fashion from his bee-keeper father, in his native village of Thriplow, south Cambridgeshire.[34]

On the Isle of Man:

> Each cottage has a 'trammon', or elderberry tree, outside the door. This is used by the 'Phynoderree' to swing in. He is a kind of faun who can bring much luck, and even helps materially in outside work.[35]

> [Fairies] like most of all to swing and play in the elder trees, and these were always thought of as fairy trees in the Isle of Man. There wasn't a house or farm that didn't have its 'tramman' tree planted by the door or in the garden 'for the fairies'. Many of them are still to be seen; the single tree will soon have grown into a

> thicket, hiding the old ruined house, but a sure sign that a house once stood there . . . When the wind was blowing the branches, it was then that the fairies were believed to be riding the tramman trees, but it was said they would desert a house or farm where the trees had been cut down. This must have happened only very rarely; no one would cut a branch of the tramman, let alone the tree itself, but if this was done the fairies grieved.[36]

> My cousin in Port St Mary, Isle of Man, wished to have some elder bushes cut down which were shading her kitchen windows. There was a long argument with the men sent to do the work – 'They' wouldn't like it ('they' being fairies) and misfortune would follow within a year. The pruning was eventually done, with our neighbour shrieking from next door.[37]

The period when elder flowers was sometimes considered to be a time of poor WEATHER. In the Basingstoke area of Hampshire this time was known as the elderbloom winter.[38] Elsewhere:

> Weather prophets say that if the weather breaks while the elder-flowers are coming out, it will be soaking wet (in Cheshire parlance, drabbly).[39]

Regardless of whether elder is considered to be malevolent or protective, most of the folk beliefs associated with it appear to be concerned with the tree's protection and preservation. Two quotations from herbalists writing in the 1940s, when wartime conditions encouraged the re-examination of Britain's herbal heritage, demonstrate the value of elder trees:

> [According to my gypsy friend] the healingest tree that on earth do grow be the elder, them sez, and take it all round I should say 'twas.[40]

> [Elder has] the unusual distinction of being useful in every part.[41]

Thus it is possible that the various folk beliefs associated with elder were due, at least in part, to efforts to protect a valuable resource. Elder is, perhaps, the wild plant most widely and diversely used in folk medicine.

> Queen of all Forest [of Dean] remedies was 'ellum blow tea' . . . The flowers were gathered in the spring and hung up to dry in

> closed paper-bags . . . in the kitchen . . . You dared not sneeze in the winter or down came the bag, a good handful was put in a jug, covered with boiling water, covered with a tea towel and left to infuse. One had to force this evil-smelling brew down one's throat willy-nilly. I loathed it, and to this day can recall the smell of cats, which emanated from it. Poultices of the mixture were used for SPRAINS, ACHES, etc., in joints, also for BOILS and 'gathered' fingers – WHITLOWS and so on. It seemed to be a universal panacea; the only use it didn't have was constipation . . . Elder berries were favoured too; they were boiled up with sugar, the resulting syrup strained, bottled and used in winter for COUGHS and COLDS . . . There is not a Forester alive over the age of 70 who does not know ellum blow tea.[42]

Similarly, a doctor who had a practice in the same area recalled that when they were taken ill older people would drink an infusion of elder flowers for a day or two before calling him. If this brought no cure they felt it was obviously a serious illness which required professional attention.[43]

Francis Bacon (1561–1626) recorded: 'they say' WARTS can be cured by rubbing them 'with a Green Elder Sticke and then burying the Sticke to rot in Mucke'.[44] Similarly:

> A 15-year-old girl, writing in 1954, says that her grandfather told her to pick a small twig of elderberry, touch her wart with it, chant the words,
>
> Wart, wart, on my knee,
> Please go, one, two, three,
>
> and put it 'down the toilet'.[45]

Other cures included:

> Elderberries when fried in mutton fat are used for boils and ULCERS.[46]

> Elder root when boiled and the water drank supposed to cure RHEUMATISM.[47]

> An infusion of elder flowers in boiling water will alleviate PILES.[48]

> A green ointment could be made from the leaves, based on mutton fat, and the creamy white flowers made Elderflower Water for the complexion. The flowers dried in the sun and stored in a paper bag

made a good remedy to break a hard cough and bring up phlegm. I always pick and dry some when they are in bloom, put the full of your fingers (one hand) in a mug, pour boiling water over and let it infuse for ten minutes. A little milk or fruit juice can be added.[49]

From my childhood in the twenties and thirties in Longbridge Deverill, Wilts . . . elderflower water to improve the complexion and whiten it.[50]

[My mother who was 94 when she died in 1987] used to collect elderflower in the spring, and dried it. In the winter if we had colds or flu, the elderflower was put in a jug covered with boiling water and put on the hob to stew. At night we were given this (strained) with sugar and a few drops of peppermint added. We were given a teacup full of this at night, and in the morning we had to drink half a cup of this cold mixture.

It was supposed to sweat out the FEVER. She used to tell me how she pulled me through PNEUMONIA by poulticing with hot flannel and sips of elderflower tea, day and night.[51]

When my three children were small and we had wintery weather (and it can be very cold up here at the foot of the Cairngorms), I made elder-flower wine, and when it was time for them coming from school I had three cups, bowl of sugar, bottle of elderflower wine and the kettle boiling, and I gave them a toddy; they never had colds or flu.[52]

Elder flowers and berries are widely collected by home wine-makers. In Victorian Surrey:

The villagers' tipple at CHRISTMAS time was home-made elder-berry wine, warmed up so that the aromatic flavour was brought out. Friends and neighbours who called to wish the compliments of the season would be sure to be offered a glass. A cheesemonger in Kingston named Pamphilon would always have hot elder wine for his customers when they came in with their Christmas orders . . . The widow of William Keen, a Farnham wheelwright, recalled how she and her husband would invite the workmen into their best room for hot elderberry wine.[53]

Apparently hot elderberry wine was being offered at Christmas time in Farnham shops well into the twenty-first century.[54]

Elderberries had various culinary uses:

> I remember a woman when I was a child [*c*.1960] in my village in Cornwall who collected elderberries to use them as currants.[55]

> [Somerset, 1940s] the juice from elderberries was used for flavouring stewed apple – very popular with people with dentures who didn't like blackberry pips 'getting under their plate'.[56]

Elder leaves have been used as a TOBACCO substitute:

> Myself, my brother and a friend always smoked elder leaves when money was not available for tailor-made cigarettes. We spent much time in the woodland of Thetford Chase, where on our regular walks we would break down, but not completely snap off, small sprigs of elder. We found that if we severed the supply of sap completely the leaves on the sprig would dry out resulting in a hot smoke. We found that if the leaves remained just slightly damp they were quite a pleasant smoke. It was obviously trial and error, sometimes they remained too wet to burn properly. We would stuff the leaves into stems of various umbellifers . . . We actually preferred these cigarettes to the tailor-made ones, but they were not available during winter.[57]

> [1950s] We smoked elder and became high.[58]

Elder twigs are characterised by their central pith which can easily be removed.

> [Colonsay] Boys aspiring to be pipers made chanters of the young branches [of elder], which are full of pith and easily bored.[59]

> Haw-blowers are made by scooping the pith out of an elder branch. Haws are blown through these.[60]

> At the beginning of the [20th] century children in parts of Devon used to make 'pop-guns' out of elder: they would force a hole through the pith, and then fashion a ram-rod out of HAZEL wood. Chewed paper would be rammed down the hollowed elder sticks, and pressed out with considerable force.[61]

> Elder . . . could be cut and cleared out and made into WHISTLES or used as pea-shooters.[62] There was another use for the Boor tree in olden times. A suitable length was cut and seasoned, then the

> white pith in the centre was scraped out, lead was then melted and poured in. When set, this made a good weapon for protection on a journey or out walking at night . . . My aunt who was born in 1894 remembered one man who had such a stick.[63]

> [South Cambridgeshire] for winter feeding one beekeeper used to make little troughs out of elderwood: he cut pieces about the thickness of a finger and five or six inches long, tapered off one end and removed the pith, and used them for replenishing the bees' honey by inserting this end in the exit hole.[64]

Less frequently recorded are uses of the pith:

> [Shropshire/Staffordshire border, 1960–71] my nan and mother made pincushions from elder pith; you would pack the pith tightly in a small bowl of water to keep it damp.[65]

> Gypsies used elder sticks to make wooden chrysanthemums – 'beautiful wooden flowers – which they sold from door-to-door.[66]

Finally:

> Fishing with elderberries beneath the float is a traditional, but now largely neglected, approach to tempt coarse fish (primarily dace and roach, but also chub). Its heyday was probably the 1950s–70s, but few anglers still try it today (including myself). It's usually associated with fishing rivers and canals and is something I've tried on and off in the past on rivers (Welland and Witham) with a bit of success, but nothing staggering. I strongly suspect they will be good bait for large carp and barbel too and shall try them this autumn!
>
> The berries look a lot like tares and large HEMP seeds (two very successful baits) when under water and this may be part of their attraction. They tend to be used by anglers that favour the 'natural bait' approach.[67]

ELECAMPANE (*Inula helenium*)

Robust perennial herb with shaggy yellow flowers, native to west and central Asia, grown in British gardens since at least AD 995, widespread and persistent on roadsides and waste ground.

Local names include:

Aligocampane in Scotland
Allecampane in Cheshire, Lancashire and Yorkshire
Hellycompane in Cornwall
Horseheal in East Anglia, where farriers and horsemen cultivated the plant and its 'bitter, aromatic leaves made an APPETITE stimulant for their charges'[1]
Scabwort, 'since it was a Norse remedy for sheep scab'[2]
SUNFLOWER in Stirlingshire
VELVET DOCK, or velvet duck, and wild sunflower, on the Isle of Wight.

> Elecampane was used by my grandmother, a late Victorian lady, for all kinds of ailments.[3]

This statement is supported by the frequent mention of elecampane, or corruptions of the word, in the doctor's speech in traditional mummers' plays. In a version of the play remembered by Thomas Hardy in 1920, when the Hardy Players were presenting a dramatised version of his novel *The Return of the Native*, the doctor exclaims:

> Yea, more; this little bottle of elecampane
> Will raise dead men to walk the earth again![4]

In Glamorgan elecampane was used to treat HYDROPHOBIA:

> There lived about sixty years ago an old woman . . . in Bridgend who cultivated elecampane in her garden. She was noted for curing hydrophobia in cattle, and farmers in the surrounding district came to her for the remedy. She made a decoction of it mixed with milk and a quantity of fowl's feathers. The other ingredients were kept as a profound secret, which she took to the grave.[5]

ELIZABETH I, Queen (1533–1603) – associated with an ASH tree in Richmond Park.

ELM (*Ulmus* spp.)

Large trees formerly characteristic of much of the countryside until devastated by Dutch elm disease from the 1960s onwards.

Clive Stace in his *New Flora of the British Isles* notes that *Ulmus* is 'an extremely difficult genus, having been interpreted in widely different ways'.[1] Thus perhaps it is surprising that collectors of plant-names, many of whom lacked an in-depth knowledge of botany, felt able to attribute names to a particular species.

Names reputedly given to English elm (*U. procera*) include:

Allom-tree in Scotland
Alme in Northamptonshire
Elem in south-west England
Ellem in Sussex
Elven in Kent and Sussex
Horse-may in Cornwall, 'coarse leaves are called horse-may, to distinguish from the small-leaved kind'
May in Devon
Owm in Yorkshire
Sugar leaves, 'given to the young leaves' in Somerset
Warwickshire-weed 'a term frequently given in books . . . but it would not be understood by the ordinary Warwickshire folk'.

Names reputedly given to wych elm (*U. glabra*) include:

Bough elm in Yorkshire
Chew-bark in Berwickshire, where 'the inner bark . . . for a certain pleasant clamminess is chewed by children'
Drunken elm in Lincolnshire
Elm-wych in Northumberland
HALSE, and hornbeam, in Somerset
Horn-birch in Surrey
QUICKEN in Warwickshire
Scotch elm in Berwickshire
Switch elm in Yorkshire
Witan elm in Shropshire
WITCH-HALSE in Somerset
WITCH-WOOD in Cumberland and Yorkshire
Wych arl in Devon
Wych halse, and wych tree, in Somerset
WYCH-WOOD in Cumberland and Yorkshire.

In Warwickshire the development of elm leaves provided guidance for the sowing of crops:

> When the elmen leaf is as big a mouse's ear,
> Then to sow BARLEY never fear;
> When the elmen leaf is as big as an ox's eye,
> Then says I, 'Hie, boys, Hie!'
> When the elm leaves are as big as a shilling,
> Plant KIDNEY BEANS, if to plant 'em you're willing;
> When the elm leaves are as big as a penny,
> You must plant kidney beans if you want to get any.[2]

On Guernsey:

> *Quànd tu veit la fieille a l'orme,*
> *Prends ta pouque et sesme ton orge.*
> When you see the elm leaf,
> Take your seed-bag and sow your barley.[3]

At one time pupils at Lichfield Cathedral School got up early on Ascension Day, and:

> put a sprig of elm on every door in the Cathedral Close. After the Ascension Day communion everyone in the procession was given sprigs of elm, and at different stations around the Close, there were readings, carrying the elm in one hand and the reading in the other. After the final reading the Dean blessed everybody and sprinkled them with holy water, and then they went back into the Cathedral where they threw the elm into the font. Nowadays LIME is used instead of elm because there isn't enough elm within or near the Close.[4]

In the 1840s:

> Common people in parts of Oxfordshire speak of the wych elm as a charm against WITCHES, and that a person under the spell of witchcraft will be cured if struck nine times with a branch of this tree.[5]

> At about the same time, an elm beside the road leading between Silsoe and Maulden in Bedfordshire was known as Beaumont's Tree, and was said to have grown from a stake thrust through the body of Beaumont, a murderer buried there. People suffering from AGUE 'would nail strands of their hair or toe nail clippings to the tree, to effect a cure'.[6]

Other remedies which used elm include:

> Many villagers [in the Upper Thames area], in cases of COLD or sore THROAT, strip off the inner bark of the young wands [of wych elm] and chew it raw, or boil it and drink the liquor. This, when cold, settles into a brown jelly that is not unpleasant to the taste. I have often taken it was a boy, preparing it according to the directions given me by my old grandmother.[7]

> Bark of elm boiled and put on a BURN cures it.[8]

Elm wood was much used for coffins: 'universally employed to encase poor humanity on its last journey to its final home'.[9]

EMETIC – BOG MYRTLE, DAFFODIL and GROUNDEL used as.

EMMENAGOGUE – see MENSTRUATION; often it is difficult to know whether a herb was being used to regulate menstruation or produce ABORTION.

EMPIRE DAY (24 May) – DAISY associated with.

ENGLISH STONECROP (*Sedum anglicum*)

Small perennial herb, with succulent, often red-tinged, leaves and pinkish white flowers, common on acidic rocks and other acidic substrates, in western, especially maritime, areas.

[Gaelic Scotland] white or pink stonecrop (*Sedum anglicum*) was considered to be a delicacy and was given the name *Biadh an t-Sionnaidh*, the prince's or lord's food.[1]

EPILEPSY – NAVELWORT used to treat.

ERYSIPELAS – treated using BLADDER CAMPION.

EVERLASTING PEA (*Lathyrus latifolius*)

Flamboyant, pink-flowered perennial (Fig. 16), native to southern Europe, cultivated as an ornamental since the fifteenth century, now widely naturalised on waste ground, railway embankments and hedgerows throughout southern England.

> [Northamptonshire] About 20 years ago I found some narrow-leaved everlasting pea, *Lathyrus sylvestris*, on the local railway embankment. I planted the seed along the roadside on my farm and now I have a mile of them. People come from miles just to see them when in flower during the first week in August.
>
> Some people call them Pharaoh's Peas. The story is that a person from the nearby village of Weebly went to Egypt and brought home some seeds which were said to come from a royal tomb in a pyramid.[1]

As everlasting pea has never been recorded for Egypt or any neighbouring countries, it would appear that this legend has no factual basis.

EVIL – BAY, BIRD-CHERRY, COCKSPUR THORN, ROWAN and YEW protect against.

EVIL EYE – 'the notion that certain people can harm other humans, animals, and even inanimate objects, simply by looking at them'[1] – associated with BEECH; GOAT WILLOW, *Caesalpinia* seeds (see SEA BEANS), ROWAN and ST JOHN'S WORT protect against.

EYEBRIGHT (*Euphrasia* spp.)

Small annual herbs, widespread on established grassland.

Eyebrights seem to have gathered very few local names:

BIRD'S-EYE in Somerset
FAIRY FLAX in Co. Donegal
Joy in Somerset
Joy-flower in Yorkshire
Peeweets in Devon.

Eyebrights are semiparasitic and obtain some of their nourishment by attaching their roots to neighbouring species. This presumably explains the 'very good cure for a disease (red water) in cows' recorded by H.C. Hart in County Donegal in 1898. This used both eyebright and slender St John's wort (*Hypericum pulchrum*), both of which were given the name rock-rue. On asking how two different plants could be given the same name and have the same herbal use, Hart's informant pointed out that both grew from the same root.[1]

> If the herb was as much used as it is neglected, it would half spoil the spectacle makers trade . . . Arnoldus de Villa Nova saith, it hath restored sight to them that have been blind a long time before.[2]

> It has been reputed good for sore EYES, but the gentlemen of the faculty have declared it does more harm than good in applications of that kind, there having been instances of persons rendered almost blind by the use of it. The highlanders do however still retain the practice of it, by making an infusion of it in milk and anointing the patient's eye with a feather dipped in it.[3]

> Infusion of leaves taken internally cures COUGHS, applied in a lotion strengthens eyes and heals sore ones . . . I have met gypsies who smoked it mixed with COLTSFOOT – and it is an ingredient of most herbal TOBACCOS – maintaining that it cured ASTHMA and CATARRH. It is widely used by gypsies for eye troubles.[4]

EYES – problems treated using BIRD'S-FOOT TREFOIL, CUCUMBER, EYEBRIGHT, GERMANDER SPEEDWELL, GREATER CELANDINE, GROUNDSEL, GROUND IVY, HEMLOCK, HOUSELEEK, MALLOW, RASPBERRY, SCARLET PIMPERNEL, water in a SYCAMORE at Clonenagh, water from TEASEL leaves, and WILD CLARY; gypsies beat newborn infants' eyes with NETTLE; ORANGE peel makes them sparkle.

F

FAIRIES – associated with DANDELION, ELDER, FOXGLOVE, GUERNSEY LILY, HAWTHORN, RAGWORT and RED CAMPION; ate BLACK POPLAR catkins; BUTTERWORT, DAISY, MARSH MARIGOLD and ST JOHN'S WORT protect against; four-leaved CLOVER enabled people to see; 'get' PARSLEY not sown on a holy day.

FAIRY FLAX (*Linum catharticum*), also known as purging flax

Inconspicuous white-flowered biennial or annual herb, widespread in dry grassland.

The few local names recorded for fairy flax include the widespread mountain-flax, fairy-lint in Berwickshire, and laverock's lint in Lanarkshire.

As its scientific and alternative common names imply, fairy flax was formerly valued as a PURGATIVE.

> [Llandudno] the old miner . . . when I questioned him . . . said, yes, indeed, their herbs 'was good for everything', and snatching up a plant or two of *Linum catharticum*, that grew near the path, 'that, now, good as Epsom salt'.[1]

This use continued on the Hebridean island of Colonsay until early in the twentieth century.[2]

In County Donegal fairy flax had the name CHICKENWEED and was considered to be 'good for SPRAINS and BRUISES',[3] while in Winchester, Hampshire, it was sold under the name of mill-mountain for an unrecorded medicinal use.[4]

FAIRY FOXGLOVE (*Erinus alpinus*)

Attractive, pink-flowered perennial herb, native to southern Europe, cultivated as an ornamental since the mid-eighteenth century and spreading to old walls and stony places.

Richard Mabey in *Flora Britannica* gives Roman wall plant, because of fairy foxglove's 'long presence on Hadrian's Wall'.[1] Similarly:

> Fairy foxglove is a small purple flower which grows intermittently on stone walls in north-east England. Local tradition says that it only grows where ROMAN soldiers have trod, and certainly it is to be found in the village of Wall (which is, of course, located near Hadrian's Wall in Northumberland).[2]

FAIRY TREE – see LONE BUSH

FALSE OAT-GRASS (*Arrhenatherum elatius*)

Coarse, unpleasant-tasting, perennial grass, common and abundant on roadsides, hedge-banks and riverbanks.

Local names, most of which appear to have been given to var. *bulbosum*, 'onion couch', a form with swollen corm-like roots:

ARNUT in Scotland
Button-grass in Cumberland
Lobbin-grass in Co. Derry
Onion-grass in Buckinghamshire
Pearl in Co. Antrim and Co. Donegal; 'from the swollen little knobs along the rootstock'
Sweet arnut in Scotland.

> As children we played a game of 'Cock or Hen' with false oat-grass. One would run their clasped fingers up the stem of the grass thus grabbing the spikelets in the hand. If one of the spikelets protruded above the others this constituted a cock and if no 'tail' was produced then it was announced as the hen.[1]

> [Thetford area, Norfolk, mid-1950s–early 1960s] when playing the game of 'cocks and hens' the spikelets could be grasped and pulled without the need to stop walking. If we were actually playing the game for points the 'puller' would guess before pulling a spikelet as to whether it would be a 'cock' or 'hen'. If s/he was correct it was one point to him/her and it was then the turn of someone else. Alternatively a person would pull a spikelet, hide it behind his/her back, and the other players would guess the sex of the hidden spikelet. Each player who guessed correctly gained a point, and then another player had a pull.[2]

> Growing up in Suffolk in the 1980s we used to take a flowering stem of grass and pull our fingers up the stem so all the seeds bunched together at the top. This would result in either a 'lump' of seeds, or a lump with a point in the middle. The game was to guess whether you would get a 'candle' (the point) or a 'cake' (the lump with no point). It was a silly game, there was no point to it really, you would play it when walking somewhere. I suppose the species was probably false oat grass, or a meadow grass, I seem to remember different species would give different likelihoods of candle or cake but we obviously didn't know the species' names.[3]

Similar games were played with YORKSHIRE FOG.

FAT HEN (*Chenopodium album*)

Widespread annual weed, common in cultivated land, formerly – since, it is said, the Stone Age – valued as a food plant. The Grauballe Man, found in a bog in Denmark and estimated to have died in about AD 300, had fat hen seeds, which formed part of his final meal, in his stomach.[1]

Local names include the widespread lamb's quarters and muckweed, and:

ALL-GOOD in Hampshire
BACON-WEED 'because it denotes rich, fat land', also beacon-weed and biacon-weed in Dorset
Confetti in Somerset
Dashpegger in Dorset
Dirt-weed in East Anglia; 'from its growth on manure heaps'
Dirty Dick in Cheshire and Wiltshire
Dirty Jack, and Dirty John in Cheshire
DOCK-FLOWER in Somerset
Dung-weed in Gloucestershire
Jack o' the Nile, John o' the Nile, John O'Neele, and Johnny O'Neele, in Shropshire
LAMB'S TONGUE in Devon and Dorset
Mails in Ayrshire
Meals in Cumberland
Meldweed in Scotland; 'cf. German *Melde*, usually for *Atriplex* spp., but also for spp. of *Chenopodium*'

Melgs in Morayshire
Midden-myles, or MIDDEN-MYLIES, in northern Scotland; 'growing on dunghills'
Milds in Cumberland, Berwickshire, Lothian and Roxburghshire and Northern Ireland
Mixen-weed in Yorkshire
Muck-hill weed in Warwickshire
MUTTON-CHOPS in Dorset
Myles in Berwickshire
PIGWEED in Hampshire and Somerset
WILD SPINACH in the Midlands.

> [A] widely used edible plant was fat hen . . . which was known in Irish as *praiseach fiáin* or 'wild spinach'. Its leaves were extensively eaten in Ireland in pre-Norman times up until spinach proper arrived to replace it.[2]

> [Colonsay] The leaves were boiled, pounded, buttered and eaten like spinach.[3]

> There used to be what I thought was a weed growing in quantities in this area, but I have heard it said that in days gone by, when green vegetables were scarce, this weed – fat hen, as it was called – was used as a vegetable.[4]
>
> In India my mother uses fat hen – *bethua saag* – in curries.[5]

Fat hen is described as 'mealy', when it is boiled a white scum rises to the surface of the water possibly discouraging novice foragers.

> Assuming *Chenopodium album* was the plant known there as 'lamb's quarters' – one of the alternative vernacular names of this species – a decoction of its stems was till relatively recently drunk in Co. Dublin for RHEUMATISM.[6]

FATIGUE – MUGWORT prevents.

FEAST SUNDAY

A day on which the eating of the first crops of a season was celebrated.

> [Histon, Cambridgeshire] the first Sunday in July is Feast Sunday; the family comes to dinner and the first bait of PEAS, CARROTS and POTATOES are planted on GOOD FRIDAY, regardless of the weather, and the first of these vegetables is always eaten on Feast Sunday, even if they are ready before. People, including my 26-year-old son, still keep these traditions up. Feast Sunday

> dates back to at least 1894, and used to be about the only time when girls who were in service were allowed home to visit their families.[1]

Elsewhere:

> [My family, who originated from the village of Whitwick, near Coalville, in Leicestershire] set early potatoes on Good Friday, they were then ready for boiling for Whitwick Wake . . . this coincided with the Church anniversary – St John the Baptist Church.[2]

FEET – DOCK leaves prevent sore; GOOD KING HENRY 'cures' sweaty; TANSY cures blistered.

FENNEL (*Foeniculum vulgare*)

Culinary herb, cultivated in Britain since Roman times, now widely naturalised, particularly near the sea.

> [Jersey] sprigs of fennel are placed in horses' harnesses to keep FLIES away. It is also used in a sauce eaten with the locally caught and popular MACKEREL.[1]

FERTILITY, human – stimulated by ASHen faggot, BLACK POPLAR, GYPSOPHILA, HAZEL, ORANGE, PERIWINKLE and a YEW at Stoke Gabriel, Devon.

FEVER – caught by picking up fallen FLOWERS; not caught by those who planted GARLIC on Good Friday; CABBAGE, CREEPING CINQUEFOIL, ELDER and TORMENTIL used to treat.

FEVERFEW (*Tanacetum parthenium*)

Perennial, white-flowered herb, native to the Balkan Peninsula, cultivated for medicinal use since late in the tenth century; now widely naturalised, mainly near habitations, throughout lowland areas, and cultivated as an ornamental. One of many plants known as BACHELOR'S BUTTONS

Other names include:

ARSE-SMART in Yorkshire
BOTHEM in Cornwall
Buncholery buttons, in Stirlingshire; 'a corruption of bachelor's buttons'
Buttons in Somerset

Devil-daisy in Somerset and Wiltshire
Featherfew in Ulster
FIELD DAISY in Somerset; 'possibly confusion with mayweed, *Matricaria* or *Tripleurospermum* spp.'
Flirt-weed in Devon
Madron in Devon
MIDSUMMER DAISY in Devon and Somerset
Nosebleed in Kent
Old-maid's scent, and stink-daisy in Somerset
Weather-vaw in Devon
Whitewort on the Isle of Wight
Yard-daisy in Somerset.

> [Norfolk] feverfew boiled and strained used to allay pain.[1]

> My mother was born in 1901 in Great Wakering, in southeast Essex . . . if we had EARACHE a red tile was heated in the oven, feverfew was placed on it and covered with a piece of flannel, and it was held to the ear.
> It's interesting that she pronounced the first e in feverfew soft, as in 'tea'.
> Although this healing use has been remembered, it seems that any connection with fever was not recognised.[2]

> Used as an emmenagogue about Strangagalwilly, in Tyrone, with the same object in view as that for which RUE is used elsewhere.[3]

In recent decades feverfew has enjoyed a high reputation as a cure for MIGRAINES and severe HEADACHES,[4] but this use does not seem to be a long-standing tradition in the British Isles.

FIBRE plants include FLAX, HEMP and NETTLE.

FIELD BINDWEED (*Convolvulus arvensis*)

Trailing or climbing perennial herb with attractive pink-and-white flowers, considered to be a troublesome weed in gardens and arable land throughout England and Wales, less common in Scotland and Ireland.

Because of its reputation as a weed, field bindweed has attracted a number of names which associate it with the DEVIL, including DEVIL'S GARTER in Somerset; the widespread DEVIL'S GUTS, 'cf. German *Teufels Nahgarn*, devil's thread' and, as deil's guts, in Scotland, devil weed, 'from its being difficult to eradicate – a piece of root less than one inch will sprout', in Sussex, and HELLWEED in Northamptonshire.

Other names include the widely used CONVOLVULUS, cornbind, withwind, withy-wind – 'equivalent to string-twist, thread-twist, cf. German *Wedewinde*' – and WITHYWINE, and:

Barbine in Shropshire
Barweed in Somerset
BEDWIND in Gloucestershire, Hampshire and Warwickshire
BELL-BIND in Cambridgeshire, Essex, Norfolk and Somerset
BELL-BINE in Cambridgeshire, Essex and Norfolk
BELLWIND in Buckinghamshire; 'cf. German *Windglocke*'
BELLWINE in Buckinghamshire
Billy-clippe in Kent
Billy-clipper in Shropshire; 'clip = to embrace'
BIND, 'pronounced with short i', in Lincolnshire
Bine, or bines, in Suffolk
BINE-LILY in Dorset
Combine in Somerset
Cornbine in Buckinghamshire
CORN-LILY in Yorkshire
CREEPING JENNY in Somerset
Drayler in Cornwall; 'Cornish draylyer, trailer'
Earwig in Somerset
Fairies' umbrella in Somerset
Fairies' wine-cups in Somerset
Fairy-umbrella in Dorset
Grandmother's petticoats in Devon
GRANNY'S NIGHTCAP in Wiltshire
Gypsy's hat in Somerset
HEDGE-BELLS on the Isle of Wight; as field bindweed is not a hedgerow plant it is probable that this is a mistake, and the name refers to hedge, or great, BINDWEED
Jack-run-i'-the-country in Yorkshire
KETTLE-SMOCKS in Wiltshire
LADY'S SMOCK in Dorset
LADY'S SUNSHADE in Somerset
LADY'S UMBRELLAS in Dorset
Lap-love in the Midlands
LILY in Hampshire and Sussex
OLD-MAN'S NIGHTCAP
Ragged shirt in Dorset
ROBIN-RUN-IN-THE-HEDGE
Sheep-bine in Essex
Shirts-and-shimmies
SUNSHADES, 'a common name'
TARE in Wiltshire
Treliw in Pembrokeshire
Way-weed in Oxfordshire
Weather-wind in Berkshire
Wheat-bine in Wiltshire
WHITE SMOCK in Devon
Wild convolvulus in Berwickshire
Willow-wind in Wiltshire
Withy in Dorset.

> In the 1950s in Invergowrie, Perthshire, bindweed was known as 'young man's death'. If you pick bindweed your boyfriend will die – seems to be associated with the rapid fading of the flower.[1]

> Shropshire . . . THUNDER-FLOWERS we always called bindweed (convolvulus) because if we picked these it would be sure to THUNDER before the day was out.[2]

It is probable that these beliefs were brought about to discourage children from gathering bindweed flowers and thereby causing damage to crops; cf. POPPY.

FIELD ERYNGO (*Eryngium campestre*)

Perennial herb, scarce in scattered, well-drained locations in southern England.

Westwood and Simpson tell how Borough Hill, near Daventry, Northamptonshire, was supposedly associated with DANES, this being proved by the prevalence thereabouts of a plant known as daneweed, which they identify as field eryngo. After visiting the Hill, Daniel Defoe noted in his *Tour Thro' the Whole Island of Great Britain* (1724–7):

> They say this was a Danish camp . . . The road hereabouts, too, being overgrown with Dane-weed, they fancy it sprang from the BLOOD of Danes slain in battle; and that if, upon a certain day of the year, you cut it, it bleeds.[1]

Anne Baker in her *Glossary of Northamptonshire Words and Phrases* listed Watling Street thistle as a name for field eryngo, and claimed, incorrectly, that 'the old ROMAN road is the only known place for this rare plant'.[2]

FIELD GENTIAN (*Gentianella campestris*)

Annual or biennial herb, in grassland and on dunes, scattered throughout the British Isles and locally common in the north.

In Orkney field gentian was valued as a TONIC, a remedy for gravel and a cure for JAUNDICE, while in the Pennines it 'served as a digestive as well as being drunk to "kill germs".'[1]

> Field gentian = sôta (meaning sweet as in nature/character) – helps digestive disorders. It is also known as rid' girse and was fed to cows that were reluctant to come into season – perhaps an aphrodisiac.[2]

In the Highlands of Scotland field gentian was used to treat a 'rickets-like disease in cattle enforcing crouching, known as the *chrùbain*, nowadays attributed to phosphorus deficiency'.[3]

FIELD MAPLE (*Acer campestre*)

Compact deciduous tree, common in England and Wales, also widely planted.

Local names include:

Box-of-matches, said to be given to the leaves in Somerset
Cat-oak in Yorkshire
CHATS given to the flowers in Yorkshire
Maplin-tree in Gloucestershire
Maser-tree, in 'early Lowland Scotch'
Spinning Jenny, 'presumably from the way in which its winged seeds spin in their flight through the air'
WHISTLE-WOOD in Clackmannanshire
Whitty-bush in Shropshire
Zigzag in Bristol.

Names given to the fruits, 'keys' include:

Boats, BOOTS-AND-SHOES, and CATS-AND-KEYS in Somerset
Hasketts in Dorset
Hooks-and-hatchets in Somerset
Ketty-keys and kit-keys in Yorkshire
LADY'S LOCKETS in Somerset
MONEY-IN-BOTH-POCKETS in Somerset
Shacklers in Devon.

In Devon late in the nineteenth century field maple was frequently known as OAK, and worn as such on Oak Apple Day.

> I have been astonished to find how constantly the Maple is called Oak. On Whit-Monday, which this year was Oak-apple Day as well (May 29th), I took an early walk into Bradley Wood. Here I met a number of children decorated with Maple, and asked them what it was for. 'It's Oak-apple Day, sir, and if you ain't got a piece of *oak-apple* they'll pinch you or sting you.' 'Will they?' I replied, 'then I must get a piece.' 'Here's a piece, sir,' said a bright lad. It was a sprig of Maple, as was all the rest they had. I said. 'This is not Oak, is it?' to which they all replied, 'It's *oak-apple*, sir.' I could give illustrations from grown people showing the same error.[2]

Similarly, in Derbyshire in the 1940s:

> Oak – this was *Acer campestre*, which had to be worn in the button-hole on 29th May (King Charles in the oak tree) on pain of being 'nettled' (*Urtica*) for non-compliance.[2]

In Nottinghamshire field maple was distinguished from oak, but could be worn as a substitute for it on 29 May:

> Some who are unable to procure it [oak], endeavour to avoid the penalty [of being stung with NETTLES] by wearing DOG OAK (maple), but the punishment is always more severe on the discovery of the imposition.[3]

FIELD SCABIOUS (*Knautia arvensis*)

Perennial with bluish-lilac flowers, on well-drained calcareous and neutral grassland, but absent from much of north-west Scotland and Northern Ireland.

Local names include the widespread BACHELOR'S BUTTONS and gypsy rose, and:

Beaver in Cornwall
BILLY BUTTONS in Somerset and Yorkshire
Blackamoor's beauty in Somerset
BLACK SOAP in Devon
BLUE BONNET in Scotland
BLUE BUTTONS in Dorset and Wiltshire
BLUE CAPS in Somerset
Blue men in Buckinghamshire
Cardies in Co. Antrim
CLOGWEED in Buckinghamshire
Coachman's buttons in Somerset
Egyptian (= gypsy) rose on the Isle of Wight
Gentleman's pincushion in Somerset

Grandmother's pincushions in Sussex
Lady cushion in Kent
Lady's hat-pins in Devon
PINS-AND-NEEDLES, purple buttons, robin's pincushion, SOLDIER'S BUTTONS, and teddy buttons, in Somerset.

Derbyshire gypsies used an infusion of field scabious leaves to strengthen the lungs and cure pleurisy.[1]

FIELD WOOD-RUSH (*Luzula campestris*)

Perennial herb, widespread in grassland, producing its sooty-black flowers around about Easter time.

Local names include:
Black caps in Berwickshire
Black-head grass in Cheshire
CHIMNEY-SWEEP in Cheshire, Lancashire, Shropshire and the West Country
CHIMNEY-SWEEPERS in Wiltshire
CROW-FEET in Yorkshire
Cuckoo-grass in Berwickshire
Davie-drap in Galloway
Easter grass in East Sussex
God's grace in Cheshire
Good Friday grass in Surrey and the West Country
GYPSY in west Wiltshire
Gypsy plant in Warwickshire
Peeseweep grass in Berwickshire; 'peeseweep is a north country name for the lapwing'
Smuts in Buckinghamshire; 'in allusion to the black appearance of the flower heads; cf. chimney-sweepers'
Sweeps in Shropshire and the West Country
Sweep's brushes in Shropshire and the West Country
Sweet bent in Ayrshire
Treacle-dabs, 'heard this name applied in Somerset . . . but only by North Country people'.

> In Cwm Einion [Ceredigion, 1950s] . . . it was known as Good Friday Grass and its flowering was the sign to put the cattle out to graze.[1]

A Shropshire name was sweep's brushes, while in Cheshire and Lancashire in the nineteenth century the names CHIMNEY-SWEEPS and CHIMNEY-SWEEPERS were used:

> When Cheshire children first see this plant in spring they repeat the following rhyme, possibly to bring them good luck:
> Chimney-sweeper, all in black,
> Go to the brook and wash your back;

Wash it clean, or wash it none;
Chimney-sweeper, have you done.[2]

FIG (*Ficus carica*)

Spreading shrubs or small trees valued for their edible fruits, native to south-west Asia, long cultivated in the British Isles and occasionally naturalised, especially beside rivers. It is sometimes claimed that Romans grew figs in Britain, but seeds found in Roman latrines were probably derived from imported dried, rather than locally grown, fruit. Fig trees at West Tarring, West Sussex, said to have been planted by St Thomas Becket (1118–70),[1] are claimed to be the oldest surviving trees in the British Isles. Alternatively the trees are descendants of those tended by St Richard of Chichester (1197–1253).[2] However the earliest written evidence for the introduction of figs dates from 1525, when Richard Pole returned from studies in Italy bringing with him trees which were planted in the garden of Lambeth Palace, London.[3]

A fig tree which grows from the south wall of St Newlyn East church, in Cornwall, is said to have grown from a STAFF carried by St Newlina, an obscure virgin martyr.

> The tradition in the village here, recorded by one of my predecessors in the 1930s, is that St Newlina, a Christian princess, planted her staff in the ground and said that this should be the site of a church. The wall of the church from which the fig grows is, however, 14th century and I have heard that fig trees were not introduced into this country until the 16th century . . . From time to time the tree has to be pruned, but by a remarkable number of coincidents some of those who have done so have met with misfortune or death.[4]

A postcard on sale in the late 1970s depicts the tree, and gives the verse:

In ancient days Newlina came,
The saint who gave this place its name.
Her staff she planted and she prayed,
'Let here a Church to God be made.'
This fig tree is her staff folks say;
Destroy it not in any way,

Upon it lies a dreadful curse,
Who plucks a leaf will need a hearse.

According to a report in the *Sunday Express* of 1 June 1958:

> Four Cornishmen have defied a 'curse of death' and lived. Warning of the 'curse' is printed beside a fig tree which grows out of the wall of the ancient parish church of St Newlyn East, near Newquay.
>
> It says death will follow within a year if any man so much as plucks one leaf from the tree.
>
> Twelve months ago four men of the village pruned the tree. One of them . . . said yesterday: 'When I was asked if I would prune the tree I said "Certainly. I'm not superstitious." But soon afterwards, when I went to fell some trees one fell on me putting me off work for three months.'
>
> Does he believe in the 'curse' now? 'Not a bit. I think it was invented to make a good yarn.'

In his *British and Foreign Trees in Cornwall* (1930) Edgar Thurston provides a photograph of the tree, but makes no mention of the legend. Since Thurston had an interest in folklore which led to the inclusion of material irrelevant to trees and shrubs in his work, perhaps this omission is significant: did he fail to collect the legend, or was it invented, rather than collected, by the parish's incumbent a few years later?

Also in Cornwall, a fig tree grows from the south-west wall of Manaccan parish church, but this tree has no legend or curse attached to it, and in August 1998 a former vicar of the parish recorded: 'I have personally cut large chunks from it and no harm has come to me, other than [that which is] natural to man!'[5] It is suggested that both fig trees were planted by the Cornish historian Richard Polwhele, vicar of Manaccan from 1794 to 1821 and of St Newlyn East from 1821 to 1838.[6]

In 1913 a description was given of three 'ATHEISTS' TOMBS' in Hertfordshire. One of these was an altar tomb on the south side of St Mary's church in Watford, which bore no inscription, but had a well-developed fig tree growing from it. According to legend a lady, or a well-known farmer, lay buried therein, and she or he asked for a fig to be placed in her or his hand; if it was true that there was another world beyond the grave, the fig would grow into a tree.[7]

> Do you know the Watford fig tree? It grows in the churchyard

> there. The story is that years ago there was an unbeliever who the local vicar kept trying to convert. When the unbeliever died he said a fig tree would grow from his grave if there was a god. I don't know if the tree is still there; I haven't been to Watford for a long time.[8]

Early in the twentieth century more than forty versions of postcards were produced of the tree, which died in the 1960s.[9]

In some parts of England figs were associated with Eastertide.

> Fig Sue is a GOOD FRIDAY drink – I don't know why only Good Friday. It is: Stew 4 oz figs in 1 pint water until tender. Rub through sieve, add one tablespoon sugar and a pinch of ground ginger. Warm 2 pints of ale, and add puree, bring to boil and serve.[10]

> My family . . . comes from a village in north Bucks, called North Marston, where I spent much of my childhood, girlhood and later. In this area it was customary to eat figs on PALM SUNDAY, always called 'Fig Sunday'. See Mark, chap. 11, verse 13.[11]

In south-west England dried figs were known as dough-figs, and a Somerset cure for EARACHE recorded in 1869 advised 'bake a bit o' dough-fig an' put un in'.[12]

The leaves, twigs and unripe fruit of fig trees produce a white latex when broken, and this latex, like that of DANDELION, has been used to treat WARTS.[13]

FIRE – BITING STONECROP, HOUSELEEK and ST JOHN'S WORT protect against.

FIRE LILY (*Cyrtanthus elatus*; syn. *Vallota speciosa*)

Bulbous herb with scarlet flowers, native to South Africa, cultivated mainly as a houseplant, but able to survive outside in warm sheltered areas. Also known as Scarborough lily:

> An incident about 200 years ago gave rise to its common name. A ship carrying some bulbs from their native South Africa to Holland was wrecked off the Yorkshire coast. Bulbs drifted ashore and were collected by the inhabitants of Scarborough.[1]

FISHERMEN – use ELDER berries as bait.

FLAX (*Linum usitatissimum*), also known as lint

Cultivated in Britain and Ireland for FIBRE since at least 1240 and, more recently, mainly for the production of linseed oil.

In common with HEMP, another fibre-plant, flax was used in LOVE DIVINATION.

> [Northeast Scotland] When the shades of evening were falling [on HALLOWE'EN] the maiden had to steal out quietly with a handful of lint-seed, and walk across the ridges of a field, sowing the seed and repeating the words:
>
> Lint-seed I saw ye,
Lint-seed I saw ye,
Lat him it's to be my lad
Come aifter and pu' me.
>
> On looking over the left shoulder she saw the apparition of him who was to be her mate crossing the ridges, as it were, in the act of pulling flax.[1]

FLEAS – BOG MYRTLE, EELGRASS and WORMWOOD deter; infest MEADOW FOXTAIL and WALL BARLEY; VIOLETS bring into house.

FLIES – BOG MYRTLE, ELDER, FENNEL, LAVENDER, MINT, NETTLE, TANSY and WALNUT deter.

FLOGGING – schoolboys used DOG ROSE gall to prevent.

FLOWER COMMUNION

Unitarians who reject the concept of the Trinity first met in London in 1774, and their church now embraces a wide range of liberal Christian beliefs.

> Eastern European Unitarians have, for the last seventy years or so, celebrated a 'flower communion', now widely used in western Europe and North America as well. In this ceremony, all attending bring cut flowers of any sort, which are gathered together in large containers at the front of the chapel at the beginning of the service. At the close of the service people file forward and take a flower, different from the one they brought, and take it home to place on their table or mantel . . . as a symbol of the larger community of the congregation of which they are a part.

> Our own congregation has a variety of this we also do once a year. Our flower arrangers make displays which are solely leaf backgrounds. Near the start of the service everyone files forward, receives a flower, and places it wherever they wish in the displays. Thus the congregation creates its own focal point in that service, while symbolically celebrating, in the vast variety of flowers purposely used, the beauty created by welcoming differences, and viewing them as a strength and not a threat.[1]

FLOWER FESTIVALS

Many churches and chapels use flower festivals, in which their buildings are extravagantly decorated with a variety of spectacular flower arrangements, as a popular means of fund-raising.

It is said that flower festivals began at Walpole St Peter, Norfolk, in 1962, but really took off in 1963 when the vicar of Flores, Nottinghamshire, and his wife organised a festival which attracted 5,000 people who came from over a wide area and enjoyed the flowers and cream teas. This festival continues with displays in the parish church and United Reformed Church chapel and ancillary events such as an art display in the local school, but due to increased competition from flower festivals elsewhere it no longer attracts such huge numbers.[1]

Similarly:

> In about 1963–5 my parish, together with others in Lincolnshire, received an impressive notice of a flower festival at Walpole St Peter, just outside Lincolnshire – a superb fenland church . . . The smart printing and detailed arrangements suggested that this event had been happening annually for some years.
>
> Two years ago we were passing near Walpole St Peter at flower festival time, called in and discovered the magnificent church, well-developed attractions and lovely displays.[2]

It appears that such events emerged from the special decorations which are prepared for patronal festivals and other special dates in the churches' year. Thus at Barnstaple, Devon, the church flower-ladies decorated the town-centre church for St Peter's Day (29 June) and 'as this church is always open and in a tourist area, many people would come into the church to see it "and the nice flower displays", and tended to place more than usual in the collection box!':

> Gradually, local people from floral art clubs joined in and from it developed the annual Flower Festival which then became detached from the Patronal Festival and moved to the time of the annual and famous Barnstaple Fair whose formal ceremonial opening took place in the town hall that was very near the church. With the town crowded for the three days of the fair, the 'fair markets' on the Friday, and the carnival on the Saturday, and the church being near the centre of it all, it brought a considerable number of people into the church and became an important fund-raising event . . . As more churches and historical buildings in the region also developed their own flower festivals, there was a heavy demand on the people from the floral-art clubs and the one in the church ceased to be an annual event and usually took place every three or four years.[3]

Although the arrangements produced at flower festivals entail painstaking planning and work, they can be disappointing; the same florists' flowers are used everywhere, and the flowers which visitors see in a Devon church in May are likely to be the same as those seen in Derbyshire in September. There is little sense of season or place. A noteworthy exception is St Peter's church in Westleton, Suffolk, which uses wildflowers, and on 29 July to 1 August 2016 celebrated its fiftieth Wildflower Festival.[4]

FLOWERING CURRANT (*Ribes sanguineum*)

Shrub with pinkish-red flowers, native to western North America, grown as an ornamental since the 1820s and now widely naturalised, often indicating the sites of former cottage gardens.

> My wife would not have the red flowering currant in the house; many years ago her mother said it would bring bad LUCK . . . (no other explanation).[1]

> [Invergordon, Ross-shire, 1950s] we were never allowed to pick or take flowering currant into the house as it was bad luck.[2]

FLOWERING RUSH (*Butomus umbellatus*)

Perennial herb with attractive pink flowers, in shallow water in canals, ditches, ponds and rivers in lowland areas, also planted as an ornamental elsewhere.

Flowering rush appears to have attracted no folklore, have no known uses, and attracted few local names: HEN-AND-CHICKENS in the Huntingdonshire fens, pride-of-the-Thames in Dorset and Somerset, and rackzen, or raxen, 'from the Old English *rixen*, rushes', in Somerset.

FLOWERING SUNDAY

In parts of Wales PALM SUNDAY is known as *Sul y Blodau*, or Flowering Sunday, and formerly it was a widespread custom to decorate graves with flowers on this day. Although the name *Sul y Blodau* occurs in Welsh literature as early as the fifteenth century, the restricting of the day to Palm Sunday appears to date from the latter half of the nineteenth century.[1] Late in the nineteenth century the custom became popular in urban parts of South Wales, where in 1896 a writer described how thousands of people would visit the cemetery in Cardiff, 'the roads thereto presenting an appearance like unto a fair'.[2] The custom, although now less important, survived at least until the 1980s, when in many parts of Wales many graves continued to receive special attention, usually by having bunches of DAFFODILS placed on them.[3]

In Radnorshire graves were decorated for EASTER Sunday. On 16 April 1870 the diarist Francis Kilvert recorded:

> When I started for Cefn y Blaen only two or three people were in the churchyard with flowers. But now the customary beautiful Easter Eve Idyll had fairly begun and people kept arriving from all parts with flowers to dress the graves. Children were coming from the town and from neighbouring villages with baskets of flowers and knives to cut holes in the turf. The roads were lively with people coming and going and the churchyard a busY scene with women and children and a few men moving about among the tombstones and kneeling down beside the green mounds flowering the graves . . . More and more people kept coming into the churchyard as they finished their day's work.[4]

Similar practices took place on local feast days in parts of Ireland.

> Dressing the graves is a custom which is practised in every parish with a few exceptions in south and mid Louth. Each parish and each churchyard has its own patron day. On that day or the Sunday following Mass was said for the dead in the parish church.

> The graves are beautifully dressed with whatever flowers are in season.
>
> A cross of flowers 2½ or 3ft high stands at the head of the grave.
>
> The grave itself is covered with moss or evergreen and outlined with BOXwood or palm [probably YEW] neatly clipped. Wreaths of flowers in blending colours – red and white, pink and white, purple and white, etc. are placed in the enclosure. People from all around come to pray for their dead and incidentally criticise the dressing.[5]

FLOWER PARADE

Between 1959 and 2013 an annual Flower Parade held at Spalding, Lincolnshire, on the first Saturday in May was a popular tourist attraction. The Parade consisted of about twenty floats elaborately decorated with tulips and other cut flowers.

Commercial bulb-growing started in the area in about 1890, when TULIPS were grown as cut-flowers, and by the 1920s the fields attracted a steady stream of tourists. In 1935, when the silver jubilee of King George V and Queen Mary coincided with tulip time, the bulb-growers planted their fields with emphasis on red, white, and 'blue' tulips, an act which brought much publicity and traffic chaos to the town. The following year an official committee was set up to sort out the traffic problems and assume overall responsibility for Tulip Time.

As the years went by the number of bulbs grown for cut flowers gradually decreased as more and more grew tulips for their bulbs. Tulips grown for their bulbs usually have their flowers removed to encourage the rapid development of new bulbs. There were fewer fields of flowering tulips for visitors to enjoy, but vast quantities of detached tulip flowers. Thus in 1959 the Spalding branch of the National Farmers Union organised the first Flower Parade which, from its start attracted large numbers of visitors.

> The initial form and steel skeleton of each float is skilfully constructed by a local blacksmith, Geoff Dodd, into the outline shape of the subject. Until 1985 the steelwork was covered with a special straw matting to form a base to receive the flowers. Polythene foam has now replaced the use of straw matting. The final stage is reached 24 hours before the actual Parade when hundreds of volunteers work into the night weaving intricate patterns in flowers.

> Each tulip head is deftly secured to the base with a wire pin until the whole float is literally covered with tulips. A single float, which may be as much as 50 feet in length, may be decorated by as many as half a million tulips, supplemented by numerous individual flower arrangements created by the local Flower Lovers' Club. Other colourful spring flowers and materials are used to complement the float's design. The result is one of indescribable beauty and it is not surprising that even visitors returning year after year still gasp with astonishment at the wonder of it all.[1]

When spring was late DAFFODILS might have to be used instead of tulips.[2] When spring was early, tulip flowers were picked and placed in cold storage until needed for the Parade.[3]

The last Parade was held in 2013, whereafter financial support from local councils was withdrawn.[4]

FLOWERS

In the 1920s children living in the poorer parts of London would readily pick up and devour any food which they found lying on the pavement: 'Waste not, want not, pick it up and eat it.' But a dropped flower was different, for 'if one of the younger children went to touch it, the rest of us dragged him back gasping "pick up a flower, pick up FEVER"'.[1] Similarly:

> It was unlucky to pick up flowers which have been dropped on the ground (brings sickness to the house!).[2]

Such beliefs probably date from the time when the urban poor only bought or had flowers when a corpse was in the house awaiting burial, so it's not surprising that juvenile minds associated flowers with illness and death.

Less common was a total prohibition of flowers indoors.

> A farmer's wife from Inkberrow [Worcestershire] (*c*.1887) visiting my mother at Aloechurch, was given a bunch of roses from our garden. But before going back to the house she contrived to drop them quietly one by one. This was noticed by my brother, who knew the reason – it brings bad luck to chickens if flowers are taken inside the house.[3]

> My mother, born in 1903, and grandmother, would not have any

> blossom of any description in the house, believing it would bring bad luck to the family.[4]

Alternatively, only flowers identified as 'flourish' or wildflowers were prohibited.

> Even before I went to primary school in Clarkstown I would be taken on walks, and, indeed, go independently on local rural walks. I used to pick flowers to take home and this is where I was told never to bring 'flourish' into the house because it was very unlucky. 'Flourish' was never clearly defined, but I associate it with flowers like ELDER or MEADOWSWEET – common factor? – white inflorescences, strong smell. I am not sure . . . My mother [born in Glasgow in 1911] always had flowers in vases in the house. My Dad grew sweet peas, CARNATIONS, etc., for this purpose – none of these were 'flourish'!
>
> It's funny how superstitions stick, I still won't wear green, nor have a green car, and I certainly wouldn't have 'flourish' in the house.[5]

> Flower unlucky indoors: LILAC, *all* wildflowers.[6]

> I was a child in Lincolnshire in the 1920s–30s . . . My mother would never let me bring any wildflowers I had picked in the meadows into the house . . . I think her insistence on keeping BUTTERCUPS, CUCKOO-FLOWERS, DAISIES, etc., in jam-jars on an outside windowsill was purely social; they were not garden flowers, so not good enough to come indoors.[7]

On the stage:

> There is a theatre superstition that you should not have live flowers on the stage. I've been told, but I don't think I believe it, that if fresh flowers fell there is a danger that the performers might slip.[8]

Sometimes it was said that flowers should not be handled by menstruating women:

> When I was a girl in the Basingstoke area of Hampshire, my mother told me I must never touch flowers during my period, or they would wilt and die![9]

See also RED AND WHITE FLOWERS.

FLOWER SERVICE

For many years until the 1970s a Flower Service was held at St Mary's church in Bridport, Dorset. According to a report in the *Bridport News* in May 1905, this event dated back to at least 1788, when Sunday Schools were first held in the town.[1] However, the Flower Services of later years were very dependent on rail transport, so it would seem that the service in its twentieth-century form evolved some time after the railway reached Bridport in 1857.

Writing in February 1985 a St Mary's Sunday School teacher recalled:

> May Sunday was kept as Sunday School Festival Day – an event which was looked forward to, as most of the girls had new dresses and hats for the occasion – usually white – they must have worn layers of warm clothing underneath as it was as cold as recent Mays. Saturday was picking day when hordes of people took the train to Powerstock, the first stop, where the fields were covered with PRIMROSES, COWSLIPS, BLUEBELLS, wild hyacinth, etc. – but cowslips were the best that travelled and retained their lovely smell. All the children brought them to church, after parading around the town to show off their bunches, which had labels with their names and addresses on them. Some 200 plus children and parents and relations filled the church during the service. The Rector then received the flowers at the chancel steps, on trays, which were then packed in cartons – sometimes twelve or more – and taken to the station to catch the 3.30 to Paddington, where they were collected and delivered the same evening to the elderly, sick and under-privileged people.
>
> Most of the recipients sent letters of thanks to the children by the following Sunday, when they were read at School. Eventually there was too much traffic on the Main Road, so the procession was shortened. Then [in 1975] unfortunately British Rail closed our branch line, so we had to send the flowers by road to Dorchester station (some 15 miles) to arrive in London the next morning to be collected by the ladies of St Stephen's, Westminster, for their old people's club. Some of the letters received then spoke of the flowers being the first wild flowers the recipients had seen . . . It was quite apparent that the flowers brought a little joy and a touch of the country to the city, but it ended in the 1970s due to people complaining about the picking of flowers.

It seems probable that other rural towns and villages organised similar events.

FLUKE (LIVER) – sheep get after eating SUNDEW; MALE FERN used to rid cattle and sheep of.

FOOD – wild plants used as include ELDER, GOOD KING HENRY, GREEN ALKANET and GROUND ELDER.

FOOL'S PARSLEY (*Aethusa cynapium*)

Annual, white-flowered herb, common on waste and cultivated land throughout much of Britain and Ireland, but absent from northern Scotland and scarce in parts of Ireland.

It is probable that many people failed to distinguish between fool's parsley and its relation COW PARSLEY. Names which have reputedly been used for fool's parsley include cow parsley in Somerset, the widespread KELK, and:

Devil's wand in Dorset
Dog-poison in Devon
False parsley in Shropshire
Lace curtains, LADY'S LACE, and pig-dock, in Somerset
RABBIT-MEAT in Yorkshire.

> Fifty-one years ago, I picked some wild flowers for my mother who was ill and my grandmother was caring for us; as I approached the house my gran said 'Don't bring those mother-die into the house, they are bad LUCK', so sadly they went into the bin. (My mother passed away a few weeks after).[1]

FOOT-AND-MOUTH DISEASE – treated using ASH, (stinking) HELLEBORE and IVY.

FORGET-ME-NOT (*Myosotis* spp., especially *M. sylvatica*)

Annual or perennial, usually blue-flowered herbs, widespread in a wide range of habitats; one species (*M. sylvatica*) commonly cultivated as an ornamental.

The name forget-me-not, of which equivalents are known in several European languages, has been used in the British Isles since the sixteenth century.[1] Two legends explain the name.

> Adam having given names to all the blooms in the Garden of

> Eden, discovered . . . that one of them had forgotten its lesson. So he renamed it forget-me-not.[2]

Alternatively, and more usually:

> A knight and his lady were walking by a fast-flowing river; the knight wishing to please his love, attempted to pick the bright blue flower growing on the bank close to the water's edge. Unfortunately he slipped and fell into the water. Throwing the flowers to his beloved, he called 'Forget me not' before being swept away and drowned.[3]

Although thousands of greetings cards depicting forget-me-nots were produced early in the twentieth century these legends do not appear to feature in oral tradition.

FORSYTHIA (*Forsythia* x *intermedia*)

Shrub of garden origin, cultivated for its early, rather vulgar, bright yellow flowers.

> In the 1950s in Wiltshire, my grandmother would not allow MIMOSA indoors because it was unlucky; she also wouldn't allow forsythia indoors, but mimosa was the one she went really mad about.[1]

FOXGLOVE (*Digitalis purpurea*)

Biennial or perennial herb (Fig. 17), common on acid soils, in hedgebanks and woodland clearings, where its pinkish purple flowers can produce spectacular displays, and on cliff-tops.

Local names include the widespread LADY'S THIMBLE, POPPY-DOCK and:

Bee-catchers in Somerset; 'when the bee is in the flower boys close the entrance to be amused by the insect's struggles', and bee-hives
Blobs, widespread; 'children pull off a flower, and with the fingers of one hand closing the mouth, and, giving the other end a slap, it burst with a noise like the word blob'
Bloody bells in Lanarkshire
Bloody fingers, widespread
BLOODY-MAN'S FINGERS in Herefordshire and Radnorshire; 'probably from the habit of children to put the red flowers on

their fingers'
Bluidy bills in Lanarkshire
Bluidy finger in northern England and Scotland
Bunch-of-grapes, BUNNY RABBITS, and BUNNY-RABBITS' MOUTHS, in Somerset
CLOTHES PEGS in Dorset
Cottagers in Co. Waterford; 'because they belong to poor people'
COVENTRY-BELLS in Dorset
Cow-bells in Devon
COW-FLOP in Cornwall, Devon and Somerset
COWSLIP, and COWSLOP in Devon
Dead-man's bellows in Berwickshire, 'i.e. pillies, male members'
DEAD-MAN'S BELLS in Dorset and Scotland
Dead-man's fingers in Inverness-shire and Ireland
Dead-man's thoombs, given to the 'tubers' in Scotland
DEAD-MEN'S BELLS in Dorset, Northumberland and Scotland
Dead-men's fingers in Ireland
Dead-men's thimbles in Somerset
Dead-woman's thimbles in Scotland
Dog-finger in Bristol
Dog's fingers in Somerset and Wales
DRAGON'S MOUTH in Sussex
Duck's mouth in Somerset
Fairies' home in Ireland
Fairies' petticoats in Cheshire and Somerset
FAIRY-BELL in Ireland, and FAIRY-BELLS in Somerset
Fairy-fingers, and fairy-gloves, both widespread
Fairy-hat in Dorset
Fairy's gloves in Cheshire
FAIRY-THIMBLES in East Anglia, Co. Derry and Co. Donegal
Fairy-weed in Ireland
Finger-cap in Somerset
Finger-hut in Devon
Finger-root in Sussex and Warwickshire
Fingers, and FINGERS-AND-THUMBS in Somerset
Finger-tips in Wiltshire
Flap-dock and FLAPPER-DOCK, flappy-dock 'most likely from the habit which children have of inflating and bursting the flower'
Flobby-dock, and flop, in Devon
Flop-a-dock in Cornwall, Devon, Somerset and Wiltshire
Flop-dock in Cornwall and Devon
Flop-docken in Yorkshire
Flop-poppy ('i.c. flowcrs which pop'), and floppy dock, in Devon
Floppy dops in Cornwall
Flop-top in Devon and Somerset
Floss-docken, and flowster-docken, in Yorkshire; 'flowster . . . to flourish or flutter in showy colours'
Folk's glove in Northumberland and Somerset, 'i.e. fairies' glove, but not all etymologists are prepared to accept this derivation'
Fox-and-leaves in Co. Donegal
Fox-docken in Yorkshire
Foxes' glove in Oxfordshire
Fox-fingers in Yorkshire

Fox-flops in Somerset; 'the plant which has grown from fox droppings'
Foxter in Scotland
Foxter-leaves in Roxburghshire
Fox-tree-leaves in Scotland and Co. Derry
Foxy in Co. Donegal
GAP-MOUTH in Somerset
Gensie pushon in Orkney
GOOSE-FLOPS in Devon and Dorset
Goose-top in Dorset
Granny-bonnet in Somerset
Green poppy, and green pop, in Cornwall
HAREBELL in Ireland
Hill-poppy in Somerset
HOLLYHOCK in Devon
King's elwand in Northumberland
Lady-glove in Shropshire
LADY'S SLIPPER in Somerset
LION'S MOUTH in Sussex
Lithmore in Co. Donegal
Lusmore in Ireland; 'i.e. large plant'
Pop-bell in Somerset
Pop-bladders in Dorset
Pop-glove in Cornwall
POP-GUNS in Cornwall and Somerset
Pop-ladders in Dorset
POPPERS, 'children pop the flower in the same way they would pop a blown-out paper bag'
POPPY, in Buckinghamshire, Cornwall and Isle of Wight; around Salisbury in Wiltshire, 'the name [poppy] is confined to *Digitalis*, *Papaver* there being known only as redweed'
Pops in Somerset
Puppy-dock in Yorkshire
RABBIT'S FLOWER in Devon
Rapper in Pembrokeshire
Scabbit-dock in Cornwall
Scotch mercury in Berwickshire
Snakes-and-ladders in Somerset
SNAPDRAGON in Devon and Somerset
SNAP-JACKS, and snaps, in Somerset
Snauper in Gloucestershire
Snoxum in the Forest of Dean, Gloucestershire; 'because of children using the flowers as crackers and exploding them by a 'snock' on the ball of the thumb'
SOLDIERS in Dorset
THIMBLE in Cumberland and Co. Waterford
Thimble-flower in Dorset and Somerset
TIGER'S MOUTH in Sussex
Turtle doves in Somerset
Virgin's fingers in Devon and Somerset
Vlops in Devon
WILD MERCURY in Berwickshire
Witches' thimble in Northumberland and Scotland
Woodwand in Devon.

According to James Britten:

> The name foxglove has, in all probability, nothing to do with Reynard, but is rather connected with the FAIRIES or little folk. The

derivation is fully borne out by other of its names; e.g. the North Country name, 'witches' thimbles', the Irish name 'FAIRY CAP', the Welsh *maneg ellyllyn* (fairies' glove), the Cheshire, 'fairies' petticoat', and the East Anglian 'fairy-thimble'.[1]

However, a belief collected from County Leitrim implies that foxgloves, rather than being fairy plants, are dangerous to fairies.

If you have a peevish child, or one that from being in good health becomes sickly, and you have reason to believe it is a fairy child, the following plan may be tried in order to ascertain whether this is the case. Take lusmore (foxglove) and squeeze the juice out. Give the child three drops on the tongue, and three in each ear. Then place it at the door of the house on a shovel (on which it should be held by someone), and swing it out of the door on the shovel three times saying: 'If you're a fairy away with you!' If it is a fairy child it will die; but if not, it will surely begin to mend.[2]

As Jeremy Harte has pointed out:

Concentrated juice of foxglove contains digitalis, which can cause death by heart spasm; about a tenth of an ounce of dry leaf is fatal for an adult, while a malnourished child would respond to a lesser dose.[3]

At least one case has been recorded, from Caernarvonshire in 1857, of a suspected changeling being killed by foxglove poisoning,[4] Thus it seems that the use of foxglove (and other ordeals to which supposed changelings were subjected) might have been an acceptable method of infanticide which enabled families to rid themselves of sickly offspring.

There are occasional records of foxgloves being considered 'unlucky', or omens of war.

[Tutbury, Staffordshire, 1950s] picking floxgloves was unlucky and they were absolutely forbidden inside a house as this gave WITCHES/the DEVIL access to the house.[5]

The summer of 1914 was a record one for foxgloves, regarding which an old [Staffordshire] man remarked, 'I don't like them, missus; they mean war. Them foxgloves is soldiers.[6]

In some areas, white foxgloves were particularly feared.

> If white flowers occur in garden foxgloves there will be a DEATH in the family.[7]

> The two most taboo flowers that were not allowed in the house were white foxgloves and LILAC.[8]

Children widely appreciated foxglove flowers as playthings.

> [Cornwall, where foxgloves is known as pop-dock]: DOCK from its large coarse leaves; pop, from the habit of children to inflate and burst the flower.[9]

> [Broadclyst, Devon] Wishing-flowers = Foxgloves, children inflated the blossom and made a wish when they burst.[10]

> [Forest of Dean, 1920s] Amusing ourselves lazily popping 'snompers'. We picked the beautiful pink foxgloves . . . then took off each flower, trapping the air with thumb and forefinger, and pushed the ends together until they'd explode with a pleasant little pop.[11]

> [Also Forest of Dean] Snomper, or snowper (rhyme with cow) – foxglove. A favourite admonition to a noisy child: 'Shut thee chops; thee bist like a bumble bee in a snowper'. A favourite occupation in summer was to trap a bee in a foxglove bell to hear it buzz angrily![12]

On Guernsey, foxglove was known as *claque*, 'derived from the children's amusement of popping or bursting (*claquer*) the flowers on the palm of the hand', and its flowering provided guidance as to when MACKEREL-fishing should start:

> *Quand tu vé epani l'claquet*
> *Met tes leines dans ten bate*
> *Et t'en vâs au macré.*
> (When you see the foxglove blossoming,
> put your fishing-tackle into your boat,
> and go off for mackerel).[13]

At Hartland in Devon foxgloves are associated with Nectan, an obscure saint to whom the parish church is dedicated. According to what appears to be a comparatively recent tradition, St Nectan and his sister arrived from Wales, and made their way towards Hartland. At Stoke they were attacked by robbers, and the saint was decapitated. However, their journey was not delayed, for Nectan picked up his head and continued. Wherever a drop of his BLOOD fell a foxglove sprang up.[14] According to the Vicar of Hartland, writing in January 1982, a Foxglove Procession is observed 'with great gusto' before morning Sung Eucharist on the Sunday nearest the patronal feast, 17 June. Although parish magazines survive from 1909, the Procession is not mentioned until 1927, when the then incumbent arranged a procession after 3 p.m. Evensong on St Nectan's Day.

In September 1998 David Ford, then Vicar of Hartland, agreed that there is no pre-twentieth-century record of the procession, and provided a different version of the legend:

> [St Nectan] came from Wales of a large family, others following him and practising their Celtic Christianity along the coast of Cornwall and north Devon (Gennys, Merewenna, Wenna, Tedda, Wynup, Endelient, etc., giving names to 24 local villages). Nectan lived here as a hermit and was beheaded by some thieves who had deprived him of some cows given as a gift. The story is that he returned to his cell carrying his head under his arm, each drop of blood became a foxglove.
>
> When we go all out to have a Procession, we set the service time and while the majority of the elder members of the congregation meet in the church, the fitter ones meet at St Nectan's Well where we sing and 'bless' the waters . . . and walk to church holding our foxglove aloft and usually trying to not look too embarrassed.
>
> Foxgloves certainly are the main flower decoration on the Sunday nearest June 17th. We have now moved the timetable of services to make the third Sunday in June a family service, so that this (much smaller than in the not so distant past) procession takes place in the presence of Sunday School members and their families.
>
> Although the procession continues, in recent years it has become much truncated; members of the congregation simply process along the nave of the church to place foxgloves in two vases placed in front of the altar, and the Well is neglected.[15]

Two other Devon churches are dedicated to Nectan. That at Welcombe, which is cared for by the same ministry team as Hartland, is abundantly decorated with foxgloves for its St Nectan's service, and a vase of foxgloves is placed at the nearby well.[16] In the south of the county, at St Nectan's, Ashcombe, in 2016 the saint's day 'was celebrated with the retelling of his legend in the school and photos of 'his' foxgloves'.[17]

In folk medicine:

> Foxglove leaves were placed in children's shoes and worn thus for a year, as a cure for scarlet fever – in Shropshire.[18]

> The *lus mor* – or soft leaves in the heart of the plant out of which fairy thimbles grow – is good for healing a CUT. The little hard thread on the back of the leaf should be pulled out and the leaf heated at the fire and applied to the cut.[19]

Foxglove provides the major British example of how traditional remedies might prove worthy of scientific investigation. In 1775 William Withering was asked for his opinion on a traditional Shropshire remedy for DROPSY. Of the twenty or so herbs the remedy contained, Withering quickly concluded that the important active ingredient was foxglove leaves. Thus, as patients for whom all other remedies had failed became available, he experimented by administering different dosages of foxglove leaves in a variety of forms. After ten years he published his results, listing 163 of his own patients and a number treated by other physicians, and, although originally used to stimulate the flow of urine, he reported that foxglove leaves had 'a power over the motion of the HEART to a degree not yet observed in any other medicine'.[20] Several of Withering's contemporaries claimed that foxglove leaves were useful for treating TUBERCULOSIS, but this was never proved.

During both world wars the foxglove was one of the 'essential species' which people were urged to collect or grow for medicinal use. During the 1914–18 war it was estimated that 20–25 tonnes of dried foxglove leaves were needed each year:

> Landowners, who often objected to foxglove leaves being collected from their land, especially after June 15, for fear of disturbing their game birds, could help . . . by instructing their head gardeners to transplant young foxgloves, or scatter their seeds in autumn, in areas to which there is easy access without disturbing any game.[21]

Again during the Second World War people were requested to collect foxglove leaves, and in 1941 Oxfordshire Women's Institutes and other volunteers, who had the support of the university's academic botanists, collected sufficient foxglove leaves to provide 350,000 doses of digitalis (enough to provide treatment for 1,000 patients for a year).[22]

Foxglove continues to be used to treat heart conditions, but at present digoxin is extracted from the woolly, or Grecian, foxglove (*Digitalis lanata*), native to eastern Europe, not the British species.

FRECKLES – FUMITORY, SILVERWEED and SUNDEW used to remove.

FRENCH MARIGOLD – see AFRICAN MARIGOLD.

FRITILLARY (*Fritillaria meleagris*)

Bulbous perennial found in damp meadows in lowland England, and widely cultivated as an ornamental. As the fritillary was not recorded as a wild plant in Britain until 1736, its status as a native must be considered unproven.

In addition to the widespread SNAKE'S HEAD, other local names include:

Bloody warrior in Berkshire; 'each one having grown from a drop of DANE'S BLOOD'
Chequered lily in Somerset
Crowcup in Buckinghamshire
DEAD-MAN'S (or MEN'S) bells in Shropshire
Death-bell in Buckinghamshire and Oxfordshire
Deith-bell in Cumberland gardens; 'from the dingy, sad colour of the bell-shaped flowers',
Doleful-bells-of-sorrow in Oxfordshire
Drooping-bell-of-Sodom in Dorset
DROOPING LILY in Cheshire
Drooping tulip in Cheshire gardens and Dorset, 'the flower hangs downwards and is much like a tulip in form'
Falfalaries in Yorkshire
FIVE-LEAVED GRASS in Oxfordshire
Frits, 'a shortened form of fritillary', in Berkshire
Froccup ('probably frog cup from its spotted flowers'), and frockup ('frog cup, or from the segments

of the perianth which hang like a frock and then turn up') in Buckinghamshire
Guinea-hen flower in Cumberland and Somerset; 'from its petals markings which are similar to those of the bird's feathers'
Jenette, a garden name in Co. Donegal
Lazarus bell in Devon and Somerset; 'seems to have been originally lazar's [leper's] bell and the flower must have been so-called from its likeness to the small bell which the lazar was bound to wear around his person, so that its tinkling might give warning of his approach'
Leopard's lily in Devon
Leper's lily in Somerset
Minety bell in Wiltshire
Mother ugly in Ireland
Pheasant lily in Cumberland
Shy widows in Warwickshire
Snake's-head lily in Buckinghamshire and Oxfordshire; 'from the shape of the flower and its spotted colour'
Solemn-bells-of-Sodom in Dorset
Toad's heads, and toad's mouth, in Warwickshire
Turkey eggs in Berkshire
Turk's head in Warwickshire; 'a corruption of turkey-hen'
Weeping widow in Northamptonshire and Staffordshire
Widow-wail in Shropshire
WILD TULIP in Berwickshire, Buckinghamshire, Northamptonshire and Warwickshire.

Fritillaries were a favourite flower in children's May garlands. In 1843 it was reported:

> This species, called by the country folk snake's head, used to flower in a meadow at Mortlake, Surrey, known from that circumstance as 'the Snake's-head Meadow', but of late years it has become very scarce, if not altogether eradicated by the ruthless hands of the village children, by whom the early showy plant was coveted as an ornament to their May garlands.[1]

> However, this was disputed by another writer who claimed that the desecration of Mortlake's fritillaries was due to the fact that 'too many *radical* botanists visited this station during the time it was in flower, such hands prove more destructive than the Mortlake children'.[2]

At about the same time:

> At Bishop's Waltham . . . the children gather the Fritillary for their May-day garlands, yet in proof of the incurious nature of the

> people of Hampshire, I could not find any one at Strathfieldsay who knew its name; some called the plants SNOWDROPS (the white variety), others DAFFODILS, whilst the rest pronounced them to be COWSLIPS![3]

Elsewhere:

> Frawcup Sunday was a Sunday in early May. Children in Thames villages, such as Haddenham, Cuddington, Dinton, Ford, Marsh, the Kimbles and in other parts, dressed in their prettiest attire and garlanded with May flowers, brought posies of the delightful little fritillary, locally known as frawcups, to the cottage doors. Older members of our village community still recall with fervent nostalgia memories of this lovely event to herald the advent of warm summer sunshine and the promise of abundance of nature's gifts.[4]

Surviving fritillary meadows are valued, protected, and visited when they are in bloom.

> [1930s] in the spring people gathered at Oaksey . . . Wiltshire to see the show of what were called locally Oaksey lilies (*Fritillaria meleagris*). They were an unforgettable sight. In the neighbouring towns, as far away as Cheltenham, street traders sold large quantities of these flowers each spring.[5]

> The last Sunday in April each year in Ducklington, Oxon, is Fritillary Sunday, when a local water meadow supporting these flowers is open to the public in aid of charity. On that day too the altar frontal in the church is embroidered with these flowers and was worked by the late owner of the field. Teas are served, there are appropriate stalls in the church and morris dancers perform. All-in-all it is a good half-day out.[6]

Ducklington's Fritillary Sunday remains popular, attracting approximately 1,000 people each year, and its programme remains little changed. Although there are better-known sites where fritillaries are more abundant, it is said that only at Ducklington can visitors wander amongst the flowers, instead of being restricted to viewing them from paths.[7]

FROSTS – cease when MULBERRY comes into leaf.

FRUIT

Make a WISH when you have the first of the season.[1]

People used to make a wish when they had the first fruit or vegetable. They always used to do that.[2]

FRUIT TREES

It was widely believed that fruit trees flowering out of season foretold misfortune or DEATH.

A blossom on the tree when the APPLES are ripe
Is a sure termination of somebody's life.[1]

In the Basingstoke district of Hampshire it was believed that WAR was foretold when fruit trees blossomed at unusual times.[2]

My grandmother told me, when I was very young, that it was unlucky take into the house, or indeed pick, a flower that was 'blooming out of season'. Quite often an odd sprig of CHERRY or PLUM will blossom in December or January because 'the DEVIL has touched it' and he wants to bring his influence into the house.[3]

[Colyford, Devon] if there are fruit and flowers on a tree, especially an apple- tree, at the same time, there will be a death in the family.[4]

In Norfolk a fruit tree which failed to produce fruit had iron nails knocked into it; 'this may have been an offering to the tree'.[5]

FUCHSIA (*Fuchsia magellanica*)

Deciduous shrub, native to Argentina and Chile, cultivated as an ornamental since 1788, now widely naturalised, especially in western coastal areas.

Local names include droppers in Cornwall, drops-of-Abel's-blood in Durham, ear-ring flower in Lincolnshire, lady's ear-drops in Devon, and tears-of-god in Co. Mayo.

Despite a record, from Merseyside in *c.*1950, of fuchsia being considered UNLUCKY and 'not to be taken into a house',[1] it appears that few people considered it to be so.

Children made dolls from fuchsia flowers:

> A memory from my childhood [1990s]: Making flower fairies from fuchsia which filled the hedgerows around west Cork where I grew up. My sister and I would trim the filaments down to just two to make two little legs with tiny shoes and then gather a small twig which we would push through the nectar-filled heart (torso) to make the arms.[2]

In Ireland:

> HONEYSUCKLE = the fuchsia (grown as a hedge, common name in Co. Antrim and Co. Down at present time) – the nectar is sucked by children.[3]
>
> Patrick O'Sullivan [aged 70+ of Birdhill, Co. Tipperary] told me . . . as a child he sucked the flower and knew it by the name of bleeding hearts.[4]
>
> Blossom of fuchsia when boiled DYES dark red.[5]
>
> [Fuchsia] is nowadays used as a quality mark symbol (as a drawing of a flowering shoot) for various produce of southern Ireland, e.g. Cork and Kerry.[6]

FUMITORY (*Fumaria* spp.)

Delicate, scrambling, pink-flowered, annual herbs, widespread on arable and waste land throughout most of the British Isles.

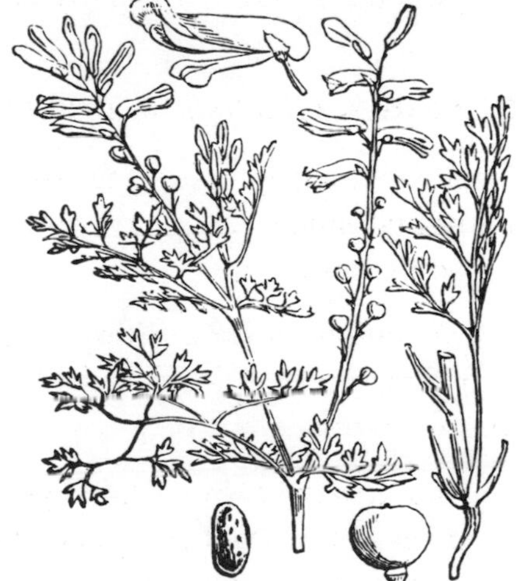

Local names include the widespread wax-dolls, 'from the doll-like appearance of its little flowers', and:

Babe-in-cradle in Somerset
Birds-on-the-bush in Dorset
Earth smoke in Sussex
Faminterry on the Isle of Man
Favourite in Plymouth
Furrow-weed in Gloucestershire
GOD'S-FINGERS-AND-THUMBS in Dorset
Hemitory in Kent

Jam tarts in Dorset
LADY'S FINGERS in Co. Dublin
LADY'S LOCKETS in Somerset
Lady's shoe in Somerset and
Wiltshire
SNAPDRAGON in
Northamptonshire.

In Wiltshire a cosmetic for removing FRECKLES was distilled from fumitory:

> If you wish to be pure and holy
> Wash your face with fevertory.[1]

In Orkney:

> The juice of this is given to children as a cure for WORMS; also in foals, but in much larger doses, of course.[2]

FUNERALS – BOX, CHRYSANTHEMUM and LILAC used; WHITE FLOWERS associated with.

G

GALINGALE (*Cyperus longus*)

Perennial sedge, native to the Channel Islands and some coastal areas, also surviving after having been planted elsewhere.

Galingale, known as *han* in Jersey-Norman-French, was used in the past in place of HEMP for cords and ROPES. It was used to make tethers for cows and halters for horses as well as matting for floors. A number of local names indicate the fact that *han* grew there, i.e. Handois and Les Hanniethe.[1]

GALL-STONES - COUCH used to treat.

GARLIC (*Allium sativum*)

Pungent, bulbous herb, unknown in the wild, but known in cultivation since *c.*3500–3000 BC.

In vampire literature and films garlic is considered to be protective against vampires and evil in general. In 1994 it was reported:

> Sotheby's [New York] is auctioning a 'vampire killing kit', with garlic powder, bible, wooden stake, moulds for silver bullets and a crucifix concealing a pistol. Most of the items, in a 19th-century mahogany case, are probably only 15 years old.[1]

Elsewhere:

> Recently I heard a colleague comment about another who he disliked that when he knew they were working together he felt the need to come wearing a string of garlic cloves and a crucifix between his shoulder blades to prevent back-stabbing. He was making it very clear that he disliked the person in question, but it seemed a very sophisticated way of doing so.

> However, about 10 years ago, a wealthy incomer to the Isle of Man sacked his gardener, also a come-over, under doubtful circumstances. In response his gateposts were wreathed with wild garlic (*Allium ursinum*, RAMSONS) by the wronged party. He got the message and was very upset by it.[2]

Since 1984 an annual Garlic Festival has been held near Newchurch on the Isle of Wight. According to the *Daily Telegraph* of 20 August 1994:

> More than 30,000 enthusiasts will head for the annual Newchurch Garlic Festival, a two-day celebration of this most pungent and versatile of vegetables. Apart from entertainment – a Garlic Queen and a Vampire King will be crowned for the occasion – there will also be the opportunity for enthusiasts to spend the weekend in a food tent serving a huge range of heavy-on-garlic dishes.

However, on 16 August 2003:

> When I arrived, mid-afternoon, the man taking money at the entrance happily reported that things were going very well with about 1,000 cars filling the Festival's car parks. The large site was packed with families enjoying what was on offer – an arena providing space for, amongst other things, falconry displays, Siberian husky racing, and displays by the Wight Diamonds Marching Band; stages for a variety of acts ranging from jazz to Punch and Judy, and a huge number of stalls provided by craftspeople, local charities and businesses.
>
> The Garlic Queen and Vampire King appeared to no longer feature in the event, and, indeed garlic was not prominent. In the Garlic Marquee there were stalls selling fresh and smoked garlic bulbs, garlic crushers, and, according to the Festival guide, 'mouth-watering delectable garlic products, from sausages, mushrooms, beer and seafood, to famous garlic fudge and ice cream'. However, on the whole it was a disappointment for most of the products were rather unimaginative and could be purchased from any medium-sized supermarket. It was a rather ordinary, pleasant event for a family outing, but didn't offer a great deal to anyone wanting to find out more about garlic.[3]

Websites currently list some thirty-five different garlic festivals in

North America, where it is said that the first such festival was held in Gilroy, California, in August 1979:

> The festival is now held annually the last full weekend in July. In 2001 over 125,000 guests attended the 23rd annual festival, which is now a three-day event put on by more than 4,000 volunteers. Between 1979 and 2001 the festival raised over $5 million to distribute to the community's non-profit charities and service organisations.[4]

Particularly in Ireland, where it was commonly cultivated until about the time of the First World War,[5] garlic was apparently widely valued as a herbal remedy, though it is probable that some records refer to wild garlic, rather than the cultivated species.

> Garlic was boiled and applied to RINGWORM to cure it.[6]

> Yellow JAUNDICE: boil two bulbs of garlic and chop them up small and boil them with a small quantity of water. Make sure no dirt enters the mixture while cooking. When the mixture is brown remove it from the saucepan and place it in a clean bowl to cool a little. Give it to the patient and allow him to drink it as hot as possible. The patient is cured after taking this mixture once. I cured myself with this cure.[7]

> Previous to the First World War garlic was used as a general cure. Slices of garlic were placed in the shoes, against the soles of the feet, as a cure for WHOOPING COUGH.[8]

Garlic planted on GOOD FRIDAY was believed to be particularly potent:

> Garlic that has been planted on Good Friday when boiled in sweet milk and given as a drench will cure any disease in people, cattle, or fowl.[9]

> If garlic is planted on Good Friday the person who plants it will not contract fever during the year.[10]

In England fewer remedies which use garlic have been recorded, but the plant is confidently recommended by herbalists:

> A cure for whooping cough, practised in Cambridge not that long ago, was to put garlic in the sufferer's socks.[11]

> Research shows that garlic can protect against some forms of HEART disease by reducing cholesterol levels, and the health food industry has sniffed a new way of making a fortune. The market in garlic supplements in Britain has doubled in the past three years, and is now [1991] worth about £10 million annually. Every day, about half a million people in the UK swallow the virtually odourless tablets.[12]

GARLICK SUNDAY (last Sunday in July or first in August) – first POTATOES dug on.

GARLIC MUSTARD (*Alliaria petiolata*), also known as hedge garlic

Short-lived, white-flowered herb, common and widespread on shaded hedge-banks, woodland edges and moist disturbed ground.

Local names include the widespread JACK-BY-THE-HEDGE, and:

Beggar-man's oatmeal in Leicestershire
Jack-by-the-hedgeside in Dorset
Jack-in-the-bush in Gloucestershire and Herefordshire
Jack-of-the-hedge in Cheshire
Jack-run-along-by-the-hedge in Wiltshire
LADY'S NEEDLEWORK, and lamb's pummy, in Somerset
Penny-hedge in Norfolk
Penny-in-the-hedge in Worcestershire
PICKPOCKET in Devon
Poor-man's mustard in Lincolnshire
Sauce-alone, and SCABS in Somerset
STICKY WILLIE in Scotland
Swarms in York.

Although once valued as an ingredient of sauces, garlic mustard has been little appreciated in recent times.

> [1550s] It is commonly used both in England and in Germany to be put in sauces in the springe of the yeare wherefore the Englishmen call it sauce alone and ye Germanes *Sauszkraut*.[1]

Allen and Hatfield record hedge garlic as being used to treat sore THROATS (in Kent?), sore GUMS and mouth ulcers in Norfolk, CRAMP in Somerset, and WOUNDS (in Kent?).[2]

GERMANDER SPEEDWELL (*Veronica chamaedrys*)

Perennial herb with attractive blue flowers, common on hedge-banks, railway embankments and woodland edges throughout the British Isles.

Local names include:
Angel's eyes in Devon
Billy-bright-eye in Ireland
Blue-eyed beauty in Staffordshire
Blue eyes in Wiltshire
Bobbies' eyes in Hampshire
CAT'S EYES in Staffordshire
Deil's flower in Scotland
Eye-of-Christ in Wales
God's eye in Devon
Mother-breaks-her-heart in Cornwall
ROBIN-RUN-THE-HEDGE in Cheshire.

> Blind-flower – *Veronica chamaedrys*; Durham (Hartlepool) where it is said by children that if you look steadily at it for an hour you will become blind.[1]

> In the Munster province of Ireland germander speedwell was known as jump-up-and-kiss-me: 'the name and its meaning seem to be so well known that if a piece of the flower be worn, the wearer is greeted with shrieks of laughter'.[2]

> In Norfolk, where germander speedwell was known as sore-eyes, 'grandmother infused the little blue flowers to make a soothing EYE-bath'.[3]

Throughout Ireland germander speedwell was used to treat JAUNDICE:

> Speed-well-blue – a herb which grows locally is believed to be a cure for jaundice.[4]

> Jaundice: get a weed which is called BLUE BELL (germander speedwell). I never heard it called anything else. You boil this weed in milk and then you give it to the person who is affected to drink.[5]

A tea made from germander speedwell was 'esteemed a valuable remedy of INDIGESTION and pains in the STOMACH' in Guernsey.[5]

GLASSWORT – see MARSH SAMPHIRE.

GLASTONBURY THORN – see HOLY THORN.

GLUE – made from BLUEBELL.

GOAT'S BEARD (*Tragopogon pratensis*)

Annual or perennial, yellow-flowered herb, common in rough grassy places in England and scattered in similar habitats elsewhere.

Most of goat's beard's local names, including the widespread go-to-bed-at-noon and Jack-go-to-bed-at-noon, refer to the fact that its flowers open early in the day and close by midday:

Jack-go-to-bed in Cornwall
JOHN-GO-TO-BED-AT-NOON in Gloucestershire, Somerset and Wiltshire
JOHNNY-GO-TO-BED-AT-NOON in Cornwall; 'the flowers open at about 4 a.m. and close at 12; fertilised flowers close earlier'
John-that-goes-to-bed-at-noon in Northamptonshire
Nap-at-noon in the Midlands, Lancashire and Shropshire
ONE-O'CLOCK in Devon and Somerset
SHEPHERD'S CLOCK in Yorkshire
SLEEPY HEAD, and TWELVE-O'CLOCK in Somerset.

Other names include JACK-BY-THE-HEDGE in East Sussex and Joseph's flower, 'probably a reminiscence of Joseph as an old man with a long beard' in Somerset and Sussex.

Closely related to the cultivated salsify (*Tragopogon porrifolius*), goat's beard was formerly used as a food. In 1660 it was recorded that:

> A very pleasant dish is made from the roots of this plant cooked in boiling water until they are tender and served with butter like parsnips, for they have a delicate flavour and yield a juice more health giving than that of parsnip or carrot. The roots are eaten raw in salads.[1]

GOAT WILLOW (*Salix caprea*), also known as pussy willow and sallow

Deciduous small tree, producing bright yellow male, and silvery grey-green female, catkins on separate trees; common in dampish ground.

Local names given to the tree include the widespread sally, saugh, and saugh-tree, and:

Black sally in Shropshire and Wiltshire
Gosling-tree in Wiltshire
Sally-withy in Herefordshire, Shropshire and Wiltshire
Sauf in Yorkshire; 'and most probably other species of willow'
Seal, or seel, in Cumberland
Willow-gull in Kent
Withey in Devon.

Names given to its catkins, especially the male ones, include the widespread goslins, and:

Cat's heads in Cambridgeshire
Cats-and-dogs, and cat's paws, in Wiltshire
Fluffy buttons in Somerset
Goose-chicken in Devon
GOSLINGS in Wiltshire
Pussy-cat in Buckinghamshire and Hampshire
PUSSY-CAT'S TAILS in Devon
Tassel-rag in Cheshire.

> The strongest condemnation of all lights on willow catkins. The soft round yellowish blossoms are considered to resemble young goslings, and are accordingly called in various [Shropshire] localities 'goosy goslins', 'gis an' gullies' or 'geese and gullies'. Whatever the name, however, the ban on the blossom is the same. No vegetable goslings may be brought into the house, for if they be, no feathered goslings will be hatched.[1]

Probably more widespread was a ban on flowering goat willow – pussy willow – indoors before PALM SUNDAY or EASTER.

> About 70 years ago in Hampshire it was thought UNLUCKY to bring palm – flowering willow – indoors before Palm Sunday.[2]

> My mother-in-law would not allow pussy willow into the house

> until after Easter. I don't know the reason for this, and she didn't seem to know either![3]

Goat willow and similar species of willow were much used as 'palm' on Palm Sunday throughout much of Europe. Throughout England it was widely known as 'palm';[4] related names included palmer in Dorset, palm-tree in Oxfordshire, Wiltshire and Yorkshire, and palm-willow in Leicestershire. On Friday 11 April 1924 the *Daily Mail* published a photograph of two young women with arms full of flowering willow, captioned 'church helpers returning after gathering palm at Ingatestone, Essex'.

> According to my mother . . . (b.1918) of Whitstable, Kent: on Palm Sunday children used to walk in procession carrying 'Palm', that is pussy willow in flower.[5]

> I was born and brought up in Surrey and from 1958 to 1964 attended Grayswood Church of England Primary School. Sallow, which we referred to as pussy willow, was collected and taken into the house and church for Palm Sunday. This was, I believe a common custom locally, and I was surprised when I saw palm leaf crosses used in other churches.[6]

> In the Durham coalfields the most important spring flower was pussy willow, or in other words the flower of sallow and its relatives. These were known as pussy willow and brought into the house on Palm Sunday, and were as important to Easter as HOLLY was to CHRISTMAS. No Easter would be complete without it.[7]

On Palm Sunday 1991 the sanctuary of Southwark Cathedral, London, was decorated with LAUREL, YEW, dried palm leaves and sallow,[8] but it appears that sallow is rarely used in English churches, which prefer to use imported dried palm crosses. However, until recently fresh sallow was utilised in churches used by congregations from other parts of Europe.

Early in the 1980s, bunches of BOX, sallow and DAFFODILS were used on Palm Sunday at the Polish Catholic Church of Christ the King in Balham, South London.[9] By 2016 this practice had been abandoned, with members of the congregation bringing a variety of 'palms', although in the church forecourt a woman sold small bunches of Alexandrian laurel, a DAFFODIL bud, and a single stem of dry willow.[10]

On 14 April 1984 – the Eve of Palm Sunday for Orthodox Christians – at a service held at the Cathedral of the Dormition of the Mother of God, in South Kensington, London, a large pile, and two jars, of fresh flowering sallow were blessed by the presiding bishop before a bunch, together with a lighted wax taper, was given to each member of the congregation. After the service the 'palm' was taken home and placed near an icon.[11] At the Cathedral's Palm Sunday liturgy in 2016, there was no evidence of sallow, and although a few members of the congregation brought small bunches of sallow twigs, these had dry, dead catkins, and had presumably been purchased, rather than gathered by those who brought them.[12]

On ST PATRICK'S DAY in County Cork:

> The children brought in a couple of bits of sally rod and the mother or grandmother stuck it into the fire and drew it out when charred. Then the sign of the Cross was made on everybody's arm with the charred end.[13]

It is difficult to ascertain if mentions of 'willow' refer to goat willow or to other species, but as goat willow when bearing its male catkins is the most conspicuous common willow, it can be assumed that many of these mentions refer to it. Around Brentwood in Essex:

> Willow and ROWAN were definitely 'good' trees and kept away WITCHES if planted near the door. Willow twigs hung on a door kept away marsh witches, as did HAZEL twigs.[14]

More enigmatically:

> Willow is, in my native East Yorkshire, the witches' tree. I stopped a puzzled Transylvanian Saxon prisoner of war from destroying one.[15]

In Herefordshire:

> The willow brings luck if brought into the house on MAY DAY, and is potent against the EVIL EYE, especially if given by a friend. It is also believed that any young animal or child struck by a willow rod, usually called a 'withy stick' or 'sally twig', will cease to grow afterwards. A woman at Pembridge said, 'I've never hit nothing with a sally twig, nor shouldn't like to either.'[16]

Despite this:

> Sally rods to beat bold children. Sally is willow, and the flexible

shoots/branches were used as canes/whips to discipline children in days not too long past.[17]

See also WILLOW.

GOITRE – GOOSEGRASS used to treat.

GOLDENROD (*Solidago canadensis* and *S.* spp.)

Robust perennial herbs, much grown as ornamentals for their bright yellow flowers, and frequently naturalised on roadsides and waste ground.

Local names include the widespread AARON'S ROD, and FAREWELL SUMMER in Somerset, FAREWELL-TO-SUMMER in Cornwall and Somerset, GOLDEN DUST, golden glow, golden wings, and WOUNDWORT ('formerly greatly esteemed as a . . . wound-herb') in Somerset.

> I took some goldenrod to a friend; she went mad, she wouldn't have it in the house at any price; she thought it UNLUCKY to have it in the garden even.[1]

There appears to be a good reason for this belief:

> Goldenrod is not a good plant to use in flower arrangements . . . Some people are highly allergic to it. Of course one never knows who is, but just to be safe it should be kept away from larger gatherings.[2]

However, in East Yorkshire:

> Cure for infected SORE places – pull goldenrod when in bloom, hang dry, boil in pan of water for 30 minutes, immerse limb in boil [*sic.*] of the solution.[3]

GOOD FRIDAY – CALVARY CLOVER, GARLIC, PARSLEY and POTATOES planted, FIG Sue drunk, rain means good HOP harvest; ORANGES rolled; SAFFRON buns eaten.

GOOD KING HENRY (*Chenopodium bonus-henricus*)

Perennial herb, an ancient introduction, found in nutrient-rich ground near farm buildings and on roadsides; native to central and eastern Europe.

Local names include the widespread FAT HEN, and:

ALL-GOOD in Hampshire; 'on account of its excellent qualities as a remedy and an esculent'
BACON-WEED in Dorset
Flowery docken in Berwickshire; 'probably floury is meant, from the mealiness of its leaves'
Good King Harry in Cambridgeshire and Yorkshire
GOOD NEIGHBOURHOOD in Gloucestershire, Oxfordshire and Wiltshire
Johnny O'Neale in Shropshire
Marcaram in Yorkshire
Margery in Lincolnshire
Markerry in Cumberland and Westmorland, where 'used as spinach'
Mercury docken in Kincardineshire
MIDDEN MYLIES in Selkirkshire
Mutton-dock in Dorset
ROMAN PLANT, the, in Lancashire
Shoemaker's heels in Radnorshire and Shropshire
Smeardock in Morayshire
Smear docken in Scotland, 'i.e. fat or grease dock; cf. German *Schmerbel*, *Schmeerwurz*, etc.'
Smeardokke in northern Scotland
Smiddy-leaves in Berwickshire; 'one of its favourite habitats . . . the vicinity of blacksmith's workshop'
WILD MERCURY in Berwickshire
WILD SPINACH in Hampshire, Isle of Wight and Somerset
Wild spinage in Berwickshire.

Although it does not seem to be highly thought of by present-day foragers, good king henry was formerly valued as a vegetable. Thus in 1660 John Ray wrote:

> The younger shoots of this plant put in boiling water and cooked for a quarter of an hour, then eaten with butter and salt, make a pleasant and health giving dish not unlike ordinary asparagus.[1]

In the twentieth century:

> Mercury – Lincolnshire spinach – or in some parts of the country they call it good king henry. Mercury is pronounced marcury in Lincolnshire and used to be quite common in gardens. My family have always grown and eaten it. I have quite a large bed of it in

> my garden; being perennial it needs little attention and no matter what the weather it comes up.
>
> . . . Mercury will fill the gap when we've finished the broccoli, kale, etc., until the peas, beans, etc. are ready. We eat it like boiled spinach, and then I like it hot or cold – also I love the flower-heads and sometimes strip the leaves off and eat it as 'poorman's asparagus'. My grandmother used to tell me because it was so deep rooted it was full of iron and minerals. For many years I've thought it was probably responsible for my good resistance to colds and infection.[2]
>
> An Isleham [Cambridgeshire] man declared in the 1930s, as a boy, he was cured of 'scurvy' or weeping ECZEMA by an old woman . . . 'She told me to ground up markery in water and drink it. That cured the scurvy and cured my sweaty FEET too.' Markery, she said, was what 'some folk call fat goose or good king henry'.[3]

In the Forest of Dean:

> My maternal grandparents (b.1856 and 1858): good king henry – infused fresh leaves and drunk – for BLADDER [troubles].[4]

GOOSEBERRY (*Ribes uva-crispa*)

Spiny shrub (Fig. 18), grown in gardens since the thirteenth century, widely naturalised in hedges and woodland.

> BABIES are found under gooseberry bushes. I was told this by several of my aunts when I was a small child (I was born in 1915), but I don't imagine that they seriously believed it themselves! They said it to avoid having to explain to me where I really came from and how.[1]
>
> In July 1975 a colleague who complained that his gooseberry bush had produced only three fruits was humorously reminded that he could hardly expect the bush to be more productive in view of the fact he had just had an addition to his family.[2]

Towards the end of 1983 expensive hand-made dolls, known as 'CABBAGE Patch Kids' were imported from the United States, where apparently children were told that babies come from the cabbage patch. The Kids had to be 'adopted' rather than bought. In December 1983

Pearl's, a shop selling miscellaneous cheap goods, in Mitcham Lane, Streatham, south London, displayed mass-produced rag-dolls in its windows, labelled 'Gooseberry bush doll, £2.50'.[3]

As the first fruit of the summer, gooseberries were often traditionally eaten at WHITSUN or at village feasts and revels. Thus, on 4 June 1830, the *Lichfield Mercury* reported:

> Lamb and gooseberry, it is well-known, are customary dishes at Whitsuntide. In this city, the usage seems to be religiously kept up. The number of lambs killed here, on Friday, was 252; and many besides were sold by country butchers who attend the market.

At Stoke-sub-Hamdon, Somerset, villagers climbed the local hill and enjoyed a Gooseberry Feast, described as a 'curious old custom' in 1875,[4] but apparently not surviving until the end of the nineteenth century.[5] At Drewsteignton, in Devon, gooseberry pasties and cream are traditionally eaten at 'Teignton Fair, held at Trinity Tide – the first Sunday after Whit Sunday'.[6] Helston in Cornwall held a Gooseberry Fair on the third Monday in July,[7] and a similar event was held at Hinton St George, Somerset, on the first Sunday after Old Midsummer's Day (5 July).[8]

During the eighteenth century the growing and exhibiting of gooseberries became a passion in the Midlands, Cheshire and Lancashire, comparable with the enthusiasm for growing LEEKS and PUMPKINS in the twentieth century. In the 1740s gooseberry clubs were formed in the Manchester area, and about a century later there were 722 varieties of gooseberry available and 171 gooseberry shows. It seems that this enthusiasm was particularly prevalent among cottage-dwelling handloom weavers. With the development of power-driven looms these weavers moved from their cottages into towns, where space for growing gooseberries was restricted. Few gooseberry shows survived the First World War. Today ten shows exist, one at Egton Bridge in Yorkshire, and about seven in Cheshire. The Egton Bridge Old Gooseberry Society, said to have been formed in 1800, had approximately 120 members in the 1980s and holds its annual show on the first Tuesday in August.[9] At present the show is curiously low key; according to the Society's president: 'Perhaps in this world of speed and high technology, the old fashioned ways of the Show have special appeal.'[10]

In Dorset:

> My grandmother used to say: 'May the skin of a gooseberry cover all your enemies'.[11]

In Ireland gooseberry thorns were used in a variety of ways to treat STYES:

> Stye: pick ten gooseberry thorns, point nine at the eye and throw away the tenth, bury the other nine and when the thorns are withered the stye will be gone.[12]
>
> If you make the sign of the cross with a gooseberry thorn on a stye for nine mornings and bury the thorn the stye will disappear.[13]
>
> Common treatment for a stye on one's eye that I saw employed by old folk in the 1920s was to prick the stye every morning with a thorn from a gooseberry bush. This vied with another commonly used cure. Have a widow touch the stye with her gold wedding ring.[14]

GOOSEGRASS (*Galium aparine*), also known as CLEAVERS.

Scrambling annual (Fig. 19), common on cultivated land, hedge-banks and disturbed soil throughout the British Isles.

Goosegrass scrambles upwards through surrounding vegetation by means of tiny recurved – Velcro-like – hooks on its stems, leaves and fruits, a feature which is responsible for many of its local names. These include the widespread beggar/s-lice, and:

Catch-grass in Cheshire
Catch-rogue in Scotland, 'growing in hedges and adhering to the clothes of those who attempt to break through them'
Clinging sweethearts in Wiltshire
Cling-rascal, and clitch-button, in Devon
Gentleman's tormentors in Suffolk
Grip-grass in Northumberland
SCRATCH-GRASS in Hertfordshire
Scratch-weed in Cambridgeshire and Northamptonshire
Snares in Yorkshire
Stick-a-back in Cheshire and Lancashire; 'it is a common amusement with children to put big pieces of the plant on each other's backs where it clings to the clothes'
Stick-buttons, and stick-donkey, in Somerset

Stickers in Scotland
Stickle-back in Cheshire and Devon
STICKY BACKS in Cumberland, Dublin and Somerset
Sticky balls in Somerset
Sticky Billy in Northern Ireland
STICKY BOBS in Dorset
Sticky buttons in Hampshire
Sticky Dick in Lancashire
Sticky grass in Scotland
Sticky Jim in Lincolnshire
Tether-grass in Northumberland.

However, the most widely used of these names at present is sticky willie (or willies), a name which is missing from nineteenth-century lists of local names and Geoffrey Grigson's *Englishman's Flora* (1955), and which appears to have first been used in Scotland, in Aberdeen[1] and Burghead, Morayshire,[2] in the 1920s, later spreading southwards.

Variations of cleavers include: clavers in Dorset, clider/s in southern England, the widespread CLIVER, and tivers in Buckinghamshire. Other names include the widespread clite/s, and:

Airess in Yorkshire
Airif in Lincolnshire
AIRIFT in Devon
Airup, and aress, in Yorkshire
BEGGAR-WEED in Northamptonshire and Nottinghamshire
Biddy-biddy in Somerset (also biddy-bids in New Zealand)
Bleedy tongues in Morayshire
Blood-tongue in Cheshire, Northumberland and Scotland
Bluid-tongue in Scotland
BOBBY BUTTONS in Somerset
BUR in Buckinghamshire and Cheshire
Bur-head in Northamptonshire
Bur-weed in Buckinghamshire, Hertfordshire and Northamptonshire
CACKEY MONKEYS in Flintshire
Claden, and claiton, in Dorset
Clapped pouch in Somerset
Claver-grass in Cumberland
Clay in Somerset
Clayton, cleden, and cleeiton, in Dorset
Cleggers in Yorkshire
Cletheren in Somerset
Clibby-clider in Devon
Climb, and clime, in Somerset
Clits in Devon
Cly in Devon, Dorset and Somerset
Clyden in Devon
Clyder/s, in Devon, Dorset, Somerset and Wiltshire; 'when collected as a herb near Harberton in the First World War, this plant was invariably called clyder; probably a variant of . . . cleavers'
Clyther in Devon, Dorset, Somerset and Wiltshire
Glider in Cornwall
Goose-cleaver in Lanarkshire
Goose-shear in Somerset

GOOSE-TONGUE in Cheshire and Somerset
Goose-weed, and goosey-gogs, in Somerset
Gosling-grass in Northamptonshire and Oxfordshire; 'given as food to young goslings'
Gosling-scrotch in Cambridgeshire, Essex and Norfolk
Gull-grass in Gloucestershire and Herefordshire; 'i.e. unfeathered gosling grass'
Haireve in Gloucestershire
Hairiff in Leicestershire
Hairough, and hairup, 'in the North'
Harris's bullets in Nottinghamshire
Hayrough 'in the North'
Hedgehogs in Somerset
Herriff in Leicestershire and Warwickshire
Hug-me-close, and huggy-me-close, in Dorset
Jack-at-the-hedge in Co. Donegal
JACK-IN-THE-HEDGE in Co. Wicklow
Jack-run-the-dyke in Northumberland
Jump-right-over-the-hedge in Lancashire
KISS-ME-QUICK in Dorset, Somerset and Wiltshire
Lizzie-in-the-hedge in Scotland
Lizzy-run-the-hedge in Berwickshire; 'when it climbs up and amidst our quick fences to the height, perhaps, of twenty feet'
Loosy-tramps in Scotland
Love, and lover's kisses, in Somerset
Lover's knots in Wiltshire
Mothers-in-law in Northamptonshire
MUTTON-CHOPS in Dorset
Pig-tail in Northamptonshire
Pimple-grass in Dorset
Rabbie-rinnie-hedge, and robin-round-the-hedge, in Ayrshire
Robin-run-by-t'-dyke, and robin-run-up-the-dyke, in Cumberland
ROBIN-RUN-IN-THE-HEDGE in Scotland and Ireland
Robin-run-the-dyke in the Lake District and Northumberland
SCURVY-GRASS in Cheshire and Yorkshire
SNAKE-WEED in Somerset
SOLDIER'S BUTTONS in Cumberland
Tongue-bleed in Suffolk and Yorkshire
Tongue-bleeder in Leicestershire
Tongue-bluiders in Northumberland
Traveller's comfort, and TRAVELLER'S EASE in Wiltshire
Turkey's food in Somerset
Who-stole-the-donkey in Somerset
Willie-run-hedge in Stirlingshire
Willy-run-the-hedge in Ireland
Witherspail in Roxburghshire.

> The sticky burrs of goosegrass, which stick to one's clothes . . . are known locally as SWEETHEARTS, and are considered certain proof that one has been with one's sweetheart.[3]

> During my childhood in Sussex/Kent in the 1920s and 1930s: if goosegrass was thrown at a girl's back and stuck there without her being aware of it, she had a sweetheart. If she took it off and dropped it, it would form the initial of her sweetheart-to-be.[4]

> Cleavers . . . clumps were made into rough balls and thrown at clothing. The number of stickers left on one's clothing indicated the number of suitors one could expect.[5]

A Guernsey name for goosegrass was *la coue* (the tail):

> This singular name originates in the common amusement of country children on All Fools' Day. They slyly stick wisps of this clinging plant on each other's backs, and then start to cry, *La coue! La coue!*[6]

As its name suggests, goosegrass was widely fed to poultry:

> [Goosegrass], chopped small, is given to GOSLINGS in this island [Wight].[7]

> We used to nibble shoots of goosegrass, and still feed the whole plant to our ducks.[8]

> During my childhood in west Dorset in the 1950s, young goosegrass was considered to be excellent food for young poultry, particularly young TURKEYS. It was gathered from the hedgerows, and chopped up using scissors.[9]

> When we used to rear turkeys under a hen we scoured the lanes for cleavers (sweethearts) as a source of early greenstuff . . . game-keepers also use it for rearing pheasants.[10]

Records of goosegrass being used in folk medicine include:

> The whole plant has been used in the spring to make a tonic drink; the fresh plant has been used in the treatment of a variety of SKIN conditions.[11]

> A tea made from the stalks and leaves will rid obstructions in the THROAT, gargled then swallowed, even goitre or CANCER.[12]

GORSE (*Ulex europaeus*), also known as FURZE or WHIN.

Spiny shrub (Fig. 20)|, producing abundant yellow, coconut-scented flowers throughout the winter months, widespread, especially on sandy heathlands.

Local names include:

Alechenagh in Co. Donegal
Brusse in Devon
Bunch-keys in Somerset
Cockles, given to the seeds, in Somerset
Crannick in Dorset
Crannock in Somerset
FINGERS-AND-THUMBS in Somerset and Wiltshire
Firr in Dublin; 'furze seems to be regarded as plural: 'That's a firr bush, when there are a lot of them together we call them furze'
French furze in Cornwall and Devon; 'to distinguish it from dwarf furze [*Ulex minor*]'
French fuzz in Cornwall, Devon and Ireland
Frez in Northamptonshire
Fur in Lincolnshire
Furra in Norfolk
Furzen in Dorset
Fuzz in Cornwall and Somerset
Gost in Herefordshire and Shropshire
Hawth in Sussex
HONEY-BOTTLE in Wiltshire
Hoth in Sussex
LING in Derbyshire and northern England
NEEDLES-AND-PINS, and PINS-AND-NEEDLES, in Somerset
Ruffet in Dorset
THUMBS-AND-FINGERS in Somerset
Vuss in the West Country
Vuz in Devon
Vuzzen in Dorset and Somerset
Whuns in Co. Antrim and Co. Down.

The flowers of gorse, like those of BROOM, are occasionally considered to be UNLUCKY when taken indoors.

> [Guernsey] May [HAWTHORN], of course, should never be brought indoors – also gorse and LILAC – both of these were considered to be UNLUCKY. As a child I was never allowed to bring any of these indoors.[1]

> [Jersey] Gorse, IVY and ARUM LILIES – none of these were ever allowed in the house.[2]

> More than 60 years ago I lived with my parents in the town of Listowel [Co. Kerry] . . . a couple of miles down the river from the town there were acres of shrub-land completely covered by . . .

> furze bushes . . . one early spring I decided to bring a bunch of them to my mother who, being a countrywoman, was very fond of wildflowers, but . . . being appreciative of the thought [she] told me to immediately take them out of the house without laying them down and never bring them in again. She told me they were extremely unlucky, and later in life I found [this] to be the view all over the country.[3]

> Some years ago, when in Fifeshire, I plucked a very fine bloom [of gorse] in a bleak season when no other wild flowers were to be seen. Meeting an elderly lady, she exclaimed on its beauty. I, thinking to please her, said, 'You can have it,' at the same time handing it to her. 'Oh,' she said 'Why did you do that? It is very unlucky to give anyone whin blossom; we shall be sure to quarrel.' I laughed and said, 'I never heard of that freit. Perhaps when one does it in ignorance it won't work.' A few days later I had the ill luck to offend the said lady. She was very angry, and gave me her opinion of me in no measured terms, ending in saying, 'That's your present of whin blossom.'[4]

> In October 1979 I was told by a London schoolteacher, then in her 20s, that in the first school she attended as a child, in Hampshire, children were afraid to touch gorse flowers because it was believed that dragons lived, or were born, in them.[5]

Despite such beliefs, widespread and much-repeated sayings associate gorse flowers with kissing and love-making.

> Kissing's out of fashion, when the whin is out of blossom.[6]

> [Yorkshire] for kissing there's room, when the gorse is in bloom.[7]

> Have you heard that you make love when the gorse is in flower? An old rambler friend told me that. But gorse is always in flower.[8]

Statements such as 'gorse may be found in bloom every month of the year'[9] frequently occur in books on plant folklore, but, in fact, the common gorse is mainly winter and spring flowering, so its flowers are difficult to find during the summer months. However, there are two other species of gorse native to the British Isles – western gorse (*Ulex gallii*) and dwarf gorse (*U. minor*) – both of which are much more sporadic in their distribution and 'mainly summer' flowering.[10]

The presence of gorse, like that of THISTLES, is said to indicate rich soil.

> [Co. Kerry] *An-t-or fe'n ainteann, an t'airgead fe 'n luachair agus an gorta fe'nbhfraoch* (Gold under furze, silver under RUSHES, and famine under heath).[11]

> A saying around here: 'Where there's BRACKEN there's gold; where there's gorse there's silver; where there's HEATHER there's poverty'.[12]

Gorse flowers were widely used as a DYE, particularly for EASTER EGGS.

> Whins . . . have a yellow blossom and woollens can be dyed by these blossoms. This is the way it is done. There is some water boiled and when it is boiling the whin blossoms are put in and they let it boil for another while. Then the woollen was put in and it was let take the dye. When it was taken out it was a beautiful yellow colour.[13]

> On Easter Sunday we collected all the weans [children] of the village and took them up to a field with a big hill on it. We made our tea after lighting a fire and got all the eggs in a big saucepan to boil them hard. Unknown to the kids we put whin blossom in the water while they were boiling and they came out a lovely yellow colour. The kids thought it was magic. Then up to the top of the hill to roll the eggs down.[14]

> [Burghead, Morayshire, 1920s and 1930s] For dyeing Easter eggs our mothers might use the outer skins of ONIONS, or whin flowers.[15]

> [North London] We used to pick gorse flowers to dye our Easter eggs.[16]

Although gorse is often considered to be little more than an invasive pest, it was formerly valued as a fuel, a food for livestock, and providing cover for game.[17]

> The Irish introduced furze . . . into Ireland to make stock-proof hedges, which it does.[18]

> Gorse was used in Westmorland for sweeping chimneys. Take a

convenient sized bush, tie a brick to the stem, climb up to the chimney (all our old cottages have ridge steps) and drop it down. Makes a glorious mess.[19]

Fuzz moots = roots of furze; we used to pull them up and take them home to burn in the kitchen range. Mother preferred them to coal as they burnt with a lovely hot clean flame.[20]

[Cawsand, Cornwall, *c.*1930] Gorse was most prolific on the cliff tops, but was always named fuzz – not furze. The dry sticks of fuzz burnt with a bright, clear, fierce flame and were prized for heating purposes and were called crinnicks or crinnix, at least that's what it sounded like. Some of the old, clustered together, cottages had a smallish deep hole running back in the wall in the kitchen usually, about a foot across and a couple of feet deep, with a cover. These were called fuzz ovens, where you burnt your crinnicks and then baked your pasty or whatever.

Sometimes a bunch of fuzz or perhaps HOLLY tied to a rope, a stone from the beach at the other end, to be dropped down the 'chimley' – we didn't have CHIMNEYS – and then hauled thro' for chimley cleaning.

Straight sticks of fuzz bush about 10–12 inches long, pointed one end were called skivers and used to hold bait in the mouths of the old wicker-work CRAB-pots, rather like kebab sticks. They were very resistant to salt water and had no pith.[21]

Bean jar [a Guernsey delicacy] was baked in the furze oven. This was a brick-lined oven in the kitchen. Furze, *Ulex europaeus*, was lighted in the oven to heat the bricks. When it was hot the food was put in.[22]

The St Fagans National History Museum has a reconstructed gorse mill from Dolwen in Clywd:

Gorse was a vital part of their [horses'] diet. It was specially grown on a large scale but had to be bruised or crushed to make it fit to eat. Small farmers bruised their gorse by hand, though water-driven mills . . . [such as the one exhibited] were common about 1800, using heavy metal spikes to crush the gorse. By 1850, however, most farmers were using lighter and cheaper hand-operated machines instead.[23]

In Ireland gorse was used to treat JAUNDICE and as a VERMICIDE.

> Furze blossoms are boiled [and] the juice used for yellow jaundice.[24]

> Whins ... are cut up and pounded. Then they are given to horses to take worms out of them.[25]

> A handful of whin blossom boiled in milk was strained and given to a child suspected of having worms.[26]

For a legend concerning ST PATRICK cursing gorse, see RUSH.

GOSLINGS – bringing GOAT WILLOW catkins indoors kill; GOOSEGRASS fed to.

GOUT – CHERRY prevents; treated using ALDER, GROUND ELDER, HORSERADISH and WILLOW.

GRAPE-HYACINTH (*Muscari* spp.)

Bulbous, blue-flowered perennials: *Muscari neglectum* native in East Anglia, and *M. armeniacum* (garden grape-hyacinth), native to south-east Europe and south-west Asia, cultivated as an ornamental in British gardens since the 1870s, and within twenty years recorded in the wild.

Two local names – CHUCKY-PIG and starch-flowers – have been recorded, both from Somerset.

> BLUEBELLS and grape-hyacinths – happiness in natural surroundings, dreadful depression in the house. When I was a young nurse, I tried to 'crack' this piece of lore, but in fact the flowers must give off some 'essence' which does make people suddenly morose and sullen . . . When I removed the flowers the atmosphere immediately lightened – and I've tested this theory for about 40 years. I've stopped now though and bluebells and grape-hyacinths stay outdoors.[1]

> My children used to enjoy squeezing grape-hyacinth flowers for the sound, and still call them squidgies.[2]

GRASS (Poaceae, formerly Gramineae)

It was widely thought that abundant grass at CHRISTMAS, foretold a large number of DEATHS in the year ahead.

[Dorset] Ev a chich'ard da look lik' a pastur' vel 'pon C'ursmas Day'll look lik' a plow'd veel avoa Medzumma Day.[1]

A green Christmas, a fat churchyard. Green meaning mild weather, absence of frost and snow. Fat churchyard meaning a full churchyard resulting from many deaths due to a mild Christmas. People preferred hard weather as frost killed germs, etc.[2]

Churchyard grass could be used to foretell the future:

[Cumbria] On ST MARK'S EVE (24 April) if a young girl picked three tufts of grass from a churchyard and put it under her pillow she would dream of her future, but she must say:

Let me know my fate whether weal or woe,
Whether my rank's to be high or low,
Whether to live single or be a bride
And the destiny my star doth provide.[3]

Children regarded grass as a readily available plaything.

As school kids (1960s/70s in a village of the Kent/East Sussex borders) we used to run our fingers up the stem of grass in full seed to pull off the grains in a little bunch, then chant 'Here's a pretty bunch of flowers . . . April showers!' as we threw the seeds all over a friend's head.[4]

I grew up near Easington Colliery in Co. Durham . . . [1970s] We would scatter grass flowers over each other, by holding the bottom of the stem and quickly moving the fingers of the other hand to the top of the stem, stripping the flowers and scattering them over your unsuspecting friend. This was accompanied by the cry of 'Shabby wedding!', as it was to imitate the throwing of rice or confetti.[5]

More widely recorded is the use of grass blades to produce 'music':

We played music on blades of grass, picking a sturdy stem the length of your thumb, then putting your other thumb up to it with the blade of grass acting like a reed between the two thumbs. By altering the angle of your thumb slightly as you blow you can change the sound a bit, or just blow hard to make a loud noise.[6]

[Cumbria, 1950s] Grasses could be held between the thumbs and blown thru' to make an orchestra of sounds if a few of us got together.[7]

Country people sucked or chewed young grass stalks for refreshment.

> [Sedgemoor, Somerset] When we were down on the moors (now known as the Levels, as our moors are lowland, liable to flooding, not high ground . . .), working at haymaking, etc., if we ran out of something to drink and the ditches were nearly empty or full of waterweeds, we would gently pull out the middle part of a big grass shoot and suck and nibble the lovely pale yellow part which had been hidden down the bottom of the plant.[8]

GRAZES – treated using COMFREY and MALLOW.

GREAT BURNET (*Sanguisorba officinalis*)

Perennial herb, locally common in damp, unimproved grassland.

Local names include: DRUMSTICKS in Gloucestershire and Somerset, maidenheads and red heads in Yorkshire, parasols in Wiltshire, and red knobs in Nottinghamshire.

> Called 'Hot weed' in parts of Brecknock and Radnor: said to cause hay to heat up if present in large amounts.[1]

Cf. RIBWORT PLANTAIN.

GREATER CELANDINE (*Chelidonium majus*)

Yellow-flowered annual herb, rarely found away from human habitation, and probably introduced for medicinal use.

Local names include:

DEVIL'S MILK in Yorkshire; 'from the acrid quality of the milky juice'
JACOB'S LADDER in Somerset
ST JOHN'S WORT in Devon
Salandine, 'pronounced salladin', in Cheshire, Cumberland and Yorkshire
Sallanders in Yorkshire
Sollendine in Co. Donegal, where 'in request for stuping sore EYES'
Swallow-wort in Cornwall and Somerset
Yellow spit in Hampshire; 'from the yellow juice which exudes from the broken stem'

The sap of greater celandine, which contains several alkaloids including chelidonin and chelerythrin,[1] was widely valued as a cure for WARTS, leading to names such as:

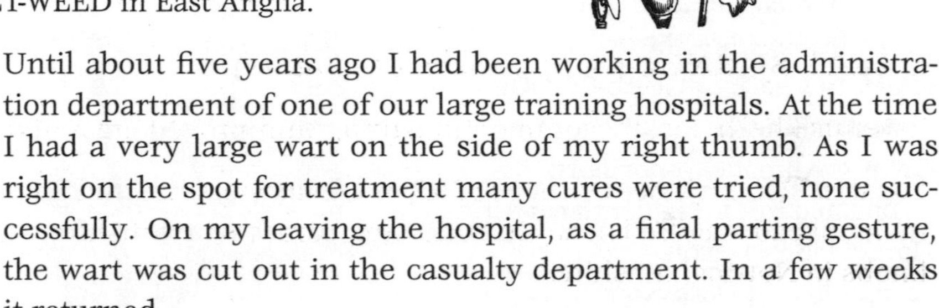

Kill-wart in North Devon
Wart-curer in Somerset
Wart-flower in Devon
Wart-plant in Somerset
WART-WEED in Suffolk
WART-WORT in Gloucestershire, Somerset and Wiltshire; 'cf. *Warzenkraut* in German'
WRET-WEED in East Anglia.

> Until about five years ago I had been working in the administration department of one of our large training hospitals. At the time I had a very large wart on the side of my right thumb. As I was right on the spot for treatment many cures were tried, none successfully. On my leaving the hospital, as a final parting gesture, the wart was cut out in the casualty department. In a few weeks it returned.
>
> I had the good fortune some weeks later to meet a man from Worcestershire, who noticed my wart and told me he could cure it. The cure was to rub the sap from the stem of the plant greater celandine on to the wart for five nights. In order that I might do that he sent me some of this plant on his return to Worcestershire. In three weeks there was no sign of the wart left, and there has been no sign of any recurrence.[2]

> In Ashford, Kent, greater celandine was much used for the treatment of warts. A stem was broken and the orange sap rubbed on the wart.[3]

Other uses included the treatment of eye complaints and, less widely, JAUNDICE.

> [Cornwall] in common with *Ranunculus repens* the greater celandine enjoys a wide reputation in making ointment for sore eyes, and is frequently called 'KENNING HERB', from 'Kenning' or 'Kennel', local names for an ulcer on the eye.[4]

> Greater celandine: the juice of the leaves passed over the eyelids for eye problems or failing sight.[5]

> An infusion of leaves and flowers of greater celandine bathed onto the eyes daily is said to reduce or even cure the development of cataracts.[6]

> The expressed juice or a decoction of the plant is in vogue with the country people of the island [of Wight] as a remedy for infantile jaundice.[7]

GREATER PLANTAIN (*Plantago major*)

Perennial herb growing on open habitats beside paths and roads.

Local names include:

Broad-leaf in Cheshire
CARL-DODDIE or CURL-DODDY in Forfarshire
Great waybreade in northern England in the sixteenth century
Hard-head in Devon
Lamb's foot in Derbyshire.
Plant ('a contraction of plantain') and ponies' tails, in Devon
RATS' TAILS in Cumberland, Norfolk and Ulster
Rat-tail dock in Lancashire
Ripple-girse in Scotland; 'a broad-leaved herb which labourers put on CUTS'
Slanlas in Co. Antrim and Co. Down
Wabret in the Scottish borders
Warba, warba-blades, and warba-leaves, in Scotland
Wayfron in Berwickshire
Wayside-bread in Wiltshire
Whybrow, or wibrow, in Cheshire.

Like BITING STONECROP and WATER FIGWORT, greater plantain was used by children to make 'fiddles'.

> [Heston, Middlesex, 1930s] a leaf [of greater plantain] was picked and the stalk partly severed to expose the tough veins, these made a violin (my grandchildren make guitars!).[1]

> Broad-leaved plantain: after making a cut across the stem with the thumb nail, the lower part of the stem could be gently pulled down leaving the veins exposed – a 'banjo'.[2]

> [North London] When I was a child in the 1950s my aunt (born 1911) taught me how to make a 'violin' out of a [greater] plantain with leaf stem stretched to make the strings, and the flower stalk as the bow. I taught this to my children (youngest aged 21 remembers this), so maybe they will teach it to their children.[3]

Names relating to such practices include angels' harps in Derby, 'when you pull the leaves apart you get the fibres showing between'.

Other children's pastimes included:

> Some of the games and pastimes we indulged in as children: Milking the cow: pulling the leaf from the stalk of the plantain to reveal the veins.[4]

> As a child in [Devon in] the 70s me and my school friends would strip the white veins from the leaf [of greater plantain] and chew it. We thought we were quite cool, because chewing gum was banned in school so we were always asked by fellow pupils where we got the gum from.[5]

Adults used greater plantain leaves to treat skin ailments, and nettle stings.

> Maternal grandparents from Loughborough Junction [South London] (grandmother was in service from age of 11) told me not to dig up broad plantains from my parents' lawn in Berkshire, as they were 'folk plasters' for healing cuts and drawing infections. Approximately 1956.[6]

> Greater plantain was a good healer; the leaf 'rough side to draw and the smooth side to heal' was bandaged on a cut or BRUISE.[7]

> As instructed by my grandmother in the 1920s my family always used the plantain leaf (*Plantago major*) for cuts and abrasions – the back or veined side of the leaf to draw out impurities and then the smooth side for healing – always worked. Granny always called it planty leaf (Hampshire colloq.).[8]

> There was another leaf, but I do not know the correct name. My father used to call it fiddle leaf. It can often be seen growing on entrances to fields . . . On one side of the leaf the surface is smooth, but the other side, well the only way I can describe it is as if there are veins on the surface. This plant is again used for

> bruising. What we used to do before applying the leaf to the affected part was to roll a half-pint empty bottle on the side of what I describe as veins, so that the moisture of the leaf would spill. This was the side for drawing out the bruise, the smooth side was for healing.[9]

> [Shetland] Greater plantain = wavverin leaf – used to treat cuts, BURNS and SORES.[10]

> My late parents came from Poland after the Second World War, and my mother occasionally used certain plants as herbal remedies . . . a poultice of the leaves of greater plantain was used for CORNS and ULCERS.[11]

> [If stung by a nettle I] crush up the leaf and put the juice on the sting, or treat it with a greater plantain leaf, or a DOCK leaf.[12]

Greater plantain seed heads were formerly valued as food for CAGE-BIRDS:

> The seedhead was stuck through the bars of a bird cage, and the birds, usually goldfinches, would peck at it.[13]

Names which refer to this use include:

BIRDSEED in Sussex, Clackmannanshire and Dumfriesshire; 'gathered when ripe and dried for putting in the cages of tame birds as winter food'

BIRD'S MEAT in Aberdeenshire

Canary-seed in Hampshire and Yorkshire; 'because canaries are so much fed upon it'

Lark's seed in Wiltshire.

Outside Europe:

> Native Americans call it Englishman's foot, because it grew wherever the newcomers walked or worked.[14]

> [Born in Oregon, brought up in California] I knew that [greater plantain] as white-man's foot.[15]

GREATER STITCHWORT (*Stellaria holostea*)

Perennial, scrambling white-flowered herb (Fig. 21), widespread in hedgerows and wood margins.

Greater stitchwort has attracted an extraordinarily large – more than

140 – local names, include the widespread BIRD'S EYE, MILKMAIDENS, PICKPOCKET, shirt buttons, star-of-Bethlehem and starwort, and:

ADDER's meat in Cornwall and Sussex
Adder's spit in Cornwall
Agworm-flower in Yorkshire; 'equivalent of snake-flower – agworm or hagworm being the North Country name for a snake or adder'
All-bones in Somerset
BACHELOR'S BUTTONS in Buckinghamshire, Dorset and Suffolk; 'referring to the button-like capsules'
Balaam's smite in Suffolk
BILLY BUTTONS in Somerset and Warwickshire
Billy White's buttons in Warwickshire
Bogey's playthings in Shropshire
BREAD-AND-CHEESE in Lancashire
Break-bones in Cheshire; from the stalks 'snapping off at the joints'
Cuckoo-meat in Buckinghamshire
Cuckoo-spit flower in Devon; 'as very often the froghopper would form bubbles in this plant, and as it coincided with the cuckoo's arrival, it was thought the cuckoo spat on it'
CUCKOO'S VICTUALS in Buckinghamshire
Cuckoo-wort in West Yorkshire
Dead-man's bones in Northumberland
Devil's corn in Shropshire
Devil's eye in Denbighshire
DEVIL'S NIGHTCAP in Dorset
Devil's shirt-buttons in Somerset
Easter bell in Devon and Ireland
EASTER-FLOWER in Sussex
EYEBRIGHT in Somerset
Grandad's shirt-buttons on the Isle of Wight
GRANNY'S NIGHTCAP in Dorset
Hagworm flower in Yorkshire; 'i.e. adder flower'
HEADACHE in Cumberland
JACK-IN-THE-BOX, and JACK-IN-THE-LANTERN, in Dorset
Jack-snaps in Somerset
Ladies-in-their-nightgowns in Plymouth
Ladies' watches in Devon
Lady's button in Suffolk
Lady's chemise in Somerset
Lady's linen in Lancashire
Lady's lint in Devon and Ireland; 'probably from the white threads at the centre of the stalks'
Lady's needlework in Somerset
LADY'S SMOCK in Cornwall
Lady strawbed in Somerset
Lady's white-hand in Dorset
Lady's white petticoat/s in Cornwall, Herefordshire and Ireland
Little-dicky-shirt-fronts in Hampshire
Little John in Somerset
Mary-at-the-cottage-gate in Somerset
MAY-FLOWER in Cumberland
May-grass in Shropshire
Milk-cans in Cheshire and Wiltshire
MILKMAIDS in Somerset

Milkpans in Wiltshire
Miller's star in Somerset and Sussex
MOON-FLOWER in Worcestershire
Moonwort in Yorkshire
MORNING STARS in Dorset
Mother Shimble's snick-needles in Wiltshire
Mother's thimble in Wiltshire
My lady's needlework in Shropshire
NANCY in Somerset
Nanny-cracker in Yorkshire; 'because the seed [pods] when ripe could be cracked between finger and thumb'
Nightingales in Wiltshire
OLD-MAN'S SHIRT in Cornwall
One-o'clock in Devon
Our Lady's needlework in Shropshire
Pigsie-flower in Cornwall, Devon and Somerset
Pisgie flower in Cornwall
PRETTY BETTY in Lancashire
Pretty Nancy in Somerset
Sailor-buttons in Hampshire
Satin-flower in Cornwall
SCURVY-GRASS in Worcestershire
SHEPHERD'S WEATHERGLASS in Lancashire
SHIMMIES, and SHIMMIES-AND-SHIRTS, in Somerset
SHIMMY-AND-SHIRTS in Dorset
Shimmy-shirts in Devon; 'shimmy = undershirt from the French *chemise*'
Sissy flower in Lancashire
Skitty-witty in Shropshire
Smell-frocks in Devon
Smock-frocks in Buckinghamshire and Devon
Smocks in Buckinghamshire
Snake-grass in Hampshire
Snake's flower in Nottingham, Wiltshire and Warwickshire
Snakeweed in Somerset
Snow in Sussex
SNOWFLAKE in Sussex
Snow-on-the-mountain in Wiltshire
Soldier's buttons in Norfolk
Starflower in Lincolnshire and Somerset
Star-grass in Yorkshire
Star-of-the-wood in Devon and Somerset
Stepmothers, sweethearts and Sweet Nancy, in Somerset
THUNDERBOLTS in Dorset
Thunderflower in Cumberland
Twinkle-star, or twinkling star, in Somerset
Wedding flowers in Gloucestershire
White bird's-eye in Radnorshire
White Bobby's eye in Hampshire
White-flower, and white-flowered grass in Wiltshire
White Sunday in Devon
Whitsuntides in Hampshire
Wild pink in Buckinghamshire.

> Piskie, Pixie or Pixy – This was the regular name for stitchwort around Plymouth some years ago. The children still say that if you gather the flowers you will be pixy-led [hopelessly lost, even in an area which you know well].[1]

In many parts of Cornwall children refuse to gather this flower, believing that a bite from an ADDER is sure to follow the act. In other districts it is held that the PIXIES or 'Piskys' hide during the daytime in the flowers of stitchwort, and that anyone gathering the flowers after sunset is sure to be 'Pixy-led' and in other ways troubled by the 'small people'.[2]

MOTHER-DIE. Form of mental torture for ultra-sensitive children, especially only children thought to be 'mothers' boys'. The child would be told to pick a greater stitchwort or its mother would die and then throw it down again or its father would die. The victims never seemed to appreciate the fallacy for a long time, even when it was pointed out by kinder-hearted children. Used as a *coup-de-grâce* for a child who somehow managed to survive physical bullying. Common in the Denby Dale [West Yorkshire] area in the 1920s and 1930s, but I haven't come across it since.[3]

[West Dorset] greater stitchwort is usually known as SNAPJACKS: children 'snap' the seed capsules.[4]

Names which refer to this practice include:

Brandy-snap in Sussex
Break-jack in Dorset
Crackers, and pop-bladders, in Shropshire
Pop-guns, and pop-jack, in Somerset
Poppers, and POPPY in Wiltshire
Snap-crackers in Essex
Snapper-flowers in Sussex
Snappers in Kent
Snaps in Wiltshire
Snapstalks in Cheshire
Snapwort in Kent.

I was an Examiner for the Naturalist Badge for Guides and Scouts in the 1930s . . . having rhymes and stories attached to plants made them easier to remember, and a lot more fun . . . greater stitchwort – we chewed the flowers of this when we got STITCH – more likely mind over matter, I think.[5]

GREATER HORSETAIL (*Equisetum telmateia*)

Colony-forming perennial herb in damp places, widespread, but scarce throughout most of Scotland.

Although it is unlikely that country folk distinguished between different horsetails, and probably regarded them all as different growth forms of a single species, three local names have reputedly been given to greater horsetail: cat's tail in Berkshire, fox-tailed asparagus in Gloucestershire, and snake-pipe in North Somerset.

> Great horsetail is uncommon hereabouts, but there is a big patch in Wychwood [Oxfordshire], and in my boyhood [*c*.1915] I remember a cottager nearby using it for scouring saucepans.[1]

Horsetails contain silica in the stems and leaves and have, therefore, been widely used for scouring:

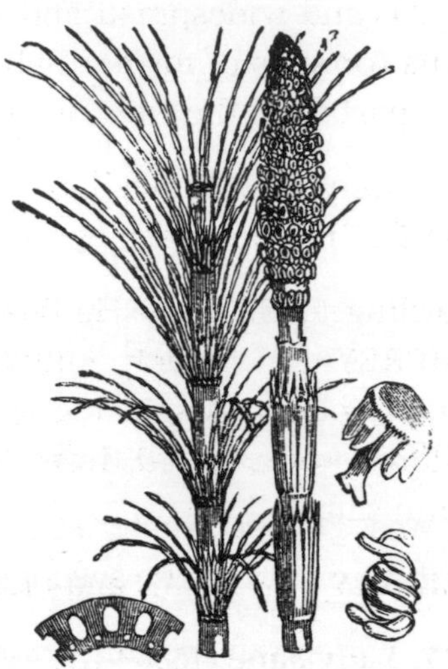

> The rough horsetail or Dutch rush, *E. hyemale*, is the most striking and, before the days of steel wool, was used for scouring pans – hence the early names of pewterwort and scrubby-grass, It was still being sold for this purpose in Austria in the 1950s.[2]

GREATER TUSSOCK-SEDGE (*Carex paniculata*)

Perennial herb, forming large clumps in damp places.

Despite its large clumps, which can give an exotic appearance to moist woodlands, greater tussock-sedge seems to have attracted little folklore, with only two local names being recorded: nat-hill in Norfolk, and shelder in Devon.

On the Isle of Wight, and no doubt elsewhere, its 'long tough culms' were used as a 'cheap though inferior substitute for straw for THATCHING'.[1]

> The dense tufts of this plant, which attain a large size in the Norfolk Fens, are cut by the peasantry and used to a certain extent in some parts of the county, as well as in Sussex, for kneeling HASSOCKS in churches. They are very durable, and have been known to last over fifty years.[2]

GREEN ALKANET (*Pentaglottis sempervirens*); formerly known as evergreen alkanet

Coarse perennial herb with attractive blue flowers, native to south-west Europe, introduced, possibly as an ornamental, or for the production of a red DYE from its roots,[1] before 1700, first recorded in the wild in 1724, and widespread and currently spreading, usually near habitations (or sites of former habitations) throughout most of the British Isles apart from the west of Ireland.

Only three local names have been recorded, all from west Somerset: BIRD'S EYE, pheasant's eye and water forget-me-not.

According to Geoffrey Halliday, in Cumbria, green alkanet was 'like COMFREY . . . widely cultivated for its medicinal properties' and therefore closely associated with houses and villages.[2]

There is one record, from London, of green alkanet flowers being used in salads.[3]

GREENFLY – AFRICAN and FRENCH MARIGOLD deter.

GREY, Lady Jane (1536?–54) – OAK trees in Bradgate Park associated with.

GROUND ELDER (*Aegopodium podagraria*), also known as gout-weed

Perennial herb, believed to have been introduced for culinary and medicinal use, now widespread in disturbed habitats.

Local names, many of which refer to the difficulty gardeners have in eradicating ground elder as a weed, include the widespread bishop's weed, GROUND ASH, ground eller and JACK-JUMP-ABOUT, and:

Ash-weed in Shropshire, Somerset and Wiltshire
Bishop's elder on the Isle of Wight
Dog-eller in Cheshire; 'from the superficial resemblance of its leaves to those of the true ELDER; here again dog undoubtedly means spurious'
Dutch elder in Wiltshire
DWARF ELDER in Hampshire and the Isle of Wight
Farmer's plague, and garden plague, in Belfast
Goat's foot in Devon
Jump-about in Oxfordshire and Warwickshire
KESH in Cumberland
Potash in Devon
Weyl-ash in Cumberland; 'i.e. wild ash'

White ash in Somerset
Wild alder in Lincolnshire, 'alder is here equivalent to elder, to the leaves of which those of *Aegopodium* bear considerable superficial resemblance'
Wild elder in Buckinghamshire
Wild esh in Cumberland.

> GOUT: drink an infusion of gout-weed twice a day . . . An old friend of mine did this successfully within the last ten years.[1]

> Ground elder (i.e. bishop's weed) almost instantly soothes nettle STINGS.[2]

> I make a quiche from the young leaves which is very popular with my family and others. People always look doubtful when I offer it, but once they've tried it they are very keen to have another slice or tartlet.[3]

GROUND IVY (*Glechoma hederacea*)

Perennial, low-growing, purple-flowered herb, widespread and common throughout lowland areas.

Local names include the widespread gill, and ROBIN-RUN-IN-THE-HEDGE, and:

Aliff in East Sussex
BIRD'S EYE, and blue runner, in Buckinghamshire
Cat's foot in northern England; 'from the shape of its leaves'
Deceivers in Essex; 'on account of its blue flowers being mistaken in early spring for VIOLETS'
Fat hen in Buckinghamshire
Gill-ale in Devon; 'according to Gerard it was formerly used in the making of ale'
Gill-creep-by-the-ground in Somerset
Gill-go-by-the-ground in Hampshire
Gill-go-to-the-ground in Somerset
Gill-run-along-the-ground in Cumberland; 'an old English name still [1922] in use'
Gill-run-the-ground in Buckinghamshire
Ground-gill in Warwickshire
Hay maiden in Dorset; 'hay means hedge and the hay maidens are plants which grow in the hedges; used for making a liquor known as hay-maiden tea'

Jin-on-the-ground in Lincolnshire
LION'S MOUTH in Sussex
MONKEY-CHOPS, and MONKEY-FLOWER in Somerset
Moulds in Rutland
Nip in Suffolk
Rat's foot, and rat's mouths, in Devon
ROBIN-RUN-THE-HEDGE in Cheshire and Derbyshire
Robin-run-up-dyke in Cumberland
Run-away-Jack in Gloucestershire
Runnidyke in Cumberland
Tunfoot, and UNDERGROUND IVY, in Somerset.

An infusion of boiling water poured over ground ivy (*Glechoma hederacea*) leaves and allowed to cool will cure STYES, a gypsy told me when I was in my teens, and it worked![1]

When my son went into the Army he did a lot of shooting, and the gun smoke made his EYES red and sore. He asked us to send something to relieve them, but we didn't know what to send, so my husband went to see a woman of gypsy origin who lived in the village to ask if she could suggest anything. She told us to put a large handful of ground ivy in a saucepan, just cover it with water, and simmer for about twenty minutes. When cool, strain and use to bathe the eyes. We sent some of this to my son and it soon cleared up the trouble. I still sometimes use this myself, as it makes a lovely smooth lotion, almost oily, and has a pleasant smell.[2]

The leaves are much used in villages to make an infusion for COUGHS, and the plant was formerly called ale-hoof or tun-hoof, because their bitter properties rendered them of use in the beer made in the old English households, before HOPS had become the common growth in our country. Even in recent times a quantity of this plant has been thrown into a vat of ALE in order to clarify it, and the ale thus prepared has been taken as a remedy for some maladies of the SKIN.[3]

Spring of the year it is the country custom (Berks and Oxon) to make an infusion of ground ivy leaves (robin-run-in-the-hedge), bairns given wineglass full before breakfast – supposed to clear skin and cool BLOOD.[4]

GROUNDSEL (*Senecio vulgaris*)

Annual yellow-flowered herb, common on waste and arable land, and in gardens.

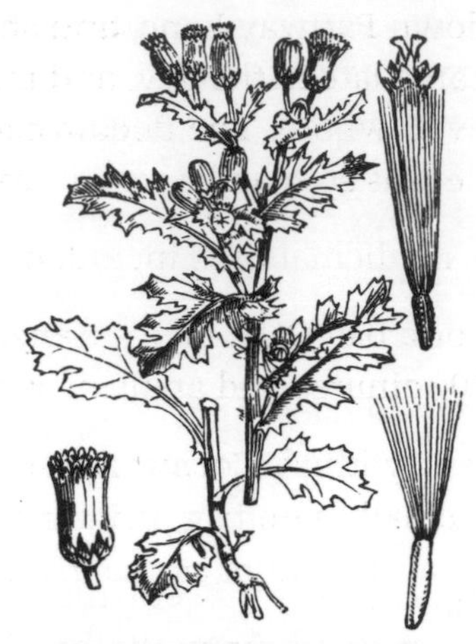

Local names include:
BIRDSEED in Yorkshire
Bogluss in Ulster
CANARY-SEED in Somerset
CANARY-WEED in Cornwall and Somerset
CHICKENWEED in Yorkshire; 'because so much used for feeding birds'
CHICKWEED in Somerset; 'well known as a valuable food for birds'
Grinsel in Wiltshire
Ground-swell, and groundwill, in Devon
Grundy-swallow in Scotland
Grunny-swally in Scotland
LADY'S FINGER in Wiltshire
Lamb's tails in Dorset
Little lie-a-bed in Somerset; 'apparently to distinguish it from the larger flowers of DANDELION and hawkbit'
Sencion, senshon, and sension, in East Anglia; 'corruption of the Latin *senecio*'
Simpson, and simson, in south and east England
Swallow-grundy in northern England
Swally in Scotland
Watery drums in Shetland
YELLOW GULL in Somerset.

Groundsel was valued in folk medicine mainly for the treatment of CONSTIPATION.

> [Cornwall] groundsel they strip upwards for an EMETIC and downwards for a cathartic.[1]

> Before castor oil had attained its popularity as a safe and efficient PURGATIVE for children it was the practice in Ireland to add a sprig or two (according to age) of groundsel to the milk, which was then boiled, strained, and given to constipated babies.[2]

> Tuesday we went to see Mr and Mrs Joby House, who used to be at Hewood. [He told us that] for constipation you boiled groundsel and lard and take that, and you will 'shit through the eye of a needle'. His sister Lucy had constipation so bad that when the doctor called in the morning he said Lucy will be dead by 5 o'clock. Mrs House went to the gypsies (Mrs Penfold) who lived

> down Partway Lane, and she told her how to cure her. The doctor came late in the day, and Lucy was running around; there was shit everywhere. The doctor had brought Lucy's death certificate, but he was so mad he tore it up and put it on the fire.[3]

Other medicinal uses included:

> For a bad CUT: take ascension, called in towns groundsel, boil for 10 minutes and apply as a poultice.[4]

> [Lostwithiel, Cornwall] for AGUE: put a handful of groundsel into a small linen bag, pricking the side next to the skin full of holes; wear it at pit of stomach and renew every two hours until well.[5]

> I can recall many things of the First World War . . . my sister had very sore EYES and the only thing that stopped the 'matter' was GRUNSELL flowers boiled then 'do' her eyes day and night with the strained liquid.[6]

As many local names suggest, groundsel was considered to be good food for CAGE-BIRDS. In London in the 1840s:

> There are no 'pitches' or stands for the sale of groundsel in the streets; but, from the best information I could acquire, there are now 1,000 itinerants selling groundsel, each person selling on an average, 18 bunches a day. We thus have 5,616,000 bunches a year, which at ½d each realise 11,700*l* – about 4s 2d per week per head of sellers of groundsel.[7]

More recently:

> Lived in Kent as a child . . . groundsel for rabbits, guinea pigs and chickens.[8]

> We used to collect groundsel to feed to our budgies.[9]

GUERNSEY LILY (*Nerine sarniensis*)

Red-flowered bulbous herb, native to coastal southern Africa, grown as an ornamental since the mid-seventeenth century.

> The Guernsey lily . . . has long been regarded as Guernsey's National Flower . . . The legend of its first appearance in the island is part of our FAIRY tradition. The story goes that when the fairy

> king won the heart of the beautiful Michele De Garis and persuaded her to go away with him to his far away kingdom, she thought of her family and how they would grieve for her. So she asked her elfin lover to let her leave some small token by which they could remember her. He gave her a bulb which she planted in the sand above Vazon Bay before embarking on the fairy craft taking her away from her island home. Later, when her distraught mother came looking for her missing daughter she found this bulb, now burst into flower – a beautiful scarlet, scentless blossom, sprinkled with fairy gold.[1]

The history of Guernsey lily, which has been associated with, or grown in, Guernsey since 1680, is confused. Four conflicting stories account for its introduction to the island: it was introduced to Guernsey by General John Lambert who, following the Restoration was held prisoner there from 1661 to 1670; bulbs washed ashore from a wrecked ship rooted on the strand and were taken into cultivation by Charles Hamilton, whose father was governor of the island from 1662 to 1670; a passenger on a ship from Japan or China who was temporarily stranded on Guernsey made a gift of bulbs to Jurat Jean de Sausmarez (1609–91); or sailors on a ship returning from China gave bulbs to an inn-keeper. It has been suggested that the oriental origin of Guernsey lily may have resulted from confusing it with the red spider lily (*Lycoris radiata*), native to China,[2] but, of course, at the time of its arrival in the British Isles, it was necessary for ships travelling from Asia to pass around the Cape of Good Hope, so bulbs of South African species could have been taken aboard there.

GUMS, sore – GARLIC MUSTARD and SAGE used to treat.

GUY FAWKES (or BONFIRE) NIGHT (5 November) – MANGOLD associated with.

GYPSOPHILA (*Gypsophila paniculata*), also known as baby's breath

Perennial herb, native to central and eastern Europe and western and central Asia, cultivated for its abundant small white flowers since the mid-eighteenth century.

Gypsophila is frequently included in wedding bouquets (and to provide bulk in cheaper bunches of flowers) probably for its decorative qualities and because it remains attractive even when dried out and dead, but according to some writers:

> The significantly named baby's breath . . . with a cloud of tiny white flowers, is a bouquet ingredient with obvious FERTILITY connotations.[1]

> *Gypsophila paniculata*, nicknamed baby's breath, is still carried to ensure a fruitful marriage.[2]

H

HAIR – BIRCH sap prevents the thinning of; CHAMOMILE and YARROW used to rinse.

HAIRY BITTERCRESS (*Cardamine hirsuta*)

Small annual herb, widespread and common on cultivated ground.

Local names include:
Jumping Jesus in the New Forest, Hampshire
Lamb's cress in Devon
Land-cress in Cheshire, Hampshire and Warwickshire
Spiky flower in Somerset
TOUCH-ME-NOT, 'shoots out its seeds when touched', in Cheshire.

Like other species of *Cardamine*, hairy bittercress was occasionally eaten:

> We ate a few weeds during the War . . . I still put hairy bittercress and GROUNDSEL in my salads.[1]

HAIRY BROME (*Bromopsis ramosa*, syn. *Bromus ramosus*)

Perennial tufted grass, frequent in shady, moist places throughout lowland areas.

> I have seen children strip the branches off the culm of *Bromus ramosus* and make a running noose of the thin end with which to catch newts. The newts submitted to having these put round their heads without any sign of fear.[1]

HALLOWE'EN – APPLES, BULRUSH, MANGOLD, PUMPKIN, SWEDE and TURNIP associated with; divination with BAY and CABBAGE attempted, straw costumes worn (see WHEAT).

HANGOVER – CABBAGE and POPPY used to treat.

HAREBELL (*Campanula rotundifolia*)

Perennial herb with delicate blue flowers, found in a wide range of infertile, well-drained habitats, but absent from much of Ireland.

Local names include the widespread LADY'S THIMBLE and THIMBLES and:

BELL-FLOWER in Devon, Dorset and Somerset
BLAVER, and BLAWORT, in Scotland
Bluebells-of-Scotland
Blue blauers, and blue blaves, in Roxburghshire
BLUE BONNET in Co. Donegal
BLUE BOTTLE in Buckinghamshire
CUCKOO in Devon
Ding-dongs in Dorset
Fairies' cups in Wiltshire
Fairies' thimbles Somerset
FAIRY-BELLS in Wiltshire
FAIRY-CUPS, and fairy-ringers, in Dorset
FAIRY-THIMBLE in Somerset and Co. Donegal
Gowk's thimles in north Scotland; 'i.e. cuckoo's thimbles'
Gowk's thumbles in north Scotland; also spelt as gowk's thummles
Gowk's thumbs in north-east Scotland
Granny's tears in Somerset
Heather-bell in Dorset and Northamptonshire
Milkort, or milkworts, given to the roots in northern Scotland
Nun-of-the-fields in Somerset
Old-man's bell in Scotland
School-bell in Wiltshire
Scotch bluebell in Argyllshire
Sheep-bells in Dorset
Sheep's bells in Devon
THUMBLE in Scotland
Tinkle-bell in Cornwall
WITCH-BELLS, or witches' bells, in Scotland
Witches' thimble in Lanarkshire and Somerset.

Many writers insist that harebell is known as BLUEBELL throughout Scotland, and cite this as an example of how English plant-names can be misleading. However, this is an exaggeration; according to a Glasgow-born botanist:

> As a young Glaswegian I learned to call *Hyacinthoides non-scripta* 'Bluebell' and many a wood in and furth of Glasgow was and is 'Bluebell Wood' to the locals. We did not know that lots of books written over many years, perhaps by mainly English authors, assured the reader that in Scotland the 'Bluebell' is *Campanula rotundifolia* (Harebell).[1]

In north-east Scotland:

> The bluebell (*Campanula rotundifolia*) was in parts of Buchan called 'the aul' man's bell', regarded with a sort of dread, and commonly left unpulled.[2]

> [Berwickshire] our children have a custom of blowing into the flower bell; and then, placing it erect on the back of the hand, they make it crack by a smart stroke with the other.[3]

HART'S TONGUE (*Asplenium*, formerly *Phyllitis, scolopendrium*)

Evergreen fern, common and widespread in dampish rocky places, but scarce in northern Scotland.

Local names include:

ADDER'S TONGUE in Devon
Burntweed in West Meath; 'used there as a remedy for BURNS'
Christ's hair in Guernsey; 'because of the single black fibrovascular bundle in the leaf-stalk'
Cow-tongue in Co. Donegal
Fox's tongue in Dublin
Fox-tongue in Co. Donegal
Horse-tongue, or hoss-tongue, in Dorset
Lamb's tongue, or lamb-tongue, leopard's tongue, and lion's tongues, in Somerset
Long leaf in Hampshire
Seaweed-fern in Surrey
SNAKE-FERN, and snake-leaves, in Somerset
Snake-tongue in Devon.

> In Devonshire the children have a graceful tale about hart's-tongue fern. It was once the pillow for the Son of Man, when He had nowhere to lay His head. In return for this service, He left two hairs of His most blessed and dear head, which the plant treasures in her ripe stems, as His legacy – two auburn hairs which children find and show.[1]

Fresh leaves of the hart's tongue are applied externally in rustic practice in the island [Wight] to bad legs! (erysipelatous erruptions) as a cooling remedy.[2]

[Co. Limerick] Burn a leaf called hart's tongue and apply it to a burn and it would cure it.[3]

[Co. Waterford] Hart's tongue fern was used as a cure for SCALDS and burns. The underside up it was laid on the scald or burn. Fresh leaves were applied when needed until the cure was complete.[4]

HARROW SCHOOL – CORNFLOWER associated with.

HASSOCKS – made from GREATER TUSSOCK-SEDGE and TUFTED HAIR-GRASS.

HATPINS – BUTCHER'S BROOM berries used to decorate.

HAWTHORN (*Crataegus monogyna*), also known (especially in Ireland) as whitethorn; flowers known as may

Small deciduous tree producing abundant white, scented flowers in early summer and red fruits, 'haws' in late summer, common and widespread, frequently planted as hedging.

Local names include the widespread haw-bush, MAY-TREE, and quick 'typically when used for hedges', and:

Aglet tree, and awglen, in Cornwall
Azzy-tree in Buckinghamshire
Blossom, given to the flowers in Devon and Somerset; 'the usual name in west Somerset'
Eglet-bloom, given to the flowers, in Devon
Haeg, or hag, in Lancashire
Hag-bush in Yorkshire
Hag-thorn, and hag-tree, in Devon and Somerset
Hathorn in Shropshire
Heg-peg bush in Gloucestershire
Hipperty-haw tree in Shropshire
Holy innocents in Wiltshire
Mahaw in Ireland
May-bush in Hampshire and Norfolk
MAY-FLOWER in Cumberland and Somerset
MOON-FLOWER in Somerset
MOTHER-DIE, given to the flowers, in Cambridgeshire, Cheshire and Yorkshire
Peggall-bush in Wiltshire

Quicks in Yorkshire
Quickset in Norfolk
Quickthorn in Lancashire and Yorkshire
Sates, given to 'young plants', in Shropshire
Scog-bush, SCROG, sgeach, and skeeog, in Ireland
SNOWFLAKE, given to the flowers, in Somerset
Spring, 'young white thorn quicks, perhaps from the usual season for planting it quick-hedges', in Suffolk
WHICKS, 'thorn plants for hedges', in Cheshire
Wick, 'a plant of hawthorn', and WICKENS, 'hedge thorns', in Yorkshire
Wicks, 'thorn plants for hedges', in Cheshire.

Names given to the fruit include the widespread cat-haws, and hag, and:

Agald in Wiltshire
Agars in Middlesex; 'we used COW PARSLEY stems as PEA-SHOOTERS, using hawthorn berries – agars – as peas'
Agarves in Sussex
Agasse in Hampshire and Sussex
Aggle-berry, 'usually applied to the berries when picked or eaten', and aggles, 'many aggles many cradles', in Devon
Aglen in Cornwall
Aglet in Devon; 'used to provide ammunition for pea-shooters'
Agogs in Berkshire
Aigarce in Sussex
Airsens in Gloucestershire
Av-en-av in Somerset; ' = half-and-half'
Bird-eagles in Cheshire; 'eagles or agles is the diminutive of hague, the more common name of the haw in Cheshire'
Birds' cherries, and BIRD'S MEAT, in Somerset

Bird's pears in Devon and Somerset
BULLS in Norfolk
Butter-haw in Norfolk and Oxfordshire
Chaws in Morayshire
Cuckoo's beads in Shropshire
Eggle-berry in Devon
Egglet, or eglet, in Cornwall and Devon
Eggs-eggs in Wiltshire
Hab-nabs in Somerset; 'a corruption of half-and-half'
Haggalans in Cornwall

Haggils in Hampshire
Haggle-berries in Devon, 'usually applied to the berries when picked or eaten'
Hails in Dorset and Somerset
Haivs in Somerset. 'we be gawin to have a hard winter, the haivs be so plenty'
Hales in Dorset and Somerset
Half-and-half in Somerset
Halves in Devon
Harsy, and harve, in Essex
Haves-and-hawses in Berkshire
Haw-berry in Cheshire and Scotland
Haw-gaw in Surrey
Haws-arglans in Cornwall
HEDGE-SPECKS in Wiltshire
Heethen-berry in Cheshire
HEG-PEGS in Gloucestershire
Hip-haw in Oxfordshire and the Lake District
Hipperty-haw in Oxfordshire and Shropshire
Hogail, and hogailes, on the Isle of Wight
Hogarve in Surrey and Sussex
Hogazel in Sussex
HOG-BERRY in Hampshire
Hoggan in Cornwall
Hog-gosse in Sussex
Hog-haghes in Hampshire
Hog-haw in southern England
Hogiles on the Isle of Wight
Johnny MacGorey in Co. Wexford
May-fruit in Yorkshire
NIPPER-NAILS in Cheshire
Orglans in Cornwall
Pigall in Hampshire, Somerset and Wiltshire
Pigaul in Hampshire
Pig-haw in eastern and southern England
Pig's ales, pig's hales, pig's hauds, pig's haws, pig's pears, and pigsy pears, in Somerset
PIXY-PEARS in Dorset and Somerset
Skiach in Morayshire; 'Gael[ic] *sciog*, a hawthorn'
Thorn-berries in Cheshire
Wind-bibber in Kent.

When the London-based Folklore Society conducted a survey of 'UNLUCKY' plants between March 1982 and October 1984, 123 (23.5 per cent) of the 524 items of information received concerned hawthorn, more than twice the number of items concerning LILAC, the second most feared plant.[1] A similar, but much smaller, survey conducted between May 2012 and April 2017 resulted in only 113 items of information being received, of which 36 per cent referred to hawthorn.[2]

> Children in Shepherd's Bush [London] who made grottoes [*c.*1920] would not use lilac or hawthorn blossom to decorate them.[3]

About thirty years later, also in Middlesex:

> Hawthorn – I used to have to put it in milk bottles on the step outside in the back garden, never allowed in house – very unlucky.[4]

Elsewhere:

> [Redhill, Surrey] When I was at school, I regret over 40 years ago, a very unpopular teacher fell downstairs. It was said that this was because she had vases of hawthorn in her classroom . . . I've no idea where we got the idea from.[5]

> My mother used to go really mad with me if I tried to take pink-flowered may into the house.[6]

> My mum gave me a ticking off for taking hawthorn into the house; she said it was very unlucky![7]

Explanations of why hawthorn blossom is considered inauspicious vary. According to a Cambridge anthropologist:

> It is considered bad luck to bring hawthorn indoors; in contrast to Christmastide greenery and Easter willow, it is a plant kept out of doors, associated with unregulated love in the fields, rather than conjugal love in the bed.[8]

Some writers have claimed that fear of hawthorn flowers is derived from memories of pre-Christian May Day celebrations which involved a May Queen being crowned with hawthorn before being ritually slaughtered.[9] However, there is no evidence to suggest that such sacrifices ever took place. Some people claim that Christ's CROWN OF THORNS was made from hawthorn, hence the tree brings misfortune.[10] More widespread is the idea that hawthorn blossoms were associated with pre-Reformation devotions to St MARY THE VIRGIN.

Superstition about May dates from the times when Catholics were persecuted for their faith.

> During the month of May – which was dedicated to the Blessed Virgin Mary – May blossoms were used to decorate the little shrines which Catholics made in their homes in her honour. If anti-Catholic officials saw May blossom being carried into a house, they recognised the household as a Catholic one and acted accordingly. Hence, to bring these flowers into a house brought 'bad luck' to the owners.[11]

However, the association of the Virgin Mary with the month of May and may blossom originated well after the Reformation, starting in Naples in the eighteenth century and spreading north-west to reach Britain and Ireland in the mid-nineteenth century.[12]

In Irish tradition hawthorns, known as LONE BUSHES, were thought to be associated with supernatural beings, or otherwise inauspicious.

> In Ireland hawthorn trees are associated with FAIRIES. I personally know a case where they had to divert a new sewer they were putting in, because there was a hawthorn in the way.[13]

> My mother, who's Irish, told me that unbaptised children were buried under isolated thorn trees.[14]

In 1866 a correspondent to the *Gentleman's Magazine* wrote:

> I have found it a popular notion among . . . country cottagers that the peculiar scent of hawthorn is 'exactly like the smell of the Great Plague of London'. This belief may have been traditionally held during the last two centuries, and have arisen from circumstances noted at the period of the Great Plague.[15]

More recently the poet Sylvia Plath (1932–63) wrote of the 'death stench of a hawthorn',[16] according to other people, may blossom smells 'of the pits where people died – from the Black Death onwards',[17] or of smallpox.[18]

Two species of hawthorn, which sometimes hybridise, are native to Britain and Ireland: the common hawthorn (*Crataegus monogyna*) and the Midland hawthorn (*C. laevigata*, formerly known as *C. oxyacanthoides*). In a letter dated 21 May 1900, the botanist Richard Paget Murray recorded:

> When in Switzerland we had plenty both of *C. monogyna* and *C. oxyacanthoides*: the latter flowering a week or two earlier than *C. monogyna*. But I often gathered a lot of *C. oxyacanthoides* for decorative purposes: and tho' in smell quite like the other when gathered, it used to absolutely *stink* of putrid flesh soon after: sometimes within half an hour. I do not remember this ever occurred with *C. monogyna*.[19]

Chemists have shown that trimethylamine, one of the first products formed when animal tissues start to decay, is present in hawthorn flowers.[20] Thus it is not surprising that our ancestors, who kept corpses in their homes for anything up to a week before burial, were fully aware of the odour of death and decay, and would not want hawthorn flowers indoors.

The misfortunes which supposedly befell people who picked or brought hawthorn blossoms indoors varied. It could 'result in a dead child',[21] 'lead to a wet summer',[22] cause the death of one's mother[23] (hawthorn blossom sometimes shared with COW PARSLEY the name mother-die), not getting a good APPLE crop 'as you would be depriving the bees of a supply of pollen',[24] or if brought indoors 'not necessarily death, but certainly illness would follow'.[25]

There is one record of hawthorn blossom being brought indoors to hasten recovery from illness:

> It is said that when there was an outbreak of typhoid in the old Post Office Street [Ayton, North Yorkshire], after eight people had died and several people in one family had had it, the doctor (Dr Megginson) found that they were all very inert and disinclined to rouse themselves; so he told the mother to get large boughs of the flowering hawthorn, and put them all over the house. This was done and the invalids soon began to rouse themselves and made efforts to recover their strength.[26]

Another instance of hawthorn being considered auspicious appears to be what was described in 1830 as an old custom in Suffolk:

> In most farm houses . . . any servant, who could bring a branch of hawthorn in full blossom on the first of May, was entitled to a dish of cream for breakfast. This custom is now disused . . . it very seldom happens that any blossoms are seen open even on Old May Day.[27]

In Ireland hawthorn could be used to protect homesteads on MAY DAY.

> Long ago on May Day the old people used [to] bring in blossoms of the whitethorn and place them on the dresser. They used [to] leave them there until that day month. They were supposed to have the power of keeping away evil.[28]

> On May Day . . . sprigs of whitethorn which have been sprinkled with blessed water procured on Holy Saturday are stuck down in village fields to prevent the 'fairies' from taking the crops, and to produce a good crop.[29]

Other practices in which hawthorn provided protection include:

> Hanging the after-birth of a calf on a thorn is preventative of fever for the COW. One was seen on a thorn by the [Hampshire] informant in 1939.[30]

> [Radnorshire] The remarkable ceremony of 'the burning bush' was long regarded as essential to the well-being of the WHEAT crop, providing a safeguard against 'smuts in the wheat'. The bush consisted of a branch of hawthorn pruned to leave four twigs at right-angles. All the men of the farm were out very early on NEW YEAR's morning, and armed with bundles of straw and a plentiful supply of beer and cyder, and carrying the bush, they visited each wheat-field in turn. In each field a fire was kindled, over which the bush was hung, to be ignited but not destroyed. The bush, or what remained of it, after all the fires had been lit, was taken to the farmhouse to be preserved carefully until the next New Year's Day . . . It is said that the custom was so widely observed that the whole countryside 'twinkled like stars' in the darkness of the early morning.[31]

> A variety of hawthorn which flowers in winter as well as at the usual time in late spring or early summer is the Glastonbury, or HOLY THORN, and in Ireland isolated hawthorn trees continue to be thought of as fairy trees or LONE BUSHES which should not be damaged or destroyed. The many superstitions which surround hawthorn have led some authors to suggest that in pre-Christian Britain there was a 'thorn cult',[32] or that the tree was associated with the supreme goddess.[33]

The saying 'Cast ne'er a clout 'til May is out' is widespread has caused some controversy: does it mean that clothing should not be discarded before may blossom, or the month of May, is out? However, most people seem to agree that it refers to the flowering of hawthorn, which varies according to local climate, rather than the month.

> Certainly it is the May or hawthorn and not the month. As a boy here in Devon it was always the way of it and as soon as May blossom showed we were told we could leave pullovers and such off. The old folk would point it out and one felt spring was here.[34]

However, a letter from Hove, East Sussex, in the *Daily Telegraph* of 5 June 1993, rejected both the month and the blossom, claiming:

> May was one Mavis (May) Dennison, a 'four-penny drab' who slept out in the Itchy Park area of Spitalfields [London] in the mid-1840s. A cunning old soak, she made it her business to spend each winter in jail, coming out reasonably well-fed and revived for the summer months, about the end of May or early June. It was from this that her contemporaries – most likely jeeringly – took up what has become our most misquoted old saw.

Newly emerging hawthorn leaves were frequently eaten by children, who knew them as BREAD-AND-CHEESE, and a variety of similar names:

BREAD-AND-CHEESE-AND-CIDER, and bread-and-cider, given to young leaves in Somerset
Bread-and-cheese bush in Dorset; 'young . . . leaves gathered and eaten by children – "bread" – and the ripe red haws "cheese"'
Bread-and-cheese tree in Somerset, 'young leaves known as bread-and-cheese eaten by children'
Butter-and-bread, CHEESE-AND-BREAD, and cheese-and-bread tree, in Yorkshire
CHUCKY-CHEESE in Devon
CUCKOO'S BREAD-AND-CHEESE in Sussex and Wiltshire
Cuckoo's bread-and-cheese tree in Leicestershire and Sussex
God's meat in Yorkshire
LADY'S MEAT in Scotland
May bread-and-cheese bush in Hampshire and Somerset
May bread-and-cheese tree in Lincolnshire and Norfolk.

Typically:

> When I was a child in Norfolk, about 20 years ago, my father would tell me that hawthorn was called bread-and-cheese because you could eat it, and so we would pick the leaves to nibble on our walks.[35]

> My grandfather [b.1897] told me that young hawthorn leaves were called bread-and-cheese when he grew up, and they used to eat them.[36]

Although they are mainly remembered as being a minor food for children, hawthorn leaves were also eaten by adults, especially during times of hardship. In 1752:

> The failure of the harvest was immediately followed by a cattle plagues which wiped out many of the local herds. The poor suffered greatly, and the Kingswood colliers, already outcast, feared

and despised, suffered more than most . . . A rumour swept through the Forest that wheat was to be exported from Bristol and this was the spark which lit smouldering unrest into open rebellion. In May 1753, the colliers, many hundred strong, marched to Bristol . . . Ravaged by hunger, they attempted to fill their empty bellies with sprigs of green hawthorn which grew along the way. This, with bitter irony, they called 'Bread and Cheese'.[37]

I grew up in Leicester . . . we used to eat the new hawthorn shoots – 'bread and cheese' – both raw and together with chopped streaky bacon as a filling for a suet rolypoly – very good![38]

I had a friend in Sandbach [Cheshire] who always collected hawthorn leaves to put in her dumplings on Sundays.[39]

There are also occasional records of hawthorn fruit – haws – being eaten.

My mother, who was born at Buxton, Derbys. in 1901, referred to the fruit of the hawthorn as 'aigie berries' and to its leaves as 'bread and cheese'. Both were eaten by children when she was young. Not very appetising, but just about edible, I would say.[40]

I am an Invernessian . . . the berries of the hawthorn tree were good to eat. They had a thick, sweet skin over a hard stone centre. They were known as 'boojuns'.[41]

Coventry 1960s and 70s – hawthorn berries were nibbled on the way home from school.[42]

An intriguing record from Berwickshire in 1853 suggests that local boys distinguished between the usually one-seeded common hawthorn, and the usually two- or three-seeded Midland hawthorn:

Boys in autumn go out in groups to gather the ripened haws, and they look out eagerly for those with double stones, which they dignify with the name of Bull-haws. Having sucked the pulp from the stone, they amuse themselves by blowing the latter at each other through their POP-GUNS, made from the hollow stalks of hemlock. Haws they believe are apt to fill the teeth with lies; for the number of 'lees' that a boy has told that day is reckoned by the number of black specks on the teeth, and the absence of specks vindicates his innocence.[43]

Outside Britain and Ireland, particularly it seems in France, hawthorn is used to treat HEART conditions. In this context the few records which we have of countrymen chewing hawthorn twigs might suggest that they were, unconsciously, self-medicating.

> I remember my father in the 1950s along with many of his male rural contemporaries, in the south of Nottinghamshire would when out walking, or working outside, often break off the last few inches of a hawthorn (*Crataegus monogyna*) twig, and placing the broken end between their teeth, chewing it, for several seconds, before replacing it with a fresh piece. They referred to this as having a bit of egg and bacon. They chewed just the end of the twig, as some people would chew a match stick or the end of a pencil.[44]

> My father . . . the son of a Lithuanian immigrant, used to take me out on his trips across the Pappert Hills, near Shotts [North Lanarkshire]. This was in the early 1960s. As we tramped across the moors he would always chew a sprig of hawthorn – perhaps it helped him stay calm.[45]

HAY FEVER – MINT prevents.

HAZEL (*Corylus avellana*)

Multistemmed (traditionally coppiced) shrub or small tree, common and widespread in woodland, hedgerows and scrubby areas; frequently planted.

Local names include:
Alse in Devon
Cobbedy-cut in Cornwall
Crack-nut, given to the nut, in Devon
HALSE in Devon, Somerset and Ireland; in Somerset 'a hazel rod is always known as a halsen stick'
Haskett in Dorset
Leemers, given to the 'ripe nuts' in Cumberland, Lancashire and Roxburghshire
Nit, 'a Scotch pronunciation of nut'
Nit-al, or nittal, in Devonshire; 'a corruption of nut-halse or

halse-nit as we Devonians term it'
Nut-arl in Devon
Nut-bush in Berwickshire
Nut-hall in Cornwall
Nuttal-bush in Devon
Nuttall in Cornwall and Devon
Victor-nut in Cornwall
WITCH-HALSE in Cornwall and Devon
Wood-nut in Yorkshire.

Names given to the male flowers (catkins) include the widespread LAMB'S TAILS, and:

CAT-O'-NINE-TAILS in Devon
Cat's tails in Northamptonshire
CHATS in Yorkshire; 'keys of ASH and MAPLE, also the hazel catkins'
Fox-tails in Somerset
Hazel-palms in Berwickshire
Kittens' tails in Dorset
Lambkins in Wiltshire
Nut-palms in Berwickshire
Nut-rags in Cheshire
Palm in Devon and Somerset
Pussies' tails in Wiltshire; 'more usually willow only'
PUSSY-CAT'S TAILS in Devon
RAG in Yorkshire.

> Some of the Highlanders, where superstition is not totally subdued, look upon the tree itself UNLUCKY, but are glad to get two nuts naturally conjoin'd, which is a good omen. These they call *Cnò-chomblaich*, and carry them as an efficacious charm against WITCHCRAFT.[1]

> [Radnorshire] it is thought . . . unlucky to take the catkins of hazel (lamb's tails) into the house, farmers holding that this will cause a bad lambing season.[2]

Cf. DAFFODIL, GOAT WILLOW and PRIMROSE.

> To ensure that the, usually, blue eyes of their newly-born babies would eventually turn brown, they [East Anglian parents] would bind a small hazel twig to the baby's back, or hang bunches of twigs in the room in which it was born.[3]

In many country areas it was believed that hazel was associated with FERTILITY.

> [Wye, Kent, 1940s] It was well-known that fresh hazel nuts are a strong APHRODISIAC.[4]

> Lots of nuts in the autumn means lots of babies in the spring.[5]

Early in the nineteenth century a curious custom was observed in some Surrey parishes:

> A ceremony not confined to this parish [All Saints, Kingston-upon-Thames] and consisting in the cracking of nuts by the whole congregation on Michaelmas Eve has scarcely fallen into disuse. The origin or meaning of this absurdity is unknown. Cracknut Sunday, in connection with the election of bailiffs, is still in the memory of many persons living.[6]

Grey squirrels, introduced from North America between 1876 and 1929 and now widespread,[7] eat immature hazel nuts before they are fit for human consumption. Before this nuts provided a useful source of protein and additional income for people living in areas where they were abundant.

> At Ashmore and other villages on Cranborne Chase [Dorset] the annual nutting expeditions were great events. The women and girls made themselves special canvas dresses and the great part of the population went off to the woods, taking their 'nammit' (noon-meat) with them. The nuts were sold to dealers for dessert and also (chiefly) for use in the DYEing industry. Often not less than £200 a year was made by the village during this season, and most families reckoned to pay their whole year's rent, if nothing more, with the proceeds. This custom has now almost come to an end. There is now little sale for the nuts to dyers, and very low prices prevailed in the years between the two great wars, so the nutting ceased to be worthwhile. During the wars prices rose again and 6d a lb for slipped nuts was obtained between 1939 and 1945. The price has now fallen to 4d a lb, and during the last decade, it is only the children and the old people who have troubled to carry on the work.[8]

It is probable that the economic value of the nut harvest stimulated taboos to discourage the unscrupulous from nutting at unreasonable times thus getting more than their share of the crop. In northern England unripe nuts were guarded by the goblins Churnmilk Peg and Melsh Dick.[9] There was a widespread, but perhaps not seriously held, belief that gathering nuts on a Sunday would attract the DEVIL's attention.

> There was also a taboo . . . against picking nuts on Sundays; if you did, the Devil would come and hold the branches down for you. This belief was sometimes deliberately used, some fifty

years ago, to stop children from spoiling their good Sunday clothes.[10]

My grandmother . . . a native of Mitcham [Surrey], used to tell of a very wicked man who went into the woods on Sunday to gather nuts. He was terrified to find that as he pulled them off the trees they came again in greater numbers.[11]

Ruth Tongue gave a song, reputedly collected by an elderly servant in Somerset, in which a girl who went nutting on Sunday took the Devil as a love:

Oh there was a maid, and a foolish young maid,
And she went a-nutting on a Sunday.
She met with a Gentleman all in black,
He took and he laid her a-down on her back,
All a-cause she went nutting on Sunday.

The outcome of this union was a baby 'which did come before the ring' and possessed horns and a tail.[12]

Nutting expeditions inevitably led to a certain amount of horse-play, so nutting, like RUSH-gathering, became a euphemism for courtship, especially that of a less restrained nature. In the well-known folksong *The Nutting Girl*:

It's of a brisk young damsel who lived down in Kent,
And she rose one May morning and she a nutting went . . .

Then a nutting we will go, a nutting we will go,
With a blue cockade all in our hats we'll cut a gallant show.

While nutting at this unseasonable time, she hears a young ploughman:

He sang so melodiously it charmed her as she stood,
She no longer had power in that lonely wood to stay,
And what few nuts that poor girl had she threw them all away . . .
So he took her some shady grove and gently laid her down.
She said 'Young man, I think I see the world go round and round.'

And so on, concluding with the warning:

For if you stay too long and hear the ploughboy sing,
Perhaps a young ploughboy you may get to nurse up in the
spring.[13]

Hazel sticks were believed to be poisonous to SNAKES.

> In July and August we were staying at Overton, near Port Eynon, Gower, where this year there happened to be an unusual number of ADDERS, one of which had bitten a sheep . . . When I saw it I suggested that the young farmer should wash it with ammonia; but he replied, 'Oh, I cure it with a poultice made of ground-ash, TANSY and hazel-leaves.' I asked, 'Why hazel-leaves?' To my amusement, he replied, 'Hazel-trees are poisonous to snakes, especially adders. In fact no creeping thing can live in or near them.'[14]

> When people began to emigrate from Ireland to foreign lands (especially USA and Australia) it was usual for them to bring a bundle of hazel rods with them to kill the snakes. One blow of the rod and the snake was no more![15]

Hazel stems being flexible and easily split had a variety of uses.

> [For Irish travellers] tents were the most common shelter of all. There were two types both constructed of bent hazel branches covered with oil-soaked bags or canvas. The smaller variety, no more than waist high at its peak, was known as a 'wattle' or 'bender' tent.[16]

Writing of Scotland, Milliken and Bridgewater record hazel being used to weave wattle walls, make wooden staples to hold thatch in place, make hurdles, walking-sticks, hoops and stakes, and creels for carrying peat and seaweed.[17]

Similar uses, and others, have been recorded throughout Britain and Ireland.

> I well remember some of the gypsies as they used to settle on Beaminster Down, which joined our farm. They were not bad, but used to cut hazel twigs from our hedges to make clothes-pegs.[18]

> When I was small gypsies used to come to our door selling orange chrysanthemums that I thought were made of hazel. I know they used to camp in a hazel coppice close to Sherborne. The wooden

> flowers were the size of a good double chrysanthemum and had wonderful swirly 'petals'.[19]

Forked hazel twigs are frequently used by dowsers.

> The *virgula divinatoria* is in high repute amongst Welsh miners; what sympathy there is between a vein of coal or lead ore and a piece of hazle [*sic.*], it would be difficult to say.[20]

> The Isles of Scilly have employed a water diviner after suffering their worst drought. Using a hazel twig, Don Wilkins, from Chacewater, Cornwall, pinpointed two water sources in 100ft deep rock to supply the 70 people on the island of Bryher.[21]

> Hazel – small forked branches are used in water divining. I have tried this and it seems to work.[22]

Examples of forked hazel twigs used by Somerset dowsers can be seen in the Pitt Rivers Museum, Oxford.

To the present day the Irish homespun is DYED with hazel.[23]

HEADACHE – CUCKOO-FLOWER and POPPY thought to produce; treated using BETONY, CHAMOMILE, CREEPING WILLOW, DOG VIOLET, FEVERFEW, HENBANE, LAVENDER, SAGE, VIOLET, WILLOW and YARROW.

HEART, conditions – DANDELION, FOXGLOVE, HAWTHORN, HEATHER, ROCK SAMPHIRE, SELFHEAL and WATERCRESS good for; GARLIC protects against disease.

HEATHER (*Calluna vulgaris*)

The name 'heather' means different things to different people. Most botanists follow Stace's *New Flora of the British Isles* and use it for *Calluna vulgaris*, whereas some members of the Heather Society, and others depending on where they learnt their plant-names, use it for what Stace calls bell heather (*Erica cinerea*).

Calluna vulgaris is a small shrub whose purplish-pink flowers clothe heaths, moorland and upland areas throughout the summer months.

Local names include the widespread grig, hadder, heath and hedder, and:

Bazzon in Cornwall and Devon
Bend in Cheshire
Black ling, and BREEAM, in Yorkshire

BROOM in Buckinghamshire, Devon, Somerset and Yorkshire; 'from its use in making brooms'
Dog-heather in Aberdeenshire
Griglans, and griglum, in Cornwall
He-heather in Northumberland and Berwickshire; 'SHE-HEATHER is *Erica tetralix* [cross-leaved heath] or *E. cinerea*'
HONEY-BOTTLE in Wiltshire
Kid in Yorkshire
Moor, 'when in blossom', in Yorkshire
Mountain mist in Somerset
Red ling in Hampshire
Satin-balls in Somerset
True heather in Nairnshire.

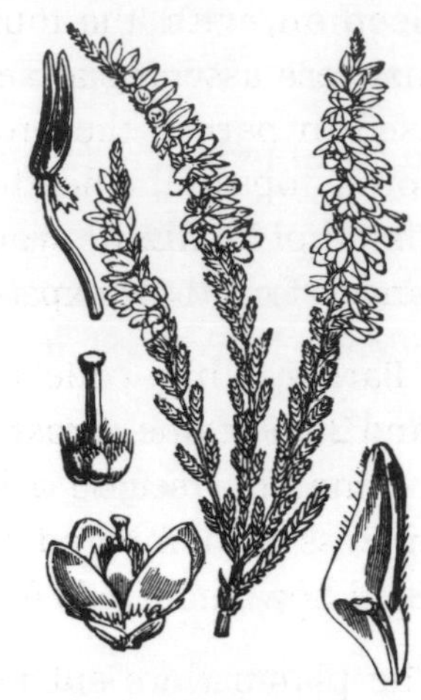

Erica cinerea, another small shrub, has purple flowers, and is common on acidic soils, but rarely, if ever, as dominant as *Calluna*.

Local names include:
Bell-ling in Yorkshire
Black heath in Hampshire
Carlin-heather in Yorkshire and Scotland; 'from Old Norse *kerling*, means witch or hag'
Cat-heather in Scotland
Crow-ling in Yorkshire
SHE-HEATHER in Northumberland and Berwickshire

Bell heather appears to have few uses and attracted little folklore, so it can be assumed that most mentions of 'heather' apply to *Calluna*.

> There's gold under BRACKEN
> Silver under GORSE
> Starvation under heather.[1]

As this saying suggests, heather grows on land which is of little agricultural use, consequently people living in these marginal areas found many uses for it.

> The tops of ling heather (*Calluna vulgaris*), *Fraoch*, produce . . . a shade of yellow DYE when boiled in water, in fact the entire plant is wonderfully versatile: it was commonly used for THATCHing houses, and even today the few Highland thatchers that remain will swear it is the best thatch in Gaeldom; it provided beds to

> sleep on, with 'the tops up and roots down' arrangement of the mattress assuring a pleasantly aromatic and sound sleep; it was used in part of the process of TANNING leather; and the fresh, young tops of the heather were (and at times still are) brewed into a kind of ale. Little wonder it is acclaimed in song and story and longed for by the expatriate Scot![2]

Anne Barker, while collecting herbal remedies in Scotland between 2007 and 2009, came across an informant in Talmine, Sutherland, who could remember seeing a heather bed being made, while a man in Wester Ross remembered how his father, a carpenter born in 1860, described how such beds were constructed:

> The plants were cut just above the roots, placed flower-side up in a box frame, and then packed in as tightly as possible until the frame was filled. It was very pleasant to sleep on, he said.[3]

In Ireland the Vikings, or in Scotland the Picts, were said to have made 'the powerfullest drink ever known', apparently a sweet ALE, from heather. When the last individuals of their race, a father and son, were captured their captors offered to spare them on condition that they would tell them how to make their ale. The father said 'Kill my son first', I am ashamed to speak in his presence.' When the son was killed, he said, 'Now, I am the only one who knows the secret; you may kill me, I will never tell it to you.' So the secret of making the ale was lost for ever. Since that time people have experimented and tried to rediscover how ale was made from heather, however, it seems that it is not possible to make a brew in which heather is the major ingredient, and it has been suggested that, in fact, the sweet Viking, or Pict, ale was brewed from honey made from heather nectar.[4]

However, Amy Stewart Fraser writing of Ballater, Aberdeenshire, in 1973, recalled:

> Mrs Leys knew all about heather ale, a popular home brew in olden times, which was made in August or September when heather was at its best. They filled a large pan with the purple flowers, covered them with water and boiled them for an hour. This was strained into a wash-tub, and ginger, hops, and golden syrup were added. Again the mixture was boiled and strained, and yeast was added when the mixture cooled. The liquid after a few days was gently poured off, leaving the barm at the bottom of the tub.[5]

Allen and Hatfield note that the three common species of heather, *Calluna vulgaris*, *Erica cinerea* and *E. tetralix*, 'are rarely if ever distinguished in the folk records', and conclude that 'as the three are readily told apart . . . this has probably been due less to taxonomic myopia than because they were assumed to share the same medicinal virtues'. They then list a wide variety of ailments which 'heather' was used to treat. Early in the eighteenth century it was considered a panacea in the Shetland Islands. In the Highlands it was valued against INSOMNIA for soothing the NERVES, and in England to treat RHEUMATISM and DIARRHOEA. But it seems that the Irish found the greatest number of medicinal uses for heather, using it to treat a weak HEART in Co. Clare, COUGHS in Co. Wicklow, ASTHMA in Co. Tipperary, rheumatism in Co. Cavan, and 'a bad STOMACH' in Co. Limerick.[6]

See also WHITE HEATHER.

HEATH-RUSH (*Juncus squarrosus*)

Tufted perennial herb, common on acidic heaths and moorland.

Local names include:

BENT in Berwickshire
BLACK BENT in northern England
Brockles in Scotland
Bruckles in Aberdeenshire; 'a word most expressive of the wire-like hardness and rigidity of the species'
Goose-corn in Cumberland
Lubba in Shetland
Rose-bent in Berwickshire
STAR in north-east Scotland; 'a bunch of stars or bruckles to redd the tobacco pipes'
Star-bent in Cumberland
Stool-bent in southern Scotland
Whirl-bent in Cumberland.

> Heath-rush = burra – the stalk (burri-stikkel) – is a popular stalk to chew while walking or working on the moorland, but it's said that if you chew too many you'll become a rabbit-mouth or get a hare-lip. Burri-stikkels were gathered to make fireside brushes well into this century. The heath rush always grows on firm dry ground, hence the saying: 'Stramp (step) fair on da burra; keep wide a da floss [soft rush].'
>
> Another use for burri-stikkels was to bind a handful tightly and use as a pot scrubber.[1]

HEDGE VERONICA (*Hebe* x *franciscana*)

Hybrid of garden origin, both parents being native to New Zealand, small shrub first raised in 1859 and widely grown in gardens as a hardy, evergreen, usually purple-blue flowered shrub, becoming naturalised mainly in coastal areas.

> [Cornwall] My family are the only people I have ever come across who have used the leaves of the plant Veronica for DIARRHOEA; peel the leaves of the plant down until the smallest pair and eat about six.
>
> Some years ago, when I lived in Plymouth, I went to an exhibition on healing plants and folklore at the local museum. One of the exhibits was Veronica which, it said, was used by Maoris for stomach troubles . . . I know my great-grandfather sailed on tea clippers, so I wondered if he had also been to New Zealand and maybe brought this idea back with him.[1]

> Children take the closed leaf tips of hedge veronica . . . the top leaf is curled back to make a sail, leaving the lower leaf as a keel, then they are floated as little boats.[2]

HELLEBORE (*Helleborus* spp.)

Green hellebore (*H. viridis*) – perennial herb, growing at wood margins, in hedgerows and in other shady places in lowland arcas; also cultivated in gardens.

Local names include:
BEAR'S FOOT in Gloucestershire
Boar's foot in Buckinghamshire
Fellin-girse in Cumberland
FELLON-GRASS in Westmorland
Green lily in Wiltshire
She-barfoot, as opposed to stinking hellebore, he-barfoot, in Warwickshire.

Stinking hellebore (*H. foetidus*) – perennial herb growing in woodland glades scrub on calcareous soils, also cultivated in, and escaping from, gardens.

Local names include the widespread bear's foot and setterwort, and:
Barfoot, 'a corruption of bear's foot' in Warwickshire
Gargut-root, 'gargut is said to be a disease incident in calves, in the treatment of which this root is doubtless employed', in Norfolk

He-barfoot, as opposed to green hellebore, she-barfoot, in Warwickshire
Setter in Norfolk (see below)
Setter-grass, 'from its use in setting or putting on a seton', in Yorkshire.

In February 1762:

> Two young children died at Fisherton Anger [Wiltshire] in a few hours after eating some bears foot, a plant recommended against WORMS. There are two sorts of this plant: 1. Two feet high, dark green leaves and whitish flowers a little purpled at the edge, now in flower – this is poisonous [stinking hellebore]: 2. A low plant not a foot high with fish-green leaves and green flowers [green hellebore]. This is good against worms.[1]

Almost a century later, on the Isle of Wight it was recorded that stinking hellebore:

> is often seen in cottage-gardens, being a rustic remedy for worms in children, but the employment of so violent a medicine has too often been followed by serious consequences, and its use is now abandoned in regular practice.[2]

In Cumbria:

> [Green hellebore] is sometimes grown by farms and has been used as a cure for cattle ailments.[3]

According to Allen and Hatfield green hellebore was used to treat 'swollen udders in cows' in Cumberland, while veterinary uses of stinking hellebore included treating FOOT-AND-MOUTH DISEASE in Leicestershire, and 'draining "bad humours" from ruminants by "settering" or "felling", involving insertion of this in open wound made in ear, dewlap or above forelegs.'[4]

HEMLOCK (*Conium maculatum*)

Erect, biennial, poisonous herb with white flowers, and under some conditions a distinctive mousey smell, believed to be an ancient introduction, now widespread and common, mostly on dampish waste ground throughout lowland areas.

Local names, many of which were also given to the similar-looking COW PARSLEY with which it is undoubtedly confused, include the widespread variant humlock, and kelk, and:

BAD-MAN'S OATMEAL in Co. Durham, Northumberland and Yorkshire; 'i.e. devil's oatmeal'
Bad-man whotmeal in Yorkshire; 'unlucky to pick (probably because it is poisonous)'
BREAK-YOUR-MOTHER'S-HEART in Dorset
BUNK in Norfolk
Caise in Yorkshire
CAKE-SEED in Devon
Cakezie, and cart-wheel, in Somerset
Dead-man's oatmeal in Northumberland
Devil's blossom in Devon
DEVIL'S FLOWER, gypsy's curtains, GYPSY FLOWER, and hare's parsley in Somerset
Hech-how in Scotland
Hever in Dorset
HONITON LACE in Devon and Somerset
Humly in Roxburghshire
Kaka in Orkney
Kakezie in Devon and Somerset
KEGGAS in Cornwall
Keish, given to the hollow stems, and 'no doubt applied to Umbelliferae generally' in North Lancashire
KESH in northern England
KEWSE in Lancashire, Lincolnshire and Yorkshire
Kexies in Somerset
Kisky in Cornwall
Koushe in Lancashire and Lincolnshire
Koushle in Lincolnshire
LACE-FLOWER, LADY'S LACE and LADY'S NEEDLEWORK, in Somerset
MOTHER-DIE in Cheshire
NOSEBLEED in Dorset
PICKPOCKET, SCABBY HANDS and STINK FLOWER, in Somerset.

A Suffolk cure for a sore EYE consisted of the leaves of true hemlock (*Conium maculatum*) chopped finely and mixed with white of egg, bay salt and red ochre. The resulting salve was applied, however, to the sound eye and not to the affected one.[1]

HEMP (*Cannabis sativa*), also known as cannabis

Annual herb, cultivated in the British Isles since early in the fourteenth century, as a source of fibre, and illegally for the production of the recreational drug known as grass, marijuana or pot; seeds used for bird-feed, and as bait by anglers.

Local names include: brunnel and BUNNLE, given to dry stalks, in Cumberland; gallow-grass in Suffolk; neck-weed, 'furnishing halters for the gibbet'; and nogs in Shropshire.

In the nineteenth century hemp seed was used in LOVE DIVINATION. In Guernsey in the 1880s:

> A vision of your future husband can . . . be obtained by the sowing of hemp-seed. The young maiden must scatter on the ground some hemp-seed, saying:
>
> Hemp-seed I sow, hemp-seed grow,
> For my true love to come and mow.
>
> Having done this she must immediately run into the house to prevent her legs being cut off by the reaper's sickle, and looking back she will see the longed-for lover mowing the hemp, which has grown so rapidly, and so mysteriously.[1]

At Wolvercote, Oxfordshire, early in the 1920s:

> Mrs Calcutt's mother was probably the last girl to try the charm of sowing hempseed . . . She, with a girl friend, went to the churchyard one CHRISTMAS EVE at midnight, carrying some hempseed, while throwing it over her left shoulder said:
>
> I sow hempseed,
> Hempseed I sow,
> He that is to be my husband,
> Come after me and mow,
> Not in his best or Sunday array,
> But in the clothes he wears every day!
>
> The friend with her was very much frightened; some people said she saw a coffin, but whatever she saw, or thought she saw, it is certain she died soon afterwards, and the people in the village evidently connected her death in some way with the visit to the churchyard, as they forbade their daughters to try this charm any more.[2]

Variations in the custom appear to have been small, the main differences being the date chosen. MIDSUMMER'S EVE seems to have been the date most widely favoured,[3] other dates included St Valentine's Eve in Derbyshire and Devon,[4] St Mark's Eve (24 April) in parts of East Anglia,[5] and St Martin's Eve (10 November) in Norfolk.[6] Hemp-seed divination on Old Midsummer Eve forms a pivotal scene in Thomas Hardy's novel *The Woodlanders*, first published in 1887. This event appears to be atypical in that his Hintock maidens sowed their seed in dark recesses in the woods, and although the village girls were the main participants, 'half the parish' turned up, young men positioning themselves so that after the sowing they would be the first to be seen by their potential lovers.

HENBANE (*Hyoscyamus niger*)

Annual or biennial herb, with yellow, purple-veined, flowers, scattered on disturbed calcareous soils mainly in southern areas.

Local names include:

HENPEN in Westmorland
HEN-PENNY in the Lake District
Hogbean in Cumberland; 'a translation of the Greek *huoskuamos*'
Hog's bean in Somerset; 'the seed is shaped like a bean, and pigs are said to eat the plant'
Loaves-of-bread, given to the seed capsules by Northamptonshire children
STINKING ROGER in Cumberland.

According to John Ray in 1660:

> The seed of Hyoscyamus placed on coal gives off a smoke with a very unpleasant smell: when passed through the mouth and nostrils by a tube it drives out small worms (vermiculi) which sometimes grow in the nostrils or the teeth. They can be caught in a basin of water so that they can be seen better.[1]

In October 1817 the rector of Wath, Yorkshire, recorded:

> Mr Faber . . . said he heard the TOOTHACHE accounted for in the following manner by a friend and he gave the account as if he believed it. He said that certain minute ephemerae of butterfly species flying about are accidentally taken into the mouth and they then make a nidus in a rotten tooth where they deposit their eggs which in the process of time are hatched and produce minute grubs which immediately begin feeding on the nerves of the tooth

> and that the remedy applied by his friend was to procure the seeds of henbane make them very dry and then set them on fire under a tin funnel, the small end of which is to be directed so that the smoke may issue against the offending tooth which will immediately kill the grubs, and that the friend had ejected several in the saliva after the operation and seen them distinct with a lens.[2]

Forty years later:

> [Henbane] is sometimes smoked like tobacco by country people as a remedy for toothache, but convulsions have occasionally followed its use in this way.[3]

However, John Gerard dismissed such cures:

> The seede [of henbane] is used by mountibancke tooth-drawers which runne about the country, for to cause woormes come foorth of mens teeth, by burning it in a chaffing dish with coles, the partie holding his mouth over the flume thereof: but some craftie companions to gaine money convey small lutestrings into the water, persuading the patient, that those small creepers came out of his mouth or other parts which he intended to ease.[4]

Vesey-FitzGerald records that gypsies commonly used henbane to cure HEADACHES and NEURALGIA.[5]

HERB ROBERT (*Geranium robertianum*)

Annual or biennial, strong-smelling herb with attractive pink (sometimes white) flowers; widespread and common in a wide range of habitats.

Local names include the widespread BIRD'S-EYE, ROBIN'S EYE, and Stinking Bob, and:

ADDER'S TONGUE in Essex and on the Isle of Wight
Angels in Dorset
Arb-rabbits, and baby's pinafore, in Devon
BACHELOR'S BUTTONS in Devon, Kent and Sussex
BILLY-BUTTONS in Buckinghamshire
BISCUIT in Devon
Biscuit-flower in Somerset
Blood-weed in Cumberland, and Co. Donegal, where 'held to be good for cattle passing blood'
Bloody Mary in Yorkshire
BOBBIES in Somerset

Bob-robert in Dorset
CANDLESTICKS in Devon and Dorset; 'the resemblance to candlesticks is striking when you see the pistils and the surrounding sepals when the petals have fallen off'
CAT'S EYES in Dorset and Hampshire
Chatterboxes in Dorset
Chinese lantern in Somerset
CRY-BABY in Devon and Somerset
CRY-BABY-CRAB in Somerset
Cuckoo's eye in Buckinghamshire and Kent
CUCKOO'S MEAT, and CUCKOO'S VICTUALS, in Buckinghamshire
Death-come-quickly in Cumberland
Dog's toe in Co. Donegal; 'in allusion to the fruit-carpels'
Doll's shoes in Somerset
Dolly's apron in Devon and Somerset
Dolly's nightcap, and dolly's pinafore, in Devon
Dragon's blood in Shropshire
DRUNKARDS in Somerset
FELLON-GRASS, and fellon-wort, in Yorkshire
Fox-flower, and fox-geranium, 'from the disagreeable scent of its leaves', in Berwickshire
Garden-gate in Buckinghamshire
Goosebill in Dorset
Granny's needles, GRANNY-THREAD-THE-NEEDLE, GYPSY-FLOWER and GYPSY'S PARSLEY, in Somerset
HEADACHE/S, in Buckinghamshire and Somerset
Hedge-lovers in Devon
HEN-AND-CHICKENS, and HOP O' MY THUMB, in Somerset
JACK-BY-THE-HEDGE in Cornwall and Somerset
Jack Horner in Devon
Jack-flower in Somerset
Jam jars in Dorset
Jenny-flower in Wiltshire
Jenny-hood in Devon and Somerset
Jenny-wren in Somerset
Joe Standley in Dorset
John Hood, and knife-and-fork, in Buckinghamshire and Somerset
KISS-ME, and KISS-ME-LOVE, in Devon
Kiss-me-love-at-the-garden-gate in Buckinghamshire
KISS-ME-QUICK in Devon, Dorset, Somerset and Wiltshire
KNIVES-AND-FORKS in Somerset
Lady Janes in Dorset
Little bachelor's buttons in Sussex
Little crane's-bill in Somerset
Little Jack, little Jan, and little Jen, and Little-John-Robin-Hood, in Devon
Little robin in Devon and Kent
Mother-thread-my-needle in Somerset
Nancy-dancy in Somerset
NIGHTINGALE in Buckinghamshire
Old-woman-threading-her-needle in Somerset
Pink bird's-eye in Buckinghamshire and Somerset
Pink pinafores in Dorset
POOR JANE in Somerset
Poor Robert, and POOR ROBIN, in

Devon and Dorset
RAGGED ROBIN Buckinghamshire and Dorset
RED-BIRD'S-EYE in Oxfordshire
Redbreast in Northamptonshire
RED RIDING-HOOD in Dorset
RED ROBIN in Cambridgeshire, Essex, Hertfordshire and Norfolk
RED WEED in Cheshire
REDSHANK in Yorkshire
Robert, Robert's bill, and ROBIN, in Devon
ROBIN HOOD in Devon and Somerset
Robin-i'-th'-hedge in Yorkshire
ROBIN REDBREAST in Northamptonshire
ROBIN'S FLOWER in Devon and Somerset
ROUND-ROBIN in Devon
Rubwort in Cheshire
Sailor's knot in Buckinghamshire
Scotch geranium in Forfarshire
SLEEPY HEAD in East Lothian
Small robin's-eye in Gloucestershire
Snapjacks in Somerset
SOLDIERS in Devon
SOLDIER'S BUTTONS in Buckinghamshire
Sparrow-birds in Devon and Somerset
Spring-flower in Dorset
Squinter-pip in Shropshire
STAR-LIGHT, STARS, stink-flower, and stinker bobs, in Somerset
Stinkin' Bobby in Cumberland
Stinking Robert in Co. Donegal
Stinking Roger in Co. Derry
Storks in Dorset
Stork's bill, and wandering Willie, in Somerset
Wild geranium in Devon, Berwickshire and Dumfriesshire
Wild pink in Gloucestershire
Wren in Cornwall
Wren-flower in Somerset
Wren's flower in Devon.

There are occasional records of herb robert being considered inauspicious because of an association with HEADACHES, SNAKES and THUNDER.

> [Crowborough, Sussex] HEADACHE-FLOWER – we smelled it like mad to see if anything would happen and it never did![1]

> Headache-plant . . . children were undecided whether it caused or cured 'an eddake'.[2]

[Hardwicke, near Gloucester] Herb robert was also called SNAKE FLOWER and was never picked because snakes would emerge from the stems.[3]

SNAKE'S FOOD, a pink flower with a red stem, we always avoided.[4]

Another plant . . . a pretty pink small flower, feathery leaves, smelt of fox musk if you touched it – we [were] told not to touch it as we always seemed to hear the rumble of thunder before our walk was over if we did.[5]

In Ireland herb robert was occasionally used in folk medicine.

People who suffered from KIDNEY trouble long ago boiled a green weed with a little pink flower on it until all the sap and juice was out of them, and the drink was allowed to cool and when cool was drunk by the person. It was always known to remove pain. The name of the weed is herb robert.[6]

Herb robert if boiled in milk and the juice given to cattle to drink will cure the MURRAIN.[7]

HERNIA – treated using ASH and BRAMBLE.

HIROSHIMA – Japanese CHERRY used to commemorate bombing of.

HOARSENESS – treated using BLACK CURRANT.

HOARY CRESS (*Lepidium draba*)

Perennial white-flowered herb, said to have been accidentally introduced via Swansea in 1802 and later via other ports, rapidly spreading throughout southern England and coastal Wales, but still scarce in Scotland and Ireland.

Local names include:

Curse-of-Thanet in Kent
Devil's cabbage in Essex and Kent
Pepperwort in Essex
WHITE WEED in Essex and Kent.

According to the *Westminster Gazette* of 6 April 1915:

The Smallholders Union omitted from their list of plants proscribed for extermination what is perhaps the most pestilent of all weeds, the whitlow-pepperwort (*Lepidium draba*), which came to us from a district not very distant from French Flanders.

> When our troops disembarked at Ramsgate after the disastrous Walcheren Expedition of 1809, the straw and other litter on which they had slept aboard ship was thrown into a chalkpit, and afterwards carted into the fields for manure by a farmer called Thompson. A huge crop of the plant, thence named 'Thompson's Curse', sprang up, spread right across England, and is now attacking the North Country. The roots of this terrible pest are many feet in length.

Alternatively, according to a Whitstable woman, born in 1918:

> Thanet weed (now hoary cress) was said to have been brought to Thanet during the 1914–18 war to feed army horses; it came with hay.[1]

HOARY PLANTAIN (*Plantago media*)

Perennial herb, with characteristic pink flowering spikes; widespread, mainly in grassland on basic soils throughout much of England.

Local names include:

Ashy poker in Wiltshire
Boots-and-stockings, CHIMNEY-SWEEPS, and COTTON-FLOWER ('from the general cottony appearance of its spike'), in Somerset
Fire-leaves, and grandmother's whiskers, in Gloucestershire (see DEVIL'S BIT SCABIOUS)
Honey-plantain in Bristol
LAMB'S EAR in Cumberland
LAMB'S TONGUE in Somerset and Sussex; 'from the shape of the leaf'
LORDS-AND-LADIES in Norfolk; 'from LOVE DIVINATION by the erect scapes'
Scent-bottles in Dorset and Somerset
SHOES-AND-STOCKINGS, and SWEEP'S BRUSH, in Somerset.

> There was a game we played with lamb-tails (flowerheads of hoary plantain) and the heads of RIBWORT PLANTAIN. But only when the plant was young. One looped the stalk around itself at the base of the head, then tugged sharply upwards. If you were successful the head would fly off.[1]

HOGWEED (*Heracleum sphondylium*)

Coarse perennial herb with white flowers, common and conspicuous in rough grassy places.

Local names include the widespread cow-parsnip ('good fodder for cows'), KECK, and RABBIT/S-MEAT, and:

Alderdraught, alderdrots, and all-rot, in Somerset

Altrot in Devon, Dorset, Somerset and Wiltshire

Arrow-rot, and bear's breech, in Somerset

Bear-skeiters in Morayshire; 'i.e. BARLEY-shooters'

Bee's nest in Somerset

BEGGAR-WEED in Bedfordshire

BILDERS in Cornwall, Devon and Somerset

BILLER/S in Devon; 'and allied plants'

Broad-kelk in Yorkshire; 'from the large leaves, kelk being an equivalent of KECK'

Bullers in Somerset

BUNDWEED in Suffolk

Buneweed in Yorkshire and Scotland

BUNNEL, or BUNNLE, in Cumberland and Lanarkshire; applied to the 'dry stalks' in the former.

Bunnen, bunnerts, and bunwort, in Yorkshire and Scotland

Caddell, and cadweed, in Devon

CAKE-SEED in Devon and Dorset

Camlicks in Suffolk

Cathaw-blow in Cumberland

Caxlies in Somerset

CLOGWEED in Gloucestershire and Somerset

Cow-belly, and cow-bumble, in Somerset

Cow-cakes in Lothian and Roxburghshire

Cow-clogweed in Gloucestershire

COW-FLOP in Cornwall

Cow-keeks in Berwickshire

Cow-keep in Fifeshire; 'another form of [cow-keeks], or maybe because cows are fond of it'

Cushia in northern England

DEVIL'S OATMEAL in Warwickshire

Dry-kesh in Cumberland

Dryland-scout in Co. Tyrone

ELTROT in south-west England

Geagles in Cornwall

Gypsy's lace/s in Somerset

HARD-HEADS in Gloucestershire

HEMLOCK in Banffshire

HORSE-PARSLEY in Somerset
Humpy-scrumples in Devon
Kecksie in Shetland
KEDLOCK in Cheshire and Lancashire
Kegga in Cornwall
KEGLUS in Cheshire
Keksi in Shetland
KESH in northern England
KESK in Cumberland
Kewse in Lancashire, Lincolnshire and Yorkshire
Kex in Warwickshire, Yorkshire and Scotland
Kishies in Cornwall
Limpenskrimps in Devon; 'before matches were introduced bundles of these were hung up near the fire, and anyone wanting a light to go into another room . . . lighted [one]'
Lisamoo, in Cornwall; 'Cornish *les-an-mogh*, pigweed'
Lump-an-scrump in Cornwall
Lumpern-scrump, or lumper-scrump, in Somerset
Madnep in Devon
MOTHER-DIE in the Midlands
Odhran in Northern Ireland
OLD-ROT in Somerset; 'i.e. eltrot'
Orn in Co. Donegal
Pig's bubble/s in Devon and Somerset; 'extensively collected as food for pigs, which are very fond of it'
Pig's cole, and pig's flop, in Devon
PIGWEED in Devon and Oxford
Piskies given to the 'dried stalks' in Cornwall
Rabbit's vittles in Somerset
Rough kesh in Cumberland
Rough kex in Cornwall
Rumpet-scrumps, and SCABBY HANDS, in Somerset
SKYTES in Scotland; 'from being used as skyters, i.e. shooters'
SNAKES' MEAT, and sweet biller, in Devon
Umplescrump in Somerset
Wippul-squip, given to 'hollow green stalks for drinking cider', in Dorset and Somerset.

HAWTHORN blossom, LILAC and devil's tobacco [hogweed] – all UNLUCKY to have in the house.[1]

The stem of hogweed was used here [Barnstaple] as a cigarette substitute, and 'boy's bacca' was a common term. Gypsies smoked them commonly.[2]

[Cornish] children are accustomed to make 'skeets' or syringes out of the living stems, and by them the plant is generally spoken of as the 'skeet-plant'.[3]

My mother back in Ireland . . . I remember this plant [hogweed] she used to gather and give it to cows for RHEUMATISM . . . she used to boil it up in water and give it to them; it kept the rheumatism at bay.[4]

HOLLY (*Ilex aquifolium*)

Evergreen small tree with glossy green leaves, small white flowers and scarlet berries, widespread and frequent in woodland and hedges throughout, and often planted as an ornamental.

Throughout the British Isles holly is the plant most widely associated with CHRISTMAS, and, indeed, in some parts of England it was simply known as 'Christmas'.

> I am nearly 70 years of age and was born and bred in Norfolk . . . Holly was never known by name by my grandfather, it was always called Christmas.[1]

Similarly, F.H. Davey records Christmas as a Cornish name for holly;[2] elsewhere the name was, 'applied to holly branches cut for Christmas, but often to the tree as well'.[3]

Local names include the widespread hollin, and:

Berry-holm, when with berries, in Somerset
CHRISTMAS TREE in Suffolk
Christ's thorn in Yorkshire; cf. *Kristtorn* in Norwegian
CROCODILE, applied to 'a small variety which grows in hedgerows, and is exceedingly bristly' in Somerset and Devon, 'i.e. low hedge holly'
Free-holly, applied to the 'smooth-leaved form' in Devon
Helver in Suffolk
HOLM in Cornwall, Dorset and Somerset
Hulm in Somerset
Hulver, and hulver-bush, in Norfolk and 'also in the north of England'
Killin in Cornwall; 'Cornish *kelen*'
POISON BERRY, applied to the fruits in Yorkshire
Prick-bush in Lincolnshire
Prick-hollin in Lincolnshire and Yorkshire
Prickly Christmas in Cornwall
Sparked holly, given to variegated forms in Somerset.

Until most posters were created using computers the association with holly and Christmas was reinforced by the fact that holly leaves are easy to draw, so whenever a notice was needed to announce a church fayre or works lunch to celebrate Christmas it was invariably decorated with sprigs of stylised holly.

Before the prevalence of central heating holly's glossy leaves and scarlet berries would remain fresh throughout the Christmas season, especially as often it was not brought in until CHRISTMAS day, or, more usually, CHRISTMAS EVE:

> UNLUCKY to bring holly into the house before Christmas Eve.[4]

> [1930s–40s] Holly must not be brought into the house before Christmas Day, or sometimes Christmas Eve. It was unlucky to bring it in before. The actual day varied from person to person.[5]

> Holly and MISTLETOE should not be brought into the house before Christmas Eve.[6]

> Rose B said that when her mother was taken ill it was in the summer, and Rose carried some holly indoors and put it up. Someone came to visit her mother, and told Rose it was all her fault that her mother was ill, as she'd brought in the holly. Rose said she was terribly terribly upset to think that she'd made her mother ill, as she was too young to think it rubbish, and she grieved about it for years and years.[7]

Occasionally holly was considered to be unsafe indoors even at Christmas:

> We decorated the tractor trailer with holly and ivy . . . and went carol singing. After three nights of singing a piece of holly fell on the trailer floor and Stewart F (aged 9) picked it up and asked, 'Please could I have it to take home?' I said, 'Yes, of course, but if you come tomorrow when we take it all down, I'll give you some more to make a trimming,' and he was so pleased, which we thought was funny for a country child. However, in a few minutes he was back, almost in tears, and threw the holly on the trailer floor saying, 'I can't have it, as Mummy says it is unlucky and will not have it indoors.'[8]

> My mother would not allow LILAC or holly in the house, neither will I, she said it was unlucky; I believed her then, and I believe her now, though I am well past three score and ten.[9]

> Lilac is supposed to be unlucky if taken indoors, also holly with red berries if taken indoors at Christmas time.[10]

Despite such prohibitions well-berried holly can be a valuable commodity in the weeks before Christmas. Thus on 24 December 1980, *The Times* reported:

> In the principal markets holly has been selling at £1.25 for a generous handful, and £5 for an armload. Supplies come more or less equally from farmers who find a useful supplementary cash crop in their hedgerows, and gypsies who gather it with or without permission where they may.
>
> Some gypsy families supplying dealers at the Western International Market, Southall, London, reckon to have made as much as £1,500 on holly alone this year . . .
>
> It seems likely that the total sales of holly this year will exceed £500,000 at wholesale prices. What retailers will take, at 20p to 30p a sprig, is anyone's guess.

Holly has several disadvantages as a plant for Christmas decorations: its leaves soon lose their attractive sheen and drop, and its twigs are often ungainly and difficult to arrange. Consequently it appears to be dropping out of favour, being to a certain extent replaced by the more shapely, leaf-less, winterberry (also known as Dutch ilex), *Ilex verticillata*, native to the United States and deciduous, so that its numerous berries appear on stems which are devoid of leaves.

In some parts, notably it seems Cornwall, a holly bush, or a large branch of holly, was used, or is, instead of the more usual conifer as a CHRISTMAS TREE:

> As a child in the late 1920s on a farm in Warwickshire, our 'Christmas tree' was always a huge bunch of holly hung from a beam – decorated with baubles as the normal tree is.[11]

> My own mother was a Cornish Bard and greatly concerned in keeping Cornish traditions alive, and we always had a holly bush instead of a pine tree for our Xmas tree; probably fir trees only started getting popular after the 1914–18 war. Two hoops pushed together and bound with holly was also known as a 'bush' and this was actually the traditional Cornish decoration.[12]

> My maternal grandparents, who lived at Mosterton and later Corscombe, Dorset, in the 1950s and early 1960s always had a large holly bush as their Christmas tree. As the children were given balloons to play with during the afternoon there were usually some tears when these touched the tree and popped.[13]

> Many Cornish families (including my family) would never dream of having a conifer as a Christmas tree – it has to be holly.[14]

The disposal of Christmas holly was subject to various restrictions:

> I am nearly 70 years of age . . . after decorations were taken down after Christmas woe betide anyone who put the holly on the fire in the house; it was another taboo of my [Norfolk] grandmother; it had to go out on the muck heap.[15]

> Burning holly in the house will burn the family fortune.[16]

> In church decorations . . . holly and IVY at Christmas . . . all out by Epiphany (12th Night) or bad luck.[17]

> I am now 75 years old. I have lived on a farm all my life . . . All decorations, holly and mistletoe had to be taken down before old Christmas day (or TWELFTH NIGHT). A saying . . . 'It must come down before old Christmas day, or the DEVIL will dance on every spray'.[18]

Although well-berried holly trees are often plundered at Christmas time, there is a widespread belief that holly trees should not be cut down or destroyed:

> Joan . . . who was born, I suspect, about 1910 and has lived her whole life in Hove and Havant . . . states that in her experience farmers find it unlucky to cut holly. To the extent that they cut their hedges around it. (I had not myself noticed that in south Hampshire their current machinery is sensitive to that – or any other consideration).[19]

> A holly tree should never be cut down . . . a farmer wanted a holly tree cutting down. He knew he shouldn't do it, so he asked one of his labourers. The labourer refused in spite of being threatened with the sack.
>
> Eventually the farmer found someone to cut the tree down, who did not believe in 'Old Wives' Tales'. This person was dead within three months, even though before cutting the tree down he was perfectly healthy.[20]

> I won't have holly trees felled unless they are dead.[21]

> Random recollections . . . mainly from South Wales . . . where holly is growing in a hedge at least one stem must be allowed to grow up freely 'to protect the hedge from LIGHTNING' (or so it was said).[22]

Possibly these restrictions on cutting down holly trees relate to their value as a winter food for farm animals. Feeding holly to CATTLE and, especially SHEEP 'during the winter is an ancient practice that doubtless goes back into prehistory', being a 'widespread, if not always documented practice up until the eighteenth century', and continuing in Derbyshire, Cumbria and Dumfries until at least the late twentieth century.[23]

In common with other berry-bearing wild trees, a good crop of holly berries foretold a hard winter:

> [According to my mother, born in Lichfield, Staffordshire, 1916] many holly berries mean a hard winter.[24]

> People used to predict a severe winter if they saw red berries in abundance on holly trees. This was because they considered nature was providing birds with enough food while the snow and frost lasted.[25]

> If a popular theory is to be believed, an abundance of holly berries heralds a hard winter . . . I have never seen so many as there are in this area this year; one bush in my garden has produced them for the first time ever.[26]

> If a holly tree is full of berries at Christmas, it is a sign of a bad winter, nature giving food to the birds.[27]

In folk medicine holly was used to treat CHILBLAINS, most commonly:

> A local cure for chilblains – thrash them with holly until they bleed.[28]

> I am now 75 years old, I have lived on a farm all my life . . . holly was considered a good remedy for chilblains by tanning the chilblains with a small twig . . . we had chilblains always in the winter time and I think one pain must have camouflaged the other, but we did it.[29]

Alternatively, in Wiltshire:

> I have often had powdered holly berries mixed with lard rubbed on my chilblains.[30]

The stiff leaves of holly were used to loosen the soot in CHIMNEYS, for example, in west Dorset:

> There were times when my mother would complain that the kitchen range would not 'draw' and the chimney would smoke, then my father, when he could put up with the condition no longer, would cut a small 'holmen' (holly) bush, attach this with the help of a bit of binder twine, to the end of a plough line, or similar length of rope, then he would weight the other end of the rope with a suitable object, a stone or brick. Now armed with this and a hammer and chisel he would ascend the stairs to the attic – there he would take a number of bricks out of the flue and drop the weight down the chimney. He then came down stairs to the kitchen, and by pulling on the rope would draw the holly bush down, bringing all the soot with it. Things would be better after that, then he would replace the bricks in the hole in the flue in the attic and all would be well.[31]

Holly was formerly used in the production of bird-lime, a sticky substance used for trapping small birds which were then eaten. On Colonsay:

> Bird-lime is the juice of the holly-bark extracted by boiling, mixed with a third part of nut-oil.[32]

HOLY THORN (*Crataegus monogyna* 'Biflora'), also known as Glastonbury thorn.

A variety of the common HAWTHORN (Fig. 22) which flowers in the winter, as well as in late April and May, each year.

What appears to be the earliest reference to the Holy thorn can be found in a lengthy anonymous poem, entitled *Here begynneth the lyfe of Joseph of Armathia*, believed to have been written early in the sixteenth century. The poet mentions that there were three thorn trees growing on Wearyall Hill, just south of Glastonbury in Somerset, which:

> Do burge and bere grene leaues at Christmas
> As fesihe as other in May when ye nightingale
> Wrestes out her notes musycall as pure glas.[1]

However, there are hints that the Thorn may have existed several centuries earlier. The Canadian botanist J.B. Phipps, who has spent much of his professional life studying hawthorns, speculates:

> Only about 16 kilometres from Glastonbury is the village of Hallatrow, whose name is a corruption of the Saxon *Helgetrev* or Holy Tree. Helgetrev already existed in 1087 as it is mentioned in the Domesday Book. Is it not possible that the holy tree of Helgetrev was a twice-flowering hawthorn and that, after its remarkable and perhaps miraculous winter-flowering characteristics had been fully appreciated and assessed, it was moved to Glastonbury Abbey?[2]

At Appleton Thorn in Cheshire, a custom known as 'Bawming the Thorn', which consisted of decorating a thorn tree which grows in the centre of the village, used to be performed each year. Local tradition asserts that a tree has stood on this site since 1125, when an offshoot of the Holy Thorn was planted there.[3] If there is any truth in this tradition, it would imply that there was a Thorn tree at Glastonbury early in the twelfth century, when the monks at its Benedictine abbey were busily accumulating their massive, but poorly authenticated, collection of relics, which was destroyed in a disastrous fire in 1184. It is quite possible that a hawthorn which flowered at about Christmas time might be added to the attractions provided to stimulate pilgrimages to the Abbey.

The *Lyfe* provides no information on the tree's origin, and does not mention the production of winter flowers. Fifteen years after its publication, four years before the suppression of Glastonbury Abbey, the Christmas flowering of the Thorn was first recorded. On 24 August 1535 Dr Layton, the visitor sent to the Abbey, wrote to Thomas Cromwell, enclosing two pieces of a tree which blossomed on Christmas Eve:

> By this bringer, my servant, I send you two Relicks: First two flowers wraped in white and black sarsnet, that in Christen Mass Even, *hora ipsa qua Christus natus fuerat*, will spring and burge and bare blossoms. *Quod expertum est*, saith the Prior of Mayden Bradley.[4]

During the reign of Elizabeth I the Thorn growing on Wearyall had two trunks:

> when a puritain exterminated one, and left the other, which was the size of a common man, to be viewed in wonder by strangers; and the blossoms thereof were esteemed such curiosities by

> people of all nations that Bristol merchants made traffick of them and exported them to foreign parts.[5]

Or, according to an earlier account:

> It had two Trunks of Bodies till the Reign of Queen Elizabeth, in whose days a Saint like Puritain, taking offence at it, hewed down the biggest of the Trunks, and had cut down the other Body in all likelyhood, had he not bin miraculously punished . . . by cutting his Leg, and one of the Chips flying up to his Head, which put out one of his Eyes. Though the Trunk cut off was separated quite from the root, excepting a little of the Bark which stuck to the rest of the Body, and laid above the Ground above thirty Years together; yet it still continued to flourish as did the other Part which was left standing; after this again, when it was quite taken away and cast into a Ditch, if flourished and budded as it used to do before. A Year after this, it was stolen away, not known by whom or whither.[6]

Later, during the reign of James I, the Thorn became popular as a garden curiosity, and the nobility, including the King's wife, Anne of Denmark, paid large sums for cuttings.[7] It is possible that this fashion saved the tree from extinction, for during the civil unrest later in the century the surviving trunk of the original tree was destroyed by a Roundhead, who 'being over zealous did cut it downe in pure devotion'.[8] In 1653 Godfrey Goodman, Bishop of Gloucester, lamented: 'The White Thorn at Glastonbury which did usually blossome on Christmas Day was cut down: yet I did not heare that the party was punished.'[9]

In 1645 John Eachard described the Thorn, then much mutilated by visitors taking bits as souvenirs, as being of the kind 'wherewith Christ was crowned'. An elaboration of this belief is that St Joseph of Arimathea brought two treasures to Glastonbury: silver containers holding the blood and sweat of Christ (which seem to have become confused or equated with the Holy Grail) and a thorn from Christ's crown of thorns, which grew and proved its holiness by flowering each year at the time of Christ's birth.[10]

Seventy years after Eachard wrote, an oral tradition collected from a Glastonbury inn-keeper claimed the Thorn had grown from a STAFF carried by Joseph of Arimathea.[11] According to tradition, the Apostles divided the world between them, with St Philip being sent to Gaul, accompanied by Joseph, who is usually thought to be an uncle of the Virgin Mary. After some years Joseph left the Apostle and with eleven others set out for Britain, arriving at Glastonbury and eventually founding the first church to be built in these islands, in AD 63.[12] When Joseph reached Glastonbury he rested on Wearyall Hill and thrust his staff into the ground, where it grew and became the original Holy Thorn. Some writers assert that it was this miracle which caused Joseph to settle in Glastonbury.

A second version of the legend relates how Joseph landed on the Welsh coast, or possibly at Barrow Bay in Somerset, but found the natives hostile. He continued his wanderings until he reached the land of King Arviragus. Although he was unable to convert the monarch he made a sufficiently good impression for land at Ynyswitrin – Glastonbury – to be granted to him and his companions. However the local people showed little enthusiasm for the new faith, and it was not until Joseph fixed his staff in the ground and prayed, whereupon it immediately blossomed, that they began to pay serious attention to the missionaries.[13] It is sometimes claimed that Joseph performed this miracle at Christmas and hence the tree has flowered on this day ever since.[14]

Some writers have asserted that there is some truth in the various legends and the Thorn did, indeed, originate from stock brought from the Holy Land, or at least a country bordering the Mediterranean. The winter flowering of the tree is explained by the suggestion that it belongs to a variety of hawthorn native to the Middle East.[15] In a leaflet produced by St John's church in Glastonbury in 1977, the then vicar of the parish firmly stated: 'Whatever the legend may say, a Thorn has been growing here for 2,000 years and it came from Palestine.' According to a 1992 study of hawthorns:

> In North Africa, flowering in late autumn and early winter is known also in populations of *C[rataegus] monogyna* that are morphologically fairly similar to the Holy Thorn.[16]

A young leafy shoot of hawthorn, labelled '*Oxyacantha autumnalis*, from Wells, Joseph of Arymathea rod', is preserved in the herbarium of the Natural History Museum in London. This specimen was included in a collection given by Robert Nicholls to the Apothecaries' Company in 1745, and formed part of 'a valuable series of plants' presented by the Company to the Museum in 1862.[17]

According to a chapbook published in 1777:

> And though the Times of superstitious Popery, in these Kingdoms, be abolished, yet Thousands of People, of different Opinions, go once a Year to see it [the Thorn], as being a most miraculous Curiosity; which also brings Foreigners beyond the Sea to behold it, at its annual Time of shewing a Wonder that is really supernatural, as being a Matter contrary to the Course of Nature, and may make us cry out with the Psalmist, 'O Lord! My God, how marvellous are thy Ways!'[18]

Also during the eighteenth century it is said that a miller trudged all the way from his home in Wales to visit the Thorn. His English vocabulary was restricted to three words, 'Staff of Joseph', but these were sufficient to ensure that he reached Glastonbury and was able to carry home a sprig from the tree.[19]

When the calendar was reformed in 1752 the Holy Thorn attracted considerable attention; people watched their trees to see if they would produce their Christmas blossoms according to the new or old calendar. In January 1753 the *Gentleman's Magazine* reported that on Christmas Eve, 24 December 1752, hundreds of people gathered in Glastonbury to see if the Thorn trees growing there would produce flowers. No flowers appeared, but when the crowds reassembled on Old Christmas Eve, 5 January 1753, the trees blossomed, confirming the onlooker's doubts about the validity of the new calendar. However, later in 1753 a correspondent to the *Magazine* stated that after reading reports of the Thorns flowering on Old Christmas Eve in a Hull newspaper, he had questioned the vicar of Glastonbury, who assured him that the trees had blossomed 'fullest and finest about Christmas Day New Style, or rather sooner'.[20]At Quainton in Buckinghamshire, a county peculiarly resistant to the calendar change,[21] over 2,000 people gathered to watch a thorn which they remembered as being a descendant of the Glastonbury tree:

> but the people finding no appearance of bud, 'twas agreed by all, that Decemb. 25 N.S. could not be Christmas-Day and accordingly refused going to church, and treating their friends on that day as usual; at length the affair became so serious, that ministers of neighbouring villages, in order to appease the people thought it prudent to give notice, that old Christmas-Day should be kept holy as before.[22]

Until early in the twentieth century people continued to visit Holy Thorns on Old Christmas Eve. In Herefordshire:

> It is believed that the Holy Thorn blossoms at twelve o'clock on Twelfth Night, the time, so they say, at which Christ was born. The blossoms are thought to open at midnight, and drop off about an hour afterwards. A piece of thorn gathered at this hour brings LUCK, if kept for the rest of the year. Formerly crowds of people went to see the thorn blossom at this time. I went myself to Wormesley in 1908; about forty people were there, and as it was quite dark and the blossom could only be seen by candle light, it was probably the warmth of the candles which made some of the little buds seem to expand. The tree had really been in bloom for several days, the season being extremely mild.[23]

Also in Herefordshire, a Thorn in the garden of Kingston Grange was annually visited by people who came from miles around, and were 'liberally supplied with cake and cider'.[24] However, such convivial gatherings sometimes degenerated into unruly behaviour, leading to some property owners destroying thorns to prevent unwelcome visitors. Near Crewkerne in Somerset, in January 1878:

> Immense crowds gathered at a cottage between Hewish and Woolmingstone to witness the supposed blooming of a 'Holy' thorn at midnight on Saturday. The weather was unfavourable and the visitors were impatient. There were buds on the plant, but they did not burst into flower as they were said to have done the previous year. The crowd started singing and then it degenerated into a quarrel and stones were thrown. The occupier of the cottage, seeing how matters stood, pulled up the thorn and took it inside, receiving a blow on the head from a stone for his pains. A free fight ensued and more will be heard of the affair in the Magistrates' Court.[25]

Similarly:

> A Holy Thorn made a brief appearance in Dorset in 1844 in the garden of a Mr Keynes of Sutton Poyntz. It was rumoured that it had grown from a cutting of the famous Glastonbury Thorn and was expected to blossom at midnight on Old Christmas Eve. 150 people turned up to see the event. Violent scenes took place, the fence was broken down and the plant so badly damaged that it died.[26]

Not surprisingly tales were told of misfortunes which befell people who attempted to destroy or damage Holy Thorns (similar misfortune ensued when people damaged LONE BUSHES in Ireland). An early attempt to destroy a tree resulted in thorns flying from it and blinding the axeman in one eye, making him 'monocular'.[27] A man who tried to cut down a tree in his garden at Clehonger in Herefordshire was less unfortunate, and was let off with a warning: 'blood flowed from the trunk of the tree and this so alarmed him that he left off at once!'[28]

Shortly before Christmas each year sprays from a Thorn growing in St John's churchyard in Glastonbury are sent to the Queen (and, formerly to the Queen Mother). In 1929 the then vicar of Glastonbury, whose sister-in-law was a lady-in-waiting to Queen Mary, sent a sprig to the Queen, reviving according to some writers a pre-Reformation practice.[29] A report in the *Western Daily Press* of 20 December 1973 stated that the custom started in Stuart times, and it is recorded that James Montague, Bishop of Bath and Wells, sent pieces of the Thorn and Glastonbury's miraculous WALNUT tree to Queen Anne, consort of King James I.[30] About ten days before Christmas children from local schools and local dignitaries gather in the churchyard and a schoolchild cuts small twigs from the Thorn, These are then sent to the Queen, who, it is said, has them placed on her breakfast table on Christmas morning. The actual ceremony varies from year to year depending on the inclination of the vicar and the headteachers.[31]

Until recently three of the several Thorn trees were pre-eminent in Glastonbury, and each had its supporters who considered it to be *the* Holy Thorn. Two of these, growing in St John's churchyard and the Abbey grounds, survive, though the latter looks unhealthy; the third, which stood, windswept, on Wearyall Hill, was vandalised in December 2010,[32] and eventually died. According to the *Independent on Sunday* of 12 December 2010, there was 'a flourishing crop of rumours as to who was to blame'. Some thought that the vandalism was a demonstration of anti-Christian feelings, others thought that it was inspired by violence at London demonstrations against student fees, and others thought people were taking revenge against the owner of the land on which the tree stood, who owed them money.[33] In April 2012 it was reported that a new tree had been planted, enclosed in a metal guard with barbed wire, and the damaged tree was recovering.[34] However, by January 2014 it seemed probable that the old tree was dead, and the younger tree, a mere twig less than a metre high, also appeared to be dead.[35] Seven months later the younger tree had completely disappeared, leaving an empty metal frame, and the old tree was clearly dead, although ribbons and other offerings were tied to its metal railings and nightlight candles had been placed nearby.[36] By December 2015 the frame which surrounded the young tree had been removed, and it appeared that no attempt was being made to replace the tree, but the railings around the old tree continued to be hung with abundant colourful ribbons.[37]

HONESTY (*Lunaria annua*)

Biennial, purple or white-flowered herb of unknown, possibly Balkan, origin, cultivated as an ornamental since at least 1570 and known in the wild before the end of the sixteenth century, now widespread on waysides and waste ground and at woodland margins. Formerly grown mainly for its translucent silvery seedpods, which were dried for use in winter flower arrangements.

Local names, nearly all of which refer to the seedpods, include:

Charity in Somerset
Devil's ha'pence in Kent
Grandmother's spectacles, and lady's lockets, in Somerset
Love-lies-bleeding in Dorset
MONEY in Devon and Somerset
MONEY-IN-BOTH-POCKETS in Dorset and Somerset; 'the flat seed-heads look like coins', or less usually money-in-every-pocket
Money-plants in Devon
Money-pockets in Somerset
MOON-FLOWER in Dorset
Old-woman's penny in Somerset
Polly-pods in Dorset
Ready-money in Shropshire
Satin-leaves in Cheshire
SHEPHERD'S PURSE in Somerset
Shillings in Dorset and Warwickshire
Silks-and-satins in Devon
Silver-leaves in Dorset.

> A bunch of dried honesty is hung inside a wardrobe for good LUCK in some Guernsey homes. I was told at a W.I. meeting in 1973 by a member that when she got married a local lady visited her and was very concerned to find that the new bride had no honesty to hang up. She brought her some straight away. My informant said that she still has this honesty in her wardrobe although she has changed house several times since then. She had made enquiries from time to time and discovered that quite lot of people

> who know the custom themselves keep honesty in their homes for good fortune. This was corroborated by other women at the meeting.[1]

But:

> I am from Yorkshire. Another odd dislike is honesty. My father is very against it, saying it brings all kinds of bad luck. He won't even have it in the garden, never mind the house. A pity, since the seedpods are so pretty, but I can't bring it in knowing it will make someone uneasy.[2]

HONEYSUCKLE (*Lonicera periclymenum*), also known as WOODBINE (or WOODBIND)

Woody twining climber with sweet-scented flowers, widespread and common in hedgerows.

Local names include the widespread HONEY-SUCK, and:

BEAR-BIND in Cheshire
BIND in Yorkshire
BINDWEED in Craven, Yorkshire
Bindwood in Cumberland and Yorkshire
Binnwood in Yorkshire
Bugle-blooms in Dorset
Caprifoy in Devon; 'cf. the old apothecaries' name *caprifolium*'
Eglantine in Yorkshire
Evening-pride in Devon
FAIRY TRUMPET, and goat-leaf, in Somerset
Goat's leaf in Devon and Dorset; 'German *Geissblatt*, whence *caprifolium*'
Gramophone-horns in Somerset
Hold-me-tight in Derbyshire
Honey-bind in Oxfordshire
Irish vine in Co. Donegal
KETTLE-SMOCKS in Somerset
LADY'S FINGERS in northern England and southern Scotland
Lamps-of-scent in Somerset
Love-bind, 'country folk have always called it love-bind denoting its resemblance to the tight embrace of lovers'; but apparently missed by collectors of plant-names
Oodbine in Oxfordshire

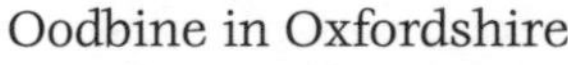

Pride-of-the-evening in Devon and Dorset
Suckle-bush in Norfolk
Suckles in Somerset
Sucklings in East Anglia
Sweet suckle in Somerset
TRUMPET-FLOWER in Somerset and Yorkshire
Widbin, 'a Scotch form of woodbine'
WITHYWIND, or WITHYWINE, in Devon
Woodbinn in Yorkshire
Woodwind in Cheshire, Gloucestershire and Shropshire.

In many areas honeysuckle was regarded as inauspicious and not allowed indoors:

> Until recently my home was in Scotland . . . Honeysuckle was never taken indoors, it was unlucky for the family, and should not be worn either.[1]

> In the early 1960s I was on holiday with my family on the Isle of Arran. I used to pick wild flowers and bring them into the hotel room where they were put in water. On one occasion, the honeysuckle was removed by the maids, and one explained to my mother that the owner (if not from Arran itself, she came from somewhere not far away) would not allow honeysuckle in the house, because on two previous occasions when that flower had been brought inside, somebody had died.[2]

> In rural Cheshire it was thought that 'honeysuckle should never be cut, because if you did you would not get a second crop of hay',[3] while in the village of Capel Hendre, Dyfed, it was said that bringing honeysuckle indoors 'gave you a sore THROAT'.[4]

In the Cambridgeshire Fens, honeysuckle was never brought into a 'home where there were young girls; it was thought to give them erotic dreams. If any was brought indoors, then a WEDDING would follow.'[5]

As its common name and many of its local names suggest, the flowers of honeysuckle were sucked for their nectar.

> As children (I was born in 1943) we used to pick honeysuckle, remove the tip at the base of the flower and suck the juice (nectar).[6]

> [Essex] honeysuckle – we used to suck the nectar out of the base of the flower.[7]

In Ireland honeysuckle was used to treat JAUNDICE:

> The bark of woodbine is a good cure for jaundice.[8]

> Jaundice is cured by Mrs Caffrey, at first she chews woodbine, and rubs it to the patient's forehead and says some prayers.[9]

HONEYSUCKLE STICK

In Sussex:

> The possession of a 'honeysuckle stick' is a guarantee of good LUCK, especially to a young man in his courtship of the lady of his choice. These sticks were 'bats' of HAZEL around which honeysuckle has entwined itself and which, when the bind has been removed, have a twisted appearance. To carry one of these when calling upon your lady love predisposes the lady in your favour and your suit will be successful.[1]

HOOD, Robin – associated with the Major OAK.

HOP (*Humulus lupulus*)

Perennial scrambling herb, native, also formerly widely cultivated mainly in Kent, Herefordshire and Worcestershire for the brewing industry, hence naturalised beyond its area of natural distribution.

In Herefordshire it was said:

> Rain on GOOD FRIDAY and EASTER DAY
> A good crop of hops, but a bad one of hay.[1]

During the Second World War:

> We used to collect NETTLES which we cooked like spinach, hop tops which were crunchy and DANDELIONS which we had in salads.[2]

In hop-growing areas dried hops are often used to decorate the bars of old-world pubs, presumably an echo of the plant's use in brewing, but sometimes said to be 'lucky'.

> In Kent they hang up hops in their houses for LUCK.[3]

Since 1986 the Hop Shop at Sevenoaks, Kent, has been providing hops for such purposes, claiming that 'garlands of hops were hung afresh every year – for good luck – in pubs and farmhouses'.[4]

In gypsy medicine:

> An ounce of hops to a pint of boiling water taken some time before meals is a good cure for loss of APPETITE. A poultice of the tops will relieve SCIATICA or LUMBAGO. An infusion of the flowers will cure WORMS in children. Put hops into a muslin bag and use the bag as a pillow and you will cure INSOMNIA.[5]

Similarly:

> According to my mother (b.1918) of Whitstable, Kent . . . dried hops under, or in, the pillow help you sleep.[6]

HORSES – BOX used as a VERMICIDE for; DOCK and JUNIPER improve their coats; IRISH SPURGE a 'grand physic for'; OAK bark used to treat sore shoulders; WHITE BRYONY used as a conditioner.

HORSE CHESTNUT (*Aesculus hippocastanum*)

Large tree, with showy white flowers and attractive shiny brown seeds – conkers – (Fig. 23) native to the Balkans, introduced early in the seventeenth century, widely planted in parklands and occasionally becoming naturalised.

Local names include:

Bongay in Suffolk
Christmas-candles, given to the flowers in Somerset
CHRISTMAS TREE in Somerset
Hobbly-flower, and hobbly-honker, in Somerset
Horse-nut in Berkshire
Horse-nut tree in Somerset
Oblionker-tree in Worcestershire
Robbers' lanterns, given to the flowers in Dorset
Roman candles in Somerset
Sheffer-tree in Oxfordshire.

Although large plantings of horse chestnut were being made at Windsor by 1699,[1] the use of its seed in the game of conkers did not become popular until much later. Thus when Britten and Holland compiled their *Dictionary of English Plant-names* (1878–86) the game seems to have been known only in a few scattered localities. Despite Britten organising boys' clubs in London, and Holland being father of a large family in Cheshire, their knowledge of the game seems scanty.

> CONQUERORS – The fruit of *Aesculus hippocastanum* L. – Ches[hire], where children thread them on strings and strike them

> one against each other. The one remaining unbroken is the 'conqueror'.[2]
>
> KONKER-TREE – *Aesculus hippocastanum* L. – Som[erset], 'A game known as Konkers is played with the fruits.'[3]

Under the name 'Oblionker' they provide additional information derived from *Notes & Queries*, in November 1878:

> Having heard this word as being in common use at Ledbury in Herefordshire, I wrote to Mr Piper, of that town – a gentleman who takes great interest in the antiquities of that county. His reply was: Oblionker is a game played by boys with horse chestnuts: each of the two contending players passes a piece of string a foot or so in length, and having a knot at the end to prevent its escape (a with of yellow willow answers equally well), through the chestnut. They then strike alternately at each other's nut whilst hung suspended, and he who succeeds in breaking that of his adversary is the winner. The first who utters the following rhyme has the right to begin:
>
> 'Obli, obli, O,
> My first go.'
>
> And on striking it is customary to say:
>
> Obli, obli, onker,
> My nut will conquer.
>
> The chestnut that has demolished the greatest number of its congeners acquires proportionate reputation, and the successes theretofore scored by the vanquished opponent are added to the achievements of the victor. Doubtless the Cymric [Welsh] boys of pre-Roman times played at oblionker.[4]

The name oblionker remained in use until at least the 1940s.

> When I went to live in Worcestershire after the last War, I found that conkers were always called Obly-Onkers, and Worcestershire children preceded the game with the solemn chant:
>
> 'Obly, obly-onker,
> My best conker,
> Obly, obly O,
> My best go!'[5]

By the second decade of the twentieth century the game seems to have become popular to the extent that in 1914 it was recorded that horse chestnut trees:

> have such an extraordinary fascination for boys in furnishing material for the game of 'conkers' (conquerors) that the value of the species as a communal tree is some districts is seriously diminished by their efforts with sticks and stones to bring down the nuts before they naturally fall.[6]

Later in the century, a correspondent in Crawley, West Sussex, asked the Natural History Museum, in London, if it was possible to have the fruit of horse-chestnuts 'doctored', but keep the tree healthy:

> The reason I ask is. Where I live we have such a tree and we are plagued each year with children throwing large sticks, stones and even house bricks to knock down the horse chestnuts. This year has been exceptionally bad. The children come over our garden fence and pull parts of the wooden structure away as missiles.

This problem can be solved by planting the variety known as 'Baumanni', 'which has double flowers which are long lasting and don't produce nuts'.[7]

Although 'conker' may be a corruption of 'conqueror', an alternative theory is that the game is a descendant of an earlier one in which snails' shells were crushed; hence 'conker' is derived from 'conch'.[8] Another earlier game was known as cobnut. In Leicestershire:

> Strings were passed through nuts, by which to use them in playing. Each player, in turn, holds his cob-nut up by the string to be 'cobbed' at by the other, and the player who first breaks his adversary's nut is the winner of the game.[9]

However, speculation about the origin of the word conker is of little or no interest to most players of the game, who are more concerned with methods by which they can obtain potential champions.

> [Ayton, North Yorkshire] Conkers: you are sure to have obtained a 'hundreder' if the chestnut was caught falling from the tree before it touched the ground.[10]

In the Northam area of Southampton in the 1930s:

> Recipes for hardening winners were varied: it might be to part bake the conker, or soak it in paraffin. Ginger Blake, Ken's friend,

> said his Dad recommended soaking them in the 'Po' overnight before baking. Ken confided in me he didn't really want to win Ginger's current prize conker.[11]

According to Jeff Cloves's *Official Conker Book*:

> Baking and vinegar are the oldest known methods of hardening your potential Conqueror and are as old as the game. In the days when everybody had a fire in their hearth, conkers were left by the fire for a few days or shoved up the chimney. With care, they wouldn't look any different from 'seasoners' [i.e. conkers which had been stored for a year]. The same applied to baking conkers in the oven. This had to be done in a low oven for about half an hour. The other method was to pickle them in vinegar.[12]

However, most players use fresh, untreated nuts, as at Shipley, West Yorkshire, in the 1940s:

> In October most boys collected horse-chestnuts or 'conkers'. The term 'conking' covered both casually looking for them and undertaking exhaustive searches at weekends. On these expeditions much expert knowledge of the peculiarities of local trees and park-keepers was amassed. Trees with pink flowers [*Aesculus carnea*, red horse-chestnut] tended to produce the inferior 'water conk', which was useless for the game of 'conkers', having nothing but fluid inside. Most park-keepers or 'parkies' objected to small boys in general, and the sport of knocking conkers off the trees with stones and sticks, or 'throwing for conks' in particular. Conkers were collected partly for their own sake; they are attractive objects and anyway small boys will collect almost everything. But they were collected ostensibly for the sake of the game. For 'conkers', large conkers had holes driven through them, and were threaded on strings. One player challenged another and perhaps claimed the right to begin by calling 'fuggy smack'. The opponent then held up his conker, dangling on its string, for the first player to aim a smack at it. They continued, aiming alternate smacks and probably watched by an excited crowd, until one conker disintegrated. The winning conker was then nominated a 'oner'. If it won again it became a 'twoer', and so on. Any score accredited to a beaten conker was passed on to the winner; thus, if a 'twoer' beat a 'fiver', it became an

'eighter'. In this way champion conkers sometimes built up large if somewhat exaggerated scores. They were carefully examined for signs of wear and tear, and maybe rethreaded to make them more secure. If the outer shell had been largely knocked off, but the kernel remained, it might be soaked in vinegar to make it harder than ever. These champion conkers could be swapped for quantities of sweets or marbles, bits of liquorice root, and so forth, but they were usually preserved by their owners. Crowds gathered to watch games in which such battered relics were concerned. As with all games there were ways of cheating at conkers. The most diabolical was to aim at your opponent's string, rather than his conker, in the hope of pulling it out of his hand and dashing it to the ground. Such treatment would obviously do no good, and one could always claim it to have been an accident. Sometimes unfair play of this kind was penalised; if the offended party called 'strings' he was awarded a free smack. There was an understandable but mistaken feeling that it was an advantage to have the smack; if your conker was cracked it would as soon break up from hitting as being hit. Sometimes, as bits of shattered conker flew everywhere, you thought you had won, only to find on looking at your string that it was your conker that was finished. Sometimes both players let go of their strings in the excitement, especially if they had become entangled; and then a dispute would arise, very serious to the players, but amusing to the crowd, over which fragments belonged to which party.

The conking season came to an end in mid-October when most boys' energies were absorbed in 'progging' [collecting material for Guy Fawkes' Night bonfires].[13]

Occasionally the game of conkers is taken up by adults who organise championships, usually to raise funds for charity. The best known of these events is the World Conker Championships. These were started at Ashton, Northamptonshire, in 1965, when a group of anglers, frustrated by the cancellation of a sea-fishing trip, took consolation in playing conkers.[14] For many years the championships were played at Ashton, but in 2013 they moved to nearby Southwick. Profits from the championships are donated to charity, and between 1965 and 2016 some £415,000 was donated, mostly to charities for the partially sighted.[15] The rules for the championships differ from those used in most playgrounds. All the conkers are supplied, ready stringed, and each player is paired with another, until one shatters the other's nut and becomes the winner. The loser drops out, and the winner, with a fresh nut, moves on to the next round. Thus players do not accumulate massive scores. In 2015 the rules stated:

1. All Conkers and Laces are supplied by the Ashton Conker Club. Laces must not be knotted further or tampered with.
2. The game will commence with a toss of a coin, the winner of the toss may elect to strike or receive.
3. A distance of no less than 8 inches (20cm) of lace must be between the knuckle and nut.
4. Each player then takes three alternate strikes at the opponent's conker.
5. Each attempted strike must be clearly aimed at the nut, no deliberate miss hits.
6. The game will be decided once one of the conkers is smashed.
7. A small piece of nut or skin remaining shall be judged out, it must be enough to mount an attack.
8. If both nuts smash at the same time then the match shall be replayed.
9. Any nut being knocked from the lace but not smashing may be re threaded and the game continued.
10. A player causing a knotting of the laces (a snag) will be noted, three snags will lead to disqualification.
11. If a game lasts for more than five minutes then play will halt and the 'five minute rule' will come into effect. Each player will be allowed up to nine further strikes at their opponent's nut, again alternating three strikes each. If neither conker has been smashed at the end of the nine strikes then the player who strikes the nut the most times during this period will be judged the winner.[16]

Elsewhere smaller championships come and go. Comparatively long-lasting events have been held at the New Inn, in Goodleigh, Devon,[17] Miskin in Mid Glamorgan,[18] and on Hampstead Heath, London, where the sixteenth annual championship was held near the bandstand in Parliament Hill Fields on 1 October 2017.[19]

It appears that the game of conkers is unknown outside Britain and Ireland, and although an annual competition is organised by *La Fédération Française de Conkers*, based in Bandiat, in the Dordogne, it was founded by British expatriates in 1991.[20] However, similar games, played with other plants, are known elsewhere. For example, in the Western Australia township of Brown Hill early in the twentieth century:

> Another game the older boys liked to play was to thread a quandong (native peach) stone on the end of a string and compete by hitting at each other's stones until one was smashed.[21]

Children inevitably found other uses for shiny fresh conkers:

> They came in very handy for making furniture for my dolls' house.
> A plump, shiny chestnut was an excellent seat for a dining-room chair, while very long pins with glass heads formed legs, and slats for the backs interwoven with wool. I was very proud of my work (age 7).[22]

Sometimes rotten conkers were used as 'stink-bombs':

> Over 40 years ago . . . we also put them [conkers] to use as stink-bombs. To explain this, the local park contained a large pond which was surrounded by horse-chestnut trees. The majority of wind-falls would land on the ground, but . . . many would end up floating on the water. As with the natural process of time, these would blacken and become brittle, allowing water to seep through and turn the centre into a putrid yellow liquid. We would gather small quantities of these fragile blackened conkers, throw on to the ground, the brittle skin cracking and releasing its odorous liquid contents to the disgust of unsuspecting passersby.[23]

Occasionally games were played with horse-chestnut leaves. In the nineteenth century around Norwich the leaf stalks were known as knuckle-bleeeders, and: 'boys try to get one another to allow them to hit them over the knuckles with the end which grows next to the branch', presumably taking it in turn until bleeding was produced.[24]

Around Chesterfield, Derbyshire, in the 1950s:

> A game was played with the leaf stalks [of horse-chestnut]. A stalk was held at each end by one person and another person would place a similarly held stalk behind it at right angles. Each would then pull in an attempt to break the other's stalk. The winner would then be challenged with another fresh stalk.[25]

Despite the attraction of conkers, horse-chestnuts are grown primarily for their attractive flowers. Chestnut Sunday is celebrated in Bushy Park, in the London Borough of Richmond, on the Sunday nearest 11 May, when its horse-chestnuts' flowers are expected to be at their best. Different websites offer different versions of the history of the event, which started in the nineteenth century, presumably after Queen Victoria had opened the Park to the public, and continued until the First World War, when it appears to have stopped, to be started after the war for what was probably its heyday in the 1930s, when bus companies ran excursions and the London Underground produced posters advertising it. After stopping again during the Second World War it was revived in 1977. The horse-chestnut trees, the flowers of which are probably already past their prime, tend to be ignored in recent years, and the main feature is a parade, which includes 'vintage bicycles, WWII military vehicles, horse rangers, marching bands and Harley Davidson motorcyclists'. Other attractions include a traditional fairground and historical re-enactments.[26]

Conkers are widely thought to deter insect pests, and, particularly, spiders.

> As a child during the early war years, when we were living on the Hants/Sussex border . . . we often used conkers for anti MOTH protection and I still do to this day, among shoes, etc.[27]

> I was advised by a friend who lives in the country and suffers enormously with spiders to put a conker in the corners of the room. This is supposed to deter spiders. I have to say I have not had a spider in my living room since I did this. Cheap remedy and worth a try.[28]

During the two world wars conkers were collected for military purposes. Early in the First World War, 1915, the need for acetone essential for the manufacture of cordite caused concern in Britain, as the usual North American supply had become erratic and costly. Chaim Weizmann, a chemist at Manchester University, was asked to investigate this problem, and initially produced a method of producing acetone from maize, but

> the shipping shortage in 1917 which forced us to restrict all unnecessary imports, introduced yet another experiment. In the autumn of that year, horse-chestnuts were plentiful, and a national collec-

> tion of them was organised for the purpose of using their starch content as a substitute for maize.

When David Lloyd George suggested to Weizmann he should receive an honour for his work, Weizmann replied that he wanted no personal honour, but spoke of his 'aspirations as to the repatriation of the Jews to the sacred land they made famous'. Lloyd George discussed these hopes with his Foreign Secretary, A.J. Balfour.

> Dr Weizmann was brought into direct contact with the Foreign Secretary. This was the beginning of an association, the out-come of which, after long examination, was the famous Balfour Declaration which became the charter of the Zionist movement [and in 1948 led to the establishment of the state of Israel].[29]

At least this is the account given by Lloyd George in his *War Memoirs*, published some twenty years later, but Geoffrey Lewis in his study of Balfour, Weizmann and Zionism states that although this story has been repeated many times, 'it was a figment of Lloyd George's imagination';[30] both Balfour and Lloyd George were sympathetic to the Zionist cause well before the outbreak of the War.

Notices in schools and scout huts read

> Collecting groups are being organised in your district. Groups of scholars and boy scouts are being organised to collect conkers. Receiving depots are being opened in most districts. All schools, Women's Voluntary Services centres, Women's Institutes, are involved. Boy Scout leaders will advise you of the nearest depot where 7s 6d per cwt is being paid for immediate delivery of the chestnuts (without their outer green husks). This collection is invaluable war work and is very urgent. Please encourage it.

Three thousand tons of conkers were taken to the factory at King's Lynn, in Norfolk, which was to manufacture the acetone, and it is estimated that several thousands of tons rotted at railway stations before they could be transported. The production process was beset by a series of problems with reliable production eventually starting in April 1918, almost too late to significantly help in the war effort.[31]

During the Second World War, in 1942, an advertisement in the

National Federation of Women's Institutes again appealed for 'scholars, Boy Scouts, etc.' to collect conkers,[32] and in 1942 it was estimated that 1,500 tons of conkers were collected.[33]

More recently, *The Times* of 28 October 1993 reported:

> Landowners may soon be growing horse chestnuts as a cash crop to sell to the international pharmaceutical industry.
>
> Forestry Commission scientists in Edinburgh have been commissioned by a German pharmaceutical company to study which strain of horse chestnut produces most aescin, a natural chemical found in conkers. It is used on the Continent for treating SPRAINS and bruising and is particularly useful in treating sports injuries.
>
> The chemical was first used by the Turks to treat bruising in horses, which, according to the commission is how horse chestnut got its name.

Other uses of conkers include the prevention of PILES:

> [Somerset, *c*.1970] old Bill, he'd had these piles for years . . . tried all manner of things, but it was just the same . . . one day this fellow got an idea about carrying a conker in his pocket, so old Bill . . . said he'd give it a go . . . He carried the conker in his pocket just like he was told, and tried to forget about it. Would you believe it, after a few days they started to go down, and soon they were all better.[34]

> Carry a conker in your pocket to prevent rheumatism.[35]

During the German Occupation of Guernsey, dried horse-chestnut leaves were used as a TOBACCO substitute.[36]

HORSERADISH (*Armoracia rusticana*)

Perennial herb with large tap roots, large leaves, and inconspicuous white flowers; of unknown origin, long cultivated in the British Isles as a medicinal herb and for the production of horseradish sauce, which traditionally accompanied roast beef; although seed-set is unknown in the British Isles, horseradish is widespread and persistent on waste-ground and sites of former gardens.

> A simple method of determining the sex of an unborn baby was for a Fenland couple to sleep with a piece of horseradish under

> each of their pillows. If the husband's horseradish turned black before his wife's, then the expected child would be a boy, and vice-versa.[1]

Some Jewish groups use horseradish as *maror*, or the bitter herb, at their Passover meal, a spring festival which commemorates the deliverance of their ancestors from captivity in Egypt.

> At the Passover Jews must drink wine and eat bitter herbs – in England that's horseradish.[2]

Thus some supermarkets sell 'Kosher Horse Radish' in the run-up to the meal.[3]

Horseradish leaves, which are superficially similar to those of DOCKS, can be used as a substitute for dock in the treatment of NETTLE stings.

> If by chance you did sting yourself with a nettle, rubbing a dock leaf on the spot helped (as kids we always called the horseradish leaf a dock leaf, which isn't strictly correct).[4]

Other medicinal uses included:

> [Fens] horseradish, applied to a CUT, would stop bleeding and draw the edges of the wound together leaving little scarring . . . Horseradish was considered an effective cure for STOMACH cramp.[5]

> Horseradish was grated and you inhaled the vapour for heavy COLDS. Horseradish root was dug up, boiled, and taken for WORMS – a common complaint in those days of insanitary cottages.[6]

> Gypsy's remedy for SCIATICA and GOUT: scrape fresh horseradish – soak in white vinegar – bathe the affected parts in liquor and use the radish as a poultice.[7]

> My now ex-husband and I lived in a Steiner community near Middlesbrough for about a year. During that time he was very depressed and often angry. He was advised by a senior member of the community to wrap horseradish leaves on his feet to draw the heat from his head. It didn't work and we divorced six months later.[8]

HOTTENTOT FIG (*Carpobrotus edulis*) and SALLY-MY-HANDSOME (*C. acinaciformis*)

Succulent procumbent herbs, introduced as ornamentals from South Africa in the late seventeenth, or early eighteenth, century; both frost sensitive, but naturalised in coastal areas, where *C. edulis* is considered invasive and a threat to native vegetation.

> The Hottentot figs . . . both garden escapes growing abundantly in the wild – can be used medicinally. The cut fleshy leaf is juicy and can be rubbed on SUNBURN for relief.[1]

HOUND'S TONGUE (*Cynoglossum officinale*)

Biennial herb, growing in disturbed dry habitats and often associated with rabbit warrens, in scattered places throughout lowland areas in the British Isles.

Local names include:

Dog's breath in Cornwall
GYPSY-FLOWERS in Gloucestershire; 'from the dark hue of its flowers'
Little burdock in Norfolk and Suffolk; referring to its fruits which, like those of burdock, bear numerous hooks
Mice in Somerset
NAVELWORT, and rats-and-mice, in Wiltshire
Scald-head in Suffolk; 'i.e. scabies, ringworm, etc.'
STICKY BUDS in Dorset; again referring to the hook-bearing fruits
STINKING ROGER in west Lancashire
Tear-coat, 'referring to the hooked fruits', in Dublin.

In September 1805 it was reported in the *Royal Cornwall Gazette* that hound's tongue, 'gathered full of sap and bruised with a hammer', would make MICE and RATS immediately leave barns and granaries.[1]

HOUSELEEK (*Sempervivum tectorum*)

Succulent perennial herb (Fig. 24), native to central and southern Europe, introduced before 1200, semi-naturalised on all walls, roofs and in graveyards, and occasionally naturalised on stabilised sand dunes and shingle.

Local names include the widespread silgreen and singreen, and:

Ayegreen in Lancashire and Westmorland
Bullock's eye in Somerset and northern England
Cyphel in northern England
Foos, fooz, and fow/s in Scotland
Foose in Northumberland and Scotland
Fouets, and fouse in Scotland
Fuets in Northumberland and Scotland
Fuit/s in Berwickshire, Fifeshire and Roxburghshire
Full in Northumberland
Fullen in northern England
Guardians-of-the-house in Scotland
Healing-blade, and HEALING LEAF, in Clackmannanshire
Hockerie-topner in Dumfriesshire
Hollick in Cheshire, Cornwall, Northamptonshire and Warwickshire
Jupiter's beard in Devon and Somerset
Mallow-rock in Somerset
Poor Jan's leaf in Devon; 'people have great faith in the healing properties of the plant'
Sel-green in Devon and Dorset
SENGREEN in Devon, Shropshire and Warwickshire
Silgren in Dorset
Simgreen in Buckinghamshire
Sinna-green in Shropshire
Sun-green in Sussex and Wiltshire
Syphel, and syphelt, in Cumberland
THUNDER-FLOWER in Staffordshire; 'from its supposed power of keeping off THUNDER'
Tourpin in Ireland
Welcome-husband-though-never-so-late in Dorset.

According to an author writing in the 1940s, houseleeks were 'at one time' deliberately placed on the roofs of old houses 'in order to keep the tiles together'.[1] However, it is more usually claimed that they were planted to protect buildings from thunder. Charlemagne (*c.*742–814) is said to have ordered houseleeks to be planted on roofs to provide protection against LIGHTNING, FIRE and SORCERY.[2] According to William Bullein in 1562:

> Old writers do call it Iovis barba, Iupiter's Bearde, and hold an Opynion supersticiously that in what house so ever it growth, no Lyghtning or Tempest can take place and doe any harme there.[3]

Similar beliefs persisted well into the twentieth century:

> Ireland shared fully in the very common European belief that the houseleek . . . protected the house from conflagration and lightning, and the growing of this plant on the roofs of thatched houses, or in specially made niches or nooks in or about the roofs or porches covered with other materials, was known in every Irish

> county. The plant is known by various names: 'houseleek' is widespread, but 'roofleek' occurs in parts of County Cork, *buachaill ti* in Galway and Mayo, *luibh a'toiteain* in west Limerick and Kerry, *toirpin* in Clare and Tipperary, and 'waxplant' in Offaly and Westmeath. Besides its virtues as a protection against fire it was also valued as a medicinal herb.[4]

> Pieces of plant-lore I learned from my teenage friends in the country during the war. . . the houseleek was planted on roofs to guard against lightning strike.[5]

> Local lore around Wymondham, where I was born, insists that one puts a houseleek on the roof to keep WITCHES and lightning away. Furthermore it will not work unless you steal them, I pinched the houseleeks I have here on the farm 37 years ago from an old thatched cottage, where it was growing in my mother's childhood (she is 89). One on the barn (north-east facing) just seems to survive and flowers spasmodically; that on my porch (south-west facing) varies tremendously, but this may be determined by whether there is a house martin's nest above it, and the third on a buttress (south-west facing) of a modern barn is growing very nicely.[6]

Sometimes houseleek were believed to provide general good fortune:

> ICE-PLANT or houseleek: if it grows on your roof you will never be entirely without money.[7]

Houseleek was much valued in herbal medicine. Allen and Hatfield suggest that houseleek and NAVELWORT were used interchangeably by country herbalists, with the former being favoured in eastern areas, where the latter is scarce.[8] It also appears that houseleek was valued to treat many of the ailments which late in the twentieth century were treated using *Aloe vera*.

> In Cornwall the leaves of the houseleek were made into a poultice for the extraction of CORNS.[9]

> The way they used to cure sore EYES was they used to have a lot of houseleek growing on their houses and they would squeeze it into a cup and rub it with a cloth into their eyes and it would cure them.[10]

> My father-in-law was brought up in Norfolk. When he was suffering from IMPETIGO a visiting gypsy woman recommended breaking off a piece of houseleek and rubbing the sores with it . . . My father-in-law (who is still alive) says the cure did work.[11]

> My grandmother's cure for EARACHE was to squeeze two or three drops of sap of houseleek directly into the ear canal. I have used this remedy successfully, with no side-effects, on all my three children, all now adults. I particularly remember going out into the garden at 2 a.m. one winter's night to gather houseleek as my son was in earache agony. He was two or three years old at the time (he's now 34). Within two minutes he was pain free and soon sound asleep.[12]

Other ailments treated using houseleek included bee STINGS in Surrey,[13] BOILS ('pushes') in Suffolk,[14] BURNS and sore LIPs in Lincolnshire,[15] RINGWORM in Cumbria,[16] SHINGLES in Shropshire,[17] and WARTS.[18]

In Ireland houseleek was used as an ABORTIFACIENT:

> My initial acquaintance with it occurred when I was about fifteen years . . . I joined a group of men of my village before a traditional type whitewashed cottage with thatched roof. The thatch terminated at either end of the cottages by 'barges' of flagstones set in lime-mortar. (In modern building terminology, this word would be 'verges'). An old man had died in this cottage the previous day and the village folk and I were attending the funeral. While we waited for the coffin to be carried to the awaiting side-car I found myself intrigued by a clump of plants growing on one of the 'barges'.
>
> 'What are the plants growing on the barge?' I asked my neighbour, an elderly man.
>
> 'That,' he said, 'is *Buachaill a'tighe*.' I remarked that I had never before heard of it. For a while he was silent. Then turning towards me he spoke in lowered tone. 'It's a strange plant, that. Now if a young girl got into "trouble" (unwanted pregnancy), her mother would take some of those plants, boil them, and give the water to her to drink. Later on, she would tell the girl to climb up on a high wall and jump down. That would make the girl alright.'
>
> When, a few years ago, I related this story to a young farmer in this locality, he answered:
>
> 'It's quite true. I saw these plants tested only a short while ago.

Walter C— came to me, and asked if I had anything to give his cow that had retained the "cleansing" (afterbirth) after calving. I made up a bottle for him with *Buachaill a' tighe*. Walter took it home and gave it to his cow. A few hours later she passed the "cleansing" and was alright.'

Buachaill a' tighe is one of several names that Irish has for *Sempervivum tectorum*. I should translate it as 'the warden of the house'.[19]

HUNGRY-GRASS

There was an Irish belief that if a certain grass was trodden on it would cause immense fatigue. Although sometimes considered to be QUAKING GRASS, hungry-grass cannot be identified with any particular species. It appears that people were particularly prone to the grass's influence when they were returning from fairs and other events at which alcohol was consumed, so it is possible that the fatigue was alcohol-induced.

> *Fairgurtha* or hungry-grass. Tufts of a peculiar grass that grows on the mountains, on which if anyone tread he immediately becomes faint and hungry and incapable of walking. People found dead in the hills are said to have had *fairgurtha*, that is, they stood on a tuft of this grass and lost the power of going on.[1]

> People used to get *feur-zorca* or get weak coming home from markets and fairs. This grass, if you walk on it you will not be able to go any further without eating something. There is *feur-zorca* at Mamore, at Slavery, and many other places.
>
> My grandfather, William McLaughlin, Carva, said that one day he and a young man were coming from Clonmany and the man got weak at this spot and he had a few cakes in his pocket and he gave them to him to eat and he got right again.[2]
>
> When I was a child [in Co. Meath] I was told by an old man in the village that if you stood on hungry-grass, if you had nothing to hand to eat, you would be rooted to the spot, and presumably starve, so I always carried a crust in my pocket.[3]

HUSSARS, 11th – known as CHERRY Pickers.

HYDRANGEA (*Hydrangea macrophylla*)

Low-growing, clump-forming shrub (Fig. 25), native to Japan, cultivated for its abundant, long-lasting flowers since the 1880s, first recorded in the wild in 1982, occasionally occurring in woodland and hedgerows and on stream banks, usually as a relic of cultivation.

> A superstition seems to have grown up in recent years that the potted hydrangeas sold by florists are UNLUCKY if brought into the house. This belief has been three times recorded in Cambridge and once in Chatteris. The blue flower seems to be more unlucky than the pink.[1]

> My grandmother's horror of hydrangeas in the house [Mitcham, Surrey, *c.*1925] is carried on by me.[2]

Hydrangea flowers, which are available in many colours and are tough and relatively long-lasting, are much-used in Peak District WELL-DRESSINGS and were formerly used to decorate floats in Jersey's BATTLE OF FLOWERS.

HYDROPHOBIA – treated using BUCK'S-HORN PLANTAIN and ELECAMPANE.

I

ILLNESS – follows IVY being brought indoors.

IMPETIGO – treated using HOUSELEEK.

IMPOTENCE – treated using ASH.

INDIGESTION – treated using BETONY, DANDELION, MINT and GERMANDER SPEEDWELL.

INFERTILITY – PENNYROYAL used to cure.

INFLUENZA – treated using YARROW.

INK – made from BILBERRY fruit.

INSECTS – deterred or killed by BALM, LAVENDER and PENNYROYAL.

INSOMNIA – HEATHER and HOP prevent.

IRISH SPURGE (*Euphorbia hyberna*)

Perennial herb growing at edges of woodland, and in hedgerows and beside streams, mainly in south-west Ireland, also in Cornwall, and rarely naturalised elsewhere.

Known in the eighteenth century as makin boy,[1] a corruption of the Irish *makkin-bwee*: '*makkin* originally meant root, but is colloquially applied to parsnip, *bwee* means yellow; *makkin-bwee* in English is therefore yellow parsnip'.

> In Galway people consider it a 'grand physic', and give it to HORSES and CATTLE, but think it too strong for human patients; nevertheless it is sometimes given, generally to the unknowing by way of a practical joke. I was told of one individual in Gort who was dosed with it a couple of years ago, and a spectator assured me that he 'ran up and down the street like a madman, and swelled so big that his friends had to bind him with hay-ropes lest he

> should burst'. The country people have a quaint notion of the way in which this medicine is to be extracted. They take about an inch of the root (in which its strongest properties lie) and scrape it into some boiling liquid, generally tea, which draws out its essence; but they firmly believe 'that if it be scraped up it will work upwards, but if you scrape it down it will work downwards, and if it is scraped both up and down it will work in every way and burst you!'[2]

In Ireland, Irish spurge was believed to be the 'most efficacious' spurge for the removal of WARTS.[3]

Also in Ireland the plant was used as a fish poison:

> I have seen the country people in the Blackwater valley crushing quantities of this plant between stones and then throwing the mass into the river to poison the salmon and trout.[4]

> Irish spurge, *Baine cavin*, is well known in Kerry and West Cork and is used to poison fish. The plant is pulled from the habitat, the river bank, and thrown into the stream or river. The white exudates (common to many spurges) contains saponins which destroy the gill tissues and prevent respiration.[5]

> In the yellow root plant there is a hole through the middle and if you bruised it white stuff like milk would come out of it, and if it went on your hands it would blister them. It was used here by old people for colouring wool for frieze coats. It was also used to catch fish. It was put into a tin box and the juice let go out into the water and it would poison the fish and they would easily be caught. They would come up to the top of the water and the boys would 'suil' them out and bring them home. It was also used for hens having the sickness.[6]

> In County Clare in Ireland Irish spurge . . . is called *bonnacheen* (phonetic, the 'ch' as in 'loch'). How it should be spelt I have no

> idea . . . The plant is collected by salmon poachers, put into sacks and then trodden on upstream from a good pool (on the River Fergus, for example, a tributary of the Shannon). The caustic juice thus released inflames the salmons' gills . . . [so that] they come gasping to the surface and are thus easily taken by net or spear. My father told me this, I have never seen it in use.[7]

> Irish spurge . . . was used for an illegal purpose in west County Cork: poaching. A sack-full of chopped spurge thrown into a stream or small river would quickly kill the fish. An elderly woman in Kilnamartyra told me that her brother was caught by the local bailiff with a salmon obtained by this method. He hit the bailiff with the salmon, made his escape and emigrated to America.[8]

Although plants have been used in this manner to poison fish in many parts of the world, especially in the tropics, there are no records of other plants being so used in the British Isles.

ITCHING POWDER – hairs in DOG ROSE fruit used as.

IVY (*Hedera* spp.)

Woody climber with glossy evergreen leaves, common and widespread, climbing on trees, old walls and rock outcrops, in hedgerows and on ground in woodland. Two species of ivy – common ivy (*H. helix*) and Irish ivy (*H. hibernica*, sometimes considered to be a subspecies of *H. helix*) are the only representatives of the essentially tropical Araliaceae family native to the British Isles. Presumably as a result of climate change, at present both are thriving particularly well, producing many succulent fruit.

Local names include the widespread ivery and ivin, and:

Bent-wood in Berwickshire
Bindwood in Scotland
Hibbin on the Isle of Man
Ivory bush in Norfolk
Pop-shells, given to the fruit in Somerset
Stone-love in Leicestershire
WOODBIND in Scotland.

Although the glossy dark green leaves of ivy were brought in to decorate homes at CHRISTMAS, in some places it was considered to be UNLUCKY:

> [According to my 73-year-old mother from Rothes, Morayshire]

> the following plants are strictly forbidden in the house . . . ivy very dangerous.[1]

> We never had ivy in the house – it was said to bring illness.[2]

> Every December I would go out and collect holly and ivy to decorate our new home. In the mid-sixties my aunt came to stay and told us that ivy indoors meant an imminent DEATH in the family and that she could only stay as a guest if it was all removed immediately. My wife has never allowed ivy indoors since.[3]

More usually ivy was forbidden indoors at any time other than Christmas, but used in Christmas decorations:

> HOLLY and ivy must not be taken in house until Christmas Eve and must be removed by January 6th.[4]

> For forty years I have used ivy, with holly, YEW and other evergreen leaves to decorate my home during Advent and Christmas.[5]

Irish ivy differs from common ivy in having larger leaves and showing little tendency to climb. In Rosneath, Dunbartonshire it was used in LOVE DIVINATION:

> In the 1940s schoolgirls took a leaf of garden Irish ivy . . . off the wall near the church, slipped it inside their blouses and sang:
>
> Ivy ivy I love you,
> In my bosom I put you,
> The first young man who
> speaks to me
> My future husband he
> will be.[6]

Ivy was widely used to remove CORNS:

> Many years ago I had a soft corn between my little toe and the one beside it. As my work at the time involved a lot of walking and standing the corn caused me a lot of discomfort. One day an old lady remarked on the fact I 'seemed to be limping.' When I told her my trouble she advised me to always keep a leaf of ivy between my toes. Since that was 50 years ago I cannot remember

how effective it was, but the old lady was sure she gave me *the* cure.[7]

Fresh [ivy] leaves steeped in boiling water and applied to corns.[8]

I've used ivy to get rid of corns. You tied it on and when you took it off you could dig out the corn.[9]

Other examples of ivy being used medicinally include:

I am in my mid-70s now. When a child of about 2 years I fell on a fire and was badly burned in my face (so my mother, R.I.P., often told me), likely a travelling woman offered to bring her ointment that would cure the burn and leave no mark; she wouldn't reveal what it was. In desperation my mother used it and it worked. T.G. I haven't a mark.

It took 30 years before I discovered the cure. When I came to live here a very old lady lived near me. I got friendly with her and discovered she worked near my home town (Kanturk) in her youth, and made, as she called it, a plaster for BURNS. She gave me instruction . . . on how to make it. This is it: You choose good green ivy leaves, wash them and dry them; boil in fresh lard enough leaves to make the lard turn a rich green colour – they crisp up on boiling. The lard can then be strained and stored in jars and will keep for years. I have seen it used with great success.[10]

An 'Ivy Cap' used to be made by joining the leaves together, and put on the heads of children who had some disease of the scalp (a sort of rash). It had good healing powers.[11]

I have a cure using the tendrils of ivy. 33 years ago [in Derbyshire] my son suffered from ECZEMA. We met the 'Grandma Gypsy' from a local gypsy group . . . one day she said she could help 'calm that skin'. She went into the undergrowth and brought back a pile of ivy tendrils with leaves. She called it Robin-run-in-the-hedge. My instructions were to boil the lot for three hours, and then leave the lot for 24 hours. I was then to wipe the 'gunge' over the affected places. This I did to please the old lady, and was surprised to find that it did indeed cool the skin.[12]

A cure for warts which I tried when I was about 6 years old (my grandmother read about it somewhere I think). I had really huge

WARTS on my hands which I loathed, obviously, and the cure was to rub the warts with the ivy leaf and then bury the leaf on a beach below the low-tide mark! Anyway I forgot all about this after a few weeks when no cure occurred, and then about two years later I was sitting in school and noticed that all the warts had disappeared more or less overnight, really bizarre.[13]

I remember (1937?) my mother using an ivy poultice on a badly suppurating WOUND – it succeeded.[14]

In Warwickshire it was recommended that ivy should be given to sick SHEEP: 'If they will not eat ivy, they are going to die.'[15] Similarly, in Yorkshire:

Ivy was offered to cattle to tempt their appetite.[16]

While in Norfolk:

An old drover in the Norwich area once told me of the miraculous healing power of ivy in relation to cattle suffering from FOOT-AND-MOUTH disease when he was a young man, in the days before it became the rule to slaughter all animals affected with this illness . . . A sick animal, in his experience, would seek out the nearest available ivy on a tree or hedge as though by instinct.[17]

Although ivy is usually assumed to be poisonous there are records of its berries being eaten:

During the German Occupation of the Channel Islands (1940–5) ivy berries were boiled and eaten.[18]

I knew of an old man who used to eat ivy berries. He used to say if they are good enough for the birds they are good enough for me. He lived until a great old age. I always thought they were poisonous, but it didn't kill him.[19]

A very widespread use of ivy was in cleaning men's suits, and, particularly the blue serge uniforms worn at one time by railwaymen, postmen and many others.

We had to pull a big bunch of coarse ivy leaves. Chop them up and stew them until soft. Keep the juice and put it in an old container. Discard the leaves. With an old clothes brush take your husband's serge suit and proceed to brush the liquid, especially [into] the lapel and neck and cuffs. Then take a clean cloth and iron it all over. It's like new. A lot cheaper than dry cleaning.[20]

My grandmother in South Africa mentioned using ivy concoction as a laundry detergent, more especially during war times; circa 1984.[21]

IVY-LEAVED CROWFOOT (*Ranunculus hederaceus*)

Low-growing, white-flowered herb, widespread in shallow water and on damp mud.

On Colonsay, ivy-leaved crowfoot pounded between stones 'was used as one of the principal ingredients in poultices for KING'S EVIL'.[1]

IVY-LEAVED TOADFLAX (*Cymbalaria muralis*)

Low-growing perennial herb with pale purple flowers; native to south central and south-east Europe, cultivated as an ornamental 'before 1602' and first recorded in the wild in 1640, now widespread on old walls and other stony places, though rarely found far away from human habitations.

Local names include the widespread CREEPING JENNY, MOTHER-OF-MILLIONS, roving sailor, wandering jew, and wandering sailor, and:

Climbing sailor in Dumfriesshire
CREEPING SAILOR in Somerset and Sussex
Creeping seefer in Kirkcudbrightshire and Wigtonshire; 'i.e. sailor'
Fleas-and-lice in Somerset
HENS-AND-CHICKENS in Kent
Lavender-snips in Hampshire
MOTHER-OF-THOUSANDS in Dorset
Mountain-creeper in Buckinghamshire
Oxford weed in Berkshire and Oxfordshire; 'from abundance on Oxford college walls'
Pedlar's basket in north-west England and Somerset
PICKPOCKET in Somerset

Rambling sailor in Cumberland, Lancashire and Somerset
Roving Jenny in Devon and the Isle of Wight
Thousand-flower in Cheshire
Travelling sailor in Hampshire
UNDERGROUND IVY and wandering jack in Somerset.

'This [ivy-leaved toadflax] is what we call wall-rabbits.'
'Why do you give it that name?'
'Because if you turn the flower upside down, and squeeze its sides, like this, it looks like a rabbit's head.'[1]

Similar names, many of which are shared with ANTIRRHINUM, seem to imply that this children's pastime is restricted to south-west England: BUNNY-RABBITS' MOUTHS, BUNNY-RABBITS, monkey-jaws and MONKEY-MOUTHS in Somerset, nanny-goat's mouth and RABBIT-FLOWER and RABBITS in Devon, and rabbits' mouths in Dorset and Somerset.

J

JACK-IN-THE-GREEN

A conical frame covered in greenery, known as Jack-in-the-Green, was a widespread feature of English MAY DAY celebrations, being particularly associated with urban chimney-sweeps.

> Throughout much of the twentieth century Jack was considered to be a manifestation of the foliate head motif which is common in medieval churches,[1] or as a rare and curious survival of an extremely ancient rite – 'the annual victim of the vegetation drama'.[2] However in the 1970s it was shown that any association between Jack-in-the-Green and foliate heads could not be justified,[3] and that Jack evolved towards the end of the eighteenth century as one of a variety of May Day begging activities.[4]

Most Jacks-in-the-Green died out before the First World War, but in the early 1980s there were a number of revivals, including that at Deptford in south-east London, which appears on 1 May, and those at Hastings in East Sussex, and Rochester in Kent, both of which continue to appear on the May Day bank holiday.

JACOBITES – SCOTS PINE associated with.

JAM – wild fruit collected for making include: BILBERRY, BRAMBLE, CLOUDBERRY and DAMSON.

JAMES ALLEN'S GIRLS' SCHOOL – CORNFLOWER associated with.

JAPANESE KNOTWEED (*Fallopia japonica*)

Vigorous, fast-growing and rapidly spreading perennial herb, native to eastern Asia, introduced as an ornamental in 1825, and first recorded in the wild in 1886, now a feared and demonised weed. In 2010 it was estimated that the cost of clearing Japanese knotweed from the Olympic site in east London was £70 million, and that the annual cost of controlling it nationally had risen to more than £150 million.[1]

Local names include: donkey rhubarb in Cornwall, German sausage in Clwyd, Hancock's curse, 'having spread from the garden of someone of that name' in Cornwall in the 1930s, Japweed, and KISS-ME-QUICK in west Cornwall.

Japanese knotweed was widely used in children's activities.

> First noted in Cumberland *c.*1940 at Cotehill by J.A. who remembers using the stems for PEA-SHOOTERS.[2]

> [Wales] Japanese knotweed was increasingly being used for making pea-shooters from the 1920s onwards . . . In some parts of North Wales this plant had no name and it was referred to generally as a form of wild rhubarb. However by the 1930s it was being referred to generally as a *cegid* or *cecs* [i.e. a plant with a hollow stem] which might indicate its acceptance into Welsh plant folklore – or at least children's folklore! As a child in North Wales in the 1950s this plant featured in our efforts at making peashooters – we called the plant *the peashooter*, a name which persists as in 'there are peashooters growing down by the river'. However, fieldwork amongst school-children in the 1980s in North Wales failed to elicit one example of such use.[3]

Similarly, in Cornwall, children used the stems as peashooters, with aglets, the fruits of HAWTHORN, being used as peas, and pipes:

> A small section was cut from the main stem immediately beneath the joint to form the base of the pipe bowl, and a short length above it. Using a nail, a hole was then bored through the 'bowl' above the joint, into which a long hollow section of a stem was fitted. As a substitute for tobacco, the dead leaves at the base of the MUGWORT (*Artemisia vulgaris*) were picked and rolled between the palms until appearing as the texture of cotton wool.

Around Veryan, also in Cornwall, children made water-squirters, known as cow-skits, from the Cornish *skyt*, meaning to squirt:

> A section of the main stem was cut to a length of about one foot or more, leaving a joint at one extremity and an open end at the other. Using a nail a hole was then drilled through the joint. A

> piece of straight leaf stem, or a twig was then bound at one end with wool or cloth and pushed into the hollow section . . . When it was found to fit snugly it was removed and the hollow section filled with water; the plunger-like twig was inserted and the water was squirted, like a syringe, through the small hole at the end.[4]

In the 1920s children in Desborough, Northamptonshire 'pulled off the maturing flowers [of Japanese knotweed], pretending they were sugar'.[5]

Elsewhere Japanese knotweed was eaten by children. In Sea Mills, Bristol, in the 1970s:

> We used to call Japanese knotweed WILD RHUBARB . . . We used to eat it on a daily basis. We snapped the stem and ate the flesh on the inside by running our teeth up it. It tasted like rhubarb hence our name for it. We would eat several sticks a week. If we were thirsty we would eat the fleshy innards of the stem as it was refreshing. Sometimes there would be fluid in the stems, presumably produced by the plant. Very rhubarby! We even took it home and dipped it in sugar.[6]

In Swansea, where Japanese knotweed was known as sally-rhubarb, 'children suck[ed] the sharp-tasting stems in high summer'.[7]

JAUNDICE – treated using BARBERRY, BOGBEAN, BROOM, BUTTERCUP, CHICORY, COWSLIP, DANDELION, FIELD GENTIAN, GARLIC, GERMANDER SPEEDWELL, GORSE, GREATER CELANDINE, HONEYSUCKLE, NETTLE and PRIMROSE.

JENNY, or JINNY, GREENTEETH – a NURSERY BOGEY associated with DUCKWEED.

JERSEY LILY (*Amaryllis belladonna*), also known as belladonna lily

Perennial bulbous herb, native to South Africa, grown for its showy pink flowers since early in the eighteenth century; occasionally surviving in the wild where thrown out or formerly cultivated.

> My mother's family were of Kent origin, and she seemed to be steeped in superstition . . . it was supposedly UNLUCKY to bring what we call pink belladonna lilies (*Amaryllis*), also known, I believe as NAKED LADIES, into the house.[1]

> Petals of Jersey lily, cover with brandy, add a little camphor and leave in a screw-top jar for about a month; it was used here . . . in the old days to treat bad CUTS by taking one leaf and putting over the wound, it stings but works.[2]

JOINTS, stiff – treated using CHICKWEED.

JOUG TREE

In Scotland trees, often SYCAMORES, used by feudal barons as their gallows were known as joug, or dool, trees.

> The Joug Tree was the local laird's gallows, for in Scotland the feudal lairds had right of life and death. By a kind of revenge, these trees became ominous to the family, and when a limb fell off a DEATH followed.[1]

JUDAS ISCARIOT – hanged himself from an ELDER tree.

JUNIPER (*Juniperus communis*)

Evergreen shrub or small tree (Fig. 26), widespread but local, occurring in a wide range of habitats including chalk downlands and heather moors.

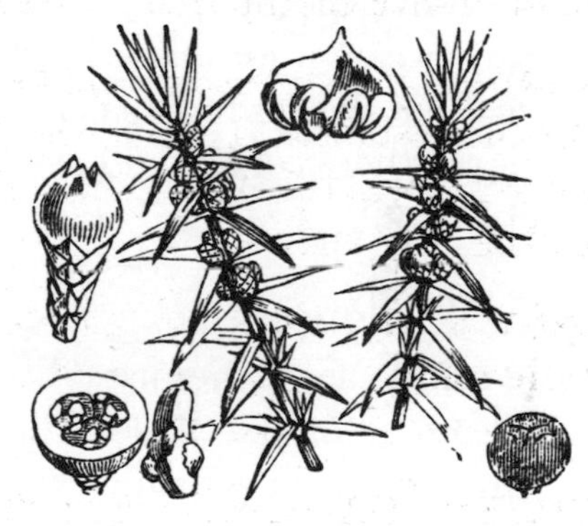

Local names include aiten in Abderdeenshire, aitnach in Banffshire and Morayshire, aitnagh in Scotland, etnagh berries given to the fruit in Angus, horse-savin in Cumberland, and melmont-berries or melmot-berries presumably applied to the fruits in Moryashire.

People said that he who cuts down a juniper would die within the year. For this reason, in many parts of Wales, aged junipers are carefully preserved, and it is customary to 'let it die of its own will', or natural death.

Twenty years ago an old farmer living in Glamorgan asserted that three DEATHS in his family followed by disaster when the 'old juniper was cut down'.[1]

Juniper was widely used as an ABORTIFACIENT, hence the Somerset name bastard-killer.[2] The court of Mary Queen of Scots (1542–87) included four ladies-in-waiting all named Mary. One of these, Mary Hamilton, became pregnant and in desperation resorted to juniper – savin. According to one of the many ballads telling of her tragedy:

> She's gane to the garden gay
> To pu' of the savin tree;
> But for a' that she could say or do,
> The babie would not die.[3]

The Pitt Rivers Museum in Oxford holds 'about five sprigs of "SAFFRON" [i.e. juniper], around 4.5 inches in length', which it acquired in 1914, with the explanatory note:

> *To prevent conception.* When a woman notices that she has missed a period she puts about the same quantity of saffron as this envelope contains into a pint jug; pours half a pint of *boiling* water on to it, covers the jug's mouth with muslin and puts a saucer on the top. She leaves the saffron to soak and, when cold, strains it through the muslin and drinks a wineglass-full for four consecutive mornings. She puts one sprig of saffron into each boot and wears it for *nine* days. The idea of this being that as the feet get hot the saffron soaks through the stocking into the foot. The sprigs here contained have been worn in the boots [of an] Oxford [woman].

More recently:

> I asked a local chemist if he had been around long enough to have heard of juniper pills – 'The Lady's Friend': typical advertisement in small print in ladies' journals: 'Late? Worried? Take Juno juniper pills.' 'Yes,' he said, he remembered them in little brown and green boxes . . . He said he hadn't seen them for five or six years; he then looked in his pharmaceutical trade book, and said, 'Yes, they're still available, Juno juniper pills.' . . . still needed even in the day of the pill.[4]

In south-west Lincolnshire:

> [Saffern] was often given by farm servants to their HORSES to make their coats shine.[5]

However, the best-known use of juniper is that of its berries to flavour gin. Almost all of the berries used in Britain today are imported from Tuscany and Eastern Europe, but in the seventeenth century berries collected in the Highlands of Scotland were exported to Holland for the production of genever gin. Later juniper became less common due to the intensification of sheep-farming, and possibly due to its wood being used as a smokeless fuel for illicit whisky stills.[6]

K

KARO (*Pittosporum crassifolium*)

Small evergreen tree, native to New Zealand, planted as a windbreak in maritime areas, occasionally self-seeding on the Isles of Scilly and in south-west England.

> [Isles of Scilly] *Pittosporum* fruits, which are sticky when ripe, are called 'pobbles' and collected by children for 'pobble fights'.[1]

KIDNEY problems – treated using BARLEY, BOGBEAN, DANDELION, HERB ROBERT, NAVELWORT, PARSLEY PIERT, PELLITORY OF THE WALL, RUE and SWEDE.

KIDNEY VETCH (*Anthyllis vulneraria*)

Variable perennial herb, of which five 'not well differentiated' subspecies are recognised in Britain and Ireland, with pale yellow to red flowers, widespread on neutral and base-rich soils.

Local names include the widespread LADY'S FINGERS, and:

Butter-fingers in Somerset
CAT'S CLAWS in Morayshire and Somerset
Crae-nebs in Northumberland; 'i.e. crow beaks'
Crawnebs in Berwickshire
Crow's freet in Gloucestershire
Dog's paise in Banffshire; 'i.e. peas'
Double lady's-fingers-and-thumbs, and double pincushion, in Wiltshire
EGGS-AND-BACON in Cornwall
FINGERS-AND-THUMBS in Devon, Dorset and Somerset
God-Almighty's fingers-and-thumbs in Devon
Granfer-grizzle, and hen-and-chickens, in Somerset
LADY'S CUSHION, and LADY'S PINCUSHION, in Wiltshire
LADY'S SLIPPERS in Warwickshire
LAMB'S FOOT in Dorset
LAMB'S TOE in Rutland and Somerset
Luck in Norfolk
PINCUSHION in Wiltshire

Twins in Yorkshire
Yellow crow's-foot in Buckinghamshire
Yellow fingers-and-thumbs in Devon.

In the Channel Islands kidney vetch leaves were used to check BLEEDING from wounds.[1]

KING'S EVIL (scrofula) – treated using IVY-LEAVED CROWFOOT and OXEYE-DAISY.

KNAPWEED (*Centaurea nigra*)

Tough, purple-flowered, herb, common in rough grassland.

Local names include the widespread HARD-HEADS and horse-knobs, and:

Arrow-head in Essex
BACHELOR'S BUTTONS in Belfast and Co. Donegal
BLACK HEAD/S in Ireland
BLACK SOAP in Devon and Gloucestershire
BLUE BOTTLE, and BLUE JACK, in Somerset
BLUE TOPS in Worcestershire
BOBBY'S BUTTONS in Somerset
Brushes in Dorset
Bull-heads, BULL-THISTLE, and bully-heads, in Somerset
BUNDS, and BUNDWEED, in East Anglia
Button-weed in Sussex
CHIMNEY-SWEEP, and chimney-sweep's brushes, in Somerset
Clobweed in Hertfordshire
Clover-knob in Nottinghamshire
Cock-head in the Midlands and northern England
CORN-BOTTLE in Devon
CORN COCKLE, and CORNFLOWER, in Somerset
DEVIL'S BIT, and devil's spit, in Somerset
Drummer-boys in Dorset
DRUMSTICKS in Somerset and Northamptonshire
Hackymore, and hairy head, in Somerset
Hardine, and hard-iron, in Cheshire, Lancashire, Nottinghamshire and Staffordshire
HORSE-BUTTON in Co. Donegal
Horse-hardhead, and horse-knap, in Devon
Horse-knob in Scotland
Horse-knops in Devon
Horse-nops in Cumberland
Horse-snap in Devon
HURT-SICKLE in Worcestershire
Iron-heart in Warwickshire
Iron-knobs in Cheshire
Iron-weed in Northamptonshire and Somerset
KNOB-WEED in Northamptonshire
KNOTGRASS in Hertfordshire
Ladies' balls in Wiltshire

LADY'S CUSHION in Kent
LOGGERHEADS in Buckinghamshire, Gloucestershire, Oxfordshire and Somerset; 'from the resemblance of the knobbed involucres to an ancient weapon so called, consisting of a ball of iron at the end of a stick'
Loggerums in Wiltshire
Matfellon in Yorkshire
Nigger-heads in Somerset
PAINT-BRUSHES in Devon
Shaving-brush in Shropshire
SWEEPS in Derbyshire
Tarbottle in Oxfordshire
Tassel in Berwickshire
Top knot in Somerset.

> In Bucks, I find that the young people still make use of the LOVE-DIVINATION by means of the knapweed . . . The florets should all be stripped off, and the rest of the flower-head placed in the bosom, If, on being withdrawn, one of these spikes, of which three should be employed, is found to have grown, that one represents the true lover.[1]

On Guernsey knapweed was known as *herbe de flon*:

> *Flon* has two different meanings. *Un flon* signifies a boil or wen on the human body; but *le flon* is a disease of cows, which causes induration of the udder after calving. To cure this, a handful of black knapweed is boiled for half an hour, and the affected part is bathed with it.[2]

L

LABOUR PAINS - RASPBERRY eases.

LABURNUM (*Laburnum anagyroides*)

Small deciduous tree, native to south central Europe, long cultivated for its ornamental yellow flowers, and formerly for hedging; occasionally escaping.

Local names include the 'fairly general' GOLDEN CHAIN, and:

Chaney ash in Cheshire
Drooping willow in Devon
Ear-rings in Cheshire
French ash in Derbyshire
French brum, i.e. broom, in Shropshire
GOLDEN DROPS in Lancashire
Golden locks in Devon
Golden rain in Dorset, Sussex and Warwickshire
Golden watch-chains in Somerset
Gold locket-and-chain in Somerset
He-broom in Fifeshire, 'perhaps meaning high broom, to distinguish it from common broom, which is of lower growth, or from low broom (*Genista tinctoria*)'
Hoburn saugh in Scotland
Lady's chain in Devon and Dorset
LADY'S FINGERS in Yorkshire
LOCKS-AND-KEYS in Devon
Pea-tree in Shropshire
Seyny-tree in Shropshire, 'the leaves are thought to resemble senna-leaves, whence the name'
Watch-guards in Cheshire
Weeping willow in Devon.

> Laburnum winter - a COLD SPELL co-inciding with the flowering of the laburnum.[1]

Laburnum hedges form a feature of the landscape in south-west Cardiganshire, where hyrbid laburnum (*L.* x *watereri*) has also been planted, usually as isolated trees. It seems that laburnum was being planted as hedges in this area in the 1840s and the practice continued until the second decade of the twentieth century. According to Arthur Chater, in his *Flora of Cardiganshire* (2010):

> There is little evidence, but much conjecture, as to when and why Laburnum was so widely planted. A common but rather fanciful idea [is] that fenceposts of Laburnum were brought from Spain as ballast in ships during the Napoleonic wars, and sprouted when put in the ground.

Other suggestions are that it was planted as a decoy for hares, which if laburnum is available 'will not browse on any other tree', or it was a 'most excellent forage' for cattle and sheep. However, Chater believes that the use of laburnum in hedges was mainly for decorative effect.[2]

LADY'S BEDSTRAW (*Galium verum*)

Perennial, yellow-flowered herb, conspicuous and widespread on dry, nutrient-poor grasslands.

Local names include:

A-hundredfald in northern England; 'presumably because of the very large number of flowers produced by the average plant'
BROOM, and BRUM, in Shropshire
CLIVER in Hertfordshire
CREEPING JENNY in Somerset
Flea-weed in Suffolk
GOLDEN DUST in Somerset
Halfsmart in Buckinghamshire; 'i.e. arsesmart, ?from use as a flea-weed'
Hundredfald in Northumberland
Joint-grass in the Midlands and northern England
Lace in Morayshire
Lady's bed/s in Aberdeenshire and Devon
Lady's golden-bedstraw in Yorkshire
LADY'S TRESSES in Somerset
MAIDEN'S HAIR in Gloucestershire and northern England
Robin-run-i't-hedge in Sheffield
Robin-run-the-hedge in Lancashire and Yorkshire
Strawbed in Devon.

Infusions of lady's bedstraw were widely used for curdling milk.

> The people in Cheshire, especially about Namptwich where the best cheese is made, do use it in their RENNET, esteeming greatly of that Cheese above other made without it.[1]

> In Arran, and some of the Western islands, the inhabitants make a strong concoction of this herb, and use it as a runnet to curdle milk: in Jura, Uist, and Lewis, &c I was inform'd they used the roots to DYE a very fine red.[2]

> Maiden's hair = lady's bedstraw – occasionally used by grandma [b.1858] to curdle milk for cheese when rennet was not available.[3]

Names which refer to this use include:

Cheese-rennet in Cumberland
Cheese-running in southern England
Curd-wort in Hertfordshire
Keeslip, 'i.e. cheeselip or rennet', in Scotland
Rennet in Cumberland, Hertfordshire and Kent; 'cf. Gaelic *lus an leasaich*, the rennet plant'
Runnet in Kent.

In 'darkest Berkshire':

> The only plant still in general demand is *Galium verum* which dried between sheets of newspaper is used for lining wardrobes and clothes-chests to deter MOTHS.[4]

LAMBING – HAZEL brought indoors leads to a poor lambing season.

LANGUAGE OF FLOWERS

During the nineteenth century numerous books were produced explaining how sentiments could be conveyed by the use of appropriate blossoms, the Language of Flowers. Although these books were popular, and their attractive illustrations ensured that they have survived rather better than many other publications of the period, it appears that they provided light-hearted amusement, rather than being actually used. Indeed, as different books often ascribed different meanings to the same flower, any attempt to convey messages by such means was liable to lead to misunderstandings.

The Language of Flowers is often said to have been of Turkish origin, introduced by Lady Mary Wortley Montagu (1689–1762). However, the method of communication by flowers and other objects which she described in her letters was a mnemonic system, which gives no meanings to flowers, and is of no relevance to the development of the Language.[1]

The Language first appears in Charlotte de La Tour's *Le Langage des Fleurs*, published in Paris in 1818. In some English publications plants have retained the meanings ascribed to them by La Tour. In others the meanings have changed, due to mistranslation, or to take into account Shakespearean and other traditions. Thus ROSEMARY, which according to La Tour meant 'your presence revives me', became associated with 'remembrance'. Other flowers had their meanings changed to take into account religious beliefs. Some British writers follow La Tour and state that PASSION FLOWER means 'faith'; others consider it to represent 'religious superstition', and in *The Catholic Language of Flowers* by the 'young ladies of Gumley House', published in London in 1861, it represented 'meditation'.

During the 1870s there was a revival of interest in the Language, stimulated by the publication in 1869 of John Ingram's *Flora Symbolica*. Other publications produced during the decade reflected the great increase in the number of ornamental plants brought into cultivation earlier in the century, and drew from a wide range of sources with the result that some flowers had several meanings ascribed to them.[2] Clearly anyone wanting to use the Language as a method of communication would need to ensure that a bouquet's recipient was consulting the right book. Indeed, when the main character in an 'erotic tale' attracted the unwanted attention of a young ballet-girl, he 'never liking to treat a woman scornfully' sent her a huge basket of flowers and a book explaining their meanings, so that she understood his love was 'elsewhere'.[3] Also important was a good appreciation of colour: is that POPPY red, in which case it means consolation, or scarlet, meaning fantastic extravagance?

There appears to be little evidence to suggest that the Language was ever much used. One rare example of its use is in Dublin in May 1868, when Rose La Touche, whose parents had forbidden her to write to John Ruskin, sent him a package of flowers. He could understand the 'rose enfolded in *erbe della Madonna*', but failed to decipher the rest.[4] However, according to a writer describing the country-house weekends which the aristocracy enjoyed before the outbreak of World War I: 'Saturdays-to-Mondays were a heaven-sent opportunity for sex . . . a note left (in collusion with the maid) beside the bottled water on the bedside table, or the placing of a code-laden flower outside a bedroom door ensured that extra-marital sex went on with ease.'[5] Also popular at that time was a series of postcards which depicted couples in an appropriate pose, a flower, and a caption, such as 'heliotrope – devotion', or 'forget-me-not – true love', allowing no scope for confusion.

Books on the Language of Flowers continue to be published with puzzling regularity; who buys them? And the Language continues to be mentioned in popular works on plant folklore:

> The white hollyhock symbolises female ambition . . . In the Language of Flowers it [HONEYSUCKLE] means 'I will not answer hastily'.[6]

LAUREL (*Prunus laurocerasus*), known in books as cherry laurel

Evergreen, white-flowered shrub (Fig. 27), native to the Balkan Peninsula, cultivated as an ornamental since early in the seventeenth century and first recorded in the wild in 1886, still widely planted and spreading as a naturalised plant.

Laurel leaves were used in LOVE DIVINATION, for 'secret letters', flavouring milk puddings, and poisoning butterflies.

> [South Cambridgeshire] a test for true love was to prick our sweetheart's name on a laurel leaf and wear it next to your heart; if the writing turned red all was well, but if it turned black, the young man loved you not.[1]

> Picked laurel leaves and with a thorn pricked our name on the back, put them inside our vests, the warmth brought out our name clearly.[2]

> [North Hampshire, 1940s] laurel leaves were good for magic writing. Gather dying yellow ones, write a message with a stick or some pointed tool, warm it (under the vest or tucked in a knicker leg), after a few minutes the words would show up.[3]

> One old lady put laurel leaves in the milk when she made cornflour moulds – 'Gave a lovely almond flavour' – I never tried it![4]

> [From my mother-in-law in Devon in the 1940s] For flavouring milk pudding or old fashioned blancmange a laurel leaf was dipped in the milk; it gave an almond flavour.[5]

> My brother in the 1950s, was a collector of moths and butterflies, which he pinned down and kept in a display case. He used to collect laurel leaves from a local hedge, crush them and put them in the bottom of a screw-top jar. Then he would put his dead moth in there and it would relax the muscles so he was able to pin it in the desired position. I remember it had a bitter almond kind of smell, and I worried about it poisoning me.[6]

The tough leaves of laurel have been much used for decorative purposes. Decorations in Devon to mark the coronation of King George VI, on 12 May 1937, involved an 'abundance of laurels' which were 'extensively used in very varied and artistic schemes'.[7] In the 1980s laurel was used to decorate London greengrocers' and, more rarely butchers' shops at CHRISTMAS.[8]

Wreaths of laurel leaves are often used in commemoration ceremonies. On the last night of the BBC Promenade Concerts, at the Royal Albert Hall, London, two promenaders place a laurel wreath on a bust of Sir Henry Wood who founded the concerts in 1895.[9] The annual Dr Johnson Commemoration held in Lichfield, Staffordshire, on or near 18 September, involves placing a similar wreath on the plinth of the author's statue in the city's Market Place.[10] On 21 October, the anniversary of the Battle of Trafalgar, each year a laurel wreath is placed on the quarter deck of Lord Nelson's ship *Victory*, now docked in Portsmouth Harbour.[11] Thus it appears that laurel wreaths have supplanted the BAY wreaths used to crown victors in classical times.

In folk medicine:

> Boil and strain laurel leaves, mix the juice with lard, and use as a cure for BURNS.[12]

> RINGWORM used to be a common complaint in the country and the cure for that was an ointment made from unsalted butter and the juice extracted from laurel leaves.[13]

LAVENDER (*Lavandula angustifolia*)

Aromatic evergreen shrub, native western Mediterranean region, cultivated in gardens since the thirteenth century, and increasingly grown as a field crop for its oil.

Lavender is reputed to be an INSECT repellent:

> Lavender bags, etc., in cupboards to keep away MOTHS.[1]

> Dried lavender to keep FLIES out of kitchen.[2]

It was also valued in domestic medicine:

> In the old days lavender was known to be good for HEADACHES and comforting the NERVES, because of its pleasant aroma; now lavender water applied to the forehead of a patient helps a lot.[3]

LEAF

> [West Sussex] if you catch a falling leaf, you will have twelve months of continued happiness.[1]

> [In my schooldays, in about 1920 it was believed that] if a falling leaf lands on your clothes you should not throw it away but keep it carefully in your satchel and it will bring you good LUCK.[2]

> [As a child in Nottinghamshire in the 1920s] to catch a falling leaf was very lucky.[3]

The leaves of deciduous trees have been considered to be symbolic of human mortality. In the words of the hymn-writer William Chalmers Smith (1824–1908):

> To all life thou givest – to both great and small;
> In all life thou livest, the true life of all;
> We blossom and flourish as leaves on the tree,
> And wither and perish – but nought changeth thee.[4]

Or, according to a south of England folk song:

> What's the life of a man any more than a leaf,
> For a man has his season and why should he grieve?
> Below in the wide world he appears fine and gay,
> Like the leaf he shall wither and soon fade away.[5]

LEEK (*Allium porrum*)

Long-cultivated vegetable, depicted in Ancient Egyptian illustrations, and known in Britain before AD 995.

The leek and the rather more ornamental and easily worn DAFFODIL are both national symbols of Wales, being worn on St David's Day (1 March) and on other occasions when such symbols are considered necessary. Judging by postcards produced early in the twentieth century the leek was then the predominant emblem, with daffodils being rarely, if ever, used. However, at present the daffodil is predominant and leeks (almost always artificial) are rarely worn.[1] Why the leek should have been chosen for this role is a matter for speculation.

> I will not presume to enter a controversy . . . by suggesting that . . . it may have become amalgamated into Druidic theology with a degree of sanctity, according to Latin writers, similar to that which rendered the leek so sacred a symbol amongst the ancient Egyptians, that to swear by these plants was considered equivalent to swearing by one of their gods, but will pass on to tell how Owen, otherwise a good antiquary, actually derives it from a prevalent Welsh custom, called *Cymhortha*, by which neighbours assemble, at seed-time, or harvest, to assist each other in completing the labour of the day; at which gathering each man contributes, by a sort of complimentary usage, a leek to the broth, which forms the dinner on the occasion; and as these leeks, he assures us, might naturally be carried in the band of the hat, he supposes the nation assumed them as a badge! . . .
>
> King James in his *Royal Apothegms* says, that it was chosen to commemorate the lamented Black Prince; but what connection subsisted between that gallant youth and the ill-scented plant, he does not inform us. Nor do the old Welsh records approach much nearer the truth. Their general testimony appears to be in favour of some battle, in which the Welsh were victorious, having been fought in a garden of leeks, from which each man gathered and wore one, to enable his countrymen to distinguish him from the enemy; to whom they had pre-determined to grant no quarter. This battle is variously stated to have occurred under the leadership of St David at the close of the fifth century, or commencement of the sixth century; or under Cadwalladr, in the year 633, when he defeated the Saxons near Hethfield, or Hatfield, in Yorkshire. It is needless to say that the idea is imaginary.[2]

It is said that in the fourteenth century Welsh archers wore green and white, the colours of leek, uniforms.[3] At Tudor and Stuart courts both the sovereign and courtiers wore leeks on St David's Day, a custom which James I considered to be 'a good and commendable fashion'.[4] In Sir Joshua Reynold's 1751 painting *Parody on the School of Athens* a Welsh musician is depicted wearing an over-large leek, while his Scottish colleague wears a THISTLE, an Englishman wears a wilted ROSE, and a Irishman wears 'a veritable meadow of shamrock'.[5] About 120 years later on St David's Day:

> In Anglesey it is the custom for boys to wear leeks up to twelve mid-day only, after that hour girls are supposed to deck themselves with the emblem of St David. Should a boy be seen without a leek in the morning, or with one after mid-day he is mercilessly pinched and the same rule applies, vice versa, to the girls.[6]

The leek forms the centrepiece of badges of the Welsh Guards, and the Royal Welch Fusiliers observed a number of leek-related rituals on St David's Day:

> The Royal Welch [Fusiliers] had its own insignia and rituals . . . St David's day, as might be expected, was particularly rich in special rites, which one hopes for [the poet Siegfried] Sassoon's sake were suspended during the First World War. It is difficult to imagine him enjoying the eating of a raw leek, one foot on the table, while the drums rolled behind him . . .
>
> [In 1917] the Quarter-master, Captain Yates was 'ingenious and on St David's Day he always miraculously had a leek for everyone's cap, however desolate the spot'.[7]

In the 1980s and 90s a favourite photograph in newspapers on 2 March showed a small girl in traditional Welsh costume stretching up to present a neatly-trimmed leek to a tall guardsman:

> Colour Sergeant Phil Atweel, of Merthyr Tydfil, receiving a leek from Joanne O'Driscoll, aged four, of Bridgend, during St David's Day celebrations yesterday hosted by the Welsh Guards at their barracks at Pirbright, Surrey.[8]

At international rugby matches supporters of rival teams deck themselves with appropriate symbols. In February 1980, for a match between England and Wales:

> The essential item of clothing is, of course, a woollen scarf with stripes of the Welsh teams colours, red and white. Other items worn by supporters included flags depicting Welsh dragons, draped over

> the shoulders; and leeks or daffodils. Some supporters wore real leeks; one particularly well equipped individual wore a red-and-white scarf, a flag, and a miner's helmet with a leek attached at the back. Some of the older men had miniature leeks made from knitting wool pinned to their lapels, and some groups of youngsters carried home-made leeks up to 3 or 4 feet high, made from white cardboard, green crêpe paper and other materials.[9]

However, thirty-five years later, when Wales played against South Africa at Twickenham, things had changed, most of the symbols were mass-produced rather than home-made, and artificial daffodils were abundant, while only three people were seen wearing leek badges – one of these being an apparently home-made brooch which incorporated both leeks and daffodils.[10]

For many years from the mid-1880s the competitive growing of giant leeks was an important pastime throughout much of north-east England. In 1893 the sixteen-member 'Pot and Glass' Leek Club at Crossgate Moor, Co. Durham, held its seventh annual show with the first prize being a pair of blankets and a picture.[11] In 1895 their show was held at Mr Lumsden's hostelry in Crossgate:

> Three leeks comprised a stand. W. Golightly took first place and a prize of £1 15s (£1.75p); W. Robson, a second, (£1 and a sheep's heart) and T. Stewart third 15 shillings (75p) and a 'beast's heart'. A special prize was awarded to Mr Robson for the best single leek in the show.[12]

Two main types of leek are commonly exhibited: long leeks which resemble the ones usually seen in greengrocers' shops, and pot leeks which are shorter and stouter. Some shows also have a class for intermediate leeks. Traditionally judges at northern leek shows considered the cubic capacity of the leeks to be of greatest importance, but the Royal Horticultural Society in its guidelines for judges has stipulated that leeks should be judged according to their condition, solidity, and uniformity (i.e. the shape of the leek and how well it compares with its companion or companions making up a stand). Most northern judges take the RHS guidelines into consideration, but continue to pay a lot of attention to the size of the leeks in a stand. As experienced growers themselves they are fully aware of the amount of effort which is necessary to produce good large leeks, whereas it is comparatively easy to produce perfect smaller specimens.

From mid September until late October newspapers in the north of England, and in other areas where leek-growing has become popular, contain reports and photographs of local leek shows. Thus, on 14 September 1974, the *Rugeley Times* reported:

> Almost £500 worth of prizes went to this year's 29 entries at the Poplars Leek Show held last weekend at the Poplars Inn, Handsacre . . . Each person submitted three leeks, all of which were grown with the use of 'recipes' formulated by the grower himself, and usually not divulged to others . . . On Monday a 'glutton's supper' was held at the Poplars, where leek soup, leek sandwiches and other food made from leeks were served to members and their families in celebration of a successful competition.
>
> Results were as follows: 1. A. Clarke, sewing machine; 2. S. Bolt, dressette; 3. C. Jessop, chest of drawers; 4. E. Fitch, quilt, blankets and sheets; 5. B. Jones, table lamp . . . 24. S. Davies, iron. The next five entrants all won £6 cash.

Further north, where shows were more common, reports tended to be shorter, often consisting of a couple of introductory sentences and a list of winners.

> Blanchland Leek Show produced an impressive display of vegetables and flowers with a mammoth leek measuring 145.1 cu ins taking top trophy. Benched by Mr P. Everitt, it was judged best in the show over his nearest rival Mr K. Heppel who, to make up for his second placing, went on to win first prize in the immediate leek class. In the flower classes the Rev. John Durnford had his prayers answered when he won first prize for his collection of flowers.[13]

In 1992 it was reported that:

> the National Pot Leek Society, with 1,000 members . . . now has branches from Scotland to Somerset . . .
>
> The Newcastle Exhibition World Open leek show, with a £1,300 first prize, is at the Northern Club, Ashington [Northumberland], on September 26–27. The event includes the Newcastle Brown Ale Heaviest ONION Challenge with an £850 first prize.[14]

More recently leek-growing has become less popular, so that in 2008 it was reported that the Ashington Leek Show had been cancelled due to the credit crunch. Rising greenhouse heating bills had hit growers. And the smoking ban was expected to affect beer sales at the event. Dick Atkinson, due to organise the show . . . said: 'People haven't got the money'.[15]

LEMON (*Citrus limon*)

Evergreen tree of unknown origin, long cultivated in tropical and subtropical regions for its sour edible fruit.

> We used to save lemon rinds after the juice had been squeezed out, put them in a jug with a bit of honey and pour boiling water over them . . . at least one lemon to a pint of water. Leave it overnight and then drink it, as much as you want, and as often as you want. You can also make it with cold water, but you have to leave it to steep longer. Spots and PIMPLES will go in a couple of days.[1]

> The best cure I know for a COUGH is some lemon and boiling water. Then breathe it through your nose.
>
> For ACNE I would also suggest a lemon. Cut it in half and with cotton wool dab the juices on, It works wonders.[2]

> My gran used to make me drink lemon juice (as tart as you can take it) with hot water [to cure a cold]. I still do and it does seem to clear my chest when I have a really hacking cough.[3]

LEMON VERBENA (*Aloysia citrodora*)

Small, frost-tender shrub with leaves that smell similar to lemon when bruised, and inconspicuous flowers. Native to temperate South America, introduced to Europe via Spain in the seventeenth or eighteenth century, occasionally grown in gardens in the more temperate parts of the Britain and Ireland.

> A Northam [Devon] woman told me that 'anyone who can make cuttings of lemon verbena grow will not die unmarried', adding with a twinkle in her eye, 'I don't think it's always true, because I can make it grow.'[1]

> Lemon-scented verbena leaves used to make a refreshing tea that cleanses the system.[2]

LESSER SPEARWORT (*Ranunculus flammula*)

Perennial, yellow-flowered, herb, common on wet soils.

Local names include:

Blisterwort in Yorkshire
Butterplate in Berwickshire
COW-GRASS in Co. Derry
CROW-FEET in Somerset
GOOSE-TONGUE in Carmarthenshire; 'with an indented leaf of a longish make, somewhat like a goose's tongue'
SNAKE'S TONGUE in Berwickshire
Spurwood in eighteenth-century Devon
WATER-BUTTERCUP in Wiltshire
WILL-FIRE in Somerset
Yellow crane in Northamptonshire.

On Colonsay, in the Inner Hebrides, lesser spearwort was used as a substitute for RENNET in cheese-making.[1] On the Channel Islands water distilled from the plant 'is said to have been the preferred local method of causing instant vomiting in cases of poisoning'.[2]

LESSER TREFOIL (*Trifolium dubium*)

Inconspicuous annual, yellow-flowered herb, common on grassy ground.

According to a survey conducted in 1893, lesser trefoil was the species which was most commonly believed to be the true SHAMROCK (Fig. 45), with 51 per cent of correspondents considering it to be the real plant.[1] A similar survey conducted in 1988 produced a similar result with 46 per cent of its participants believing it to be true shamrock.[2] In recent years plants sold as shamrock in London have invariably been lesser trefoil.[3]

LETTUCE (*Lactuca sativa*)

Annual herb of unknown origin, widely grown as a salad-vegetable.

Despite being much grown by gardeners and allotment-holders, and being an almost essential ingredient of salads in Britain and Ireland, lettuce appears to have attracted little folklore.

> Lettuces . . . were formerly believed to have magical and healing properties, including the power of arousing love . . . in medieval times they were often included in love-potions and charms. They were also said to promote child-bearing if eaten in salads by young

> women, or taken in the form of decoctions made from the juice or seeds.
>
> Some years ago, what seems to be a confused and inverted version of this last belief was recorded at Richmond in Surrey, where it was stated that too many lettuces growing in a garden would stop a young wife having children. In 1951 the *Daily Mirror* printed some letters on this subject, in one of which (published on 20 July) the writer asked whether it was true that eating the plant was bad for brides. In another letter (26 July), a woman wrote: 'After being childless for a number of years, I was advised by a specialist to eat plenty of lettuce, and give my husband some too. In less than six months, my first babby was on its way.'[1]

> Sleeplessness: eat lettuces for supper.[2]

Lettuces are offered to the dancing lions which participate in Chinese New Year celebrations in London's Soho[3] and elsewhere.

> In Hong Kong, where I come from, lettuce has the name which is like 'vegetable of life'. That's why lettuces are offered to the dancing lions at Chinese New Year celebrations. As the lion approaches the lettuce it's quiet and slothful, but once it gets the lettuce in its mouth the drums start sounding loud, and the lion revives – like it's had its batteries recharged.[4]

LICE – LOUSEWORT, ORANGE and SPINDLE kill or deter.

LIGHTNING – BAY, BITING STONECROP, ELDER, HOLLY, HOUSELEEK, OAK and ST JOHN'S WORT protect against.

LILAC (*Syringa vulgaris*)

Small deciduous small tree with sweetly scented white, mauve or purple flowers, native to south-east Europe, cultivated as an ornamental since late in the sixteenth century, becoming popular in Victorian times, spreading by suckers beyond gardens and persisting on sites of former gardens.

Local names include the widespread laylock, and:

ASH, and blue ash, in Gloucestershire
DUCKS' BILLS in Devon; 'from the shape of the flowers'
Lealock in Lincolnshire
Leelock in Norfolk

Lily-oak in Scotland
MAY in Devon
May-blossom in Ipplepen, Devon, 1909; 'at which village I find it the only name by which this plant is known'
May-bushes in Devon, 'lilac is called may-bushes, and may is restricted to *Crataegus*'
MAY-FLOWER in Cornwall, Devon and Somerset
OYSTERS, 'the name by which bunches of lilac-blossom are known', in Devon
Pipe-tree in Devon and Dorset; 'a common name in old books'
PRINCE-OF-WALES' FEATHER in Devon
PRINCE'S FEATHER in Somerset
Princey feather in Rutland
QUEEN'S FEATHER in Devon
Roman willow in Lincolnshire
Spanish ash in Gloucestershire
STARS in Somerset
White ash, given to the white-flowered form in Gloucestershire
Whitsuntide in Somerset.

Between March 1982 and October 1984 when the Folklore Society conducted its survey of plants which were believed to produce misfortune if picked or taken indoors, it was found that HAWTHORN was the most widely feared with lilac being the second.[1] An analysis of records of 'UNLUCKY plants' submitted to Plant-lore Archive between May 2012 and April 2017 revealed a similar result.[2] Many people are happy to grow lilac trees in the garden, but will not allow the flowers to be taken indoors, and it appears that white-flowered forms are more feared than mauve or purple ones.

> I remember visiting an old lady, about 1970, who had a lovely white lilac in her garden. Something she said about it being difficult for her to get into the garden made me offer to go and pick her some lilac, as it was in flower. She made all sorts of excuses to stop me, and finally said it was unlucky to bring it indoors. She was from Lancashire originally, but had lived in Worthing for very many years.[3]

> May blossom [HAWTHORN] – this is never brought into the house as it was a sign of bad luck. The same goes for white lilac, but funnily enough not mauve or other coloured lilac.[4]

> Think it was back in the mid or late 1950s. Went to get on a bus from Colchester to come home. Had visited friends and had an armful of light mauve lilac. The driver didn't want to let me on – said lilac was unlucky (nothing to do with the quite strong per-

fume). Fortunately I was able to persuade him – think it was the last bus![5]

It is possible that lilac was considered unlucky because, despite its short flowering season, it was used at FUNERALS; presumably its scent masked the odour of decaying flesh.

> My father told me people won't have lilac in the house because it was used for lining either coffins or graves, I'm not sure which.[6]

In west Dorset:

> I don't know if we have ever said for farming: 'Never buy calves when the lilac is out as it is the dearest time, and they are most likely to SCOUR.' It seems to be true, but I suppose the scour part comes in as cows are then on lush grass and the milk is richer.[7]

In Wales:

> Lilac blossoms were supposed to indicate changes in the WEATHER. If they kept closed longer than usual, fine weather might be expected. If they opened rapidly rain would fall soon. If the lilacs quickly droop and fade, a warm summer will follow. Late-flowering lilacs indicate a rainy season.[8]

LILY OF THE VALLEY (*Convallaria majalis*)

Rhizomatous herb, with fragrant white flowers, native in dry woodland, also commonly cultivated for its fragrant, white flowers.

Local names, nearly all of which were collected by A.S. Macmillan, include:

Dangle-bell, dangling bell, and fairies' bells, in Somerset
INNOCENT in Dorset
JACOB'S LADDERS in Gloucestershire
Lady's tears in Devon
Lily-confancy, linen-buttons, liricon fancy, little white-bells, and male lily ('evidently a corruption of May lily'), in Somerset
MAY-BLOSSOMS in Devon.

Like some other white fragrant flowers, such as HAWTHORN, lily of the valley is sometimes thought to cause misfortune if taken indoors.

> Lily of the valley: lucky to grow in the garden, but not to bring into house – reason not known.[1]

> It is UNLUCKY to take lily of the valley into the house.[2]

> My mother used to say that if you planted lily of the valley someone in the family would die.[3]

In Sussex:

> A very long-standing legend asserts that he [St Leonard] did actually live in St Leonard's Forest, near Horsham, and moreover that he once killed a dragon there. The battle was long and ferocious, and as a reward for Leonard's courage, Heaven granted that wild lilies of the valley would spring up wherever his BLOOD had sprinkled the earth.[4]

At Helston, in Cornwall, lily of the valley flowers are traditionally worn by participants in the annual Furry Dance, normally held on 8 May, unless that day falls on a Sunday or Monday (the local market day).

> At seven o'clock the Early Morning Dance, the first of the day, begins. There is one for the young people who, like their elders later on, dance through the narrow streets and in and out of the gardens and houses, all of them wearing lilies of the valley, the particular flower of the festival.[5]

In France lily of the valley is associated with MAY DAY,[6] and during the Second World War, when her home in Bignor, Sussex, was being used as a 'secret house' for French Resistance workers, Barbara Bertram recalled that amongst the embarrassing number of presents that the workers brought, perhaps the one she valued most was a bunch of lily of the valley which she found on her breakfast plate one May Day:

> They had been picked in France the night before, following the charming French habit of giving Our Lady's Tears on the first of her month.[7]

Lily of the valley was formerly valued for its medicinal properties as a cardiac tonic and a diuretic.[8] The antiquary Abraham de la Pryme (1671–1704) recorded in his diary:

> The flowers of the lillys of the valley, which grow in vast quantitys in these Broughton [near Brigg, Lincolnshire] woods, are now ripe and open. Here is come some men from Coronel Bierly's, that is above fifty miles off, to begg lieve to gather some. Other are come,

> some twenty, some thirty, some forty miles. There are at least gather'd in these woods yearly as many as is worth 60*l.* or 100*l.*; for when dry'd they are commonly sold for seventeen, eighteen, and nineteen shillings a pound.

In 1870 when the diary was edited for publication by the Surtees Society the editor noted that there were 'great quantities' of lily of the valley in Broughton and Manby woods, and 'people still come from a great distance to gather the flowers and take away their roots, which are medicinally valuable'.[9]

There are few recent records of lily of the valley being used as a folk remedy.

> A neighbour of ours, probably in his late 70s or early 80s, described to me how his mother . . . used to treat CUTS and abrasions, including apparently quite serious ones, successfully by covering them with fresh lily of the valley leaves held in place by a bandage.[10]

LIME (*Tilia* x *europaea*)

Large, deciduous tree, much planted in parkland.

Although widely planted, lime seems to have attracted little folklore.

> I am an Invernessian . . . the fruit of the lime tree was soft and sweet; it was known as hen's apples.[1]

> I was taught by other children in my Dorset village seventy-odd years ago to produce an ear-splitting WHISTLE from a leaf of common lime. Leaves were at their best for this purpose in June, when fully developed but flexible. I recently demonstrated to impressed grandchildren that I have not lost this art. The leaf is held taut against the lips.[2]

LIPS, sore – treated using HOUSELEEK and PRIVET.

LONE BUSH or FAIRY TREE

> Except perhaps for raths, duns, and lisses – the fairy forts of legend – nothing in Ireland is more closely associated with the fairy folk than are certain types of tree. Wherever one goes in the country one does not have far to look to see some lone thorn bush grow-

ing in a field. The thorn bush is locally reputed to be under fairy protection.[1]

Although these bushes were usually HAWTHORN, other isolated trees were sometimes considered to be lone bushes.

In Ireland [the rivals to hawthorn as lone bushes] are, in order of merit, the HAZEL, the BLACKTHORN, the bourtree - which is the English ELDER - the sally [GOAT WILLOW], the ALDER, the HOLLY, the BIRCH, the OAK - especially the twisted mountain oak - the BROOM, the SCOTS FIR; also, to my personal knowledge, in at least two instances, the ROWAN or mountain ash.[2]

Numerous accounts of lone trees were contributed to the Irish Folklore Commission's 1937–8 Schools' Folklore Scheme.

In the neighbourhood of my uncle's farm at Ballyduff, there is a lone bush growing in the middle of a big field belonging to Mrs Smythe. The people said it was not right to cut this bush. One day a man, who wanted to fence a gap in a ditch, came with his saw to cut it down.

When he got inside the bark, he heard a voice saying, 'If you go farther with it you will be sorry,' and he kept cutting away. When he had it half cut blood began to shoot out, still he cut another little bit.

Then he thought of himself and he fled leaving the saw stuck in the bleeding bush. This man died inside a week.

The bush is there still and from the cut up it is decayed, but the stump is quite sound.[3]

There is a field beyond there in Kilquiggan in which there is a rath. In the rath there was a bush (and it is said that it is UNLUCKY to touch a bush which stands alone in a rath). Well this man cut the bush to use it for fencing and when he came home after cutting it he found his horse dead in the stable.[4]

Similarly:

The house of the Irish cousins in Roscommon/Ballymote, Co. Sligo, was bought by a man, he wanted to cut down one or more thorn trees, no one would, he threatened to import Protestant labour from Derry, 70 miles north, so the locals cut the tree under protest. This annoyed the fairies, and they got into his bank balance and turned all the figures from black to red, so he went bust and left for Australia.[5]

> At the Ulster Folk and Transport Museum they have these fields they try to make look like traditional fields. So they planted a hawthorn tree in the middle of one of them, to look like a fairy thorn. A few years later they decided they didn't want a thorn tree there, so they asked the men to cut it down, but they wouldn't do so, because it was a fairy thorn. It was a different lot of men from the ones who planted it, but it does show how soon these things develop.[6]

> I have returned from a holiday in Donegal, Ireland, and spoke to an old aunt of mine, regarding 'fairy trees'. According to local tradition, one should not interfere with these trees, as it is unlucky. She told me some years ago, when her sons were out playing, they brought home some things that looked like bean pods, which they found at the bottom of the fairy tree. My aunt opened one of them and found a little fairy man inside. He was like a little doll dressed in red cap, green jacket and trousers. She was very cross with the children, so much so that they never again repeated the offence.[7]

Accounts of fairies being seen are rare; more usually lights are reported as being seen in the vicinity of lone thorns.

> I have lived in London for almost 30 years, but I did live in Glenties, Co. Donegal, where I was born, and fairy trees are very common in that part.
>
> I lived near a house that one of these trees was cut down to build and every man that lived in it went mad, but women came to no harm. I remember lights round the house at night, but no one lives there any more, so no lights appear. But these things are still going on over there.[8]

Such tales are reminiscent of English stories of hauntings which are supposed to have been put about by smugglers to discourage people from venturing near their hiding-places. It seems probable that some lone bushes originated as markers of sites where people gathered for illicit activities. Such an explanation is perhaps hinted at in a late nineteenth-century account:

There was such a tree on the lone mountain between Feakle and Gort near the mearing of Clare and Galway. When a boy my attention was directed to it by the parson of Feakle, who said it was considered a fairy bush, and pointed out the worn spot under it where they danced. The fairies were said to have left the country during the famine years (1848–52) as the grass grew on the bare spot, but they returned afterwards.[9]

LONG PURPLES, in Shakespeare's *Hamlet* – see LORDS-AND-LADIES.

LORDS-AND-LADIES (*Arum maculatum*), also known as CUCKOO-PINT

Perennial tuberous herb, often with dark purple spotted leaves, and distinctive flowers, followed by bright orange-red fruits; widespread in hedgerows and woodland, but thought to be present as the result of introduction in Scotland and the Isle of Man.

Local names include the widespread ADAM-AND-EVE ('the dark spadices represent Adam and the light ones Eve'), cows-and-calves, CUCKOO-PINTLE, ladies-and-gentlemen, LAMB'S LAKENS ('i.e. toys'), parson-in-the-pulpit, priest's pintle ('i.e. penis'), WAKE-ROBIN, and wild arum, and:

ADDER'S FOOD in Somerset and Wiltshire
ADDER'S MEAT, or ADDER'S MAIT, in Cornwall, Devon and Somerset
ADDER'S TONGUE in Cornwall and Somerset
ANGELS-AND-DEVILS in Somerset; 'the light parts of the flowers are the angels and the dark parts are the devils'
Aron in Scotland
Arrow-root on Portland, Dorset
Babe-in-the-cradle in Somerset
BLOODY FINGERS in Hampshire
BLOODY-MAN'S FINGER in Somerset ('from its lurid purple spadix') and Worcestershire ('i.e. devil's finger')
Bobbin-and-Joan in Northamptonshire
Bobbin Joan in Cornwall and Northamptonshire

BOBBINS in Buckinghamshire
Bobby-and-Joan in Northamptonshire
Bullocks in Somerset
BULLS in Dorset
Bulls-and-cows in Lincolnshire, Northamptonshire, Northumberland and Yorkshire; 'from the spadices which are sometimes dark red and sometimes pale pink or nearly white, giving an idea of male and female'
Bulls-and-wheys in Westmorland and Yorkshire; 'whey or quey, a heifer'
CALVES' FOOT in Somerset; 'from the shape of the leaf; it bears a similar name in France and Flanders'
Cocky-baby on the Isle of Wight
Cows-and-kies in North Yorkshire
Cow's parsnip in Somerset
Cuckoo-babies on the Isle of Wight
CUCKOO-COCK in Essex
CUCKOO-FLOWER in Northamptonshire
Cuckoo-lily in Cambridgeshire, where 'not allowed inside house'
Cuckoo-pint in Yorkshire
CUCKOO-SPIT in Lancashire
DEAD-MAN'S FINGERS in Worcestershire
Devils-and-angels in Dorset and Somerset
DEVIL'S BIT in Co. Dublin and Co. Tyrone
Devil's candles in Co. Kildare; 'we . . . were told it was poison and not to touch it'
Devil's ladies-and-gentlemen in Denbighshire
Dog-bobbins in Northamptonshire
Dog's dibble in Devon
Dog's spears in Somerset
Fly-catcher in Wiltshire
Frog's meat in Dorset
Gentleman's finger in Wiltshire
GENTLEMEN-AND-LADIES in Oxfordshire
Gentlemen's-and-ladies' fingers in Wiltshire
GETHSEMANE in Cheshire
Great dragon in Sussex
Hobble-gobbles in Kent
JACK-IN-THE-BOX in Buckinghamshire, Somerset and Northern Ireland
Jack-in-the-green in Somerset
Jack-in-the-pulpit in Cornwall, Leicestershire, Lincolnshire and Somerset
Jack-jump-up in Cornwall
Kings-and-queens in Devon, Co. Durham, Lincolnshire and Somerset
Kitty-come-down-the-lane-jump-up-and-kiss-me in Kent
Knights-and-ladies in Dorset and Somerset
Ladies-lords in Kent
Lady-and-larks in Berkshire
Lady-lords in Kent
LADY'S FINGER in Gloucestershire, Kent and Wiltshire
LADY'S KEYS in Kent
LADY'S SLIPPER in Wiltshire
LADY'S SMOCKS in Dorset, Hampshire and Somerset
LADY'S TRESSES in Somerset; 'it is difficult to see the reason'

Lamb-and-flag in Devon
Lamb-in-a-pulpit in Devon and Wiltshire
Lamb-in-the-pulpit in Wiltshire
Lamb-lakins in Northamptonshire
LILY in Wiltshire
Lily-grass in Sussex
Lords-and-ladies fingers, in Warwickshire
Lords-and-ladies-in-their-coach in Hampshire; 'the red ones being lords, the white ladies'
MANDRAKE in Yorkshire
Man-in-the-pulpit in Somerset
Mares-and-stallions in Berkshire
Men-and-women in Somerset
Moll-of-the-woods in Somerset and Warwickshire
Mouse-in-the-hole in Cornwall
NIGHTINGALE in Essex
Old-man's pulpit in Somerset
Ox-berry, and ozberry, in Worcestershire
Parson-and-clerk in Devon and Somerset; 'more often called parson-in-the-pulpit'
Parson-in-church in Cornwall
Parson-in-his-smock in Lincolnshire
Parson-pillycods in Yorkshire
Parson-preach-sermon in Berkshire
Parson's billycock in Somerset
Passon's pintle in Devon
Pig-lilies in Somerset
POISON-BERRY, given to the fruit in Yorkshire
Poison-fingers in Dorset
Poison-root in Wiltshire
POKERS, and preacher-in-the-pulpit, in Somerset
Priesties in Lancashire
Priest-in-the-pulpit in Somerset
Priest's pilly in Cornwall, Somerset and Westmorland
QUAKERS in Lancashire; 'when the spadices are dull-coloured'
Ramp-bobs in Lancashire
Ramp-campion in Bedfordshire
Ramps in Cumberland
RAM'S HORNS in Sussex
Ramson in Cumberland
Rat-berries, given to the fruit, in Co. Clare
RED-HOT POKER in Somerset
Schoolmaster in Sussex
Silly loons, and silly lovers, in Somerset
Small dragon in Sussex
Snakes' eggs in Essex; 'given to the fruit because so poisonous'
SNAKE'S FOOD in Devon, where applied to the fruit, Dorset and Somerset
SNAKE'S HEAD, in Dorset
SNAKE'S VICTUALS in southern England
SOLDIERS in Somerset
Soldiers-and-angels in Devon
SOLDIERS-AND-SAILORS in Somerset
Stallions, and stallions-and-mares, in Lincolnshire and Yorkshire
Standing gusses in Somerset
Starchmoor, starch-root, and starchwort, on Portland, Dorset
Sucky calves, sweet calves, and SWEETHEARTS, in Somerset
Toad's meat in Cornwall
White-and-red in Dorset
Wild lily in Devon.

> As a child I was told by a nurse-maid not to pick [lords-and-ladies], as ADDERS got their poison from eating it. It was called in North Tawton [Devon] SNAKES' MEAT. In 1939 . . . a Portuguese lady told me that as a child she had been told not to pick the plant for the same reason.[1]

> [Devon] Our parents would tell us that adders ate wild arum berries to obtain the poison for their fangs. The plant was always 'parson-in-the-pulpit' or 'snake-food' down here, and of course despite the fact it wasn't true it served to remind us that the berries are poisonous to humans.[2]

> It was believed in eastern Cambridgeshire that cuckoo pints, if brought into the house, gave TUBERCULOSIS to anyone who went near them.[3]

> [Dorset] When we were very young (and innocent) we used to say you (girls) should never touch a cuckoo pint; if you did you'd become pregnant. Where that silly idea started I do not know.[4]

Presumably the association of lords-and-ladies with PREGNANCY is suggested by the somewhat phallic-shaped spadix of the flower, which is responsible for the various 'pintle' names given to the plant, and led to it being considered to be an APHRODISIAC. In his play *Loves Metamorphosis*, published in 1601, John Lyly, who used the plant's older name of wake robin, wrote: 'They have eaten so much of wake robin, that they cannot sleep for love.'[5]

Also, in the 1940s and 1950s lords-and-ladies had phallic associations in Warwickshire:

> Cuckoo pint in seed caused great mirth amongst children as this was nicknamed dog's dick and we thought it very daring to look at one.[6]

Sometimes the leaves of lords-and-ladies have dark purple spots, and like those of EARLY PURPLE ORCHID, they are said to have been stained by Christ's BLOOD.

> There is a popular superstition in North Wales that this plant grew at the foot of our Saviour's cross, and in consequence of which the leaves became spotted.[7]

In the Cambridgeshire Fens lords-and-ladies was associated with St Withburga.

> Old Fenmen in the last century . . . held the traditional belief that when the nuns came over from Normandy to build a convent at Thetford they brought with them the wild arum or cuckoo pint. When the monks of Ely stole the body of St Withburga from East Dereham and paused, on their way back, to rest at Brandon, tradition has it that the nuns of Thetford came down to the riverside and covered the saint's body with the flowers. During the long journey down the Little Ouse of the barge bearing St Withburga several of the lily flowers fell into the river, where they threw out roots. Within an hour they had covered the banks as far as Ely with a carpet of blooms and, more remarkable still, these flowers glowed radiantly at night. Fenmen already old at the turn of the century could recall that, when a new church was consecrated in the newly formed parish of Little Ouse, the Bishop of Ely in his sermon warned his congregation against Romish superstitions and practices and stressed that there was no factual foundation to the story of St Withburga and the lilies.
>
> The pollen of the flowers does, in fact, throw off a faint light at dusk and when the Irish labourers came in large numbers to find work in the Fens during the famines in their own country during the last century, they called the lilies Fairy Lamps. The Fen lightermen had long called them Shiners.[8]

St Withburga's body on its barge covered with flowers is reminiscent of John Everett Millais' famous painting *Ophelia*, now in the Tate Britain, London. This shows Ophelia's death by drowning in Shakespeare's *Hamlet*, Act V, where the coronet which Ophelia made from WILLOW branches and wildflowers is described:

> Therewith fantastic garlands did she make
> Of crow-flowers, NETTLES, DAISIES, and long purples
> That liberal shepherds give a grosser name,
> But our cull-cold maids do dead men's fingers call them.

Although many commentators agree that long purples were EARLY PURPLE ORCHIDS, at least one Shakespearean scholar has concluded:

> Shakespeare was thinking of the wild arum or cuckoo pint . . . judging both by the appearance of the flower itself, and by the consistency with which – at least from the thirteenth to nine-

teenth centuries – it has been known, in England and elsewhere, by 'grossser' phallic names.[9]

In the sixteenth century lords-and-ladies was valued as a source of the STARCH required to stiffen the elaborate ruffs then worn by the aristocracy. In the 1560s 'tubs and other utensils necessary for the preparation of starch which are nowadays banished to the launderies, were to be seen in the most aristocratic residences. Washing, drying, hanging out and ironing were performed in the presence of nobles as today are music and arts'.[10] In 1597 John Gerard observed:

> The most pure and white starch is made from the rootes of Cuckowpint; but most hurtfull to the hands of the laundresse that hath the handling of it, for it choppeth, blistereth, and maketh hands rough and withall smarting.[11]

Soon after this time the use of starch from lords-and-ladies began to decline, due to the severe blistering of its users' hands. However, in 1797 the making of lords-and-ladies starch was rediscovered. In that year the Royal Society offered a gold medal or 30 guineas for the discovery of a method of manufacturing starch from a material not used as food for man. This was won by Mrs Jane Gibbs, of Portland, Dorset, whose

> starch, or arrowroot, as it is usually called, was prepared by her by crushing in a mortar the corms of *Arum maculatum*, stirring the mass with water, and straining off the liquor, from which the fecula was allowed to subside; this was again washed and afterwards dried. She stated, and the statement is confirmed by the then Rector of the island, that she had in her possession 2 cwt of the starch; and was ready to supply any quantity of the same, whenever required, at 11d per lb.[12]

In 1824 a visitor to Weymouth reported:

> in the island [of Portland] the roots [of lords-and-ladies] are dug in large quantities and when made into powder, many hundredweights are sold in Weymouth for starch and nourishment for invalids, and is also used in pastry, soups, puddings, etc.[13]

However, by the mid-nineteenth century only one elderly woman continued to make arrowroot.

> My informant tells me she obtains, on an average, 3lbs from a peck

> of corms; more in June, less in May. During the whole season she considers three dozen lbs to be a good average quantity to obtain; and for this she asks 1s 4d per a lb. It is highly valued by the Portlanders, who say it is good for sick people.[14]

LOUSEWORT (*Pedicularis* spp.)

Small perennial, pink-flowered herbs, common in a variety of damp habitats, including heathland and unimproved grassland, but scarce in many parts of central and south-eastern England.

Local names include:

COCK'S COMB in Scotland
Cow's wort in Nottinghamshire
Dead-man's bellows in Berwickshire and Somerset; 'i.e. pillies, male members'
Moss-crop in Sheffield
Moss-flower in Cheshire
RATTLE-BASKETS, and rattle-pods, in Somerset
RATTLE-WEED in Norfolk; 'from the rattle of seeds in the dry capsule'
Shackle-boxes in Devon
Tom's wort in Berkshire.

In Shetland lousewort was known as bee-sookies or honey-sookies, due to its 'nectar-filled flower-tubes'.[1] Similarly:

> When we were kids we picked lousewort and sucked the flower stems. I don't really know if this was wise, but I'm still alive [aged 68].[2]

Other names which suggest that children sucked the flowers for their nectar include:

Hinney flooer in Shetland
Honey-cap, and honey-cup, in Co. Donegal
HONEYSUCKLE in Hampshire and Co. Donegal
SOOKIES in Shetland
Suckies in Ayrshire
Wild honeysuckle in Co. Donegal.

Despite its name, there seems to be little evidence that lousewort was used to deter lice.

> I heard . . . that in the days before interior-sprung mattresses, etc., when sleeping on straw-filled palliasses, lousewort got its name as it was useful to have some of this in your bedding as it inhibited lice.[3]

LOVE – EARLY PURPLE ORCHID promotes, SUNDEW a 'love charm'.

LOVE DIVINATION – APPLE, ASH, BAY, CABBAGE, COWSLIP, CRAB APPLE,

DAISY, DANDELION, FLAX, GOOSEGRASS, HOARY PLANTAIN, Irish IVY, KNAPWEED, LAUREL, ORANGE, OXEYE-DAISY, PEA and YARROW used in.

LUCK, BAD – BLACKTHORN, BLUEBELL, BULRUSH, CHRISTMAS GREENERY indoors after Twelfth Night, CUCKOO-FLOWER, FLOWERING CURRANT, FOOL'S PARSLEY, FORSYTHIA, FOXGLOVE, HONESTY, HONEYSUCKLE, MADONNA LILY, MIMOSA, MONKEY PUZZLE, PRIVET, ROSEBAY WILLOWHERB, SAFFRON and YARROW associated with.

LUCK, GOOD – BITING STONECROP, four-leaved CLOVER, bark from the Coombeinteignhead CORK OAK, DAME'S VIOLET, HOLY THORN, HONESTY, HONEYSUCKLE STICK, a caught LEAF, MONEY TREE, PEA pod with nine peas, PEARLWORT, WHITE HEATHER and YEW used on PALM SUNDAY bring.

LUGHNASA, festival celebrating the start of the harvest in Ireland – BILBERRY associated with.

LUMBAGO – prevented or treated using AGRIMONY, HOP, NETTLE, NUTMEG and POTATO.

LUNGS – BLACK CURRANT 'protects'; FIELD SCABIOUS strengthens; treated using MAIDENHAIR FERN, NETTLE and SOAPWORT.

LUNGWORT (*Pulmonaria officinalis*)

Perennial herb (Fig. 28) with yellow-spotted leaves and flowers which open pink and become blue as they age; native to Europe, and apparently introduced in the sixteenth century, a commonly cultivated ornamental, becoming naturalised in lowland areas, but rarely recorded in Ireland.

Local names include the widespread Jerusalem cowslip, and:

ADAM-AND-EVE in Cambridgeshire, Cumberland and Somerset; 'on account of the two-coloured flowers'
Adam-Isaac-and-Jacob in Lincolnshire
Bedlam cowslip in Somerset; 'i.e. Bethlehem'
Beggar's basket 'in gardens' in Cheshire
Bottle-of-all-sorts in Cumberland; 'no doubt in allusion to the flowers of two different colours'
Children-of-Israel in Dorset
Cowslip of Bedlam, and Good Friday plant, in Somerset
Jerusalem seeds in Buckinghamshire, Devon and Somerset
JOSEPH-AND-MARY in Cornwall and Dorset

Josephs-and-Maries in Dorset, Hampshire and Wiltshire
Joseph's coat-of-many-colours in London
Lady Mary's tears in Dorset
LADY'S MILK-SILE in Cheshire; 'i.e. strainer'
LADY'S PINCUSHION in Cheshire and Yorkshire; 'the white spots on the leaves resembling pins'-heads on a cushion'
MOTHER MARY'S MILK in Mid Glamorgan; 'the white splashes on the leaves were supposed to have got there when the BVM was feeding the infant Jesus and there was a small "accident"'
MOUNTAIN SAGE, and sage-of-Jerusalem, in Cumberland
Soldier-and-his-wife on the Isle of Wight
SOLDIERS-AND-SAILORS in Dorset and Suffolk; 'in allusion to its two colours – red and blue'
Spotted Mary in Herefordshire and Radnorshire
Spotted virgin in Herefordshire
Thunder-and-lightning in Banffshire
Virgin Mary, in Hampshire; 'a corruption of Joseph and Mary . . . the pink flower is Joseph and the blue Mary'
Virgin Mary's cowslip in Gloucestershire, Shropshire and Worcestershire
Virgin Mary's honeysuckle in Cheshire and Shropshire
Virgin Mary's milk-drops in Monmouthshire and Wiltshire
Virgin Mary's tears in Dorset.

> Lungwort is associated with the Virgin Mary because it has blue and pink flowers – these two colours being the colours of the Virgin Mary's clothes in medieval paintings.[1]

> [Salisbury] I had an old woman weeding in my garden, and proposed to her to turn out a plant or two of it [lungwort], to which she strongly objected, and said, 'Do ee know, Sir, what they white spots be?' 'No I don't.' 'Why, they be the Virgin Mary's Milk!, so don't ee turn em out, for it would be very unlucky!'[2]

> At Osmington [Dorset] . . . there is a survival of a sweet simple, old-world piece of folklore about spotted liver-wort. The cottagers like to have it in their gardens, and call it 'Mary's Tears'. The legend is that the spots on the leaves are the marks of the tears shed by ST MARY after the crucifixion. Farther . . . her eyes were blue as the fully opened flower, and by weeping the eyelids became red as the buds.[3]

Good Friday plant, a name recorded from South Somerset, presumably refers to a similar legend.

LUST – PENNYROYAL promotes.

M

MACKEREL – fishing season starts when FOXGLOVES flower; FENNEL sauce eaten with.

MADONNA LILY (*Lilium candidum*)

Bulbous herb, believed to be native to south-east Europe and south-west Asia, long cultivated for its showy white flowers. Said to have been introduced to Britain by the Romans, and apparently first depicted in English art in a ninth-century miniature of Queen Ethelreda, founder of Ely Abbey. Known as WHITE LILY until the nineteenth century, when it was renamed Madonna lily to distinguish it from other, more recently introduced, white-flowered lilies.[1]

In common with other white flowers, such as LILAC and SNOWDROP, the Madonna lily was sometimes considered inauspicious.

> The large white lilies that grow in cottage gardens mean death if brought into the house . . . My husband was going to the solicitor to make his will – a daunting task! and as he was driving along he could smell a strong smell that reminded him of funerals; he stopped his vehicle and took a look in the back, and there was a bunch of lilies that an aunt had given me the night before – I did not want them in the house because of the smell – anyway he threw them over the hedge – and felt much happier.[2]
>
> A man came to the house and noticed a lily in the garden. Said it should not be brought into a house where there was an unmarried girl, or she would never get married.[3]

As its name suggests, the Madonna lily is a long-established emblem of St Mary the Virgin, particularly in paintings of the Annunciation – the visit of the Archangel Gabriel to Mary to announce that she will conceive and bear a son. In England Our Lady of Walsingham is usually depicted seated and holding a lily. By extension it has become a general emblem of virgin saints. St Anthony of Padua is often depicted as 'a rather "soft" young man, carrying the child Jesus and a lily'.[4]

> In Dorset the price of wheat could be foretold by examining Madonna lilies: 'a . . . calculation is made from the number of the blossoms as shown on the majority of its spikes, each blossom representing one shilling per bushel'.[5]

Madonna lilies were valued for healing CHILBLAINS and CUTS.

> When I was a child I suffered very badly from chilblains which frequently 'cracked' and became very sore. One of these on my thumb became badly infected, and many (and painful) bread poultices did nothing to reduce the infection. It was then my mother used the 'lily treatment' as she called it.
>
> A leaf from the garden Madonna lily was wrapped round the wound, and after several applications, for about a week I think, all the pus was cleared. Now was started the healing process. The reverse side of the leaf was placed on the wound, and I was soon cured.[6]

> Some of the older women of Great and Little Waltham still use the petals of Madonna lilies on bad cuts and say they heal quickly.[7]

> When we had a cut or WOUND we used to put something out of a bottle on, it was a green glass bottle; we were told it was lily leaves in brandy. This happened in 1911.[8]

MAIDENHAIR FERN (*Adiantum capillus-veneris*)

Attractive fern occurring as a native plant on damp limestone cliffs mainly on the western coasts of England, south-west Wales, Ireland and the Isle of Man; also widely cultivated and becoming naturalised on damp sheltered walls.

In common with other attractive ferns, native populations of maidenhair were threatened by the activities of nineteenth-century collectors. Also, in the 1840s on Inishmore, on the Aran Islands, valued for its medicinal properties:

> There is a great deal of this plant heedlessly destroyed. The inhabitants use it in decoction on cases of bad COLDS and complaints of the LUNGS, and for this purpose, instead of merely cutting the fronds, they pull it up root and branch.[1]

MALE FERN (*Dryopteris filix-mas*)

Common and widespread fern.

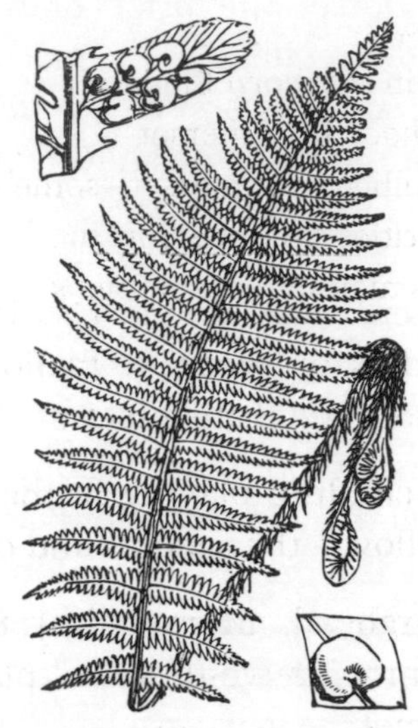

Local names include:

Basket-fern, in Cornwall, Hampshire and Co. Derry; 'from the hollow basket-like form in which the fonds grow up'
DEAD-MAN'S HAND in the Scottish borders; 'and some other ferns, from the appearance of the young fronds before they begin to open resembling a closed fist'
Fearn-brackins in Cumberland; 'and allied species'
Lucky hands in Nottinghamshire and Rutland; 'formerly this fern and its root were applied to many superstitious uses'
Meckens in Cumberland
St John's hands in Rutland
SNAKE-FERN in Hampshire.

> Male fern is used as a medicine for cattle and sheep. The roots are boiled and strained and given to animals as a cure for fluke.[1]

> Some flockmasters used to treat their liverfluked sheep with a weekly dose of 4oz of salt or a monthly treatment of male fern.[2]

During the Second World War children collected the roots of male fern 'under the direction of botanist R.W. Butcher, of the Ministry of Supply's Herb Committee, to provide a home-grown cure for tapeworm'.[3]

MALLOW (*Malva sylvestris*)

Purple-flowered, erect or spreading herb, an ancient introduction to the British Isles, now widespread on rough ground in lowland areas.

Also erroneously called MARSHMALLOW (often corrupted to marsh-mallice); local names include:

Billy buttons, applied to the 'heads' in Devon
BULL'S EYES in Devon
Flibberty-gibbet, and flower-of-an-hour, in Somerset
French mallow in Cornwall

HORSE-BUTTON in Co. Donegal
Maish-mallow in Somerset and Wiltshire; 'a corruption of marsh mallow'
Mallace in southern England
Mallow-hock in Somerset
Mash-mallice in Somerset; 'some authorities say "in error" . . . others maintain the name is correctly given, because this plant was, and still is, used in making *mashes* in poultices'
Mash-mallow in Gloucestershire
RAGS-AND-TATTERS in Dorset and Somerset
Round dock in Somerset; 'from the roundness of its leaves'
WILD GERANIUM in Somerset.

The nutlets, or 'seeds', of mallow were gathered and eaten by children throughout the British Isles.

> As children playing in the fields we would eat the 'fruit' of wild mallow – the seeds – and call them cheeses.[1]

> [Burghead, Morayshire] Biscuities were the little seed-heads from a mauve-flowered plant, which we ate, but never knew its name![2]

Names given to the fruits as a result of this practice include the wide-spread CHEESE-CAKE, and:

BREAD-AND-CHEESE in Lincolnshire and Yorkshire, 'thought by children to resemble cheeses in shape and flavour'
BREAD-AND-CHEESE-AND-CIDER, bread-and-cheese-and-kisses, and bread-and-cheesecakes, in Somerset
Butter-and-cheese in Dorset and Somerset
Cheese-cake flowers in York
Cheese-flower in Dorset
Cheese-log in Buckinghamshire
Chock-cheese in Devon
CHUCKY-CHEESE in Cornwall, Devon and Somerset
Custard-cheeses in Lincolnshire
Fairy-cheeses in Yorkshire
Frog-cheese in Oxfordshire
Ladies' cheeses in Dorset and Somerset
Loaves-of-bread in Dorset,

Middlesex and Somerset
Old-man's bread-and-cheese in Somerset
Pancake-plant in Devon, Lincolnshire and Somerset
PANCAKES in Lincolnshire
Pans-and-cakes in Somerset
Pick-cheese in Hertfordshire and Norfolk
Truckles-of-cheese in Somerset.

Mallow was also used in a wide range of folk remedies, many of which were remembered late in the twentieth century.

> [Lincolnshire] 'Marshmallow' used to be grown in gardens because it was useful in cases of BLOOD POISONING. 'Pancake' is another name that it is known by owing to its 'pie'-shaped seeds. (This is the common mallow.)[3]

> [Co. Cork] SPRAINS and stiff joints: a good remedy for these complaints was to boil marsh mallow and to rub well into the affected places.[4]

> [Co. Meath] Get marsh mallows and boil them, then bathe the place where you have RHEUMATISM and it will cure it.[5]

> My aunt, in Notts in the 1920s, used to send us out to gather the leaves of mallow; she made a wonderful ointment – but no one ever knew the recipe. This was useful for SORES, GRAZES and BRUISES.[6]

> A large colony [of common mallow] around the farmyard at Llanfaredd, Radnor – said by the owner to have been used as a leaf poultice to cure sprains of horses' legs.[7]

> [Cornwall] When we were young mother . . . bathed our eyes with the liquid from the boiled leaves of mallow, if we had any eye complaints.[8]

> An old lady in Criccieth, Caernarfon, tells me that mallow leaves, when she was a youngster over 60 years ago, were used as a remedy if one suffered from TOOTHACHE. The leaf would be pressed in the area of the mouth where the pain seemed to be at its worst.[9]

MANDRAKE (*Mandragora officinarum*)

Perennial herb with a robust tap-root, producing a rosette of leaves and dull purple flowers, followed by orange fruit; native to the Mediterranean region; occasionally cultivated in botanic gardens.

In England roots of WHITE BRYONY were commonly used as a substitute for those of mandrake.

In classical times there was a great deal of superstition surrounding the digging of mandrake roots. One of these was that the root should be attached to a starving dog to pull it up when enticed by having meat thrown towards it. It is probable that such ideas were invented by herb-gatherers who wanted to protect their livelihoods.[1]

> There used to be this chemist – well, pharmacist – in Aberystwyth . . . who had a mandrake plant in his garden. When he died, about ten years ago, he bequeathed it to the College Botanic Garden in his will. Quite a party of people gathered to remove the plant – my father was one of them – but they hesitated before digging it up, because it was supposed to be unlucky. They joked about getting a starving dog to pull it up, but the College gardener said he didn't worry, so he dug it up. He did so, and he is still alive and happy.[2]

MANGOLD (*Beta vulgaris* ssp. *vulgaris* cv.), also known as mangel wurzel

Biennial herb with swollen orange-coloured root, bred in the eighteenth century, formerly much grown for its roots which were harvested in the autumn and fed to cattle throughout the winter.

Although today's HALLOWE'EN lanterns are usually made from PUMPKINS, TURNIPS and SWEDES were formerly used for this purpose and, if obtainable, the larger roots of mangolds were probably preferred. The top of the mangold was cut off to form a lid, the inside hollowed out, and holes cut, usually to make a face, before a candle is placed inside, and a string attached so that the lantern can be carried.

In parts of the country where hallowe'en was not traditionally celebrated, mangold lanterns, known as punkies, were made for Bonfire Night, 5 November.

> In the Wiltshire village of my boyhood . . . we had a communal bonfire on the high hill that overlooks the village. We ran about with blazing besoms that had been dipped in tar. We made lanterns of hollowed-out mangolds or turnips with grotesque faces and stumps of candle placed inside.[1]

> [West Dorset, 1950s] on Guy Fawkes Night a bonfire of hedge-trimmings was lit and fireworks let off. On the day before the fire my brother and I would each make a punky by placing a piece of candle inside a hollowed-out mangold on which a face had been carved. These were proudly carried to the bonfire in the evening.[2]

At Hinton St George, Somerset, children carry punkies around the village on the last Thursday in October. According to local tradition punkies:

> were first used by the worried wives of Hinton to look for their husbands who had gone to Chiselborough fair and got drunk.
>
> The women hollowed out mangolds and made them into lanterns to search for their wayward husbands in the dark.[3]

According to an 85-year-old woman interviewed in 1988, in her childhood:

> We used to go to the big houses, knock on their doors and say 'give us a candle, give us a light' and they used to give us a penny or a piece of candlewax, which we used a lot in those days . . . At that time Punky Nights were not the organised affairs they are now – the village children got together and went off on their own.[4]

In 1982:

> We went to Hinton St George for Punky Night this year. We got there at about 6.30 p.m. There were a few people standing along the streets, so we wandered along to the village hall, where there were lots of people. The children carrying the punkies were very small; one child had a punky made from a swede, but the others all carried mangolds – the designs included dragons, primroses, and houses. They stood around for about three quarters of an hour, then a tractor pulling a decorated trailer drew up. On the trailer were a punky king and queen, and prince and princess – all very young children, and a man playing a guitar. On the front of the trailer the words of the punky ditty were written out:
>
> It's punky night tonight,
> It's punky night tonight . . .
>
> The procession was formed, led by a man dressed in a white smock and top hat who rang a bell, followed by the tractor and trailer, and the children carrying their punkies. There were so many adults

> looking after the small children that you couldn't really see the punkies. You couldn't really hear the singing either. They went all round the village, knocking on some doors, and collecting . . . Then they returned to the hall for a social gathering at which the punkies were judged.[5]

More recently the 'procession' has become simplified. People gather at the village hall, where in 2017 a storyteller told the story of Chiselborough Fair, walk around the village, probably less than a third of them carrying punkies, and stop three or four times to half-heartedly sing the punky song, before returning to the village hall, where the punkies are judged.[6]

MANURE – EELGRASS used as.

MARIGOLD (*Calendula officinalis*), known in books as pot marigold

Herb of unknown origin, introduced before 1000; formerly grown for culinary or medicinal use, now grown for its ornamental orange (or, in some cultivars, yellow) flowers; also becoming naturalised, though rarely persisting.

Local names include:

Flaming meteor in Essex
GOWLAN in Northumberland
Mally-gowl in Yorkshire
MARY-GOWLAN in Northumberland
Merry-go-rounds in Dorset
Nobody's flower in Wiltshire
SUNFLOWER in Somerset.

Marigold was used to treat MEASLES, and in Wiltshire known as measle-flower:

> The dried flowers having some local reputation as a remedy. Children, however, have an idea that they may catch the complaint from handling the plant.[1]

> [Yoxford, Suffolk] Marigold tea for measles: 1 doz. heads of marigolds in full flower, 1 pint boiling water; let it stand, give to child in wineglass three times a day.[2]

MARRIAGE – CHILEAN MYRTLE associated with.

MARSH MARIGOLD (*Caltha palustris*), also known as KINGCUP

Perennial herb (Fig. 29), producing shiny yellow flowers in springtime, widespread in moist habitats.

Local names include the widespread HORSE-BLOB, mare-blobs, may-blob, MAY-FLOWER, WATER-BLOB, and WATER-BUTTERCUP and:

BACHELOR'S BUTTONS in Devon and Somerset
Bee's rest, and big buttercup, in Somerset
BILLY-BUTTON in Somerset and Wiltshire
Billy-o'-buttons in Somerset
Blogda, bludda, and blugga, in Shetland
BOBBY'S BUTTONS in Devon and Somerset
Bog-daisy in Yorkshire
Boots in Cheshire and Shropshire
Bull-buttercup in Essex
Bull-cup in Somerset
Bulldogs in Devon
Bull-flower in Devon and Somerset
Bullrushes in Devon, Somerset and Wiltshire
BULL'S EYES in Devon, Dorset and Somerset
BULRUSH/ES in Somerset and Wiltshire; 'from some nursery legend that Moses was hidden among its large leaves'
Butter-blob in Yorkshire
BUTTERCUP in Devon and Somerset
Butter-flower in Wiltshire; 'cf. German *Butterblume*'
Carlicups in Somerset
Chirms in Northamptonshire; 'perhaps from some confusion with *Nuphar lutea*, the flowers of which resemble those of *Caltha*, cf. churn'

Claut in Wiltshire
Cow-cranes in Northamptonshire
CRAZIES in Buckinghamshire
CRAZY in Dorset and Gloucestershire
Crazy Bet/s, crazy Betsey, and crazy Betty in Dorset and Wiltshire
Crazy-lilies in Dorset
Crow-cranes in Oxfordshire
CROW-FLOWER in Bristol
CUPS-AND-SAUCERS in Somerset
Dale-cup in Somerset
DANDELION on the Herefordshire/Radnorshire border
DILL-CUP, and downscwobs, in Dorset

DRUNKARDS in Cornwall, Devon; 'because if picked and put in a vase of water they soon drink up all the water', and Somerset, 'on account of its fondness of water . . . children say if you gather them you will get drunk, or if you look at them long you will take to drink'
Fairy-bubbles in Ireland
Fiddle in Banffshire
Fire-o'-gold In Buckinghamshire
GILCUP in Dorset and Somerset
GILTY-CUP in Wiltshire
Goblin-flowers in Munster
Golden buttercup in Somerset
GOLDEN CUP, 'the usual name', and golden kingcup, in Somerset
Golden knobs in Berkshire, where 'much used in May-morning garlands', and Somerset
Golden waterlily in Dorset and Somerset
Goldicup in Cornwall
Goldilocks in Somerset
Gollan in Scotland
GOLLAND in Lancashire, Northumberland and Caithness
GOWAN in Cumberland and Scotland
Grandfather's buttons, grazies, and gypsy's money in Somerset
Halcups in Hampshire
Hobble-gobble in Buckinghamshire; 'the sound made by mud when anyone walks on it'
Horse-buttercup in Devon and Somerset; 'horse in local names of flowers is frequently used to designate a larger or coarser kind'
Horse-hooves in Shetland, presumably because of the shape of the leaves
Ivy-bells in Somerset, an unlikely name
John crane, and John Georges, in Buckinghamshire
Johnnie cranes in Northampton
King's cob in Berkshire and Hertfordshire
LIVERS in Dorset
Lockan-gowans, and locket-gowan, in Cumberland
MARIGOLD in Yorkshire
Marsh-lilies, and marshmary, in Somerset
Mary-buds in Dorset and Warwickshire
Mary's gold in Somerset
May-blubs in Wiltshire
May-bubbles, applied to the 'flowers and buds', in Somerset and Wiltshire
May-buttercup, and meadow-bout, in Lancashire
Meadow-bright in Northamptonshire
Meadow-gowan in Ayrshire
Mere-blobs in Derbyshire
Mire-blob, MOLL-BLOB, and molly-blob, in Northamptonshire
Monkey-bells in Somerset
OLD-MAN'S BUTTONS in Dorset and Somerset
Open-gowan in Cumberland; 'from its spreading flowers as opposed to the closed flowers of lockin-gowan' [globe flower, *Trollius europaeus*]

POLICEMAN'S BUTTONS in Somerset
Polyblobs in Herefordshire
Publicans in Yorkshire
Publicans-and-sinners in Oxfordshire; 'applied . . . to marsh marigold . . . and buttercups . . . when they grow together'
SOLDIER'S BUTTONS, and water-babies, in Somerset
Water-blebs in Lincolnshire
Water-blubbers in Gloucestershire
Water-bubbles in Dorset, Gloucestershire, Oxfordshire and Somerset
Water-flower in Oxfordshire
Water-geordies, and water-georgies, in Somerset
Water-goggles in Oxfordshire
Water-golland in northern England and southern England
Water-gowan in Cumberland
Water-gowlan in Yorkshire
Water-gowland in northern England and southern Scotland
WATER-LILY in Somerset and Wiltshire
Watter-blob in Yorkshire
Wild fire in Kirkcudbrightshire
Will-fire in Mearnshire
Yellow blobs in Leicestershire
Yellow boots in Cheshire
Yellow crazies in Wiltshire; 'crazies = buttercups'
Yellow cups in Wiltshire
Yellow gollan in Scotland
YELLOW GOWAN in Cumberland and Scotland
Yellow gowlan in northern England and southern Scotland
Yellow waterlily in Somerset.

There are occasional records of marsh marigold being considered UNLUCKY, no doubt because like BULRUSH it grows in potentially hazardous places.

> Locally called kingcups and considered unlucky. I have wondered since if this was due to their habitat as undoubtedly one always acquired very muddy feet getting to pick them. I could never resist them however, and stuffed them into jars to grace my playhouse. Washed my feet off with a quick paddle in the stream and hoped no one would notice![1]

More usually marsh marigolds are considered to provide PROTECTION, particularly on May Eve (30 April).

> As a 4-year-old in Portaferry, County Down (1910!) I remember seeing cottage roofs strewn with 'MAY' on the appropriate day . . . it was not HAWTHORN but marsh marigold.[2]

> [Co. Antrim] Over 30 years ago, when I was a small boy, it was a tradition here in our village to gather kingcups from the local

> meadow on 30 April in late afternoon. Then before nightfall I put a kingcup in the letterbox of each house in the village. This supposedly kept evil FAIRIES from entering the house before May Day, 1 May. Many people, especially the elderly, were very superstitious concerning this custom, and would remind me beforehand not to forget to pay them a visit. This was also a good source of pocket money as some people were very pleased to receive their May flowers as they were called. I carried on this custom into my teens.[3]

Anne O'Dowd mentions similar customs surviving or being remembered in the 1960s in counties Antrim, Armagh, Monaghan and Tyrone, and provides photographs of an elderly man in Castlebar, County Mayo, 'with his bucket of [marsh marigold] flowers for laying on doorsteps in the town on May Eve, 2014'.[4]

> In Shropshire the MAYFLOWER was the marsh marigold. My aunt's brother (Morton R. Evans, 1875–1970) for some years after World War II used to bring the flower into the house on the first day of May.[5]

On the Isle of Man:

> It was the kingcup or marsh marigold that was most sought after as a source of protection against WITCHES. One of its Manx names was *Lus y Voaldyn*, the Herb of Beltain (though it was commonly called *blughtan*), and as it was essential that it should be among the flowers gathered on May Eve, the first signs of its flowering or 'breaking' were anxiously looked for. A farmer in Kirk German, who kept a diary over a number of years in the late eighteenth century, noted down the breaking of the *blughtan* each spring. Sometimes it was late, and there was a growing note of anxiety in his April entries: 'The *blughtan* not broke yet,' 'No sign of the *blughtan* breaking.' Then at last, almost at the end of the month, 'The *blughtan* broke' – but only just in time for its flowers like small golden suns to add their brilliance to the May rites.[6]

On North Uist in the Outer Hebrides:

> Marsh marigold – hang it around the horns of your cows. They won't eat it. Make a garland and it increases the milk yield.[7]

MARSH SAMPHIRE (*Salicornia* spp.)

Annual succulent herbs with inconspicuous flowers, common and widespread in saltmarshes and other muddy saline areas in coastal districts.

Although the name glasswort is used in books, people living in coastal areas usually refer to *Salicornia* as samphire, samfer, or SAMPER, and gather it for eating.

> Sampion . . . a corruption of samphire – Chesh[ire] about Runcorn, Helsby and the neighbourhood, where it is hawked about by cart-loads for pickling.[1]

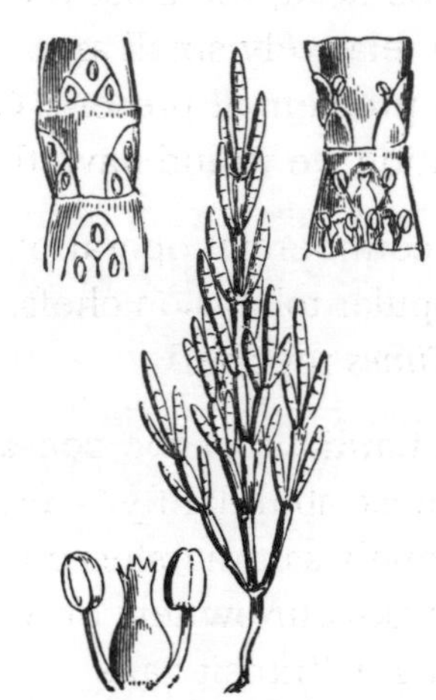

> Villagers living near the river [Humber] gather plants for their own use. The plants are either boiled and served with butter, or pickled. Marsh samphire is served in at least one local pub, but as far as I know it is not marketed in these parts.[2]

Samphire achieved wider popularity after August 1981 when it was served to guests following the wedding of the Prince of Wales and Lady Diana Spencer, and it is now available from upmarket fishmongers around the country. As the plants offered by such salesmen are usually very uniform, and available throughout the year, one assumes they are cultivated and imported, rather than collected along Britain's coast. However, samphire is still collected and sold in markets. In June 1990 a fish stall in King's Lynn market, in Norfolk, explained:

> Samphire is an edible seaweed. Wash then boil for 15 mins. Eat by pulling the green from the roots, dip in vinegar or butter, nice with cold meat and brown bread and butter. 70p per lb.[3]

This was remarkably cheap: in Guildford, Surrey, in June 1990, 'samphire sea-weed' was priced at £1.20 for ½ lb, while in Chichester, West Sussex, 'samphire grass' was offered at 75p for ¼ lb. Thirteen years later a stall in Swansea Indoor Market offered samphire at £6.30 a pound, or £13.70 a kilogram.[4]

Inevitably there are reports of the collection of wild samphire being banned. In June 1991 it was ruled at King's Lynn that under the Wildlife and Countryside Act of 1981 collectors who uprooted the plant without landowners' consent risked prosecution. However, at least one landowner was willing to support the collectors' traditional rights:

> The National Trust owns a seven-mile stretch of coast from Stiffkey to Salthouse. 'Eating samphire is an age-old tradition,' said Joe Read, the trust warden. 'It is such an abundant crop and only a relatively small amount is taken that we do not think it much of a problem at present. Clearly if it became a big commercial operation we would have the power to stop it.'[5]

Such commercial operations are feared when samphire is mentioned by popular television chefs. Following such a mention, on 9 June 2001 *The Times* reported:

> Guardians of the coastal nature reserves where samphire grows most abundantly – principally the tiny inlets that fracture the lonely salt marshes on the East Anglian coast – fear an invasion of pick-your-owners or, worse, an illicit harvest by hit squads arriving in Transit vans.

MARY, QUEEN OF SCOTS (1542–87) – COTTON THISTLE associated with.

MAT-GRASS (*Nardus stricta*)

Perennial grass widespread, and sometimes dominant, on upland moors.

Local names include BLACK BENT and wire-bent in Berwickshire, soldiers in Somerset, and white bent in Northumberland.

> I used to see doormats made . . . from flat sods of mat grass (*Nardus stricta*) in northwest Yorkshire, some sixty years ago.[1]

MAY ALTAR – COW PARSLEY and PRIMROSE used to decorate.

MAY-BIRCHERS

> In Cheshire in the earlier part of the nineteenth century were groups known as May Birchers, who . . . coming to a house after

> dark on MAY EVE, they would quietly leave a branch of some tree or shrub on the doorstep, to be discovered by the inhabitants next morning; this conveyed a message, complimentary or insulting, according to the plant chosen. The code was based on rhyme; nut for slut, PEAR if you're fair, PLUM if you're glum, BRAMBLE if you ramble, ALDER (pronounced 'owler') for a scowler, and GORSE for the whores. There was also HAWTHORN, which did not rhyme with anything but counted as a general compliment.[1]

MAY DAY (1 MAY) – DEVIL made shirts from NETTLE, in France LILY OF THE VALLEY, in Herefordshire GOAT WILLOW, and in Ireland HAWTHORN and PRIMROSES associated with, JACK-IN-THE-GREEN appears; RUNNER BEANS sown; WALNUT beaten; known as Pinch-Bum Day in Sussex and Sting Nettle Day in Devon (see NETTLE).

MAY EVE (30 April) – LOVE DIVINATION using YARROW attempted; MAY-BIRCHERS active on; MARSH MARIGOLD, PRIMROSES and ROWAN provide protection on.

MAY GARLANDS

The children's custom of preparing garlands on May was formerly widespread in England.

> The May garland is a summer emblem of very great antiquity, which has given May Day the secondary name of Garland Day. It has always varied considerably in form and shape, ranging from a simple bunch of flowers tied at the top of a long staff, or a pole wreathed with flowers, to the elaborate double-hoop garland, densely covered with spring flowers of every kind, or the less usual pyramid, also thickly covered by blossoms and usually very tall, often rising to a height of 5 or 6ft. Sometimes there is a May Doll seated in the centre of a hoop garland, or fixed upon the front of a pyramid.[1]

Children carried such garlands round the village, to outlying farms and big houses; the more wealthy members of the community being expected to make donations to the carriers. Flora Thompson in her *Lark Rise* (1939) provides an account from the carriers' point of view,[2] while an account from another vantage point is provided by Princess Alice, Duchess of Gloucester, (1901–2004). At Boughton, Northamptonshire, one of her childhood homes, four villages stood at the edge of the park:

> From each of them, on May 1, would come in turn separate groups of children, the girls garlanded and the boys holding bunches of spring flowers as they carried their May Queen on a throne smothered in blooms and leafy branches. After singing songs they would dance around the maypole and then to the delight of all, we showered them with well-polished pennies, thrown as far as possible to cause the greatest chase and scramble.[3]

Two places where garlands survived longer than most are Abbotsbury in Dorset, and Bampton in Oxfordshire.

Abbotsbury garlands were first mentioned in 1867 and more fully described in 1893, when the custom took place on 13 May and garlands were carried around the village by children of fishing families. Similar garlands were prepared in other South Dorset villages until the 1940s,[4] but only the Abbotsbury ones survived until the twenty-first century. Late in the twentieth century two garlands, one of wild flowers and the other of cultivated flowers, were carried around the village on the evening of 13 May.[5] However, although the garlands were carried by children they were prepared by adults, and in 2012 the family which had been doing this for many years decided that they would not continue.[6] Since then some attempts have been made to revive the custom. In 2015 it was reported that 'the custom dormant for some years [had] made a surprise reappearance',[7] but in 2016:

> They made a garland in Abbotsbury this year and placed it outside the community hall with a box beside it, but no one took any interest in it. It wasn't paraded or anything. It's not a community event.[8]

At Bampton the making of garlands is said to stretch back for centuries:

> The showing and judging of the garlands takes place on the Monday of the Spring Bank Holiday, at 11 a.m. In the Market Square. To make May Garlands is very easy but time consuming. All you need are two hoops which you cover with wildflowers. These wildflowers can be DANDELIONS, BUTTERCUPS, COW PARSLEY – in fact anything which is easy to find. Tie them into small bunches and attach them to the hoops with string. Fix the two hoops together so that they form a 'ball' shape. On top fix a

> bunch of flowers – years ago the top posy had to be made of red or pink PEONIES. Sit your favourite doll or teddy in the centre, find a long stick or broom handle to hook it through and two people to carry it and you have yourself a traditional May Garland.[9]

COWSLIP, CROWN IMPERIAL and FRITILLARY used in; CUCKOO-FLOWER 'usually omitted from'.

MEADOW FOXTAIL (*Alopecurus pratensis*)

Widespread and common perennial grass.

In Derbyshire in the 1950s, the flowers would be stripped off the stalk of meadow foxtail grass, leaving the stem with the floret stalks. This would be quietly twiddled into the hair of the child sitting in the desk in front. A swift yank would speedily remove all the hair attached.[1]

> Talking of Chinese haircuts brought back painful memories of my Warwickshire preparatory school in the early Forties. In my experience it was invariably Meadow Foxtail that was used. It produces flower heads in time to function as an instrument of torture at the start of summer term, deployed on heads intently watching cricket.[2]

> As a child, evacuated to Elgin, Scotland, during World War II . . . one grass we would not even pick was the meadow foxtail, *Alopecurus pratensis* – because it was said to be infested with FLEAS![3]

Cf. TIMOTHY.

MEADOWSWEET (*Filipendula ulmaria*), also widely known as queen-of-the-meadow

Perennial herb (Fig. 30) with creamy-white flowers, widespread in damp situations.

Local names include:

AIRIFT in Devon
Arrigologher in Co. Donegal
BITTERSWEET in Yorkshire
Blackin-girse in Shetland; 'from its former use as a dye'
Courtship-and-matrimony in Cumberland; 'from the scent of the flowers before and after bruising, which is thought to be typical of the two states in life'
Dropwort in Somerset
FAREWELL SUMMER in Dorset

GOAT'S BEARD in Devon and Dorset
Harif in Devon
HAYRIFF in Devon and Shropshire
HONEY-FLOWER in Devon
King-of-the-meadow in Shropshire
KISS-ME-QUICK in Somerset
Lady-o'-the-meadow in southern Scotland
Lady's ruffles in Cumberland, and Northamptonshire, where applied to 'the garden form'
Maid-of-the-meadow in Cheshire, and maids-in-the-meadow in Somerset
May-of-the-meadow in Warwickshire; 'may, i.e. maids'
Meadow-king in Shropshire and Ireland
Meadow-maid in Pembrokeshire
Meadow-queen in Perthshire and Renfrewshire
Meadow-soot in Wiltshire
Meadow-weed in Berkshire
Meaduart, and my lady's belt, in Scotland
New-mown hay, OLD-MAN'S BEARD, and QUEEN'S FEATHER, in Somerset
SUMMER FAREWELL, or summer's farewell, in Devon and Dorset
Sweet hay in Dorset and Sussex
Sweet meadow, and SWEET WILLIAM, in Dublin
TEA-FLOWER, and wild hops, in Somerset
WIRE-WEED in Hampshire; 'from the tough stems'.

Like other plants with pale rather messy flowers, meadowsweet is sometimes considered to be UNLUCKY.

> Our border matrons say that, if smelled too much, the QUEEN-OF-THE-MEADOW will cause people to take fits.[1]

> Meadowsweet is regarded as a fatal flower in Wales. There is an old story to the effect that if a person falls asleep in a room where many of these flowers are placed death is inevitable . . . it is considered quite dangerous for anyone to fall asleep in a field where it was to be found in abundance.[2]

> Eastbourne, 1941 (wartime), I was a keen collector of wildflowers at that time and brought back large armfuls of exceptionally fine sprays of 'Queen-of-the-meadow' . . . I brought it into the house where my mother and I were staying as guests, and met our hostess, a woman of fifty or more (unpleasant, petty-minded and very violent R.C. convert and very superstitious) who promptly told me not to bring it into the house. 'Why ever not?' Whereupon the woman nearly became crazy, she was screaming at me in the end. Ghastly bad luck would descend on the house if I did; I thought it

> too foolish for words and tried to persist in bringing it in. She was beside herself and screamed that she would throw out my mother and me if I did. In the end my mother intervened and begged me to leave it outside. I kept one bit secretly for my collection and left the rest outside.[3]
>
> In the west of Ireland it is deemed unlucky to bring meadowsweet into the house. My aunt was very strict about this.[4]

In Caithness meadowsweet was known as malara, or mallary, tea.

> It was considered to be unlucky to bring into the house. I never knew its correct name, but the local one used by the children was Malara Tea. How it came by that name I never knew.[5]

Further evidence for meadowsweet being considered inauspicious is provided by a name given to it by a 79-year-old Nottingham woman in October 1985: 'As children we always called meadowsweet "old man's pepper".' 'Old man' in plant-names is usually considered to be a euphemism for the Devil.

In the Retford district of Lincolnshire:

> Meadowsweet they used to smoke like TOBACCO. They dried out the flowers.[6]

Elsewhere:

> In 1952 or 1953 a friend used to walk along the canal at Lapworth near Solihull in the summer with a neighbour's daughter to gather meadowsweet to put in the toilets. The two houses built together had outside toilets built over a stream. His aunt's had two pedestals and a fireplace, although he never saw the fireplace being used. The meadowsweet was used to disguise the smell.[7]
>
> Meadowsweet. In Ireland when my grandmother's little calves had DIARRHOEA she boiled this plant and give the cool water to the little calves, which cured the diarrhoea.[8]

MEASLES – COWSLIP wine, MARIGOLD and RED DEADNETTLE used to treat.

MELANCHOLY – YARROW used to treat.

MENOPAUSE – RASPBERRY alleviates 'symptoms of'.

MENSTRUATION – FEVERFEW used to promote; menstruating women should not handle FLOWERS.

MEXICAN FLEABANE (*Erigeron karvinskianus*)

Perennial herb (Fig. 31) with abundant reddish-pink and white flowers, native to Mexico, cultivated as an ornamental since 1836, first recorded in the wild in 1860, and abundant on walls, and other rocky places, mainly near the sea in south-west England and the Channel Islands, but now rapidly spreading in other coastal districts and inland.

Like RED VALERIAN and ROSEBAY WILLOWHERB, both of which gathered many names as they rapidly spread, Mexican fleabane appears to be accumulating a growing number of local names.

David McClintock, in his *Wildflowers of Guernsey* (1975) records the name St Peter Port daisy, it having 'been first recorded from the British Isles from St Peter Port, having been known there since at least 1860, and being locally abundant already in the 1890s. It has spread over much of the island for many years now, always on walls, which it decorates delightfully.'

Other names include: Falmouth daisy, Fowey daisy, Hayle daisy, and Mexican daisy in Cornwall, wall daisy in Jersey, and fleabane daisy, under which name it was offered for sale in Battersea, London, September 2016.

MEXICAN HAT (*Kalanchoe daigremontiana*)

Succulent perennial, producing numerous plantlets at its leaf-margins, native to Madagascar, cultivated as a house-plant, rarely flowering.

Although any medicinal uses seem to be unknown in Britain, immigrants from Europe use Mexican hat to prevent or cure COLDS.

> My mother in the country in Lithuania had a big plant . . . If you had a cold you would squeeze a leaf and sniff it.[1]

> If you have a cold coming on, take and break a leaf, squeeze out the juice and put a couple of drops up the nostrils. It makes you sneeze and gets rid of the cold. Always used by my mother in the Ukraine. It doesn't help if you already have a cold.[2]

MEZEREON (*Daphne mezereum*)

Small deciduous shrub, producing attractive purplish-pink flowers which appear before the leaves, probably native in calcareous woodlands but not recorded in the wild until the mid-eighteenth century, widely cultivated as an ornamental and escaping.

Local names include mazalium in Buckinghamshire, mazell or mezell in Hampshire, mysterious plant in Derbyshire, paradise-plant in Gloucestershire and Somerset, and red-berry laurel in Yorkshire.

> In Willoughton, Lincolnshire, the berries of mazeerie (mezereon) were swallowed like pills as a cure for PILES.[1]

MICE – CAPER SPURGE, HOUND'S TONGUE, MINT and QUAKING GRASS deter.

MICHAELMAS DAY (29 September) – BRAMBLE fruit not to be eaten after; CRAB APPLE used in divination on, WILD CARROT associated with.

MICHAELMAS EVE (28 September) – HAZEL nuts cracked on.

MIDGES – ELDER deters.

MIDSUMMER'S EVE (20 June) – APPLES christened; ELDER at Rollright Stones visited; HEMP seed and ORPINE used for LOVE DIVINATION; MUGWORT most potent on.

MIGRAINE – FEVERFEW used to treat.

MILKWORT (*Polygala* spp.)

Small perennial herbs, with blue, purple, pink or white flowers, widespread in short grassland.

Local names include:

CROSS-FLOWER in Somerset

Fairy-soap in Co. Donegal; 'the "mother" (i.e. roots and leaves) . . . used as a lather by fairies'

Four-sisters in Co. Waterford; 'in allusion to the four colours – white, pink, blue and purple – of flowers on different plants'

Gang-flower in northern England

Gang-weed in Lancashire

Jack-and-the-beanstalk, MOTHER MARY'S MILK, and procession-flower ('in allusion to its use in the Rogation-tide processions'), in Somerset

ROBIN'S EYE in Hampshire; 'also applied to others, as those of the FORGET-ME-NOT'
SHEPHERD'S THYME in Wiltshire
Sisters in Co. Waterford
Waxworks in Wiltshire
Wild liquorice in Dorset.

Known as *herbe de paralysie* in Guernsey, where used to ward off or cure paralysis (see TORMENTIL).

MIMOSA (*Acacia dealbata*), also known (in Australia) as silver wattle.

Small tree with silvery-grey fernlike leaves and abundant yellow flowers, native to Australia; in the 1950s and 60s much imported from southern France as a springtime cut-flower; now cultivated as an ornamental in warm, sheltered places.

Despite its popularity as a cut-flower some people considered mimosa to be UNLUCKY.

> My mother was very superstitious about flowers: May blossom, LILAC and mimosa . . . The mimosa was considered a fore-warning of disaster, as I found out to my cost when I purchased a bunch out of my first week's earnings and it was promptly thrown into the dustbin (much to my disgust), but I never could find out why.[1]

MINT (*Mentha spicata*)

Perennial aromatic herb with pale pink, lilac or white flowers; an ancient introduction of unknown origin, commonly cultivated for culinary use, becoming naturalised when discarded or escaped.

Like other culinary herbs, mint was sometimes said to grow best where the wife was dominant.

> Mint in the garden means a woman rules the house (Luton, 1960-ish), i.e. roots are somewhat hard to eradicate once established.[1]

In addition to its culinary uses mint was valued as an air freshener and a fly deterrent.

> I've been told by many an old lady that when you leave your home to go on holiday mint should be picked from the garden and put in

> water and placed in a room. This apparently is supposed to keep the air fresh while the house is closed up. I don't know how true this is, I've never tried it.[2]

> I used to know this isolated place in Glen Luce, Wigtownshire/ Galloway, where the outside toilet was known as 'Mint Cottage', and there was a bunch of mint – which was replaced every week – placed above the door. Only ladies used it; the men had to go over the hills over the back – I never discovered where they went.[3]

> A vase of mint in the kitchen keeps away FLIES.[4]

> I remember my mother used to hang up bunches of mint to keep flies away.[5]

Occasionally mint was used to deter other pests:

> Pot of mint outside door supposedly keeps MICE away.[6]

Unsurprisingly mint had medicinal uses, mainly to treat STOMACH conditions.

> As a child [1940s] I lived for a while in the village of Wye in east Kent . . . mint tea helped belly ache.[7]

> Mint tea eases INDIGESTION.[8]

Other uses included:

> A friend in Aldershot [Hampshire] a few years ago . . . told me that for many years she suffered from HAYFEVER. One year when it was very bad a gypsy called at the house and looking around the garden said: 'You have the cure here. Pick some fresh mint every day and put it in a muslin bag. Put it in your pillow and inhale the scent during sleep. Also wear some during the day.' My friend did so and was permanently cured.[9]

MISCARRIAGE – RASPBERRY prevents; see also ABORTION.

MOLES (*Talpa europaea*) – CAPER SPURGE deters.

MOLUCCA BEANS – see SEA BEANS.

MONEY – flourishes where BAY flourishes.

MONEY TREE (*Crassula ovata*)

Succulent houseplant (Fig. 32), frequently grown in Chinese takeaways, producing pale pink flowers when pot-bound; native to South Africa.

Also known as good luck tree, tree of happiness, tree of heaven, and jade plant.

In January 1978 a woman interviewed on the BBC television programme *Nationwide* explained how her luck had dramatically improved since she had acquired a 'Money Tree' which she brought along for viewers to see. The interview caused something of a sensation, and the 'Tree' was eventually identified as the common houseplant *Crassula ovata* (also, but incorrectly, known as *C. argentea* and *C. portulacea*). By August many London florists were stocking small rooted cuttings of the plant, priced from 50p to £1.50 each.[1] Although reference books published before 1978 do not include money tree as a name for *Crassula ovata*, it appears that the name was used earlier:

> [1986] My mother has had one of those money trees . . . for about 15 years, and she's always called it a money tree.[2]

However, the name does not seem to have been widely known before it was mentioned on television. Thus television, often regarded as a destroyer of folk beliefs, can act as a super-efficient spreader of such beliefs, particularly if these are taken up in the popular press.

> Joyce Brown proved that money DOES grow on trees when she won £20,000 on the *Sun*'s bumper bingo competition.
>
> Three weeks ago her sister Maureen gave her a money-plant – told her that if she talked nicely to it she would have a win.
>
> Last week Mrs Brown, 60, of Norwich, Norfolk, won £70 at a local bingo club.
>
> 'I said thank you very much to my plant, went to bed, and the next morning my husband Neville woke up shouting "You've won on the *Sun*!"'[3]

Five months later:

> Green-fingered granny Clarice Cowell celebrated a £40,000 *Sun* Bingo win in champagne style yesterday – and said a big thank you to her lucky money plant!

> For clever Clarice believes that the house plant helped her join our list of super *Sun* Bingo winners . . .
>
> 'It's an old wives' tale, but if you talk to them they are supposed to bring you LUCK. I am always talking to mine. I was making a fuss of it only yesterday and I swear that's why I won.'[4]

Sometimes it is thought that if a money tree is given away its owner's luck goes with it.

> We bought this plant at a hospital fair two years ago, when we asked what it was, we were told it was a money tree. It would bring us good fortune, but if we ever gave the whole plant away the good luck would go with it. It's alright to give away cuttings.
>
> My wife has just won a Grand National sweepstake.[5]

MONKEY PUZZLE (*Araucaria araucana*)

Large evergreen coniferous tree, native to Chile and Argentina, introduced late in the eighteenth century and widely cultivated in parks and large gardens, also sometimes looking uncomfortable in smaller gardens.

Monkey puzzle was associated with the DEVIL and bad LUCK.

> There is a Fenland belief that if a monkey puzzle tree was planted on the edge of a graveyard it would prove an obstacle to the Devil when he tried to watch a burial. Many elderly Cambridgeshire people believe that the tree is an unlucky one.[1]

A widespread belief among children was that one should keep silent when walking under a monkey puzzle.

> In the 1930s I well remember the massive monkey puzzle which stood in Peterborough Recreation Park. As we children walked beneath its mis-shapen boughs we believed that penalties worse than death would befall us if we spoke so much as one word! I wonder if our hands clapped over our mouths had any connection with the monkey's 'speak no evil'?[2]

Similar beliefs have been recorded from Forfar, Angus in 1954,[3] Coulsdon, Surrey in 1960,[4] and Ealing, London, in 1979,[5] but have not been collected in recent years.

MONTBRETIA (*Crocosmia* spp.)

Perennial herbs, native to South Africa, introduced in the nineteenth century and cultivated for their orange flowers, becoming widely naturalised on hedgebanks and roadsides.

Two local names have been recorded from west Cornwall: bias lilies ('first brought to the area . . . about 1900 by a man called Tobias Wallis') and COCK'S COMB.

> [Co. Antrim] montbretia leaves were used as WHISTLES, different tones were produced depending on the breadth of the leaf.[1]

MOON – CABBAGES planted 'on the new moon'.

MOTHERING SUNDAY – VIOLET associated with.

MOTHS (CLOTHES) – BOG MYRTLE kills; conkers (HORSE-CHESTNUT seeds), LADY'S BEDSTRAW, LAVENDER and TANSY deter.

MOUSE-EAR HAWKWEED (*Pilosella officinarum*)

Perennial, lemon-yellow flowered herb (Fig. 33), common on well-drained soils and in short grassland throughout the British Isles.

Despite having an ungainly 'official' English name, it appears that mouse-ear hawkweed has acquired few local names – various variants of the standard name, and FELLON-HERB, recorded by Britten and Holland from east Cornwall.

> Herbal cures used in the Ashford area of Kent: my father boiled the roots of 'mouse ear' (as he called it) with chemists' liquorice to make a mixture for taking to alleviate COUGHS and COLDS. It tasted vile and we were glad to get better![1]

> We used to collect mouse-ears from the common years ago and my mother made cough mixture with it.[2]

MOUTH ULCERS – treated using SAGE.

MUGWORT (*Artemisia vulgaris*)

Coarse, aromatic, perennial herb with silvery leaves and inconspicuous flowers, common in waste places, roadside verges, and rough ground throughout lowland Britain.

Local names include:
APPLE-PIE in Cheshire
BOWLOCKS in Scotland
BULWAND in Caithness and Orkney
Council-weed, 'it always appears after the Council has been out', in Lancashire
Docko in Berkshire, where the 'dried leaves are largely smoked by the country lads'
DOG'S EARS in Pembrokeshire
FAT-HEN in Buckinghamshire
FELLON-HERB in Cornwall
Gall-wood in Scotland
Green bulwand in Shetland
GREEN GINGER in Lincolnshire and Yorkshire
Grobbies in Orkney
Migwort in Dorset
Mogford in Somerset, where 'sought by old women as "good for the inside"'
Mogrund in Somerset
Mogvurd in Somerset
Moogard in Caithness
Motherwort in Yorkshire
Muggar in Cornwall, where village boys used the flowers to make cigarettes
Muggart in Co. Donegal and Scotland
Mugger, rhyming with sugar and smoked by boys in JAPANESE KNOTWEED pipes, in Cornwall, and Scotland
MUGGERT in Cumberland, Scotland and Ireland
Muggert-kail in Morayshire
Muggins, and muggons, in Scotland
Muggurth in Ireland
Mugweed in Cheshire
Mugwood in Cumberland, Co. Durham, Shropshire, Yorkshire
Old Uncle Harry in Somerset
Sailor's tobacco in Hampshire and Hertfordshire
Silver leaf in Cornwall
Smotherwood in Lincolnshire
Will wormuth, and wormuth, in Devon; 'will = wild'
WORMWOOD in Buckinghamshire.

> [On Colonsay] The leaves when young and tender are frequently made use of by the highlanders as a pot-herb.[1]

Also on Colonsay,[2] and elsewhere mugwort was much used as a TOBACCO substitute.

> [Shortlane End School, near Truro, 1934–38] Mugwort – mugger – smoked by schoolboys.[3]

> An old lady, born *c.*1870, from Crewkerne, Somerset, called mugwort . . . moggle. She said the leaves, after being hung up to dry, made substitute tobacco. My grandfather tried this during World War II, but was not overly impressed.[4]

> [Chesterfield, Derbyshire, *c.*1930] When about 10 I and some of my pals thought we would try smoking. Our pipes were hollowed out acorns and grass for the stem. The 'tobacco' was mugwort. It was so dreadful that I was cured of smoking for ever.[5]

On the Isle of Man:

> 5 July is the day on which the Tynwald, or Manx parliament, transacts its business in the open air at St Johns, as it has done for over a thousand years . . . A well attested custom, revived by Archdeacon Kewley in 1952, is the wearing of sprigs of mugwort (*Bollan Bane* or *Bollan feailleoin*) at Tynwald . . . Some older accounts emphasise that soldiers wore mugwort when attending Tynwald, and it may be that mugwort is a typical Norse sign of loyalty to the king, or that it was the plant badge of the Kings of Man.[6]

According to Allen and Hatfield:

> The medico-magical potency with which this visually unprepossessing plant has been credited through much of Eurasia and the strikingly familiar beliefs associated with it in different regions of that landmass indeed suggest that it may be one of the oldest herbs known to mankind. Sacred to thunder gods, its power to ward off evil influences was believed to be greatest on MIDSUMMER EVE.

However, they provide few definite records of its use in folk medicine, the most widespread one being its reputation for 'easing COLDS, heavy COUGHS and especially CONSUMPTION'.[7]

More recently it has been recorded that 'as long as a piece of mugwort . . . was carried, people did not suffer from fatigue',[8] and:

> Mugwort: Chinese women in Burgess Park [Southwark] collect it on a regular basis. When asked, they say 'for the back'. I think they use it for moxibustion.[9]

MULBERRY (*Morus nigra*)

Small deciduous tree, cultivated for its edible red fruits; native to central Asia, cultivated in Britain since at least the mid-sixteenth century, and occasionally naturalised.

> In the Western Counties it is asserted that frost ceases as soon as the mulberry bursts into leaf.[1]

The mention of mulberry in the children's game 'Here we go round the mulberry bush' has stimulated a great deal of speculation, especially as mulberry is a tree, rather than a bush. Sometimes it is said that the game is derived from prisoners taking exercise in a prison yard.

> Silver medal in the gardens category at the Chelsea Flower Show has gone to two inmates of Leyhill Open Prison. The men, guarded by two burly prison officers, have clearly been enjoying themselves in the Chelsea sunshine. The Leyhill Garden. Entitled 'Here we go round the Mulberry Bush', is a representation of an old-fashioned exercise yard at HM Prison Wakefield. At its centre is a fine specimen of *Morus nigra*, the bush around which the men used to walk on their daily exercise – hence the nursery rhyme.[2]

Some years earlier, according to the Opies:

> An ingenious joke-history for the 'Mulberry Bush' was going the rounds in 1978. The knights who were intent on killing Thomas à Becket first hung their swords on a mulberry tree, still, of course, extant. They scalped the Saint singing 'this is the way we do our hair'. They washed their hands afterwards to get rid of the guilt, and said their prayers around the body.[3]

The Opies also list some of the other plants which have been mentioned in place of mulberry bush in the game: BARBERRY bush (recorded in the USA, 1882 and *c.*1990), BRAMBLE bush (recorded in 1849), GOOSEBERRY bush (known by Thomas Hardy, born in Dorset in 1840), HOLLY bush (Nottinghamshire, 1894), IVY, or ivory, bush (Norfolk, 1894) and prickly pear (*Opuntia ficus-indica*, known by T.S. Eliot, born in St Louis, Missouri, in 1888). As Roud observes, 'the sheer variety of plants mentioned in the past shows that it is dangerous to read too much significance into the mulberry bush element of the song'.[4]

However, the bramble or blackberry possibly merits further consideration.

> I have been unable to find any attempt to explain why blackberries were known as mulberries in East Anglia. It is true that there is a species of bramble with a bigger berry than most others and that is given the name of mulberry by some East Anglian countrymen even today, but I believe there is another explanation as the phrase *going a-mulberrying* was used for gathering blackberries of all sizes and species . . . Only people of substance had their mulberry trees. The ordinary people had to be content with the bramble. It is probable that the blackberry was referred to ironically as the mulberry because it was the poor man's fruit . . . One further piece of evidence does suggest that this meaning of mulberry was linked particularly with the ordinary country people. This is the action rhyme: *Here we go round the mulberry bush*. Genuine mulberries grow into trees, therefore this mulberry bush must have been a bramble.[5]

The confusion between mulberry and blackberry is long-standing. In classical Latin the same word, *morum*, was used for both, by Horace for mulberry, and by Virgil and Ovid for blackberry. In this context it is perhaps interesting to consider the various cures which were effected by crawling around, or more usually under, brambles; did these have any influence on the evolution of the game?

Not surprisingly old mulberry trees are sometimes associated with historical figures. Milton's Mulberry Tree, in the garden of Christ's College, Cambridge, is said to be associated with the poet John Milton, who entered the College in 1625.[6] While a tree growing on the twentieth-century Streatham Park housing estate in South London, is associated with Samuel Johnson, who spent much of his time at Streatham Park, then the home of Henry and Hester Thrale, during the fifteen years before Henry's death in 1781.[7]

In East Anglia mulberry was used as a 'sure cure' for DIARRHOEA:

> Boil the green leaves from the mulberry tree and drink the infusion.[8]

Tottenham cake, or Quaker tray bake, was first made by the North London baker in the 1880s as an inexpensive treat for children.

> Henry used a plain, simple recipe. The icing was flavoured and coloured pink, using mulberries from a mulberry tree growing in the Tottenham Quaker Meeting House garden. Children could buy a square for one penny or pay just half a penny for any broken pieces.

In 1901, when Tottenham Hotspurs beat Sheffield United, to become the only non-league side ever to win the Football Association cup, local bakers 'made huge trays of Tottenham cake, handing them out to children, and to this day Tottenham cake is made all around north London and served at White Hart Lane football ground at special events'.[9]

MULLEIN (*Verbascum thapsus*)

Biennial herb with silvery-felted leaves and spikes of bright yellow flowers, common and widespread on waste ground, in grassy places.

Local names include the widespread AARON'S ROD, ADAM'S FLANNEL, hag-taper and:

Aaron's flannel in Dorset
Agleaf (i.e. 'hedge-leaf') and ag-paper in Buckinghamshire
Beggar's blanket in Cumberland and Somerset
Beggar's stalk in Cumberland
Blanket-leaf in Somerset and Warwickshire; 'on account of the woolly texture of the leaf'
Blanket-mullein in Cheshire
Candlewick in Dorset and Somerset; 'useful for wicks of lamps'
CLOTE in Wiltshire
Cotswold witch in Gloucestershire
Cuddy-lugs in Northumberland; 'i.e. donkey-ears'
Duffle in Suffolk; 'i.e. a woollen cloth'
Fairies' wand in Dorset and Somerset
Flannel, and flannel-flower, in Somerset and Suffolk
Flannel-petticoats in Somerset
Flannel-plant in Cornwall, Hampshire, Isle of Wight and Kent
Fluff-weed in Norfolk and Somerset
French poppy in Devon; 'i.e. a French kind of foxglove'
Golden grain in Devon
GOLDENROD in Devon, Somerset and Fifeshire

Hag-leaf in Buckinghamshire and Somerset
Hare's beard in Dorset and Somerset
High taper in Somerset and Northern Ireland; 'probably from A.S. *hege* or *hega* = hedge, and taper, its stalks when dipped in grease being formerly used for burning'
Lady's flannel, LUNGWORT, Moses' blanket, and moth-plant, in Somerset
Mullein-dock in Norfolk
Old-man's flannel in Somerset
Our Lord's flannel, and Our Saviour's flannel, in Kent
Poor-man's blanket in Co. Donegal
Poor-man's flannel in Buckinghamshire, Gloucestershire and Somerset
Rabbit's ear in Cornwall
Rag-paper in Buckinghamshire
Shepherd's club in the Isle of Wight and Lanarkshire; 'an old country name'
Shepherd's staff in Cumberland
Snake's plover in Dorset and Wiltshire
Soldiers' blankets in Kent
Soldier's tears in Dorset
SWEETHEARTS in Somerset
VELVET DOCK in Devon and Somerset
Velvet poppy in Cornwall
Virgin Mary's candle in Co. Limerick.

In former times dried flower-stalks of mullein were used as candles. Thus when mullein and weld (*Reseda luteola*) appeared after excavations at Mount Grace Priory, in North Yorkshire, it was claimed that these had 'sprung to life' providing an insight into the daily lives of the Carthusian community which was dissolved in 1539:

> Monks would have used the weld to provide yellow DYE for their garments while the mullein would be dried and dipped into tallow to make the wicks for processional candles'.[1]

However as both species are likely to spring up on disturbed soil anywhere, the idea that the Mount Grace plants grew from sixteenth-century seed is probably no more than wishful thinking.

During the German occupation of Jersey (1940–5) dried mullein leaves were used as a substitute for TOBACCO.[2]

> [Guernsey, 1880s] for cows which were attacked by stranguary [difficulty in passing urine] the following remedy was used: Take the leaves of mullein; chop them up very fine, mix them with bran and water, then give the whole to the cow.[3]

> For BRONCHITIS and ASTHMA: the leaves of the mullein plant, dried, and put in a clay pipe and smoked like tobacco, the smoke to be inhaled.[4]

> Mullein plant . . . is boiled and the juice is strained off. The drink is used as a cure for a COLD and also CONSUMPTION and other diseases of the lungs.[5]

> I recall Bob Penfold [a gypsy] showing me mullein out on Braunton Burrows, and calling it cough flannel, the leaves of which were, I understand, used for a COUGH mixture, as good as 'ammonia and ipecac'.[6]

MUMPS – treated using PRIVET.

MURRAIN (disease in cattle) – caused by eating SUNDEW; treated using HERB ROBERT.

MYRTLE (*Myrtus communis*)

Evergreen shrub with attractive glossy leaves and abundant white flowers and purple-black fruit, native to the Mediterranean region and long cultivated as an ornamental.

> Myrtle was much esteemed in Wales, where they say that if it grows on each side of the door the blessings of love and peace will never depart from the house. To destroy myrtle is to 'kill' both love and peace. Sprigs of myrtle, with its blossoms, are not only used by brides, but in some parts of Wales they were worn in the girdle or bodice of young girls when going to their first Holy Communion. Sprigs were also placed in cradles to make babies happy.[1]

It is frequently claimed that old myrtle bushes are descendants of sprigs carried by Queen Victoria in her wedding bouquet.

> When I was a child I lived in an old house on the outskirts of Bel-

> fast, and on the south-facing wall outside my nursery there grew a fine myrtle bush which flowered freely. According to local tradition this bush grew from a sprig from Queen Victoria's wedding bouquet.[2]

> About four years ago an elderly lady came to shrub nursery for which I am propagator asking me to give a good home to a potted myrtle plant (*Myrtus communis*) as she was moving away from the area into a flat. She said it was reputedly grown from a cutting from Queen Victoria's wedding bouquet and thought we might like to propagate from it.[3]

However, an examination of press reports suggests that Victoria used only ORANGE blossom at her wedding in February 1840, and the first recorded use of myrtle by a royal bride is at the wedding of the Queen's cousin, Princess Augusta of Cambridge, in June 1843. At this, the bride's dress was decorated with myrtle, her head-dress was 'a wreath of orange flowers and myrtle', and the bride-cake decorations incorporated myrtle buds. Reports of this wedding explain the myrtle as being 'the emblematic flower of Germany'. Fifteen years later, at the marriage of Victoria's eldest daughter, also named Victoria, the bride wore a myrtle-decorated gown, myrtle 'being the bridal flower of Germany'.[4] It seems that sprigs from this myrtle were cultivated at Osborne House, the royal residence on the Isle of Wight, and thereafter myrtle seems to have been a favourite for royal brides.

> When Princess Anne married Captain Mark Phillips on 14th November 1973 . . . the princess's bouquet contained a sprig of myrtle grown from myrtle used in the wedding bouquet of Queen Victoria . . . Miss B.D. Hadow, manager of Moyses Stevens Limited, florists to the Queen and makers of Princess Anne's bouquet [commented] . . . 'We do not find that myrtle is asked for very much these days for bridal bouquets. On the few occasions that we have used it, it has nearly always been because the bride's grandmother or some other relative had some growing in the garden and wished it to be put into the bouquet for sentimental reasons.'[5]

N

NAPPY-RASH – treated using COMFREY.

NAVELWORT (*Umbilicus rupestris*), also known as pennywort

Fleshy perennial herb (Fig. 34), with circular leaves and spikes of ivory-coloured flowers, growing on walls and in rocky places mainly in western districts, but currently spreading eastwards.

Local names include:

BACHELOR'S-BUTTONS in Devon
Corn-leaves, in Worcestershire; 'applied to CORNS and WARTS'
Cows in Cornwall
CUPS-AND-SAUCERS in Devon, Dorset, Somerset and Wiltshire
CUT-FINGER in Worcestershire; 'a vulnerary'
Dimple-wort in Devon
Half-pennies-and-pennies in Devon
ICE-PLANT in Devon and Somerset
JACK-IN-THE-BUSH in Scotland
Kidney-weed, and kidney-wort, in Somerset; 'from a distant resemblance of its leaves to the outline of a KIDNEY', 'used for the kidneys and against the stone'
Lover's links in Roxburghshire
Lucky moons in Dorset
Maid-in-the-mist in southern Scotland
Milk-the-cows in Cornwall
Money-pennies, or money-penny, in Devon
NIPPLEWORT in Sussex
PANCAKES in Devon; 'from the shape of its leaves'
Penny-cake in Cornwall, Devon and Pembrokeshire
Penny-caps in Devon and Somerset
Penny-cod in Cornwall
PENNY-GRASS in Co. Fermanagh
Penny-hats in Devon
Penny leaf in Devon; also penny-leaves in Somerset and Ireland
Penny-pies in south-west England and Sussex; 'used for rubbing on CHILBLAINS'
Penny-plates in Devon
Penny-wall on the Isle of Man
PRINCE'S FEATHER in Yorkshire
Royal penny, and SUNSHADES, in Somerset.

> Among my childhood group in mid-Devon in the 1930s navelwort was quite the most favoured agent of WEATHER forecasting. One selected two large leaves, spat liberally into them, pressed them together and threw them into the air. Should they continue to adhere when striking the ground rain would ensue, should they part dry weather could be expected. Being Devon, it usually rained, but, on reflection, I incline to the view that the liberality of spittle was the main determinant of the outcome (but this, of course, was the main fun involved!).[1]

> Pennywort leaves can be softened, a piece of grass used to pierce the stem and then blown up like balloons. (I could never do this, but often saw others do it).[2]

Like HOUSELEEK navelwort had many medicinal uses. Allen and Hatfield suggest that the two were interchangeable, with navelwort being used in western areas where it is common, and houseleek being used further east.[3] In March 1854 the Phytological Club heard how navelwort was used by the 'lower classes of English people'. In Poole, Dorset, it was used to treat epilepsy, in Monmouthshire and Herefordshire to treat urinary obstructions and fits, and also in Herefordshire, and in Worcestershire, to treat corns and warts, and 'make a cooling ointment, and . . . juice is expressed and mixed with cream as a cooling lotion for sore faces and chaps in children'.[4]

> Mixed pennywort and butter (unsalted) for BURNS.[5]

> A local farmer tells me he always uses pennywort if he gets a thorn or splinter in his fingers. He peels the 'skin' from the back of the leaf before applying and 'tis drawn out in a day or two.[6]

NERVES – CENTAURY, HEATHER and LAVENDER calm; DANDELION good for.

NETTLE (*Urtica dioica*)

Perennial clump-forming herb with stinging hairs, widespread and probably increasing, particularly on nutrient-rich soils.

Local names include the widespread variant ettle, and:

Coolfaugh in Co. Donegal
Devil's leaf in Somerset
DEVIL'S PLAYTHING in Sussex
Gicksy in Somerset; 'also the stalk

of wild parsley, out of which primitive pipes were made'
Heg-beg in Scotland
Hiddgy-piddgy in Devon
Hitty-titty in Oxfordshire
Hoky-poky, or hokey-pokey, in Devon
Hop-tops in Wiltshire; 'a very common local name for the tops of young nettles, formerly gathered and boiled by country people'
Jenny-nettle on the Isle of Man
NAUGHTY-MAN'S PLAYTHING in Somerset and Sussex; 'i.e. devil's'
Stingy nettle in Devon, Northamptonshire and Oxfordshire
Tanging nettle in Yorkshire.

As Nicholas Culpeper, writing in 1652, observed: nettles 'are so well-known that they need no description at all, they may be found by feeling in the darkest night'.[1]

> Nettles in many parts of Scotland were till not many years ago used as food, and were looked upon as a wholesome diet. The young and tender leaves were gathered, boiled, then mashed . . . mixed with a little oatmeal, and reboiled for a short time. They were cooked in the same way as 'greens', which were and still are thought to possess medicinal virtues. In the north such a dish went by the name of 'nettle kail', as the dish of 'greens' went by that of 'chappit kail'. But the nettle . . . was used under the form of 'nettle ale', for the cure of JAUNDICE. The ale was prepared in the following manner: a quantity of nettle-roots was gathered thoroughly washed, and then boiled for hours in water till a strong extract was got.
>
> The extract was then treated with yeast, 'barm', fermented, 'vrocht', and bottled. A man whose mother was in the habit of making this ale lately told me he had often drunk it, and found it quite palatable.
>
> In one district at least the medicinal virtue of the nettle lay in its being 'unspoken', i.e. no one must speak to the gatherer of it,

and collected at the hour of midnight. The following is a [Kincardineshire] story . . .

'Geordie Tamson, who lived near Jollybrands on the south turnpike, not far from the toll-bar, lay sick. After weeks of treatment by the doctor, Geordie lay ill, without the least token of improvement. A "Skeely woman" from Dounies, a village not far off, was called in. She at once prescribed a supper of "nettle kail", and added that the dish must be made of "unspoken nettles", gathered at midnight.

That very night by eleven o'clock three young men, friends of Geordie's, from Cairgrassie, were on their way to the Red Kirkyard of Portlethen, where there was a fine bed of nettles . . . the nettles were gathered, carefully taken to the sick man, cooked of course, and given to him. A complete and speedy recovery followed.'[2]

Culpeper considered nettle to be a herb of Mars:

You know Mars is hot and dry, and you know as well that Winter is cold and moist; then you may know as well the reason why Nettle tops eaten in Spring consume the phlegmatic superfluities in the body of Man, that the coldness and moistiness of winter hath left behind.[3]

More recently nettles have been considered to be a health-giving food, particularly in Ireland:

Then there were the nettles, we had to take them to make soup; oh, it was lovely.[4]

When she [my mother] was a child approximately 70 years ago, she was given (or forced to take!) three meals of boiled nettles in the month of March to clear the BLOOD![5]

Eat three meals of boiled young nettles in springtime and it will keep you free of disease for the rest of the year.[6]

To boost blood iron drink water from boiled nettles, or eat young nettle leaves.[7]

My mum remembers eating boiled nettle tips in spring as a tonic.[8]

The idea that eating nettles in springtime ensured good health was more than superstition. Before the ready availability of frozen, canned and imported vegetables people experienced a period each year when fresh vegetables were scarce. The last of the winter vegetables were past their prime and early summer vegetables were not ready for harvesting, so the consumption of young nettles would undoubtedly provide nutritious variety to an otherwise impoverished diet.

According to researchers at the University of Wales, 'the oldest recorded dish, more than 8000 years ago' in the British Isles was nettle pudding. Nettle leaves, BARLEY flour, salt and water were blended together and added to stews as a form of dumpling.[9]

In Scotland nettles were considered worthy of cultivation for food:

> Nettles, indeed, were so prized for their nutritious properties that they were 'forced' as a kind of spring kale, replacing the cabbage varieties that would usually have been destroyed by the frosts by that time of year. Sir Walter Scott mentions this practice in his novel *Rob Roy* [1817], when the old gardener at Loch Leven raises early nettles under glass. The preferred variety had brown stems and dark foliage, and was sometimes 'earthed up' . . . to blanch the stalks. A second crop can be obtained later in the year by cutting down the old stems before flowering, and harvesting the regrowth.[10]

During the Second World War:

> Nettles were positively promoted as food. I know because I was a cookery demonstrator, and it was my lot to demonstrate how to make nettle and egg mould. I had to pick the nettles the day before – I hated it and still remember getting stung everywhere now. You had to boil up the nettles, place them in this mould and add hard-boiled eggs to make it look good.[11]

> [Blackpool] During the last war my Girl Guides gathered . . . nettles from the local golf course which we thought were for medicinal purposes. After the war we got a letter of thanks for the 'dried vegetables' used to supplement the diet of our merchant sailors![12]

Readers of the *Sunday Times* were encouraged to make a 'luxurious form of nettle toast':

> Toast slices of bread. On each put a pile of well-strained and seasoned nettles. Make a hole in the centre and put in a raw egg. Over the egg place a rasher of bacon and put under the griller, the fat from the bacon cooks the egg.[13]

Elsewhere:

> After the [Second World] War people in Berlin lived on nettles. I wasn't there myself, but I was told by people who were.[14]

> In the West Country, 'a decoction made from boiling nettle leaves in salt water was used . . . by farmers' wives when they needed a substitute for RENNET when making cheese'.[15]

Cornish yarg cheese is coated in nettle leaves; according to the Lynher Dairies Cheese Company's publicity in 2006:

> Cornish yarg is instantly recognisable thanks to its distinctive nettle leaf coating. The leaves are applied when the cheese is two days old. The cheese is then left to mature for 12 weeks during which time a delicious edible rind of natural moulds develops on the nettles. The mould can vary slightly in colour and quantity throughout of the life of the cheese.
>
> The leaves are picked locally by hand between May and June having reached optimum size for Cornish yarg. They . . . are frozen within 24 hours to preserve their freshness and to kill the sting!

Although 'yarg' suggests an old Cornish name, and the cheese is promoted as 'traditional':

> Yarg is in fact the surname of the original producer, Allan Gray, spelt backwards. The cheese was developed from an old recipe 20 years ago, and nettles were chosen as a coating because they were an old-fashioned way of ripening cheese.[16]

As nettle flowers start to develop the leaves become coarse and unsuitable for human consumption:

> The springtime nettle of the [New] Forest was . . . said to be unfit for eating after a certain date . . . tradition claims that on MAY DAY the DEVIL gathered nettles to make his shirts.[17]

Nettle beer was a favourite home-brewed drink.

> [Great Glen, Leicestershire, 1920s] I well remember the old zinc bath used for making herb beer, going out into the meadow gathering a huge amount of nettles, giving it a rinse, then pouring pans of hot water over it, leaving it to infuse until it cooled. She [my grandmother] then added nutmeg, allspice, raisins and yeast;

> it bubbled away like mad for a period of time. Then strained and bottled . . . I don't think it lasted very long, as there always seemed to be a brew on the go.[18]

Until the 1980s 'traditional' nettle beer – a pleasant, refreshing and slightly ginger-flavoured drink – was sold in Heysham, Lancashire, and people had memories of enjoying the beer which an old lady used to sell from her cottage in the village.[19] Such brews were widely made in rural areas, it seems probable that their alcohol content though low was sufficient to kill any germs which might be present in potentially harmful water supplies.

Before the discovery of antibiotics nettles were valued as a food to keep young TURKEYS healthy.

> In the 1930–40 period people who reared flocks of turkeys used to feed them with nettles. They put an old stocking on their hand, took a knife and went out to the fields to cut nettles; some people even cut them by the sack full. They made a pot of Indian Meal gruel (maize). Then the nettles were chopped fine. My aunt used to take a big handful in her bare hands and squeeze them (this way they don't sting) and cut them up with a sharp knife. Then they were put into the boiling gruel and stirred around. When the mixture was cooled it was thick and the nettles were cooked.[20]

> [Hertfordshire] Before pre-mixed food could be bought for day-old turkeys the growers had to make their own. Day-olds are, and always have been, the hardest things to raise, and turkeys particularly so. My husband chopped stinging nettles and mixed them with hard-boiled eggs to give them the required nourishment and vitamins until they were old enough to eat the formula he mixed in his mill.[21]

Allen and Hatfield found 311 localised records of nettles being used in folk medicine in Britain and Ireland. Of these almost two-thirds (194) related to their use as a spring tonic, and 76 records were to treat rashes. Other major uses included as a counter-irritant for rheumatic conditions (48 records) and to treat COUGHS, COLDS and LUNG troubles (29 records).[22] However, the primary concern of these researchers was the geographical rather than the historical distribution of remedies; 'where' rather than 'when'. Recently collected material suggests that nettles are mainly concerned with treating RHEUMATISM, although their use as a spring tonic is still widely remembered.

Examples of nettles being used to treat rashes include:

> Nettles boiled and the resultant liquid drunk, is a cure for 'Lily Rash'. Lily Rash is the result of juice from the daffodil family – usually making sore and itching fingers. This is usually when the skin is irritated when either picking or packing flowers during the season for market.[23]

> Tips of young stinging nettles when boiled and eaten cure rashes and allergies.[24]

> After 21 years on medication for psoriasis, my friend tried eating stinging nettles in soup or vegetables. In just over a fortnight he was cured, although the scars remain.[25]

The use of nettles to treat rheumatic and arthritic pains is extremely widespread, being recorded (possibly incorrectly as nettles are neither native nor abundant in all the countries listed) from France, Germany, Italy and Poland, India, Australia and New Zealand, and North and South America.[26] Records from Britain and Ireland include:

> I am informed by some maiden ladies living in Torquay that nettle-stings are a cure [for rheumatism]. The nettles must be applied to the affected parts.[27]

> I remember a man who pulled a bunch of nettles and got his wife to hit his back with them. He maintained it helped [his ARTHRITIS].[28]

> A friend, now over 85 years old, who came from Fife recalled being told how her father was cured of a frozen shoulder when nettles were laid on the bare skin of the shoulder area and kept on for, probably, several days. When the nettles were removed, and despite the blisters caused by them and the pain suffered, the man was cured, for all time.[29]

> Wild nettles if boiled are good for arthritis.[30]

> In the late 1980s I worked in a bedding-plant nursery in north London. Perennial nettles used to grow between two of the glass-houses. We would harvest them and rub them on our joints in order to ward off rheumatism and arthritis – as a preventative medicine.[31]

> In the late 1930s I was a nurse-midwife in the Lake District. In the village where I lived a farmer was noted for his laziness. He would lie in bed in the mornings saying his LUMBAGO was so

bad that he couldn't get up to do the milking. This chore fell to his wife and after a time she got fed up with it, so she went to the village 'know-all' for advice. The old woman told her to gather a bunch of nettles, a bunch of THYME, a few DOCK leaves, mix them together and tie them up in a bundle. With this she was to lash the area of pain for at least 10 minutes (not too hard). After some persuading the husband made himself comfortable on his tummy and the treatment commenced, however the wife got a bit over enthusiastic and the nettles fell on to private parts.

The husband leapt out of bed and dancing around the room swore he would kill the 'old hag' who had told her to do this. The story in the village was that he was never heard to complain of lumbago again and his wife was never again asked to do early morning milking. A sure cure.[32]

Other remembered remedies include:

Dry nettles and burn them in a closed room over hot embers. Breathing the pungent aroma will relieve BRONCHITIS and ASTHMA attacks.[33]

Nettle tea was made and drunk for STOMACH complaints.[34]

[Devon] A gypsy was beating his child's eyes with stinging nettles. Kid was screamin'! What the hell was 'e about? He zaid: 'You'll never see a gypsy with glasses!' No, you don't either. I couldn't believe it because the child was only just born.[35]

I cured my BLOOD PRESSURE. I was about 55, and the doctor in Tiverton zaid: 'Oh, you'll have to take blood-pressure tablets all your life.' I went out . . . and down to B: 'I got blood pressure . . .' 'What be goin' do?' I zaid: 'I shall pick some stinging nettles. Three-gallon bucket full.' And I stuffed them down with a quart of water, and pushed them down and down and boiled them. And then I strained it . . . And the water was green. I only had to take it for a month, because it was so strong. I thought it better to do it strong. I only done it for a month. I done a quarter bottle. Every morning before breakfast, drink a wine glass full. And the doctor zaid to me: 'You come in every month. You'll have to be tested every month.' Course, I went in the end of the month. He measured me, and he couldn't understand! 'What's the matter, then, Doctor?' He zaid: 'You 'aven't got no blood pressure!'[36]

There is one record, from a New Forest gypsy in 1952, of nettles being used as a contraceptive. A man should lay nettle leaves thickly as a sole inside his socks and wear them for 24 hours before engaging in intercourse. The gypsy claimed to have tried this, and proved its effectiveness.[37]

Occasionally nettles were used to deter FLIES:

> [1920s] A freshly gathered bunch of nettles hung up in the kitchen or larder was supposed to keep flies away.[38]

> Nettles were hung up to deter flies (they didn't).[39]

Nettles have a long history as a source of FIBRE, but such use appears to be more prevalent in mainland Europe than in the British Isles. It can be difficult to distinguish between FLAX and nettle fibre, and it is thought that some fabrics said to have been made from the latter were, in fact, made from the former. It appears that at one time 'nettle-cloth' was a general name for any fine fabric. Gillian Edom in her history of nettle fibre notes that 'by the nineteenth century the word "nettle-cloth" had become firmly established as one that described any textile possessing certain characteristics and possibly that made from imported ramie (*Boehmeria nivea*)'.[40]

> I have a note that the Patent Office holds over 50 patents for different ways of retting nettles, but I have never checked this. Fibre can be prepared from nettles of any age. Young shoots are reputed to yield a fibre that can be made into a very good imitation silk, and so on through the season, the strong autumn fibres being used for sacking, etc. Nettle cloth is reputed to have been made in Britain until the 1920s.[41]

Towards the end of the twentieth century there was a revival of interest in nettle fibre. In 2003 jeans made from such fibre were 'tipped to be the next big thing when they reach London shops':

> The expansion of farming nettles for their fibres is being encouraged by politicians in central Europe and the EU has been investigating the possibility of farming them since 1999.[42]

About eighteen months later, on 3 July 2004, the *Sun* reported: 'the latest UK design fad – knickers made from nettles – gives a whole new meaning to having a tingling down below!'

Nettles were also used as a dyestuff, being collected for this purpose during the Second World War. In 1942 County Herb Committees were asked to gather 100 tons of nettles; 90 tons were collected, and used for the extraction of a dark green DYE for camouflage, and the chlorophyll was used as a tonic and in other medicines.[43]

> People long ago also dyed wool. First the wool was put into water and potash and was boiled for some length so that it would make a good dye. Then a lot of nettles were pulled and they were boiled in some clean water. Then the juice was taken off the nettles, and the wool was put into this, and it was boiled again for about one hour. When the wool had taken a green colour it was taken out and it was washed to get all the pieces of nettles off it and it was dried and spun.[44]

> When I was at school during the war we used to collect things . . . LIME flowers, stinging nettles and ROSE hips. They used to dry stinging nettles, laying them out on sort of big canvas sheets. I didn't like collecting stinging nettles. On the list of clothing we had to take to school were stinging nettle proof gloves – they were terribly difficult to get.[45]

On Oak Apple Day (29 May) children who were not wearing a sprig of OAK were punished by being stung with nettles.The saying 'it's all muck and nettles' appears to be restricted to Lincolnshire.

> On Royal Oak Day children expected their fellows to wear an oak leaf . . . Any of our contemporaries who failed to display an oak leaf were accosted by others with the choice of 'Muck or nettles?' The challenger held in one hand a stick lathered in fresh cow dung and in the other a bunch of nettles, so the victim was given the choice of being either daubed or stung. This led to the expression 'It's muck or nettles' being given to any forced choice of unpalatable alternatives.[46]

> 'It were all muck and nettles after that.' This phrase was used of a social gathering in Legbourne village hall (or some similar hall). The event had started with a meal and a few informal speeches – this was regarded by the speaker as the 'easy part'. It was followed by a disco . . . this part was regarded by the speaker as 'hard work' and he left the gathering at this stage and went home.[47]

In some coastal areas such as Sussex in the 1930s and 40s,[48] around Cromer in Norfolk,[49] and South Devon, 2 May was Sting Nettle Day. Around Hove and Worthing:

> The first of May is Pinch-Bum Day,
> The second of May is Sting-Nettle Day.[50]

In South Devon:

> I was born in Brixham, and we had a little rhyme that goes:
>
> First of May Ducking Day
> Second of May Sting Nettle Day
> Third of May Petticoat Day.
>
> First of May we carried water about with us to fill our water pistols which we squirted at people (Ducking). In the evening the fire brigade went to the harbour and with their hoses washed down the monument of William Prince of Orange. After that to the Town Hall to do the same.
>
> Second of May we chased the girls with sting nettles.
>
> Third of May we chased the girls and tried to lift their dresses.[51]

> May 1st was a great day for us youngsters 50 years ago, it was Ducking Day . . . but Sting Nettle Day followed (May 2nd). The boys used to collect stinging nettles and chase all the girls trying to sting their legs; of course they hardly ever caught us and there was much laughter. The boys were not bad if they caught you, they would think twice about stinging. May 3rd was Petticoat Day . . . the boys would chase after us, trying to see what our petticoats were like.[52]

It appears that these customs died out just before or during the Second World War.[53]

The Bottle Inn, in Marshwood, on the Dorset–Devon border, hosts the annual World Nettle Eating Championships on the Saturday before the Summer Solstice. This is said to have started:

> when two old farmers were talking about the size of the nettles growing on their farms. One of them said to the other, 'I bet mine are bigger than yours; if they're not, I'll eat the buggers.'[54]

Or, according to the Bottle's preferred version:

> Nettles first came to the fore at the Bottle Inn around 1986, when

> two farmers were having an argument about who had the longest nettles on their land. The landlady . . . commented 'What makes you think you have the longest nettles, we'll have a competition open to everyone in the area and we'll see who has the longest nettles.' The Longest Stinging Nettle Competition was born.
>
> The competition had been running for three years when local hospital porter and ex-guardsman Alex Williams entered a stinging nettle 15ft 6ins long, he said at the time, 'If anyone beats that I'll eat it.' An American couple on holiday staying in the area came up with a nettle 16ft long and Alex true to his word promptly ate the nettle.

In 1997 the pub's landlord decided to hold a musical Summer Solstice celebration and talked to his customers about suitable sideshows. Alex asked for a corner of the beer garden and challenged anybody to eat more nettles than he could.[55] In 1999:

> Bridport Hospital porter Alex Williams had to swallow his pride when he failed to win back the world nettle championship at the weekend.
>
> Back after a year's absence through ill health, Mr Williams could only manage to swallow 24ft of the prickly plants in the unusual annual contest at the Bottle Inn, Marshwood.
>
> And that was a whole 10ft less than the winner – the amazing 80-year-old Terry 'Bluey' Hunt from Axminster. Runner up was last year's champ Tim Beer, also from Axminster, who swallowed 28ft of nettles.[56]

The current record is 80ft, eaten by the 2010 and 2014 winners.[57] When such lengths are mentioned people imagine that whole stems are eaten with the contestants having to chew through the fibrous stalks. In fact only the leaves are eaten. In 2001 the pub's landlord explained:

> They don't have to eat the stalks, they would be far too indigestible. The secret is to get the leaves past your lips without touching them. You must not let your mouth get dry – I sell a lot of cask-conditioned ale during the contest.[58]

However, the consumption of ale, or other drinks, leads to another problem. Any contestant needing to leave the competition to urinate is disqualified. Although termed the World Nettle Eating Championships, it appears that the event is unique and no similar championships are held elsewhere. Occasionally a 'celebrity' contestant takes part, but most of the participants live within an evening's drive of the venue. It is said that farmers and builders with their work-roughened hands do well; celebrities, whose hands tend to be otherwise, are unlikely to achieve success.

In recent years the future of the Bottle has at times seemed precarious and it has not been possible to hold the event every year; potential competitors and spectators are advised to check before travelling.

NETTLE STINGS – DOCK and HORSERADISH used to treat.

NEURALGIA – treated using ASH, CHAMOMILE, HENBANE and POPPY.

NEWTS – HAIRY BROME used to catch.

NEW YEAR'S DAY (1 January) – 'Penny for DAISY' custom took place.

NIPPLEWORT (*Lapsana communis*)

Annual, or short-lived perennial herb with inconspicuous yellow flowers, common and widespread in disturbed shady places.

Local names include:

Ballagan in Ayrshire
Bare-land leaf in Northumberland; 'where used for making a healing salve'
Bolgan-leaves in Scotland; 'bolgan is a Scotch word for a swelling that becomes a pimple'
Carpenter's apron in Warwickshire
Dock-cress in Dorset
Hasty-roger in Devon
Hasty-sargeant in Dorset and Somerset
Jack-in-a-bush in Gloucestershire
Lamb's lettuce in Somerset
Mary alone in Gloucestershire
Pig's cress in Dorset
WORMWOOD in Somerset.

According to Lightfoot, writing in 1777, 'in some parts of England the common people boil them as greens, but they have a bitter and not agreeable taste'.[1]

NITS – SPINDLE kills.

NOSEBLEED – YARROW used to treat or cause.

NURSERY BOGIES

> There is a group of spirits that seem as if they had never been feared by grown-up people but had been invented expressly to warn children off dangerous ground or undesirable activities.[1]

Such bogies include Awd Oggie and Lazy Laurence who protected APPLES, and Jenny Greenteeth who grabbed children who ventured near DUCKWEED-covered pools, and Churnmilk Peg and Melsh Dick who protected unripe HAZEL nuts.

NUTMEG (*Myristica fragrans*)

Tree, native to the Moluccas, cultivated in tropical areas for their seeds, which when grated are used as a spice.

It was widely believed that carrying a nutmeg would, like carrying a POTATO, avert RHEUMATISM, or, less commonly ensure good health.

> A nutmeg carried in the pocket wards off the rheumatics. (Firmly believed in the Doncaster area today. I have been given one for each of my pockets by an earnest well-wisher).[1]

> My father carries a nutmeg to prevent rheumatism. He's 87 and has all sorts of other things, but he doesn't have rheumatism.[2]

> Nutmeg is good for LUMBAGO, you put one in your coat pocket when you're gardening, but you miss it if you change your coat and wear one without a nutmeg in its pocket.[3]

> Some time ago I asked an elderly lady who was born in Suffolk what treasures she had in an underpocket she always wore . . . a well-worn quite small thing, which looked like a small bean, and she said that it was her nutmeg. I said, 'Why in the world do you carry that?' 'Oh,' she said, her mother gave it to her when she was about 17 years of age to always carry in her pocket to ensure good health . . . She died three years ago in Brighton aged 91.[4]

Grated nutmeg could be used to treat DIARRHOEA:

> I was constantly ill until aged about three years and my mother swore it was gran's old methods that pulled me through. For diarrhoea I was given boiled milk with a good sprinkling of freshly grated nutmeg.[5]

O

OAK (pedunculate oak, *Quercus robur*, and sessile oak, *Q. petraea)*

Large, long-lived, deciduous trees (Figs 35 and 36), common and wide-spread throughout the British Isles, formerly much valued for their timber which was used in the construction of wooden-framed and other buildings and battleships.[1]

Pedunculate oak, sometimes known as English oak, is the unofficial national tree of England, partly because its value to ship-builders made it important to the defence of the realm. It is also the official tree of the Irish Republic, having been selected in 1990 when the Irish had a stand at a horticultural show in Osaka, Japan, for which they needed a national tree, and hurriedly adopted oak.[2] In September 2006 the Conservative Party adopted a stylised oak tree as its emblem, replacing the 'freedom torch' which it had used since 1977, and said to reflect its then leader David Cameron's interest in the 'environment and quality of life'.[3]

A widespread rhyme suggests that the time at which oak and ASH trees produce their leaves can predict the WEATHER which can be expected during the approaching summer.

> [West Dorset, early 1960s] An old labourer who frequently visited my parents would recite:
>
> If the oak is out before the ash,
We shall surely have a splash.
>
> Then he would scratch his head and remark, 'but there's another way of saying that:
>
> If the ash is out before the oak,
We shall surely have a soak.
>
> He believed that both rhymes implied wet summers.[4]

However, it is generally agreed that if ash produces leaves before oak

– a rare occurrence – then the summer will be wet.

Oak before ash – splash;
Ash before oak – soak.[5]
If the oak is out before the ash
We shall only have a splash;
If the ash is out before the oak
We shall surely have a soak.[6]

The fruit of oak, acorns, were believed to protect against LIGHTNING. According to a folklorist writing in the 1960s:

> Oaks are thunder-trees, once sacred to Thor, and as such they are thought to protect against lightning and thunderbolts. It was quite commonly thought that they are never struck by lightning . . . Their branches, or their acorns, kept a house from being struck. When 'pull-down' blinds were fashionable, the bobbins at the ends of the blind-cords were usually shaped like acorns for this reason.[7]

Alternatively, according to the *Weekly Telegraph* of 5 February 1938, 'ever since the days of the DRUIDS, the acorn has been the accepted charm against lightning'. As might be expected of a newspaper article, no evidence is given to support this claim, but useful information is provided about then-current beliefs:

> Quite a number of airmen carry with them when flying an acorn . . . At an inquest on the body of a farm labourer killed by lightning, a witness testified that it was the worst storm he had ever been out in. 'But I was not frightened,' he added, 'I had acorns in my pocket.'
>
> Not only airmen but others carry this charm, among them steeplejacks. Many steeplejacks would not dream of going aloft in stormy weather without carrying an acorn.

The Cuming Museum in the London Borough of Southwark holds a collection of 'charms' presented by Edward Lovett (1852–1933) in 1916. This includes a number of cord-knobs or blind-pulls shaped like acorns, umbrella tassels 'with wooden acorns', an acorn-shaped brass pendant, and a glass perfume bottle pendant also in the shape of an acorn, all of which were said to provide protection against lightning.[8]

The twenty-ninth of May is known as Oak Apple, or Royal Oak Apple, Day, and commemorates the restoration of the monarchy in

1660. King Charles II is probably best remembered for the way in which he escaped capture after the Battle of Worcester by hiding in an oak tree. At one time he considered setting up a new order of chivalry, the Knights of the Royal Oak, but abandoned the idea thinking it might 'keep awake animosities which it was part of wisdom to lull to sleep'.[9] However the wearing of oak (or occasionally FIELD MAPLE) leaves, and, if possible, oak apples, on 29 May each year persisted until well into the twentieth century.

> When I went to school, 60 odd years ago, there was one day we used to call 'Oak Ball Day'. We were supposed to wear an oak leaf or oak ball on our coats and if we didn't the other kids used to attack us with stinging NETTLES.[10]

> [East Cowes, Isle of Wight, 1920s and 1930s] Oak Apple Day (29 May) wear a sprig to get luck – an oak apple makes you even luckier.[11]

> When I was attending Little Gaddesden Village School [Hertfordshire, *c*.1945] it was always the custom to carry a piece of oak to school to prevent being caught and stung with nettles on Oak Apple Day.[12]

In the 1950s Iona and Peter Opie found that Oak Apple Day celebrations were practised by children in a dwindling number of places, 'in parts of the north country, especially Cumberland, Westmorland, Furness, and the North Riding, and also quite commonly in the north Midlands, in a broad belt stretching from Shrewsbury to the Wash'.[13] The informal marking of the date by schoolchildren seems to have died out soon after, but Oak Apple Day traditions continued in some schools for a further twenty years or so. In the early 1960s at Audley House Prep School, in Chesterton, Oxfordshire, which was 'run by a deeply traditional eccentric', it was considered '*de rigueur* to wear a bunch of oak leaves in the lapel – or risk being labelled a traitorous poltroon', although the punishment of whipping bare legs with nettles had been discontinued.[14] About a decade later, 'a tiny country voluntary-aided grammar school with an even smaller boarding house attached' in Wem, Shropshire, celebrated Oak Apple Day at morning assembly on a Friday during summer term:

> All the boarders would rise at 4 a.m. and head into the countryside: they would gather in trailer loads of nettles and decorate the

> school stage. It would literally be covered, except for a 'safe route' for the teachers to walk along. On the table behind which the headmaster stood for assembly would be picked out in oak leaves 'Custom demands a holiday' – it may have been 'Custom requests a holiday', I'm not sure. The table would also be decorated with oak apples. I think the headmaster may have been presented with an oak apple by a senior boarder. Assistant masters would know to check whether there was actually a chair beneath the nettles where a 'chair shape' existed, sometimes the chair was not there. Many masters chose to stand rather than sit in dew-soaked leaves.
>
> It was traditional for First Year boys to wear short trousers as part of their uniform. To get to assembly all the day-pupils had to run the gauntlet of huge handfuls of nettles being held below waist height (that was a firm rule!) by boarders. Second Years and above were protected by their long trousers, but for First Years it was quite an ordeal. Wearing an oak apple guaranteed a boy (more or less!) complete immunity from having nettles held against his legs . . . I was not aware of any major complaints from parents.[15]

The oak apples worn on Oak Apple Day were not the hard, globular galls which most people know as such today, but the larger, less regular spongy galls which rapidly develop on oak trees in May and mature in June or July. The gall-wasp, *Andricus kollari*, which creates the now familiar marble galls, widely known as oak apples, was deliberately introduced from the Middle East to Devon in about 1930, so that the galls, which contain 17 per cent tannic acid, could be used for dyeing and ink-making.[16]

Elsewhere more formal Oak Apple Day events are held. At Northampton a statue of King Charles on All Saints Church has a wreath of oak leaves placed around its neck, before a service, attended by civic dignitaries, is held inside the building.[17] For many years, at least since early in the twentieth century, the town hall at Worcester was decorated with oak on Oak Apple Day, but this custom ceased in 2017, when the city council demanded payment of £160 for doing so, and banned the placing of anything on the hall's railings.[18]

In Cornwall:

> An oak branch is still placed on the top of the tower of the parish church of St Neot on Oak Apple Day. St Neot was, of course, a Royalist parish during the Civil War. The oak bough is always sup-

> plied by Lampen Farm. The bough is hoisted up the outside of the tower. When in place, the vicar says prayers. I always used prayers for the Royal Family and for the government of the day, and closed with the Lord's Prayer and a blessing. In addition, the St Neot Women's Institute holds an Oak Apple Day fair.[19]

However, the most formal of Oak Apple Day events does not take place on Oak Apple Day, but on a Thursday early in June. This is Founder's Day at the Royal Hospital, in Chelsea, London, which is said to have been held each year since the Hospital's foundation 1681 (or, according to other accounts, 1692). Presumably the current date for the event is necessitated because the Hospital's grounds host the Chelsea Flower Show late in May. On the day, the Grinling Gibbons statue of Charles II has oak boughs placed around its base, though early photographs of the event show it completely concealed by oak branches, and a more recent statue of a Pensioner has an oak twig placed in its hand. A dignitary, usually a member of the Royal Family, takes the salute as all those Pensioners who are able to do so parade past, wearing oak leaves in their lapels.[20]

Two events held on Oak Apple Day which appear to have little connection with Charles II are the Great Wishford Oak Apple Day, in Wiltshire, and the Castleton Garland Day, in Derbyshire. It appears that the Great Wishford event was originally associated with Whitsun, but later moved to 29 May.[21] It is said to assert the villagers of Great Wishford's right to 'collect dead wood all the year round and cut green boughs on Oak Apple Day – no more, no less' in Groveley Wood, on the Earl of Pembroke's estate. Early in the morning people go around the village making a great noise, and later villagers go to Groveley Wood to collect oak branches, which they bring back to decorate their houses, and a large branch decorated with ribbons is hoisted up the church's tower. Later in the day villagers travel by coach to Salisbury where four village-women dressed in early nineteenth-century costume perform two simple dances, one using oak twigs and the other 'nitches' or bundles of dry wood,[22] in the Cathedral Close. Then people process into the Cathedral for a short service in the quire which concludes with everyone shouting 'Grovely, Grovely, Grovely and all Grovely'. All the participants wear oak leaves, some with oak apples.[23] Later in the day, at 2 p.m., a procession is held through the village, finishing at a fete, which commences with the women performing their dances again.

The day concludes with a race around the village.[24]

At Castleton, in Derbyshire Peak District, a man on horseback, the King, in Stuart costume but almost covered by a conical garland of greenery and flowers, leads a procession around the village. At the end of the procession the garland, apart from its topknot, 'the Queen', is hoisted up to the top of the church's tower, while the Queen is placed on the village war memorial.[25]

In 1991 the environmental organisation Common Ground sought to promote Oak Apple Day by producing a card which carried the exhortation, 'Revive an ancient festival – wear the oak on the 29th of May', but its campaign appears to have been short-lived and little noticed.

For a while oak leaves were used to commemorate people who had died in road accidents. Thus the *Eastern Daily Press* of 18 November 1996 reported that on the previous day:

> Relatives wept openly at Norwich Cathedral as they remembered loved ones who died in road accidents around the country . . . Many wiped back tears as they laid paper oak leaves at the altar bearing the names of their late children, partners, brothers, sisters, parents and grandparents.[26]

According to the charity RoadPeace which inaugurated this event, the first such service was held in Coventry Cathedral in 1992:

> They invited all affected by such tragedies to put forward names to be read out in this act of remembrance. The names were written on oak leaves and on leaving the church people were given acorns . . . as a symbol of continuing life.[27]

The third Sunday in November is now designated as World Day of Remembrance for Road Traffic Victims, but oak leaves no longer appear to be used.

Being long-lived, and by British standards large trees, individual oaks can become the focus of legends. Gospel Oaks, or their supposed descendants, still survive in some localities, or are remembered as place-names. These trees are usually said to have served as places where the gospel was read during beating the bounds ceremonies which perambulated parish boundaries on Rogation Day. Or well-known preachers supposedly delivered sermons from under their branches. One of the most famous of these trees formerly stood near the church at Polstead, Suffolk. According to tradition, the Saxon missionary St Cedd preached under the original tree in AD 753. For many years this was thought to be no more than a 'picturesque legend', but when the tree collapsed in 1953 it was found to have over 1,400 annual growth rings, indicating that it would have been a mature 200-year-old at the time of the saint's preaching. Since about 1910 an annual service has been held near the tree, this having been initiated by the Rev. Francis Eld, Rector of the parish from 1895–1921, who had a strong interest in antiquarian matters. Since the death of the original tree, a replacement tree, which is said to have been self-sown from one of its acorns, has been used, although a report in 1958 noted that the service was then held under a SYCAMORE tree, while the young oak was 'slowly growing towards maturity'.[28]

The Major Oak in Sherwood Forest was voted as England's Tree of the Year in a poll organised by the Woodland Trust in 2014.[29] This tree – which is named after a local antiquarian Major Hayman Rooke (1723–1806) – is popularly associated with Robin Hood and his merry men who were 'said to have hatched their projects for the redistribution of wealth' beneath its branches.

> The Major Oak . . . lies half a mile north of Edwinstowe. Its age is difficult to estimate, but it is not thought to be older than the sixteenth century. Now one of the centres of Robin Hood tourist commerce, this part of Sherwood Forest owes its prominence and all its supposed detailed associations with the legend to the romantic interest in Robin which developed in the nineteenth century.[30]

In the 1990s there were worries about the state of the tree's health. According to *The Times* of 30 August 1990:

> The Major Oak . . . Robin Hood's legendary hide-out is being drenched daily with thousands of gallons of water because of fears that during the hot, dry weather it could be destroyed by fire or drought.

In 1992 it was reported that a micropropagation company had been awarded a £23,000 grant from Nottinghamshire Council to clone the tree:

> The Robin Hood legend is an important money-spinner for the authority, which runs Sherwood Forest country park where the 80ft oak with its 240ft spread is a main attraction. The plan is to plant a copy to replace the Major Oak when it dies. Dr Wright [of

> the company] and the council think that many of the one million visitors to the forest each year would buy small versions of the oak to plant at home.[31]

The history of the Boscobel Oak, in which King Charles II hid in September 1651, is complicated.

> Immediately the story of Boscobel became known people flocked to see the house and the oak, and almost at once the tree was injured by souvenir hunters removing its young boughs. The damage was so great that before 1680 the owners of Boscobel, Basil and Jane Fitzherbert, were forced to crop part of the tree and protect it with a high brick wall . . . In 1706 John Evelyn wrote that he had heard that the 'Famous Oak near White Ladys' had been killed by people hacking the boughs and bark', and six years later William Stukeley described the tree as 'almost cut away by travellers'. He also remarked that a 'thriving plant from one of its acorns' was growing 'close by the side'.[32]

It was presumably this formerly 'thriving plant' which C.A. Johns reported on in 1847, when he thought the tree was 'now rapidly following its predecessor to decay'. However, the tree survives, 'but looks rather modest for a 300-year-old'. In 2001 the Prince of Wales planted a sapling, grown from an acorn from the present tree, nearby.[33]

In the 1930s it was said that the oaks in Bradgate Park, Leicestershire, were pollarded as a sign of mourning after the execution of Lady Jane Grey in 1554.[34] In Fort William, Inverness-shire, an oak tree known as the Hanging Tree was cut down 'amid controversy and protest' in 1984–5 to make way for a new public library. The tree is believed to have been a JOUG TREE on which the local chieftain hanged wrongdoers. After the tree's destruction:

> there has been a sequence of unexplained incidents at the library, trivial in themselves and not on the face of it connected. These happenings have, however, become associated in people's minds and are now explained as the work of restless or mischievious spirits. There have been no apparitions, only a series of 'accidents'.[35]

There are occasional records of oak being used medicinally.

> DIARRHOEA: grate a ripe acorn into warm milk and give to the patient.[36]

RINGWORM: get six leaves of an oak tree, boil them and drink the water in which they were boiled.[37]

Oak bark was commonly used as a cure for sore shoulders in HORSES. The bark was boiled and the sores washed with the water.[38]

OAT (*Avena sativa*)

Annual grass, of agricultural origin, cultivated in Britain since the Iron Age, widely grown as a cereal to provide food for horses and humankind.

Oats were to be sown on *Ffair Garon* (Caron Fair) [Tregaron, Cardiganshire] on the fifteenth of March, although *Ceirch du bach* (little black oats – a primitive local type of oat) did not have to be sown until the twenty-first of the month. Adverse climatic conditions in the uplands of North Wales led to later sowing there, and the favoured time for sowing oats in those districts were *tridiau y deryn du, a dau lygad Ebrill* (the three days of the blackbird, and the two eyes of April), which are taken . . . to refer to the last three days of March and the first two of April according to the 'old calendar', i.e. from the tenth to the fifteenth of April.[1] In Northumberland:

> Upon St David's Day
> Put oats and BARLEY in the clay.[2]

Care was needed to ensure that sheaves of oats were dry before they were stacked.

> They always used to say about oats, three Sundays after it was cut. Whether it was cut on the Friday before the first Sunday or the Monday before the first Sunday, they always used to say it had to be left out for three Sundays.[3]

> Oats should have the church bells rung over them three times after they have been cut, i.e. must be left while the bells are tolled on three successive Sundays. Otherwise there will be sickness in the village.[4]

> [Newcastle-on-Clun area Shropshire] Oats should be cut and stooked in the field and not brought into the barn 'until the parson has preached over them three times'.[5]

There are few records of oat being used medicinally.

Oats are good for ECZEMA. Soak the oats in warm water and apply to the affected area. Rub the oats on the skin. The roughness of the oats gently scrapes away the dead skin, and the oil in the oats repairs the skin. I have a lot of success with this, especially when the eczema is caused by nerves.[6]

For costumes made from oat straw see WHEAT.

OAK APPLE DAY (29 May) – FIELD MAPLE, NETTLE and OAK associated with.

OLD CHRISTMAS DAY (6 January) – see APPLE.

OLD CHRISTMAS EVE (5 January) – SWEET CICELY watched on.

OLD MIDSUMMER'S EVE (4 July) – HEMP seed used in LOVE DIVINATION.

ONION (*Allium cepa*)

Bulbous herb of unknown origin, widely cultivated for culinary use.

Cut onions were believed to absorb germs, and according to some people this provided protection from infections.

> SCARLET FEVER broke out in Whitechurch [Warwickshire] in the autumn of 1915. One young mother assured me she had taken away the chance of infection by peeling some onions and burying the peelings. At Stratford an onion was often suspended in a dwelling under the idea that it would turn black if the house was infected.[1]

> On one occasion some other children in the house contracted scarlet fever, and as a result mats outside the doors were covered with raw Spanish onion, my mother claiming that any germs would be attracted to it. Anyhow, we did not get fever.[2]

> A half of onion was hung in the house every winter, a cut onion was supposed to take the germs.[3]

A Cheshire farm, although ringed by infection, escaped the disastrous outbreak of FOOT-AND-MOUTH DISEASE on British farms in 1968. The farmer's wife ascribed their immunity to the onions she laid along the windowsills and doorways of the cowsheds.[4]

> When I was living in Orkney in the early 1950s I had no refrigerator or anything like that, just cold marble in the pantry. I used to put half an onion in the pantry to keep away the germs. I used to have to tell the children not to use it as it 'collected all the bugs'.[5]

It has been stated that the inhabitants of God's Providence House in Chester were spared from the plague due to their having placed onions at the entrances to the home.[6] However, history does not support this. The last outbreaks of plague in Chester occurred in 1605 and 1647–8, but the house was not built until 1652, following extensive damage to the centre of Chester by artillery bombardment during the Civil War.

> The legend concerning the House's escape from the plague, without mentioning onions, is commonly given by both printed and two-legged guides, and can be traced back to George Batenham's *The Stranger's Companion to Chester*, published in 1821. Presumably this is inspired by the House's name. The words 'God's providence is mine inheritance' were used by Richard Boyle, Earl of Cork, to adorn buildings erected by him . . . but there does not appear to be any connection between the Earls of Cork and Chester's God's Providence House.[7]

Presumably the idea that onions absorbed germs led to the Norfolk practice of placing onions beside corpses in the 1920s and 30s:

> Raw onions were also put on a saucer beside a coffin while the body stayed in the house before burial. My mother was horrified to go into a neighbour's house to pay her respects to the dead body and find 'No onion!' I asked her what it was for – she said to keep the DEVIL away.[8]

However, other people thought that cut onions were harmful rather than protective.

> According to my grandmother who was born in Lancashire in 1897, a cut piece of onion should never be kept as it will attract all the bad from the air into it.[9]

> I've been told that during the First World War nasty people used to rub their bullets on onions, this meant that any wounds caused by them became infected with germs.[10]

The growing of mammoth onions has its enthusiasts, although o nion competitions are less common than LEEK shows, and almost every year a world-record-breaking onion is produced. The usual variety to be shown in these competition is the Kelsae onion, which in 1982 was said to have originated 'over 40 years ago in an old Scottish country estate, and was introduced to the public by the Kelso nursery-men Laing and Mather':

> Apart from its remarkable capacity to grow to an enormous size, it is a handsome, good quality onion with very high neck and shoulders, giving it a flask-like appearance.

In 1981 the 'World Record for the heaviest Kelsae onion was broken by Mr Bob Rodger of Crail, Fife, with an onion weighing 6lb 7⅝oz (2.93kg)'.[11] In 1992 Robert Holland, of Cumnock, Strathclyde exhibited an 11lb 2oz specimen, 'said to be the biggest in the world', at the National Kelsae Onion Festival in Harrogate, North Yorkshire.[12] By 2014 an 18lb 11oz winner had been produced by Tony Glover, of Moira, Leicestershire:

> It took him almost a year to grow his record-breaker using a greenhouse that is heated in winter and cooled in summer, and with lights fitted to stimulate [*sic.*] when the days are short and the weather poor. His onions are given 'nitrogen-rich food and . . . [he has] to make sure the humidity is just right.[13]

The Gloucestershire town of Newent claims to have held an annual onion fayre from late in the thirteenth century, when Welsh drovers passing through to Gloucester would purchase onions, until the twentieth, when 'the war years saw its demise'. However, in 1996 the locals decided to revive the fayre to 'celebrate local food and drink'. Currently the fair consists of the main streets being lined with stalls, mainly raising funds for local charities, a shop-window dressing competition, an onion-eating competition, an onion show, which, it is claimed is 'the only show in the country dedicated to the onion family', and a wide range of other entertainments, ranging from a dog show and fairground rides to belly-dancers.[14]

Onions were widely used in folk medicine.

> [Horseheath, Cambridgeshire] 'If an onion is eaten every morning before breakfast, all the doctors might ride on one horse' was a local saying.[15]

The onion had many uses, The inside of an onion skin placed on CUTS, and scratches acted as a type of elastoplast . . . An onion placed on a wasp or a bee STING soon took the pain away. A mixture of onions and sugar in water was a cure for WHOOPING COUGH. Rubbed on the head it was believed a cure for BALDNESS.[16]

In the late twenties and thirties onions, chopped and stewed in milk, were used to cure colds and COUGHS.[17]

My grandmother, from Derbyshire, used to cut a small onion in half and pour sugar on the halves. Then she would roast it until the sugar caramelised . . . This was given as a hot dish to family members with heavy colds. It seemed to work.[18]

Wasp sting – rub with a slice of raw onion.

Severe bruising – rub gently at regular intervals with a slice of raw onion.[19]

To cure baldness: cut and bruise an onion. Rub the sap mixed with a little honey into the bald patch, keep rubbing until the spot gets red. This concoction if properly applied would grow hair on a duck's egg.[20]

Heat a whole onion. When warm, but not hot, place in an old pair of tights and hold it to the ear as relief for EARACHE. My mother did this for my sister in the 1970s.[21]

A boiled onion was often mashed and inserted into a sock and tied around a sore THROAT as a cure until the mid-1950s.[22]

For CHILBLAINS: rub feet with half a peeled onion.[23]

[Around Chipping Ongar, Essex] RHEUMATISM may be warded off by carrying a small onion in the pocket.[24]

An old Irish remedy for PILES was the application of a poultice of boiled onions.[25]

My maternal grandmother came from the then rural area of Middlesex. Her cure-all was onion syrup made by slicing onions and covering the slices with brown sugar overnight. The resulting brown liquid was distributed to us children quite liberally at the slightest sign of a cold or sore throat.[26]

> Schoolboys recommended that an onion rubbed on hands before they were caned would alleviate the pain.[27]

> The schoolboy . . . does not fail to make use of the time-honoured charm against the sting of the cane, viz. the rubbing of an onion on the palm of the hand.[28]

Onion skins provide an easily available dye for colouring EASTER EGGS:

> In our childhood days of the 20s and 30s in and around Burghead, Moray . . . for the dyeing of Easter eggs our mothers might use the outer skins of onions, or whin [GORSE] flowers.[29]

OPIUM POPPY (*Papaver somniferum*)

Annual herb, cultivated mainly as an ornamental, with showy flowers in a variety of colours, and for its seeds which are used as food, and the production of the pain-killer morphine, also forming short-lived but conspicuous colonies in dry disturbed situations; probably native to the eastern Mediterranean area, but long cultivated in these islands with seeds being found in Bronze Age archaeological sites.

Although commercially produced laudanum – a tincture of opium – was frequently used to provide relief from pain and hunger (see, for example, Mrs Gaskell's 1848 novel, *Mary Barton*), home-made remedies derived from opium poppies were also prepared, particularly it seems in Cambridgeshire.

> Opium chewing was common, and druggists in the townships such as Wisbech and Holbeach sold great quantities of 'the stuff' across the counter, often so much as 14lb (6.4kg) a day . . . a Fenland physician writes that 'a patch of white poppies was usually found in most Fenland gardens' and that poppy-head tea was often given to children during TEETHING.[1]

> The universal pain-killer for RHEUMATISM and, indeed, for all muscular and nerve pains, was opium, freely obtainable in the poppy tea which every Fen housewife could make from the white poppies . . . she grew in her garden. She took it herself, gave it to her husband and even down to the youngest baby with teething troubles.[2]

> Sleep in children: A decoction of poppy heads. You will see poppies growing in many cottage gardens, where they are cultivated it is always for this reason – it saves the mothers a great deal of trouble. Forty years ago many of our farm labourers used to take laudanum regularly, it was easier to carry about the home decoction – an old man of nearly 80 years always had 4oz a week of the very strongest.[3]

ORANGE (*Citrus sinensis*)

Evergreen tree, with fragrant white flowers, native to southern China but long cultivated in tropical and subtropical countries for its fruit.

Occasionally orange peel was used instead of APPLE peel in LOVE DIVINATION.

> [Early 1930s] I used to toss long strips of orange peel to find out the initial letter of my future husband when I was about 5–6. The peel was cut by a grown-up of course, but we were given the peel. It always ended up as a curved letter – C, G and S – so this was a source of anxiety to me as I was affianced to my beloved sweetheart whose name began with D; no way could you produce D, so I gave up doing it. But it was popular among little girls of our group; apple peel was used too, but often broke.[1]

Orange blossom, often artificial rather than the real thing, was frequently used in WEDDING bouquets or to decorate wedding cakes.

> The orange tree, simultaneously bearing golden fruit, sweet-scented flowers and leaves – typifying fertility, through this abundance – is a traditional ingredient in love charms and marriage luck. Saracen brides wore its flowers as a sign of fecundity and crusaders are said to have carried the custom to the west.[2]

> MYRTLE was rivalled in popularity in the nineteenth century by orange blossom, which seems to have been introduced to Britain from Spain in the 1820s. Oranges are said to have been the golden apple given by Juno to Jupiter on their wedding day, and the blossoms symbolise good luck, fertility and happiness.[3]

Egg-rolling, which involved rolling eggs down slopes, was for many years a popular Easter-time activity in many parts of the British Isles.[4]

Writing of Preston, Lancashire, in 1883 the historian Anthony Hewitson recorded that on Easter Monday many thousands of children assembled in Avenham Park to 'roll their eggs, oranges, etc'.[5] Orange rolling was also held on GOOD FRIDAY, on Dunstable Downs, Bedfordshire, where in 1972 it was said that hundreds of children gathered to roll oranges down Pascombe Pit. This was said to be symbolic of the stone being rolled away from Jesus's tomb.[6]

However, a few years later:

> Good Friday: Orange-throwing for children has always been traditional on Dunstable Downs . . . but has lapsed in the last year or two after a hooligan threw an orange at the Mayor.[7]

Since the 1970s an Orange Race has been organised late in August each year in Totnes, Devon. It is said that in the 1580s Sir Francis Drake either gave a boy an orange which the boy dropped, or Sir Francis bumped into a boy carrying a basket full of oranges making him drop it. Children raced to pick up the orange, or oranges, and the race commemorates this. Several races for people of different ages take place, the contestants have to kick or toss, but not carry, their fruit along the street, the winners being those who first to reach the end of the course with their oranges 'mostly intact'.[8]

Orange peel continues to have a variety of uses. According to a letter in the *Sunday Telegraph* of 20 November 1994:

> Orange skins have a natural chemical that is more powerful than anything on the market today. Anyone infested [with head LICE] should get the skins from two fresh oranges and place them about the head.

Other uses include:

> My father uses orange peel to keep CATS out of our garden. Apparently it works![9]

> You can put orange peel into your TOBACCO moist, or to restore it. I used to do it when I was a student and I did it again recently when I found some old tobaccco.[10]

> In County Durham the girls chewed pieces of orange peel before going to a dance because it made their EYES sparkle![11]

> Orange peel is splendid for lighting fires . . . we always used it when we were young.[12]

> You can put orange peel, fabric-softener or bicarbonate of soda in the shoes overnight, which should mean fresher shoes the following morning.[13]

ORANGE LILY (*Lilium* x *hollandicum* and *L.* cvs)

Ornamental bulbous perennial, of garden origin.

On and around about 12 July each year Orange Lodges in Ulster and elsewhere march in commemoration of the defeat of King James II by King William III (William of Orange) at the Battle of the Boyne in 1690.

> Our banner poles are decorated with orange lilies and SWEET WILLIAM. If the traditional orange lily is not available substitutes such as the Peruvian lily [*Alstroemeria* cvs] or one of the new lily hybrids are used.
>
> The orange lily is traditionally the symbol of the Orange Order and this is traced to orange lilies growing at the site of the Battle of the Boyne (similar in concept to the Flanders POPPIES).[1]

However, during William's lifetime and until the mid-eighteenth century it appears that the ORANGE tree, rather than the orange lily was used by his supporters. On 12 July 1812 a traveller found Tandragee, County Armagh, to be

> a perfect orange grove. The doors and windows were decorated with garlands of the Orange lily. The bosoms and heads of the women and the hats and breasts were equally adorned with this venerated flower.[2]

Forty-five years later:

> The Twelfth celebrations in Belfast in 1857 orange lilies decorated windows, arches and buttonholes . . . Today it flourishes in cottage gardens and country graveyards, on banner-poles and in the painted decoration of lambeg drums.[3]

In the Republic of Ireland:

> There are many kinds of poultices for drawing out BOILS or SORES, but the best is the orange lily. The roots are dug out of the

ground, washed clean, and cut. Then they are boiled in water and the water into [a] substance like cornflour. Then it is applied to the boil or CUT and draws out all the matter and cleans it up.[4]

ORPINE (*Sedum telephium*)

Perennial herb with dull reddish-purple flowers in hedgebanks, and rocky places, native, and cultivated in and escaping from gardens.

Local names include:

Alpine in Cheshire
Arpent, and arpent-weed, in Hampshire
FAREWELL-TO-SUMMER in Lancashire
Harping Johnny in Norfolk
JACOB'S LADDER in Kent; 'from the regular alternate leaves'
Live-long-love-long in Somerset and Sussex
MIDSUMMER MEN in Northumberland, Somerset and Wiltshire
Orphan John in East Anglia
Orpie-leaf in Scotland
Orpies in Berwickshire
Solomon's puzzles in London.

John Aubrey (1626–97) wrote:

> Also I remember, the mayds (especially the Cooke mayds and Dayry-mayds) would stick-up in some chinkes and joists, etc., Midsommer-men, which are slips of Orpins. They placed them by Paires, sc: one for such a man, the other for such a mayd his sweet-heart, and accordingly as the Orpin did incline to, or recline from ye other, that there would be love or aversion; if either did wither, death.[1]

According to the 1853 edition of John Brand's *Popular Antiquities*:

> In one of the tracts printed about 1800 at the Cheap Repository, was one entitled Tawney Rachel, or the Fortune-Teller, said to have been written by Hannah More. Among many other superstitious practices of poor Sally Evans, one of the heroines of the piece, we

> learn that 'she would never go to bed on Midsummer Eve without sticking up in her room the well-known plant called Midsummer Men, as the bending of the leaves to the right, or to the left, would never fail to tell her whether her lover was true or false . . .'
>
> On 22nd January, 1801, a small gold ring . . . was exhibited to the Society of Antiquaries by John Topham, Esq. It had been found by the Rev. Dr. Bacon, of Wakefield, in a ploughed field near Cawood in Yorkshire, and had for a device two orpine plants joined by a true-love knot, with this motto above: '*Ma fiance velt*'; i.e. My sweetheart wills, or is desirous. The stalks of the plants were bent to each other, in a token that the parties represented by them were to come together in marriage. The motto under the ring was, '*Joye, l'amour feu*.' From the form of the letters it appeared to have been a ring of the fifteenth century.[2]

Although this passage has been much repeated, it should be treated with caution. If the ring still exists its present whereabouts are unknown, but the Ralph Harari collection contained what was considered to be a similar ring, the bezel of which was 'engraved with a device of two plants, joined by a tasselled cord'. The collection has been dispersed, but examination of the illustrations in its catalogue[3] reveals an unidentifiable plant, which is certainly not orpine.

On 11 June 1873 Francis Kilvert, having seen midsummer men plants earlier in the day, recorded in his diary that 'Mother remembers the servant maids and cottage girls sticking them up in their houses and bedrooms on Midsummer Eve, for the purpose of divining about their sweethearts.[4]

It appears that such divination was last practised, in Sussex, in the 1920s.[5]

Despite Grigson's listing of the names HEAL-ALL from Northern Ireland, and HEALING-LEAF from Scotland, it seems that orpine was rarely used in folk medicine. Allen and Hatfield suggest that it 'seems likely to have owed its medicinal uses to the learned tradition of the books', at one time being grown in cottage gardens as a WOUND herb, and, in 1767 in Sutherland, to have been used to treat the bites of mad dogs or ADDERS.[6] Wyse Jackson notes that in 1804, 'a decoction of the leaves is said to be a very serviceable and forcible DIURETIC'.[7]

OXEYE-DAISY (*Leucanthemum vulgare*)

Perennial herb with showy white, yellow-centred, flowers, widespread, and when flowering conspicuous, in grassy places.

Local names include the widespread bull-daisy ('from its large size compared with other daisies'), DOG-DAISY, HORSE-DAISY ('from its size and coarseness'), MOON-DAISY, MOONS, and:

Big daisy in Yorkshire and Berwickshire
Big kokkeloori in Scotland, where name also applied to 'mayweeds'
BILLY-BUTTONS in Shropshire
Bishop's posy in Co. Donegal
BOZZOM, or BOZZUM, on the Isle of Wight
BULL'S EYES in Cornwall, Cumberland and Somerset
BUTTER-DAISY in Devon, Dorset and Somerset
Caten-aroes in Lancashire
Cow's eyes in Cornwall
CRAZY BET in Wiltshire
Devil's daisy in Middlesex
Drummer-daisy in Somerset
Dun-daisy in Devon; 'probably a contraction of dunder-daisy'
Dunder-daisy in Devon and Somerset; 'doubtless a corruption of thunder daisy . . . in Somersetshire there is an old tradition connecting it in some way with the Thunder God'
Dundle-daisy in Somerset
Dutch morgan on the Isle of Wight
ESPIBAWN in Ireland; 'Irish *easbog ban*, white bishop'
FIELD DAISY in Devon
Fried eggs in Somerset and Wiltshire
Gadjerwraws, or gadjevraw, in Cornwall; 'Cornish *caja vras*', great daisy
Gay gorran in Co. Durham
Gay gowan in Scotland
Girt oxeye in Cumberland
GOLLAND in Northumberland
Goode in Lancashire
GOWAN in Devon and Scotland
GOWLAN in Northumberland
Grandmothers in Nottinghamshire
Great daisy in Cumberland
Gypsy-daisy in Devon, Dorset, Norfolk and Somerset
Harvest daisy in Dorset
HORSE-BLOB in Northamptonshire
Horse-gollan in Northumberland and Scotland
HORSE-GOWAN in Scotland
Horse-gowlan in Northumberland and Scotland
HORSE-PENNIES in Derbyshire
Large dicky-daisy in Cheshire
London daisy in Dorset and Somerset
Maise in Lincolnshire
Maithen in Gloucestershire and Wiltshire
Margarets in Yorkshire
Mather in Herefordshire
Mathern, maudlin, and mauthern, in Wiltshire

MAYWEED in Suffolk
MIDSUMMER DAISY in Somerset, Sussex and Warwickshire
Monnies, and MOON-FLOWER, in Somerset
Moon-pennies in Cheshire, Derbyshire and Yorkshire
Moon's eye, mother-daisy, and mowing daisy, in Somerset
Muckle-gowan in Scotland
Muckle-kokkeluri in Scotland, where also given to 'mayweeds'
Open star in Somerset
Patio-marguerite in the Glasgow area; 'a genteel older name'
Penny-daisy in Nottinghamshire
Poor-land daisy in Northamptonshire
POVERTY-WEED in Cheshire; 'as it generally flourishes on exhausted soil'
Rising sun in Somerset
Thunder-daisy in Devon, Dorset and Somerset
THUNDER-FLOWER in Dorset; 'more commonly called thunder-daisy'
White golds in Cumberland and Lancashire
White gould in Cumberland; 'as opposed to *Chrysanthemum* [now *Glebionis*] *segetum*, yellow gould'
White gowan in north-east Scotland
White gowlan in Northumberland
White gull in Cumberland
WHITE MAY in Dublin
Wild marguerite in the Glasgow area; 'a genteel older name'.

Like the smaller daisy (*Bellis perennis*), oxeye-daisy was used by children in LOVE DIVINATION.

> As children we used to make chains from dog daisies. (I don't know the correct name for this plant. They are similar to the daisy but much larger and usually found in meadows). This plant or flower was also used in a game where one would pluck a petal each time as he/she recited: 'she/he loves me, she/loves me not', until all the petals were gone.[1]

> [Somerset, 1950ish] My experience is the commonplace one of pulling petals off a flower (e.g. dog daisy) and saying, 'She loves me/She loves me not', etc.[2]

Allen and Hatfield note that despite its wide distribution oxeye-daisy has been used in folk medicine only in the Highlands of Scotland and 'limited areas' of Ireland.[3] On Colonsay it was 'esteemed as an excellent remedy for KING'S EVIL'.[4]

OXLIP (*Primula elatior*)

Perennial herb with pale yellow flowers, in woodlands in East Anglia and Buckinghamshire. 'Oxlips' found elsewhere are invariably the hybrid (*P.* x *polyantha*), between COWSLIP and PRIMROSE, sometimes known as 'false oxlip'.

> Many people in East Anglia . . . believe that the plant began its decline with the extinction of the wild boar. They say that the plant depended on the droppings of these creatures, and the only reasons they have survived in the area is because the oxlip woods, largely left to their own devices, still contain the last cases of wild boar dung![1]

P

PAINS - CORN MARIGOLD, FEVERFEW, POPPY and YARROW soothe.

PALM SUNDAY (known in Wales as FLOWERING SUNDAY, q.v.) - BOX, CYPRESS, GOAT WILLOW and YEW used; FIGS eaten.

PAMPAS GRASS (*Cortaderia selloana*)

Perennial tussock-forming grass, cultivated as an ornamental since 1848, and first recorded in the wild in 1925, native to South America; most wild plants are believed to result from plants thrown out from gardens or deliberately planted.

Local names include pompous grass in Devon, and Australian grass, feathery plume, GHOST GRASS, Queen Anne's plumes, and spear in Somerset.

According to the London *Evening Standard* of 30 November 2011, the television presenter Mariella Frostrup received 'unwanted inquiries' after placing two pampas grass plants on the balcony of her flat in Notting Hill, London. It was explained that the plants were a sign that she was seeking 'casual, adventurous sexual encounters'. Returning to the story on 2 December, the *Standard* commented on Norwegian-born Frostrup's ignorance of 'the fact pampas grass is a secret code for swinging'. 'This', it claimed, 'is one of the oldest urban, or rather suburban, myths under the sun.'

PANSY (*Viola* x *wittrockiana*)

Annual, or short-lived perennial herb, widely grown for its showy, multicoloured flowers since early in the nineteenth century, often self-seeding beneath hanging baskets and in similar situations in urban areas.

My grandmother had a pansy game or story which goes like this:

> There were these five lovely daughters and they all went to the ball in the most lovely velvety ballgowns. My grandmother would pick off the five petals and hand them to me to admire the gowns. Then she would say – but they left their old mither (she was a Scot, my gran!) all alone sitting with both her legs in one stocking! Then she would show me the two spindly plant parts from which she had pulled a little green covering. I was always fascinated that this was so, And, even now, I often pull apart a fading flower just to repeat my gran's story![1]

In May 2014 a visitor to the Natural History Museum's Wildlife Garden recalled that when she was young her mother told her that pansy was the Cinderella Plant, the two upper petals which are usually monochrome were said to be 'Cinderella's plain dress with apron', the two side petals which are more showy were the 'stepsisters' matching nice dresses', and the lower petal which is biggest and most colourful was 'stepmother's very fancy dress'.

In the 1920s 'pansy' became slang for a male homosexual, particularly one of a rather effeminate nature. Any association between the flower and homosexuality is unclear, and it is said that pansy is a corruption of the early 'nancy' which was in use towards the end of the nineteenth century, and was in turn derived from the term 'Miss Nancy' which was used in the 1820s. Occasionally pansies are still used by gay-rights campaigners. Thus in November 2007 it was reported:

> Thousands of flowers were planted last month in Liverpool to mark this year's Homotopia festival to remember hundreds of people who have been killed and attacked through homophobic violence . . . Exactly 2,000 pink and purple pansies were planted for the annual queer arts festival but they will also serve as a memorial to victims of hate crime generally and in remembrance of a homophobic-motivated murder in St Johns Gardens.
>
> The flowers will remain in the gardens until the final day of the festival on 19 November, when they will be given away to the public as a symbolic stand against hate crimes.[2]

PARALYSIS – MILKWORT and TORMENTIL ward off or cure (see the latter).

PARSLEY (*Petroselinum crispum*)

Biennial, aromatic herb, origin unknown (probably eastern Mediterranean), cultivated as a culinary herb since at least late in the tenth century, and occasionally becoming naturalised in maritime areas.

Probably the most widely used culinary herb in Britain, parsley gathered a large number of folk beliefs. In common with other herbs, such as SAGE, it was believed that it thrived where the wife was dominant.

> Where parsley stays green all the year round the wife wears the trousers.[1]

> My late husband, an officer and later Merchant Navy master . . . found it difficult to settle ashore.
>
> In his allotment, he planted parsley which came up in abundance and he was delighted, but a few days later a neighbour told him that if parsley flourished the wife was the boss in the house. He dug it up.[2]

> If parsley grows well in the garden the missus is boss.[3]

> [Fens] In the years of economic depression, boys were far more welcome than girls . . . for this reason parsley . . . was never allowed to grow too high or too thick. If it did so this signified that the wife's influence in the home was stronger than her husband's, and so her children would probably all be female.[4]

Or parsley grew best for whoever was the dominant partner in the marriage.

> Amongst people of the same background as my wife and myself it is thought that parsley should be sown by the head of the household.[5]

> [In the village of Whitwick, near Coalville, Leicestershire, it was thought] the person who wears the trousers needs to set the parsley.[6]

More rarely it was thought that people for whom parsley grows well are malevolent: parsley flourishes only when sown by a rogue,[6] or only a WITCH can grow parsley.[7]

In Somerset:

> Parsley must be planted on a holy day or the FAIRIES will get it.[8]

More usually it was recommended that parsley should be planted on GOOD FRIDAY.

> [Great Torrington, Devon] a belief prevails in this district that . . . to have your parsley all the year round it should be sown this day [Good Friday].[9]

> A north-country saying is that parsley will grow best if sown by the lady of the house before twelve noon on Good Friday. I do it and it works.[10]

> [Sussex] Good Friday is the traditional day for . . . the sowing of parsley . . . if it is sown on this holy day, not only will it sprout quickly, but it will come up curly.[11]

> We had to sow parsley on Good Friday because it had to go three times to the DEVIL before it germinated.[12]

The belief that parsley seed 'visits' the devil is widespread, the number of visits varying in different parts of the country.

> [Devon] Parsley goes three times to the Devil before it comes up.[13]

> [? South Wales] Parsley – it was said to go six times to the Devil before it germinates. It is very slow.[14]

> [Walton-le-Dale, Yorkshire] Parsley seed after being sown went seven times to the devil before it came up.[15]

> [Sussex] Some people say its roots go seven times to Hell and back before it will sprout.[16]

> There is a saying in the North Riding of Yorkshire that 'parsley seed (when it has been sown) goes nine times to the devil'.[17]

> [In Herefordshire, parsley] is said to 'go to the 'owd 'un nine times afore it comes up'.[18]

Also widespread was the belief that parsley should not be transplanted or given away.

> A poor woman near Morwenstow attributed a sort of stroke which had affected one of her children after whooping-cough to the moving of the parsley-bed; and in a neighbouring (Devonshire) parish, the parish clerk, it was believed, had been bed-ridden 'ever since the parsley-mores were moved'.[19]

[Around Ilmington, Warwickshire] parsley must not be transplanted. If it is, a member of the family in whose garden the parsley plants are set will die within the year.[20]

My father would never transplant parsley, he said it meant DEATH in the family.[21]

At the last whist drive at Holditch Mrs Harris picked some parsley, brought it up and gave it to me (as we haven't any). Phyllis Down was horrified as she said all sorts of awful things would befall me; I should have gone down to Gardners Farm and stole it; you should never have parsley given to you. Two nights later I was taken ill, so there's proof for you![22]

Children were told that BABIES came from parsley-beds.

Inquisitive children with us are usually told that babies are dug up from the parsley-bed, and sometimes it is vexatiously added that the boys are dug up from beneath a GOOSEBERRY bush.[23]

[Guernsey, 1882] the origin of babies was variously accounted for . . . they are brought over in band-boxes from England in the mail packets . . . are dug out of the parsley beds with golden spades.[24]

In traditional medicine parsley was valued for gynaecological matters.

Take plenty of parsley to recuperate quickly after childbirth. This was told to me by my mother in the 1920s/30s, as told to her by her grandmother in 1890, and practised by her mother (who had 23 surviving children) in Berkshire in the mid-nineteenth century.[25]

[London, 1982] if you want to bring on your period put a sprig of parsley inside your vagina for 12 hours – your period should start 24 hours later.[26]

[Fenland] village girls who became pregnant before marriage had faith in parsley, eating it three times daily for three weeks.[27]

[Dublin] My mother Peggy Walsh [1926–2009] would never grow parsley in the garden in case the eldest daughter (me [born 1963]) became pregnant. It was grown in copious amounts after I got married.[28]

Parsley under the armpits will dry up a mother's milk.[29]

PARSLEY PIERT (*Aphanes arvensis*)

Small, low-growing annual herb, widespread on arable and bare ground.

Local names include the widespread parsley-breakstone and:

BEAR'S FOOT in Cumberland
Bowel-hive, or bowel-hive grass, said to be an effectual remedy in the bowel-hive [inflamation of the bowels] of children', in Berwickshire
Breakstone parsley in Staffordshire
Colick-wort in Herefordshire
Parsley-brexton in Buckinghamshire
Parsley-vlix in Dorset.

> [Gypsy remedy] parsley piert . . . also known as horse wort: infusion of the dried herb is good for gravel and other BLADDER troubles.[1]

> Parsley breakstone, washed and boiled, then drunk (by my mother) for gravel in the KIDNEYS.[2]

PARSNIP (*Pastinaca sativa* ssp. *sativa*)

Perennial herb, long cultivated for its swollen roots which are eaten as a winter vegetable.

> [Guernsey] Old farmers used to say, with regard to digging the ground for their parsnip crop, that they should begin it whilst eating the bread baked for Christmas.[1]

> [Guernsey] The seedlings germinate slowly and irregularly, and it was said that parsnip goes three times to the DEVIL before it comes up.[2]

> [*c*.1940, Bray and Cookham, Berkshire] A local man used to look after the roads and lanes . . . he earned £2 a week. His wife used to make wine from parsnips which she sold to American service-men. When she died they pulled back the wallpaper in her cottage and found a pile of red pound-notes – she didn't have a licence to make wine, so she was afraid to bank her money.[3]

In the late 1990s it was reported that the landlady of the Rose and Lion pub in Bromyard, Herefordshire, paid an annual rent in parsnips. Clause 4 in the legal agreement with the pub's owners, the Wye Valley Brewery, stated, 'The licensee shall personally deliver to the licensor on the 20th of December each year a kilo of parsnips which shall have been grown by the licensee in the garden.'[4]

PASQUE FLOWER (*Pulsatilla vulgaris*)

Perennial, violet-purple flowered herb, restricted to dry calcareous grassland in central and eastern England, where it can be abundant; also cultivated as an ornamental.

Local names include: bluemony, 'i.e. blue emony or anemone' in Rutland, and COVENTRY-BELLS in Hildersham, Cambridgeshire in 1597.

Pasque flower was often considered to grow on the sites of battles against Danish invaders.

> Traditionally associated with the DANES. It is supposed to grow only where their BLOOD has been shed and is known in Hampshire as DANES' BLOOD. One of the few places where its purple bell-like flowers can be seen is on the Downs dividing Hampshire from Berkshire – curiously enough, the site of King Alfred's battlefield.[1]

> When we used to live in that area [Berkshire Downs, on the Oxfordshire border in the 1950s] we were told the story that the pasque flower grew on the sites of battles. The Danes planted these flowers where they fell in battle.[2]

> I spent part of my childhood in the 1930s in the north Chilterns where *Pulsatilla vulgaris* (pasque flower) grew above the village of Barton-le-Clay. I was always told that this plant grows only 'where Danish blood was spilt'. Indeed, one can understand how this idea developed as the Danelaw boundary was through that area and the Danish headquarters were at Bedford.[3]

The Cambridgeshire name Dane's flower presumably refers to similar beliefs.

PASSION FLOWER (*Passiflora caerulea*)

Vigorous perennial scrambler with purple flowers, and, more rarely, orange fruits; native to South America, introduced late in the seventeenth century.

Local names include:

Christ-and-the-apostles in Devon

Crown-of-thorns in Dorset

Good Friday flower, STAR-OF-BETHLEHEM, story-of-the-cross, the ten commandments, and the TWELVE DISCIPLES, all in Somerset.

> [A] flower which is traditionally associated with the CRUCIFIXION is that called the Passion-Flower . . . the Spanish friars in America first called it 'flower of the passion' (*flos passionis*), and, by adding what was wanting made it an epitome of our Saviour's Passion . . . The name was given by the superstitious in former times, who saw in the five anthers a resemblance to the five wounds received by Christ when nailed to the cross. In the triple style are seen the three nails employed; one for each of the hands, the other for the feet. In the central receptacle one can detect the pillar of the cross, and in the filaments is seen a representation of the crown of thorns on the head. The calyx was supposed to resemble the nimbus or glory, with which the sacred head is regarded as being surrounded.[1]

PASSION SUNDAY (fifth Sunday in Lent) – PEAS eaten on.

PEA (*Pisum sativum*)

Climbing annual herb, native to southern Europe, long cultivated for its immature seeds which are a popular (or at least easily prepared, when bought frozen) vegetable.

> [Cambridgeshire] farmers used to say that it would be a good year for peas if the hedges dripped on ST VALENTINE'S morn.[1]

Peas were widely used in LOVE DIVINATION:

> [Co. Cavan] if you put a pod with nine peas in it up over the door the first person who enters will have the same name as your 'future'.[2]

> [Horseheath Cambridgeshire] When peas were being 'coshed' or shelled, a keen lookout was kept for a 'cosh' holding nine peas and when a girl found one she would hang it over her doorway in the belief that she would marry a man with the same initial as the first person who passed under it.[3]

Alternatively, in West Sussex:

> If you find nine peas in the first pod you gather it bodes you good LUCK.[4]

In parts of northern England peas were traditionally eaten on Carling Sunday, which is usually considered to be Passion Sunday, the fifth Sunday in Lent.

> Peas or BEANS fried in butter with vinegar and pepper are still [1950] served in many houses on Passion or Carling Sunday. There is considerable doubt as to the meaning of the word 'carling'; in the Midlands the day is called Care Sunday, a name which is supposed to refer to the sorrow or care of Our Lord's Passion.[5]

> Here in the fastness of East Cleveland this [Carling Sunday] is still observed in the old mining communities. It, in fact, takes place on the fifth Sunday in Lent . . .
>
> Carlings are no delicacies, either. They are black peas, usually reserved for pigeon food, soaked like mushy peas and eaten with salt and vinegar.
>
> To the uninitiated they are utterly vile, tasting neither like peas nor anything else, but to the East Cleveland carling enthusiast they are something to be looked forward to.
>
> The name for the Monday which follows Carling Sunday I would not mention to your reader's delicate ears![6]

> Carling Sunday: 3 weeks before Easter, where I live now in Cumbria a kind of brown dried pea, usually given to pigeons, soaked and boiled with molasses is eaten. Not a family tradition with me, but I eat them now – OK once a year.[7]

PEAR (*Pyrus communis*)

Deciduous tree, cultivated for its sometimes fleshy fruits since at least late in the tenth century, occasionally naturalised.

> [Ireland] it is UNLUCKY to bring pear blossom into the house. It signifies a DEATH in the family.[1]

A black pear features on the Worcestershire county coat of arms, on the badge of the county cricket club, and elsewhere; and it is said that Worcestershire bowmen fighting at Agincourt in 1415 had pears emblazoned on their standard.[2] This variety, known as either Worcester Black or Black Worcester, is said to have been brought to the city of Worcester by monks from Warden Abbey in Bedfordshire, where 'the pear, or at least a pear, appears in monastic reference as far back as the thirteenth century'. Today the name Warden, or Wordon, pear is given to 'a group of large hard pears that never ripen fully and so are only any good as culinary or cooking fruit'.[3]

A particularly well-known black pear is the tree planted by the Prince of Wales (later King Edward VIII) to commemorate the opening of Cripplegate Park in Worcester in October 1932. Late in the twentieth century it was thought that the black pear was becoming endangered, so the County Council in conjunction with Pershore College ran a scheme selling young trees, with the result that several hundred new trees have been planted; 'the sight of the dark reddish fruit hanging on the trees into November is no longer such a rare sight'.[4] The Worcester trees have recently been studied by Wade Muggleton, and their DNA compared with trees from Bedfordshire, believed to be Warden pears. He found that the Cripplegate Park tree and other trees of about the same age are not true Worcester Blacks, but the trees which have been sold in recent years, which he refers to as 'the Worcester Woods Type' and some of the Bedfordshire trees are identical. Thus it appears that the legend that the Worcesters Blacks originated from Warden Abbey could be true (though of course there remains the possibility that the variety originated in Worcester and was taken to Warden).[5]

> George Broun, 10th Laird of Coalstoun [Colstoun, East Lothian], who died in or before 1524, married Marion Hay, daughter of the second Lord Hay of Yester. The dowry of this lady consisted, in part, of what has long been known as the Coalstoun Pear. Hugo de Grifford of Yester, her remote ancestor, famed for his necromantic powers, was supposed to have invested this Pear with the extraordinary virtue of securing for the family which might possess it unfailing prosperity. This Pear is preserved at Coalstoun with the care due to so singular an heirloom, which, regardless of the superstition, must be esteemed a very wonderful vegetable curiosity, having existed for more than 500 years . . .
>
> Sir George Broun of Coalstoun, Bart., married Lady Elizabeth M'Kenzie . . . and this lady is reported to have bitten a piece of the famous Pear. It being expected that some calamity would follow on such an outrage . . . Accordingly, in 1699, Sir George was constrained by the pressure of incumbrances to sell the estate; but he was fortunate in meeting with a purchaser in the person of his

> brother, Robert Broun . . . However, a much greater calamity soon befell the Laird of Coalstoun and both his sons were drowned on the 5th of May 1793, and Coalstoun passed to an heiress.[6]

Today the pear 'is held safely at Colstoun in a silver reliquary given to the family by the town of Haddington'.[7]

PEARLWORT (*Sagina procumbens*)

Tiny, mat-forming perennial herb, common and widespread in a variety of habitats including pavements and lawns.

Local names include beads (or beeds) in Wiltshire, BIRD'S EYE in Sussex, little chickweed in Somerset, and POVERTY, 'the plant is very indicative of poor land', in Norfolk.

On Colonsay pearlwort 'is said to have been one of the plants that were formerly fixed over doors for good LUCK.[1]

PEASHOOTERS – made from ANGELICA, COW PARSLEY, ELDER, JAPANESE KNOTWEED and RED VALERIAN.

PELLITORY OF THE WALL (*Parietaria judaica*)

Perennial herb with inconspicuous flowers, common on walls and other, mainly man-made, habitats in lowland areas.

Local names include: billie-beatie in Belfast, penalty-of-the-wall in Buckinghamshire, wall-sage in Warwickshire, and peniterry in Ireland, where apparently schoolboys used it to protect themselves from being beaten:

> Peniterry, peniterry, that grows
>   by the wall,
> Save me from a whipping,
> Or I'll pull you roots and all.[1]

Medicinal uses include:

> [Guernsey] A tisane made from this plant is used in cases of DIABETES.[2]

> [Gypsies use] The juice from the leaves as an ingredient of ointment that cures ULCERS, running sores and PILES . . . Infusion of

the leaves allays all BLADDER troubles. Ointment from crushed root is good for piles.[3]

> My father boils . . . pellitory of the wall and drinks the brew thus obtained. He says it's good for the KIDNEYS. It acts against kidney stones as far as I remember.[4]

PENIS – SUN SPURGE used to arouse.

PENNYROYAL (*Mentha pulegium*)

Perennial, aromatic herb (Fig. 37); native in damp grassy places, such as beside village ponds, also occasionally cultivated and escaping.

Local names include the widespread organ and organy, and:
CREEPING JENNY in Somerset
Lily-royal in southern England
Lurkey-dish in Cheshire
Orgal in Cornwall
Orgin in Dorset
Pudding-grass in Cumberland and Somerset; 'because it is used in hogs-puddings'
Pudding-herb in Whitby, Yorkshire, 'where it is used for flavouring black puddings'
Whirl-mint in Hampshire.

The *pulegium* in pennyroyal's scientific name is derived from *pulex*, Latin for flea, which implies that the plant was used as a flea repellent;[1] however, there are few records of it being used for killing INSECTS in the British Isles.

> *Pouliet* or *Poue-ye* . . . pennyroyal . . . so called at the Vale [Guernsey] and said to be efficacious in destroying vermin on children's heads. These parasites are known as *pouas*.[2]

> Pennyroyal: campers rub the leaves on the skin to use as an insect repellant.[3]

Far more abundant are records of pennyroyal being used as an ABORTIFACIENT. According to an early eighteenth-century writer, its liquid juice and oily tincture would 'provoke the terms in women, expel the birth, dead child, and afterbirth'. The 'potestates or powers' prepared from it could cure INFERTILITY in men and women, while the essence would 'much increase the seed in both sexes and strongly provoke lust'.[4] Until the First World War pennyroyal syrup was valued as an abortifacient by the poor in Salford, now part of Greater Manchester,[5] and in a North Riding sword-dance play, collected in 1926, a man-woman who is killed, simulates labour and is revived by being dosed with 'oakum-pokum pennyroyal'.[6] A nurse training in a London hospital in the 1940s found that one of her patient's records revealed how she had suffered a miscarriage after swallowing a box of pennyroyal pills.[7] About fifteen years later, pennyroyal extracts were resorted to by Halifax telephone operators. If there was a strong minty smell in the workroom, someone would be sure to say 'Hallo girls, who else has been got now?' Bottled pennyroyal was commonly used as 'a standby for girls who missed a period', and it worked in some cases, 'when it probably would have worked anyway'.[8] In more polite society:

> Cleaning out a church cupboard some years ago we found a little Boots tin that had held pennyroyal and TANSY pills 'for delicate females'.[9]

It is said that in the nineteenth century 'organ tea' was drunk at 'tea treats and lady's meetings'.

> In Cornwall . . . organ tea was made from pennyroyal and as any herbalist will tell you this herb was considered dangerous to drink during pregnancy as it provoked miscarriage . . . So were these modestly held gatherings to drink tea really a way of trying to control the size of one's family? The ladies would hardly have been unaware of the properties of the herb, but it was all highly respectable, and any ulterior motives for drinking this beverage were discretely cloaked under the social concept of a tea-party.[10]

PENNYWORT – see NAVELWORT

PENTECOST – see Whitsun.

PEONY (*Paeonia officinalis*)

Perennial herb with showy red flowers, native to Europe, believed to have been introduced before the end of the eleventh century.

Most of the recorded local names are variants, or mispronunciations, of the standard name:

Nanpie in Yorkshire
Pianet in Cheshire, Cumberland, Lincolnshire and Shropshire
Pianny-rose in Co. Antrim, Co. Donegal and Co. Down
Pie-nanny in Yorkshire
Piney, or piny, in Gloucestershire, Oxfordshire and Somerset
Piony in Cheshire
Posy in Wiltshire; 'from its size'
Whitsun balls in Bristol.

> On the death of a friend in the summer, an old lady, who was on a visit of condolence to the widow, went quietly into the garden and counted the flowers on the peonies. On her return . . . [she] said she had counted the flowers on each of the peonies in the garden, and there was an odd number on each plant, which was a sure sign of DEATH in the house before the year was out.[1]

> If there are odd numbers of flowers on a peony it is a sign of death.[2]

> [Peony] is called 'sheep-shearing rose' by many from the rough joke of filling the folds of the petals with pungent snuff or pepper at sheep-shearing feast in order to enjoy the torments of those who innocently smell it at that period.[3]

> [West Sussex] A necklace turned from the root of the peony is worn by children to prevent CONVULSIONS, and aid dentition.[4]

> Recipe for Red Peony COUGH Syrup, handed down from my grandmother . . . To every head of bloom (picked fully out) pour over 1 pint boiling water; let stand for three days; strain off. To each pint of liquid add 1 lb sugar and ½ oz of whole cloves (put in muslin) and boil for one hour. Bottle when cold.[5]

PERIOD PAINS – RASPBERRY used to treat.

PERIWINKLE (*Vinca minor*, lesser periwinkle, and *V. major*, greater periwinkle).

Perennial, evergreen, blue-flowered herbs, cultivated as ornamentals and escaping into hedgerows and woodland. Lesser periwinkle introduced before the end of the tenth century, from Europe (distribution as a native plant obscure due to extensive cultivation). Greater periwinkle introduced before the end of the sixteenth century; native to the Mediterranean region.

Local names include:
BACHELOR'S BUTTONS in Somerset
Blue betsy in Devon
BLUE BOTTLE, BLUE BUTTONS, blue shirts, and blue smock, in Somerset
Blue-eyed Mary in Somerset
Blue stars, and BLUEBELL, in Devon and Somerset
COCKLE in Devon, Gloucestershire and Somerset
Cockle-shells in Somerset
Dicky-dilver in Somerset and Suffolk
Dwinkle, and fairies' paintbrushes, in Somerset
Fairy-paintbrush, and gwean, in Cornwall
KETTLE-SMOCKS, and LITTLE STAR, in Somerset
Merry-goes in Devon
Nightshade, 'an error due to confusion'?
Old-woman's eye in Dorset
PAINTBRUSHES, and parachutes, in Somerset
Pinpatch in Sussex
SENGREEN on the Isle of Wight; 'i.e. evergreen'
TUTSAN in Suffolk; 'i.e. all-heal'
UMBRELLAS in Somerset.

A well-known rhyme which describes a bride's apparel runs:

> Something old, something new,
> Something borrowed, something blue.

Near Cheltenham, in Gloucestershire:

> The 'something blue' that the bride wears is the blue periwinkle. One informant told me that it must be worn in the garter for FERTILITY.[1]

In Wiltshire:

> Many old cottages have periwinkles along the garden banks. These were often planted when a newly-married couple took possession of the cottage, in the belief that the periwinkles would ensure a lucky and happy marriage.[2]

Similarly, in Cambridgeshire:

> Fen people believed that if a young married couple planted a patch of periwinkles in the garden of their first home their life together would be a happy one.
>
> The periwinkle is one of the flowers held by many in Cambridgeshire people to wither quickly if worn as a buttonhole by a young girl of a flirtatious nature or by an unchaste wife.[3]

Popular works of the sixteenth and seventeenth centuries mention periwinkle as a promoter of conjugal love. First published in English in *c.*1550, *The Book of Secrets*, falsely attributed to Albertus Magnus (1193–1280), stated that when periwinkle was 'beaten unto a powder with worms of the earth wrapped around it, and with an herb called *Semperviva*, in English HOUSELEEK, it induceth love between man and wife, if it be used in their meats'.[4] Similarly, in 1652, Culpeper claimed that the periwinkle was owned by Venus, and its 'leaves eaten by man and wife together causeth love between them'.[5]

In west Dorset wild periwinkles were associated with St Candida (also known as St Wite), to whom the church at Whitchurch Canonicorum is dedicated. Candida has been variously identified as the Breton princess St Gwen (or St Blanche), a monk named Witta who accompanied St Boniface, and a Saxon local holy woman who was slain by Danish invaders. Towards the end of the twentieth century the saint's shrine, which remains intact within the church, became the focus of a growing cult. It is said that in earlier times:

> After venerating the shrine, our pilgrim made his way to the saint's well, about a mile away at Morcomelake. The waters of St Wite's Well enjoyed a reputation as late as the 1930s as being 'a sovereign cure for sore eyes' . . . The wild periwinkles that carpet nearby Stonebarrow Hill every spring, are still known locally as 'St Candida's Eyes'.[6]

In nineteenth-century Oxfordshire greater periwinkle was known as CUT-FINGER, and its leaves were 'commonly applied to chapped hands, and . . . said to have healing properties'.[7]

PERUVIAN LILY (*Alstroemeria* cvs)

Used as a substitute if ORANGE LILY is not available for Orange Order parades.

PETTY SPURGE (*Euphorbia peplus*)

Small annual herb, common and widespread on cultivated and disturbed ground throughout lowland areas.

Local names include FAIRY-DELL in Dorset, MILKWEED in London, MILKWORT in Wiltshire, seven-sisters in Co. Donegal, WART-GRASS in Lincolnshire and Yorkshire, and WART-WORT in Wiltshire.

> [Andreas, Isle of Man, June 1929] *Lhuss-ny-fahnaghyn* – used for WARTS.[1]

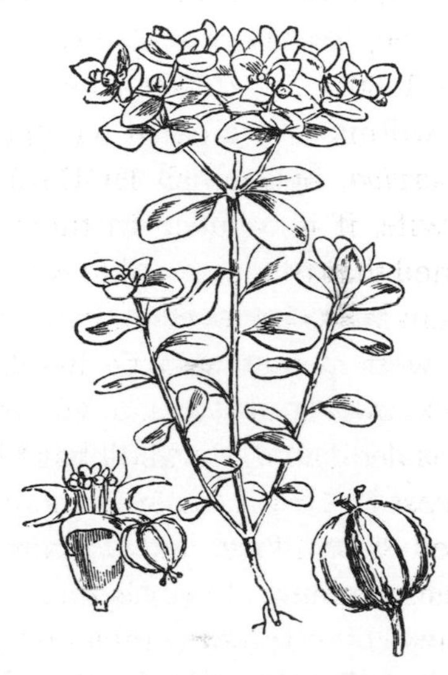

> When I was a very young boy (about 5 years old) I went to a gypsy camp close to Bodmin. There was a lady there complaining that the elderly lady had failed to 'charm' a very disfiguring wart on her face. The gypsy said she had another way provided she was given silver. WART-WEED was produced and the 'milk' was applied to the wart, and she was told to do this daily, The husband refused to pay. The gypsy then cursed him and rubbed his forehead with wart-weed. Later that day we saw the husband in Bodmin and there was a big red cross on his forehead.[2]

PIGNUT (*Conopodium majus*)

Perennial, white-flowered herb, with globose tubers, common in grassland, open woodland and hedgebanks.

Local names include:

Arnit in Northumberland, Yorkshire and Scotland

ARNUT in Northumberland, Yorkshire, Scotland and Co. Donegal

Bad-man's bread in Yorkshire; 'i.e. devil's bread'

Cain-and-Abel in Co. Durham

Cat-nut in Yorkshire

Cronies in Scotland

Cuckoo-potato in Co. Donegal
Curluns in Kirkcudbrightshire and Wigtownshire
Deil's bread, and deil's oatmeal, in Yorkshire
Earth-chestnut in Cornwall
Ernut in Roxburghshire
Ewe-yorlins in Cumberland
Fairy-potatoes in Ireland
Fare-nut, and fern-nut, in Cornwall; 'fare, a young pig, O[ld] E[nglish] *fearh*'
Gennet in Yorkshire
Gernut in northern England; 'cf. *gernotte* in Norman French'
Gourlins in Scotland
Gowlins in Inverness-shire
Ground-nut in Somerset
Grove-nut in Cornwall
Hare-nut in Dorset, Lancashire and Yorkshire
Hare-root, and heare-nut, in Dorset; 'hares are fond of the green leaves'
Hog-nut in Devon and Somerset
Hornecks in Scotland
Jack-durnals in Cumberland
Jack-jennets in Yorkshire
Jacky-journals, and jocky-jurnals, in Cumberland
Kellas, kelly, and killimore, in Cornwall; 'Cornish *keleren*'
Kipper-nut in Cumberland
Knot-girse in Scotland
Knotty meal in Inverness-shire
Loozie-arnit, lousy arnit, lousy arnuts, lucie arnit, and lucy-arnots, in Scotland
Pig's nuts in Cornwall and Somerset
SCABBY HANDS in Cumberland
Swine-bread, and truffle, in Inverness-shire
Underground nut in Cornwall and Devon
Yannut in Yorkshire
YAR-NUT in Lincolnshire and Yorkshire
Yennut (or yenut) in Yorkshire
Yernut in Cheshire and Yorkshire
Yowe-yorlings, yowe-yornals, yowe-yornut, and YOWIE-YORLIN, in Cumberland.

The roots of pignut were widely dug up and eaten by children.

> Pignut: we used to dig the roots in the First World War and eat the noodles [i.e. nodules] at the end of the root![1]

> [Wimborne St Giles, Dorset, *c.*1920s] the tubers of pignut were dug up with flints, and eaten, but we washed them in the river.[2]

> [Mid Oxfordshire, 1926–*c.*1940] Eating pignuts. These were the swollen roots of a small umbelliferae plant growing in meadowland. They were about as big as a small hazel nut and tasted similar.[3]

PIGS – fed on BRACKEN and COW PARSLEY.

PILATE, PONTIUS – associated with Fortingall YEW.

PILES – treated using BUTTERCUP, CELANDINE, CHESTNUT, conkers (HORSE-CHESTNUT seeds), MEZEREON, ONION and PELLITORY OF THE WALL.

PIMPLES – DOCK and LEMON used to treat.

PINE (*Pinus* spp.)

Evergreen, coniferous trees; one native species SCOTS PINE, and several introduced species and some of which have become naturalised, grown for timber, as windbreaks and as ornamentals.

In Guernsey:

> Pines are very UNLUCKY trees indeed. Whoever planted a row of them ran the risk that the property on which they stood might change hands, or pass from their rightful heir to a younger branch of the family. My grandmother always said that if you fell asleep under a pine tree you would not wake up.[1]

However, elsewhere:

> A remark made to me the other day by a friend who said a parent had been advised to take his ailing son for a fortnight's holiday at a spot of the Welsh coast where pines are plentiful. 'Get the child into the smell of the pines for an hour or two everyday', the serious advice ran. And he did, and the child, by all accounts, was cured.[2]

PINEAPPLE (*Ananas comosus*)

Perennial herb, originating in South America, now widely cultivated in tropical and subtropical areas for its succulent fruit.

Stone pineapples are often seen at the entrances to even quite modest homes. According to the *Miami Herald* of 4 February 1984:

> Members of the British East Indies Company brought the plants to England where they were cultivated in 17th century hot houses. Only the very rich could afford the fruit, so it became synonymous with opulence and high living. Middle-class party givers had to be

> content with renting the fruit for centerpiece displays. American sea captains gave the custom a new twist during the era of New England's great sailing ships. When a captain arrived home from the tropics he brought pineapples with him. He placed one on the newel (post) of his front stairway to let everyone know he was home and waiting for his friends to visit.

However, although pineapples are now cultivated throughout tropical Asia, the first fruits to reach Britain came from the West Indies.[1]

PINEAPPLE-WEED (*Matricaria discoidea*)

Small, annual herb (Fig. 38), smelling similar to pineapple when crushed, with unshowy yellow-green flowers, on disturbed and waste ground. Region of origin unknown, cultivated since the late eighteenth century, first recorded in the wild as an escapee from Kew Gardens in 1871, since when it has become the most widespread neophyte (species introduced and naturalised since 1500) in these islands.[1] Known in older wildflower books as rayless-mayweed, or rayless-chamomile.

Despite being a comparatively recent addition to the British and Irish flora, pineapple weed has been used to produce a yellow-orange dye in Shetland,[2] medicinally elsewhere.

> In Welsh according to the gentleman who brought it to my house it's known as *Pefelen*. The gentleman stated as a child he had several styes around the neck, and that his grandmother made a poultice out of the leaves of the plant – 'the poultice pulled like anything – but it got rid of the styes.[3]

PINK SORREL (*Oxalis articulata*)

Perennial, pink-flowered herb; native to temperate South America, introduced as an ornamental in 1870, and first recorded in the wild in 1912, naturalised on waste ground mainly in southern areas.

Although pink sorrel is the sort of cottage-garden plant which might be expected to collect a wide range of local names, only three have been recorded: SHAMROCK used on the Dorset/Somerset border in the 1950s,[1] BREAD-AND-CHEESE, and vinegar.

In common with other *Oxalis* species, pink sorrel was eaten by children:

> [I'm now in my 60s, as children my older brother and myself would] eat the stem of the pink shamrock flower. It tasted a bit sour, but was alright – when we lived at the farm we had several plants outside the front door.[2]

> Our dear neighbour is 89 . . . she is a real country woman . . . as a child she ate leaves, called bread-and-cheese, or vinegar . . . from a plant with pink flowers.[3]

PIXIES – BURDOCK and GREATER STITCHWORT associated with.

PLAGUE – BUTTERBUR and PRIMROSE associated with.

PLANTA GENISTA

The *planta genista* which provided the name of the Plantagenet dynasty, rulers of England from 1154 to 1399, is usually identified as BROOM. It is said that Geoffrey of Anjou, father of Henry II, the first Plantagenet king, selected the plant as his emblem.

> A sprig of genista was adopted by Gefroi, Duke of Anjou . . . [who gathered the wild flower] when passing through a rocky pathway; he saw on either side bushes of yellow broom clinging with firm grasp to the huge stones, or upholding the crumbling soil: 'And thus' (said he) 'shall that golden plant ever be my cognizance firmly rooted amid rocks, and yet upholding that which is ready to fall. I will bear it in my crest, amid battlefields if need be, at tournaments, and when dispensing justice.' Thus saying, the warrior broke off a branch, and fixing it in triumph on his cap, returned to his castle.[1]

PLEURISY – FIELD SCABIOUS used to treat.

PLUM (*Prunus domestica*)

Deciduous tree grown for its succulent fruits, native to south-west Asia, but cultivated since late in the tenth century, and frequently naturalised.

APPLE seedlings produce fruit which rarely have any culinary value,[1] but the fruit of plum seedlings are usually pleasantly edible, hence there are a large number of local varieties, often unnamed, both cultivated in gardens and semi-naturalised in hedgerows.[2]

> 'A good WHEAT year, a fine plum year.' This is a prevailing saying in N. Notts, and one which I have heard from many persons this year [1887], the crops of both being very good.[3]
>
> Should an earwig get into one's ear, it was believed it could best be got out by holding a ripe plum over the other ear.[4]

PNEUMONIA – ELDER used to treat.

POACHERS – use IRISH SPURGE to poison fish.

POINSETTIA (*Euphorbia pulcherrima*)

Native to Central America where it grows as an unattractive lanky shrub, the plants which are familiar today have been carefully cultivated using controlled lighting to produce colourful, usually red, bracts. The plant was first introduced to the USA in 1825 by Joel Roberts Poinsett, and by the early 1900s had become a popular CHRISTMAS houseplant. About fifty years later its popularity spread to Europe, where it is sold in great quantities during the Christmas period; apparently becoming less popular in recent years.

A story, read at a carol service in a Brixton, London, church in December 1998, explains how poinsettias became associated with Christmas:

> Pepita was a young girl who lived in a small village in Mexico. Every Christmas Eve as everyone in the village gathered in the church for the Mass, every family would bring a gift for the Christ Child. Pepita had nothing to bring – her family was very poor. Her cousin Pedro attempted to console her. He told her, 'I am certain that even the most humble gift, given with love, will be acceptable in God's eyes.'
>
> Pepita was determined to bring something, so she found some flowering weeds along the side of the road. She made them into a bouquet, but she was embarrassed at the thought of bringing a bunch of weeds to lay before the manger. But it was the only gift that she had to offer.
>
> The church was almost full by the time she got there and Pepita

held the bouquet close to her side hoping that the others would not notice it. She hesitated before walking down the aisle, but she remembered the words of her cousin. She prayed, 'Oh God, this may not be the grandest of gifts, but may it be the gift given with the most love.'

Her spirits began to lift and she walked down the aisle holding the bouquet proudly in front of her. By the time she got to the front of the church her eyes were brimming with tears of love as she knelt and laid her bouquet in front of the nativity scene.

Suddenly, the bouquet of weeds burst into blossoms of brilliant red. And everyone there was certain that they had seen a Christmas miracle. From that day on the bright red flowers were known as the *Flores de Noche Buena*, or Flowers of the Holy Night. However, you know them as poinsettias.

POOKA (Irish supernatural being) – contaminated BRAMBLE fruit.

POP-GUNS – made from ELDER; HAWTHORN seeds used for ammunition in.

POPPY (*Papaver rhoeas*)

Annual herb with conspicuous scarlet flowers, widespread on arable and disturbed lands in lowland areas; much sown in 2014 to mark the centenary of the outbreak of the First World War.

Local names include the widespread copper rose and RED WEED, and:

BULL'S EYES in Somerset
Butterfly-ladies in Dorset
CANKER, 'from its red colour and its detriment to arable land', in Norfolk and Suffolk
CANKER-ROSE in East Anglia; cf. the Dutch *kanker-bloemen*
Cheese-bowls in Somerset
Chesbow in Scotland
Cockeno, in Berwickshire; 'evidently from *coch*, the Celtic for scarlet and hence the name is probably coeval with the early inhabitation of the district'
Cock-rose, also given to 'any wild poppy or red flower', in Yorkshire and Scotland
COCK'S COMB in Berwickshire
Cock's head in Scotland
Collinhood in Lothian and Roxburghshire
Cop-rose, 'from its red rose like flower and the cop or button-shape of its capsule'
CORN COCKLE in the West Country
CORNFLOWER in Devon
Corn poppy in Cornwall
Corn rose in Dorset and Somerset

Cup-rose in northern England, 'and probably [applied to] other species'
Cusk in Warwickshire
Devil's tongue in Cornwall
Dog rose, 'general' in Dublin
EYEBRIGHT in Somerset
Fireflout in Somerset and Northumberland
Golliwogs, also golliwogs' heads, applied to the seed heads, in Somerset
Guy in Suffolk
HOGWEED in East Anglia
Old-woman's petticoats, and paradise lily, in Somerset
Pepperboxes in Somerset, presumably referring to the seed capsules.
Poison poppy in north Buckinghamshire
Pope in south Northamptonshire, where 'going a poping is commonly used as a term for weeding poppies'
Poppet in Warwickshire
POPPLE in Yorkshire
Red cap, red cup, red dolly, red huntsman, and red nap in Somerset
Red petticoat in Kent
Red poppy, 'as opposed to poppy, *Digitalis purpurea*' [FOXFGLOVE] in Devon
Red rags in Dorset
RED SOLDIERS in Somerset
SCABBY HANDS in Dublin
SLEEPY HEAD in Somerset
SOLDIERS in Norfolk, Northamptonshire, and Wiltshire, 'a field of these is supposed to resemble an army of Redcoats'
Soldiers' blood, 'perhaps thinking of Flanders fields'
Summer poppy in Somerset
Wild maws, and yedwark, in Derbyshire.

It was widely thought that the picking of poppy flowers would cause THUNDER.

> About Wooler [Northumberland] it was wont to be called the THUNDER-FLOWER or Lightnings; and children were afraid to pluck the flower, for, if perchance, the petals fell off in the act, the gatherer became more liable to be struck with LIGHTNING, nor was the risk small, for the deciduousness of the petals is almost proverbial.[1]

Names which relate to this belief include thunderball in Warwickshire, THUNDERBOLT in Devon, Cheshire and Westmorland, thunderbowt in Shropshire, thunder-cup in Berwickshire, thunner-flower in west Cumberland, and thunnor cups in Northumberland.

It was also believed that poppies could cause EARACHE or HEADACHE.

> The popular name for field poppies, as well as cultivated ones, in this district [Worksop, Nottinghamshire] is 'earaches' . . . It is said that if they are gathered and put to the ear a violent attack of earache will be the result.[2]

> The red poppies that grow in cornfields in Ireland are in the counties of Carlow, Wexford, Wicklow and Waterford called 'Headaches', and are particularly obnoxious to females, the more so to young unmarried women, who have a horror of touching or being touched by them. The flower is sometimes used with logwood and copperas to DYE wool and yarn black, but otherwise the weed is considered poisonous.[3]

According to a gypsy woman born in Somerset in 1941:

> The red poppy that grew in the pea fields and gave us all headaches when we picked close to them.[4]

The name earaches is also recorded from Derbyshire, while names which relate to the headache belief include the unlocalised HEADACHE-FLOWER, and heead-vahk, 'because it gives you one' in Yorkshire.

Alternatively, Gabrielle Hatfield suggests that the name headache-flower relates to the Norfolk practice of chewing poppy seeds to treat HANGOVERS.[5]

It is probable that such beliefs were partly propagated to discourage children from gathering the attractive flowers and thereby damaging crops. However, another belief, that poppies can cause blindness, may contain an element of truth. According to a veteran of the First World War, who described lengthy marches through fields of poppies, the flowers was 'so bright they affected our eyes – we could see red for days'.[6] Similarly, in nineteenth-century London tailors found that:

> Of all colours scarlet, such as used for regimentals, is the most blinding . . . there's more military tailors blind than any others.[7]

In North Marston, Buckinghamshire, poppies were known as blind-eyes and children were cautioned by their parents 'not to gather it, for it will blind your eyes'.[8]

> Blind-eyes or blind-eye have also been recorded as names for poppy in Norfolk and Northamptonshire and County Tyrone, while the name blindy-buffs was recorded in Yorkshire. In south-

> west Wiltshire the name blind-man was given to poppies which were 'locally supposed to cause blindness if looked at too long'.[9]

Elsewhere sniffing poppies was thought to induce SLEEP.[10] In Thetford, Norfolk, in the late 1960s:

> My mother never let me keep flowers, especially poppies, in my bedroom, because they would make me drowsy; they 'sucked out the oxygen'.[11]

However, despite these superstitions, children collected poppy flowers to make dolls.

> The dollies I remember from the cornfields . . . were the little ballerina poppy dollies that our mother taught us to make from the scarlet poppies . . . Tiny features were marked with a pin or a pencil point on the green seed capsule which wore its own natural flat ribbed cap; a few stamens were removed in front of the face, the remaining ones forming an impressive Elizabethan-type ruff. The petals were turned down and tied tightly at the waist with cotton or a blade of grass. A length of stalk was pushed through between the petals for arms, and, if fussy, another piece could be pushed up into the waist to make a second leg.[12]

> [We were perpetually on the move during my childhood, but 50 years ago, probably in Staffordshire] we made Chinese-men poppy dolls. The petals were turned back. Two pieces of stalk were used, one as a cross-piece for the arms, and another to add to the still attached stalk to make a second leg. A piece of wool gave a waist to the poppy petal dress, and a Chinese face drawn on the exposed seed box.[13]

> When my children were at school in Craigavan, Northern Ireland, *c.*1969, they used to say to me 'Do you want to see the fairies dance?' They would pull a flower – either field poppy or lesser bindweed – and pull the petals back over the stem to make a 'skirt', and holding the stem between thumb and forefinger just behind the flowerheads (within the skirt) move the finger to and fro to cause the flowerhead to partially rotate and return – this caused the 'fairy' to dance.[14]

Since early in the nineteenth century poppies have been associated with those who died in battle:

> The red poppies which followed the ploughing of the field of Waterloo after the Duke of Wellington's victory were said to have sprung from the BLOOD of the troops who fell during the engagement.[15]

In the twentieth century the poppy became a symbol of remembrance of those who died in the First World War and subsequent wars. During the second battle of Ypres, in May 1915, a Canadian doctor, Colonel John Macrae, wrote:

> In Flanders fields the poppies blow,
> Between the crosses, row on row
> That mark our place; and in the sky
> The larks still bravely singing fly
> Scarce heard amid the guns below.
>
> We are the dead. Short days ago
> We lived, felt dawn, saw sunset glow,
> Loved and were loved, and now we lie
> In Flanders fields.
>
> Take up our quarrel with the foe;
> To you from failing hand we throw
> The torch; be yours to hold it high,
> If ye break faith with us who die
> We shall not sleep, though poppies grow
> In Flanders fields.

Published in *Punch* in December 1915, the poem caught the eye of Moina Michael, an American woman, who was inspired to produce a poem which expanded the poppy motif:

> We cherish too, the poppy red
> That grows on fields where valor led,
> It seems to signal to the skies
> The blood of heroes never dies.
> But lends a lustre to the red
> Of the flower that blooms above the dead
> In Flanders fields.

Miss Michael took to wearing a poppy to 'keep faith in those who died', and in November 1918 a French woman, Madame Guerin, decided to get poppies manufactured in France and sell them with any profits being used to help people returning to war-devastated areas. Thus runs the 'official' history, but other things were happening elsewhere. In June 1918 when members of Papton Adult School visited the Derbyshire home of the socialist, philosopher and (to use the terminology of later times) gay-rights campaigner, Edward Carpenter, he spoke on 'War and Reconstruction':

> literature was distributed along with red poppies. The Millthorpe poppies were symbols of a grassroots patriotism, for the mighty wartime state had overlooked the wounded men discharged from the forces, and two left-wing organisations had taken on their welfare: the National Association of Discharged Soldiers and Sailors (N.A.D.S.S.) and the National Federation of Discharged and Demobilised Soldiers and Sailors (N.F.D.D.S.).[16]

When the British Legion organised its first Poppy Day, 1921, it used poppies imported from France. This event raised £106,000 and the Legion thought it should find a British source to supply its poppies. By June 1922 Major George Howson had established a poppy factory, employing five disabled men, in east London. Initially only lapel poppies were made, but in 1924 wreath making was added to the factory's activities, the first large wreath being laid by the Prince of Wales on 11 November 1924. By 1925 Howson's workforce had grown to fifty, making it necessary to search for larger premises, which were eventually found at Richmond, Surrey. Until 1975 only ex-servicemen, with preference being given to the disabled, were employed, but since then the pool of potential recruits has expanded to include ex-servicewomen and the widows and disabled dependants of ex-servicemen. In July 1989 the factory provided work for 130 employees, 80 per cent of whom were disabled. An additional 45 or so workers, who were aged, severely disabled, or in poor health, assembled poppies in their own homes. Each year the factory produced approximately 36 million poppies (which were sold by some 250,000 volunteers), 75,000 wreaths, and 250,000 remembrance crosses. The 1988 Poppy Appeal raised over £10 million to help ex-service personnel and their dependants.[17] According to the *Independent* of 11 November 2006 the Legion expected to sell a record 37 million poppies that year. Ten years later it was reported that 40 million poppies would be sold.[18]

Since 1933 white poppies have also been worn during the run-up to Remembrance Sunday. Initially these were promoted by the Co-operative Women's Guild as a 'pledge to peace that wars must never happen again'. It was intended to extend 'the narrow nationalistic and militaristic view of Remembrance to remembering all the dead in wars, irrespective of nationality, civilians as well as those in the armed forces'. The use of white poppies reached its peak in 1938, but they continued to be used sporadically until 1980, when they were adopted by the Peace Pledge Union. The Union organises an alternative wreath-laying at the Cenotaph in Whitehall, London, at which a wreath of white poppies is used. In 1988 the Union expected to sell 50,000 poppies,[19] while a record 110,000 were sold in 2015.[20]

In the 1990s black poppies were produced to commemorate black and Asian people who died in war,[21] and for some years from 2011 purple poppies, 'remembering the animal victims of war', were produced by Animal Aid. More recently such poppies have been abandoned, it being thought that people might consider animals to be 'valiant servants of people', rather than innocent victims of war, and superseded by a purple-paw badge which commemorates all animals which are 'victims of human exploitation'.[22]

For several decades from the early 1880s the area around the Norfolk villages of Overstrand and Sidestrand was known as Poppyland. This was initially promoted by the writer Clement Scott, who published an article entitled 'Poppyland' in the *Daily Telegraph* of 30 August 1883. Scott seems to have been particularly moved by the sight of the tower of the former church of St Michael, Sidestrand, the remainder of which had been moved to safety and re-erected further inland in 1881. Around the tower remained the graves of villagers with poppies growing from them, 'a garden of sleep' which became the subject of his verse. As an influential drama critic, Scott's enthusiasm and writings encouraged various well-known people to visit the villages, and where they went others, inevitably, followed. Scott's articles were gathered in a series of books, starting with *Poppyland Papers* in 1886. Postcards, china – 'from candle holders to full tea services' – and a perfume and soap called Poppyland Bouquet were produced. The Great Eastern Railway enthusiastically took up the theme and produced posters encouraging the public to visit north Norfolk. By the outbreak of the First World War the allure of Poppyland had faded. Scott died in 1904, regretting the change that his publications had brought to two formerly quiet villages. The tower eventually toppled into the sea in 1916 and the surrounding graveyard gradually slipped away. Souvenirs continued to be sold between the two world wars, and then interest declined, until the early 1980s, when a Poppyland centenary walk was held in August 1983 and the production of a BBC film in 1985 revived interest.[23]

Allen and Hatfield note that poppies have been used as a soporific in the Isle of Man and the Scottish Lowlands, but 'feature in the folk records much more often in the PAINkilling role'. Some of the ailments they were used to treat included EARACHE in Somerset and County Tipperary, NEURALGIA in Montgomeryshire and County Wicklow, and TOOTHACHE in Sussex and parts of Ireland.[24]

POTATO (*Solanum tuberosum*)

Perennial herb, cultivated for its edible tubers in the Andes for over 7,000 years, introduced to Europe late in the sixteenth century and by 1800 becoming a staple food of the poor.

On Looe (or St George's) Island off the south coast of Cornwall:

> We had been told that it was traditional to plant early potatoes on Boxing Day. As it was now February it seemed matter of urgency to sow them as soon as possible, and I was determined to get started at once.[1]

Elsewhere planting was delayed to later in the year. In Ireland:

> Early potatoes: it was believed that this crop should be sowed before ST PATICK'S DAY.[2]

> Potatoes should be planted . . . about St Patrick's Day, when he 'turns up the warm side of the stone'.[3]

However, the most common date for planting potatoes was GOOD FRIDAY, which can fall on any date between 20 March and 23 April, when at least token plantings were made. Explanations of why this should be done varied. From the practical point of view it was a day when farm labourers had time off so they were able to get on with garden work. In Rutland and south-west Lincolnshire before the Second World War Good Friday was 'believed to be symbolical of the Resurrection after three days, thus promoting the growth of the buried tubers'.[4] In County Derry it was thought that 'because it was a Holy Day the potatoes would be better'.[5]

> We always plant new potatoes on Good Friday to ensure a good

crop, a local tradition in Sussex which we have always adhered to.[6]

[Cumbria] My grandfather always planted his potatoes on Good Friday, and insisted it had to be done then to get a good crop (he always did).[7]

In County Mayo seed potatoes 'are dressed during planting, with a pinch of salt and human excrement', in Kerry 'a piece of CYPRESS is stuck into the ridge on planting day, and on harvesting a branch of the same was burnt'.[8]

In 1836 it was reported that in County Mayo:

The digging of potatoes for daily use commences after Garlick Sunday [the first Sunday in August], but not to such an extent as to prevent a great rise in price.[9]

Elsewhere:

Garland Sunday, being the last in July, is in Galway the day on which the first digging is permitted. In Kerry the date is 7 July, the day of the local patron saint. In Cork, some potatoes, however few, should always be dug on 29 June. In Mayo, in every home the end of the potato harvest is celebrated by a feast. In Tipperary, when new potatoes first appear on the table, it is usual for each to say to the other: 'May we all be alive and happy this time twelve months.'[10]

Garlick or GARLAND SUNDAY are two of several recent names given to the ancient Celtic festival of Lughnasa, which was formerly celebrated to welcome the first fruits of the harvest. In many parts of Ireland the day was known as the Sunday of the New Potatoes.[11]

Wishes were often made when the first new potatoes were eaten (cf. FRUIT).

There is an old saying that you'll get what you wish for when you are sitting down to your first dinner of new potatoes.[12]

It is a custom to wish when eating the first new potato.[13]

Being readily available, potatoes were used in children's toy-making and games. In Shetland:

Tattie craa – Choose a nicely rounded potato, then collect some stout wing feathers from a large bird, such as the herring gull or crow. Insert the points of the feathers (using 3 or 4) slightly angled and splayed out into the end of the potato, then throw it high in the air, and as it descends it twirls around very fast making a loud whirring noise. The positioning of the feathers and the number used can be experimented with, adding interest to the game.

Staelin swine – A player sits on a stool with a cloth in his hand, and four to six potatoes lie before him on the floor. The second player approaches and recites:

A'm come fae da high laands,
Gyaan ta da low laands,
Seekin swine, geese and gaeslins.
We lost a peerie aalie pootie,
An me midder pat me ower ta see
If he wis among your eens.

The seated player then replies: 'Look du, du's welcome' whereupon the other starts to turn over the potatoes, examining them carefully, and saying: 'Hit's no dis een, an no dis een.' The aim of the game is to get one of the potatoes in his hand and escape before the other player touches him with the cloth.[14]

Potatoes were used in folk medicine, and were believed to prevent RHEUMATISM and CRAMP.

I knew a clergyman in the North of England, a graduate of Oxford, who used to carry in his trousers pocket, and recommend to others, a potato as a cure for rheumatism.[15]

[Somerset/Devon] A potato carried until it gets hard in the pocket of the patient, is firmly believed as a cure for rheumatism. It is supposed to 'draw the iron out of the blood': too much iron, and consequent stiffness, being the root of the complaint.[16]

[Co. Offaly] A cure for rheumatic pains, SCIATICA and LUMBAGO, is to carry in your pocket a Champion potato which you stole from your neighbour's pit.[17]

[Twickenham, London] My dad used to carry a potato in his pocket for years, about 1970s, for rheumatism.[18]

A potato in the bed helps do away with cramp.[19]

Cure for leg cramps . . . a large raw potato in your bed near your legs.[20]

Occasionally a potato might be used to provide more general protection, or even to attract good luck.

> In May 1978, when discussing the football pools with a group of middle-aged women working in South Kensington, they mentioned some of the 'lucky charms' which they hoped would bring them luck. These included short lengths of string, pressed four-leaved CLOVERS, and tiny withered potatoes, all of which they carried around in their purses.[21]

Other medical uses included:

> [Co. Leitrim] A sore THROAT was cured by roasting potatoes and putting them into a stocking, and tying them around the neck.[22]

> [Co. Dublin] Sliced potato to be rubbed on the CHILBLAIN every night for three times and in a short time the chilblain will be cured.[23]

> I have been assured by an elderly neighbour, a lady, that a certain cure for leg ULCERS, and other hard to heal lesions is obtained by simply grating a portion of raw potato over the sore. She has used the treatment herself in the past with complete success.[24]

> In my family we have long used raw potato on our potential BRUISES to great effect. The potato must be cut and used raw and rubbed on the place immediately. Some years ago I slammed a door on my upper lip when leaving the flat . . . I returned immediately to the flat for a potato, cut it and then rubbed it on the lip – no bruise appeared.[25]

> [Ireland] To get rid of a WART, cut a potato in half, rub one half on the wart, plant the two halves together.[26]

Nonmedical uses included:

> [Horseheath, Cambridgeshire] For laundry, water in which potatoes have been boiled was frequently used as a substitute for the expensive and perhaps better known flour STARCH.[27]

My father had to drive his car in the rain, but the windscreen wipers had stopped. A neighbour came out with half a potato and rubbed it over the outside of the windscreen to help disperse the water droplets. Apparently the starch helps achieve this.[28]

POULTRY – bringing DAFFODIL and PRIMROSE indoors leads to bad luck with; DOCK seeds fed to; TANSY used to treat 'pip' in.

PREGNANCY – caused by touching LORDS-AND-LADIES; RASPBERRY used during.

PRIMROSE (*Primula vulgaris*)

Perennial herb (Fig. 39) producing yellow flowers in springtime, common in a wide range of habitats, typically hedgerows and open woodland.

Local names include:
Buckie-faalie in Caithness
Butter-rose in Devon
Darling-of-April, early rose, and EASTER ROSE, in Somerset
Fearkan in Co. Donegal
First rose in Caithness and Somerset
Golden rose, and GOLDEN STARS, in Somerset
GOSLINGS in Co. Tipperary; 'as they were in flower when goslings hatched'
Green jackets in Devon
Key-flower in Somerset
LENT-ROSE in Devon
May-flooer in Shetland
MAY-FLOWER in Shetland and Co. Sligo
May-spink in Aberdeenshire
Pimerose in Cheshire
Pimmirose in Shropshire
Pimrose, 'a very common mispronunciation', and primerole, in Somerset
Primerose in Berwickshire and Cumberland
Simmeren in Yorkshire
WHITE MAY in Co. Tyrone
White summer in Dublin and Co. Tyrone.

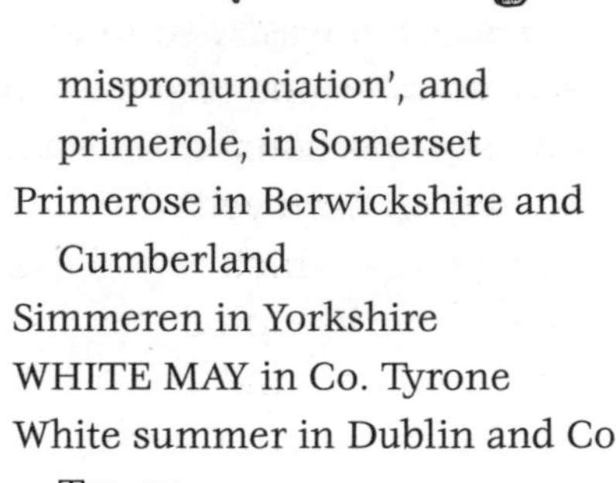

Although primrose flowers are typically pale yellow, they exhibit considerable diversity from pale pink to dowdy purple. Such plants, which appear to be increasing in abundance, presumably result from hybridisation with cultivated polyanthus, and have been collected and transplanted in gardens. These include: beef-and-greens, 'a variety having a red and white green calyx', in Yorkshire BUTTER-AND-EGGS, 'having a double calyx growing one out of the other', in Somerset.[1]

Primroses like DAFFODILS were associated with POULTRY-keeping.

> In East Norfolk some old women are still found who believe that if a less number of primrose than thirteen be brought into a house on the first occasion of bringing in, so many eggs only will each hen or goose hatch that season. When recently admitted into deacon's orders, my gravity was sorely tried by being called in to settle a quarrel between two old women, arising from one of them having given one primrose to her neighbour's child, for the purpose of making her hens hatch but one chicken out of each set of eggs. And it was seriously maintained that the charm had been successful.[2]

> I have always heard that it was unlucky to take primroses indoors before April, particularly if one had fowl hatching![3]

> [From an elderly Payhembury, Devon woman] Don't bring one or two primroses into the house. Bring a whole bunch. Otherwise your chicken and ducklings will die. [She] was highly sceptical about [this]. She had reared chickens, ducks and geese all her working life, and had never paid any attention to it.[4]

> 'Of course you had to bring at least thirteen primroses into the house. Do you bring less, it were no use; it didn't serve. Thirteen was the number or more. It didn't matter if you had more; but you dursn't have less.' The rationale of the custom was then immediately clear as one of the old people pointed out; thirteen was the number traditional to a clutch of eggs placed under a hen during the spring. Each yellow primrose was, therefore, the analogue of a young chick which would eventually emerge from the egg. If one grants that like produces like – an unquestioned assumption of the primitive mind either in Britain or in Borneo – it is folly then to bring in fewer primroses than you hope to have healthy young chicks.[5]

There are also hints that primroses were connected with the health and well-being of humans. In the mid-nineteenth century it was remarked that no primroses grew at Cockfield in Suffolk.

> Nor, it is said do they thrive when planted, though they are numerous in all surrounding villages, which do not apparently differ from Cockfield in soil. The village legends says here, too, they

were once plentiful, but when Cockfield was depopulated by the PLAGUE, they also caught the infection and died, nor have they flourished since that time.[6]

An old Cheshire name for a primrose was PAIGLE. If it bloomed in winter it was an omen of DEATH.[7]

[Lake District, *c.*1930s] a single primrose brought into a house is a sign of a death in the family.[8]

On the Isle of Man:

MAY EVE, as its old Manx name *Oie Voaldyn* indicates, was the eve of the festival of Beltaine. As elsewhere in Europe, yellow blooms and green branches were used to decorate and protect the house and cattle. Primroses (*sumark*) and MARSH MARIGOLDS (*Lus vuigh ny boaldyn, Lus airh ny lheannaugh, bluightyn*) were the favoured flowers. This use gave the name Mayflowers – also shared in Man with the lady's smock [CUCKOO-FLOWER] – to the latter. Primroses and marsh marigolds still appear in jars of water on May Day, even on office windowsills and shop counters in the towns.[9]

Similarly, in Ireland:

[Co. Meath] Though it is dying out now, but still surviving in this district, a custom was to scatter primrose or COWSLIPS outside the front door on May Day.[10]

[Co. Donegal] In years gone by people used to throw a primrose at the byre door [on May Eve] so that the FAIRIES would not take away the milk from the COWS for the year.[11]

Elsewhere in Ireland this custom seems to have become Christianised:

[Co. Sligo] Primroses, or mayflowers, used to decorate the MAY ALTAR, dedicated to Our Blessed Mary.[12]

In East Yorkshire primroses were associated with St John of Beverley.

Since 1929 a Patronal Festival in honour of St John of Beverley has been held at St John's Well at Harpham, the saint's birthplace.

Children from the village pick primroses in the nearby woods and place bunches of them on the saint's tomb during the Patronal Festival at Beverley Minster.[13]

In recent years, probably because primroses have usually finished flowering by St John's feast day, 7 May, wildflowers have been used instead of primroses, though a rather faded, undated photograph of St John's tomb in Harpham parish church has the caption: 'A token of appreciation for each generation of children who from your village each year have picked primroses to place on the tomb of St John of Beverley in Beverley Minster.'[14]

In 1883 the Primrose League was formed in memory and in support of the political ideas of Benjamin Disraeli (1804–81), who served as prime minister in 1868 and 1874–80. Primroses are said to have been Disraeli's favourite flower, and the League promoted the wearing of them on 19 April, the anniversary of his death.

> We have had an opportunity of seeing how flower-lore originates and grows, for we recently added to our calendar in connection with the late Lord Beaconsfield [Disraeli's title from 1876], a new festival, under the title of Primrose Day . . . we shall have this day [Primrose Day] commemorated, as Royal OAK Day has been, for many years, Ladies will wear primroses in their head-dresses, and gentlemen sport them in button-holes, and wreaths will still be consecrated at the Earl's shrine.[15]

At its peak in 1910 the League had two million members, but the enthusiasm shown for Primrose Day in the decades after Disraeli's death was not maintained, and today it passes virtually unnoticed. According to *The Times* of 19 April 1990:

> Tories steeped in the 'One Nation' tradition who thought the pendulum had swung back their way have suffered a setback: an acute shortage of primroses in Primrose Day . . . Alas, members of the Primrose League, including about 30 Tory MPs, who had hoped to be sporting buttonholes today are finding them in short supply. The early spring has meant that in most places, Disraeli's favourite flower has long since bloomed and wilted.

Though the League disbanded in 2004, on 19 April 2014 a 'small delegation' gathered at Disraeli's statue in Parliament Square, London, but it appears that their activities did not involve primroses.[16]

An occupation which has fascinated generations of children is the attempted production of pink or red primroses. James Britten recorded in 1869 that around High Wycombe in Buckinghamshire it was commonly believed that 'spring flowers' – primroses with bright purplish flowers – could be produced by planting a primrose in cow dung, Polyanthuses could be obtained by placing COWSLIP root upside down in soot.[17] Similarly:

> [West Dorset, 1950s] When I was about nine or ten, I remember my father telling me that if you planted a primrose upside down it would produce red flowers. I tried doing this but the plant invariably died, so I never produced any red flowers.[18]

Such beliefs were investigated in a paper 'On the variation and instability of the coloration of flowers of the primrose (*Primula vulgaris*) and cowslip (*P. veris*)'. According to this, the idea that the flower-colour of primroses and cowslips could be changed by such methods as described by Britten had 'long been held by practically all country people throughout Britain'. The evidence assembled in the study varied: some people transplanted pink-flowered wild primroses into their gardens and were surprised when these later produced yellow flowers; others found primroses retained their colour no matter how and where they were transplanted, and others claimed to have induced plants to change their flower-colour from yellow to pink. The author concluded that it was possible to make a normal primrose plant produce pink flowers by planting it in rich, well-manured soil,[19] but his evidence seems weak. The belief was undoubtedly stimulated by the fact that pink-flowered primroses do occur quite frequently, presumably the result of hybridisation between wild primroses and garden polyanthuses. It is noteworthy that primrose plants which bear pink or reddish flowers also tend to produce several flowers on each stalk, thus indicating a polyanthus as a parent. Such plants tend to colonise areas where the soil had been disturbed, where a ditch has been dug or a hedge repaired, leading people to conclude that during this work a primrose root had been inadvertently upturned, thus providing 'proof' for the belief.

The 'primrose path', defined in the *Oxford English Dictionary* as 'a path of pleasure', but often used to imply a way to a rather illusory or uncertain pleasure, seems to have first appeared in Shakespeare's *Hamlet* I. iii: 50 (1602):

> Doe not as some ungracious Pastors doe,
> Shew me the steepe and thorny way to Heauen;

Whilst like a puft and recklesse Libertine
Himselfe the Primrose path of dalliance treads.

Similarly in English traditional songs primroses are frequently associated with courtship.

As I walked out one midsummer morning
For to view the fields and to take the air,
Down by the banks of the sweet prim-e-roses
There I beheld a most lovely fair.[20]

And when we rose from the green mossy bank
In the meadows we wandered away;
I placed my love on a primrosy bank
And I picked a handful of may.[21]

In folk medicine:

[Co. Longford] The primrose . . . cure for yellow JAUNDICE. Boil the roots in water and preserve the liquid in bottle and take a wine glass full each morning.[22]

A gypsy cure for SKIN complaints on the face: take three primrose leaves and boil them in a pint of water, drink the water.[23]

I am 87 years old now. Mother used to make a healing ointment, also a mixture for curing RINGWORM. The ointment was made with pork lard and primrose leaves.[24]

A lady who lived near here had the 'BURN Cure' . . . I don't know the recipe, but primrose was one of the plants used, and she had to dig for the roots in winter-time. It was an ointment made with beef or mutton suet, and was very good. It was used to treat even quite bad cases of burns and SCALDS and was always successful.[25]

Primrose pie seems to have been eaten only by the impoverished.

From my mother who was 93 years old last week: Her mother [on the Isle of Wight] used to cook primroses in a pie, due to low income, and the children (my mother and aunt) would be sent to pick the primroses from the woods/hedgerows.[26]

PRIMROSE-PEERLESS (*Narcissus* x *medioluteus*)

Bulbous perennial, hybrid between bunch-flowered daffodil (*N. tazetta*) and pheasant's eye daffodil (*N. poeticus*), both native to southern Europe; long cultivated as an ornamental, surviving on sites of old gardens, where previously planted or bulbs were discarded; first recorded in the wild in the mid-eighteenth century.

Local names include DAFFY-DOWN-DILLY in Dublin, sweet nancy in West Yorkshire and given to the double-flowered variety in Somerset, and WHIT-SUNDAY or White Sundays in Devon.

Until about 1920, when it was exterminated by heavy grazing, an extensive colony, covering about half an acre, grew in a field between the vicarage and the church at Churchill in North Somerset. An elderly man still living in 1965 could recall how he used to pick bunches of the flowers and sell them at Weston-super-Mare market.[1]

According to Ruth Tongue:

> A Crusader came home to Churchill after years of heat and bloodshed in the Holy Land. He had gone away rich – he came home poor, but he had brought his beloved wife a carefully cherished present – two bulbs of the Primrose Peerless. She had always loved rare flowers.
>
> Alas, when he reached Churchill the primroses on her grave were blooming for the fourth time. In despair he flung the precious bulbs over the churchyard wall and, falling beside his lady's grave, died of a broken heart.
>
> Throughout the centuries the bulbs have grown and flourished and kept his memory alive.[2]

PRIVET (*Ligustrum ovalifolium*)

Evergreen shrub, extensively used for garden hedging (sometimes known simply as 'hedge'), surviving on the sites of old gardens or where garden waste has been dumped; native to Japan, cultivated since the mid-nineteenth century.

The white, strongly scented flowers of privet were often banned from being taken indoors.

> I remember in about 1963, when we were living in South Shields,

that my mother would nor allow privet flowers in the house.[1]

[Lancashire] When I was very much younger, I brought a sprig of flowering privet home when my grandmother was visiting. As soon she saw it she looked horrified and recounted the story of her mother telling her it would bring bad luck after my grandmother herself brought it home as a little girl. Shortly after, their house burnt down (luckily no one was injured)![2]

In London during the first decade of the twentieth century privet was associated with DIPHTHERIA.

When I was in hospital I heard much gossip about various sources, previously thought innocuous, from which you could now 'catch the fever': privet leaves, putting an iron key in your mouth, passing a smelly drain without a handkerchief to clap over your mouth and nose.[3]

Children place privet leaves between their thumbs to produce 'strange noises'.[4]

Records of privet being used in folk medicine are rare.

To cure sore LIPS get some privet leaves and chew them and let the juice flow over the sore lip. The pain will be intense but the lip will undoubtedly get better. None of the juice should be swallowed.[5]

[Wiltshire cure for MUMPS] Put privet berries in a pan with water to cover. Boil till the juice is out of the berries and tip the juice into a small bottle, together with cream from top of the milk. Leave berries in a jar with enough water to cover. Do not take till it has cooled for at least three hours. Dose: teaspoon of juice and one berry, once daily and after food.[6]

PROTECTION – MARSH MARIGOLD and ROWAN provide.

PSORIASIS – NETTLE used to treat.

PUMPKIN (*Cucurbita maxima*)

Large, sprawling annual herb, native and long cultivated in Central America, said to have been introduced to Europe during Tudor times, cultivated for its large, globular, orange fruits.

Until recent decades pumpkins entered British consciousness mainly as providing the fruit which Cinderella's fairy godmother transformed into a golden coach. The popular story-book and pantomime versions of this tale are derived from Charles Perrault's *Histoires ou Contes du temps passé*, first published in 1697.

> Her Godmother, who was a Fairy said, 'You want to go to the Ball, isn't that it?'
>
> 'Yes,' sighed Cinderella.
>
> 'Well, if you're a good girl, I shall send you,' said her Godmother.
>
> She took her into her own room and told her, 'Go into the garden and bring me a pumpkin.'
>
> Cinderella went straight to the garden and picked the finest she could find, and took it to her Godmother, without the least idea how a pumpkin could help her go to the Ball. Her Godmother scooped it out to a hollow skin, then tapped it with her wand, and the pumpkin was instantly turned into a beautiful gilded coach.[1]

More recently pumpkins have become increasingly familiar. From the 1980s onwards HALLOWE'EN, the celebration of which had previously been mainly restricted to Ireland and the northern part of the United Kingdom, has become increasingly popular elsewhere. This extension of the celebration seems to have been the result of the spread of American culture and the opportunity to create another commercially exploitable festival. During the mid and late 1980s pumpkins began to appear in greengrocers' shops in October, and pumpkin lanterns began to be made in homes. Cheap plastic toys, often in the shape of pumpkins and imported from the Far East, appeared in shops. The popular home-makers' magazine *Family Circle* in its November 1990 issue included a 'Hallowe'en Special', which included instructions on how to make a pumpkin cake (i.e. a cake decorated to look like a pumpkin), pumpkin pie, and a pumpkin jack o'lantern. Even when pumpkins are not included in Hallowe'en decorations, orange, the colour of pumpkins, has become one of the three colours – together with ghosts' white and witches' black – associated with the festival.[2]

According to a report in *The Times* of 13 October 2004: 'there are now about five main growers who plant a total of 1,000 acres of pumpkins each year for a business which is worth £25 million a year at retail.' Since that time the production and sale of pumpkins has undoubtedly increased, and as most of the fruits are only used for making

Hallowe'en lanterns, with the flesh being discarded, there is growing concern about the amount of edible material which is wasted. Thus in 2014 children attending Windmill and Larkrise primary schools, in Oxford, shaped spooky faces and used the leftover pulp and seeds to make 'pumpkin scones, roast pumpkin seeds and jewellery out of leftovers'.[3]

The rapidly growing fruits have a fascination which verges on obsession for the enthusiasts. Giant pumpkin competitions attracted competitors in scattered villages during the latter part of the twentieth century, but have become less popular. For many years from 1967 the Greyhound Pumpkin Club, named after a pub in the village of Broughton, in Hampshire, held an annual show in October. The pumpkin show was the climax of a season that commenced with a competition for the heaviest stick of RHUBARB in June to a competition for the longest blanch LEEK in December.[4] Other places which hold, or have held, pumpkin shows include Barrington, Somerset, where 'more than twenty of the bulging vegetables were on show' in 1994,[5] and Haydon Bridge, Northumberland, where, at its annual pumpkin and marrow show in 2002, the winning pumpkin weighed 285lb.[6]

Although many of the smaller pumpkin clubs and shows have died out, giant pumpkins continue to fascinate and are considered newsworthy each autumn. On 12 February 2013 *The Times* reported that the seed company Thompson and Morgan were seeking a grower for a seed of The Freak II, 'the first pumpkin to weigh more than 2,000lbs', for which they had paid €200 (£171).[7] Three years later it was reported that Thompson and Morgan had paid £1,250 for the 'world's most expensive pumpkin seed'.[8] This was cultivated at the Royal Horticultural Society's garden at Hyde Hall, in Essex, and in December the Society was able to proudly announce that it had produced the 'UK's heaviest out-door grown pumpkin', which weighed 605kg (1,333lb 13oz) and had a circumference of 4.5m (15ft).[9] Since 1998 the annual weigh-off of giant pumpkins has taken place at the Jubilee Sailing Trust Autumn Pumpkin Festival, at Netley, near Southampton, Hampshire, on a Saturday early in October. Here, on 14 October 2017, Ian and Stuart Paton of Lymington broke the British record by producing a 2269.4 lb (1029.38 kg) pumpkin.[10]

PURGATIVE – ALDER BUCKTHORN, FAIRY FLAX and GROUNDSEL used as.

PURPLE MOOR-GRASS (*Molinia caerulea*)

Tufted perennial grass, widespread, and sometimes dominant on damp heaths and moors, and in open woodland.

Local names include:

Blawing garss in Lanarkshire
Flying bent in Northumberland and southern Scotland
Kabe, and KEER, in Co. Donegal.

> Fishermen in the isle of Skie make ROPES for their nets of this grass, which they find by experience will bear water well without rotting.[1]

> Purple moor-grass: referred to as disco grass in Rhayader, Radnor area. When you walk through the tussocks you stagger about and look as if in a disco, I presume. Also still cut as horse hay in mid and west Wales in September.[2]

Q

QUAKING GRASS (*Briza media*)

Attractive small perennial grass, common in unimproved grassland, particularly on base-rich soils.

Local names include the widespread LADY'S HAIR, QUAKERS and totter-grass, and:

Beetle-grass in Suffolk
COW-QUAKES in Berwickshire, Derbyshire, Shropshire and Somerset
Datheron-duck in Scotland
Dawther in Kent; 'i.e. dodder'
Didder-grass in Lancashire
Diddery-docks in Co. Durham
Dithering grass in Lancashire; 'dithering is a North Country word for shivering'
Dithery-dicks in Yorkshire
Dithery-dother in Northumberland
Dodder-grass in Cumberland and Kent; [dodder] 'has the same meaning as dother or dither, words of northern dialects used to denote trembling'
Doddering dickies in north-east England
Doddering dillies in Co. Durham, Shropshire and Montgomeryshire
Doddering grass in Lancaster, Cumberland and Co. Durham
Dodderin' grass, and dodderin' Nancy, in Cumberland
Doddering jockies in Yorkshire
Doddle-grass in Somerset and Wiltshire
Dothering dickies in Northumberland
Dothering dicks in northern England
Dothering docks in Berwickshire
Dothering ducks in northern England
Dothering grass in west Lancashire
Dotherin' grass ('dothering means trembling'), and dotherin' Nancy in Cumberland
Dudder-grass in Norfolk
Earthquakes in Northamptonshire
Fairy-grass in Co. Donegal, Co. Limerick and Dublin
Flax-seed in Dublin
Hay-shakers in Cheshire
Horse-and-chariots in Somerset
Ladies' hair in Berwickshire
Ladies' hand in Northumberland
Lady-shakes in Yorkshire

MAIDENHAIR in Suffolk
MAIDEN'S HAIR on the Isle of Wight and in Norfolk
MEADOW-FLAX in Kirkcudbrightshire; 'where it is used medicinally'
OOMAN'S TONGUE in Berkshire; 'because of the motion caused by the lightest breeze, so they are always on the move'
Quack-ducks in Hertfordshire
Quaker-grass in Northamptonshire, Worcestershire and Yorkshire
Quakers-and-shakers in Cheshire
Quiver-grass, 'presumably Hampshire'
RATTLE-BASKETS in Somerset
RATTLE-GRASS on the Isle of Wight
Ring-all-the-bells-of-London, 'said to have been often heard in England and Ireland'
Shak-and-trammel in Scotland
Shake-ladies in Lancashire
Shakers in Cheshire, Shropshire and Wiltshire
Shakie-tremlie in Forfarshire
Shaking grass in Buckinghamshire, Devon, Shropshire and Somerset
Shakky-trem'el in Morayshire
Shaky-grass in Buckinghamshire, and Dublin
Shickle-shackles in Somerset
Shiver-grass in Hampshire
Shivering grass in Warwickshire
Shivering Jimmy in Sussex
Shiver-shakes in Somerset
Shivery bibs in Berkshire
Shivery grass in Dublin
Shivery shakeries, or shivery

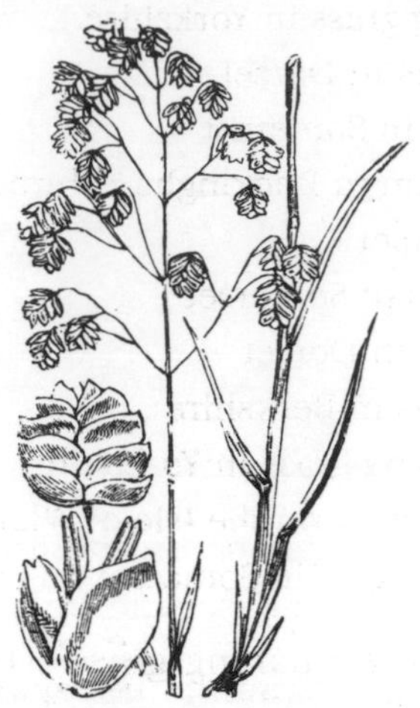

shakes, in Somerset and Wiltshire
Sillar shakle, or siller tassels, in southern Scotland
SILVER BELLS in Devon
Silver ginglers in Roxburghshire
Silver shakers in Dumfriesshire, Kirkcudbrightshire and Roxburghshire
Silver shekels in Somerset
Suisilk in Nottinghamshire
Swaggering grass in Lancashire
Thrimlin jockies in Yorkshire
Toad-shakers in Somerset
Totter-bobs in Lincolnshire
Tottering grass in Cambridgeshire and Yorkshire
Totter-robin in south-west Lincolnshire
Tottery grass in north Hampshire
Totty grass in Surrey
Trembling grass in Lancashire, Suffolk, Yorkshire and Berwickshire
Trembling jock in Yorkshire
Wag-a-wams, and wag-a-wands in Somerset

Waggering grass in Yorkshire
Wag-wafers in Dorset
Wag-want in Somerset
Wag-wanton in Buckinghamshire and Dorset
Wag-winds in Somerset
Wagwonts in Dorset
Water-wags in Berkshire
Whacker/ing-grass in Yorkshire
Wibbly-wobbly on the Isle of Wight
Wigger-waggers in Somerset
Wiggle-waggle on the Isle of Wight
Wiggle-waggle-wantons in Berkshire
Wiggle-wantoms in Somerset
Wiggle-wants in Wiltshire
Wiggle-woggles on Portland, Dorset, where considered 'unlucky to have in the house'
Wiggy wantons in Somerset
Wig-wag-wanton in Buckinghamshire.

In Yorkshire quaking grass – trimmling jock – was believed to deter MICE:

> A trimmling jock i' t' house
> And you weeant hev a mouse.

> Dried in bunches, with its brown seeds on a tall stem, it was commonly stuck on the mantel-shelf, as believed to be obnoxious to mice.[1]

In the Cambridgeshire Fens:

> It was an old belief that it [maidenhair or quaking grass] grew only in places where a young woman had drowned herself – usually on account of some unhappy love affair. In W.H. Barrett's [born 1891] boyhood the old people used to declare that they noticed maidenhair grass flourished particularly well wherever a corpse, dragged from the river, had been laid on the bank. He himself noticed, when minding cows as a young boy, that the beasts would never eat or trample on maidenhair grass.[2]

The mysterious plant known in Ireland as HUNGRY GRASS, *fairgurtha* or *fairgorta*, has sometimes been identified as quaking grass:

> *Fear gorta(ch)*: quaking grass, a mountain grass supposed to cause hunger-weakness when trodden on, hence a violent hunger, abnormal craving for food, diabetes . . . it is prevented by carrying food and cured by a grain of OATS . . . it occurs on Mt Brandon and in mountains near Omeath between two cairns which peasants will not pass without carrying food.[3]

QUINSY – SELFHEAL and TORMENTIL cure.

R

RABBITS, pet – fed on DANDELIONS; when sick fed on SHEPHERD'S PURSE

RADISH (*Raphanus sativus*)

Annual herb of unknown, assumed Mediterranean, origin, cultivated in Britain and Ireland since at least the tenth century for its red-skinned fleshy root, a popular salad vegetable.

> A curious annual custom, dating from time immemorial, was celebrated at Levens Hall . . . near Kendal . . . known as the Radish-feast . . . [it] is attended by the mayor and corporation of Kendal and most of the gentry . . . who partake of radishes and oatbread and butter . . . and . . . athletic competitions.[1]

This event was formerly held on 12 May when the nearby village of Milnthorpe held its fair. The right to hold the fair was granted in 1280, and it is believed that the feast dates from late in the seventeenth century, when it was said that the owner of Levens Hall wanted an entertainment which would attract greater attention than that provided by his neighbour at Dallam Tower.[2]

> After the repast came the 'colting' of new visitors. The neophytes were brought into a ring, or 'haltered' . . . when they were required in turn to stand upon one leg and drink what was called 'constable', a weighty glass filled with half a pint of 'Morocco' [a strong ale brewed at the Hall]. This had to be emptied at one draught, pledging at the time the ancient house in the words 'Luck to Levens as long as the Kent flows'. Declining or failing to accomplish this feat, the forfeiture of a shilling was required for the benefit of the under gardeners, and well they deserved it, for it is said that it took a full day's work for four men to clean the radishes eaten at this festival, which were conveyed to the tables in wheelbarrow loads.[3]

It appears that the Feast ceased 'well before' the Second World War'.[4] Elsewhere radish feasts were associated with parish elections. In Oxfordshire:

> The annual meeting for the election of churchwardens . . . will be held in the vestry of the Parish Church on Easter Tuesday . . . The Radish-feast will be held at the New Inn, New Street, immediately after.[5]

At Andover, in Hampshire:

> The [radish] feast was usually held on the day of the election of officers, and the person who supplied it was chosen by ballot.[6]

RAGWORT (*Senecio jacobaea*, also known as *Jacobaea vulgaris*), formerly known as ragweed

Perennial herb with showy yellow flowers, common in grassy places.

Local names include:

Agreen in Cumberland
Balcairean in Ireland
Beaweed in Scotland
Bennel in Ireland
BINDWEED, and binweed, in Scotland
BIRDSEED in Dublin
Bogluss bwee, in Co. Donegal; 'bogluss, i.e. bugloss, a name for several widely different plants with rough leaves'
Boholawn in Ireland; 'Irish *buadhghallan*'
Booin, bouin, and bowens, in Cumberland
Bowlochs in Wigtownshire
BOWLOCKS in Scotland; 'Gaelic *buaghallan*'
Buchalan in Co. Tipperary, 'a shortening of the Gaelic *buchalan bui*, or yellow boy'
Buckles-bwee in Co. Donegal
BUNDWEED in Suffolk and Scotland
Bunkieshaun in Co. Tyrone
BUNNEL, and BUNNLE, in Upper Clydesdale
Canker-weed in East Anglia; 'applied in CANCER'

Cheedle-dock in Cheshire
Cow-foot in Shropshire
Cradle-dock in Cheshire
CROWFOOT in Shropshire
Curly doddies in Scotland
Cushag on the Isle of Man
Daisies in Somerset
Devildums in Dorset
Dog-stalk in Yorkshire
Dog-standard/s, or dog-standers, in Worcestershire and northern England
Ellshinders in Berwickshire
Fairy-horse in Ireland, or fairy's horse, in Co. Donegal, 'well-known to be the plant on which FAIRIES ride through the air at night'
Feanut in Lancashire
Fizz-gigs in Berwickshire
Flea-nit, or flea-nut, in Cheshire
Fleenurt in Lancashire
Fly-dod in Cheshire; 'generally covered with a dusky yellow fly, which accounts for the first part of the name'
Goosegan in Co. Clare, 'pronounced goosegawn, probably a corruption of the Gaelic *geosadan*'
GOWAN in Scotland; 'historically sometimes, but name normally applied to CORN MARIGOLD'
GRUNSEL, and grundswaith, or grundswathe, in Cumberland
GULD in Scotland; 'historically sometimes'
GYPSY in Somerset
HARVEST FLOWER in Cornwall
Haygreen in Devon
James's weed, and James's wort, in Shropshire
KADLE-DOCK in Cheshire
KEDLOCK in Lancashire
Keedle-dock in Cheshire and Lancashire
KETTLE-DOCK, and mare-fart, in Cheshire
Mare's fart in Cheshire and Shropshire
MUGGERT, and muggets, in Cumberland
RAGGED JACK in Devon and Yorkshire
Scattle-dock in Lancashire
Scrape-clean in Lincolnshire
Seggrom, seggrums, and SEGGY, in Yorkshire
Sleepy dose in Banffshire
Stinking alisander in Northumberland and Stirlingshire
STINKING BILLY in Lincolnshire
Stinking Davie in Fifeshire
Stinking Nanny in Nottinghamshire
Stinking weed in southern Scotland
SUMMER FAREWELL in Devon and Gloucestershire
TANSY, weebie, and WEEBO, in Scotland
WILD CHRYSANTHEMUM in Devon
Yack-yard in Lincolnshire
Yallers in Somerset
Yark-rod in Lincolnshire
Yellow boy in Co. Donegal
Yellow daisy in Dorset
Yellow ell-shinders in Berwickshire
YELLOW TOP in Northumberland
YELLOW WEED in Oxfordshire and Berwickshire.

In Scotland there was a belief that WITCHES and fairies travelled on ragwort stalks.

> Tell how wi' you on ragweed nags
> They skim the muirs an' dizzy crags.[1]
>
> On auld broom-besoms, and ragweed naigs,
> They flew owre burns, hills, and craigs.[2]

Elsewhere:

> It is regarded as UNLUCKY to take into your hands the ragwort plant, also known as *boholàun*. In rural Ireland you were likely to be reminded not to strike a beast with a *boholàun* as this action would bring misfortune to the beast. There is an old saying around here: 'Don't call it a weed though a weed it may be, 'tis the horse of the fairies, the *boholàun buidhe*.'[3]
>
> I come from the Isle of Man. My father told me that ragwort was a magical plant; if you pulled it you had to apologise for it, or else the fairies might get you. He wasn't a Manxman himself, but he might have got it from an old Manxman.[4]

Also on the Isle:

> Manx National Flower – *Cushag* (Manx Gaelic), *Senecio jacobaea*, ragwort, dog's standards – no great antecedents. Reputedly a sarcastic Victorian Governor-General said it must be, there is so much of it in the fields.[5]

After the Battle of Culloden in 1746 the victorious English are said to have renamed the attractive garden flower SWEET WILLIAM in honour of their leader, William, Duke of Cumberland. The defeated Scots retaliated by naming the obnoxious weed ragwort Stinking William.[6]

Alternatively:

> In my native Highlands masses of the common ragwort grows in fields . . . it is usually referred to in the North as STINKING WILLIE, partly on account of its unpleasant smell, but more for the fact that it sprang up everywhere that William, Duke of Cumberland, otherwise known as the Butcher, had been when he perpetrated the massacre after the Battle of Culloden in 1746, and Stinking Willie it has remained. The seeds were supposed to have come from the fodder provided for the Butcher's horses.[7]

In the Hebrides ragwort was used in the children's game *Goid a' Chruin*, recorded in Gaelic in 1953:

> We played it during the interval at school when I was a little boy. You had always to be on a level piece of ground . . . we had an excellent stretch of green sward a short distance from the school, and we gathered there. And it was always in autumn that this game was played because we had a sort of plant that was very common then, we call it *buaghallan* in Gaelic. And that was plucked out by the roots, and two rows were set up on the green; boys and girls were in each row. *An Crun* [the Crown] was one of those *buaghallans*, one of those plants. It was thrown there on neutral territory between the two ranks and the person who was successful in breaking through and taking up the crown and right down . . . round, circling their opponents, and right round and coming back, without being caught, that person was successful. But he always had to give the next, his companion, a chance, and he himself just looked on. But if he was caught with the *buaghallan*, with the crown, by his opponents he had to go outside the camp, and he was called a *cnoimhags* [maggot!]. Now the side that had more failures (or more *cnoimhags*) was the side that lost, and the row that had less, of course, was the one that was successful.[8]

RAIN – WHITEBEAM and WHITE POPLAR predict.

RAMSONS (*Allium ursinum*), also known as wild garlic

Bulbous herb, strongly smelling of garlic and producing attractive white flowers in late spring, common and conspicuous in dampish woodlands throughout much of Britain and Ireland.

Local names include:

Badger's flowers in Wiltshire
Brandy-bottles in Dorset
Devil's posy in Shropshire; 'cf. *Tufelschnoblech*, devil's garlic, Switzerland
Gypsy onion in southern England
Gypsy's gibbles in Somerset
Gypsy's onions in southern England
IRON-FLOWER in Somerset
MAY-FLOWER in Lancashire
Moly in Devon and Somerset
Onion-stinkers in Somerset
Rames on the Gower Peninsula
Ramp in Dorset and Somerset; also RAMPS in northern England, Scotland and Ireland
Ramsden on the Isle of Wight

RAMSEY in Cornwall, Devon and Norfolk
Ram's horns in Gloucestershire
Rams in Yorkshire
Rommy, and roms, in Yorkshire
Rosems in Staffordshire and Yorkshire
Sives in Dorset
SNAKE-FLOWER in Dorset and Somerset
Snake-plant, and SNAKE'S FOOD, in Somerset
Stink-plant in Lincolnshire
Stinking Jenny, stinking lilies, and water-leek, in Somerset
Wild leek in Berwickshire.

Despite mention by writers on foraging for wild foods,[1] and the consequent sale of ramsons in farmers' markets and similar places,[2] there appears to be little evidence that it was gathered as food in the British Isles.

> According to my mother (b.1918) of Whitstable, Kent, the little new buds of ramsons are delicious to eat.[3]

More widespread was the use of ramsons in folk medicine. According to a seventeenth-century proverb:

> Eat LEEKS in Lide [March] and ramsins in May,
> And all the yeare after physitians may play.[4]

> [Reminiscences of her aunt, licensee of the Halfway House, between Douglas and Peel, Isle of Man.] In the 1930s the pub had a garden/orchard. *Allium ursinum* grew at one end of the orchard. Bulbs were dug when they first sprouted in spring, washed and dried in the sun on a clean tea-towel. Then packed into a wide-mouthed glass jar with dark brown sugar. Light Jamaica rum was poured over and the whole was stored in a cool dark place – e.g. wardrobe – until the following winter, when it was used for chesty colds and COUGHS.[5]

> Ramsons: leaves soaked in milk for bad tummy.[6]

My father always swore by wild garlic to treat nettle STINGS . . . that was in North Wales in the last 10–15 years.[7]

I know a horseman locally who fed the leaves of *Allium ursinum* to his ponies to expel parasitic WORMS; whether it actually got rid of them or just gave the ponies a boost generally that masked the worm burden, I can't tell.[8]

RASHES – treated using BROAD BEAN, CHICKWEED, COMFREY, NETTLE and WATERCRESS.

RASPBERRY (*Rubus idaeus*)

Erect spiny shrub, with white flowers and fleshy red fruits, native in wood margins, hedge-banks, where it rarely produces a worthwhile amount of fruit, also widely cultivated for its fruit, and escaping into the wild.

Raspberry leaves were valued to treat gynaecological conditions.

Raspberry leaves for easing labour pains. A preparation made from these leaves is sold in the Bradford herbalists' shops under the title 'Mother's Friend'.[1]

Raspberry leaf made a great muscle relaxant. I used to take it for period pains – it worked better than paracetamol or any conventional analgesic.[2]

Raspberry leaves for a tea to help with symptoms of menopause, from paternal grandfather, approximately 1958, Oxfordshire.

[Sheffield] Raspberry leaf tea: 1 oz. raspberry leaves (wild or cultivated) to two cups of water.

Dose: 2 cups a day during the last month of PREGNANCY to give an easy delivery.[3]

Also mentioned in the Sheffield area for preventing MISCARRIAGE, increasing mother's milk, for painful menstruation, DIARRHOEA, and as an eye-wash for sore EYES. After being soaked to the skin, to avoid a COLD take raspberry leaf tea and go to bed.[4]

Many years ago when I kept dairy goats . . . at kidding time I fed them raspberry leaves to ease birthing.[5]

RATS – HOUND'S TONGUE and WHITE BRYONY deter.

RED AND WHITE FLOWERS

The belief that red and white flowers together in a vase cause misfortune or DEATH is widespread, but apparently not very old. According to the *Sunlight Almanac* (1896) dreaming of red and white flowers foretold death,[1] but the belief in its current form seems to have evolved in the first half of the twentieth century.

> I have just retired but started nursing during the war. I found that red and white flowers in the same vase made some patients uneasy; they would mutter 'Red and white, someone will die.' If the colours were separated into a vase of red and one of white this was acceptable.[2]

> The family of a friend of mine went to visit their mother in hospital in Galashiels. When they got there they were very upset to see a vase of red and white flowers placed on the locker beside her bed. They complained to the sister, and told her that they would rather be told that their mother was going to die, instead of having them dropping hints. They thought that the red and white flowers indicated that the hospital had given up hope with their mother.[3]

> I was a nurse in the 60s and was told, on pain of being dismissed, never put white and red flowers together as it signified death with the blood and bandages. I was amazed that such a superstition was upheld in a hospital, but I obeyed the rules. Still do![4]

> I spent most of my working life as a regional tutor in the NHS covering all the specialities of medicine from dentistry to chiropody in hospitals all over NE Thames Region. Back in the 1970s, all bouquets of red and white cut flowers were quickly discarded by nurses. Apparently, having red and white flowers on display in the hospital wards was a sign of bad luck and superstitious older folk associated the red and white blooms with an imminent death; something to do with blood and bandages during the First World War, I believe. By the 1980s all cut flowers had been banned from bedside tables on grounds of hygiene, although flowering pot plants were sometimes tolerated, so long as one avoided arrangements of red and white cyclamen![5]

However, red and white flowers are not always considered to be inauspicious:

At WHITSUN many Anglican churches decorate with red and white flowers – the symbolism of the fire and the wind of the Holy Spirit.[6]

Red and white are the colours of the City [of London], so when there are events at the Guildhall there are always masses of red and white flowers.[7]

RED CAMPION (*Silene dioica*)

Short-lived perennial herb, with pale pink to bright red flowers, common, mostly in semi-shaded places, but also on exposed cliff-tops in Britain, less frequent in Ireland.

Local names include the widespread BACHELOR'S BUTTONS and CUCKOO-FLOWER, and:

ADAM'S FLANNEL in Oxfordshire
Adder's flowers in Devon, Hertfordshire and Somerset
Billy-buttons in Essex
BIRD'S EYE in Devon and Somerset
Bobby-hood in Somerset; 'more common Robin Hood'
Bob-robin in Cornwall, Somerset and Wiltshire
Brid een in Cheshire; 'i.e. birds' eyes'
BULL-RATTLE in Buckinghamshire
BULL'S EYES in Devon and Somerset
Butcher's blood in Somerset
Cancer in Scotland
Cock-robin in Cornwall, Devon and Somerset
CROW-FLOWERS in Somerset
CUCKOO in Devon, Nottinghamshire and Somerset

CUCKOO-PINT in Northamptonshire and Somerset
DEVIL'S FLOWER 'about Liverpool'
Dolly-winter in Cornwall
DRUNKARDS in Somerset

FINGERS-AND-THUMBS in Somerset
Flea-bites in Cornwall
GEUKY-FLOWER in Devon and Somerset; 'i.e. cuckoo-flower'
God's stocking in Lancashire
Goose-and-gander in Somerset
Gramfer (or granfer) Jan in Dorset, Somerset and Wiltshire
Gramfer-greygles, 'sometimes distinguished as red gramfer-greygles', and GRANFER-GRIGGLES, in Dorset
GYPSY-FLOWER in Somerset
HARD-HEAD in Lancashire
HEADACHES in Cumberland
JACK-BY-THE-HEDGE in Sussex
JACK-IN-THE-HEDGE in Somerset and Sussex
JACK-IN-THE-LANTERN in Dorset
Jandy-crowders in Devon; 'jandy might be intended for Johnny, while crowder means a fiddle'
Jan granfer in Somerset and Wiltshire
Johnny-woods in Dorset
KETTLE-SMOCKS in Somerset
KISS-ME-QUICK, 'sometimes is added "for mother's coming!"', in Lancashire
Large robin's-eyes in Gloucestershire
Little-red-riding-hood in Devon and Somerset
Lousy beds in Cumberland
Lousy betty in Lancashire
Lousy soldier's-buttons in Lancashire; 'an . . . old nurse says she would not have gathered one, as a child, for anything, because they are so often covered with small insects'
Mary Janes in Somerset
Mintdrop in Northumberland
Plum-pudding in Essex, Somerset and Suffolk
POOR JANE in Somerset
POOR ROBIN in Cornwall, Devon and Somerset
Puddens in Somerset
Rabbit's rose in Devon
RAGGED JACK in Dorset
RAGGED ROBIN in Somerset; 'more usually . . . known as Robin Hood'
Red bachelor's-buttons in Suffolk
RED BIRD'S-EYE in Radnorshire
RED BUTCHER in Gloucestershire
Red Jack in Cheshire
Red Jane in Somerset
Red mintdrops in Northumberland
RED RIDING-HOOD in Devon, Dorset and Somerset
RED ROBIN in Cornwall, Devon, Dorset, Somerset and Wiltshire
RED SOLDIERS in northern England
Red wolf in Somerset; 'very commonly called red riding-hood, and this is possibly the explanation of the wolf'
ROBIN in Devon and on the Isle of Wight
Robin- (or robin's) flower in Cornwall and Devon
ROBIN HOOD in Devon, Dorset and Somerset
Robin-i'-the-hedge in Yorkshire
ROBIN REDBREAST in Cornwall and Devon

ROBIN-RUN-IN-THE-HEDGE in Dorset
ROBIN'S EYE in Devon and Somerset
Robins in Somerset; 'more often called Robin Hood'
Rob roys in Somerset
Rose-campion in Gloucestershire and Yorkshire
ROUND-ROBIN in Devon and Kent, 'to distinguish it from the ragged robin [*Silene flos-cuculi*]'
Sarah Janes in Devon
Scalded apple in Shropshire
SOLDIERS in Cheshire and northern England
SOLDIER'S BUTTONS in Lancashire
Soldier's fleas in Guernsey
SWEET WILLIAM in Lancashire, and Shetland, 'where the flowers are larger than usual and often a deep magenta' [sometimes distinguished as a separate subspecies, ssp. *zetlandica*]
Tidy robins on the Isle of Wight
WAKE-ROBIN in Yorkshire
Water-poppies in Lincolnshire
Wild geranium in Cheshire.

In nineteenth-century Cumberland red campion was known as mother-dee (i.e. mother-die):

> There is a superstition amongst Cumberland children that if they pick the flower some misfortune will happen to their parents.[1]

Similar beliefs persisted in the area well into the following century. The name fadder-dies was recorded in 1937,[2] and:

> Egremont, Cumbria, 1960s-70s – pink campion known as mother-and-father-die. We as children never picked them.[3]

More recently red campion has been associated with FAIRIES, SNAKES and THUNDER.

> [Isle of Man] red campion, being the fairies' flower, *Blaa ny Ferrishn*, was UNLUCKY and should not be picked.[4]

> *Blodwyn Neidr* (snake flower) – red campion – used to be one of my favourite flowers, but my grandmother was convinced that I would be attacked by a snake if I brought it into the house.[5]

> In Cnwch Coch, near Aberystwyth, about 1921–31, we called red campion *Blod Trane* (*Blodyn Taranau*: Thunder Flower) as it was believed that a storm of THUNDER and LIGHTNING would follow if it was picked.[6]

RED CLOVER (*Trifolium pratense*)

Native perennial herb with pinkish-red flowers, common in grassy places; also widely cultivated for fodder or as a green manure.

Local names include the widespread CLAVER, COW-GRASS and HONEY-SUCK, and:

Bee-bread in Kent
Broad-clover in Dorset and Isle of Wight
BROAD-GRASS in Dorset
Clover-rose in Devon
Cow-cloos in northern Scotland
Fairy-pops in Dorset; 'pops = sweets . . . refer[s] to the sweetness which children extract by sucking the flowers'
HONEY-STALKS in Somerset
Honey-suckers in Somerset
HONEYSUCKLE in Dublin and Co. Tyrone
Horse-shamrock in Dublin
King's crown in Somerset
Knap in Dorset
Lady's posies, and little honeysuckle, in Somerset
Marl-grass in Somerset and Wiltshire
Plyvens, given to the flowers, in Scotland
Red cushions in Somerset
Sickers in Northamptonshire
Sleeping Maggie in Northumberland
SOOKIES, soukie clover, soukie soo, soukies, and souks, in Scotland
SUCK-BOTTLE, given to the flowers, in Northamptonshire
SUCKLERS, given to the flowerheads, in Northumberland
Suckling in East Anglia
Sugar-bosses, or sugar-busses, in Somerset
Sugar plums in Buckinghamshire.

As many of these names suggest, red clover flowers were widely sucked by children to extract their nectar.

> Red clover: sweet flowers picked by us as children [I am now 43], called honey-suck.[1]

> Red clover . . . I'm glad to see that children still take the flower-heads of this plant, tug out the flowers and suck the nectar from their base.[2]

> [Easington Colliery, Co. Durham, 1970s] As children we would pick pink clover flowers and pull the little petals and suck the nectar from the bottom of the petals. The sweetness was more imaginary than real, but it was a common thing to do.[3]

In 1988 7 per cent of the specimens obtained in response to an appeal for examples of 'true SHAMROCK' were found to be red clover.[4]

> Montgomery Red Clover, selected from the wild in Mont., was important commercially as a fodder plant for hill cattle before the days of plant breeding; a Montgomeryshire Red Clover Society was formed in 1920, and the strain is still grown in New Zealand, where it is known as 'Monty Red' and from whence it is exported to Europe.[5]

RED DEADNETTLE (*Lamium purpureum*)

Small, annual herb with pinkish purple flowers, common and widespread on disturbed and waste ground.

Local names include:

Bad-man's posey in Cumberland
Bad-man's posies in Northumberland
BEE-NETTLE in Somerset
Blackman's posies in Cumberland
Bumble-bee flower in Somerset
DEAF-NETTLE in Yorkshire
DEE-NETTLE in Cumberland, Worcestershire and Yorkshire
Dog-nettle in Cheshire
DUMB-NETTLE in Somerset
DUNCH-NETTLE in Dorset and Wiltshire
French nettle in Shropshire
LAMB'S EARS in Somerset
LAMB'S TONGUE in Caithness
Purple dead-nettle, and RABBIT-MEAT, in Shropshire
Red archangel in Lancashire
TORMENTIL in Shropshire.

Although there are numerous records of children sucking nectar from flowers of WHITE DEADNETTLE, it seems as if they rarely bothered with red deadnettle's smaller flowers.

> [South Yorkshire, 1950s and 60s] sweet nettle – mainly pink – remove each little floret and suck out a tiny bit of sweetness (nectar).[1]

In County Meath:

> When children have MEASLES they used to boil the red deadnettle roots in sweet milk and give it to them to drink so as to bring out the measles.[2]

REDSHANK (*Persicaria maculosa*), also known as spotted persicaria

Herb, with reddish stems, pink flowers, and in exposed situations with reddish-brown blotches on its leaves; common and widespread on disturbed and waste ground.

Local names include:
Alice in Norfolk
Arse-smart in Lincolnshire and Wiltshire
Crab's claw in Dorset and Somerset
Devil's pinch/es in Dorset
FAT HEN in Somerset
Flooering soorik in Shetland
LAMBS TONGUES in Devon
LAVENDER in Somerset
Lover's pride in Sussex
Pig-grass in Lincolnshire
PINCUSHION in Somerset
Red joints in Dorset
Red knees in Cheshire and Lancashire
Red legs in Somerset
Red weed in Cheshire
Sauch weed in Ayrshire; 'sauch = willow'
Saucy Alice, and smart-weed, in Norfolk
Stone-weed in Somerset and Wiltshire
WILLOW-WEED in Yorkshire
Yellowin girse in Shetland.

Various legends told how redshank leaves acquired their characteristic reddish markings. In Gaelic-speaking areas of Scotland the plant was known as *Am boinnefola* (the blood-spot) or *Lus chrann ceusaidh* (herb of the tree of CRUCIFIXION); 'the legend being that this plant grew at the foot of the Cross'.[1]

> A few old folk may still be found in west Cornwall who believe that the spotted persicaria grew at the foot of the Cross and that the dark markings on the leaves are due to drops of BLOOD falling from our Saviour's body.[2]

> [Co. Cork] There is a flower called the redshank and it is said that it got its colour when Our Saviour was dying on the cross. That flower was near the cross and a drop of blood fell on it and that is how it got its name and colour.[3]

Elsewhere redshank was associated with ST MARY THE VIRGIN, an unknown murderess, or the DEVIL.

> The Oxonian, however, says that the Virgin was wont of old to use its leaves for the manufacture of a valuable ointment, but that on one occasion she sought it in vain. Finding it afterwards, when the need had passed away, she condemned it, and gave it the rank of an ordinary weed. This is expressed in the local rhyme:

> She could not find in time of need,
> And so she pinched it for a weed.
>
> The mark on the leaf is the impress of the Virgin's finger, and persicaria is now the only weed that is not useful for something.[4]

Presumably the names useless recorded from Scotland, pinchweed recorded from Northumberland and Oxfordshire, and Virgin Mary's pinch recorded in Berkshire, refer to this legend.

> *Herbe traitresse* . . . this name originates in a Guernsey legend to the effect that a woman who had committed a murder wiped her blood-stained fingers on the leaves of the plant, which betrayed her, and led to her detection. Ever since then the leaves have been marked in the centre by a dark spot.[5]

> Talking to a man from Norfolk . . . he told me that in East Anglia they'd known it [redshank] as devil's arse-wipe.[6]

> About 10 years ago at the Royal Welsh Show I was manning a small plot of arable weeds . . . when I noticed an elderly farmer inspecting the crop. I struck up a conversation . . . as he was going he turned and said to me pointing at redshank with a smile, 'Do you know what we used to call that one then?' I said no, and he replied 'Arsewipe' (because of the brown blotch on the leaf).[7]

RED VALERIAN (*Centranthus ruber*), known in older wildflower books as spur-valerian

Perennial herb (Fig. 40), producing flamboyant flowers in a variety of pinks and red, and white, native to the Mediterranean area, grown as an ornamental in British and Irish gardens since late in the sixteenth century and first recorded in the wild in 1763, now widely naturalised on sea-cliffs, consolidated shingle, walls and quarries, and apparently adapting to other, less well-drained, habitats.

It has been suggested that although red valerian spread rapidlyknowledge of its name spread more slowly, thus necessitating the invention of new names.[1] Local names include:

American lilac in Devon
Bouncing Bess in Devon and Dorset
Bouncing Betsy in Devon and Dorset
Bovisand soldier, the plant being 'very abundant' in Bovisand, in Devon
Cat-bed in Lincolnshire

TOP 26. Juniper, Easedale Tarn, Cumbria

ABOVE LEFT 27. Laurel wreath on Holocaust Memorial, Peterborough, Cambridgeshire, 30 January 2016

RIGHT 28. Lungwort, Beverley, East Riding of Yorkshire

ABOVE 29. Marsh marigold, South West Coast Path, Cornwall

ABOVE RIGHT 30. Meadowsweet, Burwash, East Sussex

LEFT 31. Mexican fleabane, Salisbury, Wiltshire

ABOVE 32. Money tree, South London Botanical Institute, Lambeth, London

ABOVE 33. Mouse-ear hawkweed, Chard, Somerset

RIGHT 34. Navelwort, Bideford, Devon

BELOW 35. Oak wreath decorating statue of King Charles II, All Saints Church, Northampton, 29 May 2015

TOP 36. Oak Apple Day, Salisbury Cathedral, 29 May 2017

LEFT 37. Pennyroyal, Royal College of Physicians Garden, Regent's Park, Camden, London

ABOVE 38. Pineapple-weed, Burgess Park, Southwark, London

TOP 39. Primrose, Gillingham, Dorset

TOP RIGHT 40. Red valerian, Crewkerne, Somerset

ABOVE 41. Ribwort plantain, Bedford

RIGHT 42. Rosebay willowherb, Stoney Middleton, Derbyshire

ABOVE LEFT
43. Royal fern, Grasmere, Cumbria

ABOVE RIGHT
44. Rye grass, Brixton, London

LEFT 45. Shamrock, card posted in Kentish Town, London, St Patrick's Day, 17 March 2013

OPPOSITE PAGE ABOVE LEFT
46. Snowdrop, Wandsworth Common, London

OPPOSITE PAGE ABOVE RIGHT
47. Spear thistle, Edensor, Derbyshire

OPPOSITE PAGE LEFT
48. St John's wort, Chatham, Kent

OPPOSITE PAGE RIGHT
49. White bryony, Sutton, Essex

LEFT 50. White deadnettle, Gillingham, Dorset

ABOVE 51. White heather *(Calluna vulgaris)*, card posted in Elgin, 1911

BELOW LEFT 52. Yarrow, Northampton

BELOW 53. Yellow corydalis, York

Cat's wee in Devon
Convict grass on Portland, Dorset
Delicate Bess, given to the white-flowered form, in Devon
DRUNKARDS, 'because the flower-heads sway about in the wind'
Drunkard's nose in Somerset
Drunken sailor in Devon
Drunken willies in Somerset
Drunken willy, in Devon
Drunkits in Somerset
Fox's brush in Huntingdonshire, Somerset and Northern Ireland
German laylock, 'i.e. lilac', in Lincolnshire
GOOD NEIGHBOURHOOD in Gloucestershire, Oxfordshire and Wiltshire
Good neighbours in Gloucestershire, Oxfordshire and Wiltshire, and 'commonly used' in west Somerset
Ground laylock in Lincolnshire
Gypsy maids, in Martock, Somerset, but 'more commonly known in that part of the county as kiss-me-quick'
Kiss-behind-the-pantry-door in Somerset
Kissing-kind on Portland, Dorset
KISS-ME, and KISS-ME-LOVE, in Devon
KISS-ME-QUICK in Devon, Somerset, Wiltshire and on Portland, Dorset
Kiss-me-quick-mother's-coming in Devon
Kiss the garden-door, but 'more commonly known as kiss-me-quick' in mid Dorset
Ladies' needlework in Cornwall, and, as LADY'S NEEDLEWORK, in Somerset and Worcestershire
Maids-of-Fowey in Cornwall
MIDSUMMER MEN in Somerset
Modest Mary in Buckinghamshire
Old-woman's needlework in Somerset
Pins' heads in Somerset
Pretty baby in Co. Donegal
Pretty Betsy in Dorset, Essex, Sussex, Co. Derry and Co. Donegal
PRETTY BETTY in Sussex
Pride-of-Fowey in Cornwall
PRINCE-OF-WALES' FEATHER on Portland, Dorset
Quiet neighbours in Wiltshire
Red money in Somerset
Saucy Bet in Cornwall
Scarlet lightning in Devon and Huntingdonshire
Setwall in 'Berkshire or Oxfordshire'
Soldier boys in Somerset

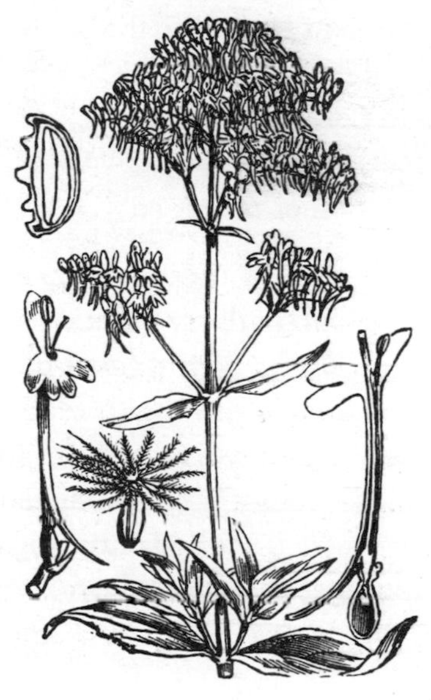

Sweet Betsy in Dorset, Kent and Somerset
Sweet Betty in Dorset and Somerset
Sweet Mary in Buckinghamshire
Ventnor pride, 'probably because this plant is very prolific in the Ventnor area', on the Isle of Wight in *c.*1940
Wall-lilac in Somerset.

On Portland, Dorset, 'dried valerian stems [were] used as PEASHOOTERS with ripe vetch seeds'.[2]

RED WATER (disease in cattle) – EYEBRIGHT and slender St John's wort cure.

REED (*Phragmites australis*)

Tall perennial grass, abundant and conspicuous, in moist situations and shallow water.

Local names include:

Bennels, and bog-reed, in Berwickshire
Daddymore in Berkshire
Ditch-reed in Shropshire
Doudle in Roxburghshire; 'found, partially decayed, in morasses, of which the children in the south of Scotland make a sort of musical instrument similar to the oaten pipe of the ancients'
GOSS in Cornwall
Pole-reed in Somerset; 'may be a corruption of pool-reed'
Pull-spear in southern England
Quell-, or quill-, rods in Co. Donegal; 'used by weavers under this name'
Spire in Devon
Star-reed in Scotland; 'with emphasis on reed, hence sometimes streed'
Stower in Orkney
Streeds in Morayshire
WATER-GRASS in Somerset
Wheel-rod in Co. Donegal.

Being abundant, rapidly growing, and robust, reed was an important resource, particularly valued as a thatching material. According to Milliken and Bridgewater, most of the few Scottish buildings that are still thatched are covered with reed, which is more resistant to attack from mice and other vermin than straw thatch, and has the advantage of being less heavy.[1] Wyse Jackson states that an (undated) survey of almost 1,000 thatched buildings in eleven counties in Ireland found that reed had been used on 39 per cent of them; other thatching materials being at present OAT and rye straw, though in the past WHEAT or BARLEY straw, reed, and, more rarely RUSHES were used.[2] Today much, probably most, of the 'water reed' used as THATCH is imported.

The biggest reed producer in Britain was the Tayreed Company, which harvested 100,000 bundles a year from reedbeds originally planted in the 1780s in an attempt to prevent river-bank erosion,[3] but in March 2005 it was reported that they were ceasing to trade, being unable to compete with cheaper reed imported 'especially from eastern Europe'.[4] Beckett and Bull in their *Flora of Norfolk* (1999) record that reedbeds continued to be managed to provide 'the raw material for Norfolk thatch for roofs, etc'.[5]

Other uses, recorded in the nineteenth century, include 'laths for partitions of rooms', reeds for Scottish bagpipes and Northumbrian small-pipes,[6] fences, and by artists for making reed-pens for sketching.[7]

More recently:

> [Mid Oxfordshire, 1930s] A game played by the boys . . . was bows and arrows. The bow was made from a stout briar. The arrows were made from the previous year's reed stem with a tip of about 3 inches of ELDER stuck in the end for weight and direction. The arrows were capable of travelling 100 yards.[8]

> [11-year-old girl] We make duck noises with reeds. Put them between your thumbs and blow to make the duck's quack. It drives my dog mad.[9]

RENNET – substitutes include BUTTERWORT, LADY'S BEDSTRAW, LESSER SPEARWORT and NETTLE.

REPUBLICANISM, in Ireland – ARUM LILY associated with.

RESTHARROW (*Ononis repens*)

Perennial pink-flowered herb, widespread in rough grassy places, especially near the sea, throughout much of lowland Britain and southern Ireland.

Local names include the widespread CAMMICK and LAND-WHIN, and:

CAMMOCK on southern England
CAT-WHIN in Somerset and Yorkshire
Cornets in Dorset
Crammick in Somerset
Donkey-weed in Lincolnshire

Dumb-cammock in Somerset
Fin, and finweed in Northamptonshire
FURZE in Wiltshire
GOOSEBERRY-PIE in Somerset
GOSS in Wiltshire; 'in this district GORSE . . . is always fuzz'
Ground furze in Somerset
Hard-hack in Essex
Harrow-rest in Lincolnshire
Hen-gorse in northern England
Horse-breath in Somerset and Worcestershire; 'it has been suggested that this name is due to the harder breathing of horses as they endeavour to plough through this plant'
Kemmick, and kramics, in Somerset
Lady-whin in Scotland
Lewte in Somerset
Liquorice-plant in Somerset
Liquory-stick in Roxburghshire
Pink shoes-and-stockings in Sussex
RAMSEY, and RAMSONS, in Devon
Rassels in Suffolk
Rust-burn in Yorkshire
Sid-fast, and sit-fast, in Scotland; 'from the tenacity with which its roots cling to the ground, rendering the plant difficult to eradicate'
Stainch in northern England
STINKING TAM, and stinking tommy, in Northumberland; 'the smell of the root is very disagreeable'
WHIN in Northamptonshire
Wild liquorice in northern England and Scotland.

> The roots are very sweet, and, when young, have the flavour of liquorice. The writer was informed by some workmen . . . that they and their fellow labourers were accustomed to suck the juice from these roots in order to assuage the thirst induced by hard toil under summer sun. The young shoots are also sweet and succulent, and in some country places they are boiled and eaten.[1]

> *Reglisse* or *Reclisse*, restharrow . . . old [Guernsey] people still remember as children eating the roots of this plant, which are said to taste very much like liquorice root. The same thing is done in the north of England, where restharrow is called wild liquorice.[2]

RHEUMATISM – prevented using ALDER, conkers (HORSE-CHESTNUT seeds), DOCK, DOG ROSE gall, NUTMEG and POTATO; treated using BIRCH wine, BOGBEAN, BROOM, BURDOCK, (creeping) BUTTERCUP, CELERY, CHICKWEED, FAT HEN, HEATHER, MALLOW, NETTLE, ONION, OPIUM POPPY, ROWAN, ROYAL FERN, SUN SPURGE, WATERCRESS, WHITE BRYONY and WOOD SAGE; in cattle 'kept at bay' by HOGWEED.

RHUBARB (*Rheum* x *rhabarbarum*)

Persistent perennial herb, the parents of which are believed to be of Siberian origin, cultivated for its laxative properties in Britain and Ireland since the late sixteenth century, and for its edible leaf-stalks since the eighteenth century.

> A slice of rhubarb placed at the bottom of the 'dibbed' hole will prevent club root in brassicas [i.e. CABBAGES].[1]

According to a letter in the *Daily Telegraph* of 22 January 2014:

> In the small burgh of Darvel in Ayrshire it was common up until the early Seventies for newlyweds to be given a root of rhubarb as a WEDDING present.

Georgina Boyes in a brief survey of colour in 'mock-obscene' riddles includes:

> What's long and thin and covered in skin
> Red in parts, stuck in tarts?
> Answer (which is usually chanted as part of the rhyme): Rhubarb![2]

In 1992 'greetings' cards with this riddle were available on sale in London.

In folk medicine rhubarb and spirit of nitre was used to cure COLDS,[3] and stewed rhubarb to treat CONSTIPATION.[4]

RIBWORT PLANTAIN (*Plantago lanceolata*)

Ubiquitous perennial herb (Fig. 41), abundant in grassy places throughout Britain and Ireland; flowers superficially dull, but surprisingly attractive when closely observed.

Local names include the widespread cocks-and-hens, fightee-cocks, HARD-HEADS, rib-grass, and SOLDIERS, and:

Baskets in Wiltshire
Black bent in Buckinghamshire
BLACK BOYS in Wiltshire
Black heads in Dublin
Blackie-tops in Somerset
Black jacks, applied to the 'heads' in Shropshire
Black man in Dorset
Black men in Dorset and Somerset
Blacksmiths in Somerset
BOBBIES, and BOBBINS, in Dorset
CANARY-WEED in Yorkshire

CARL-DODDIE in Scotland and Co. Donegal
CAT'S TAILS in Somerset
CHIMNEY-SWEEPERS in Northamptonshire, Warwickshire and Worcestershire
CHIMNEY-SWEEPS in Northamptonshire, Somerset and Warwickshire
Cock-grass in Devon and 'the only name used by farmers' in west Somerset
Cocks in Co. Armagh, Belfast and Co. Donegal
CONKERS, and conqueror-flowers, in Somerset
Cruk-slanish in Co. Donegal
Curl-doddies in Forfarshire and Co. Donegal
DEVIL-AND-ANGELS, and donkey's ears, in Somerset
Fechters in Scotland; 'i.e. fighters'
Frenchmen's head in Ireland
GYPSY in Somerset
Headman in Perthshire
Jack straws in Yorkshire
Kemps in northern England and Scotland; 'kemp is to fight in northern dialect'
Kempseed in Selkirkshire
LADY'S MANTLE in Dorset
LAMB'S TAIL, and LAMB'S TONGUE, in Somerset
LORDS-AND-LADIES in Norfolk
Men-of-war in Somerset
Monkey-heads in Hampshire
Niggers in Somerset
Niggers' heads in Devon
Pash-leaf in Pembrokeshire
RATS' TAILS in Cumberland
Red-hot poker in Somerset
Ripple-grass in Ireland; 'a common name throughout the country'
Ripplin-grass in Lanarkshire
Sodgers in Scotland
Soldier's tappie in Forfarshire
SWEEP'S BRUSH in Somerset
Swords-and-spears in Dorset
Violin-strings in Somerset.

On the Shetland Islands ribwort plantain was known as Johnsmas-flooer:

> The local name refers to an ancient Scandinavian custom which also had a place in the folklore of Orkney and Faeroe. We cannot do better than quote from Hoeg (1941): 'In these parts of the country [south-west and west Norway], and only here, it has been (and partly still is) common to foretell the future by means of that plant, by taking one or two flowering spikes of it on St John's Eve, picking off the stamens, and keeping the spike over night, usually under the pillow: If new stamens were developed in the morning (as will generally be the case), certain wishes would be fulfilled, mostly concerning matters of live and death, or love. Similar customs and names of the plant are known from the Faroes and the

> Shetlands, and must consequently be assumed to be very old, no doubt from before the year 1468.' This custom was still providing amusement in Shetland as late as the 1920s.[1]

Throughout the British Isles ribwort plantain was used in children's games.

> Children everywhere play with the flower-stems of ribwort plantain, striking one against the other until the loser's breaks, like CONKERS. Plymouth kids call this game 'Kings'.[2]

> As children we played a game called 'Fighting Cocks' with the long-stemmed seed heads of ribwort plantain. Two children each took a 'cock' and they kept hitting them against each other until the head was knocked off one.[3]

> [Aberdeen, 1930s] CARL DODDIES is the name of a game played by children with *Plantago lanceolata*. The name has extended from the game to the plant itself. During the 45 rebellion supporters of Charles, the Young Pretender, were known as Carls, and the supporters of King George were known as Doddies – Doddie being the local name for George. Children emulated their parents and took sides, hence the game's name.[4]

> My brother and I used to play a game in which you try to knock the flowerhead off a plantain on the way home from school. We called it fox-and-geese; the loser – the one whose flowerhead was knocked off was the 'goose'. That was in Warwickshire, the Stratford-on-Avon area. We thought it was rather a girlish game.[5]

This pastime, which was also known as Blackmen, Cocks and Hens, Hard Heads, Knights, and, in parts of North Wales, *ceiliogod* (cocks) or *taid a nain* (grandfather and grandmother), has a long history.

> In the historical poem *Histoire de Guillaume le Maréchal*, written soon after 1219, the story is told how the boy William Marshal, later to become Earl of Pembroke and Regent of England, but then not ten years old, was detained as a hostage in the king's camp, while Stephen was besieging Newbury. One day the boy picked out the plantains (*les chevaliers*) from the cut grass strewn on the floor of the tent, and challenged the king, '*Beau sire chiers, volez joer as chevaliers?*' The challenge being accepted, William laid half the 'knights' on the king's lap, and asked who was to have the first

stroke. 'You' said the king, holding out his knight, which the small boy promptly beheaded, greatly to his own delight. King Stephen (strictly in accordance with the rules of the game) then held out another plantain, but the game was interrupted. It matters not whether the story is apocryphal; as early as the thirteenth century a poet has shown himself to be familiar with the game.[6]

In another childhood game:

The stem and flowering heads of the plantain were used as a missile, by somehow knotting and flicking the ripe seed-head.[7]

Friends who lived in Dumfriesshire many years ago reminded me that a game was played in this way: 'Take a long, strong stalk of plantain, loop it over the head of the flower, pull it quite tightly so that the head shoots off. The winner is the one whose plantain head travelled furthest.'[8]

I was born in 1948 in Ferndown, before it was built up, and our playground was the fields, common and forestry around our home . . . Ribwort plantain (bootlace): tie in a knot and then quickly shoot the end off.[9]

On a walk with a friend and her son near Leeds a couple of years ago, he picked some of those plants with the long stalks and big seed heads (I've no idea what they're called, but it was the same plant we used as kids for the same game of knotting the stem and shooting the head off). But this kid accompanied the knotting and shooting with the song 'Miss Molly had a dolly, but her head popped off.'[10]

We also picked the plantain flowers, wrapped the long stalk around in a loop and popped the head off, saying, 'Grandmother, Grandmother jumped out of bed.'[11]

According to Mont Abbott, a farm labourer born at Entsone, Oxfordshire, in 1902, on summer evenings young men would cycle to neighbouring villages to view the girls who

was dotted about like daisies . . . We foreigners 'ud pull up on our bikes within winking or aiming distance, picking the Daisy we fancied from the bunch by looping the pliant stem of a plantain into a capapult and pinging her bare neck with its . . . seedhead.[12]

Similarly:

> In Herefordshire in the 1870s my uncle attracted my aunt by firing a plantain flowerhead at her. It must have worked, they got married.[13]

Ribwort plantain leaves were used in folk medicine, mainly to stop BLEEDING:

> [Co. Kerry] When a person got a cut and was bleeding profusely, he pulled some *slanlus* (rib grass) leaves, chewed them in his mouth and then placed the pulp on the wound. This always stopped the flow of blood.[14]

> Rib grass is chewed and put on a wound to stop bleeding.[15]

In recent years they have also been used to treat nettle STINGS, but this remedy does not appear to be indigenous, but seems to have been brought in by people from other parts of the world. Typically:

> When I got stung by a nettle a woman from Australia told me to rub ribwort plantain on it.[16]

Ribwort plantain leaves dry more slowly than those of grass, so when hay was being made it was important to make sure that they were dry, or else the resulting hay stack might overheat.

> When building a hay rick in rather showery unsettled weather, old Durrant would casually mention 'There's a fair bit of fire grass hereabouts Gaffer' and again the Manager knew that the rick was in danger of beginning to overheat. 'Better put a chimney in, Durrant,' he would say. Fire grass was the name given by the older men to the ribwort or narrow-leaved plantain, which holds quite a bit of moisture in its leaves.[17]

Other names which refer to this problem include fire-leaf in Somerset, and FIREWEED in Shropshire.

RICKETS – BRAMBLE and ROYAL FERN used to treat.

RINGWORM – treated using ASH, GARLIC, HOUSELEEK, LAUREL, OAK and PRIMROSE; WHITE BRYONY cures and causes.

ROCK SAMPHIRE (*Crithmum maritimum*)

Perennial herb, leaves said to smell of furniture polish when crushed, found in rock crevices, on sea-walls and stabilised shingle, common around the south and west coasts of Britain and Ireland, extending north to Suffolk and south-west Scotland.

Local names include:
Camphire in Cumberland
Creevereegh in Co. Donegal
Greirig in north-west Donegal, where 'gathered and sold under this (its Irish) name'
Passper in Scotland, 'from French *perce-pierre*, pierce rock'
ST PETER'S HERB in Somerset
SAMPER on the Isle of Wight
Shamsher in Cornwall.

Britten and Holland also list rock-semper, and semper, from East Holderness, Yorkshire,[1] and Grigson lists the former name from Northumberland,[2] both of which counties are far beyond the plant's natural distribution.

According to Rosemary Parslow and Ian Bennalick, writing of the Isles of Scilly:

> There is an account by Robert Heath in 1750 of the preservation of samphire for pickling by covering it in strong brine and then pickling in vinegar – when it was 'sent in small Casks to distant Parts for Presents . . . In his 'Observations', Borlase (1756) comments that the 'Sampier – is the best and largest kind (far superior to the Cornish)'. There has been a recent resurgence in interest in using the plant both as a pickle and as a herb.[3]

Elsewhere:

> [Colonsay] Samphire is much sought after for pickling sometimes at the risk of human life (men being suspended from the rocks by ropes).[4]

[Co. Donegal] Semper grows in the cliffs by the sea . . . eaten in its raw state relieves pain in the HEART.[5]

ROGATION SUNDAY (fifth Sunday after Easter) – CHERRY orchards blessed on.

ROMANS – BUTCHER'S BROOM, FAIRY FOXGLOVE, FIELD ERYNGO, WILD CLARY, WINTER ACONITE and YELLOW CORYDALIS associated with.

ROPE – made from PURPLE MOOR-GRASS.

ROSE (*Rosa* cvs)

Thorny shrubs, long and widely cultivated in many forms for their showy, fragrant flowers.

The red rose as a symbol of England is occasionally worn on St George's Day, it is also associated with love, and the Labour Party. Such roses are not naturally available on St George's Day (23 April), so imported flowers have to be used, and the English, unlike the Irish, Scots and Welsh, have not had to assert their nationality. Consequently red roses are not commonly worn on the country's patron saint's day, but in 1986, when the Labour Party adopted the red rose as its symbol, there was unease in the City of London and elsewhere.

> The colour of the rose for St George's Day became an important issue at Common Council [of the City of London] last week, as the Court strove to prove that it was non-political.
>
> A motion was tabled by Mr Brian Boreham that all Common Councilmen should wear a red rose provided by the Corporation for their meeting on St George's Day. However, Mr Norman Harding said that he would be voting against the motion since the red rose was now an emblem of a political party. His reasoning was that the Court should remain apolitical.[1]

> The Common Council can't make up its mind what colour rose should be worn on St George's Day – but the City of London branch of the Royal Society of St George has no doubts . . . Mr John Minshull Fogg, Hon. Secretary and chairman-designate of the branch, said this week, 'The red rose is the rose of England, and that's what we shall be wearing. We always have done and always will.'[2]

> It has been my custom, in previous years, to wear a red rose on April 23 – a custom shared with many of my fellow countrymen and women. However, I am a Civil Servant and as such must show no tendency to favour any individual political party.
>
> Given the rose has become the symbol of one of our major political parties I wonder if any of your correspondents could suggest what, if anything, would be appropriate wear on St George's Day.[3]

Following the Labour Party's defeat in the 1992 general election its use of the red rose as a symbol became less flamboyant, but it continues to appear on election literature.

The Royal Northumberland Fusiliers wore a red and white rose on St George's Day.

> St George's Day was their regimental day, when all ranks wore a red and white rose. This tradition is now continued by all battalions of the Royal Regiment of Fusiliers. The roses are thought to symbolise the unity of the houses of Lancaster and York.[4]

The first of August has been designated as Yorkshire Day, 'an event established by the Yorkshire Ridings Society in 1975 to act as a focal point for county pride, because on this day in 1759 soldiers from the Yorkshire regiments who had fought in the Battle of Minden in Germany picked white roses, the county emblem, from nearby fields as a tribute to fallen comrades'.[5]

In recent years white roses have been used to commemorate those whose lives have been cut short as a result of terrorism or some other tragedy. In the United Kingdom this practice seems to have become prominent following the murder of the Yorkshire Member of Parliament Jo Cox, on 16 June 2016. On 21 June *The Times* reported:

> Parliament was united yesterday. United in grief, a white rose pinned to the chest of every politician in memory of this daughter of Yorkshire, taken before her time.

A white rose and a red rose – the Labour Party symbol – were attached to the seat in the House of Commons which Ms Cox usually occupied. On 22 June, which would have been the MP's 42nd birthday, a memorial meeting was held in Trafalgar Square, with many of those attending carrying single white roses.[6] Similar roses were worn by members of parliOUT, the UK Parliament's LGBTIQ group, the following Saturday at the annual London Pride Parade.[7] Seventy-one people were killed by a fire at Grenfell Tower, Kensington, on 16 June 2017, and some mourners attending a memorial service six months later at St Paul's Cathedral carried white roses.[8]

As symbols of love, red roses are in great demand on ST VALENTINE'S DAY.

> [1991] The annual Valentine red roses spree was in full swing again yesterday. And prices of our favourite bouquet went through the roof as florists cashed in on love . . .
>
> Shop owner Ian Duncan blamed worldwide demand. 'The trouble is that Valentine's Day falls on the same day everywhere, unlike occasions like Mother's Day,' he added. In Birmingham, where some shops were charging £3 a bloom angry wholesalers said their price was only 65p–75p. 'We're fed up with it,' said one. 'It happens every year.'
>
> At Sally's, in Dulwich, south-east London, a dozen red roses had soared 300 per cent from £9 to £36. The assistant was frank. 'It's Valentine's now, so everything has gone right up,' she said.[9]

Most of the roses sold on Valentine's Day are imported from Colombia. In 1994 it was reported that although 450 rose-growers on the plains around Bogotá had suffered their most severe frost for many years, this year they expected to export more than 375 million flowers. Valentine's Day roses account for between 30 and 70 per cent of a grower's annual income.

> For Valentine's Day, staff levels can quadruple, as farms work 24 hours a day to meet the demand. Teams of workers will pick and pack by hand 150,000 stems in an eight-hour shift. After 24 hours in cold storage, the flowers are then transported to the airport in locked, refrigerated lorries.[10]

In recent years, possibly because red roses have become so expensive, their use on Valentine's Day appears to be less dominant and a wider range of flowers, red blooms predominating, are sold. The purchasers of Valentine's Day flowers are mainly male; few women buy flowers for their male sweethearts, but since 2016 unattractive multi-coloured 'rainbow' roses, presumably appropriate for people in same-sex relationships, have been available.[11]

Since Roman times rose petals have been showered on important people, or strewed on religious occasions. On 12 June 1928 the *Daily Mail* reported on the wedding of the daughter of the singer Dame Clara Butt:

> A shower of thousands of rose petals greeted Miss Joy Kennerley Rumford . . . when she married Major Claude Cross at Holy Trinity, Brompton.
>
> These petals were softly tinted tissue paper and been supplied by the Disabled Men's Industries for most of the big society weddings. Princess Mary and the Duchess of York ordered them for their weddings . . .
>
> The petals are made in pink shading to white and orange shading to cream. Boxes of separate colours can be had for 1s 6d a thousand petals, but most brides choose mixed colours.

At St Albans, in Hertfordshire, roses are placed on the shrine of St Alban, England's first Christian martyr, during a festival held late in June each year. For much of the twentieth century the festival was a Rose Service, held on the Sunday nearest Alban's feast day, 22 June. In the 1980s Sunday-school children from throughout the diocese gathered in St Albans Abbey to take part in a simple, joyful service, during which the entire congregation, led by the bishop with his crozier decorated with roses, filed up the aisle to place home-grown roses on the saint's shrine.[12]

In recent years the event has become the Alban Pilgrimage. In the morning there is a procession, consisting of giants representing Alban and people associated with his legend, clergy, choir, band and dignitaries, after which there is a service in the Abbey. In the afternoon a Festival Evensong is held, at the conclusion of which the clergy and choir lead the congregation to place (or, it seems more often, to throw) roses on the shrine. Whereas home-grown roses were used in the past, today long-stemmed red roses are sold in the Abbey between the two services, and it is these which are most often thrown on the shrine; very few people bring their own roses.[13]

In Unitarian chapels roses are sometimes used in children's naming ceremonies. Unitarians reject the doctrine of original sin, and therefore most of them consider it unnecessary to use water in naming or christening ceremonies to wash away sin.

> We prefer the symbolism of the rose, chosen because of its beauty and its thorniness. The words generally spoken are some version of:
>
> [Name] we give you this rose today.
>
> We have taken the thorns off it for this occasion.
>
> But we know that, even if we would, we cannot remove the thorns from your life. Therefore we hope that your life, like this rose, will be beautiful in spite of the thorns.
>
> We then stroke the child's face with the rose petals, not necessarily with any deep theological meaning, other than connection with life itself, but more for the pleasure of it. If the child is old enough (four months on) they generally try to eat it, which is alright too.[14]

Occasionally roses are required in lieu of rent. According to legend in 1381 a fine of one red rose a year was imposed on Lady Constance Knollys, who built a footbridge connecting two of her properties in Seething Lane, London, without first obtaining the Lord Mayor's permission. The custom came to an end in the seventeenth century, but was revived in 1924, with the presentation of the rose to the Mayor taking place on, or near, 24 June each year.[15]

> I was born in 1922 in the village of Newburn-on-Tyne [Northumberland]. Most of the local activities, buildings, etc., were connected to the Percys and the Dukedom of Northumberland. In the village was a row of cottages, called Duke's Cottages – four small, one larger. My mother's Aunt Martha lived in the larger. This comprised of a front porch, sitting room, breakfast-cum-kitchen-cum-dining room, three bedrooms, and gardens front and back. For this lavish accommodation she paid a token floral rent, if I remember correctly one red rose, in June every year. She lived there until 1959.[16]

By 2002 this rent was forgotten, but according to an account in the *North Mail* of 8 August 1956:

> The Duke of Northumberland walked into Newburn Urban Council chamber yesterday to collect some rent; and after smelling it, he turned to the councillors and said, 'It's lovely.'
>
> The rent was only a rose – payment for a strip of land owned by the Duke and leased by the council for a rose park.
>
> This strip was part of the lands granted to the Duke's descend-

ants in 1329. But until two years ago it had remained derelict.

Then the council decided to turn the weed patch into a beauty spot. They asked the Duke to rent them the site.

The cost, decreed the Duke, would be one red rose, payable each year at the time of the council's August meeting . . .

'This is the loveliest and most extraordinary rent I've ever received.' said the Duke.

In 1995 it was reported that Mary Cornelius-Reid was providing a 99-year lease for the plot of ground on which Naomi House Children's Hospice, at Sutton Scotney, Hampshire, was to be built, with the rent being set at one dozen red roses a year.[17]

Rose bushes are commonly planted in memorial gardens at crematoria and cemeteries, where they look rather dull and straggly for most of the year but colourful for a few weeks in the summer. In 2002 the City of London Cemetery and Crematorium reported:

> At this time of year, our Memorial Gardens (which cover 32 acres) are full of roses in bloom, and many people choose to dedicate rose bushes to their departed loved ones . . . The Memorial Gardens has approximately 600 rose beds containing around 20,000 roses. Every year we inspect each rose and in July or August make a list of those that need replacing. The replacements number between 3,500 and 4,000 . . . We use approximately 18 varieties of rose in the Memorial Gardens.[18]

ROSEBAY WILLOWHERB (*Chamerion angustifolium*)

Perennial herb (Fig. 42), producing flamboyant purple-pink flowers; common, widespread and often forming dense colonies on disturbed ground, including cleared woodland, burnt ground and tracksides.

Until the mid-nineteenth century rosebay willowherb was a rare plant. In 1798 William Curtis noted:

> In the third [1724] edition of Ray's *Synopsis*, this plant is said to have been growing wild near Alton in Hampshire [Curtis's birthplace]: in confirmation of this, I have myself found it in a wild unfrequented wood near the same place . . . Mr Hudson in his *Flora Anglica* [1762] mentions it growing in Maize Hill, beyond Greenwich.[1]

About seventy year later, the Worcester botanist Edwin Lees observed:

> Rosebay willowherb has become numerous in several parts of the Vale of Severn, and promises to spread, incited to take possession of new-made roads and embankments.[2]

Why this rapid expansion of the plant's distribution took place is a matter for speculation. The most usual explanation is that it spread along the developing railway network, alternatively perhaps the expansion resulted from the introduction of a more vigorous adaptable form from overseas.[3]

It seems as though this rapid spread led to the plant accumulating a large number of alternative names, the most widespread of these being fireweed:

> Rampant in Northumberland on areas of rough ground, especially commons and railway embankments. It is a moot point whether it is called fireweed because it spreads so rapidly and its flowers are purplish pink, or because in the days of steam engines it was often ignited by flying sparks from the engines.[4]

> Rosebay willowherb referred to as FIREWEED in Scotland – it is one of the first flowering plants to return after a heath fire.[5]

Owing to its rapid colonisation of burnt ground, rosebay willowherb became conspicuously abundant on London bombsites during the Second World War, leading to such names as bombweed, bomb-site weed, LONDON PRIDE, and London's ruin, and it was even rumoured that rosebay willowherb seeds were actually put into German bombs.[6] In 2002 the charity Plantlife designated rosebay willowherb as London's 'county flower'.[7]

Outside London:

> Clydebank was bombed during the last war and one of the casualties was the Singer Sewing Machine factory. On the bombsite a profusion of rosebay willowherb sprang up, which locals of that vintage now call Singer Weed.[8]

> Westwood lily – my wife learnt this name in Sheffield [25–30 years ago]; it is a purely local name and refers to a district in their village (High Green) which us called Westwood and where these flowers grew in profusion.[9]

Rosebay willowherb's leaves somewhat resemble those of some narrow-leaved willows. Thus it has been given names such as:

Blooming sally in Cumberland and Ireland
Blooming willie in Co. Tyrone
Frenchaloo, or French willow, in Warwickshire
Flowering withy in Berkshire
French saugh in Lanarkshire.

Other names include:

Blood-vine in Hampshire
CAT'S EYES in Shropshire
Dog's parasol in mid Wales
Plum jam in Dorset
Railway chrysanthemum in Derbyshire, Lancashire and South Yorkshire
Red buffer in Shropshire
Romping molly in Yorkshire.

There are occasional records of rosebay willowherb being associated with bad LUCK.

> [Shropshire] Rosebay willowherb – never picked – was called MOTHER-DIE.[10]

> [Macclesfield area, Cheshire, 1940s] Rosebay willowherb should not be picked, otherwise a THUNDER-storm will ensue, or, more horrifically, your mother will die. I think we called this plant THUNDER-FLOWER.[11]

Although books on foraging mention rosebay willowherb,[12] and it has been used as a tea elsewhere in the world,[13] there seems to be no tradition of it being used as food in the British Isles; indeed one writer on wild foods states 'it is far too bitter to enjoy as any kind of vegetable'.[14] However, the seeds were sometimes used as a TOBACCO substitute – 'but it was a very hot and strong smoke'.[15]

ROSEMARY (*Rosmarinus officinalis* or, alternatively, *Salvia rosmarinus*)

Fragrant, evergreen shrub, native to the Mediterranean area, cultivated in British and Irish gardens as a culinary herb and ornamental since at least late in the fourteenth century, occasionally becoming naturalised, mainly on old walls, in southern England.

> [Stockleigh Pomeroy, Devon] if you had a rosemary bush growing near the house no WITCH could harm you.[1]

Alternatively:

> You always had to plant rosemary in your garden, so that you wouldn't be short of friends.[2]

> In Hampshire rosemary was known as friendship bush. Every house had one; I can't remember seeing a house that didn't have one.[3]

> I have always referred to rosemary as the plant of friendship.[4]

In common with other culinary herbs, such as PARSLEY and SAGE, rosemary was thought to grow best where the wife was dominant.

> There is also a saying in Yorkshire that rosemary will not grow in the garden of a house unless the woman is master.[5]

> A woman came to ask me with many apologies whether I would plant some rosemary cuttings; she did not wish to be rude, and Mrs Wist had said it was a very delicate thing to ask anyone to strike rosemary cuttings, but she did want a bush so badly. I said I did not mind doing it for her, why should it be such a delicate thing to ask anyone. She replied, 'Don't you know it only strikes in the house where the mistress is master, and if it won't grow here, it won't grow anywhere.' So I got 14 shoots off the branch, 11 grew, three didn't. My husband was so interested he went to look at the cuttings every day, and said the three that did not grow were evidently where he got a look in.[6]

> Some people consider it UNLUCKY to grow rosemary. I imagine it may be because if it did well it meant that 'the woman wore the trousers'.[7]

Various legends associate rosemary with the Holy Family's flight to Egypt.

> When I was five or less I heard that rosemary flowers, which were formerly white, became blue after the Virgin Mary stopped to do some washing on her way to Egypt and hung her robe to dry on a rosemary bush.[8]

Rosemary is widely associated with remembrance.

> When I was at school our teacher always gave us a piece of rosemary – for remembrance – before exams.[9]

> In the Cathedral [Derby] they sprinkle us with rosemary branches dipped in water in remembrance of our baptismal vows.[10]

This association is mainly due to a passage in Shakespeare's *Hamlet*, 'there's rosemary, that's for remembrance'. Consequently people participating in the Shakespeare Birthday Celebrations, in Stratford-upon-Avon, Warwickshire, on 23 April each year wear, or carry, twigs of rosemary.

> Rosemary, DAFFODILS and PANSIES (see *Hamlet*, Act IV, Scene V) feature in the Shakespeare Birthday Celebrations every year, worn as buttonholes or carried as posies in the procession through the town which ends at Holy Trinity Church. Here all the flowers are put in the chancel round Shakespeare's tomb . . . The celebratory procession was started about 1898–1900 by the headmaster of King Edward VI Grammar School, where Shakespeare is said to have been a pupil. The boys took evergreens including rosemary to the church on Shakespeare's birthday, and this became an annual event. It is now an international celebration of a great dramatist.[11]

In 1994 most of the participants in the Birthday Celebration procession wore sprigs of rosemary, but the flowers carried and placed around the memorial in Holy Trinity Church seemed to be readily available seasonal or florists' flowers, which had no particular association with the bard.[12] In 2016, the 400th anniversary of Shakespeare's birth, not only people in the procession, but also most of the many onlookers wore sprigs of rosemary, with the result that later in the day the streets along which the procession walked were delightfully scented with crushed, dropped twigs.[13]

In addition to its culinary use, rosemary was also valued as a hair rinse. Typically:

> Rosemary leaves in the bath revives, and strained off, having been in hot water, make a nice rinse for hair and leaves it shiny.[14]

> There was an old rosemary bush in the garden and small twigs were broken off and put into a saucepan to boil up in water that was used to wash my hair.[15]

ROWAN (*Sorbus aucuparia*), also known as MOUNTAIN ASH

Small tree with white flowers and orange-red fruits, widespread in woods, moorland, and rocky places; also widely cultivated.

Local names include the widespread QUICKEN, quicken-tree, wicken, wicken-tree, and wiggen, and:

Caers in Cornwall

Care, or cares, and care-tree, in Cornwall and Devon, 'Cornish *kerdhyn*, cf. Irish *caorthann*'

Cayer in Pembrokeshire

Chit-chat in Wiltshire

Cock-drink, or cock-drunks, in Cumberland; 'the fruit is reputed to possess the property of intoxicating fowls'

Creaghan in Co. Donegal

Cuirn on the Isle of Man

Dog-berry, referring to the fruit, in Cheshire and Cumberland

Kair, kear, and kearn, in Devon

KEER in Cornwall and Devon

Kitty-keys in Somerset

QUICKBEAM in southern England and Ireland

Quickbeam-tree in Hampshire, Somerset and Sussex

Quicken-wood in Lincolnshire and Yorkshire

Rantry in Scotland

Ranty-berries, applied to the fruit in Co. Antrim and Co. Down

Rodden-tree, roddin, and roddin-tree, in Scotland

Roden-, or rodin-, tree in Scotland, where fruit known as rodens or rodins

Royne-tree in Yorkshire

Shepherd's friend in Dorset

Sip-sap in Lancashire and Yorkshire

Twickband in Hampshire

Twick-bine in Devon

Whicken, and WHISTLE-WOOD, in Yorkshire

Whitten-tree in Shropshire and Ireland

Whitty in Radnorshire and Shropshire

Whitty-tree in western England

Wickey in Cheshire, Derbyshire, Co. Durham and Shropshire

Wigan in Yorkshire

Wiggen-tree, and wiggin, in Cumberland, Westmorland and Wales

Wiggy in Co. Durham

Wilchen, or wilchin, in Northamptonshire and Worcestershire
Witch-beam in Devon
Witchen in Northamptonshire and Worcestershire
Witch-wicken in Lincolnshire
Witch-wood in Cumberland, Co. Durham, Northumberland and Yorkshire; 'from its supposed efficacy against witchcraft'
Withen in Lancashire
Withy in Herefordshire and Shropshire
Witty-tree in Worcestershire
Wychen in Cheshire.

Throughout the British Isles, but especially in Ireland and the Highlands and Islands of Scotland, rowan was valued for its protective powers.

> Whitty-tree . . . in Herefordshire they are not uncommon; and they used, when I was a boy [1630s], to make pinnes for the yoakes of their oxen of them, believing it had the vertue to preserve them from being fore-spoken, as they call it; and they used to plant one by their dwelling-house, believing it to preserve from witches and evill eyes.[1]

> [Scots] believe that any small part of this tree carried about with them will be a sovereign charm against all the dire effects of witchcraft. Their cattle also, as well as themselves, are supposed to be preserved by it from evil; for the dairy-maid will not forget to drive them to the shealings or summer pastures with a rod of the rowan-tree, which she carefully lays up over the door of the sheal boothy, or summer-house, and drives them home again with the same. In Strathspey they make, for the same purpose, on the first day of May, a hoop of the wood of this tree, and in the evening and morning cause all the sheep and lambs to pass through it.[2]

In 1945 the new owner of a croft in north-west Scotland was warned against destroying a clump of rowans which obscured the view from her kitchen window:

> 'Ach – but you must no' be cutting them all down, whateffer!' she exclaimed. 'You must be leaving one . . . because a rowan tree near the house keeps evil spirits away. It's good to have a rowan tree in your garden, and if you hang a wee sprigie of it over the byre door, your beasts will be well too, and bad luck willna come to themselves.'[3]

Recent examples of similar beliefs include:

> A rowan tree is good to have growing in the garden, or nearby, because it wards off the influence of WITCHES.[4]

> The rowan . . . is said to be the home of good FAIRIES. I have one growing in my garden, given to me by my mother after I got married.[5]

> I remember from my childhood in the twenties and thirties in Longbridge Deverill, Wilts, mountain ash – with coal – picked and worn in cap to keep bad fairies away.[6]

> [Newcastle-on-Clun area, Shropshire] Rowan tree, known as witty tree (witches' tree) was said to protect people from EVIL.[7] Country people never burnt it in their homes. As late as the 1950s one farmer was agitated when his neighbour put rowan branches on a bonfire and warned that terrible things would happen to them. Their relationship was never the same again![7]

In New Zealand Scottish settlers and their descendants planted rowan trees near their homes.

> In the face of different work, climate and seasonal pattern of the Southern Hemisphere . . . rowan trees [are] still planted at the entrance to most modest suburban [Dunedin] villas to avert the evil eye, witches, or other potency of ill fortune . . . A (related?) custom of Otago children was to soak red rowan berries in water and then sprinkle the fluid around doors and the like as some sort of preventative of evil.[8]

The Pitt Rivers Museum, Oxford, holds examples of rowan used for protection. Three young twigs each tied into a simple knot were acquired by the Museum in 1893, and bear the label:

> Rowan tree loops, protective against witches. Two were placed on the railings of Dr Alexander's house, Castleton, Yorks, the third on a gateway before the church porch. They were placed by a horseman who turned his horse thrice before setting each loop.

Homes, crops, and cattle were particularly at risk on May Eve.

> [Driney, Co. Leitrim] The first smoke from a chimney on May morning is apt to be used by witches for bringing bad luck on a

house. This may be guarded against in the following way: On May eve, get a bunch of rowan leaves, and tie it up the chimney to dry, then on May morn, light this, and let that be the first smoke to go out of the chimney; for witches can do nothing with it.[9]

[Co. Galway] May Eve they stick a piece of mountain ash in their crops, that way the fairies would not take the luck of the crops.[10]

[Co. Cavan] On May Eve the farmer cuts rowan berry in the shape of a ring and ties it to the cow's tail with a red string. It is an old belief that butter would be taken off the milk if rowan berry was not tied to cows' tails.[11]

[Co. Wexford] cows going out on May morning are struck with a quickerberry switch, which prevents any person putting any evil on them or taking their profit or butter.[12]

Rowan could provide protection when people attempted to rescue others from the fairy realms. John Rhys described how a captive dancing in a fairy ring could be rescued. Two or more strong men should hold a long rowan pole so that one of its ends rested in the middle of the circle. When the invisible captive is felt to grasp the pole the men should pull with all their strength; the fairies will not be able to intervene because of their aversion to rowan wood.[13] Similarly, in a Highland folktale published in 1823, a man successfully rescued his brother from a *shian*, or fairy hill, after being advised: 'Return to the *shian* in a year and a day from the time you lost him, fasten a rowan cross to your clothing and enter boldly, and in the name of the Highest claim your brother.'[14]

It is often asserted that the rowan's protective qualities are due to its 'red berries – there is no better colour against evil.'[15] However, rowan flowers being white, scented, and messy, have all the characteristics of inauspicious flowers such as those of HAWTHORN and MEADOWSWEET, and in Lancashire, at least, rowan was considered unlucky.

[Accrington, *c.*1940] 'Some people won't have rowan in the garden. I knew one woman who refused to buy a house which she liked in every other way, because it had two rowans.'[16]

70 Accrington schoolchildren (11–14 years old) were asked about 'unlucky' plants; according to one child it is 'unlucky to transplant mountain ash, or bring cuttings into the house'.[17]

The name poison-berries has been recorded for rowan fruits in Northumberland, Somerset and Yorkshire, but Wyse Jackson observes:

> Rowan berries (*caora*) were one of the most significant fruits of ancient Ireland . . . [they] have been used for a wide variety of culinary purposes, such as jams and jellies, or fermented and distilled to make an alcoholic spirit.[18]

The *aucuparia* in rowan's scientific name is explained as being derived from the Latin *avis* (bird) and *anceps* (fowler), 'because wildfowlers used the fruits as bait for their nets'.[19] Coles, in his *Adam in Eden* (1657) records fowler's service as a name for rowan: 'boyes and fowlers use the berries as baits to catch blackbirds &c'.[20] Two names are recorded from Cumberland, cock-drunks and hen-drunks, because 'the fruit is reputed to possess the property of intoxicating fowls.'[21]

There appear to be few records of rowan being used in folk medicine.

> [Ireland] an infusion of the leaves is a popular remedy for RHEUMATISM (an oz. to one pint); dose, one wineglassful. The leaves, when burned and inhaled are said to be useful in ASTHMA.[22]

> My late parents came from Poland after the Second World War, and my mother occasionally used . . . herbal remedies . . . Frost-nipped rowan berries (the frost removes the sourness) were infused in vodka and after six months or so this was used as a medicine for STOMACH ache.[23]

> Some gypsies I stayed with for a while used ground up [mountain] ash berries, dried first of course, as a flour for small cakes for unwell children.[24]

ROYAL FERN (*Osmunda regalis*)

Large fern (Fig. 43), found in damp habitats, mainly in western areas, but cultivated in gardens and escaping elsewhere.

Other names include the 'fairly general' flowering fern, and:

Ash-leaf fern in Cornwall
Bog-onion in Cumberland, where 'considered a specific for RICKETS in children', and Ireland
King-fern in Co. Donegal
King-o'-the-ferns in Devon
Royal bracken in Scotland
Tree-fern in Wales.

In the Lake District, where the plant is known by the name of bog onion, the caudex is still used as an outward application for SPRAINS and BRUISES: it is beaten and covered with cold water, and allowed to remain thus during the night; in the morning a thick starchy fluid is the result, which is used to bathe the parts affected.[1]

[Co. Galway] Bog onion resembles a fern in appearance but its root is somewhat the same as an onion bulb The root is converted into a juicy substance and used as a rub for RHEUMATISM or SCIATICA; it is often found to be a complete cure. First the root is cut into slices and then pounded into a mash. It is then put into a bottle or corked vessel and water supplied in proportion to the size of the root. It is then left to set for about two days until it forms a thick juicy substance.

The bog onion flowers at night in the month of June, but there is some mystery attached to it, because at the approach of daylight the flower disappears. It is known to be there by leaving seeds after it.[2]

[Co. Clare] A poultice of bog-onion is a cure for a sprained limb. The onion is pounded and pressed to the sprain, and is then covered with a bandage.[3]

RUE (*Ruta graveolens*)

Perennial small shrub, with bluish-green leaves, yellow flowers and a distinctive odour, native to the eastern Mediterranean area, said to have been introduced by the Romans, little grown after their departure and re-introduced in the mid-sixteenth century.

Formerly known as herb-of-grace, corrupted to herb-grass in Somerset, and herby-grass in Sheffield, rue also has the name teardrops recorded in Berkshire.

As a symbol of repentance and sorrow rue could be used to bless or curse, help or harm.

> [Herefordshire, early nineteenth century] nosegays of rue, enclosing a piece of half-eaten bread and butter, were dropt in the church porch by a deserted female, to denote an unhappy wedding.[1]

> It's only a few years since a young girl went to Cusop [Herefordshire], to the wedding of a young man who had jilted her, waiting in the church porch till the bridegroom came out, she threw a handful of rue at him, saying 'May you rue this day as long as you live!' . . . the curse would come true, because the rue was taken direct from the plant to the churchyard, and thrown 'between holy and unholy ground' . . . if there was any difficulty in obtaining it for this spiteful purpose, rue-fern [WALL-RUE], the leaves of which resemble it, might be used; it must be found growing on the churchyard wall, and gathered directly from thence.[2]

Rue was valued to treat a variety of ailments.

> [Horseheath, Cambridgeshire] Rue tea was prescribed for improving one's APPETITE.[3]

> A former resident of Tule, Staffordshire, remembers having been given a tea made from rue for the relief of COUGHS and COLDS.[4]

> My mother – who died last year at the age of 88 – when she was small she was considered 'sickly' . . . Perhaps because of this each spring my grandmother would make rue tea. She picked rue leaves from the garden and poured boiling water over them; when the infusion was cold it was given to mum to drink. Grannie considered this to 'clear the BLOOD'.[5]

> Rue is supposedly good for the KIDNEYS, but I am not clear how this should be used.[6]

RUNNER BEAN (*Phaseolus coccineus*), also known as kidney bean

Perennial climbing herb with scarlet flowers, native to Central America, introduced to Spain late in the fifteenth century and known to have been cultivated in England as an ornamental in the mid-seventeenth century; now grown, as an annual, for its edible immature fruit.

In different places different days in late April or early May were recommended as being the time for sowing runner beans. The earliest of these days were associated with the feast of ST GEORGE (23 April).

> [Devon] It was unwise to plant kidney beans until after George Nympton Revel [on the Wednesday after the last Sunday in April, last held in *c.*1939].[1]

> They always used to say that kidney beans should be planted on the day of Hinton St George [Somerset] Fair. That used to be held on the third Thursday in April.[2]

Other dates included MAY DAY in the Ashford area of Kent,[3] and the exceptionally late 23 May at Lower Quinton, Warwickshire.[4] In Hertfordshire in the late 1950s 14 May was known as Runner Bean Day,[5] while in South Somerset:

> May 6 was known . . . as Kidney Bean Day, and it was believed that, if you did not plant your beans then, they would not flourish.[6]

Alternatively, other plants could provide guidance.

> Never plant kidney beans until the buds of the HAWTHORN have opened.[7]

> [Warwickshire] do not plant your kidney beans until ELM leaves are as big as sixpences, or they will be killed by frost.[8]

RUSH (*Juncus* spp.)

Perennial herbs, common in damp open places.

A legend recorded in 1937 in County Galway explains why the tips of rushes become brown and withered.

> One night as St Patrick went to bed he warned his servant that if he talked in his sleep he might be impolite. The servant was warned to listen to all what the Saint said.
>
> After sleeping for a while the Saint shouted: 'Bad luck to Ireland!' The listening boy responded: 'If so, let it be on the tips of the rushes!' After sleeping a little longer the Saint shouted again: 'Bad luck to Ireland!' The boy answered: 'If so, let it be on the highest part of the white cows!' After another short sleep the Saint shouted: 'I'll say again what I've said twice already: "Bad luck on

> Ireland!"' The boy answered: 'If so let it be on the bottom of the furze!' On waking the Saint asked his servant if he had said anything during his sleep, and, if so, what. The boy replied that he had said 'Bad luck to Ireland' three times. 'And what did you say?' asked the Saint. The boy explained what he had said. Ever since the tips of the rushes have been withered, the tips of the horns of white cows have been black, and the lower parts of GORSE bushes have been withered, and every priest should have a boy serving him at Mass.[1]

Alternatively:

> St Patrick got a piece of a dog to eat in a house one day. When he found out what it was he cursed the place. When he had it cursed he was sorry, but he couldn't take back the curse . . . without putting it on some other thing, so he put it on the tops of the rushes.[2]

In Ireland crosses are made for St Brigid's Day, 1 February. Although these crosses can be made from a variety of materials, including straw, sedge or bent grass, hay, wood, goose quills, tin, wire, cardboard and cloth,[3] it appears that rushes were originally used, and crosses sold in Irish craft shops are usually made of these.[4] Brigid is said to have been an abbess who lived from *c.*450 to 523,[5] but few facts are known about her life, and she seems to have absorbed many of the attributes of a pre-Christian deity.

> One day St Brigid was passing by an old shed and she heard a moaning cry. She entered the shed and beheld a dying man, She went over to him and spoke to him about God. He would not listen to her. She tried in vain to bring him to Our Lord, but he would not listen. Finally St Brigid went out and formed a cross of rushes. She returned to the man, and when he saw the cross he was moved to sorrow. He made his confession and received the last sacrament. Such was the origin of St Brigid's cross.[6]

In impoverished damp moorland areas rushes were an important resource. Writing of Ireland, Wyse Jackson notes they were used for 'thatch for houses and corn stacks, matting, hats and caps, for making and stuffing saddles and mattresses, making ropes, rattles, fishing nets, baskets, bracelets, belts, whips and rush candles'.[7] In North Wales late in the eighteenth century:

> The church is a humble gothic structure, the floor covered with rushes . . . The practice is almost universal through Wales. The floors are without pavement, and as straw is scarce, quantities of dried rushes are laid thick over the floor, for sake of warmth and cleanliness. The houses, few in number, are primarily mud cottages with rush-clad roofs; and, not being white washed, wear an aspect little inviting to the passing traveller . . . The hut consisted of one room . . . the floor was of native soil . . . a few bundles of rushes thrown down for a bed.[8]

Rush-lights, usually made from soft rush (*Juncus effusus*) seem to have been widely used throughout Britain and Ireland.

> Until about 60 years ago [*c.*1880] the people of Ireland had to provide their own light. They had no oil lamps or no light of any kind except the rush candles made by themselves. Each household made bundles of rush candles at a time and used them as required. The man or woman of the house brought in a bundle of rushes. He then peeled each rush to a small strip from one end to the other. He treated each of the rushes in the same way. When he had this done he prepared a vessel of fat. This vessel was boat-shaped and he poured the heated grease into it. Then he dipped the rushes into the grease and drew them through it until every bit of the rush was covered with fat. Then he would allow the rushes to dry and afterwards they were again dipped, and so on until they were thick enough for use. They were taken and stored up for the winter months.[9]

About a hundred years earlier Gilbert White provided an account of how rush-lights were prepared in Hampshire by 'decayed labourers, women and children', adding that most country people had little need for such lights 'because they rise and go to bed by daylight'. It appears that bacon fat was most often used, though in areas where bacon was not much eaten 'especially by the sea-side, the coarser animal oils will come in very cheap'. Gertrude Jekyll, in *Old West Surrey*, 1904, mentions a ninety-year-old friend who showed her how rush-lights were made, recalling that her mother said 'mutton fat's the best; it dries hardest'.[10] Wyse Jackson, who prepared rush-lights from *Juncus effusus* and beef fat, found that his light burnt 'well and steadily for about 40 minutes', but provided 'very poor light . . . barely enough to do more than faintly illuminate the room'.[11]

The strewing of rushes on church floors was formerly widespread, and frequently mentioned in churchwardens' accounts.

> S. Mary-at-Hill, London:
> 1493 – For three bundles of rushes for new pews, 3d.
> 1504 – Paid for 2 berdens rysshes for strewying newe pewes, 3d.
> St Margaret's, Westminster:
> 1544 – Paid for rushes against the Dedication Day, 1s 5d.
> Kirkham, Lancashire:
> 1604 – Rushes to strew the Church cost this year 9s 6d.[12]

Rush-strewing continues at a few churches, though it is the carrying of rushes to the church – rush-bearing – that is usually more important. In Cumbria rush-bearings take place at Ambleside, Grasmere, Musgrave, and Warcop. Of these, the Grasmere bearing, held on the Saturday nearest St Oswald's Day (5 August), is said to be the first recorded. Grasmere churchwardens' accounts record that in 1680 one shilling was spent on 'ale bestowed on those who brought rushes and repaired the Church'.[13] The church remained primitive with an unpaved floor until 1840, so until then the strewing of rushes was necessary to try and improve conditions for worshippers. In 1789 it was recalled:

> About the latter end of September, a number of young women and girls (generally the whole parish) go together to the tops of the hills to gather rushes. These they carry to the church, headed by one of the smartest girls in the company. She who leads the procession is styled the Queen, and carries in her hand a large garland, and the rest usually have nosegays. The Queen then goes and places her garland upon the pulpit, where it remains till after the next Sunday. The rest then strew their rushes upon the bottom of the pews, and at the church door they are met by a fiddler . . . the evening is spent in all kinds of rustic merriment.

It appears that throughout much of the nineteenth-century rush-bearing took place in July, with children preparing 'garlands of such wild flowers as the beautiful valley produces, for an evening procession'. In 1885 the date was moved to the Saturday nearest St Oswald's Day, 5 August, and since 2002 it has moved again to the last Saturday in the school year,[14] usually the second Saturday in July. In the morning volunteers, mostly women and children, gather in St Oswald's Church to prepare a number of traditional bearings, which eventually 'look as if they are made entirely of rushes, but are in fact created by winding the rushes around strong and practical materials like wood, wire netting and string'. These bearings are further decorated with flowers before being taken to the village school. At 3.30 p.m. an informal procession of people carrying bearings, led by a young man carrying a cross covered in yellow CHRYSANTHEMUMS, leaves the school and progresses to the village green, where the rush-bearing hymn is sung (without enthusiasm), before continuing to the church, which has its floor covered with a deep layer of rushes. A short service is held after which a variety of entertainments take place in the churchyard. The bearings are placed around the church and remain there for about a week.[15]

Ambleside's rush-bearing takes place on the first Saturday in July, while Warcop holds its festival on St Peter's Day, 29 June, unless this falls on a Sunday, in which case the event is held on Saturday 28 June. The National Trust owns a highly idealised painting of Grasmere's rushbearing, by Frank Bramley early in the twentieth century, and a mural in St Mary's Church, Ambleside depicts its 1944 festival.

At the Cumbrian rush-bearings young girls carry token rushes on a sheet, elsewhere, in Lancashire and Yorkshire, the focus is an elaborate rush-cart. The first mention of rushes being transported by cart is in 1617, but the custom as known today seems to have become popular in the mid-eighteenth century, and as communities became more industrialised the carts became more elaborate.[16] These were usually two-wheeled vehicles with a tall pyramid of rushes on them, pulled by young men. Although the need for rushes as church floor-covering diminished, the making of carts continued usually for the local wakes, but by the end of the nineteenth century rush-carts had died out in most places. Occasional revivals took place early in the twentieth century, and since then a number of persistent revivals have taken place. At Saddleworth in Greater Manchester the local morris side initiated a revival in 1975, and at Sowerby Bridge in West Yorkshire a revival took place in 1976 to mark the Queen's Silver Jubilee. Both events continue to thrive. The Sowerby Bridge cart processes on a lengthy route around the town and surrounding villages on the first complete weekend in September, making token presentations of a bunch of rushes and a bunch of BULRUSH at various churches, and followed by a large number of morris sides (fifteen in 2016).[17]

The flexible stems of rushes were widely used in children's pastimes.

> [Dorset, *c.*1930] making little green baskets (only girls did this) from soft rush.[18]

> Rushes we picked, wound around and made a boat.[19]

> [1940s] I used to make little 'braided' chains from the young pliant stems of *Juncus effusus* . . . Two stems were tied at one end and by repeatedly laying one strand over the other a very simple chain could be produced.[20]

> Soft rush . . . a friend in Kilrush told me once how her grandmother (R.I.P.) taught her to make rush hats by 'weaving' several rushes together to end up with a product like a witch's hat.[21]

> [Durham coalfields, 1950s] I was shown by my father how to make a 'whip'. This was made from the common field rush. The whip looked like an African fly whisk with loose head and plaited handle. It seemed the making of it was the important part of its use, for it had no known purpose.[22]

The collections of the National Museum of Ireland include approximately forty playthings made from rushes, with rattles being the most frequent, but also including pinkeen (little fish) nets, miscellaneous 'toys', and a swimming aid.[23]

RYE GRASS (*Lolium perenne*)

Perennial native grass (Fig. 44), abundant, and widely cultivated to provide monotonous grazing for dairy cattle.

Local names include:

Crap in Suffolk and Sussex
Crap-grass in Sussex
Devon eaver, 'used more especially amongst Somerset farmers'
Eaver in Cornwall, Devon and Somerset; 'French *ivraie*, drunkenness, from the intoxicating qualities of darnel [*Lolium tremulentum*], an allied species'
Eaver-grass, and eever, in Devon
Ever in Cornwall and Devon
Ever-grass, every, and every-grass, in Dorset
Evor in Devon; 'but the term is often used . . . for any grass seeds

other than clover'
HARDHEADS in Somerset
Hayver in Cornwall
Heaver/s in Cornwall and Devon
PICKPOCKET in Devon
Ray in Wiltshire
Seeds, or sids, in Oxfordshire and Sussex
White nonsuch, given to the seed in Norfolk.

[Leicestershire, 1940s] rye grass – tinker-tailor grass – taking off the flowering glumes to count off 'tinker, tailor, soldier, sailor, rich man, poor man, beggarman, thief' – ending at the occupation of one's future husband.[1]

[Retford, Lincolnshire] We used to count the rye grass – we'd say 'He loves me, he doesn't, he would, if he could, but he can't' to the top. Then we'd say 'What house? Little house? big house? pig stye? barn?' We'll get married: 'Coach, carriage, wheelbarrow, muck cart'. 'What'll you marry in? Silk, satin, muslin, rags'. 'How many children will you have?' – then you'd count up the stalk. Oh, I've missed one. 'What sort of man will he be – rich man, poor man, beggar man, thief.'[2]

[Desborough, Northamptonshire, 1920s] We used to go up rye grass spikelets saying 'He loves me, he loves me not'.[3]

Names which relate to similar pastimes include the unlocalised LOVE-ME-LOVE-ME-NOT, and:
Aye-no-bent in Gloucestershire
Does-my-mother-want-me in Somerset
Soldiers-sailors-tinkers-tailors in Wiltshire
Tinker-grass in Yorkshire
What's-your-sweetheart in Sussex
Yes-or-no in Somerset.

S

SAFFRON (*Crocus sativus*)

Autumn-flowering crocus of unknown origin, formerly grown in the British Isles for its stigmas and styles which are used as a spice.

Saffron is recorded as being cultivated around Walden in Essex in the time of Edward III (1327–77), and by the beginning of the sixteenth century it was so extensively grown locally that the town began to be known by its current name of Saffron Walden. Eventually the importation of less expensive saffron made local production uneconomical, and by 1790 the growing of saffron had 'virtually ceased' in the area.[1]

Saffron (with DODDER sometimes being substituted for the real thing) cake was a local delicacy in Cornwall.

> The fishermen of both north and south Cornwall believe that saffron brings BAD LUCK, and that saffron cake carried in a boat spoils the chance of a catch.[2]

In Cornwall and Devon, saffron buns and clotted cream were eaten on GOOD FRIDAY.[3]

In some parts of England JUNIPER was known as saffron.

SAGE (*Salvia officinalis*)

Perennial evergreen herb, native to southern Europe, long cultivated in the British Isles as a culinary herb.

In common with other herbs, such as PARSLEY, sage was thought to grow best either for the dominant partner in a marriage, or if the wife was dominant.

> [Bishops Nympton, Devon] after a wedding the bride and bridegroom must each plant a small sage bush brandise-wise. The size

> to which the sage bushes grow will show which will be the ruler in the house; it will of course be the planter of the larger of the two bushes.[1]

> [In Plymouth it is said that where sage] flourishes well, it denotes that the mistress is head of the household.[2]

> In Bucks it is not only maintained that the wife rules where sage grows vigorously – a notion elsewhere attached to the ROSEMARY – but a farmer recently informed me that the same plant would thrive or decline as the master's business prospered or failed. He asserted that it was perfectly true, for at one time when he was doing badly, the sage began to wither, but as soon as the tide turned the plant began to thrive again.[3]

> As the sage bush flourishes so does the family.[4]

In addition to its culinary uses sage was also used for cleaning teeth and the treatment of gum problems.

> Teeth: clean with fresh sage leaves.[5]

> A gypsy 'toothpaste' in the 1940s and 1950s was chopped sage and salt in an equal parts mix, rubbed on the teeth with Irish linen.[6]

> A sage leaf with its thick vein removed will relieve sore GUMS if put between the denture and the gum. Leaves steeped in water then used as a gargle will relieve sore THROATS.[7]

> Wash and crush a sage leaf to put on mouth ulcers, like a little poultice. It will ease the pain, disinfect, soothe and very quickly heal.[8]

Other medicinal uses include:

> Boil sage leaves, drink water for ARTHRITIS.[9]

> In 1950 I was in extreme pain because my milk would not dry up, and my baby was eight months old . . . We were living in rooms, and the elderly landlady was concerned for me, but many years earlier had experienced the same painful condition when she had a stillborn baby, and at the time she and her husband kept a hardware shop and used to buy pegs from the local gypsies. One old gypsy woman saw what her problem was and suggested she sleep with a bunch of fresh sage under each armpit!! It worked for her,

and as she had sage in her garden, she offered some to me. At that stage I was desperate to try anything, so kindly accepted her offer, and passed a very uncomfortable night trying the cure. I was astonished to find that the following morning the milk had disappeared and I was free of the frightful pain and discomfort.[10]

[During the First World War] HEADACHES, etc. – sage tea: wash the leaves and boil them 10–15 minutes, leave to cool, not too bad to take; we certainly were cured.[11]

Sage tea – taken to help CANCER – a few fresh leaves daily in hot water.[12]

ST ALBAN – ROSE used to commemorate.

ST ALDHELM'S DAY (25 May) – WELL-DRESSING at Frome, Somerset.

ST ANTHONY OF PADUA – MADONNA LILY associated with.

ST BRIDE (BRIDGET) – tenuously associated with a BLACK POPLAR at Aston-on-Clun, Shropshire.

ST BRIGID'S DAY (1 February) – RUSH crosses made; straw costumes worn (see WHEAT).

ST CANDIDA (WITE) – PERIWINKLE associated with.

ST COLUMCILLE'S DAY (15 June) – TURNIP should be planted before.

ST CONGAR – YEW at Congresbury, Somerset, associated with.

ST DAVID'S DAY (1 March) – DAFFODIL and LEEK worn on; barley and OAT sown on.

ST FINTAN – SYCAMORE at Clonenagh, Co. Laois, associated with.

ST FRANKIN'S DAYS

In 1894 Sabine Baring-Gould found in Devon that:

In the Taw valley, at Eggesford, Burrington, etc., there exists a saying that the 19th, 20th, or 21st May, or three days near that time, are 'Francismass' or 'St Frankin's days', and that then comes a frost that does much injury to the blossom of APPLES. The story relative to this frost varies slightly. According to one version, there was a brewer, name of Frankan, who found that cider ran his ale so hard that he

> vowed his soul to the Devil on the condition that he would send three frosty nights in May to cut off the apple-blossom annually.
>
> The other version of the story is that the brewers of North Devon entered into compact with the Evil One, and promised to put deleterious matter in their ale on condition that the Devil should help them by killing the blossom of the apple-trees. Accordingly, whenever these May frosts come, we know that his Majesty is fulfilling his part of the contract, because the brewers have fulfilled theirs by adulterating their beer. According to this version, St Frankin is a euphemism for Satan.[1]

Thirteen years later in the south of the county:

> My gardener, on being told to put some bedding plants from the greenhouse into the open to harden, said it would not be well to do so 'until Franklin nights were on'. To my inquiry, he answered that he did not know who St Franklin was, but people thereabouts never thought the cherries or mazards were safe from frosts until St Franklin nights were over, and these nights were the 19th, 20th and 21st May. On these nights we had severe frost at [Newton] St Cyres, and the potatoes were much cut.[2]

ST GEORGE'S DAY (23 April) – DANDELIONS gathered by wine-makers; red ROSES worn, RUNNER BEANS sown on.

ST JAMES THE GREAT'S DAY (25 July) – APPLES christened.

ST JAMES THE LESS'S DAY (1 May) – APPLES christened.

ST JOHN OF BEVERLEY – PRIMROSES used to commemorate.

ST JOHN'S DAY (24 June) – ST JOHN'S WORT associated with; STOCKS treated to produce double flowers on.

ST JOHN'S EVE (23 June) – RIBWORT PLANTAIN used in divination.

ST JOHN THE BAPTIST – ate YELLOW RATTLE.

ST JOHN'S WORT (perforate St John's wort, *Hypericum perforatum* and other *Hypericum* spp.)

Perforate St John's wort: perennial (Fig. 48), yellow-flowered herb, common in dry grassy places in England and Wales, less frequent in Scotland and Ireland.

Local names include:

AMBER in Kent (and USA)
John's wort in Somerset
Penny-john in Norfolk
Rosin-rose in Yorkshire
TOUCH-AND-HEAL in Co. Antrim and Co. Down
Yellow star-of-Bethlehem in Sussex.

Species of St John's wort were formerly valued for providing protection. The name *Hypericum* was originally used by the Greeks for a plant which was placed above religious figures with the purpose of warding off evil spirits.[1] It is not known whether or not this plant was, in fact, a species of what is now known as *Hypericum*, but there are certainly many records of the genus being used in Britain for such purposes. An early thirteenth-century work on the life of St Hugh of Lincoln contains an account of a woman who was tormented by a 'licentious demon' in the shape of a young man. After much suffering she was approached by another spirit who advised her to take a certain plant, hide it in her bosom, and scatter it around her house. The demon lover found the plant 'disgusting and stinking' and was unable to enter the house while the plant remained in place. The woman considered the plant, which was known as '*Ypericon* in Greek, and in Latin either the perforated plant or St John's wort', to be her sole defence against the demon, but later she found simple piety provided adequate protection. A monk to whom she showed the plant in turn showed it to a young couple in Essex, who were subsequently cured and protected from attacks from demons with whom they had been seen to speak. The writer of the life noted that physicians regarded St John's wort to bc 'a sovereign remedy against poison – even if this is due to the bite of a poisonous animal'. Therefore it was 'ridiculous to suppose that a bodily remedy for snakebite should not by God's mercy have been effective against the assaults of the ancient Serpent'.[2] A twentieth-century writer states that St John's wort is 'one of the most beneficent of magical herbs, protecting equally against FAIRIES and the DEVIL', and quotes a couplet spoken by a demon lover who was unable to approach a girl carrying the plant:

> If you would be true love of mine
> Throw away John's Wort and Vebein [VERVAIN].[3]

It is probable that women who were assaulted by demon lovers would, today, be diagnosed as suffering from a form of DEPRESSION, so perhaps it is unsurprising that in recent years St John's wort has been used as an antidepressant:

> The plant contains a red pigment, hypericin, which is thought to account for its antidepressant and tranquillising powers. In one trial of 3,250 depressed patients, 80 per cent showed improvement when taking *Hypericum*. A smaller trial demonstrated that it was particularly effective against seasonal affective disorder (SAD), the gloom that can descend once the evenings draw in . . . The modern preparation of St John's wort was first marketed in 1989, but by 1996 its sales were running at £41 million a year.[4]

In the eighteenth century:

> The superstitious in Scotland carry this plant [perforate St John's wort] about with them as a charm against the dire effects of WITCHCRAFT and enchantment. They also cure, or fancy they cure, their ropey milk, which they suppose to be under some malignant influence by putting this herb into it and milking afresh upon it.[5]

The English translation of a Gaelic incantation collected from a Hebridean cottar in the nineteenth century runs:

> St John's wort, St John's wort,
> My envy whosoever has thee,
> I will pluck thee with my right hand,
> I will preserve thee with my left hand,
> Whoso findeth thee in the cattlefold,
> Shall never be without kine.[6]

Also in the Hebrides:

> St John's wort is one of the few plants still cherished by the people to ward away second-sight, enchantment, witchcraft, EVIL EYE and DEATH, and to ensure peace and plenty in the house, increase and prosperity in the fold, and growth and fruition in the field. The plant is secretly secured to the bodices of the women, and in the vests of the men, under the left armpit. St John's wort, however, is effective only when the plant is accidentally found. When this occurs the joy of the finder is great.[7]

In the Forest of Dean:

> [According to my grandparents, b.1856 and 1858] St John's wort was brought into the house and tied into bunches, hung in the windows – I believe it was supposed to prevent LIGHTNING striking the house.[8]

St John's wort's protective qualities were particularly valued at Midsummer. According to a folklorist writing in 1977 about the Feast of St John the Baptist, or Midsummer's Day:

> the saint's own golden flower, St John's wort – which is quite clearly a sun-symbol – was brought indoors to promote good fortune and protect the house from FIRE.[9]

Whether St John's wort is 'clearly a sun-symbol' is debatable, but it was certainly widely used to decorate and protect homes on MIDSUMMER'S EVE. According to John Stowe, writing of London in 1603, but possibly describing much earlier practices:

> On the vigil of St John the Baptist, and on St Peter and Paul the apostles, every man's door being shadowed with green BIRCH, long FENNEL, ST JOHN'S WORT, ORPIN[E], white lilies and such like, garnished upon with garlands of beautiful flowers, had also lamps of glass with oil burning in them all night . . . which make a goodly show, namely in New Fish street, Thames street, etc.[10]

In Cornwall:

> [Midsummer was] the time for lighting Midsummer Bonfires, an ancient custom rescued from extinction by the Old Cornwall Movement in 1929 and still flourishing . . . At St Cleer the fire is crowned with a witch's broom and hat, a sickle with a handle of newly cut OAK is thrown into the flames and wreaths of St John's wort are hung around the village – all this was traditionally said to banish witches.[11]

At these celebrations the *Arlodhes an Blejyow* (Lady of the Flowers) casts a bunch of plants tied with coloured ribbons into the flames. These plants are chosen to represent both 'good' and 'bad' herbs, and a booklet produced by the Federation of Old Cornwall Societies in 1977 lists thirty-six suitable species. The list of good plants is headed by St John's wort, known in Cornish as *losow sen Jowan*.[12]

In Wales:

> [On St John's Eve] it was custom in many parts of the country to place over the doors of houses sprigs of St John's wort or, if this were not available, the common MUGWORT: the intention was to purify the house from evil spirits. St John's wort gathered at noon on St John's Day was thought to be good for several complaints and if dug at midnight on the Eve of St John the roots were good for driving the devil and witches away.[13]

In Ulster:

> [On St John's Eve] the flower of St John's wort was brought into the house as protection against the evil eye.[14]

Also on Midsummer's Eve, St John's wort was used to foretell the future.

> In Hertfordshire and elsewhere in England, girls would use St John's wort to test their chances of matrimony. On Midsummer's Eve they would pluck a piece of the plant; if it appeared fresh the following morning the prospects were good.[15]

In the mid-nineteenth century St John's wort was used in Wales to predict life expectancy. A piece of the plant was gathered for each person in the house, cleaned 'free from dust and fly', and each piece named after a member of the household before being hung on a rafter. In the morning the pieces were examined; those whose pieces had withered most were expected to die soonest.[16]

In Aberdeenshire it was believed that if one slept with a plant of St John's wort under one's pillow on St John's Eve, the saint would appear in a dream, give his blessing, and prevent one from dying during the following year.[17]

Writers on folklore mention St John's wort as being valued as a medicinal herb but they are unspecific about which illnesses it was used to treat. According to Ruth Tongue:

> There are a few herbs which are almost universal specifics. St John's wort is one; an infusion of the leaves will cure CATARRH, grow hair, heal CUTS and make a poultice for SPRAINS. An ointment made from it is good for BURNS.[18]

If she was correct, it is surprising that no information on the plant having any medicinal use has been contributed to the Plant-lore Archive project.

ST JOSEPH OF ARIMATHEA – HOLY THORN associated with.

ST LEONARD – LILY OF THE VALLEY grew from his BLOOD.

ST MARK'S EVE (24 April) – GRASS and HEMP seed used in divination.

ST MARTIN'S EVE (10 November) – HEMP seed used for LOVE DIVINATION.

ST MARY THE VIRGIN – COW PARSLEY, DEVIL'S-BIT SCABIOUS, HAWTHORN, LUNGWORT, MADONNA LILY, REDSHANK and SEA BEANS associated with; see also MAY ALTAR.

ST MICHAEL'S DAY (29 September) – see MICHAELMAS DAY.

ST MOALRUDHA – BUTTERWORT associated with.

ST NECTAN – FOXGLOVE associated with.

ST NEWLINA – FIG tree at St Newlyn East associated with.

ST PATRICK – BUTTERWORT associated with; cursed RUSH (and gorse).

ST PATRICK'S DAY (17 March) – BLACK MEDICK, GOAT WILLOW and SHAMROCK associated with; POTATOES sown on or before.

ST PETER'S DAY (29 June) – APPLES christened on; Barnstaple church decorated with flowers (see FLOWER FESTIVALS); RUSH-bearing at Warcop.

ST RICHARD OF CHICHESTER – tended FIG trees at West Tarring.

ST STEPHEN'S DAY (26 December) – straw costumes worn (see WHEAT).

ST SWITHIN'S DAY (15 July) – APPLES christened.

ST THOMAS BECKET – planted FIG trees at West Tarring.

ST VALENTINE'S DAY (14 February) – BROAD BEANS planted; PEA crop good if 'hedges dripped'; red ROSES associated with

ST VALENTINE'S EVE (13 February) – HEMP seed used for love divination.

ST VITUS' DANCE (Sydenham's chorea) – WOOD SAGE used to cure.

ST WITHBURGA – LORDS-AND-LADIES associated with.

SALLY-MY-HANDSOME – see HOTTENTOT FIG.

SCALDS – HART'S TONGUE and PRIMROSE used to treat.

SCARLET FEVER – FOXGLOVE cures.

SCARLET PIMPERNEL (*Anagallis arvensis*)

Annual low-growing herb producing red flowers, widespread on arable land in lowland areas.

Local names include:
Adder's eyes in Hertfordshire
BIRD'S EYE in Buckinghamshire, Oxfordshire and Wiltshire
Bird's tongue in Norfolk
CHICKWEED, and chiuk, in Somerset
Crimson pirate in London
CRY-BABY, and CRY-BABY-CRAB, in Somerset
Drops-of-blood in Wiltshire
EYEBRIGHT in Cornwall
Ladybird, and laughter-bringer, in Somerset
Little Jane in Devon
Mother Redcap in Warwickshire
Nupinole in Wiltshire; 'no doubt due to progressive corruption of pimpernel'
OLD MAN in Wiltshire
Old-man's friend, and old-man's glass-eye, in Somerset
ORANGE LILY in Dumfriesshire
Owl's eyes in Somerset
Peepers in Yorkshire
Pheasant's eye, pumpernal, and ragged jack, in Somerset
Red bird's-eye in Devon
Red weed in Dorset
Shepherd red-eye in Yorkshire
Shepherd's calendar in Devon
Shepherd's daylight in Somerset
SHEPHERD'S DELIGHT in Lincolnshire
Shepherd's joy in Dorset and Somerset
SNAP-JACKS, and STAR-OF-BETHLEHEM, in Somerset
Tom pimpernel, or tom pimpernowl, in Yorkshire
Wind-pipe in Somerset; 'possibly a corruption of wink-a-peep or wink-and-peep'
Wink-a-peep in Cheshire, Shropshire and Staffordshire.

Since scarlet pimpernel flowers tend to close in cloudy weather, they were thought to predict changes in the weather, leading to such names as the widespread poor-man's weather-glass, and weather-glass, and:

Change-of-the-weather, and farmer's weather-glass, in Somerset
Grandfather's weather-glass in Devon
Little peep-bo, and little peeper, in Somerset; 'no doubt from the way in which it opens or closes its petals according to the weather'
Old-man's weather-glass in Somerset
Ploughman's weather-glass in Wiltshire
Shepherd's glass in Norfolk and Rutland
Shepherd's warning in Lincolnshire
SHEPHERD'S WEATHERGLASS in Yorkshire
SUNFLOWER in Cumberland
Weather-flower, and weather-teller, in Somerset.

Alternatively, the flowers, which supposedly close in the afternoon, could be used to indicate the time, stimulating such names as:

SHEPHERD'S CLOCK in Buckinghamshire and Gloucestershire; 'from its closing its flowers at a certain time, i.e. about 2 p.m.'
Shepherd's dial in Middlesex
Shepherd's watch in Cambridgeshire, Essex and Norfolk
TWELVE-O'CLOCK in Somerset.

In Devon in the early 1950s:

> A lady used to walk the roads and lanes, always with a bicycle with a basket on the front into which she would put hedgerow and field wildflowers. Scarlet pimpernel, she told me, was better than cold tea for sore EYES.[1]

SCIATICA – HOP, HORSERADISH, POTATO and ROYAL FERN prevent or cure.

SCOTS LOVAGE (*Ligusticum scoticum*)

Perennial herb, frequent around the coast of Scotland and the coast of Northern Ireland.

Local names include lovage, Scotch parsley, sea-parsley, shemis, shunas and siunas, the last three derived from the Gaelic *sionnas* or *shunnis*.

According to Lightfoot, writing in 1777:

> Frequent in the Western islands of Jura, Isla, Iona and Skye, in which last it is call'd by the name of *shunis*, and is sometimes eaten raw as a sallad or boil'd as greens.[1]

SCOTS PINE (*Pinus sylvestris*)

Evergreen tree, native to the Scottish Highlands, but much planted, and naturalised elsewhere, particularly on sandy soils.

Local names include:

Bay-lambs, given to the male flowers in Yorkshire
Deal-tree in East Anglia and Northamptonshire; 'the fir that mainly produces the deal timber of commerce'
Eye-glasses in Somerset'; 'it is not easy to see the connection'
Firren-dales, or firren-deals, in Norfolk
Keyball in Devon
Sticky tree in Somerset.

Names given to the cones include:

Berk-apple, or birk-apple, in Yorkshire; 'and probably cones of other species'
CUCKOO in Essex and Yorkshire
Deal-apple in eastern England and Northamptonshire
Fir-top in Scotland
OYSTERS in Devon; 'the scales of which, with the seeds, nearly enough resemble oyster-shells'
Pur-apple in Northamptonshire
Sheep in Yorkshire
Tory-tops in Co. Cork.

Being a conspicuous, evergreen and easily recognisable tree, Scots pines were planted as landmarks, for example indicating drovers' routes, and farms which were willing to offer accommodation for them and their cattle.[1] In the Cotswolds, Scots pines were said to have been planted as an expression of Jacobite sympathies, and indicated places where fugitive Jacobites could find safe harbour.[2]

In Scotland in the 1820s:

> Splinters of the wood used by the Highlanders in the place of candles, the younger individuals of a family holding them in turn.[3]

SCOUR (diarrhoea in cattle) – YELLOW BARTSIA causes; most prevalent when LILAC in flower; treated using COMFREY and SHEPHERD'S PURSE.

SCROFULA – see KING'S EVIL.

SCURVY – CRANBERRY, SCURVY-GRASS and WATERCRESS cures.

SCURVY-GRASS (*Cochlearia* spp.)

Annual to perennial herbs with glossy dark green leaves and white flowers, mostly restricted to maritime areas, but spreading inland along regularly salted roads.

Local names include: BADMAN'S OATMEAL in Co. Durham, screebie- (or screevie-)grass in Scotland, and scrooby-grass in northern England and Scotland.

> [Shetland islanders] have much scurvy-Grass; God so ordering it in his wise Providence that *Juxta venenum nascitur Antidotum*, that seeing the SCURVY is a common Disease of the Countrey, they should have the Remedy at hand.[1]

SEA BEANS, also known as Molucca beans

The seeds of approximately fifty tropical plants have been recorded on the shores of western Europe, having been carried by the Gulf Stream from Caribbean islands and north-east South America.[1] The transatlantic origin of such seeds was recognised by European scientists as early as 1670,[2] but for inhabitants of remote areas on the western coasts of the British Isles the origins of the seeds remained unknown. The three seeds which attracted greatest attention were those of two members of the pea family, Leguminosae: *Caesalpinia bonbuc* and *Entada gigas*, and a member of the bindweed family, Convolvulaceae: *Merremia discoidesperma*.

The largest and most frequently found of these is the heart- or kidney-shaped seed of *Entada*, with which 'superstitions are associated' in western Ireland,[3] and which 'are washed ashore on the Isles of Scilly, where the children call them lucky beans'.[4]

Writing of Cornwall in 1602, Richard Carew noted:

> The sea strond is strowed with . . . certain Nuts, somewhat resembling a sheepes kidney . . . the outside consisteth of a hard darke coloured rinde. the inner part, of a kernel voyd of any taste, but not so of virtue, especially for women trauayling in childBIRTH, if at least, old wiues tales may deserve any credit.[5]

The Cuming Museum in the London Borough of Southwark holds a collection of objects relating to London superstitions assembled by Edward Lovett, who served on the Folk-lore Society's council from 1903 to 1920. Many of the objects were used to prevent illness and misfortune, and amongst them is an *Entada* seed – a 'Lucky Bean' – which in north-west London was supposed to ensure good fortune. The fresh, shiny appearance of the seed suggests that it has not undergone lengthy immersion in water, and it is probable that it was brought home by a traveller who acquired it in the tropics.[6]

At about the same time as Lovett was assembling his collection, imported beans, possibly *Entada* seeds, were being sold by quack doctors. A South Shields, County Durham, lodging-house keeper, who had several quacks (or 'crocuses' as they were commonly known) among her lodgers, remembered:

> Another would buy a pound of foreign beans from the chemist for a shillin' – like little hard pebbles they were – and sell them as lucky beans or magic beans at one and six each. If they put a thread through to wear round your neck, then it cost ye two bob. Proper bloody frauds they were.[7]

In the 1940s *Entada* seeds were searched for along the Pembrokeshire coast, for it was believed that such seeds would bestow good luck on their finder if they were worn on his (or, more usually, her) person. Many people made their seeds into brooches, lockets or similar trinkets for this purpose.[8] More recently *Entada* seeds have been offered for sale as 'Lucky Sea Beans' at seaside resorts, where they form part of the stock of shops selling shell ornaments, exotic seashells and similar items.[9] Like most, probably all, of the shells, the beans have been imported from tropical areas. Occasionally the name given to the seeds varies according to the expected interests of potential buyers. Thus 'lucky bingo sea beans' were offered at 10p each in Blackpool, Lancashire in 1986, and 'lucky folklore sea beans' were offered, also at 10p each, at Ambleside, Cumbria, in 1988.

It appears that *Caesalpinia* and *Merremia* seeds were valued only in the Highlands and Islands of Scotland. *Merremia* seeds, which have a characteristic cross-marking, were used as amulets to ease childbirth. In 1891 Lieutenant-Colonel Feilden sent such a seed, which he had acquired about twenty years earlier, to the Royal Botanic Gardens, Kew. This seed's Gaelic name meant Mary's Bean, and, in Roman

Catholic communities in the Outer Hebrides, it was believed that if the seed was clenched in the hand of a woman in labour it would ensure an easy delivery. Such seeds could become treasured heirlooms, and Feilden's example had come from a North Uist woman who stated that it had formerly belonged to her grandmother.[10]

According to Alexander Carmichael, the great collector of Hebridean folklore:

> *Arna Moire*, kidney of Mary, *tearna Moire*, saving of Mary: This is a square thick Atlantic nut, sometimes found indented along and across, the indentations forming a natural cross on the nut. It is occasionally mounted in silver and hung round the neck as a talisman. Every nurse has one which she places in the hand of the woman to increase her faith and distract her attention. It was consecrated on the altar and much venerated.[11]

In January 1893 a *Merremia* seed was sent by the Revd Alexander Stewart to a meeting of the Society of Antiquaries of Scotland:

> I send you a specimen of a kind of amulet very highly prized by the people of the three Uists – North Uist, Benbecula, and South Uist – which is known locally as *Aire Moire* (Virgin) Mary's kidney, It is really a kind of bean occasionally picked up on the shores of the Outer Hebrides . . .
>
> It is considered all the more valuable and sacred if, as in this specimen, there is something like a cross on one side of it. Midwives use it as a charm to alleviate the pains of parturition. Very often also a small hole is drilled through either end and through these holes a string is passed and looped, so that it may be hung round the neck of children when they are TEETHING, or suffering under any infantile ailments.
>
> It is most in request amongst Catholics, as its local name implies; but Protestants also sometimes use it. It is oftenest met with in South Uist and the Island of Barra, where at least three-fourths of the people are Roman Catholics. Canary-coloured specimens are sometimes got, almost white, and these are very highly prized. These amulets are greatly valued, and it is not easy for outsiders to get specimens.[12]

As *Merremia* seeds are characteristically brownish-black, Stewart's statement about yellow or almost white seeds is confusing, and it is probable that these light-coloured seeds were, in fact, *Caesalpinia*, rather than *Merremia*. *Caesalpinia* seeds are round and light greyish-yellow in colour, their surface being acorn-like with hair-like cracks. In 1703 Martin Martin gave information on the then current uses of *Caesalpinia* seeds. They were worn around children's necks to protect their wearers from WITCHCRAFT and the EVIL EYE, and were believed to change from yellow to black if any evil was intended. Martin claimed to have observed this change, but was unable to suggest any explanation.[13] It is possible that seeds worn by children who were neglected and unwashed might become darker in colour and that such children were suspected of being bewitched.

Caesalpinia seeds were also used to protect CATTLE, and Martin was told by Malcolm Campbell, Steward of Harris:

> that some weeks before my arrival there, all his cows gave blood instead of milk for several days together; one of the neighbours told his wife that this must be witchcraft, and it would be easy to remove it, if she would put the white nut, called the Virgin Mary's Nut, and lay it in the pail into which she was to milk the cows. This advice she presently followed, and having milked one cow into the pail with the nut in it, the milk was all blood, and the nut changed its colour to dark brown; she used the nut again, and all the cows gave pure good milk which they ascribe to the virtue of the nut.[14]

SEA BEET (*Beta vulgaris* ssp. *maritima*), also known as WILD SPINACH

Perennial herb with dark green shiny leaves, common and widespread around the coasts of England, Ireland and Wales, less frequent in Scotland; the parent of BEETROOT, MANGOLD and foliage beet (spinach beet and Swiss chard).

Sea beet has long been gathered for food:

> [Isle of Wight, 1856] boiled instead of greens, the sea beet is much relished by the poorer classes.[1]

> [Isles of Scilly] young leaves of sea beet are collected and boiled as spinach.[2]

> According to my mother (b.1918) of Whitstable, Kent, we ate wild spinach, but then we grew it in our garden as it was cleaner there.

It is very hardy and self-sows all the time. (But her children hated it as it was stringy and tough, and made a bad feel on our teeth, like RHUBARB).[3]

SEA BINDWEED (*Calystegia soldanella*)

Perennial herb with attractive pink flowers, scattered on sand dunes and consolidated shingle along the coast, but absent from northern Scotland.

A plant associated with the Stuarts is the sea bindweed . . . In 1745 Prince Charles Edward landed on the Island of Eriskay, and from his pocket he scattered seeds of this white striped pink convolvulus, which he had gathered while waiting to embark to France. These seeds grew and seeded themselves in turn and are still found growing at this spot and nowhere else in the Outer Hebrides.[1]

SEA CAMPION (*Silene uniflora*)

Procumbent, perennial, white-flowered herb, common in rocky maritime habitats.

Local names include balloons in Somerset, buggie-flower in Shetland, dead (pronounced deed)-na's grief in Northumberland, THIMBLE in Berwickshire, white snapjacks in Somerset, and witches' thimbles in Northumberland.

From an elderly friend in Porthnockie, about 30 miles from Burghead: DEADMAN'S BELLS = sea campion . . . it was untouchable, never picked and never brought into the house; she thinks the reason for this ban was that in that area of steep cliffs it grew on rocky ledges, highly dangerous for children.

A friend who used to live in Buckie knew sea campion as Devil's Hatties; it grew in a dangerous area called 'The Back o' the Head', i.e. the headland of Burghead.[1]

Despite this, sea campion flowers were used in childhood pastimes.

Sea campion has five different Jersey French Norman names, three of which indicate its use in a children's game. The calyx is turned inside out and the petals removed together with all but two of the stamens, these being retained to represent the arms of a washerwoman hanging the washing on the line. This is sometimes called a crinolined lady, and heated exchanges have taken place over this point.[2]

Sea campion flower turned inside out makes a little ballerina.[3]

SEDATIVE – BALM used as.

SEASONAL AFFECTIVE DISORDER – ST JOHN'S WORT effective against.

SELFHEAL (*Prunella vulgaris*)

Small perennial herb with purple (occasionally pinkish) flowers, common and widespread in grassy places.

Local names include the widespread heart-o'-the-earth and PRINCE'S FEATHER, and:

All-heal in Cheshire, Somerset and Yorkshire
Blue curls in Devon and Dorset; also USA
Brownwort in Somerset; 'from its being supposed to cure the disease called in German *die braune*, a kind of QUINSY'
Brunnels in Somerset
Bumblebees in Yorkshire
Caravaun-beg in Somerset
Carpenter-grass in Cheshire 'considered very efficacious in curing CUTS'
Carpenter's herb in Gloucestershire and Somerset; 'good for cuts'
Fly-flowers in Gloucestershire
HEARTSEASE in Co. Donegal and Dublin; 'an infusion of the plant is highly esteemed for the HEART'
Hook-herb in Somerset
Keanadha-hassog in Co. Donegal
LADY'S SLIPPER in Hampshire
London bottles in Ayrshire
PICKPOCKET in Essex
Proud carpenter in Cheshire
Scotch gramfer-griggles in Dorset
SOLDIER'S BUTTONS in Somerset; an unlikely name
TEA in Caithness
TOUCH-AND-HEAL in Co. Antrim and Co. Down.

[North Hampshire] nursemaids will warn their little charges 'not to pick black-man flowers' (*Prunella vulgaris*) telling that the plant belongs to the DEVIL, who is exceedingly annoyed when it is gathered, and will certainly appear in the night to carry off the child who has so angered him.[1]

Despite its name there are few records of selfheal being used as medicine.

> [Colonsay] a popular remedy for chest ailments, it [selfheal] was collected in summer, tied in bundles, and hung up to the kitchen roof to dry for winter use. The plants were boiled in milk and strained before using; butter was added.[2]

> If you are anxious to get rid of a COUGH, go out to the field and gather some little purple plants called selfheal. Put them in water and boil them. Then drink the juice which has boiled out.[3]

> The minerac herb [selfheal] cures the minerac [a mysterious wasting] disease . . . Nine pieces are got (saying the name of the person you require it for) washed clean and rubbed until a froth is produced from it. The froth is mixed with water and turns it green. The person that needs it drinks it three mornings in succession and blesses himself each time. He is not allowed meat, eggs or much butter when taking it.[4]

SERVICE-TREE (*Sorbus domestica*)

Deciduous tree, with white flowers and greenish-brown fruit, native, but fewer than fifty trees known, restricted to two localities in England and one in Wales.

In August 1853 the Worcestershire Naturalists' Field Club visited a service-tree growing in Wyre Forest, then believed to be the only wild representative of its species in the British Isles:

> There was an undoubted feeling of superstitious protection attached to the tree, whose fruit was commonly said, by foresters living in the vicinity, 'to keep out the WITCH from habitations', and for this reason they hung up the hard fruit, which would remain a long time without decaying. The tree is commonly called by the foresters the Whitty, or Whitten, pear; perhaps from the old English word *witten*, to know, meaning the *wise tree*. They distinguish it from the mountain ash [ROWAN], which they commonly call Witchen, and though a protective power is attributed to that tree, yet the 'Whitty Pear', they say is 'stronger'.[1]

Despite this the tree was 'maliciously destroyed' by fire in 1862.[2]

SHALLOT (*Allium ascalonicum*)

Perennial herb, a cultivated variety of ONION grown for its edible bulbs.

It was widely said that shallots should be planted on the shortest day (21 December) and harvested on the longest (21 June).

> My late father worked on the land . . . a keen gardener, everything had to be done at the proper times. His shallots were planted upon the shortest day to be harvested on the longest day.[1]

However, it appears that this is more frequently a remembered saying, rather than a regular practice.

> You should plant your shallots on the shortest day, and harvest them on the longest one. We did this once, and they kept really well.[2]

> [Ashford area, Kent] shallots planted shortest day, harvested longest day (though we usually planted ours in February).[3]

SHAMROCK

Although the harp is Ireland's official emblem, the shamrock is an important symbol for the Irish, both at home and overseas. Shamrock is used to promote a wide range of Irish products and organisations, including garden peat, the National Tourism Development Authority and the national airline. Throughout the struggles for Irish independence the shamrock motif was utilised both by those wanting an independent state and by those who wanted to maintain Ireland within the United Kingdom.

The word shamrock is believed to have derived from the Irish *seamroge* – 'little CLOVER' – and first appeared in print, as shamrote, in 1571. However the wearing of shamrock on ST PATRICK'S DAY was not recorded until 1681, and the legend about Patrick using its leaves to demonstrate the Holy Trinity was first recorded in 1726.

The earliest references to shamrock describe its use as food. Thus in 1571 Edmund Campion describing the diet of the Irish wrote:

> Shamrotes, watercresses, and other herbs they feed upon: oate mele and butter they cramme together.[1]

Later writers, including Richard Stanihurst,[2] and Edmund Spencer, equated shamrock with WATERCRESS, thus the latter, who was probably describing Ireland as he had observed it in about 1582, wrote:

> Ere one yeare and a halfe they were broughte to such wretchedness as that anye stonye harte would have rued the same. Out of every corner of the woods and glinnes they came creeping foorthe upon theyr handes, for theyr legs could not beare them . . . they did eate of dead carrions . . . and yf they founde a plotte of watercresses or sham-rotes there they flocked as to feast for the time.[3]

In 1597 the herbalist John Gerard stated that shamrock was a species of clover.[4] From the illustrations he provides and his description of 'common meadow trefoil' it is apparent that he was referring to the two agriculturally important species, RED and WHITE CLOVER, and it seems that either of these is the plant to which Campion referred. However, as early as 1570, Matthias de l'Obel recorded that a form of cake made from meadow trefoil was eaten by the Irish:

> The Meadow Trefoil . . . with a purple flower called Purple Trefoil, and with a whitish flower White Trefoil . . . and there is nothing better known, or more frequent than either or more useful for the fattening whether of kine or of beasts of burden. Nor is it from any other than this that the mere Irish, scorning all other delights and spurs of the palate, grind the meal for their cakes and loaves which they knead with butter, and thrust into their groaning bellies, when, as sometimes happens, they are vexed and high maddened with three days' hunger.[5]

Later writers who mentioned clover being used for food include Henry Mundy, who in his *Commentarii de aere vitali, Esculentis ac Potulentis*, claimed that the Irish 'nourish themselves with their shamrock (which is purple clover) are swift of foot and of nimble strength'.[6] Mundy's work, in which he strongly advocated a vegetarian diet, was published in 1680 and enjoyed sufficient popularity for six editions to be produced before the end of the century. Subsequent writers who made similar statements, presumably derived from the *Commentarii*, included John Ray in his *Historia Plantarum* (1686), and Linnaeus, who in his *Flora Lapponica* (1737) stated:

> The swift and agile Irish nourish themselves with their shamrock which is Purple Trefoil: for they make from the flowers of this

> plant, breathing a honey odour, a bread which is more pleasant than that made from the Spurrey.[7]

It has been claimed that the first association of St Patrick with shamrock occurs on a copper coin minted in Kilkenny 'for the use of the Confederate Irish in the wars of Charles I'. This coin is said to depict 'St Patrick, with mitre and crozier . . . displaying a trefoil to the assembled people'.[8] Although no such coin appears to have been produced during the reign of Charles I, a halfpenny piece minted in the 1670s accurately fits this description.[9]

By 1681 or thereabouts shamrock was being worn on St Patrick's Day; according to the English traveller Thomas Dinely:

> The 17th day of March yeerly is St Patrick's, an immoveable feast when ye Irish of all stations and conducions wear crosses in their hats, some of pins, and the vulgar superstitiously wear shamroges, 3-leaved grass, which they likewise eat (they say) to cause sweet breath.[10]

Since that time the wearing of St Patrick's crosses has declined, while the popularity of shamrock has grown. What was once a practice restricted to the 'vulgar' has now become widespread both among the Irish and their descendants overseas, whereas by early in the twentieth century crosses were worn only by girls and small children and have now been completely abandoned.[11]

The establishment of shamrock as a symbol of Ireland and the Irish seems to have been complete by the end of the seventeenth century. In 1689, when James Farewell published his *Irish Hudibras*, 'a coarse satire on everything Irish', he referred to Ireland as 'Shamroghshire'. The *Hudibras* was said by its author to have been 'taken from the Sixth Book of Virgil's Aeneid and adapted to present times', and shamrock takes the place of the golden bough which eased Aeneas' passage through the Underworld.[12]

Almost thirty years later the association of shamrock with the Trinity was first recorded by Caleb Threlkeld:

> This plant is worn by people in their Hats upon the 17 Day of March yearly (which is called St Patrick's Day). It being Current Tradition, that by this Three Leafed Grass, he emblematically set forth to them the Mystery of the Holy Trinity. However that be, when they wet their Seamaroge, they often commit Excess in

> Liquor, which is not a right keeping of a Day to the Lord: Error generally leading to Debauchery.[13]

The actual identity of St Patrick's shamrock has been the subject of much debate, involving various species of clover, BLACK MEDICK, WOOD SORREL and WATERCRESS. Irish tradition is unhelpfully vague. Colgan, who made extensive studies of the plant in the 1890s, was informed:

> First of all the mystic plant is not a clover, in the next place it never flowers, and finally it refuses to grow on alien soil.[14]

A century later:

> A half-Irish friend, in her late 50s, last night solemnly (and quite sincerely) made the statement that 'Shamrock will not grow in England.' Her mother, living in London, repeatedly brought back plants – they inevitably died, even when planted at the site of origin in pots with native soil and transported while apparently growing healthily. When pressed to say what the 'clover' was, she said it was never seen to flower and looked just like 'ordinary' clover.[15]

Or, as a woman who described herself as a 'Northern Irish colleen' wrote:

> There are no spots on shamrock, it is pure green. The leaf is small and it grows in bunches between grass on Irish soil. Shamrock must not be picked before St Patrick's Day, otherwise it dies off until the same time next year . . . Most Christian Irish people wear shamrock in their lapels with pride on St Patrick's Day.[16]

In the 1950s schoolchildren who were unable to select true shamrock for St Patrick's Day and appeared wearing 'clover' were castigated by their companions and accused of not being truly Irish.[17]

Although watercress is widely eaten, the fact that its leaves typically have more than three leaflets suggests that it cannot be shamrock. It seems that the equating of shamrock with watercress resulted from a misreading of Campion's 1571 *Historie of Ireland*. However, it is worth noting that a well on the Commons of Duleek, in County Mayo, was known as Shamrock Well, and until the late 1940s watercress, remembered as being the finest in the district, was gathered from it.[18]

The suggestion that wood sorrel was the real shamrock was promoted by the English botanist James Bicheno, who seems to have had his attention drawn to the subject during a lengthy tour of Ireland in 1829. While he admitted that white clover was the plant then considered to be shamrock, he argued that wood sorrel was the plant originally given this name.[19]

Although this suggestion has not been generally accepted, it merits examination. Firstly there is the similarity between the Irish names: *seamsoge* for wood sorrel, and *seamroge* for clover. Secondly, wood sorrel can be identified with the edible plant mentioned by the earliest authors who write about shamrock. BREAD-AND-CHEESE and many similar names demonstrate how wood sorrel was nibbled as a delicacy by children living in rural areas.[20] Furthermore the plant's sour taste would certainly freshen the mouth, even if it did not, as Dinely's informants claimed, 'cause sweet breath'. Another early passage which probably refers to wood sorrel or watercress rather than clover, is found in Fynes Moryson's *Itinerary* (1617), based on observations made in 1599:

> They willingly eat the herbe Schamrocke being of sharp taste which as they run and are chased to and fro they snatch like beasts out of the ditches.[21]

Clovers do not have a sharp taste, and the ditch habitat seems more suited to wood sorrel, and especially watercress.

Bicheno explained that as Irish woodlands were destroyed, wood sorrel became scarce, and was subsequently replaced as a St Patrick's Day emblem by the clover which thrived in the newly created pastures. Perhaps a simpler explanation is that wood sorrel has a major disadvantage when worn as a buttonhole or pinned to a hat. Its leaves rapidly wilt, so the more robust leaves of clover would be more suitable for this purpose.

Flowering shamrock is rarely depicted on St Patrick's Day greetings cards, but when it is, wood sorrel is the plant usually shown.[22] In North America where potted 'shamrock' plants appear in supermarkets before St Patrick's Day, species of *Oxalis*, the same genus as wood sorrel, are most frequently offered.[23] However, during the last 150 years or so the shamrock sold in Ireland and Britain has almost always been a species of clover or medick.

Apart from the instances already mentioned, there seem to be few records of clover being used as a food by humans in Britain and Ireland.

Although its flowers can be sucked to extract their nectar, clover was not used on the same scale as wood sorrel was nibbled by country children. Though the Irish traveller Dervla Murphy writes of eating stewed clover – 'the very same as clover at home' – in Afghanistan and Pakistan,[24] it seems probable that Murphy confused clover with a related pot-herb, such as fenugreek (*Trigonella foenum-graecum*). Other mentions of clover being eaten are unconvincing. John Lightfoot wrote that when corn was scarce poor people in Ireland would make a wholesome and nutritious bread from the powdered flowers of red and white clovers.[25] No recipe is known for this bread, and it appears that Lightfoot, usually a reliable recorder of plant-lore, was in this instance relying on Linnaeus's *Flora Lapponica* instead of oral informants.

In 1893 Nathaniel Colgan, a clerk with the Dublin constabulary, published his findings on the identity of shamrock. He had requested people to send him rooted bits of shamrock which he grew on in his garden, When the plants were mature enough for accurate identification it was found that four species were represented. White clover and lesser trefoil were the plants most often considered to be shamrock, while red clover and black medick were less frequently favoured as such.[26] Colgan also bought shamrock from three Dublin hawkers, each of whom assured him that she was selling the authentic plant. He planted his purchases and when they were identifiable found he had three different species: lesser trefoil, red clover and white clover.[27] Eventually he concluded that lesser trefoil was probably the true shamrock. This was the plant which decorated greeting cards said to contain the genuine Irish plant.

Several years before Colgan's investigations James Britten had examined bunches of shamrock sold in London, and he too found that they most frequently were of lesser trefoil.[28] The harp-shaped wreath placed on the Guards Division Memorial in London on St Patrick's Day 1977 was composed of this species,[29] and in recent years it appears to be the only species sold as shamrock in London.[30] In Dublin lesser trefoil is planted in St Patrick's Park, on the supposed site of St Patrick's well.[31]

In 1988 Colgan's survey was repeated on a much larger scale. As a result of appeals in the media, Charles Nelson of the National Botanic Gardens, Glasnevin, received 221 plants which grew to maturity, and on examination proved to be much the same mixture of species as those sent to Colgan. In 1893 Colgan found that lesser trefoil was the plant most commonly found to be shamrock, being submitted by 51 per cent of his contributors; in 1988 this remained the species most frequently submitted, being sent in by 46 per cent of Nelson's contributors. The second most popular candidate was white clover, sent in by 34 per cent of the 1893, and 35 per cent of the 1988, correspondents. As Nelson concluded, 'little significant change evidently has taken place during almost one century in the folk concept of the shamrock'.[32]

A custom which took place towards the close of St Patrick's Day was drowning the shamrock. The wearer would remove his shamrock and place it in his final drink of the evening; when the toasts had been drunk the shamrock would be removed from the bottom of the glass and thrown over the left shoulder.[33]

Although some Irish people still try to gather their own wild shamrock for St Patrick's Day, at least in towns most of what is worn is commercially produced.[34] Stringent legislation in the United States, whereby the importation of living plants is carefully restricted, has ensured that the amount of shamrock grown for export as living bunches is now negligible. In 1980 it was reported that most commercial growers plant small areas, the size of their plantings apparently being limited by the amount of labour available to harvest and pack the plants.

> A grower in the Kanturk area has been growing the crop for the past eight years on about $^{1}/_{20}$ acre. Seed is collected from selected plants in their plots. Selection is based on appearance, vigour and freedom from purpling. Seed is sown in early July in beds outdoors. The plants are planted in September at 6 in x 6 in into the ground from which a crop of early potatoes has been taken. No fertiliser is added. Plants are lifted early in March and the roots washed. They are sold to an agent in Mallow who collects himself.
>
> The gross value of the crop is about £200. About 70% of the plants reach the required size. The main drawbacks are heavy labour requirements at harvesting, variable weather during the growing season, and diseases. Downy mildew (*Peronospora trifoliorum*) can be troublesome in damp seasons.[35]

St Patrick's Day celebrations at which shamrock is worn are held in Irish embassies around the world, and in Britain a member of the Royal Family (for many years from 1966 Queen Elizabeth the Queen Mother) present shamrock to the Irish Guards.[36]

SHEEP – put to graze on uplands when BILBERRY and CLOUDBERRY come into leaf; especially fond of COTTON GRASS; fed HOLLY; tempted with IVY when sick; BOG ASPHODEL, BUTTERWORT and SUNDEW harm.

SHEEP'S BIT (*Jasione montana*)

Perennial herb with attractive blue flowers, common on acidic substrates mainly in western areas.

Local names include:
BACHELOR'S-BUTTONS in Dumfriesshire
BLUE BONNETS in Cumberland, Devon, Dorset and Somerset
BLUE CAP in Yorkshire
IRON-FLOWER in Cheshire
POLICEMAN'S BUTTONS, and TAILOR'S NEEDLES, in Cornwall.

The picking of the flowers of sheep's bit is supposed to give one WARTS.[1]

SHEPHERD'S PURSE (*Capsella bursa-pastoris*)

Annual herb with insignificant white flowers, widespread and common on disturbed ground.

Local names include:
BAD-MAN'S OATMEAL in Co. Durham; 'i.e. devil's oatmeal'
Blind-weed in Yorkshire
Case-weed in northern England; 'from its little purse-like capsules'
Casewort, and churchyard elder, in Somerset
Clappedepouch in northern England; 'meaning clap or rattle pouch, a name that alludes to the licensed begging of lepers, who stood at the crossways with a bell and a clapper'
Cow-pecks in Wiltshire
FAT HEN in Gloucestershire and Yorkshire 
Gentleman's purse in Somerset
Guns in Somerset; 'no doubt owing to the explosive dispersal of the seeds; the name is sometimes

given to other plants with explosive fruits'

HENS-AND-CHICKENS in Somerset

Lady's pouches in Aberdeenshire

Lady's purses in East Anglia and Scotland

Money-bags in Somerset

NAUGHTY-MAN'S PLAYTHING in Birmingham; 'i.e. the devil's plaything'

OLD-WOMAN'S BONNET in Somerset

Pepper-and-salt in Middlesex; 'from its pungent taste when bitten

PICKPOCKET, widespread; 'children gather it and repeat "Pick pocket, penny nail. Put the rogue in the jail"', or 'from their impoverishing of the land of the farmer'

Pickpocket-to-London in Yorkshire

PICK-PURSE in Somerset; 'from its robbing the farmer by stealing the goodness of his land'

Poor-man's parmacetty in Somerset; parmacetty = spermaceti

Poor-man's purse in Somerset

Poverty-purse in Lincolnshire

Rifle-the-ladies'-purses in Banffshire

Shepherd's pedler in Wiltshire

Shepherd's pouch in Hertfordshire and Somerset

Shepherd's scrip in Somerset

Stony-in-the-wall in Lincolnshire

Tacker-weed in Somerset

Toywort in northern England

Ward-seed in Devon

Witches' pouches in Elgin and Morayshire.

> [Yorkshire] on finding a root of shepherd's purse . . . open a seed vessel. If the seed is yellow you will be rich, but if green you will be poor.[1]

More widespread was the belief that picking shepherd's purse would in some way harm, or upset, the picker's mother.

> [Berwickshire] children have a sort of game with the seed-pouch. They hold it out to their companions, inviting them 'to take a haud o' that'. It immediately cracks, and then follows a triumphant shout – 'You've broken your mother's back.'[2]

> In Middlesex, schoolboys offer to their uninitiated companions a plant of the shepherd's purse, and request them to pluck off one of the heart-shaped seed-pods, which done, they exclaim, 'You've picked your mother's heart out!' This was practised in Chelsea in my own schooldays, and, as a Lancashire name for the plant is 'Mother's-heart', it seems likely that the custom is widely extended.[3]

Other names which relate to this practice included pick-your-mother's-heart-out in Warwickshire, and pluck-your-mother's-heart-out in Oxfordshire.

> In the Invergowrie area of Perthshire in the 1950s: if you pick the shepherd's purse your mother will die – associated with the plucking of the heart-shaped capsules.[4]

In folk medicine shepherd's purse was valued as a cure for DIARRHOEA and high BLOOD PRESSURE.

> A cure for scour in cattle and diarrhoea in human beings. When a young calf was bought at the mart she always gave it some tea made from shepherd's purse, and it would be better in an hour. They were nearly always affected by scour because of a change in their diet when they were newly bought, She used the root sometimes, but mostly the leaves, or as she called them the 'leafs'.[5]

> The best herbal remedy is one I have used on my pet RABBITS for over 50 years. Some shepherd's purse (*Capsella bursa-pastoris*) is very effective if given to a rabbit with diarrhoea or a tummy upset. In the last few years I had an adult rabbit that was refusing to eat and looking quite poorly. It refused to eat any shepherd's purse, so I broke some into small pieces and forced it into its mouth. It gradually swallowed some and then ate some. In two hours it had started to recover and in the next few days went on to make a full recovery. It lived for several more years after this until it died of old age.[6]

> Shepherd's purse is a tonic for rabbits; they're very fond of it.[7]

> We . . . nibbled on parts of shepherd's purse and these parts of shepherd's purse were given to older people to lower their blood pressure.[8]

SHINGLES – HOUSELEEK used to treat.

SHIPWRECK – foretold when BURNET ROSE flowers out of season.

SHROVE TUESDAY – CHRISTMAS GREENERY kept to provide fuel for cooking pancakes.

SICKNESS – CUCKOO-FLOWER and YARROW cause.

SILVERWEED (*Potentilla anserina*)

Perennial herb with silvery leaves and yellow flowers, widespread and common in a wide range of habitats.

Local names include the widespread GOOSEGRASS and goose-tansy, and:

Blithran in Co. Donegal; 'no doubt a form of the original Irish name . . . *brisclan*'
Brisken in Co. Donegal
BUTTERCUP in Buckinghamshire
CAT'S TAILS in Somerset
Dog's tansy in Scotland
Fair-days in Berwickshire and Northumberland
FAIR-GRASS in Forfarshire and Roxburghshire
Fern-buttercup in Wiltshire
Fish-bones in Somerset
GOLDEN FLOWER, golden sovereigns, and LAMB'S EARS in Somerset
Mackerel-weed in Cornwall
Moor-grass in Scotland; 'cf. Faroese *murgras*'
PRINCE'S FEATHERS in Lancashire and Somerset
Scented buttercup in Devon
Silver feather in Oxfordshire
Silver fern in Oxfordshire, Somerset and Wiltshire; 'has fern-like silvery foliage'
Silver grass in Wiltshire
Swine's beads in Orkney
Swine's murriks in Shetland; 'murrik, i.e. root'
TANSY in Cumberland, Northamptonshire and Yorkshire
TRAVELLER'S EASE in Warwickshire' 'so called because applied to galled feet'.

Britten and Holland considered 'midsummer silver' – 'a little herb, which continues all the Year of a bright Ash colour' – mentioned by Aubrey 1718, to be silverweed. According to Aubrey in Lingfield, Surrey, 'the inhabitants are very fond of ghirlands or garlands made of Midsummer Silver . . . and have crowded the church and their own houses with them'. However, by 1808 the custom was forgotten, though 'the Midsummer Silver is common here'[1]

In 1670 John Ray recorded that children around Settle in the West Riding of Yorkshire dug up silverweed roots and ate them.[2] Such practices seem to have been widespread until comparatively recent times. According to a writer born in Strood, Kent, in 1806:

> Children in rural districts of England sometimes lay them [silverweed roots] over a brisk fire and then eat them. They are very small, but to some of us in childhood they seemed as pleasant as the fruit of the CHESTNUT.[3]

Particularly during times of scarcity silverweed roots were valued as food in Scotland. On Colonsay:

> The roots were gathered and eaten raw and also boiled like potatoes. The local value, in former times, attached to this article of food may be realised from the fact that it was termed *an seachdamh aran* (the seventh bread).[4]

Alexander Carmichael noted:

> The root [of silverweed, *brisgein*] was much used throughout the Highlands and Islands before the POTATO was introduced. It was cultivated and grew to a considerable size. As certain places are noted for the cultivation of potato, so certain places are noted for the cultivation of silverweed. One of these was Lag nan Tanchasg in Paible, North Uist, where a man could sustain himself on a square of ground of his own length. In dividing *morfhearann*, common ground, the people lotted their land for *brisgein* much as they lotted their fishing-banks at sea and their fish on shore. The poorer people exchanged *brisgein* with the richer for corn and meal, quantity for quantity and quality for quality. The *brisgein* was sometimes boiled in pots, sometimes roasted on stoves, and sometimes dried and ground into meal for bread and porridge. It was considered palatable and nutritious.[5]

Names which appear to relate to this practice include:

BREAD-AND-BUTTER, and BREAD-AND-CHEESE, in Somerset

Marsh corn in Counties Antrim, Down and Tyrone; 'the root is roasted and eaten'

Mascorns in Morayshire; 'once important as a famine food . . . now ignored'

Mascrop in Scotland

Mashcorns, and mashy-corns, in Co.Antrim and Co. Down

Moss-corns in Morayshire and Selkirkshire
Moss-crops in Scotland.

Silverweed was also valued as a cosmetic:

> A friend, whose early home was a Highland manse, has described . . . how eagerly the plant was gathered in summertime by the female part of the household, and steeped in buttermilk to remove FRECKLES and brownness which the sun had brought to the fair cheek.[6]

SKIN CONDITIONS – treated using BANANA, BOGBEAN, BOG MYRTLE, BUTTERCUP, CELANDINE, GOOSEGRASS, GROUND IVY, PRIMROSE and WATERCRESS.

SKULLCAP (*Scutellaria galericulata*)

Perennial, purple-flowered herb, locally common in damp places including canal-sides, riverbanks and marshes.

> Bunches of skullcap used to be placed alongside the tea urn at Pilkingtons Glassworks [?Liverpool], as many men added it to their drink.[1]

SLEEP – DOG ROSE gall and POPPY induce.

SLENDER SPEEDWELL (*Veronica filiformis*)

Small perennial herb, producing abundant blue flowers, native to the Caucasus and Turkey, cultivated as an ornamental since early in the nineteenth century, during the twentieth century becoming increasingly naturalised and widespread in lawns, churchyards, and damp grassy places.

> Shropshire names include everlasting sin, and sin: 'said to be so called because "it is so attractive and so prevalent"'.[1]

SMALLPOX – treated using APPLE; HAWTHORN said to smell of.

SNAPDRAGON – see ANTIRRHINUM.

SNAKES – picking CUCKOO-FLOWER leads to attacks by; emerge from HERB ROBERT stems; HAZEL poisons.

SNOWDROP (*Galanthus nivalis*)

Bulbous herb (Fig. 46), producing white flowers in early spring; formerly considered to be native, known in cultivation in the British Isles late in the sixteenth century, and first recorded in the wild in 1778, now widely naturalised in woodlands and churchyards and on riverbanks England and Wales, less common in north-west Scotland and Ireland.

Local names include:

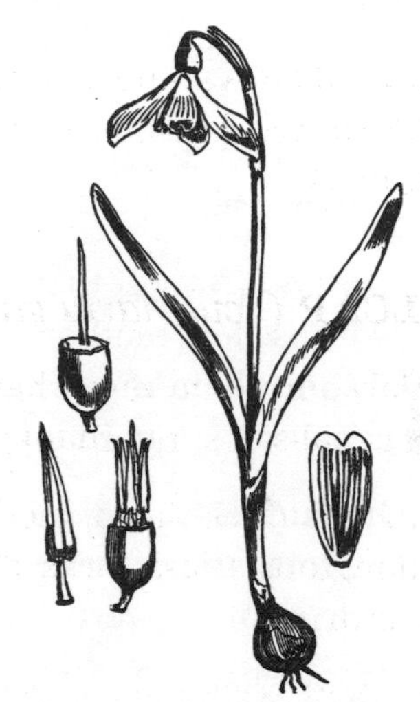

Candlemas-bells in Essex, Gloucestershire and Wiltshire
Death's flower, dew-drops, dingle-bell, drooping bell, drooping heads, DROOPING LILY, and Eve's tear, in Somerset
Fair maids in Hampshire, Lincolnshire and Norfolk
February fair-maids in Somerset and Westmorland
Mary's tapers in Somerset
MORNING STARS in Somerset; 'probably in confusion with star of Bethlehem, *Ornithogalum*'
Naked maiden, and pierce-snow, in Somerset
Shame-faced maiden in Wiltshire
Snow-dropper in Buckinghamshire and Gloucestershire
Snow-piercer, white bells, white cup, and white queen, in Somerset.

Despite attracting a surprisingly large number of enthusiasts – galanthophiles – some of whom collect varieties distinguished mainly by minor features of the flower's inner petals, snowdrops are widely associated with DEATH and considered to be UNLUCKY.

> In a London flowershop today – January 29, 1931 – I asked for some snowdrops. The assistant replied: 'No, sir, we are not allowed to sell them.' I expressed surprise, and was told that Mr – (presumably the proprietor of the shop) thinks them unlucky.[1]

> I am a District Nursing Sister working in Lancashire, and the following story was related to me by an elderly patient in a farmhouse.

For many years her mother had refused to have snowdrops in the house, even though they grew profusely in the orchard. Girls from east Lancashire towns were often employed to help in the large house and would ask permission to pick the flowers to take home on their day off. This they were allowed to do so long as they left them in water on the doorstep.

After the old lady's death there was an occasion when a wedding party announced their intention, at short notice, of arriving to pay respects to an ailing relative. Snowdrops were brought in to decorate the tables, it being early in the year and no other flowers available so easily. Within three months the bridegroom was dead, and, needless to say, snowdrops have never since been brought into the house.[2]

In Scotland [Argyll] they won't allow HAWTHORN indoors. And snowdrops. My mother would let me pick snowdrops, but she always put them in a jar on the windowsill outside. She wouldn't let us have them in indoors.[3]

By the late 1940s my mother was a widow in Edinburgh and we had a Scots family, some of whom lived in Glasgow in a tenement flat. We, in our poverty, would struggle to find something to take (on the bus) to the 'Glasgow Aunts', maybe something from the garden such as green beans in season, since we were proud of our garden and they had no garden. One cold early spring day we scoured the garden for the first snowdrops and gathered a sacrificial bunch of the best we could find. We arrived triumphantly at the flat, only to have a severe reaction from my excitable aunt. 'Oooooh, Angel's tears! Don't bring those in here!' I think we left them on the doorstep and took them home again when we left.[4]

Do not take snowdrops into the house – unlucky – so when as a child [1940s] I picked them for my mother to sell in the market, I sat outside and bunched them with IVY leaves surrounding them and tied each bunch with bits of wool. I usually got 6d a bunch for them.[5]

SNUFF – SPIGNEL used to scent.

SOAP – BRACKEN used in the manufacture of; SOAPWORT used instead of.

SOAPWORT (*Saponaria officinalis*)

Perennial, pink-flowered herb, native to mainland Europe, but extensively naturalised so native range unknown; long cultivated and escaping to form conspicuous colonies, mostly in dampish situations.

Local names include:

BOUNCING BETT in Devon and Dorset; 'especially applied to the double-flowered form'
Farewell-summer in Monmouthshire, 'from its flowering in the months of August and September'
Farewell-to-summer in Somerset
Gill-run-by-the-street in Kent and Sussex
Hedge-pink in Hampshire
Stewed gooseberries in Somerset.

Both the scientific name *Saponaria* (from the Latin *sapo*, soap) and the English name refer to the former use of soapwort roots as a substitute for soap. In 1931 when Admiral Sir Herbert Fetherstonhaugh inherited Uppark, in West Sussex, he ordered the faded, ragged Italian brocade curtains, which had been hanging since 1740, to be thrown away, but his wife thought otherwise. She consulted two aunts who in turn sought advice from a herbalist, who brought along a trug full of soapwort:

> The soapwort plants were placed in old-fashioned bags of muslin and put to boil in a huge cauldron in Uppark's still room . . . The liquid was allowed to cool a little before the great experiment began. Then the first curtains, removed from the Prince Regent's famous four-poster, were dunked in the soapwort and at once the water turned as black as ink . . . the curtains came out of the cauldron looking like shredded bundles of pink seaweed. Yet as they dried, it was obvious that they were marvellously clean and the rose-coloured fabric bloomed anew.[1]

Soapwort also had some medicinal uses.

> The chief use to which the herb was put in Ireland was the treatment of inflammation of the LUNGS.[2]

> [Gypsy remedy] A decoction of the root applied to a bruise, or to a black eye, will quickly get rid of the discoloration: slices of the freshly dug root laid on the place have the same effect but are slower in action.[3]

> The Jersey-French-Norman name of *des mains jointes* comes from the plant's jointed rhizomes, which were used to heal wounds in cattle.[4]

SOIL – BRACKEN, CHICKWEED and THISTLE indicate fertile; CORN SPURREY indicates poor.

SOLOMON'S SEAL (*Polygonatum multiflorum* and *P.* spp.)

Perennial rhizomatous herbs, with arching stems producing pendent greenish-white flowers; *Polygonatum multiflorum* native in open woodland and hedgebanks in central England, but also, like the garden Solomon's seal (*P.* x *hybridum*), widely planted and becoming naturalised.

Local names include the widespread David's harp ('curving like the neck of a harp'), and Jacob's ladder (cf. the old apothecaries' name *scala coeli*, ladder of heaven), and:

Job's tears in Cornwall
LADY'S LOCKETS in Somerset; 'from the hanging flowers'
Lily-of-the-mountain in Warwickshire; 'to distinguish it from LILY OF THE VALLEY'
Sealwort in Somerset
Sow's teats in Dorset
Vagabond's friend in the Lake District
Zolomon's zale in Berkshire.

> [Gypsy remedy] an ointment made from the leaves and applied to a bruised or black eye will quickly get rid of the discoloration.[1]

> My mother, who was born in the second half of the nineteenth century, told me when I was a little boy of the wonderful efficacy of solomon's seal for drawing out the blackness of a BRUISE. This apparently traditional remedy . . . was used once on some

member of my family – I have forgotten who it was – and a local doctor who was called in to the case was most impressed and made a note of the remedy. As I remember it, the washed rhizome was grated and the pulp was bandaged on the bruise as a cold poultice.[2]

SORCERY – see WITCHES.

SORES – treated using ALDER, BUTTERBUR, CARROT, CELERY, DOCK, GOLDENROD, GREATER PLANTAIN, MALLOW, ORANGE LILY and (Turkish) VALERIAN.

SORREL (*Rumex acetosa*)

Perennial herb, widespread and common in grassy places.

Local names include the widespread GREEN SAUCE, and sour dock ('sometimes extended to *R. acetosella* [sheep's sorrel]'), sour docken, sour grass, and:

Bitterdabs in northern England
Bog-sorrel in Co. Donegal
BREAD-AND-CHEESE in Devon
Brown sugar in Somerset
CELERY in Buckinghamshire; 'eaten by children'
Cock-sorrel in Yorkshire
Crow-sorrel in Co. Donegal
CUCKOO'S MEAT in Cheshire
Dock-seed in Devon
DONKEY'S OATS in Devon
GREEN SNOB in Warwickshire
GREEN SORREL in Buckinghamshire
GYPSY'S BACCA, and GYPSY'S TOBACCO, in Somerset
Hunters, given to the seeds, in Somerset
Lammie sourocks – Roxburghshire
London green-sauce in Lancashire
REDSHANK in Roxburghshire
Red sour-leek in Northern Ireland
Salary in Oxfordshire
Salery ('leaves . . . eaten by children'), and sallet ('simply a pronunciation of salad'), in Buckinghamshire
Sarock in Scotland
Smart-ass in Shropshire

SOLDIERS in Somerset; 'for the reddish colour of its stems, petals and sepals'
Sooracks, or sooraks, in Scotland
Soorik in Shetland
Soorocks in Scotland
Sorrow in southern England
Sourack in Scotland
Sourdabs in northern England
Sour dockling in north-east England and Berwickshire
Sour dogs in Dorset
Sour dooks in Cumberland
Sour gob in Lancashire
Sour grab in Cornwall, Devon and Somerset
Sour grass in Norfolk and Yorkshire
Sour grobs on the Gower Peninsula
Sour leaves in Somerset
Sour leek in Roxburghshire and Belfast
Sourlick in Roxburghshire and Northern Ireland
SOUROCKS in Scotland and Ireland
Sour sabs in Devon, 'sabs = sauce?'
Sour salves in Devon
SOUR SAPS in Cornwall and Devon
Sour sodge in Buckinghamshire
Sour sog in Cornwall
Sour sops in Cornwall and Devon
Sour suds in Devon
Sow-sorrel in Hertfordshire
TEA in Somerset
Tom Thumb's thousand fingers in Kent
Vinegar-leaf in Bedfordshire and Derbyshire
WILD RHUBARB in Lancashire
Zour-docks, zour-grabs, zour-zabs, and zour-zob, in Devon.

As many of its names imply, sorrel was frequently eaten by children for its refreshingly sour-tasting leaves.

> [Dorset, *c.*1930] common sorrel was, of course, eaten – 'but it'll make 'ee bad if 'ee eats too much'.[1]

> [Invergordon, Ross-shire, 1950s] as kids we used to chew a leaf growing on the ground, which we called sourey souracks.[2]

More rarely sorrel was gathered by adults for culinary use. In north-west England it was frequently used in BISTORT puddings, and people of Polish extraction gathered the leaves to make soups.[3]

> As children in wartime we used to search spring meadows for the first leaves of sorrel – called sour docks in Berwickshire – to cure our spots. We sucked the sour juice and spat out the chewed leaves.[4]

SOUTHERNWOOD (*Artemisia abrotanum*)

Aromatic shrub of unknown origin, long cultivated in British gardens, rarely flowering, and occasionally found on tips and waste ground.

Local names include the widespread old man, and:

Apple-riennie, apple-ringie, and apple-ringin, in Scotland
Boy's love in Devon, Dorset and Somerset; 'it is said . . . that its ashes were formerly made into an ointment and used by young men to promote the growth of the beard'
KISS-ME-QUICK in Somerset
Kiss-me-quick-and-go in Devon
Lad-love-lass in Yorkshire
Lad-savour in Lancashire
LONDON PRIDE in Wiltshire
Maiden's delight in Cornwall; 'more often called boy's love'
Maiden's ruin in Cheshire and Devon
Maid's love in Northamptonshire
Mother-wood in Lincolnshire
Nobby-old-man, 'so called by [London] costermongers'
OLD-MAN'S BEARD in Dorset and Wiltshire
Old-man's love in Northumberland
Oud-man in Yorkshire
Overenyie in Aberdeenshire
Sloven-wood in East Anglia
Smelling wood in Oxfordshire
Suttenwood in Essex.

As some of its local names seem to suggest, southernwood was associated with courtship.

> A form of proposal of marriage by an inarticulate young Fenman was provided by the plant Southernwood . . . known in the Fens as OLD MAN or Lad's Love. The youth . . . would cut some sprigs of the plant and put them in his buttonhole before setting out with the village lads on a spring or early summer evening stroll. Presently, leaving his companions, he would wander along the lanes, where he would find little groups of giggling girls, and would pass them by ostentatiously sniffing at his buttonhole to show that his thoughts were turned to matrimony. If the girls went by unheeding, he knew he was unlucky, but if they turned and came slowly back towards him he knew that his herbal decoration had not gone unnoticed. After a show of hesitation he then removed the buttonhole and handed it to the girl of his choice. If she spurned him, she probably threw his offering to the ground and might even smack the bold suitor on his face. If he was acceptable, however, she would inhale the pungent scent of the Lad's Love and, after

> some teasing from her companions, would put her arm through his and the pair would set off on their first courting stroll.[1]
>
> Before the First World War the [Cambridgeshire] lads used to wear Lad's Love (Southernwood) as a buttonhole to attract females, or sprigs of WHEAT were used for the same purpose.[2]

SOWTHISTLE (*Sonchus asper*, prickly sowthistle, and *S. oleraceus*, smooth sowthistle)

Annual, yellow-flowered herbs, stems and leaves producing white latex when broken; common and widespread on waste and disturbed ground.

Local names include the widespread milk-thistle, and:

DINDLE in Norfolk
Dog's thistle in Somerset
Hare's lettuce in Devon
Milk-thrissel in Co. Donegal, where 'used for feeding cows'
Milkweed in Somerset
Milkwort in Dorset
Milky dashels in Cornwall, Devon and Somerset
Milky dassel in Cornwall and Devon
Milky dicel in Devon and Somerset; 'known to every schoolboy as the food of his pet rabbit'
Milky dickle/s in Devon and Somerset
Milky disels, 'second word clearly pronounced dye-zels', in Devon
Milky disle in Cornwall and Devon
Milky tassel in Cornwall
Mulk-vissel in Devon
Rabbit's victuals in Somerset
SOWBREAD in Kent
Sow-dingle in Lincolnshire, where also applied to 'other plants much dissimilar in appearance'
Sow-flower in Wiltshire
Sow-thristle in Berwickshire
Swine's thistle in Co. Donegal; 'good food for pigs'
Swine-thistle in northern England and Scotland
Swinnies in Berwickshire
Virgin's milk in Somerset.

In common with other latex-producing plants, such as FIG and PETTY SPURGE, sowthistle was believed to cure WARTS.

> [Co. Meath, sowthistle] is a great cure for warts . . . cut the thistle and get the milk out of it and put it on the wart.[1]

> For at least three generations in my family we have used sowthistle to remove warts. That's in Bromley, Kent.[2]

SPEAR THISTLE (*Cirsium vulgare*)

Spiny, biennial herb (Fig. 47), with attractive pale purple flowers; widespread and common in grassy areas, considered to be a 'noxious weed'.

Local names include the widespread boar-thistle, and:

Bell-thistle in Warwickshire
Bird-thistle, 'because goldfinches and other birds feed on the seeds; or perhaps a form of bur thistle', and blue thistle, in Worcestershire
Bow-fistle, or boar-thistle, in Cheshire
Buck-thistle in Yorkshire.
Bull-distle, or BULL-THISTLE, in Somerset
BUR given to the seed head in Yorkshire
Bur-thistle in northern England and Scotland
CUCKOO-BUTTONS in Somerset
DISLE in Cornwall
Horse-dashel in Cornwall and Devon
Horse-thistle in Devon and Somerset
Quat-vessel in Hampshire
Row-dashle in Devon; 'i.e. rough thistle'
Scotchman in Dublin.

> [Ulster] Scotch thistle (*Cirsium vulgare*) – pull off the florets and their young pappuses and peel of the spiny bractlets, and you found a nice little nutty core.[1]

> My maternal uncles, when the thistle – *Cirsium vulgare* – was flowering, used to split and cut through the flowerheads, pull all the flowers out, and eat the succulent bases.[2]

Presumably the name cheese, given to the receptacle of the flower in Dumfriesshire, refers to this practice.

SPIDERS – conkers (HORSE-CHESTNUT seeds) q.v.

SPIGNEL (*Meum athamanticum*)

Perennial aromatic herb, with white or pinkish flowers, in mountain grassland from northern England and North Wales to central Scotland.

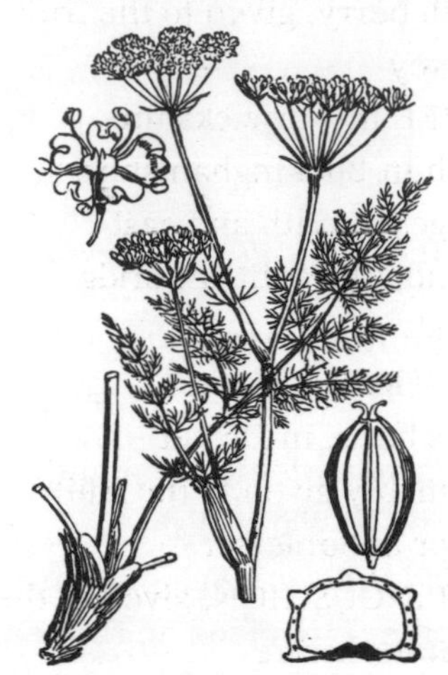

Local names include:
Badminnie in Scotland; 'from the circular seeds'
Baldmoney in Westmorland and Yorkshire
Balmoney in Yorkshire
Bawd-ringie in Perthshire
Highland micken in Stirlingshire
Houk in Northumberland
Moiken in Perthshire
Muilcoinn in Scotland.

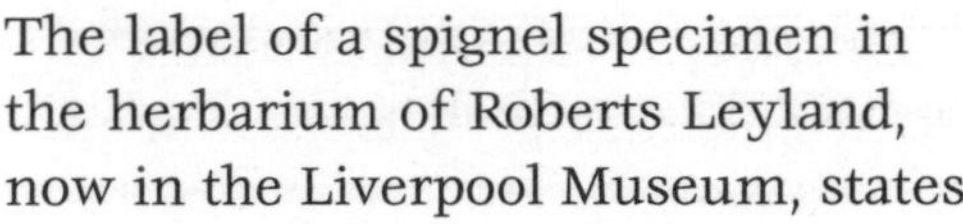

The label of a spignel specimen in the herbarium of Roberts Leyland, now in the Liverpool Museum, states:

> [Collected] in the 2nd lane after you pass the Booth Wood Inn on the road from Ripponden to Oldham, July 1837 . . . the plant is however nearly destroyed, not by the rapacity of Botanists but by Snuff takers in the neighbourhood of the place where it grows who dig up the roots for the purpose of scenting their snuff.[1]

Geoffrey Halliday, writing of Cumbria, notes:

> The fruits . . . are extremely pungent, smelling of curry powder. The equally strong-smelling roots of 'The Westmorland Herb' were once sold in London as a sexual stimulant.[2]

SPINDLE (*Euonymus europaeus*)

Deciduous shrub or small tree, widespread on base-rich soils, and widely planted elsewhere, inconspicuous throughout most of the year, but producing bright pink fruit which split open to reveal orange seeds in the autumn.

Local names include the widespread skewer-tree and skewer-wood, and:
Bitch-wood in Worcestershire
Cat-tree in Buckinghamshire
Catty-tree in Shropshire
Cat-wood, and death-alder ('where

it is thought unlucky to bring into the house'), in Buckinghamshire
Dog's timber in Devon
Dog-timber in Somerset
Dog-tooth-berry, given to the fruit in Surrey
DOG-TREE in Warwickshire
Foul rush in Buckinghamshire
Gatteridge in south and east England; 'the fruit gatteridge berries'
GATTER-TREE in Kent
Hot-cross buns, in Dorset; presumably given to the fruit
Ivy-flower in Somerset
LADY'S PINCUSHION, given to the fruit in Berkshire
Louse-berries, given to the fruit in Gloucestershire and Warwickshire
Needle-tree in Lancashire
Peg-wood in Devon
PINCUSHION, given to the fruit in Gloucestershire and Warwickshire
Pincushion-shrub in Buckinghamshire
Popcorns, given to the fruit in Somerset

Prick-wood in Cumberland, Somerset and Sussex
Skiver in Somerset and Wiltshire
Skiver-berries, given to the fruit in Somerset
Skiver-tree in Somerset and Wiltshire
Skiver-wood in Devon, Dorset and the Isle of Wight
Spindle-wood in Gloucestershire
Stink-wood on the Isle of Wight
WITCH-WOOD in Suffolk
Yewnanimous in Co. Donegal; 'a gardener's corruption'.

The name spindle was first used by William Turner (1509?–68) who stated that although he had seen the tree many times he had not been able to find an English name for it. He chose this name because in the Netherlands the tree was called *spilboome* and was used for making spindles. Many of spindle's local names hark back to similar uses. According to Grigson (1987): 'spindles, skewers (pricks and skivers) and pegs have given it most of its names, and gypsies still make skewers out of the hard, white, tough wood'.[1] Richard Mabey notes that although spindle wood was suitable for making spindles, it did not seem to be 'especially favoured' for this purpose in Britain.[2] However, writing of the Plymouth area in 1880, Archer Briggs observed:

> The wood of this shrub is that most commonly employed for butchers' skewers . . . its beautifully coloured fruit would be seen more frequently if bushes were not so sought out and cut down for the purpose of skewer-making.[3]

Spindle fruits were valued as a pesticide, according to Theophrastus (*c.*372–*c.*287 BC), 'this shrub is hurtfull to all things . . . the fruite heerof killeth'.[4] John Evelyn (1620–1706) reported that baked and powdered spindle berries were sprinkled on the heads of small boys to kill nits and lice.[5] Isabel Wyatt, writing of Huish Episcopi, Somerset, early in the 1930s, noted that villagers 'scattered their houses with the powdered leaves of spindle-berry, a natural insecticide'.[6]

SPOTS – SORREL used to cure.

SPRAINS – treated using COMFREY, ELDER, FAIRY FLAX, HORSE-CHESTNUT, MALLOW, ROYAL FERN, ST JOHN'S WORT and WATER FIGWORT.

SPRINGBEAUTY (*Claytonia perfoliata*)

Annual white-flowered herb, native to western North America, cultivated since the 1790s, and since the 1930s becoming increasingly naturalised on open sandy ground in East Anglia and elsewhere.

> Some 20 years ago, children on the Isles of Scilly ate leaves of *Claytonia perfoliata*, and called them water weed.[1]

In recent years cultivated springbeauty has been frequently sold under a variety of names as a salad vegetable in farmers' markets.[2]

SPURGE LAUREL (*Daphne laureola*)

Evergreen, green-flowered, inconspicuous shrub in deciduous woodland, most common in south-east England, probably present only as the result of introductions elsewhere.

Local names include copse-laurel on the Isle of Wight, fox-poison in Lincolnshire, laurel-wood in Gloucestershire, sturdy lowries in Co. Durham, and wood-laurel in Gloucestershire, Hampshire, the Isle of Wight and Somerset.

> The Rev. G.E. Smith tells me that the spurge laurel is collected

> in large quantity from the woods of Sussex, by persons who go at stated periods round the country for the purpose, and supply the markets at Portsmouth and Chichester, where it sold as HORSE medicine, but he was unable to ascertain in what manner or for what diseases it was employed.[1]
>
> The acrid bark is in some counties used as a blister, and the still more acrimonious roots are employed to alleviate the TOOTH-ACHE; but they should be applied with caution.[2]

STAFF (walking stick) – BUTTERWORT provided; FIG tree at St Newlyn East, HOLY THORN, and YEW at Congresbury grew from.

STAG'S-HORN SUMACH (*Rhus typhina*)

Deciduous small tree, producing inconspicuous flowers followed by crimson fruiting-heads and brilliantly coloured autumn leaves; native to eastern North America, cultivated since the early seventeenth century, and extending by suckers beyond gardens to railway embankments and similar habitats.

Local name include: cobber's wax and monkey-tree in Wiltshire, and red-hot poker in Somerset.

> In the Bromsgrove area [Worcestershire] there is a belief that the stag's-horn sumach, if growing in the garden of a house is sure to bring marital strife to the family living in the house. This was related to me by an old gardener who died in 1973 full of years. He lived until he was about 90 years of age . . .
>
> Notwithstanding more than 50 years of collecting such information, I have not encountered this belief elsewhere, but it was firmly held by other gardeners of my informant's generation.[1]

STARCH – prepared from LORDS-AND-LADIES and POTATO.

STAR OF BETHLEHEM (*Ornithogalum umbellatum*)

Bulbous herb, with white, green-striped flowers, native to southern Europe, cultivated in the British Isles since the mid-sixteenth century, first recorded in the wild in 1772, and now widespread, and becoming increasingly common, in grassland in lowland England, Scotland and Wales; little recorded in Ireland.

Local names include:

ANGEL'S TEARS, and (the) apostles, in Somerset
SHAME-FACED MAIDEN in WILTSHIRE
SNOWFLAKE in Devon and Wiltshire
STAR-FLOWER in Buckinghamshire
Starry eyes in Somerset
Stars in Berkshire
Stars-and-garters in Wiltshire
SUNFLOWER in Devon.

Like GOAT'S BEARD, star of Bethlehem flowers are said to close at midday; names which seem to relate to this characteristic include:

Betty-go-to-bed-at-noon, and JOHN-GO-TO-BED-AT-NOON, in Shropshire
JOHNNY-GO-TO-BED-AT-NOON in Norfolk
MORNING STAR in Hampshire
Nap-at-noon in Shropshire (and North America)
Noon-peepers in Wiltshire
ONE-O'CLOCK in Devon
Open-and-shut in Wiltshire
Peep-o'day in Shropshire
Six-o'clock-flower in the Midlands
Six-o'clocks in Buckinghamshire
Six-o'clock sleepers in Lincolnshire
Sleepy Dick in Lancashire
Ten-o'clocks in Somerset
Twelve-o'clock in Cornwall, Dorset, Oxfordshire and Somerset.
Wake-at-noon on the Isle of Wight and in Wiltshire.

> Tiny children used to search for star of Bethlehem flowers. As children in Sunday School we were told that lilies alluded to by Jesus when warning his disciples against worrying about material provision: 'Consider the lilies of the field, how they grow; they toil not, neither do they spin; and yet I say unto you that even Solomon in all his glory was not arrayed like one of these.' Apparently stars of Bethlehem grew wild on slopes and plains of Palestine . . . and were called lilies. This description captured the imagination and interest of small, young children, and when returning home they would walk and play in the nearby meadows seeking this flower – a treasure and a prize in their view if they were lucky enough to find one. To my knowledge there were only half a dozen or so of these plants in one meadow.[1]

STINGS – treated using BETONY, CHICKWEED, DOCK, GREATER PLANTAIN, GROUND ELDER, HOUSELEEK, ONION, RAMSONS and RIBWORT PLANTAIN.

STINKING IRIS (*Iris foetidissima*), also known as GLADDON.

Clump-forming evergreen herb, with dowdy yellowish-purple flowers followed by capsules which split revealing succulent orange seeds which can persist throughout the winter; native in woodland and hedgebanks on dry, often calcareous, soils, in southern England, naturalised elsewhere, and more widespread, due to being grown as a winter ornamental.

Local names include:

Bloody bones in Dorset and Co. Durham; '?from its use in fractures'
Blue devil in Somerset
Blue seggin in Ayrshire; 'i.e. sedges'
Dagger-flower in Devon and Somerset; 'from the leaves'
Dragon-flower in Devon; 'possibly a corruption of dagger-flower'
Dragon's tongue in Kent
Field lily in Dorset
FLAG in Devon
Gladden in Cornwall
Glading root in Co. Tyrone; 'the medicinal root'
Gladwyn in Somerset
Lever in Cornwall
Poison berry/berries in Devon and Somerset
Roast-beef plant in Cornwall, Devon and Dorset; 'the smell of the bruised leaves though very disagreeable, has been likened by many to that of roast beef'
Snake's fiddles on the Isle of Wight; '? vittles, victuals'
SNAKE'S FOOD in Devon and Somerset
Snake's meat, and snake's poison, in Devon
Spurge-wort in Dorset, 'because the iuyce of it purgeth'
Windmills, 'a schoolchildren's name', in Somerset.

> Having a purging qualitie . . . the country people of Sommersetshire have good experience who use to drink the decoction of the roote. Others do take the infusion thereof in ale and such like, wherewith they purge themselves, and that unto very good purpose and effect.[1]

STITCH – treated using GREATER STITCHWORT.

STOCK (*Matthiola incana*)

Short-lived perennial with greyish leaves, and sweetly scented white, pink or purple flowers, cultivated as an ornamental since the sixteenth century, naturalised (or possibly native) on sea-cliffs in southern England.

Local names include:

Gilliflower in Devon and Gloucestershire
Ginger-flower in Somerset
Jillofer in Dorset
Jiloffer, and jiloffer-stock, in Somerset
Lady whit-smock, given to white-flowered variety, in Somerset
Linnigas, sinnegar, Whitsun gilaffer (gilawfer or gilofer), and zinegar, in Somerset.

Double-flowered forms of stock were particularly admired.

In Alderney it was believed that if two single stock flowers were tied together exactly at noon on 24 June (the festival of St Jean [John]), plants grown from their seed would be double-flowered.

STOMACH PAINS AND PROBLEMS – treated using BALM, BETONY, BILBERRY, BOGBEAN, CHAMOMILE, DANDELION, GERMANDER SPEEDWELL, HEATHER, HORSERADISH, MINT, NETTLE, RAMSONS, ROWAN, wild THYME and TORMENTIL; in cattle treated using YARROW.

STUART, CHARLES EDWARD (Bonnie Prince Charlie, 1720–1788) – SEA BINDWEED associated with.

STYES – treated using APPLE, (creeping) BUTTERCUP, DOCK, GOOSEBERRRY, GROUND IVY and TREE MALLOW.

SUNBURN – BLACK BRYONY, DOCK, HOTTENTOT FIG (and SALLY-MY-HANDSOME), SUNDEW and TORMENTIL used to ease.

SUNDEW (*Drosera rotundifolia*)

Perennial insectivorous herb, with reddish leaves covered in glistening hairs and small white flowers; common on damp acidic heathland, often associated with sphagnum moss.

Local names include:

Fly-catcher in Devon
Fly-trap in Somerset
Iles in Cornwall; 'by eating which sheep were supposed to get iles or FLUKE'
LONDON PRIDE in Somerset
Moor-gloom in Yorkshire
MOOR-GRASS in Cumberland and Yorkshire
Oil-plant in Co. Donegal
SOLDIERS-AND-SAILORS ('it is not easy to reason'), and STICKY BACKS, in Somerset.

Like BOG ASPHODEL and BUTTERWORT which grow in similar peaty habitats, sundew was believed to make cattle and sheep sicken.

> The name red-rot, by which it is distinguished in some of our rural districts, on account of its supposed share in the injurious effects experienced by sheep which feed on pastures such as it loves, but of which it is most probably innocent.[1]

> [Scotland] Sundew . . . has, according to one interpretation of the Gaelic, a somewhat uncomplimentary name – *Lus na Feàrnaich*, said to mean 'the plant of *earnach*', a disease in cattle, sometimes identified as MURRAIN, and reputedly caused by eating this poisonous plant.[2]

On the Isle of Man:

> Sticky-leaved common sundew – *Lus ny Greih*, *Lus yn eiyrts* or *Lus y ghruiaghtys* – was used as a love charm. Traditionally it was surreptitiously slipped into the clothing of the person who was to be attracted. When the plant was on display in the Manx Museum (1964–1983) it usually vanished to become a signal between teenagers, rather than a charm of power.[3]

On Colonsay:

> Some ladies mix the juice with milk so as to make an innocent and safe application to remove FRECKLES and SUNBURNS.[4]

SUN SPURGE (*Euphorbia helioscopia*)

Annual herb with yellowish inflorescences, common in cultivated and disturbed soils in lowland areas.

Local names include the widespread kirnstaff, and wart-weed, and:

Cat's milk in Worcestershire
Churn-staff in Cheshire, Cumberland and Lancashire; 'from its straight stem spreading

into a flat top'
Deil's apple-trees in Clackmannanshire; 'and other species'
Deil's kirn-staff in western Scotland
DEVIL'S MILK in Middlesex and Worcestershire; 'from the acrid quality of the milky juice'
FAIRY-DELL in Dorset
Little giddie in Clackmannanshire
Little good in Berwickshire
Little goodie, or little guid, in Northumberland and Scotland; 'i.e. the Littlegude, the Devil'
Little gweedie in Aberdeenshire
Mad-woman's milk, and mamma's milk, in Buckinghamshire
MILKWEED in Essex, Hertfordshire and East Anglia
MILKWORT in Essex; 'and other species'
Mouse-milk in Yorkshire
Pig's milk in Dublin
Potatoes-in-the-dish in Dorset
Ret-weed in Norfolk; 'the acrid juice is believed to be efficacious in taking away warts'
Saturday night's pepper, and Saturday's pepper, in Wiltshire
Virgin Mary's nipple in Devon
WART-GERSE in Cumberland
WART-GRASS in Cumberland and Derbyshire
WART-WORT in Shropshire and Yorkshire
Wet-weed in Norfolk
WHITLOW-GRASS in Lincolnshire; 'probably from its local use'
WRET-WEED in Norfolk, Suffolk and Scotland.

A new resident on the Isle of Man approached me recently with a request for a botanical name which he had been told Port St Mary fishermen used (as he delicately phrased it) 'to rub on themselves to get themselves a bit excited' . . . [the plant was sun spurge], in Manx *Lus y Bwoid Mooar*, which Manx speakers had interpreted to me as 'the plant of the big knobs'. This I accepted as a reference either to its well-known use as a WART cure, or to the flower's structure. The name properly means . . . 'the plant of the big penis' . . . the milky juice of sun spurge was applied to the human penis, and promptly produced considerable swelling. If this caused excessive discomfort the organ could be dipped in milk (soured milk was usually recommended).[1]

Although all spurges produce white latex when broken, it appears that sun spurge was the one most favoured for the removal of warts.

> [Orkney] The white, milk-like fluid in the hollow stem was applied to warts to remove them, hence the local name warty-girse.[2]

> [Co. Longford] The seven-sisters is a green branchy plant with seven stems on each stalk . . . When a stem is broken away a milky fluid appears on the stalk. This if rubbed on warts is a cure.[3]

> [1930s, mid Oxfordshire] The latex in the stems and leaves of sun spurge was used on warts. The latex when set was supposed to stifle the wart by not allowing it to breathe. I have cured my own warts by using this method.[4]

In Northumberland:

> [My 78-year-old cousin says] Sun spurge . . . was boiled up, the resultant yellow-green liquid was drained off and was drunk with an equal quantity of water, night and morning, to relieve aches and pains of a RHEUMATIC nature.[5]

SUNSTROKE – treated using WALNUT.

SWEDE (*Brassica napus* ssp. *rapifera*), known as rutabaga in North America

Biennial herb with swollen, pale orange, tap root, believed to have originated in Scandinavia or Russia, cultivated in these islands since about 1700, as a root vegetable and as fodder for sheep.

It is apparent that in different parts of Britain and Ireland the names swede, TURNIP, and neep have been used to refer to different plants.

> I'm from Liverpool. Throughout my childhood swedes were known as turnips. I never heard the word swede until supermarkets arrived. We used to make HALLOWE'EN lanterns with these large turnips as children. We would hollow out the flesh with a knife and cut slits to make a face. Then we would put candle stumps or nightlights inside and replace the top. This is an Irish custom and may have arrived here with Irish immigrants. It was commonly done in England in the 1950s and 60s.[1]

> [Co. Armagh] Hallowe'en was a big event for children in my

> childhood (1950s). Naturally we didn't celebrate Guy Fawkes Night, so it all happened on Hallowe'en – bonfires, trick or treat (though I think we called this something different), etc. We made our lanterns from large swedes, not PUMPKINS, and the story is that it was Irish people emigrating to America who took the custom over there. Finding that pumpkins were bigger and better they were used for lanterns instead – and finally the custom came back to England from America – or so I heard![2]

Like other commonly grown vegetables, swedes were used as folk remedies.

> [Tiverton, Devon] Swedes – for a COUGH – grate as finely as possible. Boil with dark brown sugar. Strain.[3]

> Cure for WHOOPING COUGH – slice a swede, cover each slice with brown sugar, leave until it is a syrup, then give a spoonful.[4]

> For KIDNEY stones, drink water from boiled swedes.[5]

SWEET CICELY (*Myrrhis odorata*)

Perennial herb, with fern-like leaves, white flowers and seedpods that taste of aniseed; introduced from mountains in central and southern Europe, first recorded in the wild in 1777, common on roadsides, rough grassy places mainly in northern areas; cultivated elsewhere.

Local names include:

Anise in Co. Durham
Annaseed in Cumberland
COW-WEED in Yorkshire
Myrrh in Cumberland and Aberdeenshire
ROMAN PLANT in Lancashire
Sweet bracken in Cumberland; 'from its aromatic fragrance, and the resemblance of its leaves to those of fern or bracken'
Sweet cis in Yorkshire
Sweet fern in Essex
Sweet humlick, or sweet humlock, in Berwickshire
Sweets in northern England
Switch in Lancashire
Wild anise in Cumberland.

Typically, in Cumbria:

> It occurs by walls, in damp gills and along roadsides, but usually close to houses and farms, reflecting its earlier culinary and medicinal use. It is still used occasionally to sweeten sour fruit.[1]

Knowledge of sweet cicely as a sweetener is widespread, possibly as a result of sugar rationing during the Second World War.

> Sweet cicely . . . abundant on roadsides in East Lothian, formerly prized for its sweetness and used for sweetening RHUBARB.[2]

> I have always used sweet cicely in cooked APPLE dishes, a habit I suppose I picked up from my mother or grandmother.[3]

On the Isle of Man:

> On Old Christmas Eve . . . a watch was, and sometimes still is, kept for the flowering of the myrrh (sweet cicely). According to tradition this blooming lasts but an hour. In many years the first leaf buds can be found, and sometimes indeed there are flowers, but both are very soon cut by frost.[4]

In Suffolk:

> I have made an infusion (tea) of fresh sweet cicely for my family's COUGHS for many years.[5]

Sweet cicely seedpods are much used in Derbyshire WELL-DRESSINGS.[6]

SWEET FLAG (*Acorus calamus*)

Perennial herb with sweetly scented rhizome, of uncertain origin, known in Britain since the sixteenth century, formerly used as a strewing-herb, widespread in shallow water in canals, streams and lakes in England, scarce elsewhere.

Local names include GLADDON and sweet seg in Norfolk, and sweet rush in southern England.

> [Merseyside] local anglers used to pull up and chew the roots of sweet flag, a plant of the water's edge.[1]

SWEET WILLIAM (*Dianthus barbatus*)

Short-lived perennial herb, native to southern Europe, cultivated in gardens in these islands since the sixteenth century for its attractive, usually reddish, fragrant flowers, and increasingly popular as a cut-flower.

Local names include:

Blooming-down in Somerset
Lady's tuft in Yorkshire
London bobs in Lancashire
LONDON PRIDE in Northamptonshire and Yorkshire
London tuft in Lincolnshire, Norfolk, Northamptonshire and Suffolk
Pink beauty, and pretty willie, in Somerset.

According to Richard Prior, 'sweet William is so called from Fr[ench] *oeillet*, L[atin] *ocellus*, a little eye, corrupted to Willy, and thence to William'.[1] However, others associate the name with William the Conqueror, St William of Rochester, King William III, or William, Duke of Cumberland.

On 21 September 1984 a correspondent to the *Daily Mirror* wrote:

> You say that the flowers called sweet william were probably named after William the Conqueror. I was led to believe they were named after William, Duke of Cumberland, who defeated the Scots at the battle of Culloden in 1746.
>
> Due to the atrocities committed by the English after the battle, he became known to the Scots as 'The Butcher of Culloden'. And when the English renamed a pretty flower sweet william in his honour, the Scots renamed their most obnoxious, smelly weed [RAGWORT] stinking billy.

Similarly:

> Sweet william – known as STINKING BILLY in Scotland, because of the defeat of the Scots at Culloden by the Duke of Cumberland (Billy) – loathed.[2]

The *Daily Mirror* published further correspondence about sweet william on 20 May 1991, when a correspondent from Rochester, Kent, wrote:

> I wonder whether I have solved the old mystery of where the name sweet william originates. I have been told it comes from St William of Rochester.

To which the letters-page compilers replied: 'no one knows, but all the Old Adams we know say it was found originally by William the Conqueror'.

In the mythology of the Orange Lodges of Ulster, sweet william is associated with King William III. When the Belfast lodges march on 12 July each year to commemorate William's victory at the Battle of the Boyne in 1690, their banner-poles are decorated with 'ORANGE LILIES and sweet william'.[3]

SWELLINGS – treated using COLUMBINE.

SYCAMORE (*Acer pseudoplatanus*)

Large deciduous tree, introduced from Europe in the sixteenth century, now extensively naturalised.

Local names include:

Cockie-bendie, given to the 'large buds' in Renfrewshire
Dog-maple in Devon
Faddy-tree 'associated with Furry, or Faddy, Day at Helston', Cornwall
MAPLE in Cumbria
MAY, 'branches carried into towns on MAY DAY'; MAY-TREE, in Cornwall
Seggy in Yorkshire
Segumber in western England
Succamore in Shropshire
Tulip-tree in Wiltshire; 'the smell or taste of the young shoots is supposed by children to resemble that of the tulip'.

Local names given to the winged fruit include:

Butterflies in Somerset
CATS-AND-KEYS in Devon
CHATS in Norfolk, Suffolk and Yorkshire
Dragonflies in Dorset
Fly-angels, fly-aways, and flying angels, in Somerset
KNIVES-AND-FORKS in Kent
Lady's keys in Wiltshire
Pigeons, and wings, in Somerset.

An ancient sycamore at Clonenagh, County Laois, is associated with the abbot St Fintan (d. 603).

> The tree is outside a graveyard on the main Dublin–Limerick road . . . the site of St Fintan's Monastery, which was one of the most famous monastic schools in Eire. Tradition says the tree grew on the site of St Fintan's Well, which was desecrated by allowing farm animals to drink in it, and the well was miraculously transferred to Cromogue about three miles away, and the site of another church or abbey of St Fintan.
>
> Water is always in [a cavity in] the tree, even during the driest summer weather, or prolonged drought, and is thought to come

> from a spring beneath the tree through pores in the tree. In recent times con-celebrated Mass at the site has been offered to honour St Fintan, patron of this parish, and in memory of the dead interred in the seven graveyards in the vicinity of the tree.
>
> The water was considered to give relief to people afflicted with weak EYES, and cure other bodily ills, as was also the water in the well at Cromogue.[1]

In Scotland feudal lords used sycamores, known as dool or JOUG TREES, as their gallows.

> The most remarkable sycamores in Scotland are those which are called 'Dool trees'. They were used by the most powerful barons in the west of Scotland, for hanging their enemies and refractory vassals on, and were for these reasons called dool or grief trees. Of these there are three yet standing [in 1847], the most memorable being one near the fine old castle of Cassilis, one of the seats of the Marquis of Ailsa, on the bank of the River Doon. It is not so remarkable for its girth of stem, as for its wide spreading branches and luxuriant foliage, among which twenty or thirty men could be easily concealed. It was used by the family of Kennedy, who were the most powerful barons in the west of Scotland, for the purpose above-mentioned. The last occasion was about 200 years ago, when Sir John Fau, of Dunbar, was hanged on it, for having made an attempt, in the disguise of a gypsy, to carry off the then Countess of Cassilis, who was the daughter of the Earl of Haddington, and to whom he had been betrothed prior to his going abroad to travel. Having been detained for some years a prisoner in Spain, he was supposed to be dead, and in his absence the lady married John, Earl of Cassilis. It is said that the lady witnessed the execution of her former lover from her bedroom window.[2]

The tree was blown down in January 1938.[3]

In Wiltshire, 'sycamores had slightly unlucky associations, perhaps because they were also sometimes known as hanging trees'.[4]

At Aldenham in Hertfordshire three sycamore trees grew from the grave of William Hutchinson (d. 1697) and his wife Margaret (d. 1706). It is said that William declared his disbelief in the Resurrection, and ordered a heavy stone tomb enclosed within iron railings; if a tree grew from the tomb future generations would know that there was a life after death. This tradition appears to have been first collected early in the twentieth century; postcards depicting the grave were issued as early as 1907,[5] but the legend associated with them does not seem to have been recorded until 1913.[6] The trees were recorded as still present 'athough much shorter as they have . . . been lopped' in 2001,[7] but by 2015 only an ivy-covered stump remained.[8] At Tewin, also in Hertfordshire, is the grave of Lady Anne Grimston (d. 1713) which has an ASH tree, said to have seven distinct stems, and a sycamore with a similar number of stems emerging from it. The dying Lady Anne is said to have gathered her friends around her and told them that: 'If there is any truth in the Word of God, may seven trees grow from my grave',[9] or, 'if indeed there is life hereafter, trees will render my tomb asunder'. However, 'all records of Lady Anne bear testimony to her piety and her belief in the Resurrection', thus it seems 'that there is no truth in the story, which does not seem to have been heard of until around 1840'.[10.] Perhaps it is noteworthy that both of these graves were enclosed within substantial railings, which protected them from mowers and grazing animals, thus encouraging the growth of trees upon them.

In many areas children made WHISTLES from sycamore twigs. In St Nicholas, Pembrokeshire, in the late 1920s:

> Spring arrived to the music of hedgerow whistles. First efforts began at the end of March when buds of hedge sycamore were pushing out pale green spikes, tinged with pink, covered with pale silky hairs, though best results came later when the tinted veined leaves as well as flowers were fully formed. Then the shoots, thick as walking sticks, were cut off and trimmed into six-inch lengths, scored round with a penknife an inch from the end. The sugary wand was soaked with spittle, and then, holding it rigid in a handkerchief, first attempts were made to loosen the bark. It had to come off – snickingly – in one piece or one had to begin again. Then with a little channel to let in the air, and the bark replaced and pulled over the tapering mouthpiece, the whistle was ready to join the birds – though unlike those music-makers, sycamore whistles became gurgley with spit.[11]

Names associated with such practices include peweep-tree in Cornwall, where 'whistles – "peweeps" [were] made from twigs on May Day', WHISTLE-TREE, also in Cornwall, and WHISTLE-WOOD in Clackmannanshire.

It seems that at one time sycamore featured in Cornish MAY DAY celebrations, but at present it is only used at Helston, and then not on 1 May, but on Flora Day, 8 May, best known for its Furry Dance. On this day:

> Very early in the morning youths go out into the neighbouring woods and gather sycamore. They return at 8.30 a.m., and, waving the branches above their heads perambulate the town, stopping at places of vantage to sing the Hal-an-Tow song.[12]

T

TAMARISK (*Tamarix gallica*)

Shrub, or small tree with a feathery appearance and pale pink flowers; native to south-west Europe, cultivated in the British Isles since the sixteenth century, mainly in coastal areas, and becoming naturalised in southern England and the Channel Islands.

Two local names, BRUMMEL and CYPRESS, have been recorded from Cornwall.

In Guernsey tamarisk, known in Guernésias as *chipre* or *saunier*, was used to make the bottoms of crab pots, as its wood 'long resists the action of sea water'.[1] Also in Guernsey, water-diviners made their divining rods from tamarisk or HAZEL.[2]

TANGERINE (*Citrus reticulata*)

Evergreen tree, probably originating in subtropical China or Indochina, widely cultivated in warm temperate and tropical areas for its juicy fruit.

Varieties of tangerine are associated with CHINESE NEW YEAR celebrations in Hong Kong and elsewhere, including Britain. At England's best-known celebrations, in Soho, London, tangerines are offered to the lions which parade around the area performing outside shops and restaurants. It is said that the tangerine is of particular significance at New Year celebrations because of the similarity of its name to the Chinese word for 'blessing' or 'fortune'.[1]

TANNING – BOG MYRTLE, HEATHER and TORMENTIL used for.

TANSY (*Tanacetum vulgare*)

Perennial aromatic herb with showy yellow flowers, an ancient introduction (or perhaps native) on roadsides and other rough grassland, formerly cultivated as a medicinal or culinary, herb and still cultivated as an ornamental.

Local names include the widespread ginger, and:

BACHELOR'S BUTTONS in Somerset
Bitter buttons in Morayshire; 'from the shape of the flower, and the bitter taste of the whole plant'
BUTTONS in Yorkshire
Golden buttons in Devon
GOOSEGRASS in Wiltshire
GREEN GINGER in Cheshire
Herb in Cornwall
Hindheel in northern England
Parsley-fern in Devon and Somerset
Scented daisies in Somerset
Scented fern in Devon and Somerset; 'from its fern-like leaves and strong smell'
Stinking elshander in Perthshire
STINKING TAM in Scotland
STINKING WILLIE in Sutherland
Traveller's rest, in Wiltshire; 'leaves are supposed to cure blistered FEET'
WEEBO in Scotland
Yellow buttons in Somerset and Scotland.

Tansy was widely valued as a cure for intestinal WORMS.[1]

> A cure for thread worms: boil the flowers or foliage of tansy weed and drink the infusion. Dose: a wineglass full each morning.[2]

Other uses include:

> Occasionally known as yellow buttons – used as a MOTH repellent, and placed round the home it can discourage FLIES and other insects.[3]

> Many years ago an old lady told me that when she was a girl in the Shropshire countryside people used to use tansy as a floor covering. This was used to keep the air pleasant. The houses had bare stone floors.[4]

Young CHICKENS are subject to a disease known as 'the pip'. The chickens pick up a worm especially in wet weather. The worm lodges in the windpipe and will eat it away. The disease can be prevented by cutting finely the tansy leaf and giving it in their ordinary food. The 'pip' is the most prevalent and fatal disease of young chickens and TURKEYS also.[5]

Married couples anxious to start a family would eat salads containing tansy. The plant grew on banks and meadows on the upland fringes of the Littleport Fens [Cambridgeshire] and . . . children were sent long distances to gather the leaves. It was said that where there were wild rabbits there was sure to be tansy, since these animals are noted for the large families they produce, the plant must have the same effect on human beings. On the other hand, many unmarried Fen girls who became pregnant chewed tansy leaves to procure a MISCARRIAGE.[6]

TAPEWORM – MALE FERN kills.

TEA – substitutes included AGRIMONY and BLACKTHORN.

TEASEL (*Dipsacus fullonum*)

Robust biennial herb, with prickly stems and leaves, and pink-purple flowers, widespread on rough ground throughout England and much of Wales, less common elsewhere.

Local names include:

ADAM'S FLANNEL in Leicestershire
Barber's brush in Essex, Dorset, Somerset and Wiltshire
BESOM in Cornwall
BRUSHES in Lincolnshire
Brushes-and-combs in Dorset and Somerset
Bull-rush in Somerset
Buttons in Dorset
Church-brooms in Essex; 'from the resemblance of the flower-heads in shape to the turk's-head brooms used for sweeping high places'
Clothes-brush in Somerset and Wiltshire
Clothier's brush in Cumberland; 'where it is grown in gardens'
COCK'S COMB in Dorset
Comb-and-brush in Wiltshire.

The paired leaves of teasel clasp its stem to create cups which collect rainwater in which small insects drown, leading to the suggestion that the plant is insectivorous. Recent research has shown that although depriving plants of these insects does not influence their stature, plants which retain their insects tend to produce a greater quantity of seeds.[1] Rainwater from teasels was believed to cure WARTS and to be good for EYES.

> [Cambridgeshire, 1660] Rain water lying stagnant in the bases of the leaves of this plant is recommended for removing warts if the hands are washed in it several times. And from this it has perhaps acquired the name of Bath of Venus.[2]

> [West Sussex] For weakness in my eyes I have been assured that the best application would be the water that is found in the hollow cup of the teasel.[3]

> [Somerset, 1914–39] The rain water held at the base of teasel leaves (basin of Venus) was thought to be a good remedy for sore eyes.[4]

> Fuller's teasel (*D. sativus*) continues to be grown for raising the nap on cloth in the Taunton area of Somerset, and was formerly cultivated more widely.[5]

> During my childhood [1920s] they used to grow teasels near Ilton. Some schoolboys were employed to harvest these during the summer holidays. The boys, aged 13, were paid five shillings for a month's work. There were one or two fields where they always grew teasels.[6]

TEETH – SAGE used to clean.

TEETHING troubles – BITTERSWEET, OPIUM POPPY and *Merremia discoidesperma* (a SEA BEAN) used to treat.

THATCH – BRACKEN, EELGRASS, GREATER TUSSOCK-SEDGE, HEATHER and REED used for.

THIRST – BROAD BEAN cures.

THISTLE (*Cirsium* spp.)

Spiny herbs, considered to be agricultural weeds.

In central Ireland thistles were thought to indicate fertile soil.

> It is good land where thistles grow. Old people tell a story of a blind man who went to buy a farm. 'Tie that horse to a thistle,' he said to

> the son. 'I don't see any thistles,' said the son. 'Oh,' said the old man, 'we'll go home, son, I won't buy this land it's too poor and bad.'[1]

Despite this, thistles are weeds which are difficult to eradicate. In Devon it was recommended:

> Speed them in May
> They are up the next day.
> Speed them in June
> They will come again soon.
> Speed them in July
> Then they soon will die.[2]

In Cornwall:

> Cut dashels (thistles) in June
> – it's a month too soon.
> Cut in July
> they are sure to die.[3]

The thistle has long been an emblem of Scotland, although the history of the emblem and the identity of the species intended are confused. An article 'What is the Scottish Thistle?' published in 1981 concluded:

> the more obvious candidate, in its strong prickliness is C[*irsium*] *vulgare* [SPEAR THISTLE], the species we choose as the Scottish Thistle.[4]

Similarly in Northern Ireland spear thistle was known as 'Scotch thistle'.[5]

> The thistle as an emblem of Scotland has early but spurious associations. Achaius, an unrecognisable, if non-existent, king of the Picts did not found the Order of the Thistle, the early history of which is unclear . . . In the 11th century did an invading Dane step barefooted on a thistle and howl to alarm the Scots? This is a good tale, often told from at least as early as 1829 . . . without any stated original source. Can it be anything but pure myth?[6]

The first artefacts which might represent thistle decorations are the 'thistle-headed pins' and Norse thistle brooches of the eighth to tenth centuries in Scotland.

> However, the heads of the pins are not decorated in detail as thistles, while brooches lodged in both the Hunterian and

> Kelvingrove Museums bear only the slightest superficial resemblance to thistle heads.[7]

It appears that the association between Scotland and thistles cannot be traced back beyond the late fifteenth century.

> In a poem written by William of Dunbar, 'The Thrissel and the Rose', in honour of the marriage of Margaret Tudor to James IV in 1503, shows that the thistle was a Scottish emblem by the early 16th century at the latest. Queen Margaret (of Denmark, died 1486) possessed a bed or table covering embroidered with thistles and her husband James III issued thistled coins. According to Innes (1959) the Earl of Orkney and Caithness is known to have been a Knight of the Order of the Thistle by 1470.[8]

Some twentieth-century authors have given the name Scottish or Scotch thistle to the decorative COTTON THISTLE.[9] As this species is believed to be an introduction to Britain,[10] it seems unlikely that it was found in Scotland as early as the fifteenth century, although if it was present it may well have been cultivated in royal gardens, thus becoming associated with the monarchy and the state. It appears that cotton thistle first became associated with Scotland when King George IV visited the country in 1822.

> Soon after the King's visit to Edinburgh, Scotland, some seeds were presented to the botanic garden at Bury St Edmunds by a relation of the Bishop of London, who received them as seeds of the identical thistle, or kind of thistle, carried in the processions that attended His Majesty in Scotland; these developed into *Onopordum acanthium*.[11]

The King's visit to Edinburgh stimulated the invention of a great amount of Scottish pageantry, so it is probable that cotton thistle replaced the spear thistle – a despised weed – as a national emblem at that time.

THORN-APPLE (*Datura stramonium*)

Annual herb with white flowers and spiny seedpods, of unknown, possibly North American, origin, grown for medicinal use since the sixteenth century, occasionally appearing on disturbed ground in lowland areas, rare in Scotland and Ireland.

> This poisonous plant, very infrequently found 'in the wild' in the [Channel] Islands, was grown, and the leaves and stems dried and smoked like tobacco as a well-used remedy for ASTHMA.[1]

THROAT – sore, induced by bringing HONEYSUCKLE indoors; treated using BLACK CURRANT, BRAMBLE, DEVIL'S-BIT SCABIOUS, ELM, GARLIC MUSTARD, ONION, POTATO, SAGE, and WOOD AVENS; GOOSEGRASS used to remove obstructions.

THUNDER – BAY, BITING STONECROP and HOUSELEEK protect against; picking FIELD BINDWEED, POPPY, RED CAMPION and ROSEBAY WILLOWHERB will produce.

THYME (*Thymus* spp.)

Dwarf fragrant shrubs with purple flowers.

Garden thyme (*T. vulgaris*) native to the western Mediterranean region, long cultivated as a culinary herb.

Wild thyme (*T. polytrichus*) widespread on free-draining short turf and in rocky places.

In addition to the widespread mother-thyme and shepherd's thyme, local names for wild thyme included:

Bank-thyme in Berkshire
Horse-thyme in Northamptonshire
Mother-of-thyme in Cumberland, Worcestershire and Ireland
Old-mother-thyme in Somerset
Tae-girse in Shetland; 'i.e. tea grass'.

> To bring a sprig of SHEPHERD'S THYME, as wild Thyme is called, into the house is thought very UNLUCKY, as by doing so you bring DEATH or severe illness to some member of your family. My informant tells me that she was charged with hastening the death of her own sister in this way, and as the neighbours and family more than once accused her of this great crime, it preyed upon her mind till it made her almost ill herself.[1]

> Gypsies regard this plant as very unlucky, and do not bring it into their wagons or tents. But it may be used out of doors as a cure for WHOOPING COUGH, boiled in water with a little vinegar added and drunk cold.[2]

> As a boy [1940s and 50s] I was shown ways of helping and healing via herbs by a woman wise in the methods . . . She used wild thyme a great deal to help people with bronchial and STOMACH problems.[3]

> [Newcastle-on-Clun area, Shropshire] Dried thyme, one tablespoon full in a cup of boiling water, allow to cool and strain. Relieves phlegm and stimulates urine output.[4]

> Some 30 years ago I was told by a South Country traveller that an infusion of wild thyme rubbed into the hair would preserve the natural colour. Well, I'm 74 today and I haven't got a grey hair in my thatch, which has remained thick and brown.[5]

TIMOTHY (*Phleum pratense*)

Widespread and common perennial grass.

Local names include CHIMNEY-SWEEP in Somerset, RATS' TAILS in Cheshire, and soldier's feathers in Berwickshire.

> [Leicestershire, 1940s] Timothy grass – a game played by boys on girls: the flowering parts were stripped off leaving a rough end about 1–1.5 inches long, these were surreptitiously twisted into the girl's hair and pulled – ouch! If done in class the victim would get into trouble for shouting![1]

> When I was out for a walk with my boyfriend at the weekend, he picked some timothy . . . he took a flowerhead and twisted it in my hair – it didn't half hurt. He learnt it at school.[2]

Cf. MEADOW FOXTAIL.

TOBACCO – substitutes include ANGELICA, BLACKTHORN, COLTSFOOT, ELDER, HORSE-CHESTNUT, MEADOWSWEET, MUGWORT, MULLEIN, ROSEBAY WILLOWHERB and TRAVELLER'S JOY; EYEBRIGHT used in 'most herbal tobaccos'; DOCK leaves and ORANGE peel used to keep moist.

TOILET PAPER – TREE MALLOW leaves used instead of.

TOMATO (*Solanum lycopersicum*)

Annual herb, native to Central and South America, introduced to Europe in the sixteenth century, and widely cultivated for its juicy, usually red, edible fruit.

The Tomatina festival, during which people pelt each other with tomatoes, has been held in Buñol, Spain, since 1945 and is frequently reported, with photographs, in the British press.[1] This custom has been copied in Sutamarchàn in Colombia and Guangdong in China,[2] and, according to the *Evening Standard* of 31 August 1999 a tomato fight, 'London's answer to Spain's annual Tomatina Festival', had been held for the second consecutive year at the Redback Tavern, in Acton. It is not known if the event was repeated in 2000, but in recent years the pub, which was popular with young Australian and New Zealand travellers, has experienced closures, name changes and on 1 January 2017 a disastrous fire.

AFRICAN MARIGOLDS planted to deter pests; in 1999 'sacred Arabic scripts' appeared in a tomato fruit (see AUBERGINE).

TONIC – CENTAURY, DANDELION, FIELD GENTIAN, TORMENTIL and WORMWOOD used as.

TOOTHACHE – CHAMOMILE, DOG ROSE gall, HENBANE, MALLOW, POPPY, SPURGE LAUREL and WILLOW used to treat.

TORMENTIL (*Potentilla erecta*)

Low-growing yellow-flowered herb, widespread and common, especially on acidic soils.

Local names include the widespread FIVE-LEAVED GRASS, and:

Aert-bark, given to the roots in Shetland

BISCUIT in Co. Antrim and Co. Down

Blood-root in Northumberland and Scotland; 'from the red colour of its roots, whence it was used in cases of DYSENTERY'

Earth-bark in Shetland; 'used as a substitute for oak bark in tanning'

Ewe-daisy in Northumberland
FIVE-FINGER BLOSSOM in Suffolk
FIVE-FINGER GRASS in Gloucestershire and Isle of Wight
FIVE-FINGERS in Suffolk
Flesh-and-blood in Berwickshire; 'obviously derived from the disease it is administered to cure, viz. dysentery'
Hill-bark in Orkney
Nyamman in Co. Derry and Co. Tyrone
Nyammany in Co. Donegal; 'a decoction of this whole plant . . . stops DIARRHOEA in man or beast'
Sheep's knapperty in Dublin
Shepherd's knot in Edinburgh and Northumberland
Shepherd's root in Berwickshire
SNAKE'S HEAD, STAR, and STAR-FLOWER, in Wiltshire
Tarment in Wiltshire
Thormantle in Devon; 'excellent as a medicine in FEVERS'
Tormenting root in Counties Antrim, Cavan, and Down
Wild mother-of-thousands in Cornwall.

> Tormentil root . . . is an ordinary medicine. Two girls inquired of one of my friends at Stratford-by-Bow [London] for a 'pennorth of tormentel'. The next week they came for more, and were asked what they wanted it for. After much hesitation and nervousness, one of the girls said that the other, her sister, had been jilted by her young man. She had consulted an old woman who was 'wise', and this old woman told her to get a bunch of 'tormentel' root and burn it at midnight on a Friday. This would so worry and discomfort the 'young man' that he would return to his sweetheart. My druggist friend told me that they came for three successive weeks, and then stopped. He does not know whether they succeeded, or gave it up as a bad job, but he thinks they won![1]

Tormentil was valued for treating diarrhoea and a number of other illnesses.

> In the Lammermuirs the root is called shepherd's knot, and is used, boiled in milk, for the cure of diarrhoea.[2]

> [Co. Cavan cure for diarrhoea] Boil 'tormenting root' and drink the juice and eat the root.[3]

> [Glen Gyle, Perthshire] When over exposure to the hot sun can catch the unwary, painful SUNBURN can be treated with a cooled lotion of tormentil steeped in boiling water.[4]

> In the 1860s Duncan Deyell, a 17-year-old from the Westside of Shetland, became very ill while at sea and had to be put ashore in Southern Ireland. He saw a doctor there who told him he had rheumatic fever, and by the time the lad got back to Shetland he was unable to walk. He couldn't even sit on a horse, and his sister brought him home lying across a pony's back. A doctor was called again, but he said there was nothing he could do for the young man and gave him only six months to live. However a relative called Mary Fraser wouldn't accept this, and sent some members of the family to dig up aert-root, which she boiled, then gave the stock to the patient. I don't know how long the treatment lasted, but the lad was well and back at sea again in a few months, and continued to work on board ship for ten years. He then retired from the sea . . . and lived until he was 74 . . . I heard this story from his grandson.
>
> My mother, who was born in 1889, and lived to the grand age of 93, said that as a child she was given aert-bark boiled in milk quite regularly, as a TONIC or for any STOMACH upset.[5]

In Guernsey tormentil was known as *esquinancée*, and 'valued as a remedy for QUINSY'. It also shared with milkwort (*Polygala* spp.) the name *herbe de paralysie*:

> The Rev. R.H. Tourtel tells me that the country people recognise two different plants as effective in warding off or curing paralysis – (1) the milkwort, which is used in the case of men, and (2) the tormentil which is used in the case of women. An old woman, who had lost the faculty of speech through an attack or paralysis, recovered it again after drinking a decoction of tormentil. I have reason to believe, however, that this distinction is not universal; because some years ago an old man at St Martin's showed me some tormentil which he had gathered, and which he said he was taking regularly as a 'tea' to avert the danger of a paralytic stroke.[6]

In areas where trees are scarce or absent, tormentil roots were used for TANNING leather. In the mid-nineteenth century shoes worn on Eigg in the Inner Hebrides were 'altogether the production of the island':

> There were few trees, and, of course, no bark to spare . . . but the islanders find a substitute in the astringent root of the *Tormentilla*

> *erecta*, which they dig for the purpose among the heath at no inconsiderable expense of time and trouble . . . it took a man nearly a day to gather roots for a single infusion.[7]

> [Tormentil] is generally used for tanning their nets by fishermen in the Western Isles, who call it *Cairt-Lair.*[8]

Trees were scarce throughout much of Ireland, and in 1727 the Irish Parliament awarded £200 to William Maple for his discovery that leather could be tanned using tormentil roots. His findings were published in a pamphlet entitled *A Method of Tanning without Bark* in 1729. This contained an illustration of a tormentil plant, and is thought to be the first botanical illustration to be published in Ireland.[9] Later in eighteenth-century Ireland tree bark was so scarce that 'some tanners resorted to a more humble source of tannin, the roots of tormentil, to encourage which the Royal Dublin Society offered premiums in 1750'.[10]

TRAVELLER'S JOY (*Clematis vitalba*), also known as old-man's beard and wild clematis

Rampant climber with woody stems which can create an exotic appearance in woodland on chalky soils, having ivory-coloured flowers which are followed by feathery seeds, widespread in southern England, introduced and apparently spreading elsewhere.

Local names include the widespread BEDWINE, HONESTY and VIRGIN'S BOWER, and:

Bear-bine in Kent
Bedwind in Dorset and Wiltshire
Beggar-brushes in Buckinghamshire
Bellywind in Hampshire
BETHWIND, or BETHWINE, in Gloucestershire, Hampshire, the Isle of Wight and Sussex
Binder in Hampshire
Blind-man's buff in Somerset; 'more commonly known as old-man's beard'
Bushy beards in Devon
By-the-wind in Wiltshire
Climbers in Kent
CROCODILE in Kent
Daddy-man's beard in Somerset
Daddy's whiskers in Devon, Somerset and Wiltshire
DEVIL'S GUTS in Dorset and Somerset
Father Christmas in Warwickshire
Father-time in Somerset
Grandfather's whiskers in Cornwall and Somerset
Granfy's beard in Somerset
Grey beard in Hampshire and Wiltshire
Hag-rope in Somerset

HALF-WOOD in Gloucestershire
Hedge-feathers in Yorkshire
Jewel-guts in Devon
Lady's bower in Gloucestershire and Somerset; 'more often called virgin's bower'
LOVE-ENTANGLED in Devon
MAIDENHAIR in Buckinghamshire
Maiden's honesty in Wiltshire
OLD MAN in Sussex
Old-man's woozard in Buckinghamshire
Poor-man's friend in Somerset
Pother-wind in Sussex
Robin Hood's fetter in Cumberland, 'in gardens'
SHEPHERD'S DELIGHT in Dorset
Silver bush in Jersey
Skipping ropes in Wiltshire
Snow-in-harvest in Northamptonshire and Somerset
TUZZY-MUZZY in Gloucestershire
Willow-wind in Gloucestershire
Withy-vine, and WITHYWIND, in Wiltshire
WITHYWINE in Gloucestershire, Somerset and Wiltshire
Wold-man's beard in Dorset.

On 13 November 1875 the *Hertfordshire Mercury* reported that 'a fire [was] supposed to have been caused by a boy who had been smoking a butt of bullbine [traveller's joy] near some straw'. Similarly:

> [As a boy in Bedfordshire in the late 1930s] I well remember . . . smoking Old Man's Beard, white flaxen weed the real name of which escapes me (it is however worth noting that it has no hallucinatory properties whatever, indeed all it does is make you cough a lot).[1]

> [East Surrey, 1950s] Smoking whiffy-wood: using short lengths of the dry traveller's joy for 'a smoke'. I can only remember the bitter and acrid taste we had to endure.[2]

Local names which refer to this practice include:

BOY'S BACCA in Somerset
GYPSY'S BACCA in Somerset
GYPSY'S TOBACCO in Dorset
Smoking cane in Devon, Hertfordshire and Somerset
Tom-bacca in Sussex
Whiffy-cane, 'young boys of about 8–13 years old used to smoke [it]', in Somerset in the 1950s.

Jewel guts = wild clematis, used in Beer [Devon] to make the bottom of crab pots, being both flexible and hard wearing.[3]

[Stratford-upon-Avon, Warwickshire] Rings made by twisting into a loop the twining stems of traveller's joy . . . were placed round a child's neck as a cure for CONVULSIONS.[4]

TREE MALLOW (*Malva arborea*, formerly known as *Lavatera arborea*)

Erect herb with woody lower stems, and pinkish-purple flowers, conspicuous and believed to be native to some western coastal areas, but introduced elsewhere.

Local names include: Bass-mallow, on Bass Rock in the Firth of Forth in the seventeenth century, where 'its leaves would have been used as a bandage or poultice by the garrison of the fortress', HOLLYHOCK in Dublin, and sunsets in Somerset.

[In Cornwall tree mallow] Sometimes called 'ku-tree', because a fomentation of the leaves is used to cure a 'ku', or 'kennal' or 'STYE' or ulcer in the eye. It is generally believed that there are male and female plants, the leaves of which must be used on those of the opposite human sex.[1]

[Dorset] MARSHMALLOW – tree mallow: grows almost like a small tree, up to 7–8ft high, with a thick tough stem and very soft downy leaves which were boiled and the resulting water, whilst still as warm as bearable, was used to bathe WOUNDS, or help draw infection to the surface. After a good spell of bathing some of the leaves were used as a poultice.[2]

[Isles of Scilly] MALLOW seeds are called 'CHEESES'. We have common, tree, Cretan and dwarf mallows – preference is given to the tree, probably because of size rather than any other reason. The cheeses are eaten green after the remains of the sepals are

> removed. They taste rather like English cob nuts. They are used in salads, but I found I needed to chop them – otherwise they were suspected of being curled up caterpillars!
>
> The leaves of mallow (preferably *Lavatera arborea*) are boiled until the resultant mush looks like well boiled spinach. It is used as hot as the patient can stand on swellings of joints – the poultice to be renewed as required, a number of times over a couple of days. This actually works. I carried out instructions on a swollen ankle – not from any act of faith, but I did it not to hurt the feelings of a dear friend, so that I could say truthfully, 'Oh, I tried it, but it didn't work' – but it did![3]

On Jersey, tree mallow leaves were used in place of toilet paper.

> [Tree mallow] was much cultivated in cottage gardens on the coast. This seemed strange until an elderly Jersey gentleman delicately pointed out that, in the past, the privy tended to be at the bottom of the garden and the tree mallows were strategically placed because of their leaves. Children used to eat the fruits, *des p'tits pains*.[4]

TRINITY-TIDE (Sunday after Whit Sunday) – GOOSEBERRIES eaten.

TUBERCULOSIS (consumption) – ELDER and LORDS-AND-LADIES associated with; FOXGLOVE supposed to cure.

TUFTED HAIR-GRASS (*Deschampsia cespitosa*)

Tussock-forming perennial grass common and widespread in dampish rough grassland.

Local names include:

Bent-grass in Cumberland
Bull-faces in Cheshire, Cumberland, Yorkshire and Berwickshire
Bull-front in Cumberland, Northumberland and Yorkshire
Bull's forehead in northern England
Bull-toppin in Cumberland
Carnation-grass in Gloucestershire
Hassock-grass in Yorkshire
Iron-grass in Shropshire
Nation-grass in Somerset and Wiltshire; 'probably a corruption of carnation grass'
Sniggle-grass, and snizzle-grass, in Shropshire
Zedge-mocks in Dorset.

Tufted hair-grass was formerly used to make church HASSOCKS.

> The Church hassocks were easy to make, though tough to dig up and they were never, as far as I know, covered with anything . . . The late Rev. R. Kettlewell in 1938, writing his history of the Parish of Great Ayton [North Yorkshire], found an early eighteenth-century Church charge of 2d each for 'Bull-front kneelers'. Within the last decade an old man of Lealholm Roman Catholic Church said: 'You owt ti hev a hassock: there's nowt mair comfortabler ti kneel on than owd bull-front. Before we had a chetch here, there was two or three of t'owd hands at Ugthorpe had bull-fronts i' their pews. I haven't seen one in use for forty years or mair.'[1]

TULIP (*Tulipa gesneriana*)

Bulbous herb, commonly cultivated for its showy, but rather short-lived, flowers since late in the sixteenth century, also shortly persistent in the 'wild' where bulbs have been thrown out, or planted.

In common with some other ornamentals, such as auricula (*Primula auricula*), pinks and CARNATIONS, the competitive showing of tulips had its enthusiasts, and during the early part of the nineteenth century there were 'possibly as many as 200 tulip shows' held annually throughout the British Isles. However, since 1936 only one has survived, the Wakefield and North of England Tulip Society, which was founded in 1836, or probably earlier, and whose 'members have always been mainly working class', including 'an extraordinary number of shoemakers'.[1]

See also FLOWER PARADE

TURKEYS – DANDELION considered good for; NETTLE and WHITE DEADNETTLE fed to; TANSY cures 'pip' in.

TURNIP (*Brassica rapa* ssp. *rapa*)

Perennial herb, with swollen, white tap root, long cultivated as a root vegetable, its leaves providing a tasty and useful springtime alternative to cabbage.

It is evident that in different parts of these islands people use the names neep, SWEDE and turnip in different ways, so that what some people call turnip is what other people call swede and vice-versa.

In common with other crops, local festivities provided guidance as to when the various stages in the cultivation of turnips would take place.

> Turnips, as far as I can remember should be sowed before the 15th June (locally the Feast of St Columcille).[1]

> Turnips should be singled by Crewkerne Fair (4 and 5 September).[2]

At HALLOWE'EN turnip lanterns were made in parts of Ireland, northern England and Scotland.

> [Co. Monaghan] Lamps out of turnips are made now only at Hallowe'en. The lamps were made by scooping the centre out of the turnip, and by cutting holes in the front for windows. Then a candle or wick is put inside which shows the light.[3]

> Turnip lanterns are still made in this area [Corbridge]. Indeed, I assumed it was fairly common throughout the country. Certainly our children made them until they grew too old for that sort of thing, but, as far as I know, it was only on Hallowe'en, and I have not seen them around on Guy Fawkes Night.[4]

> As kids in the 1980s growing up in Aberdeen we had never seen a PUMPKIN. At Hallowe'en we were given turnips to carve. They were so hard that my parents would use a hammer and chisel – this would involve a great deal of swearing. My father was so fire safety conscious we were not allowed candles, but used mini torches inside the neepie lanterns.[5]

In the Shetland Islands:

> Ducking for APPLES was not a common practice . . . However a substitute was found in slices of turnip cut into letters of the alphabet, which were put into a tub of water to be retrieved in the mouth by lads and lasses. It was great fun trying to get the initials of your heart's desire, and should you end up with strange letters, it was always exciting trying to work out who the Fates had in store for you.[6]

In folk medicine turnips were valued for treating COLDS and COUGHS.

> [Used by my father-in-law's grandmother, around Melton Constable, Norfolk, about 60 years ago]. Take a large white turnip,

> wash it and slice it about $^1/_8$ to $^1/_{16}$ inches thick. Lay the slices around the sides of a dish . . . sprinkle each slice with brown sugar (demerara). A liquid will in time . . . run into the centre of the dish. Drink this to cure the cold.[7]

> When I was a small child living in North Wales [round about 1979] my brother and I were extremely ill with WHOOPING COUGH . . . One evening the doorbell rang and our mother opened the door to a gypsy selling this and that; she heard us whooping in the house and told my mother to use this family remedy to heal our coughs: it consisted of a ripe turnip, cut in half, enough of the inside should be scraped out and filled with a sugar, wrapped in foil, and left overnight, a syrup would form, this should be fed to the patient like a medicine every hour. She said this would make us better (I do not remember the timescale). It did![8]

TUTSAN (*Hypericum androsaemum*)

Small, yellow-flowered shrub, common in open woodland and on shady hedgebanks, also widely cultivated as an ornamental.

Local names include:

All-clean in Somerset
AMBER in Kent; 'because of its smell'
Butter-nut, given to the fruit in Co. Derry; where also applied to 'other larger sorts [of *Hypericum*] grown in shrubberies'
Devil's berries in Cornwall
Park-leaves in Somerset
Stitson in Devon; 'a corruption of tutsan'
Sweet amber in Sussex; 'probably from its resinous odour'
Sweet leaf in Devon
Tipsen-leaves in Buckinghamshire and Devon
Tipsy in Devon and Somerset
Tipsy-leaf in Devon
Titsum in Devon, Somerset and Sussex
Titsy-leaf in Devon
Titzen in Cornwall
TOUCH-AND-HEAL in Buckinghamshire
Touchan-leaves in Hampshire
Touch-leaf in Wales
Treacle-leaf in Cumberland.

> [Hampshire] tutsan . . . is known as 'touchen' or 'touched' leaves, and its glossy berries which turn from green to red are said to be stained with the BLOOD of the DANES.[1]

During the nineteenth century tutsan leaves were commonly placed in bibles.

> In England, and also on the Channel coast of France, the dry leaves, for their scent (likened to ambergris, so the names 'Amber' and 'Sweet Amber') for good LUCK, were put between the pages of prayer books and the Bible.[2]

Similarly, Britten and Holland (1886) record the 'still-prevailing custom, taught by mothers to their children, of placing its leaves – under the name of touch-leaf or touching-leaf – between the leaves of their bibles'.[3]

It has often been suggested that the name tutsan is derived from the French *toute-saine*, meaning 'all wholesome',[4] or 'all-heal',[5] suggesting that it was formerly valued as a medicinal herb. However, there appears to be little evidence that it was ever important as a folk remedy in Britain and Ireland.

> [Co. Leitrim] To prevent a mark: touch-and-heal leaf. This grows in fields; you will find an odd one at river's edge. There is a berry on it, first it's red, then turns black. The ointment is made with lard and the leaves mixed together, the leaves being pounded first.[6]

TWELFTH NIGHT (8 January) – APPLE trees wassailed; HOLLY and other CHRISTMAS GREENERY must be removed from indoors.

TYNWALD DAY (5 July) – MUGWORT worn on.

U

ULCERS – treated using CABBAGE, COMFREY, ELDER, GREATER PLANTAIN, PELLITORY OF THE WALL, POTATO and WATER FIGWORT, on cow's udders treated using DWARF ELDER.

UNITARIANS – celebrate FLOWER COMMUNION; use ROSES in naming ceremonies.

UNLUCKY – CACTI, CINERARIA, CUCKOO-FLOWER, ELDER, FRUIT TREES flowering out of season, FUCHSIA, GOLDENROD, HAZEL, HAWTHORN, HOGWEED, HOLLY, HYDRANGEA, IVY, JERSEY LILY, LILAC, LILY OF THE VALLEY, LONE BUSHES, MARSH MARIGOLD, MEADOWSWEET, PEAR, PINE, RAGWORT, RED CAMPION, ROSEMARY, SNOWDROP, SPINDLE, YELLOW IRIS, WHITE FLOWERS and wild THYME considered to be.

URINARY CONDITIONS – treated using ANNUAL KNAWEL and CELERY; see also BLADDER.

V

VALENTINE'S EVE (13 February) – BAY used for divination on.

VALERIAN (*Valeriana officinalis*)

Perennial, tall-growing herb with pale pink flowers, widespread in a wide range of damp habitats.

Local names include:

All-cure in Oxfordshire
Black elder in Cornwall
Bolaira in Co. Donegal
Cat's love in Yorkshire; 'since cats like the scent of the dry root'
Cat-trail, given to the root, in Yorkshire; 'attractive to CATS, and used for trailing or enticing them into traps laid where they infest'
CUT-FINGER, and cut-finger leaf, in Wiltshire
Drunken slots in Somerset
Falery in Derbyshire
Fathery-ham in Somerset
Filaera in northern England and Northern Ireland
HEAL-ALL in Cornwall and Oxfordshire
Lady's needlework in Berkshire
Maid's conceit in Co. Derry
Phillaira in Co. Donegal; 'corruption of valerian'
POOR-MAN'S PEPPER in Wiltshire
Valara in Cumberland and Co. Donegal
Villera in Co. Antrim and Co. Down.

> [In West Sussex valerian] was always called cut-leaf and I well remember being told to rub the leaf into CUTS and scratches.[1]

> [He told me Turkish valerian, *V. phu* is] called God's hand leaf . . . He explained that the leaves were wrapped round such things as an infected SORE on a finger and kept in place overnight.[2]

VERMICIDE – BISTORT, BOG MYRTLE, BOX, BRACKEN, GORSE, HELLEBORE and MALE FERN used as.

VERRUCA – treated using BANANA.

VERVAIN (*Verbena officinalis*)

Inconspicuous perennial herb with pink flowers, said to have been widely cultivated in medieval gardens, now found on open ground and waysides in southern England, and less frequently elsewhere.

Local names include berbine in Kent, Juno's tears and pigeon-grass in Berkshire, and simpler's joy in Cornwall, Somerset and Warwickshire.

In the mid-nineteenth century:

> Many people now living can remember how general a practice it was some years since, to hang a piece of vervain around the neck of a child to avert infection; some believing it to be an amulet or charm, others thinking it a herb of powerful properties. Besides this, it was taken medicinally, or worn to cure existing disease, and thus deemed efficacious in 30 different complaints, in some of which it was particularly recommended that it should be tied round the neck with a white ribbon.[1]
>
> [West Sussex] vervain dried leaves 'worn in a black silk bag', are recommended as a cure for weakly children.[2]

Since then the use of vervain seems to have been forgotten, and nothing concerning it has been contributed to the Plant-lore Archive project.

VICTORIA, Queen (1819–1901) – drank BIRCH sap to prevent hair thinning; reputedly carried MYRTLE in her wedding bouquet; popularised WHITE HEATHER.

VIOLET (*Viola odorata*)

Small perennial herb, producing bluish-purple or white, scented flowers early in the year; native throughout much of England on roadsides and hedge-banks, and in open woodlands; cultivated and becoming naturalised elsewhere.

Local names include:

Blue jackets in Somerset
English violet in Scotland
Gypsy violet in Somerset
Horse-violet, 'given to the red or pink variety' in Dorset and Somerset
Hump-backs in Somerset; 'presumably from the way the stalk bends near the flower'
Miss modesty in Dorset

Miss scenty, and modest maiden, in Somerset

Vilip in Devon.

> One violet is UNLUCKY, but a bunch of violets is not. Another 'Granny' saying . . . Granny was raised in East Anglia.[1]

> Violets will bring FLEAS or vermin into house.[2]

> I was born in 1914 . . . occasionally we would eat violet seeds – not the green pod – when they were white.[3]

It seems that violets were considered to be particularly appropriate as gifts on Mothering Sunday.

> Violets (*Viola odorata*) to be picked on Mothering Sunday (when had a special place where the rarer white and purplish red violets grew). Kent/Sussex, 1920–30s.[4]

> I have memories of attending a Mothering Sunday service at Mosterton [Dorset] church when I was a child, sometime in the 1950s. All the mothers present were given a mass-produced card with pictures of violets printed on it. Rather strange as it was a country parish, and long before the days when the picking of wildflowers was discouraged, but my mother kept the card in the drawer where she kept family photographs and the like until she died.[5]

Violets had a number of medicinal uses.

> Boil wild violets and drink the juice and it would cure a pain in the head.[6]

> I have used violet leaves for extracting thorns and bad [a wound which has become septic, or infected, or inflamed]. The back of the leaves for drawing and the front for healing. I put five leaves on top of each other. They are gentle and clean the area with no pain.[7]

Maud Grieve recorded in 1931 that in recent years fresh violet leaves had been used both externally and internally to treat CANCER, but:

> A continuous daily supply of fresh leaves is necessary and a considerable quantity is required. It is recorded that during the nine weeks that a nurseryman supplied a patient suffering from cancer of the colon – which was cured at the end of this period – a violet

bed covering six rods of ground was almost entirely stripped of its foliage.[8]

According to a gypsy remedy recorded in 1944:

> A poultice of the leaves [of violet] steeped in boiling water is good for cancerous growths: an infusion of the leaves will aid internal cancers, and, I have been told, will even cure them.[9]

Similarly:

> Recipe for infusion of violet leaves for use in cases of cancer: Take a handful of fresh green violet leaves, and pour about a pint of boiling water on them. Cover them, and let them stand for about 12 hours (until water is green), then strain off the liquid and dip into it a piece of lint. Warm a sufficient quantity of the liquid. Put on the wet lint hot wherever the malady is. Cover the lint with oil-skin or thin mackintosh. Change the lint when dry or cold. An infusion should be made fresh about every alternate day . . . Originated, I think, in Maidstone, it has been in the family many years.[10]

W

WALKING STICKS – made from ASH, BLACKTHORN and CABBAGE.

WALL BARLEY (*Hordeum murinum*)

Annual grass, widespread and common on waste ground throughout most of England, scattered elsewhere.

Despite its English name, wall barley is not associated with walls, presumably the name results from equating the Latin *murinum*, mouse-grey (pale brownish grey), with *murale*, of, or on, walls.

Local names include:

Caterpillars, and flea-grass, in London
Jack-go-up-your-arm in Suffolk
PUSSIES, and squirrel-tail grass, in Kent
Way-bent in Cambridgeshire.

Wall barley was used in a variety of rather unpleasant children's games.

> [*c*.1940] if placed under the sleave of the dress or jumper and you swung your arms backwards and forward it would slowly climb up. We used to have races with these.[1]

> [*c*.1940] we used to bombard each other with grass darts which we said were FLEA-ridden. They stuck very well to woollen clothes when thrown, and as children we used to get very ratty if we got one in our jumper without realising it for some time, as everyone said they were full of fleas![2]

> Wall barley – used as darts. Nottingham, 1950s and 60s.[3]

WALL RUE (*Asplenium ruta-muraria*)

Inconspicuous fern, widespread on mortar and calcareous rocks.

Local names include LADY'S HAIR in Somerset, and rue-fern in Devon.

When RUE was unavailable, wall rue could be used as a substitute by a jilted girl at the marriage of her former young man.

WALNUT (*Juglans regia*)

Deciduous tree, thought to be native to south-west and central Asia, and possibly introduced by the Romans, widely planted primarily for its wood, and increasingly becoming naturalised.

> A falling walnut tree is the herald of calamity. A great walnut tree on the lawn of my old [Berwickshire] home keeled over one tea-time in 1939.[1]

A widespread saying is:

> A woman, a dog, and a walnut tree,
> The more you beat them, the better they be.[2]

This rhyme was discussed in the newsletter *Plant-lore Notes & News* between February 2002 and March 2003, without any satisfactory conclusion being reached. According to a report from Hertfordshire, it is necessary to 'splash' or beat walnut trees on MAY DAY to make the sap rise. As the walnut is not native to England, it 'needs to be beaten stir up the sap and make it rise, or else there will be no walnuts'.[3]

In Norfolk:

> I have a walnut tree in my garden that was grown by my father from a nut at the beginning of the 1940s . . . Though fruitful, however, the tree was not prolific, and my father, still living locally, showed us how to improve the crop. The mildest of men, he would slash it annually with his stick, normally prefacing his actions with: 'A dog, a woman, a walnut tree . . .' He wouldn't have the heart to do it to a dog, and I'd like to see him try beating my mother, but on the walnut it worked. Since he died, I have continued the practice. This autumn thrashing is the only arboriculture our walnut gets, but it flourishes . . . It is the easiest way to get the

nuts off and at the same time make the tree shed dead wood of the previous year.[4]

Karen Russell, Technical Secretary of the Walnut Club, suggested that beating 'breaks the tips of the branching, this causing more new growth, the more flowers and hence, fruit'. However, she knew of no growers who beat their trees except to bring the nuts down.[5] Alternatively, according to Bob Flowerdew in his *Organic Bible* (1998), if you damage the bark of walnut trees their timber will have a burred grain, and thus be more valuable.[6]

The varied uses made of walnut include:

> My grandfather (b.1865) always planted a walnut tree in his horses' field because FLIES don't go under a walnut.[7]

> [Somerset, 1940s] walnut juice – [from] the green casing of walnuts – stained the skin brown so we could look 'foreign'. Unfortunately it didn't wear off, neither could it be washed off. I remember older girls tried it on their legs instead of stockings – bit blotchy though.[8]

> Boiled walnut shells make a very successful DYE for wool, popular because it does not require a mordant (to fix the colour).[9]

Before the First World War, Somerset roadmakers could be seen 'squatting by the roadside, breaking stones with a hammer and wearing spectacles made out of half walnut shells, with a small hole drilled in the middle and tied round the head with cord'. These were later replaced by glass spectacles covered with a wire mesh.[10]

> Leaves of the walnut tree – wet them and place on the head for sunstroke.[11]

WAR – FOXGLOVE and FRUIT TREES flowering out of season foretell.

WARTS – caused by picking SHEEP'S BIT flowers; removed using APPLE, ASH, BROAD BEAN, CAPER SPURGE, DANDELION, ELDER, FIG, GREATER CELANDINE, HOUSELEEK, IRISH SPURGE, IVY, NAVELWORT, PETTY SPURGE, POTATO, SOWTHISTLE, SUN SPURGE, water at the base of TEASEL leaves (and, it seems, any plant with white sap).

WATER – HAZEL and TAMARISK used in divination of.

WATERCRESS (*Nasturtium officinale*, syn. *Rorippa nasturtium-aquaticum*, and *N.* x *sterile*, syn. *R.* x *sterilis*)

Perennial white-flowered herbs, widespread in and beside shallow streams, and in other damp places. Wild watercress (*N. officinale*) can have brownish-purple leaves; sterile watercress (*N.* x *sterilis*), the commonly cultivated plant, has leaves which stay green throughout the year.

Local names include:

Beller in Co. Derry; 'biller or beller is the common name . . . among the lower classes'
BILDERS in East Anglia and Ireland
BILLER in Co. Derry and Co. Donegal
Brook-lime in Buckinghamshire; applied to 'the large form . . . considered quite distinct from true watercress'
Carpenter's chips in Gloucestershire
CARSONS in Scotland, 'from carse, the old spelling of cress'
Crash in Lancashire
Creese, 'a very common name', and kerse, in Somerset
Long tails in Dorset
Rib in East Anglia
Tang-tongues, or teng-tongues, in Scotland; 'being pungent to the taste'
Tongue-grass in Ireland
Water-crashes in Yorkshire
WATER-GRASS in Ireland; 'it is called about the street by the abusive name of water-grass'
Well-girse, and well-grass, in Scotland
Well-karse, or well-kerse, in Northumberland and Scotland
Wild skirret in Scotland.

Early mentions of SHAMROCK refer to an edible plant which both Standihurst in 1577 and Spenser in 1633 equate with watercress.[1]

> Watercress, or watergrass, could be gathered from a clean, running stream, cooked and used instead of CABBAGE. It could also be eaten raw with bread and butter . . . some people call it St Patrick's cabbage as it needed no dressing.[2]

The cooking of watercress seems to have been an uncommon practice; most gatherers ate it raw, and there was a widespread belief that watercress, like pork, should not be eaten when there is no letter R in the name of the month. Two explanations have been given for this: wild watercress is in flower and therefore unsuitable for eating during the summer months,[3] or, as water levels are likely to be low and streams sluggish during the summer, watercress might be unclean then.[4]

In addition to having various medicinal uses, especially in Ireland, it appears that watercress was considered to be generally health-giving.

> A 'simple' person down here [Devon] was said to have 'never ate his watercress', and thus it was thought that the plant, which is common enough even today in the wild, was one which gave intelligence, rather like fish.[5]

> A Hertfordshire watercress-grower used to live next door to me, he used to say 'There's a baby in every bunch of Hertfordshire watercress'.[6]

> Watercress rubbed in the SKIN takes away RASH and other skin blemishes.[7]

> Watercress eaten raw is good for HEART disease.[8]

> RHEUMATISM: it is said to eat watercress was used as a cure.[9]

> [Isle of Mull] Watercress was eaten as a cure for SCURVY. It was a common disease after the '45, when people lived on shellfish and salt meat. They came from miles away for the cress.[10]

> Poultice for a hurt sustained doing manual work, such as an over-tired arm, apply a bunch of watercress to the skin and lay a slice of bread softened in hot water on afflicted part. Bandage and leave overnight. Hurt will disappear.[11]

WATER FIGWORT (*Scrophularia auriculata*, syn, *S. aquatica*)

Perennial herb widespread at margins of lakes and waterways in low-land areas.

Local names include the widespread STINKING ROGER, water-betony, and:

Angler's flower in Somerset
Babes-in-the-cradle in Wiltshire
Bishop's leaves in Somerset and Yorkshire
Black doctor in Yorkshire
Brown net in Devon and Somerset
Brownwort in Cornwall and Somerset
Brunnet in Devon
Crooool, and cresset, in Wiltshire
Fiddlesticks in Wiltshire
Fiddlewood in Yorkshire

Huntsman's cap in Cornwall; 'from the shape of the corolla'
Poor-man's salve in Devon
Rose-noble in Co. Durham and Co. Donegal
Scaw-dower in Cornwall; 'Cornish *scaw dowr*, water elder, from the smell'
Soap-leaves in Somerset
SQUEAKERS in Devon
Stinking Christopher in Cumberland
Venus-in-her-car in Somerset.

A Devon name for water figwort was crowdy-kit:

> An interesting word, coming from the Welsh for Fiddle . . . this plant is known as 'Fiddles', and 'Fiddle-wood' in some places, 'so called because the stems are by children stripped of their leaves and scraped across each other fiddle-fashion, when they produce a squeaking noise.'[1]

Cf. BITING STONECROP.

> [Co. Donegal] Pounded and bruised is good for slays and SPRAINS.[2]

> [Cornwall] The leaves of this plant were formerly held in high repute as an application for ULCERS.[3]

> I was recently told that the old people in Ipplepen, near Newton Abbot, use 'Water Betony' or water figwort for curing ulcers and CUTS externally. Several people still living had it applied by their parents and say it was very effective.[4]

WEATHER – ELDER blooming coincides with poor weather; the emergence of ASH and OAK leaves foretell the summer weather ahead; flowering LILAC 'supposed to indicate changes'; NAVELWORT and WHITE POPLAR used to forecast.

WEDDINGS – follow HONEYSUCKLE being brought indoors; GYPSOPHILA, MYRTLE, ORANGE blossom, WHITE HEATHER and YELLOW IRIS used; RHUBARB root given as a gift; straw costumes worn (see WHEAT).

WELL-DRESSING

In the Peak District of Derbyshire a thriving tradition is the creation of pictures, composed mainly of plant material, on screens to decorate village wells during the summer months. Well-dressing traditionally started at Tissington on Ascension Day, and continued until the Late Summer Bank Holiday, at the end of August, when wells at Eyam and Wormhill are dressed.

Lack of information on the early history of well-dressing has, inevitably, led to speculation about its origin. According to one popular theory:

> Springs and wells have always been venerated, from exceedingly remote times onwards, because water is a basic necessity of life . . . Wells were honoured with religious ceremonies and dances, and decorated with flowers and green branches at the greater festivals. When Christianity came, water-worship, as such, was strictly forbidden, but most of the ancient and well-loved springs were purged of their pagan associations, purified, and rededicated to the Blessed Virgin Mary or to one of the Saints . . . well-dressing in Derbyshire is a relic of this ancient form of worship, though it is hardly necessary to say that in no part of that county has it continued uninterrupted since pagan times.[1]

Some writers claim that the Tissington dressings started in 1350 as a thanksgiving that the village, apparently due to the purity of its water, had not been ravaged by the Black Death. Alternatively the dressings started in 1615, when Tissington's wells continued to produce water even though the surrounding countryside was drought-stricken.[2]

It appears that well-dressing in its present form evolved in the late eighteenth or early nineteenth century.

> A visitor to Tissington, writing in 1758 said: 'We saw the spring adorned with garlands; in one of these was a tablet inscribed with rhymes.' If there had been pictures, surely he would have said. But when Ebenezer Rhodes went there in 1818 he found 'newly gathered flowers disposed in various devices', some arranged on boards 'cut to the figures intended to be represented, and covered with moist clay, into which stems of flowers are inserted'.[3]

These 1818 dressings are very similar to those which appear today.

> Wooden trays, *c*.2.5 cm deep and up to 3.7 m in length, are filled with moist clay onto which is pressed natural, mostly plant, materials to form vivid mosaic-like pictures usually depicting religious themes or ecclesiastical buildings. Materials commonly used include: petals, sepals of HYDRANGEA, fruits of ALDER . . . various seeds, PARSLEY leaves and lichens. When the picture is completed the tray is placed in an erect position at a village well, the well remaining dressed for about a week.[4]

Obviously the plant materials used vary according to when the dressings take place. Hydrangea flowers, which are much used late in the summer, are unavailable when Tissington dresses its wells.

> The hydrangea must surely be considered as the most important of all. Its availability throughout a long season, together with a wide colour range, provides dressers with petals for skies, robes, pattern and backgrounds. Subtle colour changes in the petals also take place from day to day during the life of the screen.[5]

Lichens which lose no colour and do not deteriorate, are also widely used.[6]

Local rules in different villages, or followed by different dressers, lay down which materials are, or are not, acceptable. In some villages only natural plant material is allowed. At the other extreme, at Wirksworth, where a competition is held for the best screens:

> Faces, hands and feet are permitted modelled in relief . . . a watch was allowed as the 'face' of Big Ben, and there have been such things used as actual leather belts, bowie knives, and knitted 'waves'.[7]

Usually flowers are stripped of their petals and each petal is pressed individually into the damp clay as the screen is covered, but:

> Barlow has 'gone it alone' with 'whole flower' dressing. Each bloom is cut to half-an-inch of stalk, which is pushed into a knife-slit in the clay. This has to be much thicker – three inches – than for petalling, the extra weight making it necessary to dress screens in position . . . Although it would be foolish to say the screens are more effective than petalled ones, they are very attractive, with deeper colour saturation, though less precise in outline.[8]

Although religious themes are usually favoured, local or national anniversaries are also featured. At Tissington in 1900:

> one well had a medallion portrait of the Queen, whose birthday fell on Ascension Day that year, and another had a view of Windsor Castle. This last was worked out almost entirely in elder catkins and grey lichens, and was very effective.[9]

Other anniversaries which have inspired dressings included the 150th anniversary of the founding of the Royal Society for the Prevention of Cruelty to Animals, at Bakewell in 1990;[10] the 1,400th anniversary of St Augustine becoming first Archbishop of Canterbury at Tissington in 1997;[11] and the centenary of the birth of the prolific children's writer Enid Blyton, at Ashford-in-the-Water again in 1997,[12] and the bicentenary of St John the Baptist church, Buxton, at Buxton in 2012.[13]

Despite the tremendous amount of painstaking work and the many varied skills involved – Tissington, with a population of 159 recorded in the 2011 census, dresses six wells – the number of places which dress wells continues to expand. In 1974 the Peak National Park Office listed 11 towns and villages in its events leaflet; a similar leaflet produced in 2015 listed 78 dressings, starting on 4 May and finishing on 19 September.

Well-dressings have spread far outside the custom's traditional boundaries.

> They have this pretty wishing-well at Upwey [Dorset], at the source of the River Wey, which they dress. They didn't dress wells in Dorset, well-dressing being a northern thing, but they make these pictures using flowers and leaves and beans – they use beans for the shoes.[14]

Since 1985 St Aldhelm's well, at Frome, Somerset, has been dressed on the saint's day, 25 May. Originally this dressing was done in the traditional Peak District way, but in more recent years things have changed. In 2015 moist clay and natural materials were no longer used, having been replaced by a drawn and coloured-in picture with materials such as shredded polythene bags being used to create vivid areas of colour.[15]

Further afield, Tom Shaw, who was born in Derby in 1916 and migrated to Australia in 1956, produced three well-dressings in Perth, Western Australia in 1985.[16]

WHEAT (*Triticum aestivum*)

Annual grass, originating in south-west Asia, now extensively cultivated as a cereal in temperate areas.

> A green [i.e. mild] Christmas means a light wheat sheaf.[1]

In Nottinghamshire a good wheat year was said to coincide with a good PLUM year,[2] while in Dorset the price of wheat could be foretold by counting the number of flowers on MADONNA LILIES.[3]

Bread made from wheat features in the Eucharist, Holy Communion, or Mass, in most churches, where it may be the 'body of Christ' or simply be part of a symbolic meal. Because of this, wheat is used as a symbol of the eucharist, and, because bread is also such an important food, wheat sheaves, or specially baked loaves of bread in the shape of sheaves, are commonly used to decorate churches at their Harvest Festival services.

According to a hand-written history of the church of St Margaret of Antioch, Sea Palling, Norfolk, displayed in the church in 1989:

> In years gone by, corn from one of the sheaves with which Palling Church was decorated for Harvest Thanksgiving was germinated in a bowl and placed on the communion table at Easter. How and when this originated is not known, but for many years it was in abeyance until it was revived at Easter 1952, when wheat from the Harvest Thanksgiving of 1951 was used.

Although this custom lapsed for a few years it was revived in about 2000:

> We do still grow the previous year's Harvest wheat for Easter . . . We plant it about three to four weeks before Easter so it is about six inches high on Easter Sunday.[4]

Similarly, a visitor attending an Easter service at the Ukrainian Catholic Cathedral of the Holy Family in Exile, in London's Mayfair, recorded:

> The bishop, priests and acolytes are shining in pure white vestments; the church alive with lilies and pots of new green wheat – loveliest symbol of the Resurrection.[5]

At St Sava Serbian Orthodox Church in Notting Hill, London, at Christmas 2017, the icons around the church each had a small pot of wheat seedlings, an APPLE, an ORANGE, some OAK leaves, some WALNUTS and some HAZEL nuts placed on tables below them, and it was explained that the wheat seedlings, symbolic of 'new life' were only used at major festivals.[6]

At present the wheat commonly grown in the British Isles is bread wheat (*T. aestivum*). Rivet wheat (*T. turgidum*) was formerly cultivated in Ireland and the north and west of the United Kingdom, but is now rarely grown, usually for cattle feed. Some forms of rivet wheat tend to produce multi-headed ears and have become the subject of legend. In the late nineteenth and early twentieth centuries it was claimed that these multi-headed forms had grown from seeds found in ancient Egyptian tombs.

> It would appear that grains of wheat, usually blackened in some way to suggest age, have been sold to tourists by unscrupulous local guides as wheat from the tombs or mummy wheat. Such wheat, being relatively fresh, often germinated if sown.

However, wheat seeds are not long-lived and do not remain viable for much over twenty-five years.[7] Several specimens labelled as 'Egyptian wheat', 'Mummy wheat', or 'Pharaoh's wheat' found their way into the collections of the Natural History Museum, London, during the late nineteenth century.

A field of rivet wheat formerly existed in the parish of Llanllwchaearn, Powys:

> This was told to me about 30 years ago by an Anglican clergyman in Tunbridge Wells, the husband of a descendant of the family involved . . .
>
> This was at the time when preachers had to be licenced by Church and State. There was this unlicenced preacher who had been out to a local centre – presumably a secret meeting – where he had preached on the text 'I will bless the Lord at all times' – it's the first verse of one of the psalms, I forget which one. Even when things go wrong you should praise the Lord.
>
> During the following week he walked to the local market. When he returned home he found his house on fire, his father dead in the kitchen, killed by soldiers, and his wife and son missing. He stood in the kitchen and said 'I will bless the Lord at all times,' reaffirming his belief.
>
> This had three results. The wind reversed direction; the soldiers

> had only fired the house at one end, leaving the wind to do the rest, but the wind changing direction put the fire out. He discovered his wife and son were safe. They had crossed a river, which they knew of but the soldiers did not. The soldiers actually burnt and destroyed the wheat crop, but when the BARLEY crop came up it had five heads on each stalk. They interpreted this as the barley with the wheat returned four-fold.[8]

An alternative version of this legend is provided in Robin Gwyndaf's *Welsh Folk Tales*, where the preacher is identified as Henry Williams (1624–84):

> Following the Restoration of the Monarchy in 1660 he spent a total of nine years in prison. During those years his furniture and farm stock were either stolen or destroyed and his house was burnt to the ground.
>
> When everything seemed to have been lost and his family were on the verge of starvation, fate intervened. Wheat sowed in a field near the house grew prolifically and astonished the whole country. From that time Henry Williams and his family suffered no more poverty . . . The field where the marvellous wheat had grown is called *Cae'r Fendith* ('the field of blessing') to this day.[9]

Two multi-branched ears of wheat, not barley as the first version of the legend suggests, which are said to have come from the field and preserved by Williams's descendants, are in the St Fagans National History Museum.

The gathering of stray ears of wheat left in fields after the crop had been harvested was an important activity for many cottagers until the early years of the twentieth century, and was briefly revived during the Second World War.

> Every harvest time after the cutting of the corn and the stacking of the sheaves, women and children would don aprons with large pockets almost the size of the apron, and glean the ears of corn left on the stubble. Sacks were then filled and taken home to feed hens throughout the winter. I gleaned the corn fields many times with my grandmother, stuffing my apron and chewing the grains of corn as we worked.[10]

In Ireland, men wearing straw costumes, or at least with straw hats or masks, appear at WEDDINGS, on ST BRIGID'S DAY, at

HALOWE'EN, as CHRISTMAS mummers or rhymers, and as wrenboys on St Stephen's Day (26 December).[11] Writers usually fail to record which cereal provides the straw for these disguises, and it is possible that OAT is most frequently used.

Estyn Evans writing in 1957 noted:

> One of the strangest traditional features of Irish weddings is the visitation of 'strawboys', that is youths wearing straw masks who attend as uninvited guests and disport themselves at the dance which normally follows a wedding. I am told that they may still be seen occasionally in Co. Cavan and parts of the west.[12]

Almost sixty years later O'Dowd observed that strawboys could still occasionally be seen at weddings in parts of Sligo, Mayo and Clare.[13] Wrenboys also continue to appear:

> People dress in some sort of straw costumes in west Kerry to celebrate what is known locally as 'the wran', i.e. the wren. This celebration takes places on St Stephen's day, the day after Christmas day. People dress in costume and parade down the street singing and performing. Money is collected also. This custom is still strong in west Kerry, and across parts of Munster, but probably long gone in other parts of the country.[14]

Even if the tradition of wearing straw disguise had died out in most places in the twentieth century, it is being revived and continues thanks to local enthusiasts.

In England two customs which use straw costumes, neither ancient and both using wheat straw, thrive. At Whittlesey (or Whittlesea) in the Cambridgeshire fens, a straw bear parades around the town on the second weekend in January. Originally associated with Plough Monday (the first Monday after Epiphany, 6 January), straw bears were first recorded in the area in the second half of the nineteenth century, dying out during the first decade of the twentieth,[15] being revived in 1980, and continuing to flourish.[16] During the nineteenth century one or more bears and their attendants toured the town expecting to receive donations of cash from householders, so it is probable that they appeared mainly at times when employment was scarce. The revival has become the first major morris dance festival of the year, with some 37 teams participating in the 2017 event. It is possible to spend all day being entertained by a variety of dancers without actually seeing the bear.[17]

Carshalton's Straw Jack first appeared in 2004, and has become a well-established event held on a Saturday early in September each year. Jack, a man inside a frame (like those used by JACK-IN-THE-GREEN) covered in miniature sheaves of wheat, parades around the town accompanied by drummers and other musicians (but no morris dancers), making lengthy stops at various pubs, before ending up at The Hope at about 5 p.m. At nightfall the Jack is burnt and people are invited to take home pieces of straw 'for luck'. The event is said to celebrate the end of the harvest (though it is unlikely that any cereals have been harvested in the area for many decades), and despite its recent origin it is apparent that some of those taking part believe that they are participating in an authentic pagan festival.[18]

WHISKY – made from BARLEY; flavoured using BITTER VETCH.

WHISTLES – made from EASTERN GLADIOLUS, ELDER, LIME, SYCAMORE, WHITE DEADNETTLE and WILLOW.

WHITEBEAM (*Sorbus aria* agg.)

A complex group of species of small trees, some of extremely restricted distribution,[1] others widely planted, characterised by leaves with white undersurfaces, and red fruit.

Local names include:

Chess-apple in Lancashire and Westmorland
Hen-apple in Morayshire
Hoar-withy in Hampshire
Iron-pear in Wiltshire; 'Iron-pear-tree Farm near Devizes is supposed to take its name from the tree growing freely thereabouts'
MULBERRY in Aberdeenshire
QUICKBEAM in Hertfordshire
Whip-crop in Hampshire and the Isle of Wight; 'the long straight and very tough shoots . . . are cut for whip-handles by waggoners'
White rice in Hampshire and the Isle of Wight; 'rice or rise is a Hants term . . . for brushwood or underwood, this shrub is called white rice from the silvery colour of the undersides of the leaves'
Whitten, and whitten-beam, in Hampshire
Widbin pear-tree in Buckinghamshire
Wild pear-tree in Derbyshire.

> The leaves of the whitebeam slightly turned and giving a glimpse of their silvery underside predicts RAIN.[2]

> Two old ladies told me that their father permitted them to eat the young leaves of whitebeam. These had an almond-like flavour. However, they were permitted to eat only seven at a time as the leaves contained traces of a deadly posion.[3]

See also DEVON WHITEBEAM.

WHITE BRYONY (*Bryonia dioica*)

Perennial scrambling herb (Fig. 49), from a parsnip-like tap-root, producing greenish-white flowers and scarlet berries. The only member of the Cucumber family (Cucurbitaceae) native to Britain, where it is mainly restricted to England, but absent from the south-west, and introduced elsewhere.

Several names for white bryony were recorded from outside the area of its natural distribution, suggesting, if they are correct, that it was once widely cultivated. Local names include the widespread wild vine (cf. French *vigne sauvage*), and:

Ache in Cornwall
Canterbury Jack in Kent
Cow's lick in Norfolk; 'given as a horse and cow medicine'
Dead creepers in Lancashire
Death-warrant in Dorset
Elphany in northern England
Hedge-grape in Worcestershire
HOP in Gloucestershire
JACK-IN-THE-HEDGE in Hampshire
MURREN in Hampshire, Norfolk and Yorkshire
Night-bonnets in Dorset
Poisoning berries in Yorkshire
Row-berry in Somerset
SNAKE-BERRY in Suffolk
Tetter-berry in Hampshire; 'i.e. for cure against RINGWORM', or 'children have an idea that the juice of the fruit will, if it touches the skin, cause tetter [ringworm]'
Vine on the Isle of Wight
Wild cucumber in Warwickshire
Wild hop on the Isle of Wight and in Yorkshire
Woman-drake in Lincolnshire, 'where *Tamus communis* is mandrake'.

White bryony roots were widely known and used as MANDRAKE. Writing of Ireland (where bryony is not a native plant) in 1726, Caleb Threlkeld noted:

> Out of the Root knavish Impostures form Shapes which they style Mandrakes to deceive the Vulgar.[1]

Sometimes such 'mandrakes' would be partially hollowed out and have grass seeds, or grains of corn planted in the cavity, so that on germination they gave the appearance of hair. Thus Hans Sloane (1669–1753) had in his 'Collection of Vegetables and Vegetable Substances' 'a mandrakes beard . . . corn putt into the root of white bryony and thence sprouting'.

> In December 1908, a man employed in digging a neglected garden half a mile from Stratford upon Avon, cut a large root of bryony through with his spade. He called it mandrake, and ceased to work at once, saying it was 'awful bad luck'. Before the week was out, he fell down some steps and broke his neck.[2]

> A 60-year-old gardener in Cambridge, asked to dig up some bryony roots, said 'That's mandrake, that is; my old dad would never touch it; said it might scream horribly if you did' . . .

Also in Cambridgeshire:

> Farmers whose barns and outhouses were overrun by RATS would send one of their men on the task of digging up 'mandrake' roots, which were crushed and put into rat holes to drive the vermin away . . .
>
> A bryony root, as does one of mandrake, often resembles the trunk, legs and thighs of a human being. W.H. Barrett [b.1891] remembers old Fenmen digging up roots, selecting those most human in shape, washing them carefully and putting on their marks – few of the older generation could read or write. On their visits to the local inn the men took their roots to join others arranged on the taproom mantelshelf ready to be judged in a competition for which each entrant paid a small fee. On Saturday night the landlord's wife would be called in to judge the exhibits, a prize being awarded to the root which most resembled the female figure. The 'Venus Nights' were popular with both landlord and customers, because the entrance fees were spent on beer and tobacco.
>
> After the prize had been awarded the winning root stayed on the shelf until it was ousted by a finer specimen. Even then it was not discarded, for if it was suspended by a string from the rafters of a sow's sty it was reckoned that more piglets would be produced. When the root was dry and shrivelled it was placed among

> the savings kept in an old stocking hidden under the mattress as a guarantee that the hoard would increase.[3]

White bryony roots were valued as a conditioner for HORSES:

> My father-in-law, now around 70 years of age, says that the powdered root of white bryony was added to the food ration of working horses on a farm where he worked in north Norfolk. The roots were placed in the household oven, at cool heat enough to dry them. Just a pinch was used once a day, and he still maintains 'it put a shine into their coats'.[4]

> In the village of Ascott-under-Wychwood [Oxfordshire] somewhere in the mid-1930s, I watched a groom preparing a mash for his hunter and adding shavings from something which looked like a dried-up parsnip hanging on the wall. On enquiring as to its identity, I received the reply 'Mandrake, the best physic there is for 'osses.' On smelling and tasting a shaving I realised it was white bryony.[5]

In the nineteenth century white bryony was used to treat RHEUMATISM:

> [White bryony root] is very acrid in its properties, and is often scraped and applied to the limb affected with rheumatism, when it causes a stinging sensation in the skin, similar to that produced by the nettle.[6]

WHITE CLOVER (*Trifolium repens*)

Low-growing, perennial herb with white flowers which are much visited by bees; widespread and common in grassy places.

Local names include the widespread HONEYSUCKLE, and:

Baa-lambs in Dorset
Bobby-roses in Cornwall
BROAD-GRASS in Dorset
Bubby-roses in Cornwall
CURL-DODDY in Orkney; 'i.e. curly head'
HONEY-STALKS in Warwickshire
LAMB-SUCKINGS in Cumberland and Yorkshire
Mull in Cornwall
Mutton-grass in Devon
Mutton-rose in Cornwall
Pussy-foot in Somerset
Quiller in Cornwall
Sheep's gowan in Berwickshire
Skally-grass, 'so called by the

vulgar', in Ireland
Smara, and smoora, in Orkney; 'Icelandic *smari*'
Sucklers in Northumberland
Suckling-clover in Suffolk
Three-leaved grass in Cornwall
White sookies, given to the 'flowering heads' in Cornwall.

As many of these names suggest, white clover flowers were sucked to extract their nectar.

> Names my friends and I used in our childhood days of the 20s and 30s . . . around Burghead, Moray . . . white clover = milkies (we sucked them).[1]

> [Berkshire/Wiltshire border] As children in the 1950s we used to suck the juice from [white] clover flowers.[2]

White clover was sometimes used as SHAMROCK:

> Shamrock on St Patrick's Day . . . before 1920 we (as children) gathered what we thought to be shamrock to send to an aunt in England – believing (as most people did) that it grew only in Ireland. We looked for a clover with small neat trefoil leaves – certainly almost always *Trifolium repens*.[3]

WHITE DEADNETTLE (*Lamium album*)

Perennial, white-flowered herb (Fig. 50), widespread on rough ground and waysides in most lowland areas.

Local names include the widespread white nettle, and:
BEE-NETTLE in Leicestershire
Black-beetle poison in Somerset
DEAF-NETTLE in Lincolnshire and Yorkshire
DEE-NETTLE in Northumberland
DUMB-NETTLE in Oxfordshire and Somerset
Dunch in Wiltshire
DUNCH-NETTLE in Dorset and Somerset
Honey-bee in Devon
HONEY-FLOWER in Somerset
HONEYSUCKLE in Wiltshire

SHOES-AND-STOCKINGS in Somerset
SNAKE FLOWER in Cambridgeshire, Essex and Norfolk
SUCK-BOTTLE in Northamptonshire
Suckie-sue in Berwickshire
White sting-nettle in Devon.

The flowers of white deadnettle attracted a number of nursery tales. A Somerset name for the plant was Adam-and-Eve-in-the-bower: 'if you turn the flower upside down, beneath the white upper lip of the corolla, Adam and Eve, the black and gold stamens lie side by side like two human figures'.[1] In Norfolk in the 1940s:

> Cinderella's slippers were found in white deadnettle flowers . . . if you look inside the white flowers you can find two black slippers.[2]

In Lancashire in the 1970s:

> The white flowers of deadnettle are always in twos. This is because they are fairy shoes that have been left outside their house. Pick a deadnettle flower, turn it upside down – they are pointed pixie shoes.[3]

In Berwickshire in the mid-nineteenth century white deadnettle leaves 'shorn into bits', are sometimes given to young TURKEYS.[4]

William Curtis in his *Flora Londonensis* (1775–98) recorded that 'boys make WHISTLES of the stalks' [of white deadnettle].[5] In Cambridgeshire in 1950s:

> The girls made whistles too, but they used the stems of white nettles which were soft to cut. It was easy to make the mouthpiece by cutting off a stem slantwise and shaving a flat on top and it was not hard to make the vent. Never the less, many attempts had to be made before an instrument was produced that would whistle.[6]

There is also one record, from Kent, 1947–50s, of whistles being made from the dried stems of white deadnettle: 'cut to a suitable length with razor/sharp knife, slot near the end, whistle'.[7]

A more widespread childhood activity was sucking the base of a white nettle flower to extract the nectar. Typically:

> I used to suck the nectar from white deadnettle flowers when I was at school [*c*.1975] in Kent.[8]

WHITE FLOWERS

Many white flowers, including COW PARSLEY, HAWTHORN and SNOWDROP, are considered to be inauspicious. Sometimes this belief is extended to include all white flowers.

> People dislike having white flowers in the house. In Long Compton [Warwickshire] some people would not enter a church which had been decorated with white flowers. It is bad luck if the first flowers of the year to be brought into the house are white. Some say that white flowers portend DEATH.[1]

> My 73-year-old mother from Rothes in Morayshire reports the following plants as strictly forbidden in the house: Any white flowers on their own, especially snowdrops, BLACKTHORN, may blossom, CHRYSANTHEMUMS – these were associated with FUNERALS and therefore death.
>
> Associated with this was the refusal ever to put RED AND WHITE FLOWERS on their own in a vase.[2]

> When I was a child (in rural Norfolk) any white flowers were known as 'funeral flowers', and we were not allowed to bring them indoors.[3]

WHITE HEATHER

Before it was popularised by Queen Victoria it seems that the idea that white heather (Fig. 51) – white-flowered forms of HEATHER, *Calluna vulgaris*, or cultivated heaths, *Erica* spp./cvs – is lucky was a little-known Highland belief. In September 1855 she recorded:

> Our dear Victoria was this day engaged to Prince Frederick William of Prussia . . . during our ride up Craig-na-Ban this afternoon he picked a piece of white heather (the emblem of 'good LUCK'), which he gave to her; this enabled him to make an allusion to his hopes and wishes.

No earlier record of the belief can be found, and this was not published until 1868, in the Queen's *Leaves from the Journal of Our Life in the Highlands.*[1]

In 1862, when the Queen met Princess Alexandra of Denmark, the future wife of the Prince of Wales, her

> heart warmed towards the exquisite creature . . . She spoke kindly to her and presented her with a sprig of heather picked by the Prince at Balmoral, saying she hoped it would bring her luck.[2]

When out for a drive on 9 September 1872, her servant John Brown

> espied a piece of white heather, and jumped off to pick it. No Highlander would pass by without picking it, for it was considered to bring good luck.[3]

Other Victorian references to white heather include those 'in connection with the weddings of Princess Helena on April 28, 1882, Princess Beatrice on July 22, 1885, and Princess Mary on July 8, 1893, when white heather was included in the bouquets of the bride or bridesmaids'.[4]

Explanations of this belief are vague and varied. It has been suggested that white heather is lucky because, unlike plants which produce normal-coloured flowers, it escaped from blood spilt on ancient battlefields.[5] Another explanation which can be found on the Internet is 'an old, old tale and a sad one', a 'Celtic' legend about Oscar, a handsome warrior-hero, and his lover Malvina. One day a messenger brought Malvina a bunch of purple heather, Oscar's last token of love before he was slain. Malvina's tears fell onto the heather turning it white:

> Ever afterwards, as she sorrowfully wandered the moors, crying for her dead lover, those of her tears that dripped on to the heather instantly turned it white. 'Although this is a symbol of my sorrow', she declared, 'may the white heather bring good fortune to all who find it.'

However, Charles Nelson in his study of white heather, notes that this legend was not mentioned by early writers, such as David McClintock, suggesting that it might be an invented late-twentieth century 'tradition'.[6]

McClintock, an eminent member of the Heather Society, reported in 1970:

> I have been told that the belief dates back to Mary, Queen of Scots, or, inevitably, to Prince Charles, the Young Pretender. But not a shred of evidence have I been able to find in support. White heather is the badge of certain clans, but hardly a convenient one to produce for a posse of men at any time of the year. One of them is the Clan Macpherson, because of a story in post-

> Culloden times of Cluny, who attributed his escape on one occasion from searchers to the fact that he had been sleeping on a clump of white heather . . . There is a similar story of the Clan Ranald, which dates back to 1544, when a battle was said to have been won because the MacDonalds stuck white heather in their bonnets.[7]

There are records of white heather being considered to be unlucky:

> My grandfather (a Scottish Royalist) always said that white heather was unlucky because of its connection with the banishment of Bonny Prince Charles.[8]

However, such beliefs are rare; white heather, and objects decorated with it, remained popular until late in the twentieth century, and were sold in Scottish gift-shops, used at weddings, depicted on postcards, and sold by gypsies in London, who often used dried white statice (*Limonium*) as a substitute. The gypsy women who sold white heather tended to be vague and unimaginative when asked what sort of good luck it is supposed to bring: 'makes your hair grow'.[9]

At an all-girls school in rural New South Wales, Australia, in the 1960s:

> The deputy headmistress . . . always had an annual consignment of 'lucky white heather' sent out from Scotland, or so she said, and of course why would we not believe it . . . An individual small sprig of white heather was given to every girl sitting her final exams . . . I believed it always helped.[10]

In 1970:

> Nowadays, white heather is an industry. There are white heather farms north and south. The heathers they grow differ: white heather is in demand on Burns' Night in January, when you will find none in flower on the moors. A tree heath, which originally came from Portugal, is grown commercially in the south-west because there it starts to flower before Christmas, and the trade is supplied from there, I have seen the same heather being hawked in London early in the year. Why people choose to grow this and not one of the superb hardy winter ones, I cannot say, but there it is. The normal white heather – the one that flowers in late summer – is commercially always ling – that is the one that the gypsies sell.[11]

More recently interest in white heather has declined and it is rarely, if ever, sold.[12]

WHITE POPLAR (*Populus alba*)

Large deciduous tree, with grey bark and leaves with distinctive white undersurfaces, suckering to produce thickets; said to have been introduced from Holland in the sixteenth century; widely planted and established in lowland areas, less common in Ireland.

Local names include:
Abbey in Somerset
Abel in Suffolk
Arbale in Somerset
ASPEN in Hertfordshire
Lady-poplar, and silver poplar, in Somerset
White back ('in allusion to the white underside of the leaves'), and white bark, in Norfolk.

> The [white poplar] tree may frequently be noticed turning up the white surface of its leaves during the huffling winds which we often experience in summer, and this is a pretty sure indication of approaching RAIN.
>
> 'I think, there will be rain' a little girl was overheard to say 'for the WEATHER tree is showing its white lining'.[1]

WHITE WATERLILY (*Nymphaea alba*)

Perennial herb, with floating leaves and white flowers; widespread in lakes, large ditches and slow-moving rivers.

Local names include:
Alau in Cornwall
BOBBINS in Buckinghamshire
Cambie-leaf in northern Scotland
Can-dock in Nottinghamshire, Somerset and Warwickshire
Can-leaves in Leicestershire and Warwickshire
Flatter-dock in Cheshire; 'flatter probably from the floating leaf'
Lough-lily in Co. Donegal
Paps in Caithness-shire; 'eaten, with the petals'
Poached egg in Buckinghamshire
Swan-amongst-the-flowers in Dorset and Wiltshire
Water-bells in 'Northern Counties'
WATER-BLOB in Northamptonshire and Yorkshire
Water-can in Leicestershire and Warwickshire; 'from the half-unfolded leaves floating on water, being supposed to resemble cans'
Water-rose in Somerset.

> A black DYE, for dyeing wool and yarn, is obtained from the large roots which are cut up and boiled.[1]

> A native of Islay . . . referring to dyeing black with *ruamalach* [white water lily roots] says that when she had a little logwood at hand she used it along with the *ruamalach*, and it improved the colour, but when she had no logwood, she just used *ruamalach*.[2]

WHITLOWS – treated using BITTERSWEET and ELDER.

WHITLOWGRASS (*Erophila verna* and *E.* spp.)

Inconspicuous, white-flowered annual herb, growing on dry open ground, walls and in pavement cracks throughout most of lowland Britain.

> Farmers used to use the flowering of whitlowgrass as a sign to sow spring BARLEY. I found it growing in the steps of the lychgate of Woolhope Church, Herefordshire – how useful for checking every Sunday![1]

Despite its name there is little evidence that whitlowgrass was used in folk medicine. Allen and Hatfield provide a record of it being used thus in Essex in the 1920s, but consider its 'folk credentials questionable' and it 'may well have been inspired by the reading of herbals'.[2] Similarly Wyse Jackson notes 'no ethobotanical uses of this plant have been noted in Ireland'.[3]

WHIT SUNDAY – churches decorated with BIRCH on; GOOSEBERRIES eaten.

WHOOPING COUGH – DOG ROSE gall prevents; BRAMBLE, BROAD BEAN, CHIVES, GARLIC, ONION, SWEDE and TURNIP used to treat.

WILD CARROT (*Daucus carota* ssp. *carota*)

Biennial white-flowered herb, growing on rough grassy, well-drained ground throughout lowland areas, but largely restricted to coastal areas in the north.

Local names include:

Bird's nest in Cumberland
Cax in Dorset
Crow's nest in Bedfordshire
Curran-petris in Scotland
ELTROT in Hampshire
FIDDLE in Lincolnshire
HILL-TROT in the New Forest, Hampshire

KEGGAS in Cornwall
KEX, and PIG'S PARSLEY, in Dorset
Rantipole in Hampshire and Wiltshire; 'so called from its bunch of leaves'.

White-flowered members of the Apiaceae (Carrot Family) confuse all but keen wildflower enthusiasts, but it appears that wild carrot, like COW PARSLEY, was associated with causing the death of one's mother:

> I loved picking flowers when I was a child (before the Second World War). I was a Londoner, but have always loved the country-side. My mother was horrified one day when I took [indoors] some flowers from the wild carrot – so pretty and delicate. She called them BREAK-YOUR-MOTHER'S-HEART. I had to take them out of the house. I have no idea why she called them that.[1]

Most of the flowers of wild carrot are white but at the centre of each flower-head there is usually a solitary dark crimson flower. Although the name Queen Anne's lace is most usually given to cow parsley in the British Isles, in North America, where wild carrot has become naturalised, it is the latter that is commonly given the name. According to legend, the Queen pricked her finger while making lace, and 'the center floret of the flower represents a drop from the Queen's blood'.[2]

In the Hebrides:

> On this day [MICHAELMAS] wild carrots are presented by the girls in a house to male visitors. An old woman named Campbell describes how 'All week before St Michael's Day we gathered wild carrots, and each hid our store on the machair. On St Michael's Day we took them up, and we girls had a great day cooking and eating them and dancing and singing. The boys had their own fun. They used to try and find and steal our carrots. We had always to give some to the first person we met after pulling them up, and also to the first person we met coming into the house when we got home.'[3]

WILD CLARY (*Salvia verbenaca*)

Perennial aromatic herb, in open grassland on base-rich soils, mainly in southern England.

Local names include blue beard in Somerset and eyeseed in Lincolnshire. The last presumably refers to the plant's use to treat EYE conditions, to which two unlocalised names also refer:

Christ's eye, 'most blasphemously called Christ's eye, because it cures diseases of the eyes'.

Clear-eye, 'old herbalists considered this one of the most efficacious of herbs in any complaint of the eye'.

> [Cotswolds] clary, or wild sage, is locally supposed to be a legacy of the ROMAN occupation of Britain. The soldiers dropped the seeds as they marched across the country. In proof of this, country people will point to the fact that it frequently flourishes along the old Roman roads.[1]

WILDE, Oscar (1854–1900) – green CARNATION associated with.

WILD PANSY (*Viola arvensis* and *V. tricolor*)

Field pansy (*V. arvensis*) an annual herb with yellow or cream, often violet-marked petals, widespread on arable and waste ground.

Wild pansy (*V. tricolor*), annual to perennial herb with usually purple, sometimes yellow-blotched flowers, widespread on arable and disturbed ground.

Although collectors of plant-names have allocated these names to one or other species, it is unlikely that any name referred solely to one species. Local names include the widespread HEARTEASE, love-and-idleness, love-in-idleness, and pink-eyed John, and:

Beedy's eyes, biddy's eyes ('biddy means chick'), and BIRD'S-EYE, in Somerset
Bleeding heart in Hampshire
BOUNCING BETT in Somerset
Buttery-entry in Derbyshire
Cat's faces in Somerset, Sussex and Aberdeenshire
Coach-horses, EYEBRIGHT, and funny face in Somerset
Gentleman-tailors in Dorset
GENTLEMEN-AND-LADIES in Somerset
Godfathers-and-godmothers in Co. Durham and Somerset
Heart's pansy in Devon
HORSE-VIOLET in Somerset
Jack-behind-the-garden-gate in Suffolk
Johnny-run-the-street, and kiss-and-look-up, in Somerset
Kiss-at-the-garden-gate in Suffolk
Kiss-behind-the-garden-gate in Warwickshire
KISS-ME in Lincolnshire and Sussex
Kiss-me-behind-the-garden-gate in Devon and Norfolk
Kiss-me-love in Dorset

KISS-ME-LOVE-AT-THE-GARDEN-GATE in Devon
Kiss-me-over-the-garden-gate in Norfolk
KISS-ME-QUICK in Cornwall
Kitty-run-the-street in Kent, Somerset and Wiltshire
Lark's eyes in Somerset
Leap-up-and-kiss-me in Hampshire, Somerset and Sussex
Little-knock-a-nidles in Somerset
Look-up-and-kiss-me in Cornwall
Love-a-li-do in Wiltshire
Love-and-idleness in Warwickshire
Love-in-idle, and love-in-vain, in Somerset
Lovenidolds in Wiltshire; 'a corruption of love-in-idleness'
Love-true in Northamptonshire
Loving idols in Wiltshire
Meet-her-at-the-entry-kiss-her-in-the-buttery, or meet-her-i'-th'-entry-kiss-her-i'-th'-buttery, in Lincolnshire
Monkey's face in Sussex
NEEDLES-AND-PINS in Dorset
Nuffin-idles in Wiltshire
Nuffy-nidles in Berkshire
Old-men's faces in Oxfordshire
Pinken-eyed John in Bedfordshire
Pinkenny John in Northamptonshire
Pink-o'-my-John in Leicestershire
SHOES-AND-STOCKINGS in Somerset
Three-faces-under-a-hood in Northamptonshire
Three-faces-under-one-hood in Somerset
Tittle-my-fancy in Norfolk
Trinity-violet in Yorkshire
Two-faces-under-the-sun in Shropshire.

> [Fillongley, Warwickshire] for HEART trouble: collect the heads of the wild pansy (called heart-ease), boil them, and drink a wine glass full of the infusion every morning.[1]

WILD SERVICE-TREE (*Sorbus torminalis*)

Deciduous tree producing white flowers and brown fruit, in woods and hedgerows in England and Wales.

Local names include:

Chequer-tree in Kent and Sussex
Chequer-wood in Kent, 'farm-labourers use it in preference to other wood to make flails for threshing corn'
Lezzory, or lizzory, in Gloucestershire
Serbs, given to the fruit, in Sussex
Shir in Surrey.

> The fruit is well known in Sussex by the name of chequers, from its speckled appearance, and sold both there and in this island [Wight]

> in the shops and public markets, tied up in bunches, principally to children. At Ryde they go under the name of sorbus-berries, but are not much in request.[1]

> Some people squeeze or sieve the pulp out and eat only that, but the fruit is best consumed in its entirety . . . the taste is something like dried apricots or tamarinds and has been described as sharpish, softly mealy and agreeably acid, of an exceedingly pleasant acid flavour . . . English botanists writing in the late eighteenth and early nineteenth centuries, seem on the whole familiar with wild services as a dessert fruit and speak of it being widely on sale . . . The principal demand, especially in the nineteenth century, for the raw fruit appears to have come from children . . . My own father who lived as a boy on a farm on the edge of Epping Forest in Essex in the early years of this [twentieth] century has often recounted how all the rural children knew where wild services grew, although they were very scarce trees, and great enthusiasm and energy was displayed every autumn in obtaining the fruit which were known as sarves, sarvers, or sarvies.[2]

It has been suggested that the name Chequers, sometimes given to inns in southern England, derives from the wild service, or chequer, tree, though at present most, if not all, signs of these inns depict chess-boards.

> At the Chequers in Smarden there is a wild service growing in the rear courtyard and evidence to show that the inn was named after the tree is given in D.C. Maynard's *Old Inns of Kent* (1925): 'Mr Mills, a local archaeologist, who has lived in Smarden for over 84 years . . . informed me that the origin of the Chequers sign is not that generally accepted – the early form of ready reckoner – for he could well remember when a boy seeing the sign of the inn garlanded in the autumn of the year with the fruit of the chequer tree'.[3]

WILLOW (*Salix* spp.)

Deciduous trees, most common in dampish situations (also small, low-growing shrubs, not considered here).

The use of willow as an emblem of grief seems to have originated in Psalm 137:

> By the rivers of Babylon we sat down and wept
> when we remembered Zion.
> There on the willow trees
> we hung up our harps,
> for there those who carried us off
> demanded music and singing,
> and our captors called on us to be merry:
> 'Sing us one of the songs of Zion.'
> How could we sing the Lord's song
> in a foreign land?

Recent writers on biblical plants note that the psalmist's tree was poplar rather than willow, but willow continues as a symbol of grief.

> The 'willows' on which the exiled Israelites hung their musical instruments . . . were Euphrates poplar [*Populus euphratica*] and not the weeping willow (*Salix babylonica*), in spite of its epithet, since it is usually considered to have originated in China.[1]

Throughout the sixteenth and seventeenth centuries willow became particularly associated with the grief felt by forsaken lovers. In the folksong 'The Seeds of Love', thought to have been written in the seventeenth century,[2] the forsaken one laments:

> For in June there's the red rose bud,
> And that's the flower for me;
> But I oftentimes have snatched for the red rose-bud
> And gained but the willow-tree.
> Oh the willow-tree will twist,
> And the willow-tree will twine,
> And I wish I were in that young man's arms,
> Where he once had the heart of mine.[3]

The idea that a rejected lover should wear a wreath or hat of willow persisted for several centuries:

> Willow caps were presented to all people who were disappointed in love. It is customary in the present day for villagers in Wales to ask a rejected suitor on the morning of his sweet-heart's marriage to another man, 'Where is your willow cap?' or 'We must make

you a willow cap.' The same applied to a spinster whose lover discards her for another girl.[4]

One of England's most beautiful folk carols associates willow with ST MARY and her son. *The Bitter Withy* explains why (particularly pollarded) willow trees rot and become hollow comparatively quickly. Jesus asks his Mother for permission to play at ball. Mary grants permission, but entreats him to keep out of mischief. He meets three high-born children, greets them, and asks them to play with him. They reply that as 'lords' and ladies' sons, born in bower and hall', they cannot play with a 'poor maid's child, born in an ox's stall'. Christ makes a bridge of sunbeams and walks over it, the other children chase after him, but the bridge fails to support them, they fall and are drowned. Their distressed mothers protest to Mary, who punishes her son:

> So Mary mild fetched home her child,
> And laid him across her knee,
> And with a handful of willow twigs
> She gave him slashes three.

Whereupon Christ curses the willow:

> Ah bitter withy, ah bitter withy,
> You have caused me to start,
> The willow must be the very first tree
> To perish at the heart.[5]

In his *De Materia Medica*, written in AD 77, Dioscorides noted that a decoction of white-willow leaves was an 'excellent fomentation for ye GOUT'.[6] This decoction was also applied externally to treat such conditions as HEADACHE, TOOTHACHE and EARACHE, and there are occasional records of willow twigs being chewed to alleviate pain.

> I am nearly 70 years of age and was born and bred in Norfolk . . . my father, if he had a 'skullache' as he called it, would often chew a new growth willow twig, like a cigarette in the mouth.[7]

> This was a great area for growing osier (withies) [*Salix viminalis*] and if anyone had a headache they chewed a sprig of withy.[8]

In 1827 a French chemist isolated from MEADOWSWEET a chemical which was later found to be present in the sap and bark of willows and given the name salicin. From this was derived salicylic acid, and towards the end of the century acetylsalicylic acid, which is more commonly known as the analgesic aspirin.[9]

Children made WHISTLES out of young willow twigs.

> After the small branches are cut to the proper form the bark is notched round with a knife, it is then beat on the knee with the knife haft, and the following lines are repeated:
>
> Sip sap, sip sap,
> Willie, Willie, Whitecap.[10]

See also CRACK WILLOW, CREEPING WILLOW and GOAT WILLOW.

WINCHESTER COLLEGE – CORNFLOWER associated with.

WIND (FLATULENCE) – YARROW used to treat.

WINE – made from BIRCH sap and DANDELION.

WINTER ACONITE (*Eranthis hyemalis*)

Small perennial herb, producing yellow flowers early in the year; native to southern Europe, cultivated as an ornamental probably since the sixteenth century, becoming naturalised in open woodland and parks, mainly in eastern areas (absent as a wild plant in Ireland).

Surprisingly few local names have been recorded: choirboys ('from the ruffs which surround the flowers') and New-year's gift in Essex, and Christmas-rose and devil's wort in Somerset.

When the six-year-old Dorothy L. Sayers moved to her new home at Bluntisham Rectory in the Fens in January 1897:

> As the fly turned into the drive she cried out with astonishment, 'Look, auntie, look! The ground is all yellow, like the sun.'
>
> This sudden splash of gold remained in her memory all her life. The ground was carpeted with early flowering aconites. Later her father told her the legend that these flowers grew in England only where ROMAN soldiers have shed their BLOOD, and Bluntisham contains the outworks of a Roman camp.[1]

WISHES – made when FOXGLOVE flower popped, first FRUIT of year eaten.

WITCHES/WITCHCRAFT – BIRD-CHERRY, ELDER and FOXGLOVE associated with; BAY, BLACKTHORN, BUTTERWORT, *Caesalpinia* seeds (see SEA BEANS), CAPER SPURGE, ELDER, ELM, GOAT WILLOW, double HAZEL nuts, HOUSELEEK, MARSH MARIGOLD, RAGWORT, ROSEMARY, ROWAN, ST JOHN'S WORT and SERVICE-TREE protect against; can grow PARSLEY; poison BRAMBLE fruit.

WOOD

BEECHwood fires are bright and clear,
If the logs are kept a year;
CHESTNUT only good they say,
If for long it's laid away;
Make a fire of ELDER tree;
Death within your house shall be;
  But ASH new or ash old
  Is fit for Queen with crown of gold.
BIRCH and FIR logs burn too fast,
Blaze up bright and do not last;
ELMwood burns like churchyard mould –
E'en the very flames are cold;
  But Ash green or Ash brown
  Is fit for Queen with golden crown.
POPLAR gives a bitter smoke,
Fills your eyes and makes you choke;
APPLE wood will scent your room
With an incense-like perfume.
OAK logs, if dry and old,
Keep away the winter's cold;
  But Ash wet or Ash dry
  A King shall warm his slippers by.[1]

The superstition that wood should be touched when something which tempts misfortune is said is widespread, but not particularly old, having first been recorded in its current form in 1877.[2]

> I remember how superstitious people were during the last war; you were for ever touching wood, hoping things would turn out.[3]

Sussex children rubbed BRUISES and other minor injuries on wood to speed healing:

Rub it on wood
Sure to come good.[4]

In Yorkshire:

If the palm of either hand itches – 'rub it on wood it's sure to be good.'

If it's the right hand, it means that you'll be paying money, but the left hand is for receiving.[5]

WOOD, Sir Henry (1869–1944) – LAUREL used to commemorate.

WOOD ANEMONE (*Anemone nemorosa*)

Small perennial herb, producing springtime white flowers; widespread in woodlands and other dampish sheltered habitats.

Local names include the widespread CUCKOO-FLOWER, enemy, and windflower, and:

BILLY-BUTTONS in Lancashire
BOW-BELLS in Worcestershire
BREAD-AND-CHEESE, and BREAD-AND-CHEESE-AND-CIDER in Dorset; ?confusion with WOOD SORREL
Candlemas-caps, and chimney-smocks, in Somerset
COWSLIP in Scotland; 'the popular name in the north'
CUCKOO in Somerset and Wiltshire
Darn-grass in Aberdeenshire, Kincardineshire and Morayshire; 'gives rise to a disease called darn or blackwater and also DYSENTERY among . . . cattle which eat it'
Drops-of-snow in Sussex
EASTER-FLOWER in Devon and Dorset
Emony in Devon, Lincolnshire and Somerset
EVENING-TWILIGHT, and fairies' windflower, in Dorset
GRANNY'S NIGHTCAP in Somerset, Warwickshsire and Wiltshire
GRANNY-THREAD-THE-NEEDLE in Somerset
Jack-o'-lantern, or jack-o'-lanthorn, in Dorset
Lady's chemise, and lady's milk-cans, in Somerset
LADY'S NIGHTCAP in Gloucestershire and Herefordshire
LADY'S PETTICOAT in Wiltshire
Lady's purse in Dorset
Lady's shimmy ('i.e. chemise'), and MILKMAIDS, in Somerset

Moggie-nightgown in Derbyshire; 'a moggie is a mouse, not a cat'
Moll-o'-the-woods in Dorset and Warwickshire
MOON-FLOWER in Worcestershire
NANCY in Dorset
Nedcullion in Co. Donegal
Nemmy in the West Country
Piss-the-bed in Aberdeenshire
SHAME-FACED MAIDEN in Wiltshire
Shoes-and-stockings, and SILVER BELLS, in Somerset
Smell-foxes in Hampshire and Somerset; 'a large colony . . . can fill the air with a sharp musky smell'
Smell-smock in Hertfordshire
SNAKE-FLOWER in Devon, Lincolnshire and Somerset
Snakes-and-adders, and SOLDIER'S BUTTONS, in Somerset
Snake's eyes in Dorset
STAR-OF-BETHLEHEM in Somerset, and Co. Donegal
White soldiers in Buckinghamshire
Wild jessamine in Dumfriesshire
Wind-plant in Lincolnshire; 'pronounced as if rhyming to bind'
Wooden bettys in Lancashire
Woolly heads in Somerset.

> One of my earliest recollections of Staffordshire is the wood anemone, which grows in very great profusion round Stanton. The natives gave it the name of THUNDERBOLT, and explained to me very carefully that I must on no account pluck it, if I did, it would certainly bring on a THUNDERstorm, and without a doubt I would be struck . . . By chance one day I went to an outlying farm, where unknown to me a wedding feast was in progress. The door was opened by someone with a smiling face, which suddenly changed to a face with alarm written on it, because I had a wood anemone in my buttonhole! I was the bringer of bad LUCK to the wedding.[1]

WOOD AVENS (*Geum urbanum*) once commonly known as herb bennet

Perennial yellow-flowered herb, widespread and common in a wide range of habitats.

Local names include:
Black bobs in Yorkshire
Blessed herb in Somerset
Gold star in Dorset
HEMLOCK in Somerset
Ram's foot, and ram's-foot root, in Devon
Wild orange-blossom in Kent
Yellow strawberry in Somerset.

In Worcestershire early in the twentieth century wood avens was 'still used in country places to put in home-made wines, and, in spring, put into ale, it is said to prevent it turning sour'[1]

> The crushed root is used [by English gypsies] as a cure for DIARRHOEA, and a little in boiling water relieves sore THROATS.[2]

WOOD SAGE (*Teucrium scorodonia*)

Perennial herb with pale greenish flowers; widespread on hedgebanks, moorland, wood margins, and a wide range of other habitats.

Local names include the widespread MOUNTAIN SAGE, and:

Ambroise in Jersey
BREAD-AND-AND-CHEESE in Somerset
Gypsy-baccy, and gypsy's sage, in Dorset
Hart's ease, 'eaten by sick deer', in Surrey
Rock-mint in Somerset
SAGE in Dublin.

In the eighteenth century Jersey people used wood sage instead of HOPS for brewing,[1] and elsewhere it was used medicinally:

> Around Dursley neighbourhood [Gloucestershire] the leaves are gathered in spring by country folk, dried and stored to make an infusion which is drunk like tea as a cure for RHEUMATISM.[2]

> [Clwyd, *c.*1930] An old lady recently deceased cured two cases of St Vitus' dance with a brew of wood sage, after doctors had given them up.[3]

WOOD SORREL (*Oxalis acetosella*)

Small perennial herb, producing white flowers in springtime, common in damp shady situations.

Local names include:

ALLELUIA in Dorset, Somerset and Wales; 'from its blossoming between Easter and Whitsuntide, the season at which the psalms were sung which end with that word'
Alleluia-plant in Dorset
CUCKOO-FLOWER in Buckinghamshire; 'sometimes confined to the blossom, the leaves being cuckoo's meat'
CUPS-AND-SAUCERS in Somerset
EVENING-TWILIGHT in Dorset
FAIRY'S BELL/S in Devon and Lancashire
Good luck in Somerset
Gowk's clover in Berwickshire and Northumberland
GREEN SNOB in Warwickshire
Hearts in Northumberland
Lady's clover in Perthshire
Laverocks in Yorkshire
RABBIT'S FLOWER in Devon
Sleeping beauty in Dorset
Sleepy clover in Dorset and Oxfordshire
SMELL-SMOCK in Hampshire
Stubwort in Somerset; 'an old name having reference to its growth about the stubs of trees'
Three-leaved laverocks in Yorkshire
Whitsun flower in Dorset
Wild clover in Lincolnshire
Woman's nightcap, and wood-ash, in Somerset.

James Bicheno, writing in 1831, identified wood sorrel as the true SHAMROCK,[1] while Lady Wilkinson claimed that the name alleluja was given to it because of

> the veneration paid to the plant, for even among the DRUIDS it was an emblem of the mysterious Three in One, which they claimed as their own peculiar secret, and endeavoured to illustrate in every possible particular of their worship. And their reverence for the plant was doubtless increased by the fact that each leaflet of the trifid leaf, is marked by a pale crescent, the emblem of the moon, and another of their sacred symbols.[2]

Regardless of such speculations wood sorrel was widely nibbled by children:

> As a child in Crowborough [Sussex] . . . wood sorrel we called egg-and-cheese – presumably from its sharp taste when chewed, but I think vinegar would be more appropriate![3]

> Wood sorrel was known to us boys as cuckoo's bread-and-cheese, and we would sample a few leaves and flowers for their astringent taste.[4]

> [Ballinahinch, Co. Tipperary] leaves of wood sorrel . . . were chewed for their enjoyable bitter-sweet taste.[5]

Names relating to this use include the widespread BREAD-AND-CHEESE, CUCKOO-MEAT, CUCKOO-SORREL and GREEN SORREL, and:

Bird's bread-and-cheese in Devon
Bird's cheese-and-bread in Cumberland
Bread-and-cheese-and-cider, and BREAD-AND-MILK, in Somerset
BUTTER-AND-CHEESE in Devon and Somerset
BUTTER-AND-EGGS in Somerset
CHEESE-AND-BREAD in the Lake District
CUCKOO-BREAD in Devon, Dorset and Somerset, 'cf. Cornish *bara-an-gok*'
Cuckoo bread-and-cheese in Cumberland
Cuckoo-cheese in Devon
Cuckoo cheese-and-bread in Cumberland
Cuckoo's sour in Shropshire
Cuckoo's victuals in Buckinghamshire
Fox's meat in Perthshire; 'its acid taste resembling that of SOUROCK, *Rumex acetosa*'
God-almighty's bread-and-cheese in Devon and Somerset
Gowk-meat in Scotland
GREEN SORREL in Devon
Hare's meat in Cornwall and Somerset
Lady-cakes in Dumfriesshire
Lady's meat in Clackmannanshire
RABBIT-MEAT in Devon
RABBIT'S FOOD in Lancashire
RABBIT'S MEAT in Cornwall, Devon and Somerset
Salt cellar in Dorset; 'from its acid flavour when eaten by children'
Sheep-soorag in Caithness
Sheep's sorrel in Dorset, East Anglia and Co. Donegal
Sookie-sooriks in Scotland
SORREL in Cheshire
Sour clover in Berwickshire
SOUR GRASS in Devon and Yorkshire
Sourocks in Co. Donegal
SOUR SABS in Devon
Sour sally in Co. Donegal
Sour sap, and SOUR SUDS, in Devon
Sour trefoil in Somerset
Wild sorrel in Lancashire
Woodsour in Yorkshire.

WORMS, INTESTINAL – BOG MYRTLE, BRACKEN, FUMITORY, HELLEBORE, HOP, HORSERADISH, RAMSONS, TANSY and WORMWOOD used to expel; see also VERMICIDES.

WORMWOOD (*Artemisia absinthium*)

Aromatic shrub, with greyish-silver leaves and inconspicuous flowers, native to temperate Eurasia, grown in British gardens for medicine and flavouring since before 1200, continuing to be cultivated by herb enthusiasts, and occasionally found in the wild.

Also known as mingwort and MUGWORT in northern England, wermout in Pembrokeshire, wermud in Northumberland, wormit in Devon and northern England, and wormod in northern England.

According to John Ray (1627–1705):

> Those who go through the countryside . . . and by chance come upon a nasty tasting BEER can improve and render it more pleasant to the palate and digestion by adding an infusion of wormwood. For the bitterness removes acidity even better than sugar.[1]

Remembered uses include fresh material to eliminate FLEAS,[2] and:

> I was brought up in Anglesey . . . wormwood grew in abundance . . . boiling water poured on and then drunk as a TONIC and to give one an APPETITE after illness.[3]

> Wormwood was used for an infusion to deal with threadWORMS, or where this was suspected, i.e. if the patient was scratching at the rear end or squirming about instead of sitting good and quiet, as was expected of my generation.[4]

WOUNDS – treated using BETONY, CARROT, GARLIC MUSTARD, IVY, MADONNA LILY, TREE MALLOW and VIOLET.

Y

YARROW (*Achillea millefolium*), also known as milfoil

Perennial herb (Fig. 52) with white or pink flowers and fern-like leaves, widespread and common in grassy places throughout the British Isles.

Local names include the widespread thousand-leaf, and:

Angel-flower in Somerset
Arrow-root in Suffolk
Badman's plaything in Lincolnshire
Bunch-o'-daisies in Dorset
Camil, CAMMICK, and CAMMOCK, in Devon
Devil's nettle in Cheshire; 'children draw the leaves across their faces which leaves a tingling sensation'
DEVIL'S PLAYTHING in Lincolnshire
DOG-DAISY in Belfast
Doggie's brose in Scotland
Farrows in Cornwall
Ginger-leaf in Berkshire
GOOSE-TONGUE in Somerset; 'cf. the German *Gansezunge*'
Green arrow in Suffolk; 'a corruption of green yarrow'
Hemmin'-and-sewin' in Hampshire
Hundred-leaved-grass in Berwickshire
LADY'S FINGERS in Dublin
LADY'S LACE on the Gloucestershire/Oxfordshire border
Meal-and-folly in Orkney
Melancholy in Orkney and Shetland; 'dried milfoil was infused and drunk as a tea to dispel melancholy'
Moleery-fea in Caithness
Old-man's mustard in Lincolnshire; 'i.e. devil's mustard'
SNAKE'S BIT, and snake's grass in Dorset

Sneezewort, and sneezing, in Gloucestershire
Stanch-girse in Scotland; 'from the styptic properties of the plant'
Staunch-weed in the West Country
Stench-girse in Scotland
Sweet nuts in Dorset
TANSY in Cheshire; 'from the finely cut leaves resembling those of the true tansy'
Thousand-leaf-grass in Staffordshire
Thousand-leaved clover in Berwickshire
TRAVELLER'S EASE in Wiltshire
Wild pepper in Berwickshire
WOUNDWORT in Somerset; 'formerly used as a vulnerary'
Yallow in Lancashire
Yarra in Devon and Scotland
Yarra-grass in Essex
Yarrel in Suffolk
Yarroway in Norfolk.

Like some other white flowers such as COW PARSLEY and HAWTHORN, yarrow was sometimes considered to be inauspicious.

> Yarrow – known as MOTHER-DIE or Fever-plant: unlucky to pick or bring into house, it is thought to cause SICKNESS.[1]

> [Essex, 1950s] When I was a child we small fry used to call it [yarrow] BREAK-YOUR-MOTHER'S-HEART. According to our folklore that's what would happen if you picked it.[2]

Yarrow was used in a wide variety of ways in LOVE DIVINATION.

> [Co. Donegal] The boys and girls cut a square sod in which grows yarrow . . . and put it under their pillow, if they have not spoken between the time of cutting the sod and going to sleep they will dream of their sweetheart. The sod ought to be of a certain size, but what the size should be seems uncertain. The custom is said to have been introduced into the country by the Scotch settlers.[3]

> [Aberdeenshire] Lasses used to take it and put it in their breasts as a charm, repeating this rhyme:
>
> Eerie, eerie, I do pluck,
> And in my bosom I do put;
> The first young man that speaks to me,
> The same shall my true lover be.[4]

> [South Devon] If a maiden wants to know who her be going to marry, her must go to the churchyard at midnight, and pluck a bit o' yarra off the grave of a young man. I knawed a woman what did it, and her told us all about it, when us was maidens. Her went up to the churchyard and her found a bit o' yarra on a young man's

grave, and as the church clock struck twelve her picked 'un and as her picked 'un her saith:

Yarra, yarra, I seeks thee yarra,
And now I have thee found.
I prays to the gude Lord Jesus
As I plucked 'ee from the ground.

Then, her said, her took 'un home, and when her got to bed her put the yarra in her right stocking, and tied 'un to her left leg, and her got into bed backwards, and as her got she saith:

Good night to thee yarra
Good night to thee yarra
Good night to thee yarra.

And again three times:

Gude night, purty yarra,
I pray thee sweet yarra,
Tell me by the marra
Who shall me true love be.

Then my old friend said to me in rather awed tones: 'He come to her in the night, and he saith "I be thee own true love Jan".' And first her married Jan Scoble, and then her married Jan Wakeham.[5]

On May Eve a girl should look for the yarrow plant. Nine leaves were taken off and the following verse was to be said:

Yarrow for yarrow, if yarrow you be
By this time tomorrow
My true love to see
The colour of his hair
The clothes he does wear
The first words he will speak
When he comes to court me.

The leaves should then be put under a pillow and the girl was supposed to dream of a future husband.[6]

A girl who already had a lover could use yarrow to find out if he was faithful:

In Suffolk . . . a leaf is placed in the nose, with the intention of making it bleed, while the following lines are recited.

Green 'arrow, green 'arrow, you wears a white blow,
If my love love me, my nose will bleed now,

> If my love don't love me, it on't bleed a drop,
> If my love do love me 'twill bleed every drop.[7]

Yarrow was reported to have been used to either stop or provoke NOSEBLEEDS, and, here again, a variety of methods were employed.

> [South Lincolnshire] Nosebleeding, to stop this, smell the flower of the yarrow; called locally 'Nosebleed'.[8]

> [My grandparents, b.1856 and 1858] put yarrow in boiling water and then over the nose for nosebleed.[9]

> I have learned, living in south Scotland, that yarrow was used for nosebleed, but in the north a handkerchief dipped in cold water and placed over the bridge of the nose was the usual cure.[10]

> The common people in order to cure the HEADACHE, do sometimes thrust a leaf of it up their nostrils, to make their nose bleed; an old practice which gave rise to one of its English names [nosebleed].[11]

Other medicinal uses included:

> [Co. Mayo] They found a great cure in yarrow for consumptive people. First of all they pulled it and cut it into small little bits. Then they boiled it in a saucepan and then they used to drink the juice and it cured that severe ailment.[12]

> [Great Yarmouth, Norfolk] For BRONCHITIS: take the flowers of yarrow, stew and drink when strained.[13]

> In late summer my mother would send us out into the local pastures to gather stalks and flower heads of yarrow. These would be wrapped in newspaper, tied with string, and put outside (under cover) to dry thoroughly. Then in the winter days when colds, flu and COUGHS were threatening, the yarrow would be broken down, put into a jug and infused with boiling water and left on the hob to keep warm. It would be served up as a medicine to all the family as required – a noxious brew, but it put paid to all colds, etc. Mum said it was the quinine in the yarrow which effected the cure, but I have my doubts.[14]

> I can remember . . . looking for a yarrow plant which my grandmother infused on the hob of our old fashioned range, back in the

1930s – her cure for ACHES and PAINS.[15]

> As a child of about 10 (present age 74) we were staying with friends of the family in Hampshire. My older sister and I developed colds and the grandmother of my friends picked sprays of yarrow (milfoil), putting it inside our handkerchiefs which we were told to inhale. She . . . advised us to permanently keep some of the flowers in our handkerchief sachets to ward off future colds.[16]

> Hot yarrow tea, made from the leaves and flower-stalks chopped and dried, for COLDS, achy backs or wind.[17]

> [Halifax area, 1930s] Yarrow, which we brewed like tea, then sieved, made a beneficial rinse for HAIR after a shampoo.[18]

> I was told by an old Scillonian that yarrow was picked and hung in the kitchen to dry . . . During the winter it would be used – the dried leaves boiled up and the resultant liquid used as a cow drench for CATTLE with STOMACH problems.[19]

YELLOW BARTSIA (*Parentucellia viscosa*)

Annual, yellow-flowered herb of damp grassy places, often near the sea, most common in Cornwall, scattered elsewhere.

> In 1958 when on holiday in Cornwall, I did some botanical recording. I got a farmer's permission to examine his paddock, and he came and asked me if I knew what made his cows SCOUR when in this field. He pointed to *Parentucellia viscosa* and said 'I thought it might be this; I don't know what you call 'e, but us call 'e ARSE-SMART.'[1]

YELLOW CORYDALIS (*Pseudofumaria lutea*)

Perennial herb (Fig. 53) with fern-like leaves and yellow flowers, native to the southern Alps, cultivated in the British Isles as an ornamental since late in the sixteenth century, first recorded in the wild about 200 years later, now widely and increasingly naturalised on old or neglected walls and stony places throughout most of England and Wales, less common in Scotland and Ireland.

Local names include:

FINGERS-AND-THUMBS in Dorset
Gypsy fern in Sussex
Italian weed in the Craven and Wharfedale areas of Yorkshire, because it was said 'to follow the ROMANS'
Kitty Barnard, in Glyndyfrdwy, Clwyd, 'after a Victorian lady who potted up the plant for sale from her garden'
LADY'S PINCUSHION in Devon
Laurel-carpet in Lancashire; 'as it seems to be one of the few flowers able to grow underneath laurel bushes'
MOTHER-OF-THOUSANDS in Devon and Somerset
PINCUSHION in Devon
Poppers in Wiltshire
Portsmouth weed in the Portsmouth area of Hampshire, where 'it seems to pop up everywhere'.

YELLOW HORNED-POPPY (*Glaucium flavum*)

Perennial herb with greyish leaves, and yellow flowers, conspicuous on shingle beaches and other coastal habitats; absent from much of Scotland and Ireland.

Local names include gold watches in Dorset, horn-poppy in Devon, Indian poppy in Somerset, and squat in southern England.

> Yellow horned-poppy – if flower is touched WARTS develop – very true, I know from personal experience – can't get rid of wart on one finger.[1]

According to James Newton, writing in 1698:

> Horned poppy with a yellow flower, vulgarly called in Hampshire and Dorsetshire, Squatmore, or Bruiseroot (as I was there informed) where they use it against BRUISES external and internal.[2]

YELLOW IRIS (*Iris pseudacorus*)

Perennial, yellow-flowered herb, widespread in a wide variety of moist habitats.

Local names include the widespread flag, or yellow flag, leavers, and seg, and:

BUTTER-AND-EGGS in Buckinghamshire, Northamptonshire and Oxfordshire
Cheeper, or cheiper, in Roxburghshire; 'so called because children

make shrill noises with its leaves'
Crane-bill in Scotland
Cucumbers, applied to the seedpods which 'when green bear a close resemblance to young cucumbers'
Daggers in Devon, Lancashire and Somerset; 'from the leaves'
Dragon-flower in Devon and Somerset; 'possibly a corruption of dagger-flower'
DUCKS' BILLS in Cornwall and Somerset
Eyeteeth in Lancashire
Flaggan in Belfast
Flagger/s in Somerset, Co. Donegal and Dublin
Flaggon in Co. Donegal; 'it is to the leaves that the name is applied, not the flowers'
Flagon/s in Co. Derry and Co. Fermanagh
Flag-plant in Lincolnshire
Fligger in East Anglia
Flower-de-luce in Sheffield
HYACINTH in Somerset
Jacob's sword in Aberdeenshire
LADY'S SLIPPER in Suffolk
Laister in Cornwall; 'Cornish *elester*'
Lavers in Cornwall and Dorset; 'O[ld] E[nglish] *laefer*, not uncommon in place-names'
Leathers, and lever-blossom, in Somerset
Levers in Dorset
Levver in Devon, also levvers in Dorset and Somerset
LIVERS in Dorset
Maiken in Lancashire
Mekkins in Cumberland, Lancashire and Yorkshire
Old sow in Norfolk
Pond-lily in Devon
Queen-of-the-marshes, and QUEEN-OF-THE-MEADOW, in Somerset
Saggan in Co. Donegal and Co. Kerry
Saggons in Co. Donegal
Seag in Cumberland
Segge, the, in Berwickshire
Seggen, seggin, or seggins, in Cumberland, Yorkshire, Scotland and Northern Ireland
Segs in Suffolk
Shalder, or shelder, applied to the 'roots', in Devon and Somerset
Skeg in Yorkshire; 'an old English name'
SOLDIERS-AND-SAILORS in Dorset
Swan-bill in Somerset
Sword-flower in Dorset
Sword-grass in Northumberland, Wiltshire and Scotland
Trinity-plant in Dorset
Water-flag in Berwickshire
WATER-LILY in Devon, Dorset, Somerset and Co. Donegal
Water-skegs in Scotland
Yellow devils in Somerset
Yellow sedge in Berwickshire.

There are occasional records of yellow iris being considered UNLUCKY, presumably because, like BULRUSH, it grows in potentially perilous places.

> My grandmother would not have yellow irises inside the house.[1]

Shetland children used yellow iris in their games.

> Another amusement of mine, long ago [1940s], was to make 'sailing boats' from the leaves of *Iris pseudacorus*. A long leaf was selected and a small lengthwise slit made in the leaf perhaps about half-way along its length. The tip of the leaf was then bent up and over until its apex would be pushed a little way through the slit, thus forming a 'sail' and a 'keel' at the same time. The little boats were either sailed down a burn or released into the sea during an off-shore wind. In the lightest of winds they covered small distances until they were lost from sight. We called them seggie boats.[2]

> Yellow iris = seg, seggie flooer, dug's lug – it produces a blue-grey and dark green DYE, and children use the leaves of the plant to make little sailing boats. It's also believed that anyone who bites a seg will develop an impediment of speech, such as a stammer.[3]

In Guernsey, yellow iris was 'formerly a favourite plant for strewing in front of a bride on her way to her WEDDING ceremony,[4] while during the German occupation of Jersey (1940–5) well-roasted yellow iris seeds were used as a COFFEE substitute.[5]

In Arran, and some other of the Western isles, the roots are used to dye black; and in Jura they are boil'd with copperas to make INK.[6]

YELLOW RATTLE (*Rhinanthus minor*)

Yellow-flowered annual herb, producing seedpods which rattle when shaken, a root-parasite on grasses and therefore encouraged by people wanting to create wildflower meadows; widespread, and apparently increasing, in meadows, road verges, and other grassy habitats.

Local names include the widespread PENNY-GRASS, and rattle, and:

Bladder-seed in Cornwall
Bull's peas in Co. Donegal
CLOCK on the 'Scotch border'
COCK'S COMB in Gloucestershire, Shropshire and Somerset; 'the plant bears an equivalent name in many of the countries in Europe'
Cow-wheat in Cumberland and Devon
Doggins in Scotland
Dog's pennies in Shetland
Dog's siller in Scotland
Fiddle-cases, applied to the 'dry capsules' on the Isle of Wight
Gowk's shillings (or shillins) in

Lanarkshire; 'from the seeds'
Gowk's sixpences in Berwickshire, Northumberland and Roxburghshire
Hay-rattle in Lancashire
Henny-penny in Yorkshire
HENPEN/S in Cumberland, Northumberland, Somerset and Yorkshire
HEN-PENNY in northern England
HORSE-PENNIES in Derbyshire, Lancashire and Yorkshire
LAMB'S TONGUE in Somerset
MONEY in Buckinghamshire, Northamptonshire and Somerset; 'from the rattling of the seeds in the pouch or pod'
Money-grass in Leicestershire
Monkey-plant in Dorset
Pence in Northamptonshire
Penny-girse in Shetland
Penny-rattle in Somerset and Sussex
Penny-weed in the Midlands
Pepperbox in Somerset
POVERTY, and POVERTY-WEED, in Somerset; 'a partial parasite fastening its suckers on the roots of grass and other plants growing near and robbing them of their sap'
Purses in Somerset
RATTLE-BAGS in Devon and Dorset
Rattle-baskets in Somerset and Wiltshire
Rattle-box in Devon, Shropshire and Ireland
Rattle-caps in Somerset
RATTLE-GRASS in Hertfordshire, Somerset and Ireland
Rattle-jack in Lincolnshire
Rattle-penny in Dorset and northern England
Rattles in Yorkshire
Rattle-traps in Dorset
Rochlis in Herefordshire; '?Flemish *rochel*, death-rattle; rochlis is death-rattle in dialects of Herefordshire and Pembrokeshire'
Rottle-penny in Dorset; 'from its dry calices rattling and the shape of its round flat capsules'
Shackle-bags in Devon, Dorset and Somerset; 'shackle means rattle'
Shackle-basket Dorset and Somerset
Shackle-box in Dorset
Shackle-caps in Somerset
Shackles in Devon
Shekel-basket in Dorset
SHEPHERD'S PURSE in Cumberland and Somerset
Snaffles in Kent
Wild musk in Berkshire.

In nineteenth-century Buckinghamshire yellow rattle was known as locusts (pronounced locus) as it was 'locally supposed' to have provided food for St John the Baptist when he was preaching in the Judaean wilderness.[1]

In Cornwall:

When the yellow rattle is in flower hay is said to be ready for cutting.[2]

YEW (*Taxus baccata*)

Evergreen, long-lived native tree, widely planted in parks, gardens and churchyards, both as solitary trees and for hedging.

Local names include:

Bow-tree, in Berkshire; 'used for bow-making – a common direction from a local would be "Go down the road past the bow-tree. . ."'
Cup-of-wine in Somerset; presumably referring to the fruits
Ife in Suffolk
View, vewe, or view, in Cheshire, Derbyshire, Lancashire and Yorkshire.

Names which refer to the fleshy reddish-pink arils which partially enclose dark seeds include:

Snat-berries in Northamptonshire; 'from the sliminess'
Snodder-gills in Hampshire
Snodgog in Kent
Snots in Lincolnshire, Somerset and Wiltshire
Snotter-berries in Somerset
Snottergall in Berkshire and Wiltshire
Snottle-berries in Yorkshire
Snotty gogs in Hampshire and Sussex.

> In Ireland yew is frequently used as palm on PALM SUNDAY, which to many Irish-speakers was known as *Domhnach an Iuir* (Yew Sunday).[1]

> Yew tree branches are used in Catholic church ceremonies on Palm Sunday and afterwards distributed to the congregation. It may be worn and afterwards some is placed in the dwelling house and byres to bring good luck.

> Yew is usually burned to make the ash for ASH WEDNESDAY ceremonies.[2]

> For Palm branches on Palm Sunday, pieces of the yew tree were used; it grew near churchyards or where there was a landlord's estate. It used to be collected, taken to church and distributed at Mass. It was always the yew and people called it Palm. Some people liked to get a good branch as they put some in the cow byre. We didn't put it in the out-buildings; we put it at the side of a picture in the kitchen. Now people bring it to the church and it is blessed there, this year it was outside in the grounds. While I and some others still bring yew, others bring whatever tree or shrub is handy, sometimes CYPRESS.[3]

Yews have stimulated people's imagination to the extent that some unusual claims have been made for them. In 1998 the botanist David Bellamy wrote:

> We also know that ever since people arrived in force upon these shores they have been in the habit of planting yew trees in acts of sanctification, close to where they eventually hoped to be laid to rest.[4]

Similarly the label on a yew tree in the Royal Botanic Gardens, Kew, in 1993, read:

> The DRUIDS regarded yew as sacred and planted it close to their temples. As early Christians often built their churches on these consecrated sites, the association of yew trees with churchyards was perpetuated.

And, according to a report in *The Times* of 19 August 1993:

> A yew tree in the village churchyard at Coldwaltham, near Pulborough, West Sussex, has been confirmed as the oldest tree in England . . . probably planted around 1000 BC by Druids.

Needless to say, such statements lack the sort of evidence that one would expect from serious scientists. Current scholarship suggests so little is known about the Druids that most of what has been written about them simply reflects what the writers would like them to have been, rather than what they actually were. Much of what has been written about yew is probably best forgotten or ignored.

Probably the most sane account of yew in the British Isles is Robert Bevan-Jones's study published in 2002. After studying the evidence accumulated over many years, and 'from many quarters', he concludes that ancient yew trees 'frequently exceed a thousand years growth',[5] and suggests that some of the oldest yews found in English and Welsh churchyards mark the sites of the hermitages or cells of early saints, but were planted by the cells, rather than the cells being built beside an existing holy tree.[6] However, Cornwall, the English county in which memory of early saints persists strongest in church dedications, lacks ancient yews.

The reason for yews being planted in churchyards has never been satisfactorily explained. In 1307 Edward I decreed that yews should be so planted to protect churches from gale damage. Robert Turner, writing in 1644, suggested that yew absorbed the vapours produced by putrefaction.[7] Writing of Somerset in 1791, John Collinson thought that yew trees were preserved in churchyards because their evergreen foliage was 'beautifully emblematic of the resurrection of the body'.[8] Others have suggested that yew trees were planted in churchyards to provide wood for making bows,[9] or, being poisonous to livestock, they were planted in churchyards so that farmers did not allow their cattle to desecrate consecrated ground,[10] or they warded off evil spirits.[11] More imaginatively:

> Yew tree is distinctly red and white, especially when the trunk is freshly cut. The heartwood is red, the sapwood . . . white. The colours were used to symbolise the blood and body of Christ.[12]

Other explanations which associate yew with Christianity include that provided on a sign on the gate to Wilmington churchyard, in East Sussex, which refers to a churchyard tree that is estimated to be about 1,600 years old: 'For those of the Christian faith a yew tree is symbolic of Christian Resurrection as it has the ability of regenerate by sending down a shoot from high up which then takes root in a crevice near the base of the tree, thus giving birth to new life.'[13] Or, according to a label in the herbarium of the Natural History Museum, London: 'Palm . . . planted in churchyards for the use of the branches on Palm Sunday and for making bows'.[14]

Recently the idea that yew wards off EVIL has led to it being used in topping out ceremonies. These events, which mark the completion of a building – 'a logical counterpart of the foundation stone ceremony'[15] – were revived, or largely created, in the 1960s. In its early form the ceremony involved placing a leafy branch on a high point of the finished building. More recently it seems that a branch of yew must be used. When the new Sadler's Wells Theatre, London, was completed on 20 November 1997 a photograph published on the following day in the *Independent* showed 'building worker J.S. Hunda Singh praying at yesterday's topping-out ceremony . . . With him is Fr Victor Stock, rector of St Mary-le-Bow, who is holding a piece of yew to be mixed with concrete to ward off evil spirits'. In January1999 the *Railway Magazine* reported:

> Celebrity steeplejack and steam fanatic Fred Dibnah has performed the topping-out ceremony for the National Railway Museum's new workshop in the traditional manner – with a sprig

of yew, used by builders for centuries to ward off evil spirits.

While in 2003, when the London Borough of Barnet's new arts building was completed:

> In keeping with the tradition of planting a yew tree at the highest point of the building for luck, a giant yew branch was swung around the theatre space on a crane to bring good luck to future productions.[16]

Despite their supposed association with early saints there are very few gospel yews - trees which sheltered preaching saints or under which the gospel was read when a parish's bounds were beaten. Most 'gospel' trees are OAKS. Bevan-Jones mentions without explanation or comment a gospel yew 'on the Bromyard to Ledbury road',[17] probably referring to the tree at Castle Froome, Herefordshire. According to a website which mentions this tree:

> There are various gospel yews, oaks and ashes scattered about the country, where the local parsons used to hold occasional services.[18]

While this is true of oak, it seems to be an exaggeration as far as yew and ASH are concerned. The Castle Frome tree was described in 1896 as a 'vulnerable yew of great antiquity, which has been known, time out of mind, as the Gospel Yew,'[19] but there is no record of religious services being held under it. No other gospel yews are known.[20]

An extremely old yew growing in the churchyard at Fortingall in Perthshire is said to be 'incontestably the most ancient specimen of vegetation in Europe'. This is possible; less probable is the belief that as a baby Pontius Pilate had been suckled under this tree when his father was a Roman legionary,[21] or alternatively he was said to have been born near the tree.[22] Neither legend is true; Pilate served as Roman governor of Judaea from AD 26 to 36, approximately thirty-five years before the Roman army ventured into Scotland. Inevitably the age of the tree has led to it becoming hollow and falling apart.

> Local tradition has it that funeral processions passed through the arch made by the ancient tree. Today only a little of the shell of this ancient hulk remains, but some parts have regrown, and several newer trunks are standing in a circle with a vast hollow centre.[23]

In November 2015 it was widely reported that the Fortingall tree, which is male, was changing gender and producing fruit.[24]

A yew tree at Congresbury, Somerset, was associated with St Congar, who is believed to have been an eighth-century hermit. In 1992:

> It is said that St Congar wished for a yew tree to provide shade; he planted his STAFF in the earth and on the following day it put forth leaves and grew into a wide-spreading tree. A portion of the ancient yew in the churchyard is still known locally as St Congar's walking stick.[25]

However, by 1998 the remains of the yew were enclosed within a beech tree: 'the old pieces of yew spill out on the ground below the spreading beech branches',[26] and by 2011 not even these remains could be found.[27]

At Ambergate, in Derbyshire:

> A gang of 18th century charcoal burners anticipated the current clamour for workplace creches by hollowing out the bough of a yew tree to make a cradle, The tree . . . which inspired the nursery rhyme 'Rock-a-Bye-Baby', is to be preserved.[28]

At Stoke Gabriel, in Devon:

> The churchyard . . . has an imposing old yew, in good condition, with some FERTILITY legends invested in it. For instance, if you are male, and walk backwards around it, or female and walk forwards, fertility is assured. Another superstition promises that wishes come true if you walk around the yew seven times. Certainly the area beneath the tree is weed free, possibly indicating the large numbers of the credulous regularly performing these rites. The tree is female, 45ft in height, 17ft in girth.[29]

Although other parts of yew are poisonous, the fleshy pinkish-red aril which partially surrounds the poisonous seed was eaten by children, and, perhaps, used in jam making.[30]

> [1597] When I was young and went to schoole, divers of my schoole fellowes and likewise myself did eate our fils of the berries of this tree . . . without any hurt at all, and not one time, but many times.[31]

> Some kids ate the red flesh of the yew berries despite the actual seeds being noxious.[32]

> [I am 88] as far as plant names are concerned the one that stands out is snot-gobbles referring to the berries of the yew tree. Not very elegant but truly descriptive of the berries I well remember tasting as a child.[33]

> [Wye, Kent, 1940s] We knew that yew seeds were poisonous, but we would eat, for its sweetness, the sticky red covering, which was known as red snot.[34]

In recent decades taxol, a chemical which occurs in yew has been found to be effective against some forms of CANCER.

> The autumnal pruning of yew trees in the grounds of a Hampshire church could aid cancer victims.
>
> Strange as it may sound, the clippings taken from 102 trees at St Mary's Church in the village of Hook with Warsash, will be the basis for a new anti-cancer drug.[35]

A popular photograph in August newspapers shows the 150-yard-long, 40-foot tall, yew hedge on the Bathurst Estate in Gloucestershire, having its annual trim: 'the cuttings are sold to drug companies, which extract from them a key ingredient of docetaxel, a chemotherapy drug used for breast, ovarian and lung cancer'.[36]

Yorkshire Day (1 August) – white ROSE associated with.

YORKSHIRE FOG (*Holcus lanatus*)

Perennial grass, widespread and common throughout the British Isles.

It is unlikely that the 'folk' distinguished between Yorkshire fog and the closely related and very similar creeping soft-grass (*H. mollis*), names which were probably shared by them include:

Dart-grass in northern England
Midge-grass in Berwickshire
Pig-wick in Yorkshire
Pluff-grass in Morayshire
PUSSY-CAT'S TAILS in Sussex
Whin-wrack in Berwickshire; 'because it is found to occupy places where whins [GORSE] have been removed'
White-topped grass in Cumberland.

Like FALSE OAT-GRASS, Yorkshire fog was used in the children's game 'Cock or hen'.

I'm originally from the Midlands (Wolverhampton). We used to play 'Tree or Bush' ('Cock or Hen') with Yorkshire fog. Basically you'd pull off a grass flowerhead, grasp it at the base, say to someone 'Tree or bush' ('Cock or hen'). Then after they had guessed you would quickly slide your fingers upwards forcing all the flowers to the top. The resulting shape would resemble either a tree or a bush.[1]

YUCCA (*Yucca gigantea*)

'Tree' native to Central America, small plants frequently cultivated for their foliage as pot plants in homes, and hotel and office reception areas. In the 1980s yucca featured in a 'dreadful contamination' urban legend.

This was told to us in good faith by one of our employees. A friend of a cousin had bought a yucca from a well-known retailer. The plant, despite care, died.

She returned it in exchange for tokens. The plant was analysed and a dead male tarantula found in the pot.

Two experts arrived at the house stating that where there was a dead male, there will be a female with offspring.

The search duly revealed the female and eight babies inside the duvet. All the bed linen and also the bed were replaced free of charge, but the retailer insisted on a secrecy agreement, agreeing to no disclosure.[1]

Notes and Bibliography

Notes

N & Q refers to the periodical *Notes* and *Queries* (1849–)
Pers. obs. (personal observation)

The Collection of Plant-lore in Britain and Ireland

1. Now in the library of the Natural History Museum, London.
2. *N & Q*, 4 ser. 6: 230, 1870.
3. See Vickery, 1978.
4. See Ayres, 2015.
5. The Irish Folklore Commissions archives can be found in the National Folklore Collection at University College Dublin; the Welsh Folk Museum has now become the St Fagans National Museum of History; the School of Scottish Studies is housed in the University of Edinburgh's Department of Celtic and Scottish Studies, and the Ulster Folk Museum has become the Ulster Folk and Transport Museum.
6. See Widdowson, 2010: 129, and 2016: 259.
7. Harvard University's Richard Evans Schultes (1915–2001) who was generally considered to have been the twentieth century's greatest ethnobotanist, declared, 'The British Isles have no ethnobotany.'
8. Vickery, 1985.
9. Since the cessation of the newsletter an attempt has been made to collect information via the website www.plant-lore.com
10. The Flora Britannica project stimulated a number of mostly short-lived, but useful local initiatives: *Plants, People, Places*, produced by the Natural History Centre, Liverpool Museum (10 issues, 1993–8); *Flora Facts and Fables*, produced by Grace Corne of Sisland, Norfolk (47 issues, 1994–2006); *Plant Matters* - A Flora Britannica project for North Wales, produced by Ann Macfarlane of Shotton, Clwyd (3 issues, 1996–7, curtailed by the compiler's ill health and subsequent death), and *Flora Sheffielder*, of which it seems only one issue was produced by Ian Rotherham and Janet Alton in 2002.
11. Elizabeth Howard, Royal Botanic Gardens, Kew, October 2017.

THE FLORA

ADDER
1. Friend, 1884: 476.
2. Friend, 1882: 3.
3. Cinderford, Gloucestershire, November 1993.
4. Grigson, 1987: 209.
5. Britten & Holland, 1886: 5.
6. Macmillan, 1922: 8.

ADDER'S TONGUE
1. Britten, 1881: 182.
2. Vesey-FitzGerald, 1944: 22.

AFRICAN MARIGOLD and FRENCH MARIGOLD
1. Campbell-Culver, 2001: 110.
2. Didcot, Oxfordshire, February 1991.
3. Barnstaple, Devon, July 1992.
4. New Longton, Lancashire, June 1993.
5. Pers. obs., September 2008.

AGRIMONY
1. Macmillan, 1922: 269.
2. Charlbury, Oxfordshire, January 1991.
3. Parsons MSS, 1952.
4. Archer, 1990: xiii.

5. Cinderford, Gloucestershire, November 1991.

ALDER

1. Porteous, 1973: 3.
2. Britten & Holland, 1886: 19.
3. Gomme, 1884: 134.
4. Thorncombe, Dorset, April 1990.
5. Cefn Coed, Mid Glamorgan, April 2011.
6. Gullane, East Lothian, February 1997.
7. Evelyn, 1664: 38; it is said that 'much of Venice is built on alder piles' (see, for example, http://www.woodlandtrust.org.uk/learn/british-trees/native-trees/alder/, accessed December 2014).
8. *Phytologist*, ser. 2, 2: 143, 1857.
9. Thorncombe, Dorset, January 2004.

ALDER BUCKTHORN

1. Vesey-FitzGerald, 1944: 23.
2. Bromfield, 1856: 23.

ALEXANDERS

1. Letter from Etchingham, East Sussex, *The Times*, 7 May 1988.
2. Plymouth, Devon, January 1993.
3. Whitstable, Kent, January 2012.

ALEXANDRIAN LAUREL

1. Pers. obs., Polish Catholic Church of Christ the King, Balham, London, 20 March 2016, and the Russian Orthodox Cathedral of the Dormition of the Mother of God, Kensington, London, 20 April 2016.

ALPINE MEADOW-RUE

1. Lerwick, Shetland, March 1994.

AMPHIBIOUS BISTORT

1. Scott & Palmer, 1987: 113.

ANGELICA

1. Barnstaple, Devon, March 1991.
2. Barnstaple, Devon, September 1992.
3. Lerwick, Shetland, March 1994.

ANNUAL KNAWEL

1. Maloney, 1919: 38.

ANTIRRHINUM

1. Ö Danachair, 1970: 25.

APPLE

1. Clifford & King, 2006: 14.
2. *N & Q*, 4 ser. 10: 408, 1872.
3. Chope, 1929: 125.
4. Reading, Berkshire, February 1987.
5. Fleet, Hampshire, March 1993.
6. *N & Q*, 1 ser. 8: 512, 1853.
7. Carre, 1975: 12.
8. Johns, 1847: 303.
9. Hutton, 1996: 46.
10. Simpson, 1973: 102.
11. *West Sussex Gazette*, 26 December 1966; cited in Simpson, *op. cit.*: 103.
12. Tilehurst, Berkshire, February 1987.
13. Hutton, 1996: 48.
14. Patten, 1974: 7.
15. *Western Gazette*, 25 January 1974.
16. Sheffield, April 1993.
17. Pers. obs.; this event was not repeated in 2018.
18. Brand, 1853: 346.
19. Aubrey, 1881: 96.
20. Leather, 1912: 104.
21. *N & Q*, 3 ser. 8: 146, 1865.
22. *Ibid.*, 8 ser. 10: 112, 1896.
23. *Ibid.*, 11 ser. 10: 152, 1914.
24. *Ibid.*, 11 ser. 10: 152, 1914.
25. Hornchurch, Essex, August 1992.
26. *N & Q*, 2 ser. 1: 386, 1856.
27. Peter, 1915: 132.
28. Clifford & King, 2006: 13.
29. Pers. obs.
30. IFCSS MSS 350: 397, Co. Cork.
31. Acomb, North Yorkshire, August 1989.
32. Hunt, 1881: 388.
33. Williams, 1987: 98.
34. Opie, 1959: 273–4.
35. Deane & Shaw, 1975: 53.
36. *N & Q*, 4 ser. 6: 340, 1870.
37. Udal, 1922: 251.
38. Stevens Cox, 1971: 9.
39. Kensington, London, November 1991.
40. Gutch, 1912: 40.
41. Briggs, 1976: 262.
42. Tongue, 1970: 155.
43. Smith, 1996.
44. Briggs & Tongue, 1965: 44.
45. Briggs, 1976: 9.
46. Brown, 2016: 7
47. Simpson, 1982: 5.

48. Thorncombe, Dorset, autumn 1974.
49. Letter from Streatham, London, *Sunday Times*, 21 December 1958.
50. Lisburn, Co. Antrim, March 1986.
51. Histon, Cambridgeshire, January 1989.
52. Royal Botanic Gardens, Kew, Surrey, September 2014.
53. Letter from Five Ashes, Sussex, *The Times*, 1 March 1929.
54. Letter from Middle Winnersh, Berkshire, *TV Times*, 23 December 1989.

ARUM LILY

1. Plymouth, Devon, June 1983.
2. Lewisham, London, April 1986.
3. Truro, Cornwall, October 1996.
4. Sadborow, Dorset, December 1982.
5. Loftus, 1994: 86.
6. *Ibid.*: 92.
7. Personal obs., Dublin and Belfast, 13–16 April 2017.

ASH

1. Jackson, 1873: 14.
2. Forby, 1830: 406.
3. Larne, Co. Antrim, October 1993.
4. Buczacki, 2002: 196.
5. Opie & Tatem, 1989: 7.
6. *Ibid.*
7. Udal, 1922: 254.
8. Thorncombe, Dorset, June 1976.
9. Opie & Tatem, 1989: 7.
10. Hart, 1898: 373.
11. Britten & Holland, 1886: 170.
12. Poole, 1877: 6.
13. Hutton, 1996: 40.
14. Legg, 1986: 54.
15. Willey, 1983: 40.
16. Hutton, 1996: 41.
17. Pers. obs., 6 January 2009.
18. www.thorncombe-village-trust.co.uk/page24.ntml, accessed December 2017.
19. Opie, 1959: 240.
20. St Ervan, Cornwall, February 1992.
21. Natural History Museum, London, March 2004.
22. Simpson & Roud, 2000: 12.
23. Email, June 2013.
24. Evelyn, 1706, 1: 92.
25. Latham, 1878: 40.
26. Opie & Tatem, 1989: 7.
27. Richards, 1979: 13.
28. Hole, 1937: 12.
29. Whitlock, 1976: 167.
30. Plot, 1686: 222.
31. White, 1822, 1: 344.
32. Ffennell, 1898: 333.
33. *op. cit.*: 334.
34. Kew, Surrey, February 1994.
35. Daingean, Co. Offaly, January 1985.
36. Larne, Co. Antrim, October 1993.
37. Dorchester, Dorset, February 1992.
38. Radford, 1998: 5.
39. IFCSS MSS 750: 242, Co. Longford.
40. Ballymote, Co. Sligo, May 1994.
41. Robertsbridge, East Sussex, June 2003.
42. Alexander, 2010: 4.
43. Boulger, 1906: 71.
44. Pers. obs., Grasmere, Cumbria, May 2015; Upper Town, Tenbury Wells, Worcestershire, November 2015.
45. Yoxall, Staffordshire, January 2016.

ASPEN

1. Lightfoot, 1777: 617.
2. Carmichael, 1928: 104.

ASTROLOGICAL BOTANY

1. Addison, 1985: 194.
2. Jones, A.E., 1980: 59.
3. Thomas, 1971, especially chapters 10–12.

AUBERGINE

1. *Independent*, 28 March 1990.
2. *Guardian*, 28 March 1990.
3. *Fortean Times*, 55: 5,1990.
4. *Independent*, 16 September 1999.

AUTUMN GENTIAN

1. Tait, 1947: 81.

BALM

1. Berkhamsted, Hertfordshire, August 2004.
2. Oker, Derbyshire, April 1997.
3. Marquand, 1906: 45.
4. Cinderford, Gloucestershire, November 1993.

BANANA

1. Opie, 1959: 336.

2. Barnes, London, October 1979.
3. Great Yarmouth, Norfolk, April 2003.
4. Deptford, London, May 2002.
5. East Ham, London, April 2004.
6. Streatham, London, September 2006.
7. Letter from Newtownabbey, Co. Antrim, *Arthritis Today*, no. 131, Winter 2006.
8. Letter from Cambridge, *ibid.*
9. Letter from Devon, in *Sunday Mirror*, 27 December 1998, where it got the response: 'bananas are rich in potassium which is essential to the movement of nerve and muscle tissue, so they may well have helped your cramps.'

BARBERRY

1. Davey, 1909: 17.
2. Clonlara, Co. Clare, June 2002.

BARLEY

1. Salisbury, Wiltshire, February 1989.
2. Letter from Wickhambrook, Suffolk, *Farmers Weekly*, 10 January 1964.
3. Vaughan Williams & Lloyd, 1968: 116.
4. Lomax & Kennedy, 1971.
5. Portland, Dorset, April 1991.
6. Childwall, Liverpool, April 2013.

BARREN GROUND

1. *Gentleman's Magazine*, 74: 1194, 1804.
2. Westwood, 1985: 127.
3. Bergamar, n.d.: 7.
4. Hythe, Kent, June 1973 and November 1987.
5. Armagh, September 1985.

BATH ASPARAGUS

1. Bath, Somerset, May 2016.

BATTLE OF FLOWERS

1. Lake, n.d.

BAY

1. Boase, 1976: 118.
2. Oxford, January 1991.
3. Oxshott, Surrey, February 1998.
4. Chope, 1938: 359.
5. Roggenburg, Switzerland, February 1995.

BEECH

1. Palmer, 1973: 79.
2. Macmillan, 1922: 24
3. Britten & Holland MSS.
4. Farnborough, Kent, January 1993.
5. Natural History Museum, London, June 2011.
6. Sidmouth, Devon, October 1991.
7. Oban, Argyllshire, October 1990.
8. St Marychurch, Devon, August 2011.
9. Barnstaple, Devon, August 1992.
10. Sisland, Norfolk, August 2007; this is the only record of this belief, and if it is true it is extremely unlikely that any of the trees survive.

BEETROOT

1. Barnstaple, Devon, May 1991.

BENT

1. Stevenage, Hertfordshire, January 1993.

BERMUDA BUTTERCUP

1. St Mary's, Isles of Scilly, September 1992.
2. Carbis Bay, Cornwall, September 1995.
3. Parslow & Bennalick, 2017: 181.

BETONY

1. Allen & Hatfield, 2004: 212.
2. Thomson, 1925: 160.

BILBERRY

1. Court, 1967: 42.
2. Pers. obs., April 2007.
3. IFCSS MSS 575: 382.
4. MacNeill, 1962: 20.
5. Hillaby, 1983: 96.
6. Lightfoot, 1777: 201.
7. Bromley, Kent, April 1991.
8. London, SE1, February 2016.

BINDWEED

1. Hammer, 2008: 80.
2. Hampstead, London, September 1987.
3. St Mary's, Isles of Scilly, November 1992.
4. Whitstable, Kent, January 2012.
5. Email, July 2016.
6. Solihull, West Midlands, April 1991.

BIRCH

1. Wright, 1936: 157.
2. Burne, 1883: 350.
3. Quoted in Pegg, 1981: 34.

4. Frome, Somerset, March 1994.
5. Pers. obs., 25 May 2015; birch is also used to decorate Lutheran churches in Berlin at Pentecost (pers. obs., June 2017).
6. *Phytologist* 1: 780, 1843.
7. Grosmont, Gwent, November 1994.
8. Milner, 1992: 20.
9. Milliken & Bridgewater, 2004: 57.

BIRD-CHERRY
1. Gregor, 1889: 41.
2. London, April 2005.
3. Milliken & Bridgewater, 2004: 154.

BIRD'S-FOOT TREFOIL
1. Allen & Hatfield, 2004: 160, 353.

BISTORT
1. Email, July 2016.
2. Girton, Cambridge, September 1985.
3. Clappersgate, Cumbria, October 1985.
4. Halifax, West Yorkshire, September 1996.
5. 'Dock Pudding', information sheet obtained from Hebden Bridge Tourist Information Centre, West Yorkshire, April 2011.
6. Smith, 1989: 11.
7. Short, 1983: 124.

BITING STONECROP
1. Trevelyan, 1909: 95.
2. Ó Danachair, 1970: 25.
3. Friend, 1882: 18.

BITTERSWEET
1. Quoted in Halliday, 1997: 349.
2. Taylor MSS.
3. Bloom, 1930: 25.
4. Hart, 1898: 206.

BITTER VETCH
1. McNeill, 1910: 114.
2. Macmillan, 1922: 203.
3. Britten & Holland, 1886: 310.
4. *Ibid.*: 208.

BLACK BRYONY
1. Bromfield, 1856: 507.

BLACK CURRANT
1. Calpe, Alicante, Spain, December 1991.
2. Lichfield Wildlife Group, Staffordshire, September 2014.
3. Grosmont, Gwent, November 1994.

BLACK MEDICK
1. Ballycastle, Co. Antrim, January 1991.

BLACK POPLAR
1. Milner, 1992: 22.
2. Sutton, Surrey, August 1993.
3. Shuel, 1985: 39.
4. Sykes, 1977: 71.
5. Box, 2003.
6. *Ibid.*: 21.
7. Shuel, 1985: 40.
8. Milner MSS, 1991–2.
9. Box, 2003: 15.
10. *Ibid.*: 22.

BLACK SPLEENWORT
1. Marquand, 1906: 38.

BLACKTHORN
1. Wolsingham, Co. Durham, May 1982.
2. Newcastle-under-Lyme, Staffordshire, March 1983.
3. Angarrack, Cornwall, February 1989.
4. Hemel Hempstead, Hertfordshire, August 2004.
5. Bromfield, 1856: 141.
6. St Osyth, Essex, February 1989.
7. Letchworth, Hertfordshire, May 2001.
8. Johnston, 1853: 57.
9. Phelps, 1977: 175.
10. Boys, 1792: 403.
11. Charles Wanostrocht, Honorary Curator and Archivist, Sandwich Guildhall Museum, March 1994.
12. Gilmore & Oalcz, 1993: 10.
13. Pers. obs., Dublin and Belfast, May 1994; it appears (Dublin, November 2014–April 2017) that these are no longer being produced.
14. Lenamore, Co. Longford, April 1991.
15. Plymouth, Devon, May 1986.
16. Callington, Cornwall, October 1996.
17. Aldbury, Hertfordshire, February 1998.
18. St Marychurch, Devon, August 2011.
19. IFC MSS, 462: 305, 1937–8.
20. Freethy, 1985: 5.

BLADDER CAMPION

1. Pratt, 1857, 2: 157; bladder campion is cultivated for culinary use in Cyprus [Paltres, 1994] and in Italy shoots from wild plants are collected and said to be in great demand by gourmets [Laghetti & Perrino, 1994].

BLUEBELL

1. Chope, 1933: 122.
2. Stetchworth, Cambridgeshire, December 1991.
3. Ballymote, Co. Sligo, May 1994.
4. Aldershot, Hampshire, April 1994.
5. Freethy, 1985: 88.

BOG ASPHODEL

1. Hart, 1898: 371.
2. IFCSS MSS 589: 62, Co. Clare.
3. Worthing, West Sussex, September 1982.
4. Thurlby, Lincolnshire, November 1996.
5. Stabursvik, 1959.
6. Malone *et al.*, 1992: 100.

BOGBEAN

1. Atkinson, 2003: 243.
2. Beckwith, 1978: 211.
3. Burravoe, Shetland, March 1994.
4. IFCSS MSS 313: 213.
5. IFCSS MSS 1121: 354.
6. Omagh, Co. Tyrone, March 1986.
7. St Fagans, South Glamorgan, February 1991.
8. Lerwick, Shetland, March 1994.
9. Orpington, Kent, January 2013.
10. Lindsay, 1856: 372.

BOG MYRTLE

1. Reynolds, 1993: 22.
2. Grigson, 1987: 243.
3. Evans, 1800: 149.
4. *N & Q*, 3 ser. 4: 311, 1863.
5. Lightfoot, 1777: 614.
6. SSS MSS SA 1969/28/A12.
7. Hart, 1898: 376.
8. Dixon, 1890: 111.
9. Edlin, 1949: 115. In North America two related species, *Myrica californica*, bayberry, and *M. cerifera*, known as candle-berry, tallow-shrub or wax-myrtle, are used for making candles which are sold at heritage sites and similar places.
10. Milliken & Bridgewater, 2004: 105.
11. Hart, 1898: 384.
12. McNeill, 1910: 167.
13. SSS MSS SA1970/164/A2.

BOX

1. Britten & Holland, 1886: 78.
2. Macmillan, 1922.
3. Paston, Cambridgeshire, November 1993.
4. Payhembury, Devon, December 2003.
5. *Daily Telegraph*, 1 December 1868; quoted in *N & Q*, 4 ser. 6: 496, 1870.
6. Pers. obs., 8 April 1979.
7. Shirley, Surrey, October 1996.
8. Smith, 1994: 242.
9. Tywyn, Gwynedd, May 2013.
10. Pers. obs., 27 March 1983. In more recent years Alexandrian laurel seems to have replaced Palm Sunday box.
11. Grieve, 1931: 121.
12. Pimperne, Dorset, January 1992.

BRACKEN

1. Smith, 2005: 114.
2. Latham, 1878: 31.
3. Smith. 2005b: 114.
4. *Sunday Express*, 17 June 1979.
5. Mac Coitir, 2006: 167.
6. Smith, 2005: 114.
7. Gibson, 1853: 152.
8. Paddington, London, May 1989.
9. Bath, Somerset, January 1991.
10. Elworthy, 1888: 529.
11. Britten & Holland, 1886: 180.
12. Ewen & Prime, 1975: 64.
13. Brand, 1813, 1: 251.
14. Rymer, 1976: 172.
15. Vesey-FitzGerald, 1944: 23.
16. Thompson, 1925: 161.
17. Spray, 2004: 7.
18. Rymer, 1976.
19. Lightfoot, 1777: 658.
20. IFCSS MSS 200: 300.
21. Lightfoot, 1777: 659.
22. Sykes, 1987: 242.
23. *Phytologist* 1: 263, 1842.
24. Thorncombe, Dorset, January 2004.

25. Balham, London, September 2005.
26. IFCSS MSS 593: 115.
27. Hillaby, 1983: 96.
28. Charmouth, Dorset, January 1994; but, according to Keller & Prance, 2015: 8, in Missiones, Argentina, 'farmers identify compacted and degraded soils by means of observations on the presence of *Pteridium aquilinum*'.

BRAMBLE
1. Leather, 1912: 21.
2. Kavanagh, 1975: 64.
3. Cranbrook, Kent, July 1983.
4. Barnes Common, London, July 2005.
5. Rowling, 1976: 101.
6. Harrow-on-the-Hill, Middlesex, October 2004.
7. Stoke Bishop, Somerset, December 1982.
8. Norton Fitzwarren, Somerset, July 1983.
9. Bath, Somerset, January 1988.
10. St Marychurch, Devon, August 2011.
11. East Tuddenham, Norfolk, October 1984.
12. Sampford Brett, Somerset, October 1993.
13. IFCSS MSS 800: 113.
14. Lenamore, County Longford, April 1991.
15. Ashreigney, Devon, July 1983.
16. Whalton, Northumberland, October 1984.
17. Sunderland, Tyne and Wear, August 2001.
18. Email, June 2015.
19. Britten & Holland, 1886: 46.
20. Brown, 1953: 217.
21. Patten, 1974: 15.
22. Oxford, August 1987.
23. Eastwood, n.d.: 80.
24. Whalton, Northumberland, October 1984.
25. *London Evening Standard*, 8 October 1957.
26. Opie & Tatem, 1989: 37.
27. Trevelyan, 1909: 320.
28. Leather, 1912: 82.
29. Simpson, 1976: 108.
30. Raven, 1978: 51.
31. K. Palmer, 1976: 114.
32. Udal, 1922: 255.
33. Deane & Shaw, 1975: 135.
34. IFC MSS 782: 257, 1941.
35. IFCSS MSS 450: 163.
36. Brompton Cemetery, London, June 2010.
37. Orpington, Kent, October 2007.
38. Northamptonshire, June 2008.
39. Tamworth, Staffordshire, June 2012.
40. Fleet, Hampshire, March 1993.
41. Tooting, London, June 2002.

BRASSICACEAE
1. Oxshott, Surrey, January 1997.
2. Minehead, Somerset, November 1993.

BROAD BEAN
1. SLF MSS, West Butterwick, Humberside, September 1970.
2. Stewart,1987: 98.
3. Barrett, 1967: 97.
4. West Stow, Suffolk, January 1991.
5. Sandiway, Cheshire, October 2004.
6. *N & Q*, 4 ser. 1: 361, 1868.
7. Wiltshire, 1975: 113.
8. Wicken, Cambridgeshire, March 1993.
9. Ryde, Isle of Wight, November 1988.
10. Much Hadham, Hertfordshire, 2002.
11. Taylor MSS, Lincolnshire.
12. Taylor MSS, Sproughton, Suffolk.
13. Bramley, Surrey, August 2002.
14. Thorncombe, Dorset, April 1982.

BROOM
1. Latham, 1878: 52.
2. Edinburgh, March 1984.
3. *Folk-lore*, 36: 257, 1925.
4. Steele MSS, 1978: 38.

BUCK'S-HORN PLANTAIN
1. Newton MSS.
2. Trevelyan, 1909: 313.

BULRUSH
1. Five Ashes, East Sussex, April 1983.
2. Borrowash, Derbyshire, January 2007.
3. Weymouth, Dorset, December 2008.
4. Bristol, November 2010.

BURDOCK
1. Davey, 1909: 261.

2. Yafforth, North Yorkshire, January 1990.
3. Hole, 1976: 39.
4. Darwin, 1996: 75.
5. Anon., 1967: 92.
6. Hole, 1976: 40.
7. Vesey-FitzGerald, 1944: 23.
8. Union Mills, Isle of Man, February 1995.

BURNET ROSE
1. Trevelyan, 1909: 99.

BUTCHER'S BROOM
1. Bromfield, 1856: 509.
2. Belfast, February 1991.
3. *The Royal Borough of Kensington and Chelsea, Local Biodiversity Plan 2007 to 2011. Draft for consultation*: 51.

BUTTERBUR
1. St Ervan, Cornwall, February 1992.
2. Blackburn, Lancashire, April 1994.
3. Cotherstone, Co. Durham, April 1994.

BUTTERCUP
1. *N & Q*, 5 ser. 5: 364, 1876.
2. Minchinhampton, Gloucestershire, January 1991.
3. Langtoft, Humberside, July 1985.
4. Pinner, Middlesex, May 2001; pastime also recorded from New Zealand [University of Otago, October 2013] and, no doubt, known elsewhere.
5. Orpington, Kent, February 2007.
6. Hewins, 1985: 69.
7. IFCSS MSS 375: 93, Co. Cork.
8. Streatham, London, February 1992.
9. Davey, 1909: 10.
10. Cong, Co. Mayo, January 1992.
11. St Ervan, Cornwall, February 1992.
12. Glynn, Co. Antrim, February 1992.
13. Aylesbury, Buckinghamshire, November 1996.
14. Howkins, 2006: 42.
15. Alton, Hampshire, June 1993.
16. Rushall, Herefordshire, November 1999.

BUTTERWORT
1. McNeill, 1910: 105.
2. Salisbury, Wiltshire, November 1985.
3. Grigson, 1987: 312.
4. Britten & Holland, 1886: 163.

CABBAGE
1. IFCSS MSS 812: 155, Co. Offaly.
2. Pollock, 1960: 62. It appears that similar practices were known in the United States, where a postcard on sale in *c.* 1910 depicts four uprooted cabbages with the words 'O, is my true love tall and grand? O, is my sweetheart bonny?'
3. Lerwick, Shetland, March 1994.
4. Barnstaple, Devon, May 1991.
5. Horsted Keynes, West Sussex, February 1991.
6. Kavanagh, 1975: 21.
7. London, SE1, October 2010.
8. IFCSS MSS 880: 331, Co. Wexford.
9. UCL EFS MSS M3, Hauxton, Cambridge, October 1963.
10. *Sunday Telegraph*, 31 August 2003.
11. Logan, 1965: 52.
12. Lambeth Horticultural Society, London, November 2015.
13. Barnstaple, Devon, May 1991.
14. Bognor Regis, West Sussex, March 1997.
15. Whitstable, Kent, March 2011.
16. IFCSS 500: 90, Co. Limerick.
17. Pratt, 1857: 21.
18. Sisland, Norfolk, April 2001.
19. Parker & Stevens Cox, 1974.
20. Anon., 1985.

CACTI
1. Porter, 1969: 135.
2. Deane & Shaw, 1975: 135.
3. Golders Green, London, May 1982.
4. SLF MSS, Sheffield, January 1970.

CALVARY CLOVER
1. Chope, 1931: 124.
2. *N & Q*, 8 ser. 12: 26, 1897.

CAMPION
1. Felmersham, Bedfordshire, March 1993.

CAPER SPURGE
1. Le Sueur, 1984: 91.
2. Little Paxton, Cambridgeshire, February 1998.
3. Itchen, Hampshire, June 1993.
4. Brixton, London, June 2004.

5. *The Garden*, May 2012: 23.
6. Royal Botanic Gardens, Kew, September 2014.
7. Somerton, Somerset, May 2015.
8. *Garden Answers*, June 1999; presumably 'courtesans', rather than 'courtiers', was intended.
9. Minehead, Somerset, October 1993.

CARNATION

1. Campbell-Culver, 2001: 13.
2. Davies, 2000: 96.
3. Christ Church, Oxford, April 2005.
4. blog.sarahlaurance.com/2008/06.oxford-rituals.htm; accessed September 2016.
5. www.thestudentroom.co.uk/showthread.php?t=400903; accessed September 2016.
6. McKenna, 2004: 226.
7. Swiss Cottage, London, February 2007.

CARROT

1. Didcot, Oxfordshire, February 1991.
2. Muchelney, Somerset, January 2007.
3. Lichfield, Staffordshire, September 2014.
4. Horley, Surrey, January 1999.
5. Larne, Co. Antrim, February 1992.

CEDAR

1. Evershed, 1877: 40.
2. Wilks, 1972: 133.

CELANDINE

1. Marquand, 1906: 45.
2. McNeill, 1910: 96.
3. Vesey-FitzGerald, 1944: 27.
4. Wheeler, 2006: 22.
5. Driffield, Humberside, July 1985.

CELERY

1. Sung by a group of drunken men on a train between Basingstoke and Woking, 3 March 2007.
2. IFCSS MSS 919: 10, Co. Wicklow.
3. SLF MSS, West Butterwick, Humberside, March 1970.
4. Taylor MSS, Norwich, Norfolk.
5. Lafont, 1984: 83.

CELERY-LEAVED BUTTERCUP

1. Lightfoot, 1777: 291.

CENTAURY

1. Britten & Holland, 1886: 202.
2. Charlbury, Oxfordshire, February 1991.
3. St Mary's, Isles of Scilly, September 1992.
4 Vesey-Fitzgerald, 1944: 23.

CHAMOMILE

1. IFCSS MSS 1128: 26, Co. Cork.
2. Eastbourne, East Sussex, February 2007.
3. Quinton, West Midlands, April 1993.

CHARLOCK

1. Burdy, 1792: 122.
2. Tighe, 1802: 483.
3. IFCSS MSS 1112: 19, Co. Donegal.
4. Lucas, 1960b: 31.
5. Drury, 1984: 49.

CHERRY

1. Keyte & Parrott, 1992: 440.
2. *The Times*, 21 May 2004.
3. Margaret Gale Collection.
4. Pers. obs., Streatham, London, 5 August 1984.
5. Pers. obs., 2008 and 2015.
6. Letter from London, W1, *Country Life*, 30 December 1971.
7. Letter from Torquay, Devon, *The Times*, 18 September 2014.

CHESTNUT

1. Weobley, Herefordshire, August 1998.
2. Vesey-FitzGerald, 1944: 23.
3. Melksham, Wiltshire, April 1990.

CHICKWEED

1. Lerwick, Shetland, March 1994.
2. Johnston, 1853: 43.
3. Spence, 1914: 100.
4. IFCSS MSS 790: 161, Co. Dublin.
5. Boat-of-Garten, Inverness-shire, November 1991.
6. Storrington, West Sussex, October 2015.
7. Mayhew, 1851: 153.
8. Wandsworth Common, London, March 1008.
9. Childwall, Liverpool, April 2013.
10. For example, Mabey, 1972; Phillips, 1983.

11. Andover, Hampshire, December 2013.

CHICORY
1. Vesey-FitzGerald, 1944: 23.

CHILEAN MYRTLE
1. St Martin, Guernsey, April 2002.

CHIVES
1. Uphill, Avon, January 1993.

CHRISTMAS GREENERY
1. Hole, 1976: 50.
2. Cambridge, November 1985.
3. Potten End, Hertfordshire, August 2004.
4. Hereford, April 2012.
5. IFCSS MSS 775: 122, Co. Kildare.
6. McKelvie MSS, 1963: 176.

CHRISTMAS TREE
1. Miles, 1912: 265.
2. *Ibid.*: 265.
3. Hutton, 1996: 114.
4. Combermere & Knollys, 1866: 419.
5. Hutton, 1996: 114.
6. Miles, 1912: 264.
7. Hole, 1976: 54.
8. *The Times*, 23 October 2010
9. Pers. obs., December 2008.
10. *Chard & Ilminster News*, 10 December 2014.
11. Pers. obs., December 2016.
12. Leaflet produced by the Hospice, December 2016; the collected trees are converted into chippings which are used in the Hospice's grounds [Nick Harris, Facilities Support Co-ordinator, November 2017].
13. *Sussex Express*, 30 December 2016.

CHRYSANTHEMUM
1. Davies, 2006: 96.
2. Pers. obs., East London Cemetery and Crematorium, January 2007.
3. Streatham, London, April 1983.
4. Mettlach, Germany, August 2004.
5. Pers. obs., 4 June 2015 and 25 May 2016.

CINERARIA
1. Lerwick, Shetland, March 1994.

CLAN BADGES
1. Phillips, 1825: 13.
2. Biggar, 2000.

CLOUDBERRY
1. MacGregor, 1935: 16.
2. Gardiner, 1843: 470.
3. Penrith, Cumbria, May 2016.

CLOVER
1. *Gospelles of Dystaues*, part 2, xv, 1507, cited in Opie & Tatem, 1989: 88.
2. Melton, 1620: 46.
3. Trevelyan, 1909: 95.
4. Edinburgh, March 1984.
5. Maynooth, Co. Kildare, February 1991.
6. Parsons MSS, 1952.
7. Letter from Woore, Cheshire, *Farmers Weekly*, 25 June 1976.
8. Stirlingshire, April 2005.
9. Hardy, 1895: 142.
10. Hunt, 1881: 107.
11. *Ibid.*: 109.
12. O'Sullivan, 1966: 225; tale recorded Dunquin, Co. Kerry, 1936.
13. MacNicholas *et al.*, 1990: 84.
14. McNicholas, 1992: 210.
15. O'Sullivan, 1966: 280.
16. Clark, 1882: 83.
17. SSS MSS SA1976.196A7.
18. Ballaghadereen, Co. Roscommon, October 1984.
19. *Observer Magazine*, 4 August 1974.
20. Wenis, 1990: 24.
21. Marston Montgomery, Derbyshire, March 1983.
22. Streatham, London, May 1991.
23. Pers. obs., promoting the National Lottery, Balham, London, June 1995; Berlin Lotto, Berlin, Germany, May 2015, and the Spanish National Lottery, Segovia, March 2016 and Madrid, June 2018.
24. *N & Q*, 1 ser. 10: 321, 1854.

COCKSPUR THORN
1. Edmonton, London, May 2016.

COCONUT
1. Quoted in Beavis, 2013: 25.
2. Harrow, Middlesex, May 2015.
3. Streatham, London, December 1989.

4. Pers. obs.
5. Gupta, 1971: 35.
6. *It's My Story*, BBC Radio 4, 4 December 2006.

COLTSFOOT
1. Pratt, 1857, 2: 59.
2. Langtoft, Humberside, July 1985.
3. Portland, Dorset, March 1991.
4. Nottingham, May 2014.
5. Mcneill, 1910: 137.
6. Saddleworth, Greater Manchester, 1998.

COLUMBINE
1. IFCSS 1075: 139, Co. Donegal.

COMFREY
1. Tooting, London, August 1987.
2. Holbeach, Lincolnshire, January 2004.
3. Barton-under-Needwood, Staffordshire, May 2002.
4. Girton, Cambridge, June 1986.
5. Omagh, Co. Tyrone, October 1986.
6. Windermere, Cumbria, November 1988.
7. Oban, Argyll, October 1990.
8. Portesham, Dorset, January 2007.
9. Wandsworth U3A, London, June 2014.

CORIANDER
1. Leamington Spa, Warwickshire, January 1993.

CORK OAK
1. Felmersham, Bedfordshire, April 1993.
2. Tooting, London, August 2015.

CORN CLEAVERS
1. Dony, 1953: 359.

CORN COCKLE
1. Fosbroke, 1821: 73.

CORNFLOWER
1. Arthur R. Chandler, Honorary Archivist, Alleyn's School, January 1994.
2. Letter from Llandrindod Wells, Powys, *This England*, Spring 1988.
3. Harrow School, January 1994.
4. Campbell-Preston, 2006: 52; this tradition appears to have died out. On 24 June 2017, when a one-day match between Eton and Harrow was being played at Lord's, no one going into the ground was wearing flowers of any kind [pers. obs.].
5. *The Trusty Servant*, 114: 34, November 2012.
6. Letter from Kilmersdon, Somerset, *This England*, Winter 1988; there appears to be a discrepancy here as the Thames Embankment was not constructed until the reign of King George III's grand-daughter, Victoria.
7. Letter from Buxton, Derbyshire, *This England*, Spring 1988.

CORN MARIGOLD
1. McNeill, 1910: 136.

CORN SPURREY
1. Lerwick, Shetland, March 1994.
2. Addlestone, Surrey, September 1997.
3. Grigson, 1987: 92.

CORPUS CHRISTI
1. Revd Anthony Whale, Cathedral of Our Lady and St Philip Howard, Arundel, West Sussex, March 1990.
2. Anon., n.d. (1): 23.
3. Pers. obs., 1980–2017, and Mrs Stella Smart, chief 'flower lady', March 1978 and August 1980. In recent years gerberas (*Gerbera* cv.) have also been used to provide points of vivid colour.
4. Revd Jeremy Fairhead, All Saints, September 1987; pers. obs., 28 May 2016.
5. Fr John Gilling, St Mary's, October 1987; pers. obs., 18 June 2017, when the twigs included ash, a variegated form of kohuhu (*Pittosporum tenuifolium),* lime, a yellow form of Mexican orange (*Choisya* ternata) and oak; and conversation with a member of St Mary's congregation.

COTTON GRASS
1. Lerwick, Shetland, March 1994.
2. Aultvaich, Inverness-shire, April 2012.
3. Milliken & Bridgewater, 2004: 159.
4. *Ibid*.: 98.
5. Johnston, 1853: 204.

COTTON THISTLE

1. *Pharmaceutical Journal*, 12 June 1875: 997.
2. Gent & Wilson, 2012: 299.

COUCH

1. Mabey, 1996: 395.
2. Vesey-FitzGerald, 1944: 24.

COW PARSLEY

1. St Bride's-Super-Ely, South Glamorgan, October 1982.
2. Market Drayton, Shropshire, March 1983.
3. Camberwell, London, February 1998.
4. London, E1, December 2003.
5. Witham, Essex, May 1983.
6. Wimbledon, London, November 1983.
7. Stowmarket, Suffolk, September 1985.
8. Corbridge, Northumberland, January 1983.
9. Wicken, Cambridgeshire, March 1993.
10. Hackney, London, February 1998.
11. Edgware, Middlesex, November 1994.
12. Welling, Kent, November 1995.
13. Natural History Museum, London, June 2015.
14. Dublin, May 1993.
15. East Finchley, London, March 1998.
16. Helensburgh, Dunbartonshire, February 1991.
17. Macmillan, 1922: 200.
18. Honor Oak, London, April 2016.
19. Grigson, 1987: 209.
20. East Bergholt, Suffolk, February 1991.
21. Sunninghill, Berkshire, February 1998.
22. Hillingdon, Middlesex, March 2001.
23. Cinderford, Gloucestershire, November 1993.

COWSLIP

1. Hole, 1937: 48.
2. Leather, 1912: 63.
3. Thorncombe, Dorset, June 1976.
4. Trevelyan, 1909: 97.
5. Haberton, Devon, October 1992.
6. IFCSS MSS 825: 123, Co. Laois.
7. Wicken, Cambridgeshire, March 1993.
8. Wickham, Hampshire, July 1996.
9. South Collingham, Nottinghamshire, March 1992.
10. Vickery, 2013: 52.
11. *Plant Life*, 74: 44, 2016.
12. Jekyll, 1908: chap. 5.
13. Shipston-on-Stour, Warwickshire, September 1993.
14. Quinton, West Midlands, April 1993.
15. Portland, Dorset, April 1991.

CRAB APPLE

1. Brand, 1853.
2. Streatham, London, May 1991.
3. Dunsford, 1978: 209.
4. Shepherdswell, Kent, October 1979.
5. Barrow-in-Furness, Cumbria, August 2004.

CRACK WILLOW

1. Great Bookham, Surrey, October 1979.

CRANBERRY

1. Lamplugh, Cumbria, July 1997.
2. Britten & Holland MSS.

CREEPING CINQUEFOIL

1. Malew, Isle of Man; Manx Folklife Survey.
2. Liverpool, March 1998; creeping cinquefoil still thrives at the locality described.

CREEPING THISTLE

1. Bloom, 1930: 245.

CREEPING WILLOW

1. Hart, 1898: 370.

CRESTED DOG'S-TAIL

1. Fowler, 1909: 296.
2. Conqueror, 1970: 145; practice also known near Truro, Cornwall, 1934–8 [St Day, Cornwall, January 1994].
3. Wyse Jackson, 2014: 277.
4. O'Dowd, 2015: 469.
5. *Ibid.*: 443.

CROWBERRY

1. Hart, 1898: 377.
2. Lightfoot, 1777: 613.

CROWN IMPERIAL

1. Hole MSS.
2. Campbell-Culver, 2001: 127.
3. Donald, 1973: 31.

CUCKOO-FLOWER
1. Brize Norton, Oxfordshire, August 1992.
2. Taunton, Somerset, April 1994.
3. Bratton, Wiltshire, January 2007.
4. Hole, 1976: 131.

CUCUMBER
1. De Garis, 1975: 122.
2. Harrow-on-the-Hill, Middlesex, October 2004.
3. Corbridge, Northumberland, January 1993.

CYCLAMEN
1. K'Eogh, 1735: 115.

CYPRESS
1. Colwyn Bay, Clwyd, June 1992.
2. Pers. obs., Dublin, May 1993 and 1994.

DAFFODIL
1. Leather, 1912: 17.
2. Chope, 1932: 154.
3. Isle of Man, spring 1982; Manx Folklife Survey.
4. Britten & Holland, 1886: 542.
5. Boase, 1976: 115.
6. Tulse Hill, London, March 1979.
7. Pimperne, Dorset, January 1992.
8. fovantbadges.com/fovant-remembers-the-anzac-day-centenary/, accessed December 2018.
9. *The Times*, 12 January 1993.
10. Hole, 1950: 45.
11. *Balham and Tooting Guardian*, 5 April 1990.
12. Dublin, August 1994.
13. *Daily Cardinal*, 8 March 2000.
14. *Australian Senior*, August 2004.
15. Gullane, East Lothian, February 1997.
16. Childwall, Liverpool, April 2013.
17. St Mary's, Isles of Scilly, September 1992.
18. Hart, 1898: 372.

DAISY
1. *N & Q*, 2 ser. 3: 343, 1857.
2. Wicken, Cambridgeshire, April 1993.
3. Trevelyan, 1909: 97.
4. Llanuwchllyn, Gwynedd, April 1991.
5. Honor Oak, London, April 2016.
6. Rambouillet, France, March 1995; similar practices recorded from Germany [Royal Botanic Gardens, Kew, September 2014] and Magadan, Russia [Sutton, Surrey, September 2014].
7. Hyson Green, Nottingham, October 1985.
8. Honor Oak, London, April 2016.
9. Briggs, 1976: 87.
10. Anon [?1910].
11. Bridport, Dorset, February 1985.
12. Letter from Stratford-upon-Avon, Warwickshire, *This England*, Winter 1988.
13. Chamberlain, 1990: 164.
14. North Shields, North Tyneside, September 1996.
15. Threlkeld, 1726: 23.
16. McNeill, 1910: 134.
17. Rawlence, 1914: 84.

DAME'S VIOLET
1. Paston, Cambridgeshire, November 1993.

DAMSON
1. Childwall, Liverpool, April 2013.
2. Longford, Shropshire, April 1997.
3. Clifford & King, 2006: 127.
4. 'Lyth Valley damson blossom', postcard produced by Dane Stone Cards, purchased 2014.
5. Parsons MSS.

DANDELION
1. St Albans, Hertfordshire, November 1979.
2. Apples, Switzerland, February 1983.
3. Belfast, February 1991.
4. Leiston, Suffolk, July 2011.
5. Daingean, Co. Offaly, January 1985.
6. Merthyr Tydfil, Mid Glamorgan, October 2000.
7. Llanuwchllyn, Gwynedd, April 1991.
8. Worcester, October 1991.
9. South Kensington, London, May 1979.
10. Stevenage, Hertfordshire, January 1993.
11. Walthamstow, London, October 2015.
12. Streatham, London, October 2011.
13. IFCSS MSS 550: 274, Co. Tipperary.
14. IFCSS MSS 313: 310a, Co. Cork.

15. IFCSS MSS 200: 73, Co. Leitrim.
16. IFCSS MSS 450: 162, Co. Kerry.
17. IFCSS MSS 589: 205, Co. Clare.
18. Mac Coitir, 2006: 199.
19. IFCSS MSS 717: 27, Co. Meath.
20. IFCSS MSS notebooks 442c, Co. Kerry.
21. IFCSS MSS 50: 295, Co. Galway; use also recorded from Porthnockie, Morayshire [Edinburgh, December 1991].
22. Langtoft, Humberside, July 1985.
23. Llanbedr, Gwynedd, February 1998.
24. St Osyth, Essex, February 1989.
25. Portland, Dorset, April 1991.
26. Aultvaich, Inverness-shire, April 2012.
27. IFCSS MSS 500: 295, Co. Galway.
28. Piltown, Co. Tipperary, April 1991.
29. Edinburgh, December 1991.
30. Tamworth, Staffordshire, June 2012.
31. Atkins, 1986: 37.
32. Wimbledon, London, November 1983.
33. Southampton, Hampshire, November 1996.

DEATH
1. Deane & Shaw, 1975: 135.
2. Radford, 1961: 268.

DEVIL'S BIT SCABIOUS
1. Friend, 1884: 50.
2. Wright, 1898, 1: 842.
3. Grigson, 1987: 362.
4. *N & Q*, ser. 9, 4: 98, 1899.
5. Tongue, 1965: 36.
6. Allen & Hatfield, 2004: 276.
7. Deane & Shaw, 1975: 124.
8. *Gardeners Chronicle*, 11 August 1870: 738.

DEVON WHITEBEAM
1. Britten & Holland, 1886: 194.
2. Cann, 2012: 41.

DOCK
1. Vickery, 2006: 18.
2. Dobwalls, Cornwall, January 1985.
3. Email, June 2015.
4. Botesdale, Suffolk, February 1998.
5. Harrogate, North Yorkshire, May 2002.
6. Natural History Museum, London, June 2015.
7. University of Natural Resources and Life Sciences, Vienna, September 2012.
8. Natural History Museum, London, October 2012.
9. Palace Road, Nature Garden, Brixton, London, July 2014.
10. Tooting Common, London, August 2011.
11. South London Botanical Institute, June 2011.
12. St Clair, 1971: 58.
13. C.E., 1951: 13.
14. Thorncombe, Dorset, September 1977.
15. Basingstoke, Hampshire, March 2004.
16. Thorncombe, Dorset, April 1986.
17. Tiverton, Devon, February 1991.
18. Colwyn Bay, Clwyd, June 1992.
19. SLF MSS, Stannington, South Yorkshire, September 1970.
20. Mardu, Shropshire, November 2004.
21. Pimperne, Dorset, January 1992.
22. South Kensington, London, February 1994.
23. Rodmell, East Sussex, December 1992.
24. Lerwick, Shetland, March 1994.

DOCTRINE OF SIGNATURES
1. Coles, 1657: 186.
2. *Ibid.*: 31.
3. Pers. obs., Streatham, London, September 1993.
4. See Lownes, 1940.
5. Hatfield, 1999: 90.

DODDER
1. Marquand, 1906: 41.
2. *Newsletter of the Camborne-Redruth Natural History Society* 7: 8, 1990.
3. Cuddy, 1991: 59.

DOG ROSE
1. West Stow, Suffolk, September 2002.
2. Botanical Society of the British Isles, Annual Exhibition Meeting, London, November 1991; other people present remembered the practice from Sutton, Surrey in *c*.1936–9; County Durham; Blackpool, Lancashire in the 1950s; and Dorset in the 1970s.
3. Solihull, West Midlands, March 1991.

4. Farnborough, Kent, January 1993.
5. Corbridge, Northumberland, February 1993.
6. Westleton, Suffolk, March 1998.
7. Ayres, 2015: 82.
8. *Journal of Botany*, 78: 269, 1940.
9. Heworth, Tyne and Wear, December 1985.
10. Wideopen, Tyne and Wear, November 1985.
11. *Villager*, Salehurst and Robertsbridge, July 2014.
12. *The Times*, 3 October 1966; reprinted in *Garden Journal* (New York Botanic Garden), 17: 39, 1967.
13. Letter from J.P. Farrow, Senior Packaging Buyer, Winthrop Laboratories, January 1986.
14. Latham, 1878: 38.
15. Rudkin, 1936: 28.
16. Burne, 1883: 194.
17. Dartnell & Goddard, 1894: 23.
18. Trevelyan, 1909: 98.
19. Baker, 1854: 78.

DOG'S MERCURY
1. Lightfoot, 1777: 621.
2. East Dulwich, London, April 2016.

DOG VIOLET
1. Luton, Bedfordshire, January 1997.
2. IFCSS MSS 500: 74, Co. Limerick.

DUCKWEED
1. *N & Q*, 10 ser. 1: 365, 1904.
2. Woolton, Merseyside, December 1980.
3. Great Meols, Cheshire, November 1980.
4. *Plants, People, Places*, 4: 2, 1994.
5. Kensington, London, November 1979.
6. Irby, Merseyside, November 1980.
7. Bebington, Merseyside, November 1980.
8. Vickery, 1983: 249.

DWARF CORNEL
1. Lightfoot, 1777: 12.
2. Marren, 1996: 453.

DWARF ELDER
1. Aubrey, 1847: 50.
2. Asberg & Stearn, 1973: 40.
3. Britten & Holland, 1886: 51.
4. *Country Life*, 113: 290, 30 January 1953.
5. Parkinson, 1640: 210.
6. Bromfield, 1856: 231.
7. Synott, 1979: 37.

DYER'S GREENWEED
1. Britten & Holland, 1886: 26.
2. Knapp, 1829: 76.
3. Nicholson, 1861: 238.

EARLY PURPLE ORCHID
1. *Quarterly Review*, July 1863: 231.
2. Radstock, Avon, March 1982.
3. Newcastle-under-Lyme, Staffordshire, March 1983.
4. K'Eogh, 1753: 49.
5. Gregor, 1874: 106.
6. Kinahan, 1881: 117.
7. Larne, Co. Antrim, January 1992.

EASTER LILY
1. Purley, Surrey, April 1983.
2. *Parish Magazine of Llandrindod and Cefnllys with Diserth* [Powys], 338, March 2003.

EASTERN GLADIOLUS
1. Lousley, 1971: 276.

EELGRASS
1. Scott & Palmer, 1987: 339.
2. Lerwick, Shetland, March 1994.
3. Lindsay, 1856: 374.
4. Mabberley, 1997: 768.
5. Wyllie-Echeverria & Cox, 1999; Prendergast, 2002.

ELDER
1. Parslow & Bennalick, 2017: 407.
2. Withering, 1776: 186.
3. Driffield, Humberside, March 1985.
4. Anon., 1916: 425.
5. Scotton, Lincolnshire, October 1996.
6. Information collected from Oxfordshire Women's Institute groups by Mary Standley Smith, 1950s.
7. Ashreigney, Devon, July 1983.
8. Union Mills, Isle of Man, October 1993.
9. Bow Street, Dyfed, October 1984.
10. Kill Village, Co. Kildare, October 1984.

11. Skibbereen, Co. Cork, January 1993.
12. Longford, Shropshire, April 1997.
13. Mitcham, Surrey, May 1986.
14. Evans, 1895: 20.
15. Heanley, 1901: 55.
16. Whitwick, Leicestershire, August 1983.
17. Ponsanooth, Cornwall, November 1993.
18. South Kensington, London, March 2004.
19. Hardy, 1895: 325.
20. Trevelyan, 1909: 103.
21. Webster, 1978: 342.
22. McClintock, 1987: 33.
23. Bracknell, Berkshire, August 1984.
24. Lenamore, Co. Longford, April 1991.
25. Parkstone, Dorset, June 1991.
26. Holbeach, Lincolnshire, January 2003.
27. Yafforth, North Yorkshire, January 1990.
28. Nicolson, 1978: 74.
29. *N & Q*, 11 ser. 12: 489, 1915.
30. Peter, 1915: 123.
31. Horseheath, Cambridgeshire, April 1991.
32. St Oswyth, Essex, February 1989.
33. Hexham, Northumberland, June 1988.
34. Girton, Cambridge, May 1988.
35. *Daily News*, 27 January 1926.
36. Killip, 1975: 35.
37. London, E1, November 1996.
38. Maida Hill, London, December 1982.
39. Hole, 1937: 49.
40. Quelch, 1941: 78.
41. Ranson, 1949: 55.
42. Cinderford, Gloucestershire, November 1993; in April 2015 a member of the Hertfordshire Herb Group, meeting in Welwyn Garden City, mentioned that she uses elder leaves to treat constipation.
43. Viney Hill, Gloucestershire, November 1993.
44. Bacon, 1631: 258.
45. Opie, 1959: 315.
46. IFCSS MSS 414: 43, Co. Clare.
47. IFCSS MSS 700: 35, Co. Meath.
48. Horsted Keynes, West Sussex, February 1991.
49. Lenamore, Co. Longford, April 1991.
50. Albury, Hertfordshire, February 1998.
51. Hill, Worcestershire, October 1991.
52. Boat-of-Garten, Inverness-shire, November 1991.
53. Alexander, 2004: 113, 117.
54. Ruskin Park, London, October 2017.
55. Chepstow, Monmouthshire, June 2011.
56. St Marychurch, Devon, August 2011.
57. West Stow, Suffolk, November 1992.
58. Weobley, Herefordshire, August 1998.
59. McNeill, 1910: 130.
60. IFCSS MSS 700: 338, Co. Meath.
61. Lafonte, 1984: 35.
62. Merthyr Tydfil, Mid Glamorgan, October 2000.
63. Lenamore, Co, Longford, April 1991.
64. Parsons MSS, 1952.
65. Westminster, London, June 2011.
66. Smith-Bendell, 2009: 232.
67. Chesham, Buckinghamshire, April 2009.

ELECAMPANE

1. Baker, 1996: 56.
2. *Ibid.*
3. Ryde, Isle of Wight, November 1988.
4. Stevens Cox, 1970: 450.
5. Trevelyan, 1909: 314.

ELM

1. Stace, 2010: 279.
2. R. Palmer, 1976: 62.
3. De Garis, 1975: 121.
4. Lichfield, Staffordshire, January 2015.
5. *Phytologist*, ser. 2, 2: 167, 1857.
6. *Folk-lore*, 56: 307, 1945.
7. Williams, 1922: 275.
8. IFCSS MSS 190: 115, Co. Leitrim.
9. Amphlett & Rea, 1909: 323.

ENGLISH STONECROP

1. Bennett, 1991: 57.

EVERLASTING PEA

1. Rushton, Northamptonshire, July 1985; site visited August 1992 and plant identified as broad-leaved everlasting pea, *Lathyrus latifolius*; still thriving there 2015.

EVIL EYE
1. Roud, 2003: 177.

EYEBRIGHT
1. Hart, 1898: 381, 284.
2. Culpeper, 1653; see www.bibliomania.com/2/1/66/113/frameset.html, accessed December 2017.
3. Lightfoot, 1777: 323.
4. Vesey-Fitzgerald, 1944: 24.

FAIRY FLAX
1. Lees, 1850: 871.
2. McNeill, 1910: 108.
3. Hart, 1898: 370.
4. Britten & Holland, 1886: 336.

FAIRY FOXGLOVE
1. Mabey, 1996: 337.
2. Hexham, Northumberland, June 1990.

FALSE OAT-GRASS
1. Jacobs, 2002: 3.
2. West Stow, Suffolk, September 2002.
3. Colchester, Essex, March 2003; although this pastime seems to have been most frequently recorded from East Anglia, in September 2014 members of the Lichfield, Staffordshire, Wildlife Group recalled doing it elsewhere.

FAT HEN
1. Uí Chonchubhair, 1995: 62.
2. Mac Coitir, 2006: 5.
3. McNeill, 1910: 161.
4. Mauden, Bedfordshire, April 1993.
5. Sutton Community Farm, Wallington, Surrey, July 2017; fat hen is occasionally sold by greengrocers who cater mainly for customers from the Indian Subcontinent [pers. obs., Tooting, London, July 2017].
6. Allen & Hatfield, 2004: 90.

FEAST SUNDAY
1. Histon, Cambridgeshire, January 1989.
2. Hornchurch, Essex, August 1992.

FENNEL
1. St Saviour, Jersey, May 1993.

FEVERFEW
1. Taylor MSS, Attleborough, Norfolk.
2. Southend-on-Sea, Essex, February 1999.
3. Hart, 1898: 375; presumably the mention of rue refers to its use as an abortifacient: 'rue can't be bought in large quantities in France, because it can be used to produce abortions' [Rambouillet, France, September 2011].
4. For example Oxshott, Surrey, February 1998; Wormshill, Kent, May 2003, and Barking, Essex, August 2004.

FIELD BINDWEED
1. Stevenage, Hertfordshire, May 1982.
2. Bessacar, South Yorkshire, April 1984.

FIELD ERYNGO
1. Westwood & Simpson, 2005: 528.
2. Baker, 1854; cited in Britten & Holland, 1886: 485.

FIELD GENTIAN
1. Allen & Hatfield, 2004: 195.
2. Lerwick, Shetland, March 1994.
3. Allen & Hatfield, 2004: 352.

FIELD MAPLE
1. Friend, 1882: 41.
2. Downley, Buckinghamshire, February 1995.
3. E., 1884: 382.

FIELD SCABIOUS
1. Thompson, 1925: 163.

FIELD WOOD-RUSH
1. Chater, 2010: 778.
2. Britten & Holland, 1886: 102.

FIG
1. A postcard of the 'Fig Garden, Worthing', posted in 1913, shows a tree labelled 'This FIG TREE is the Oldest in England planted by THOMAS BECKET'.
2. Vince, 1979: 26.
3. Campbell-Culver, 2001: 24.
4. Letter from the Revd Peter Denny, St Newlyn East, January 1978.
5. Letter from the Venerable Trevor McCabe, Archdeacon of Cornwall, August 1998.
6. Manaccan, Cornwall, April 1998.
7. *N & Q*, 11 ser., 8: 425, 1913.

8. Lewisham, London, April 1986.
9. Westwood & Simpson, 2005: 353.
10. Clappersgate, Cumbria, October 1985.
11. Bow, London, April 1990; Mark 11: 12–14 relates how Christ noticed a fig tree, went to see if it was fruiting, and on finding it sterile – 'for it was not the season for figs' – cursed it.
12. Wright, 1905, 2: 135; the word dough-fig was still current in west Dorset in the 1960s, when dried figs appeared at Christmas time.
13. Hatfield, 1998: 10,19; Whitlock, 1976: 164; also known in Italy [Brompton Cemetery, London, July 2010].

FIRE LILY
1. *Dalesman* 39, 12: 945, March 1978.

FLAX
1. Gregor, 1874: 103.

FLOWER COMMUNION
1. Hampstead, London, October 1993.

FLOWER FESTIVALS
1. Email, February 2007.
2. Sleaford, Lincolnshire, February 2007.
3. Goodleigh, Devon, February 2007.
4. https://www.eastlife.co.uk/events/westleton-wild-flower-festival-2016/; accessed November 2016.

FLOWERING CURRANT
1. Shavington, Cheshire, March 1983.
2. Email, September 2013.

FLOWERING SUNDAY
1. Owen, 1978: 80.
2. *Bye-gones*, 9 September 1896, quoted in Hole, 1976: 74.
3. St Fagans, South Glamorgan, April 1983.
4. Plomer, 1977: 30.
5. IFCSS MSS 675: 93, Co. Louth.

FLOWER PARADE
1. Simpson, 1987.
2. *The Times*, 28 April 1986.
3. *Ibid.*, 25 April 1991.
4. *The Garden*, May 2013: 16.

FLOWERS
1. Gamble, 1979: 94.
2. Aberdovey, Gwynedd, July 1983.
3. Partridge, 1917: 311.
4. Macclesfield, Cheshire, April 1982.
5. Bridge of Weir, Renfrewshire, November 2015.
6. Caernarvon, Gwynedd, March 1993.
7. Wheatley, Oxfordshire, February 2007.
8. Paddington, London, July 1990.
9. Maida Hill, London, December 1982.

FLOWER SERVICE
1. Quoted in Udal, 1922: 42.

FOOL'S PARSLEY
1. Shavington, Cheshire, March 1983.

FORGET-ME-NOT
1. Browning, 1952: 149.
2. *Ibid.*: 94.
3. Addison, 1985: 98.

FORSYTHIA
1. Winchmore Hill, London, May 1984.

FOXGLOVE
1. *Science Gossip* 1, February 1870: 43.
2. Duncan, 1896: 163.
3. Harte, 2004: 119.
4. Sikes, 1880: 57.
5. Stevenage, Hertfordshire, May 1982.
6. Hodson, 1917: 452.
7. Holbeach, Lincolnshire, January 2003.
8. Itvaich, Inverness-shire, April 2012.
9. Britten & Holland, 1886: 153; name also recorded from Devon and Somerset.
10. Laver, 1995: 320.
11. Foley, 1974: 18.
12. Cinderford, Gloucestershire, November 1993.
13. Marquand, 1906: 39.
14. Dunsford, 1981: 176.
15. Pers. obs., 19 June 2016.
16. Pers. obs., 20 June 2016.
17. Rev. Stephen West, Bishopsteignton, Devon, June 2016.
18. Haynes, Bedfordshire, August 1984.
19. IFCSS MSS 1128: 26, Co. Cork.
20. Withering, 1785.
21. Ayres, 2015: 30, 31.
22. Summers, 2015: 154.

FRITILLARY
1. *Phytologist* 1: 580, 1843.
2. *Ibid.* 1: 814, 1843; 'radical botanists' in this context presumably means people who collected roots of plants, possibly for cultivation in their gardens.
3. Bromfield, 1850: 965.
4. Weston Turville, Buckinghamshire, June 1987.
5. St Albans, Hertfordshire, October 1996.
6. Abbeymead, Gloucester, August 1997.
7. Pers. obs., 23 April 2017.

FRUIT
1. Edgware, Middlesex, March 1977.
2. Great Bedwyn, Wiltshire, January 1991.

FRUIT TREES
1. *N & Q*, 9 ser 12: 133, 1903.
2. Maida Hill, London, March 1978.
3. Towcester, Northamptonshire, August 1982.
4. Smith, 2004: 305.
5. Taylor MSS, Mattishall, Norfolk.

FUCHSIA
1. Bromborough, Merseyside, November 1990.
2. Hertfordshire Herb Group, Welwyn Garden City, April 2015; pastime also known in Aberdeen in the 1980s [Walthamstow, London, October 2015].
3. Holywood, Co. Down, December 1991.
4. Clonlara, Co. Clare, June 2002.
5. IFCSS MSS 1112: 453, Co. Donegal.
6. Clonlara, Co. Clare, June 2002.

FUMITORY
1. Dartnell & Goddard, 1894: 55.
2. Spence, 1914: 101.

GALINGALE
1. St Saviour, Jersey, May 1993.

GARLIC
1. *The Times*, 10 January 1994.
2. Douglas, Isle of Man, April 1992.
3. Pers. obs.
4. Ontario, Oregon, March 2002.
5. McBride, 1991: 83.
6. IFCSS MSS 98: 154, Co. Mayo.
7. IFCSS MSS 790: 36, Co. Dublin.
8. Larne, Co. Antrim, January 1992.
9. IFCSS MSS 990: 71, Co. Cavan.
10. IFCSS MSS 232: 29, Co. Roscommon.
11. Girton, Cambridge, October 1985.
12. *The Times*, 6 March 1991.

GARLIC MUSTARD
1. Turner, 1551–6; quoted in Britten & Holland, 1886: 415.
2. Allen & Hatfield, 2004: 117.

GERMANDER SPEEDWELL
1. Britten & Holland, 1886: 50.
2. Britten & Holland MSS.
3. Hatfield, 1994: 37.
4. IFCSS MSS 440: 348, Co. Kerry.
5. IFCSS MSS 290: 159, Co. Offaly.
6. Marquand, 1906: 42.

GOAT'S BEARD
1. Ewen & Prime, 1975: 118.

GOAT WILLOW
1. Burne, 1883: 250.
2. Kirkby-in-Furness, Cumbria, February 1998.
3. Great Bookham, Surrey, October 1979.
4. Britten & Holland, 1886: 366; Grigson, 1987: 258.
5. Upminster, Essex, April 2011.
6. Kidwelly, Carmarthenshire, September 2012.
7. Appleshaw, Hampshire, January 2014.
8. Pers. obs., 24 March 1991.
9. Pers. obs., 27 March 1983.
10. Pers. obs., 24 April 2016.
11. Pers. obs.
12. Pers. obs., 24 April 2016.
13. IFCSS MSS 325: 11.
14. Yafforth, North Yorkshire, January 1990.
15. South Stainley, North Yorkshire, March 1992.
16. Leather, 1912: 19.
17. National Botanic Garden, Glasnevin, Dublin, August 2014.

GOLDENROD
1. Allenton, Derbyshire, March 1983.
2. Westminster Quaker Meeting, London, September 2012.
3. SLF MSS, Aldbrough, East Yorkshire, April 1972.

GOOD KING HENRY
1. Ewen & Prime, 1975: 78.
2. Washingborough, Lincolnshire, March 1994.
3. Porter, 1974: 47.
4. Cinderford, Gloucestershire, November 1993.

GOOSEBERRY
1. Reading, Berkshire, February 1987.
2. Streatham, London, September 1992.
3. Pers. obs.
4. 'News from the Past', *Pulman's Weekly News*, 15 July 1975.
5. In December 1975 neither the headmaster of the local school nor the parish priest was able to find any memories of the Feast.
6. Warkleigh, Devon, April 1975.
7. Palmer & Lloyd. 1972: 168.
8. Watson, 1920: 276.
9. Smith, 1989: 109.
10. Information displayed at the 2017 Egton Bridge Show.
11. Thorncombe, Dorset, April 1978.
12. IFCSS MSS 212: 370, Co. Leitrim.
13. IFCSS MSS 800: 53, Co. Offaly.
14. Larne, Co. Antrim, October 1993.

GOOSEGRASS
1. Oxshott, Surrey, November 2002.
2. Edinburgh, October 1991.
3. Driffield, Humberside, July 1985.
4. Farnham, Surrey, December 1985.
5. Mardu, Shropshire, November 2004.
6. Marquand, 1906: 42.
7. Bromfield, 1856: 240.
8. Muchelney, Somerset, January 2007.
9. Streatham, London, December 1983.
10. Charmouth, Dorset, January 1994.
11. Hatfield, 1999: 58.
12. Mardu, Shropshire, November 2004.

GORSE
1. St Peter Port, Guernsey, April 1984.
2. St Aubin, Jersey, April 1984.
3. Ballybunion, Co. Kerry, October 1984.
4. *Weekly Scotsman*, Christmas 1898.
5. Streatham, London, February 1993.
6. Ballycastle, Co. Antrim, January 1991.
7. Tooting, London, March 1999.
8. Essex, March 2003.
9. Baker, 1996: 64.
10. Stace, 2010: 185.
11. Lucas, 1960a: 186.
12. Newton Rigg, Cumbria, September 1988.
13. IFCSS MSS 212: 61, Co. Leitrim.
14. Castlerock, Co. Derry, February 1989.
15. Edinburgh, October 1991.
16. Balham, London, July 2004.
17. See Lucas, 1960a and 1979, and Harris, 1992.
18. South Stainley, North Yorkshire, March 1992.
19. Appleby-in-Westmorland, Cumbria, October 1996; practice also recorded from Glynn, Co. Antrim, February 1992, and no doubt known elsewhere.
20. Martinstown, Dorset, May 1991.
21. Callington, Cornwall, October 1996.
22. St Martin's, Guernsey, April 2002.
23. William, 1991: 24.
24. IFCSS MSS 500: 76, Co. Limerick.
25. IFCSS MSS 212: 61, Co. Leitrim.
26. Glynn, Co. Antrim, February 1992.

GRAPE-HYACINTH
1. Paston, Cambridgeshire, November 1993.
2. Wallasey, Merseyside, September 1996.

GRASS
1. Udal, 1922: 267.
2. Daingean, Co. Offaly, January 1985.
3. Metheringham, Lincolnshire, April 1994.
4. Bristol, September 2013.
5. Email, April 2016.
6. Old Basing, Hampshire, September 2012.
7. Email, January 2013.
8. Muchelney, Somerset, January 2007.

GREAT BURNET
1. Llandrindod Wells, Powys, September 1991.

GREATER CELANDINE
1. Clapham *et al.*, 1962: 102.
2. Letter from Highgate, London, *Hornsey Journal*, 17 August 1956.
3. Bexhill-on-Sea, East Sussex, February 1991.

4. Davey, 1909: 23.
5. Mardu, Shropshire, November 2004.
6. Whitstable, Kent, January 2012.
7. Bromfield, 1856: 26.

GREATER PLANTAIN
1. St Ervan, Cornwall, February 1992.
2. Lee, London, April 1993.
3. Email, January 2011.
4. Holmwood, Surrey, November 2009.
5. Mitcham Common, Surrey, June 2015; also Blackheath, London, June 2009.
6. Horniman Museum, London, September 2013.
7. Lenamore, Co. Longford, April 1991.
8. Binfield Heath, Oxfordshire, February 1998.
9. Cefyn Coed, Mid Glamorgan, August 2011.
10. Lerwick, Shetland, March 1994.
11. Bromley, Kent, April 1991.
12. Peabody Woods, London, June 2011; practice widespread in Europe: Estonia [Tooting Common, London, August 2011], Italy [Brockwell Park, London, October 2016], and Poland [Royal Botanic Gardens, Kew, Surrey, September 2011].
13. Lenamore, Co. Longford, April 1991.
14. Uí Chonchubhair, 1995: 204.
15. Stockwell, London, June 2010.

GREATER STITCHWORT
1. Friend, 1882: 44.
2. Davey, 1909: 73.
3. SLF MSS, Welwyn Garden City, Hertfordshire, April 1976.
4. Thorncombe, Dorset, January 1993.
5. Chard Junction, Somerset, May 1985.

GREATER HORSETAIL
1. Charlbury, Oxfordshire, February 1991.
2. Mabey, 1996: 13.

GREATER TUSSOCK-SEDGE
1. Bromfield, 1856: 553.
2. Jackson, 1871: 173.

GREEN ALKANET
1. Mabey, 1996: 309.
2. Halliday, 1997: 355.
3. Peckham, London, July 2012.

GROUND ELDER
1. UCL EFS MSS, Kilburn, London, September 1963.
2. Hurlford, Strathclyde, October 1996.
3. Kensington, London, March 2007.

GROUND IVY
1. Tadley, Hampshire, February 1998; ground ivy also used to treat styes at Longbridge Deverill, Wiltshire in the 1920s and 30s [Aldbury, Hertfordshire, February 1998].
2. Shipston-on-Stour, Warwickshire, September 1993.
3. Pratt, 1857, 2: 64.
4. Taylor MSS.

GROUNDSEL
1. Hunt, 1930: 416.
2. Moloney, 1919: 30.
3. Thorncombe, Dorset, March 1993.
4. Taylor MSS, Baconsthorpe, Norfolk.
5. Deane & Shaw, 1975: 123.
6. Pocklington, East Yorkshire, December 1996.
7. Mayhew, 1851: 155.
8. Peckham, London, July 2012.
9. Blunsdon, Wiltshire, April 2012.

GUERNSEY LILY
1. De Garis, 1975: 120.
2. De Sausmarez, 1970.

GYPSOPHILA
1. Baker, 1977: 77.
2. Bloxham & Picken, 1990: 82.

HAIRY BITTERCRESS
1. Mulcheney, Somerset, January 2007.

HAIRY BROME
1. Syston, Leicestershire, January 1991.

HAREBELL
1. Dickson, 1991: 158. Although 'bluebell' is frequently cited as an example of how English names can cause confusion, the scientific names given to the plant known as bluebell have been unstable; now known as *Hyacinthoides non-scripta,* names used during the twentieth century included *Scilla non-scripta, Endymion non-scriptus*, and *Scilla nutans*.

2. Gregor, 1881: 148; aul' man (old man) is a euphemism for the Devil.
3. Johnston, 1853: 135.

HART'S TONGUE
1. Marson, 1904.
2. Bromfield, 1856: 634.
3. IFCSS MSS 500: 74, Co. Limerick.
4. IFCSS MSS 650: 128, G. Waterford.

HAWTHORN
1. Vickery, 1985.
2. www.plant-lore.com/news/survey-of-unlucky-plants/ accessed September 2017.
3. East Bedfont, Middlesex, September 1978; for an account of the children's begging custom of making grottoes see Vickery & Vickery, 1977.
4. Email, March 2017.
5. Anonymous telephone call, July 1983.
6. Tooting Common, London, October 1998.
7. Elgin, Morayshire, June 2013.
8. Goody, 1993: 256.
9. Maple, 1971: 31.
10. Mickleover, Derbyshire, March 1983; Exeter, Devon, July 1984; Shingle Street, Suffolk, February 1998; East Sheen, London, February 1998.
11. Letter from East Grinstead, East Sussex, *Sunday Express*, 16 May 1982.
12. Warner, 1978: 281; Father James, Guardian of the Franciscan Priory, Portishead, Avon, June 1982.
13. Tooting Common, London, March 1999.
14. South East London Folklore Society, June 2009.
15. Gomme, 1884: 206.
16. http://inthebloodstream.wordpress.com/2009/01//21.whitsun-sylvia-plath/,accessed June 2010.
17. Camden Town, London, February 1998.
18. Tadley, Hampshire, February 1998.
19. Allen, 1980: 119.
20. Challenger, 1955: 266.
21. Blurton, Staffordshire, March 1983, from 'gran who lived in Wiltshire'.
22. Hartshay Hill, Derbyshire, March 1983.
23. Chiswick, London, July 1983.
24. Newcastle-under-Lyme, Staffordshire, March 1983; from a farmer in rural Cheshire.
25. Witham, Essex, May 1983.
26. Dickinson MSS, 1974: 41.
27. Forby, 1830: 426.
28. IFCSS MSS 413: 35, Co. Kerry.
29. IFCSS MSS 825: 139, Co. Laois.
30. Heather, 1940: 406.
31. Howse, 1949: 206.
32. Cornish, 1941.
33. Graves, 1948.
34. Barnstaple, Devon, September 1992.
35. Natural History Museum, London, October 2012.
36. Lichfield, Staffordshire, September 2014.
37. Lindegaard, 1978: 8.
38. Fleet, Hampshire, April 1994.
39. Orpington, Kent, March 2011.
40. Bath, Avon, January 1983.
41. Solihull, West Midlands, March 1991.
42. London, SE1, February 2016.
43. Johnston, 1853: 78.
44. Email, May 2016.
45. Whitstable, Kent, January 2017.

HAZEL
1. Lightfoot, 1777: 587.
2. Howse, 1949: 207.
3. Porter, 1974: 19.
4. Alton, Hampshire, June 1993.
5. Charmouth, Dorset, January 1994.
6. Biden, 1852: 58.
7. Buczacki, 2002: 462.
8. Dacombe, 1951: 44.
9. Briggs, 1976: 285.
10. Simpson, 1973: 65.
11. Emslie, 1915: 161.
12. Tongue, 1967: 54.
13. Purslow, 1972: 1.
14. *Folk-lore* 7: 89, 1896.
15. IFC MSS 462: 310, Co. Carlow, 1937–8.
16. Gmelch & Kroup, 1978: 18.
17. Milliken & Bridgewater, 2004.
18. Thorncombe, Dorset, August 2003.
19. Orpington, Kent, March 2015.
20. Evans, 1800: 404.
21. *The Times*, 23 April 1992.

22. Ludlow, Shropshire, March 2012.
23. IFCSS MSS 1112: 358, Co. Donegal.

HEATHER
1. Brompton Cemetery, London, October 2015.
2. Bennett, 1991; 58.
3. Barker, 2011: 57.
4. Nelson, 2000b.
5. Fraser, 1973: 78.
6. Allen & Hatfield, 2004: 122.

HEATH-RUSH
1. Lerwick, Shetland, March 1994.

HEDGE VERONICA
1. Calpe, Alicante, Spain, November 1991.
2. St Mary's, Isles of Scilly, September 1992.

HELLEBORE
1. *Wiltshire Family History Society Journal*, 46: 6, 1992.
2. Bromfield,1856: 15.
3. Halliday, 1997: 118.
4. Allen & Hatfield, 2004: 352.

HEMLOCK
1. Porter, 1974: 43.

HEMP
1. Stevens Cox, 1971: 10.
2. Parker, 1923: 324.
3. Wright, 1940: 12.
4. Wright, 1938: 152.
5. *Ibid.*: 187.
6. Baker, 1974: 6.

HENBANE
1. Ewen & Prime, 1975: 75.
2. Aitkin, 1944: 128.
3. Pratt, 1857, 1: 128.
4. Gerard, 1597: 284.
5. Vesey-FitzGerald, 1944: 25.

HERB ROBERT
1. Pershore, Worcestershire, October 1991.
2. Charlbury, Oxfordshire, February 1991.
3. Newcastle-under-Lyme, Staffordshire, March 1983.
4. Portland, Dorset, March 1991.
5. Eardiston, Worcestershire, April 2012.
6. IFCSS MSS 975: 27, Co. Cavan.
7. IFCSS MSS 575: 354, Co. Tipperary.

HOARY CRESS
1. Upminster, Essex, April 2011.

HOARY PLANTAIN
1. Cliftonville, Kent, January 2012.

HOGWEED
1. Leek, Staffordshire, March 1983.
2. Barnstaple, Devon, September 1992.
3. Davey, 1909: 220.
4. Wandsworth Common, London, March 1998.

HOLLY
1. Two Locks, Gwent, March 1993.
2. Davey, 1909: 105.
3. Grigson, 1987: 115.
4. St-Bride's-Super-Ely, Glamorgan, October 1982.
5. Longford, Shropshire, April 1997.
6. Sandiway, Cheshire, October 2004.
7. Thorncombe, Dorset, December 1982.
8. Thorncombe, Dorset, January 1983.
9. Holywell, Clwyd, March 1994.
10. Jedburgh, Roxburghshire, June 1994.
11. Mordiford, Herefordshire, December 1991.
12. Calpe, Alicante, Spain, December 1991.
13. Streatham, London, May 1992.
14. St Wenn, Cornwall, November 2003.
15. Two Locks, Gwent, March 1992.
16. Plymouth, Devon, February 1998.
17. Harrow on the Hill, Middlesex, October 2004.
18. Tregaer, Monmouthshire, October 2013.
19. Havant, Hampshire, August 1982.
20. Boundary, Staffordshire, March 1983.
21. South Stainley, North Yorkshire, March 1992.
22. Merthyr Tydfil, Mid Glamorgan, October 2000.
23. Mabey, 1996: 246.
24. Stratton, Dorset, September 1983.
25. Daingean, Co. Offaly, January 1985.
26. Letter from Reigate, Surrey, *The Times*, 9 July 1990; the winter of 1990–

91 was cold with heavy snow in Wales and parts of England 6–7 December.
27. Maynooth, Co. Kildare, February 1991.
28. Great Plumstead, Norfolk, October 1989.
29. Tregaer, Monmouthshire, October 2013.
30. Whitlock, 1976: 167.
31. Studley, 1988: 23.
32. McNeill, 1910: 110.

HOLY THORN
1. Anon., 1520.
2. Phipps, 2003: 16.
3. Hole, 1976: 26.
4. Batten, 1881: 116.
5. Collinson, 1791: 265.
6. Rawlinson, 1722: 109.
7. Collinson, 1791: 265.
8. Taylor, 1649: 6.
9. Rawlinson, 1722: 301.
10. Hole, 1965: 39.
11. Rawlinson, 1722: 1.
12. Hole, 1965: 35.
13. Anon., n.d. (2): 6 & 23.
14. Wilks, 1972: 98.
15. Batten, 1881: 125.
16. Christensen, 1992: 111.
17. Vickery, 1991: 81.
18. Anon., *The Holy Disciple, or, The History of Joseph of Arimathea*, Glasgow, 1777 (other versions of this chapbook, are provisionally dated from 1710 to 1805); quoted in Stout, 2007: 40.
19. Bett, 1952: 139.
20. *Gentleman's Magazine*, 1753: 578; it seems as if in 1752/3 Holy Thorns flowered steadily throughout much of the winter, as they continue to do today.
21. Stout, 2007: 38.
22. *Gentleman's Magazine*, 1753: 49.
23. Leather, 1912: 17.
24. *Ibid.*
25. *Pulman's Weekly News*, 10 January 1978.
26. Waring, 1977: 68.
27. Howell, 1640: 86.
28. Leather, 1912: 17.
29. Anon., 1977.
30. Rawlinson, 1722: 112.
31. See Bowman, 2006 and www.plant-lore.com/news.thorn-cutting-ceremony-glastonbury/ for an account of the 2015 event.
32. *Daily Mirror*, *The Times*, *Western Daily Press*, 10 December 2010.
33. Simpson, 2011: 8.
34. *Western Daily Press*, 4 April 2012.
35. Pers. obs., 8 January 2014; see www.plant-lore.com/news/holy-thorn-update-2/
36. Pers. obs., 6 August 2014.
37. Pers. obs., 16 December 2015.

HONESTY
1. De Garis, 1975: 119.
2. Stetchworth, Cambridgeshire, December 1991.

HONEYSUCKLE
1. Apples, Switzerland, February 1983.
2. St Andrews, Fife, January 2013.
3. Newcastle-under-Lyme, Staffordshire, March 1983.
4. September 1983.
5. Porter, 1969: 45.
6. Clevedon, Avon, March 1993.
7. Southwark, London, March 2012.
8. IFCSS MSS 190: 167, Co. Leitrim.
9. IFCSS MSS 800: 122, Co. Cavan.

HONEYSUCKLE STICK
1. Williams, 1944: 59.

HOP
1. Leather, 1912: 245
2. South Croydon, Surrey, May 2014.
3. Kew Gardens, Surrey, April 1999.
4. hopshop.co.uk/hops/hops-for-decoration/, accessed March 2017.
5. Vesey-FitzGerald, 1944: 25.
6. Upminster, Essex, March 2011.

HORSE CHESTNUT
1. Hadfield, 1957: 162.
2. Britten & Holland, 1886: 116.
3. *Ibid.*: 548.
4. *N & Q* 5 ser. 10: 378, 1878; presumably Piper was unaware that horse chestnuts did not arrive in Britain until early in the seventeenth century

(the volume also contains other correspondents' speculation about the origin of the word 'oblionker').
5. Sidmouth, Devon, October 1991. At the Hampstead Heath Conker Championships, London, first held in 2001, spectators are urged to chant 'Obly, obly-onker, let's play conkers' [pers. obs., 2 October 2017].
6. Bean, 1914, 1: 170.
7. *Ibid.*
8. Hadfield, 1957: 392.
9. Evans, 1881: 126; see also Gomme 1894: 71.
10. Dickinson MSS, 1974: 38.
11. Sharman, 1977: 60; 'po' = chamber-pot, Ginger soaked his conkers in urine.
12. Cloves, 1 993: 76.
13. Ogden, 1978: 71.
14. *Independent*, 10 October 1990.
15. https://www.worldconkerchampionships.com/, accessed September 2017.
16. 2015 World Conker Championships programme.
17. Vickery, 1995: 193; this championship started in the 1960s and continued until the 1980s, when it came to an end, to be briefly revived in 2003 and 2004 [Julia and Justin Gardner, owners of the New Inn, September 2017].
18. *South Wales Echo*, 9 November 1981.
19. Pers. obs.
20. Vickery, 2010: 87, 191.
21. Moore, 1989: 21; quandong = *Santalum acuminatum*.
22. Letter from Norwich, Norfolk, *The Times*, 27 October 1989.
23. Stoke Newington, London, August 2000.
24. Britten & Holland, 1886: 292.
25. Llandrindod Wells, Powys, September 1991.
26. www.fbhp.org.uk/bushy-park/cs/intro.html.php; accessed September 2017.
27. Wookey, Somerset, September 2012.
28. Battersea, London, August 2016.
29. Lloyd George, 1938: 349.
30. Lewis, 2009: 115.
31. Ayres, 2015: 85.
32. Kitchen, 1990: 69.
33. Ayres, 2015: 84.
34. Palmer, K., 1976: 113.
35. Kensington, London, 1979.
36. McClintock, 1975: 99.

HORSERADISH

1. Porter, 1969: 12.
2. Streatham, London, October 1996.
3. Pers. obs., Clapham, London, 1 April 2001.
4. St Osyth, Essex, February 1989.
5. Porter, 1958: 118.
6. Cinderford, Gloucestershire, November 1993.
7. Taylor MSS.
8. Tower Hamlets Cemetery Park, London, May 2016.

HOTTENTOT FIG

1. St Mary's, Isles of Scilly, November 1992.

HOUND'S TONGUE

1. Pickering, 1995: 20.

HOUSELEEK

1. Brimble, 1944: 73.
2. Phillips, 2012: 465.
3. Grigson, 1987: 183.
4. Ó Danachair, 1970: 25.
5. Alnwick, Northumberland, March 1998.
6. Hempstead, Norfolk, January 2003.
7. Rushmere St Andrew, Suffolk, February 1989.
8. Allen & Hatfield, 2004: 134.
9. Davey, 1909: 193.
10. IFCSS MSS 50: 298, Co. Galway.
11. Bexhill-on-Sea, East Sussex, February 1991.
12. Liversedge, West Yorkshire, April 2004.
13. Salmon, 1931: 317.
14. Botesdale, Suffolk, September 2011.
15. Rudkin, 1936: 27.
16. Armathwaite, Cumbria, October 1988.
17. Newcastle-on-Clun, Cheshire, October 2004.

18. Chiddingstone Causeway, Kent, April 1997; Harrogate, North Yorkshire, April 2009.
19. Kiltimagh, Co. Mayo, April 1983.

HUNGRY-GRASS
1. Kinahan, 1881: 109.
2. IFCSS MSS 1112: 390, Co. Donegal.
3. East Sheen, Surrey, April 2012.

HYDRANGEA
1. Porter, 1969: 45.
2. Paston, Cambridgeshire, November 1993.

IRISH SPURGE
1. NHM MSS, herb. John Blackstone, 1712–53.
2. Hart, 1873: 339.
3. Allen & Hatfield, 2004: 170.
4. NHM MSS, herb. H.J. Ryden, fl.1860s.
5. Dublin, March 1992.
6. IFCSS MSS 450: 90, Co. Kerry.
7. South Rauceby, Lincolnshire, March 1997.
8. Mallow, Co. Cork, May 2016.

IVY
1. Stanton-on-the-Wolds, Nottinghamshire, January 1983.
2. Barrow-in-Furness, Cumbria, August 2004.
3. Ware, Hertfordshire, November 2013.
4. Boundary, Staffordshire, March 1983.
5. Childwall, Liverpool, October 2013.
6. Helensburgh, Dunbartonshire, February 1991.
7. Larne, Co. Antrim, January 1993.
8. Cinderford, Gloucestershire, November 1993.
9. Earls Court, London, June 2010.
10. Ballyclough, Co. Cork, October 1990.
11. Lenamore, Co. Longford, April 1991.
12. St Mary's, Isles of Scilly, September 1992.
13. Inverness, April 2007.
14. Mordiford, Herefordshire, May 1993.
15. Wharton MSS 1974: 196.
16. Addingham Moorside, West Yorkshire, April 1993.
17. E.A. E[llis], Green medicine, *Eastern Daily Press*; undated cutting, January 1975.
18. Bonnard, 1993: 26.
19. Corfe Castle, Dorset, October 2011.
20. Castlerock, Co. Derry, February 1989.
21. Herne Hill, London, May 2013.

IVY-LEAVED CROWFOOT
1. McNeill, 1910: 95.

IVY-LEAVED TOADFLAX
1. Abbotsbury, Dorset, May 1983.

JACK-IN-THE-GREEN
1. Raglan, 1939.
2. James, 1961: 288.
3. Basford, 1978.
4. Judge, 2000.

JAPANESE KNOTWEED
1. Mabey, 2010: 4, 265.
2. Halliday, 1997: 167.
3. Cardiff, January 1994.
4. Tangye, 2008: 46.
5. Robertsbridge, East Sussex, June 2003.
6. Bristol, July 2002.
7. Mabey, 1996: 108.

JERSEY LILY
1. Nicholson, Victoria, Australia, July 1983.
2. St Lawrence, Jersey, April 1993.

JOUG TREE
1. Briggs, 1971: 541.

JUNIPER
1. Trevelyan, 1909: 105.
2. Grigson, 1987: 24.
3. Child, 1889: 387.
4. Great Bedwyn, Wiltshire, July 1993.
5. Wright, 1905, 5: 200.
6. Milliken & Bridgewater, 2004: 62.

KARO
1. Woodnewton, Northamptonshire, June 1992.

KIDNEY VETCH
1. Bonnard, 1993: 23.

KNAPWEED
1. Friend, 1884: 14.
2. Marquand, 1906: 41.

LABURNUM
1. Great Bookham, Surrey, October 1979; probably referring to Hampshire.
2. Chater, 2010: 306.

LADY'S BEDSTRAW
1. 1597: 968.
2. Lightfoot, 1777: 116.
3. Cinderford, Gloucestershire, November 1993.
4. Oxford, December 1993.

LANGUAGE OF FLOWERS
1. Halsband, 1965: 387–9, 464–5.
2. Elliott, 1984: 63.
3. Wilde, 1999: 161.
4. Mahood, 2008: 164; *erbe della Madonna* is IVY-LEAVED TOADFLAX, which Ruskin regarded as his personal emblem.
5. Nicolson, 2006: 84.
6. Addison, 1985: 135,137.

LAUREL
1. Parsons MSS, 1952.
2. Plymstock, Devon, January 1993.
3. Orpington, Kent, February 2007.
4. Plymouth, Devon, January 1993.
5. Tilehurst, Berkshire, February 1987.
6. Yatton, Somerset, November 2016.
7. *Hartland Times* 156: 39, 2007.
8. Pers. obs., Chelsea, Streatham and Tooting, London, December 1983.
9. Pers. obs., 11 September 1993.
10. Pers. obs., September 2014.
11. Boase, 1976: 175.
12. IFCSS MSS 575: 324, Co. Tipperary.
13. Glynn, Co. Antrim, February 1992.

LAVENDER
1. Harrow-on-the-Hill, Middlesex, October 2004.
2. Bromley, Kent, March 1997.
3. Boat-of-Garten, Inverness-shire, November 1991.

LEAF
1. Latham, 1878: 9.
2. Great Bookham, Surrey, November 1979.
3. Oban, Argyllshire, October 1990.
4. *New English Hymnal*, 1986, no. 377.
5. Palmer, 1979: 216.

LEEK
1. Pers. obs., Wales versus South Africa rugby match, Twickenham, London, 17 October 2015, and St David's Day, Cardiff, 2017.
2. Wilkinson, 1858: 137.
3. www.historic-uk.com/HistoryUK/HistoryofWales/The-Leek-National-emblem-of-the-Welsh/, accessed March 2017.
4. Caption to portrait of Philip Proger (1585–1644), in the National Museum Cardiff: 'Proger is shown holding a leek, and the portrait is the earliest known of a Welshman doing so'.
5. Nelson, 1991: 50; painting now in the National Gallery of Ireland, Dublin.
6. *N & Q*, 5 ser. 7: 206, 1877.
7. Moorcroft Wilson, 1998: 190; 334.
8. *The Times*, 2 March 1989; see also *ibid.*, 2 March 1991.
9. Pers. obs., 16 February 1980.
10. Pers. obs., 17 October 2015.
11. Calderbank, 1984: 11.
12. *Op.cit.*: 12.
13. *Hexham Courant*, 21 September 1990.
14. *Op. cit.*, 12 September 1992.
15. The *Sun*, 15 August 2008.

LEMON
1. Chard, Somerset, March 1991.
2. Beaconsfield, Buckinghamshire, November 2003.
3. Lewes, East Sussex, November 2011.

LEMON VERBENA
1. Chope, 1929: 126.
2. Plymouth, April 1993.

LESSER SPEARWORT
1. McNeill, 1910: 96.
2. Bonnard, 1993: 13.

LESSER TREFOIL
1. Colgan, 1893.
2. Nelson, 1990.
3. Pers. obs., 2016–17.

LETTUCE
1. Radford, 1961: 217.
2. Taylor MSS, Ashby, Norfolk.
3. Pers. obs., 1971–2016.

4. Natural History Museum, London, March 2004.

LILAC
1. Vickery, 1985.
2. www.plant-lore.com/survey-of-unlucky-plants/, accessed June 2017.
3. Worthing, West Sussex, February 1982.
4. Bristol, January 2013.
5. Clacton-on-Sea, Essex, January 2013.
6. Gorleston, Norfolk, April 1991; elsewhere, 'an old lady I know in Budapest says that you shouldn't give anyone lilac; it's only used for funerals' [Natural History Museum, London, June 2003].
7. Thorncombe, Dorset, May 1982.
8. Trevelyan, 1909: 96.

LILY OF THE VALLEY
1. Driffield, Humberside, March 1985.
2. Dunkineely, Co. Donegal, February 1986.
3. Tooting, London, February 2001.
4. Simpson, 1973: 34.
5. Hole, 1976: 75.
6. Griffin-Kremer, 2011 & 2015.
7. Bertram, 1995: 23.
8. Grieve, 1931: 481.
9. Jackson, 1870: 137.
10. Girton, Cambridge, August 1989.

LIME
1. Solihull, West Midlands, March 1991.
2. Sidmouth, Devon, December 1992.

LONE BUSH
1. Mac Manus, 1973: 51.
2. *Ibid.*: 1973: 52.
3. IFCSS MSS 717: 103, Co. Meath.
4. IFCSS MSS 1020: 243, Co. Cavan.
5. South Stainley, North Yorkshire, March 1992.
6. Queen's University, Belfast, September 1997.
7. Bracknell, Berkshire, August 1984; when visited in 1991 the tree was found to be a hawthorn.
8. London, September 1984.
9. Kinahan, 1888: 266.

LORDS-AND-LADIES
1. Brown, 1952: 298.
2. Barnstaple, Devon, May 1991.
3. Porter, 1969: 42.
4. Thorncombe, Dorset, March 1982.
5. Grigson, 1987: 430.
6. Aldershot, Hampshire, April 1994; name also recorded from Welshpool, Powys, August 1994.
7. Bromfield, 1856: 529.
8. Porter, 1969: 41.
9. Wentersdorf, 1978: 417.
10. Prime, 1960: 37.
11. Gerard, 1597: 686.
12. *Phytologist*, 4: 1030, 1853.
13. Prime, 1960: 50.
14. *Phytologist*, 4: 1031, 1853.

LOUSEWORT
1. Tait, 1947: 79.
2. Lerwick, Shetland, July 2011.
3. Honor Oak, London, July 2016.

LUNGWORT
1. Bratton, Wiltshire, April 1983.
2. Britten & Holland, 1886: 481.
3. Udal, 1922: 17.

MADONNA LILY
1. Campbell-Culver, 2001: 49.
2. Brize Norton, Oxfordshire, August 1992.
3. Swansea, West Glamorgan, April 1984.
4. Attwater, 1970: 51.
5. Udal, 1922: 257.
6. Worcester, January 1998.
7. Little Waltham, Essex, January 1978.
8. Stowmarket, Suffolk, February 1989.

MAIDENHAIR FERN
1. Ogilby, 1845: 349.

MALE FERN
1. ICSSS MSS 232: 129, Co. Roscommon.
2. Barrington, 1984: 50.
3. Lockton & Whild, 2015: 139.

MALLOW
1. Christchurch, Dorset, May 1991.
2. Edinburgh, December 1991; practice also recorded from Holland, 'we eat them and they look like little Dutch cheeses' [Natural History Museum, London, August 1993].
3. Rudkin, 1936: 27.

4. IFCSS 350: 182.
5. IFCSS MSS 700: 121.
6. Oban, Argyll, October 1990.
7. Llandrindod Wells, Powys, September 1991.
8. Calpe, Alicante, Spain, November 1991.
9. Caernarfon, Gwynedd, April 1995.

MANDRAKE

1. Stearn, 1976: 290.
2. Kew, Surrey, July 1984.

MANGOLD

1. Whitlock, 1978: 145; Whitlock was born in 1914.
2. Vickery, 1978: 156.
3. *Pulman's Weekly News*, 31 October 1972.
4. *Chard & Ilminster News*, 9 November 1988.
5. Thorncombe, Dorset, November 1982.
6. Pers. obs., 2015 and 2017.

MARIGOLD

1. Dartnell & Goddard, 1894: 101.
2. Taylor MSS.

MARSH MARIGOLD

1. Mardu Clun, Shropshire, October 2004.
2. Charlbury, Oxfordshire, January 1991.
3. Carnlough, Co. Antrim, January 1989.
4. O'Dowd, 2015: 33, 44.
5. Bristol, January 1999.
6. Killip, 1975: 173.
7. Milliken & Bridgewater, 2004: 157.

MARSH SAMPHIRE

1. Britten & Holland, 1886: 414.
2. Hull, Humberside, May 1988.
3. Pers. obs.
4. Pers. obs.
5. *The Times*. 25 June 1991.

MAT-GRASS

1. Teulon-Porter, 1956: 91.

MAY BIRCHERS

1. Simpson, 1976: 148.

MAY GARLANDS

1. Hole, 1975: 58.
2. Thompson, 1939: Chapter 13; see Judge, 1993 for an appraisal of the accuracy of Thomson's work.
3. Gloucester, 1983: 34.
4. Robson, 1993.
5. Vickery, 1995: 236; 2010: 162.
6. Peter Robson, April 2014.
7. *FLS News* 77: 4, 2015.
8. Peter Robson, May 2016.
9. http://www.bamptonoxon.co.uk/pages.asp?id=69&pageName=Annual Events; accessed November 2009.

MEADOW FOXTAIL

1. Llandrindod Wells, Powys, September 1991.
2. Winchester, Hampshire, September 1991.
3. Stevenage, Hertfordshire, January 1993.

MEADOWSWEET

1. Johnston, 1853: 59.
2. Trevelyan, 1909: 96.
3. Kensington, London, January 1983.
4. Camberwell, London, May 2014.
5. Anon. note, January 2012.
6. Cottam MSS: 50.
7. Lichfield, Staffordshire, January 2015.
8. Anon. note, 1998.

MEXICAN HAT

1. Streatham, London, June 2012.
2. Sydenham, London, January 2011.

MEZEREON

1. Rudkin, 1936: 26.

MIMOSA

1. Bayswater, London, July 1983; see also forsythia in the present volume.

MINT

1. Leamington Spa, Warwickshire, January 1993.
2. Churchdown, Gloucestershire, January 1988.
3. Canterbury, Kent, November 1993.
4. Rushmere St Andrew, Suffolk, February 1989.
5. Richmond, Surrey, May 2003; also recorded from New Zealand [University of Otago, October 2013].

6. Cheslyn Hay, Staffordshire, October 2014.
7. Alton, Hampshire, June 1993.
8. Tulse Hill, London, October 2012.
9. UCL EFS MSS M22, St Leonard's on Sea, East Sussex, October 1963.

MONEY TREE
1. Pers. obs.
2. St Paul's School, Barnes, London, February 1986.
3. *Sun*, 23 August 1982.
4. *Ibid*., 24 January 1983.
5. London Natural History Society, April 1983.

MONKEY PUZZLE
1. Porter, 1969: 63.
2. Peterborough, Cambridgeshire, June 1981.
3. Opie, 1959: 218.
4. Opie & Tatem, 1989: 260.
5. Natural History Museum, London, October 1979.

MONTBRETIA
1. Larne, Co. Antrim, October 1993.

MOUSE-EAR HAWKWEED
1. Bexhill-on-Sea, East Sussex, February 1991.
2. Itchen, Hampshire, June 1993.

MUGWORT
1. Lightfoot, 1777: 469.
2. McNeill, 1910: 137.
3. St Day, Cornwall, January 1994.
4. Winchester, Hampshire, April 1996.
5. Rugby, Warwickshire, February 1998.
6. Garrad, 1984: 76.
7. Allen & Hatfield, 2004: 297.
8. Chiddingstone Causeway, Kent, June 1997.
9. Peckham, London, June 2012.

MULBERRY
1. Thiselton-Dyer, 1889: 120.
2. *The Times*, 20 May 1992.
3. Opie, 1985: 291.
4. Roud, 2010: 268.
5. Evans, 1969: 13.
6. Pers. obs., May 2016.
7. Pers. obs., December 2015; the bulk of this tree fell on 18 August 2017, leaving 'a small ... trunk and it is hollow and soft inside'. Later in the month most of the remaining tree was removed leaving only its stump and a few leafy twigs (pers. obs. September 2017).
8. Taylor MSS, East Harling, Norfolk.
9. Information sheet prepared by Janet Thompson, September 2013.

MULLEIN
1. *Daily Telegraph*, 26 June 1996.
2. Shirreffs, 2015: 84.
3. Stevens Cox, 1971: 6.
4. IFC MSS 36: 252, Co. Laois, c.1930.
5. IFCSS MSS 660: 347, Co. Louth.
6. Barnstaple, Devon, May 1991; in Alsace, France, 'mullein makes a fantastic syrup to treat coughs' [email, August 2017].

MYRTLE
1. Trevelyan, 1909: 105.
2. Broomborough, Merseyside, November 1990.
3. Verrall, 1991: 28.
4. *The Times*, 10 February 1840, 11 February 1840, 29 June 1843; I am indebted to Charles Nelson for sharing his analysis of royal wedding flowers.
5. Baker, 1974: 28.

NAVELWORT
1. Salisbury, Wiltshire, January 1992.
2. Minehead, Somerset, November 1993.
3. Allen & Hatfield, 2004: 134.
4. *Phytologist* 5: 135, 1854.
5. Llanuwchllyn, Gwynedd, April 1991.
6. St Ervan, Cornwall, January 1992.

NETTLE
1. Culpeper, 1652: 155; for a fuller account of nettle-lore see Vickery, 2008.
2. Gregor, 1884: 377.
3. Culpeper, 1884: 156.
4. Castlerock, Co. Derry, February 1989.
5. Danesfort, Co. Kilkenny, April 1991.
6. Ballymote, Co. Sligo, May 1994.
7. Llanuwchllyn, Gwynedd, April 1991.
8. Little Sandhurst, Berkshire, August 2003.

9. *The Times*, 14 September 2007.
10. Milliken & Bridgewater, 2004: 37.
11. Kentish Town, London, September 1995.
12. *Plants, Places, People* 7: 10, 1995.
13. Turner, 1961: 208.
14. Girton, Cambridge, June 1996.
15. Radford, 1998: 6.
16. *Daily Telegraph*, 14 April 2003.
17. Boase, 1976: 115.
18. Leicester, September 1993.
19. Streatham, London, 1992.
20. Lenamore, Co. Longford, April 1991.
21. Great Bedwyn, Wiltshire, January 2002.
22. Allen & Hatfield, 2004: 85.
23. St Mary's, Isles of Scilly, September 1992.
24. Headcorn, Kent, January 1993.
25. Letter from Killarney, Co. Kerry, *Arthritis Today* 131, Winter 2006.
26. Randall, 2003: 19.
27. Chope, 1935: 138.
28. Letter from Girvan, Ayrshire, *Daily Record*, 6 June 1985.
29. Edinburgh, January 1992.
30. Clonmel, Co. Tipperary, February 1993.
31. Tulse Hill, London, 1996.
32. Metherington, Lincolnshire, April 1994.
33. Langtoft, Humberside, July 1985.
34. Histon, Cambridge, January 1989.
35. Smith, 2004: 303.
36. *Ibid.*: 304.
37. MacPherson MSS.
38. Larne, Co. Antrim, November 1991.
39. Woodnewton, Northamptonshire, June 1992.
40. Edom, 2010: 21.
41. Addlestone, Surrey, February 1998.
42. London *Evening Standard*, 30 January 2003.
43. Ranson, 1949: 84.
44. IFCSS 212: 79, Co. Leitrim.
45. Chiswick, London, February 1991.
46. SLFS MSS, from Boston, Lincolnshire, March 1971.
47. Barton-upon-Humber, Humberside, February 1992.
48. Simpson, 2002: 118.
49. Chandler, 1993: 11.
50. Simpson, 2002: 118.
51. Paignton, Devon, November 1984.
52. Brixham, Devon, November 1984.
53. Attempts to collect memories since 1984 have been unsuccessful.
54. Thorncombe, Dorset, July 1998.
55. http://www.thebottleinn.co.uk/nettlespage.htm, accessed January 2008.
56. *Bridport & Lyme Regis News*, 25 June 1999.
57. https://en.wikipedia.org/wiki/The-Bottle-Inn, accessed December 2017.
58. *Bridport & Lyme Regis News*, 22 June 2001.

NIPPLEWORT

1. Lightfoot, 1777: 445.

NURSERY BOGIES

1. Briggs, 1976: 313.

NUTMEG

1. SLF MSS, Doncaster, South Yorkshire, April 1968.
2. Kew Gardens, Surrey, January 1999.
3. Brixton, London, June 2013.
4. Taylor MSS, Debenham, Suffolk, April 1925.
5. Anon., April 2002.

OAK

1. Mills, 2013 provides an excellent overview of oak in the culture of the British Isles, with well illustrated accounts of many historic trees.
2. Vickery, 2004: 59.
3. www.plant-lore.com/plantofthemonth/plants-and-politics/, accessed May 2017.
4. Vickery, 1978; 157.
5. Letter from Enmore, Somerset, *The Times*, 20 March 1990.
6. Tickhill, South Yorkshire, February 1998.
7. Radford, 1961: 253.
8. The Museum is currently (June 2017) closed; images of the relevant objects can be seen on www.heritage.southwark.gov.uk/search/

lovett%2520acorn, accessed June 2017.
9. Yallop, 1984: 29.
10. Letter from Church Gresley, Staffordshire, *Daily Mirror*, 1 November 1973.
11. Ryde, Isle of Wight, November 1988.
12. Hemel Hempstead, Hertfordshire, August 2004.
13. Opie, 1959: 263.
14. Seven Sisters, London, May 2009.
15. Nottingham, May 2009.
16. Darlington & Hirons, 1975: 151.
17. Pers. obs., 29 May 2015, when the service was poorly attended, and the celebrant ranted about the evils of regicide, which he regarded as the most heinous of crimes.
18. www.facebook.com/groups/Traditionalcustomsandceremonies/, accessed June 2017.
19. The Revd E.G. Allsop, former vicar of St Neot, November 1989.
20. Pers. obs., press reports and the Royal Hospital website.
21. Ross, 1987: 10.
22. Ross, *op. cit.*; pers. obs., 2017.
23. There is confusion about the identity of oak apples; people who wear them wear the real thing, but the banner carried in the procession, which has been in use since at least 1987, depicts marble galls [pers. obs., 29 May 2017].
24. Pers. obs., 2017.
25. Roud, 2006: 199.
26. In November 2017 no mention was found of a service in Norwich Cathedral.
27. Letter from Rita Hare, Administrator, RoadPeace, 20 August 2003.
28. www.plant-lore.com/plantofthemonth/polstead-gospel-oak/, accessed July 2017; Bill Wigglesworth, Polstead churchwarden and local history recorder, July 2012.
29. *The Times*, 14 November 2014.
30. Holt, 1983: pl. 10/11.
31. *The Times*, 21 May 1992.
32. Weaver, 1987: 29.
33. Mills, 2013: 169–73.
34. Palmer, 1985: 17.
35. Douglas, 1989; Fort William Inverness, October 1993.
36. Taylor MSS, Woolverstone, Suffolk.
37. IFCSS MSS 800: 219, Co. Offaly.
38. IFCSS MSS 1075: 135, Co. Donegal.

OAT

1. Williams-Davies, 1983: 229.
2. Balfour, 1904: 175.
3. Sadborow, Dorset, March 1975.
4. Woodstock, Oxfordshire, January 1983.
5. Sandiway, Cheshire, October 2004.
6. Glamorgan, May 2003.

ONION

1. Bloom, 1930: 245.
2. UCL EFS MSS M13, Carshalton Beeches, Surrey, October 1963.
3. Hill, Worcestershire, October 1991.
4. Baker, 1996: 115.
5. Canterbury, Kent, November 1993.
6. Jeacock, 1982.
7. Peter Bamford, Local Studies Librarian, Cheshire County Council Archives and Local Studies, Chester, November 1993.
8. Ifield, West Sussex, March 1997.
9. St Andrews, Fife, September 1988.
10. West Wimbledon, London, November 2003.
11. Larkcom, 1982: 58.
12. *The Times*, 22 September 1992.
13. *Metro*, 19 September 2014.
14. 2017 Newent Onion Fayre and Food Village Event Day Programme 2017; pers. obs. 9 September 2017.
15. Parsons MSS, 1952.
16. St Osyth, Essex, February 1989.
17. Ryde, Isle of Wight, November 1988.
18. Childwall, Liverpool, August 2011.
19. Stockport, Greater Manchester, April 1991.
20. IFCSS MSS 175: 313, Co. Sligo.
21. Liss, Hampshire, June 2013.
22. Holbeach, Lincolnshire, January 2004.
23. Llanuwchllyn, Gwynedd, April 1991.
24. Smith, 1959: 414.
25. Glynn, Co. Antrim, February 1992.
26. Taunton, Somerset, January 1992.

27. Opie, 1959: 375.
28. Gutch, 1912: 31.
29. Edinburgh, October 1991.

OPIUM POPPY

1. Crompton, 2003: 344.
2. Porter, 1969: 85.
3. Taylor MSS, Norwich, Norfolk.

ORANGE

1. Great Bedwyn, Wiltshire, November 1988.
2. Baker, 1977: 78.
3. Bloxham & Picken, 1990: 82.
4. Newall, 1971: 335; Hutton, 1996: 202.
5. Quoted on a display board in the Avenham Park Pavilion, October 2017.
6. Palmer & Lloyd, 1972: 137.
7. SLF MSS, Wood Green, London, March 1977.
8. www.telegraph.co.uk/only-in-britain/francis-drake-inspired-totnes-orange-races/, accessed May 2017.
9. Wallington, Surrey, September 2004.
10. South London Botanical Institute, March 1996.
11. Lewis, 2001: 156.
12. Girton, Cambridge, September 2001.
13. *The Times*, Home Forum, 18 March 2004.

ORANGE LILY

1. Belfast, May 1994.
2. Loftus, 1994: 16.
3. *Ibid.*
4. IFCSS MSS 1100: 248, Co. Donegal.

ORPINE

1. Aubrey, 1881: 25.
2. Brand, 1853: 329.
3. Boardman & Scarisbrick, 1977.
4. Plomer, 1977: 234.
5. Simpson, 1973: 123.
6. Allen & Hatfield, 2004: 138.
7. Wyse Jackson, 2014: 530.

OXEYE-DAISY

1. Daingean, Co. Offaly, January 1985.
2. Leamington Spa, Warwickshire, January 1993.
3. Allen & Hatfield, 2004: 305.
4. McNeill, 1910: 136.

OXLIP

1. Stevenage, Hertfordshire, January 1993.

PANSY

1. Long Melford, Suffolk, November 1993.
2. *Pink Paper*, 15 November 2007.

PARSLEY

1. Wimbledon, London, November 1983.
2. Letter from Barry, South Glamorgan, in *Daily Mirror*, 7 June 1989.
3. Weobley, Herefordshire, August 1998.
4. Porter, 1958: 113.
5. Purley, Surrey, January 1978.
6. Hornchurch, Essex, August 1992.
7. Letter from London, NW1, in *Daily Mirror*, 26 May 1962.
8. Waltham Abbey, Essex, March 1991.
9. Amery, 1905: 114.
10. Letter from Orpington, Kent, in *Daily Mirror*, 26 May 1962.
11. Simpson, 1973: 113.
12. Hamworthy, Dorset, May 1991.
13. Chope, 1932: 155.
14. Merthyr Tydfil, Mid Glamorgan, October 2000.
15. *N & Q*, 4 ser. 6: 211, 1870.
16. Simpson, 1973: 113.
17. *N & Q*, 6 ser. 11: 467, 1885.
18. Leather, 1912: 21.
19. King, 1877: 90.
20. *Folk-lore* 24: 240, 1913.
21. Hemel Hempstead, Hertfordshire, August 2004.
22. Thorncombe, Dorset, November 1984.
23. *N & Q*, 4 ser. 9: 35, 1872.
24. Stevens Cox, 1971: 7.
25. Farnham, Surrey, December 1985.
26. Opie & Tatem, 1989: 299.
27. Porter, 1958: 113.
28. Natural History Museum, London, June 2015.
29. St Martin, Guernsey, April 2002.

PARSLEY PIERT

1. Vesey-FitzGerald, 1944: 27.
2. Quinton, West Midlands, April 1993.

PARSNIP
1. *Report and Transactions of the Guernsey Society for Natural Science* 2: 276, 1893.
2. De Garis, 1975: 121.
3. Fulham, London, February 2012.
4. *Western Daily Press*, 21 December 1996; see also *The Times*, 4 January 1997.

PASQUE FLOWER
1. Boase, 1976: 115.
2. Anonymous telephone call, April 1991.
3. Hillingdon, Middlesex, February 1998.

PASSION FLOWER
1. Friend, 1884: 192.

PEA
1. Parsons MSS, 1952.
2. Jones, 1908: 323.
3. Parsons MSS, 1952.
4. Latham, 1878: 9.
5. Hole, 1950: 42.
6. Letter from Boosbeck, Cleveland, *The Times*, 3 April 1985.
7. Townhead, Cumbria, August 1989.

PEAR
1. SLF MSS, Birmingham, September 1977.
2. Muggleton, 2016: 40.
3. *Ibid.*, 2017: 40.
4. *Ibid.*, 2016: 40.
5. *Ibid.*, 2017: 40.
6. *Scottish Antiquary* 5: 181, 1891.
7. Information from 'Cameron, of The Colstoun Team', September 2017.

PEARLWORT
1. McNeill, 1910: 105.

PELLITORY OF THE WALL
1. Britten & Holland, 1886: 374.
2. Marquand, 1906: 44.
3. Vesey-FitzGerald, 1944: 27.
4. Skibbereen, Co. Cork, April 1994.

PENNYROYAL
1. Stearn, 2002: 251.
2. Marquand, 1906: 45.
3. Hillside Garden Club, Eastcombe, Gloucestershire, December 2013.
4. Salmon, 1710: 846.
5. Roberts, 1971: 100.
6. Helm, 1981: 26.
7. Shepherdswell, Kent, October 1979.
8. Leeds, October 1981.
9. Barton, Cambridgeshire, 1998.
10. Menheniot, Cornwall, 1991.

PEONY
1. *N & Q*, 4 ser. 12: 469, 1873.
2. Taylor MSS.
3. *N & Q*, 5 ser. 9: 405, 1878.
4. Latham, 1878: 44.
5. Charlwood, Surrey, April 2004.

PERIWINKLE
1. Hammond MSS, 1970: 28.
2. Whitlock, 1976: 163.
3. Porter, 1969: 47.
4. Best & Brightman, 1973: 8.
5. Culpeper, 1652: 170.
6. Waters, 1987.
7. Britten & Holland, 1886: 139; name also recorded from Dorset [Macmillan, 1922: 76].

PETTY SPURGE
1. NHM MSS, herb. C.I. Paton.
2. Truro, Cornwall, December 1993; sample of plant received and identified.

PIGNUT
1. Didcot, Oxfordshire, February 1991.
2. Sidmouth, Devon, October 1991
3. Letchworth, Hertfordshire, May 2001.

PINE
1. De Garis, 1975: 117.
2. *Times Weekly Review*, 31 May 1956.

PINEAPPLE
1. Raphael, 1990: xxxii.

PINEAPPLE-WEED
1. Stace & Crawley, 2015: 387.
2. Lerwick, Shetland, March 1994.
3. Bethel, Gwynedd, March 1999; sample of plant received and identified.

PINK SORREL
1. Milborne Port, Somerset, September 2011.
2. Tooting, London, January 2013.
3. Leiston, Suffolk, July 2011.

PLANTA GENISTA
1. Friend, 1884: 390.

PLUM
1. Brown, 2016: 212.
2. For a brisk survey of some local varieties of plum see Clifford & King, 2006: 329.
3. *N & Q*, 7 ser. 4: 485, 1887.
4. Parsons MSS, 1952.

POPPY
1. Johnston, 1853: 30.
2. *N & Q*, 5 ser. 9: 488, 1878.
3. *Ibid.*, 3 ser. 8: 319, 1865.
4. Smith-Bendell, 2009: 207.
5. Hatfield, 1994: 40.
6. Hewins, 1981: 147.
7. Quennell, 1984: 167.
8. Ward, 1891: 119.
9. Macmillan, 1922: 28.
10. Wormshill, Kent, May 2003.
11. Natural History Museum, London, December 2004.
12. Hersom, 1973: 79.
13. St Mary's, Isles of Scilly, September 1992.
14. Barton upon Humber, Humberside, February 1992; practice also recorded from Alsace, France [email, August 2017], Hungary [Borough, London, June 2009] and north-east Italy [London University of the Arts, October 2012].
15. Thiselton-Dyer, 1889: 15.
16. Rowbottom, 2008: 395.
17. Royal British Legion Poppy Factory, December 1989.
18. *Daily Star*, 11 November 2016.
19. PPU leaflets, 1988; Seed, 1988.
20. Hill, 2016.
21. www.plant-lore.com/news/white-poppies/, accessed December 2017.
22. https://www.animalaidshop.org.uk/accessories/the-purple-poppy, accessed December 2016.
23. Stibbons & Cleveland, 1990.
24. Allen & Hatfield, 2004: 78.

POTATO
1. Atkins, 1986: 23.
2. Daingean, Co. Offaly, January 1985.
3. Lisburn, Co. Antrim, March 1986.
4. Frome, Somerset, June 1978.
5. Castlerock, Co. Derry, February 1989.
6. Little Paxton, Cambridgeshire, February 1998.
7. Barrow-in-Furness, Cumbria, August 2004.
8. Salaman, 1949: 117.
9. Danaher, 1972: 165.
10. Salaman, 1949: 117.
11. Ó Suilleabhain, 1967: 68.
12. IFCSS MSS 232: 269, Co. Roscommon.
13. IFCSS MSS 800: 14, Co. Offaly.
14. Lerwick, Shetland, March 1994.
15. *N & Q*, 8 ser. 9: 396, 1896.
16. Whistler, 1908: 89.
17. IFCSS MSS 812: 110.
18. Raynes Park, London, August 2015.
19. Rushmere St Andrew, Suffolk, February 1989.
20. Thornton Heath, Surrey, August 2015.
21. Streatham, London, October 1991.
22. IFCSS MSS 212: 10, Co. Leitrim.
23. IFCSS MSS 790: 34, Co. Dublin.
24. Glynn, Co. Antrim, February 1992.
25. Kensington, London, March 2007.
26. Natural History Museum, London, October 2013.
27. Parsons MSS, 1952.
28. Email, July 2016.

PRIMROSE
1. It seems likely that Macmillan, 1922: 45, was mistaken when recording this name, and it refers to a form with a double corolla (petals) rather than a double calyx (sepals).
2. *N & Q*, 1 ser. 7: 201, 1853.
3. Cappamore, Co. Limerick, October 1984.
4. Bath, Avon, December 2003.
5. Evans, 1971: 68.
6. *N & Q*, 1 ser. 7: 201, 1853.
7. Hole, 1937: 47.
8. Metheringham, Lincolnshire, April 1994.
9. Garrad, 1984: 76.
10. IFCSS MSS 700: 186.
11. IFCSS MSS 1112: 441.
12. Ballymote, Co. Sligo, May 1994.
13. Burton Agnes, Humberside, November 1988.

14. Pers. obs., May 2017, when the tomb of St John was decorated with local flowers including cow parsley and perennial cornflower (*Centaurea montana*). During the Festival of St John of Beverley service held in the Minster, posies of flowers from Harpham were placed on the tomb. Although these flowers were described as 'wild' they included forget-me-not and lily of the valley, both of which were probably gathered from gardens.
15. Friend, 1884: 7.
16. www.plant-lore.com/news/primrose-day-revived, accessed March 2017. On 19 April 2017 neither Disraeli's grave in the churchyard of St Michael and All Angels, Hughenden, nor his memorial inside the church, was decorated with primroses, though a basket of primrose plants had been placed near a statuette of him in Hughenden Manor; the Disraeli statue in Parliament Square, Westminster, was undecorated.
17. Britten, 1869: 122.
18. Streatham, London, April 1991.
19. Christy, 1928.
20. Copper, 1971: 219.
21. Karpeles, 1987: 64.
22. IFCSS MSS 740: 293.
23. Corscombe, Dorset, March 1975.
24. Stowmarket, Suffolk, February 1989.
25. Lenamore, Co. Longford, April 1991.
26. Storrington, West Sussex, October 2015.

PRIMROSE-PEERLESS
1. Churchill, Avon, June 1978.
2. Tongue, 1965: 202.

PRIVET
1. Ealing, London, May 1982.
2. Swaythling, Hampshire, April 2009.
3. Rolph, 1980: 56.
4. Tooting Common, London, July 2003.
5. IFCSS MSS 775: 83.
6. Macpherson MSS.

PUMPKIN
1. Philip, 1989: 11.
2. Pers. obs., Uttoxeter, Staffordshire, and London, October 1991, and elsewhere during subsequent years.
3. *Oxford Guardian* (Didcot edition), 30 October 2014.
4. Jack Orrell, chairman of the Greyhound Pumpkin Club, November 1990; in December 2009 Jim McDonald of the Grey Hound reported that the Club no longer existed.
5. *Chard and Ilminster News*, 19 October 1994; this event ceased in about 2001 [Barrington, Somerset, December 2009].
6. *Hexham Courant*, 13 September 2002.
7. According to an earlier report in *The Times*, on 22 January 2013, the seed cost £200.
8. *The Garden*, June 2016.
9. *Ibid.*, December 2016.
10. Pers. obs.

PURPLE MOOR-GRASS
1. Lightfoot, 1777: 96.
2. Llandrindod Wells, Powys, September 1991.

QUAKING GRASS
1. Gutch, 1901: 61.
2. Porter, 1969: 45.
3. Dinneen, 1927: 436.

RADISH
1. *N & Q*, 5 ser. 8: 248, 1877.
2. Wilson, 1940.
3. Curwen, 1898: 36.
4. C.H. Bagot, Agent, Levens Hall Estate, November 1993.
5. *N & Q*, 1 ser. 5: 610, 1852.
6. *N & Q*, 5 ser. 8: 355, 1877.

RAGWORT
1. Robert Burns, 'Address to the Deil', verse 9, 1785.
2. Henderson, 1856: 59.
3. Ballymote, Co. Sligo, May 1994.
4. Natural History Museum, London, December 1995.
5. Manx Museum, June 1994.
6. Letter from Upholland, Lancashire, *Daily Mirror*, 21 September 1984.
7. Evesham. Worcestershire, January 1982,
8. Bennett, 1991: 59.

RAMSONS
1. e.g. Phillips, 1983: 22; Mabey, 2012: 308.
2. e.g. Balham, London, April 2014.
3. Upminster, Essex, April 2011; although being a keen botanist interested in all aspects of natural history, the informant, Mary Smith, does not appear to have personally sampled ramsons buds.
4. Aubrey, 1847: 51.
5. Contributed to the Manx Folklife Survey, October 1991.
6. Newcastle-on-Clun, Shropshire, November 2004.
7. London, May 2011.
8. Tebay, Cumbria, January 2017.

RASPBERRY
1. McKelvie MSS, 1963: 273.
2. Cliftonville, Kent, January 2012; also Ballymote, Co. Sligo, May 1994.
3. Horniman Museum, London, September 2013.
4. Chapelthorpe, Selkirkshire, April 2002.
5. Steele MSS, 1978: 81.

RED AND WHITE FLOWERS
1. Opie & Tatem, 1989: 164.
2. Penicuik, Midlothian, April 1982.
3. Notting Hill Gate, London, May 1985.
4. Email, January 2013.
5. Upminster, Essex, July 2016.
6. Catford, London, September 1984; and pers. obs. elsewhere since.
7. Fleet Street, London, August 1983.

RED CAMPION
1. Britten & Holland, 1886: 342.
2. Murray, 1937: 180.
3. Workington, Cumbria, May 2015.
4. Garrad, 1984: 79.
5. Bow Street, Dyfed, March 1984.
6. Garth, Gwynedd, April 1984.

RED CLOVER
1. Christchurch, Dorset, June 1991.
2. Stoke, Plymouth, January 1993.
3. Email, April 2016
4. Nelson, 1990: 34.
5. Trueman *et al.* 1955: 139.

RED DEADNETTLE
1. Penge, London, September 2012.
2. IFCSS MSS 717: 352.

REDSHANK
1. Cameron, 1883: 61.
2. Davey, 1909: 389.
3. IFCSS MSS 375: 90.
4. Friend, 1884: 6.
5. Marquand, 1906: 42.
6. London, N1, February 1997.
7. Llanbadarn, Ceredigion, January 2017.

RED VALERIAN
1. Vickery, 2010: 114–15.
2. Portland, Dorset, April 1991.

REED
1. Milliken & Bridgewater, 2004: 87.
2. Wyse Jackson, 2014: 436.
3. Milliken & Bridgewater, 2004: 85.
4. www.telegraph.co.uk/news/uknews/486639/Reed-beds-reach-end-of-an-era-html/, accessed November 2017.
5. Beckett & Bull, 1999: 253.
6. *Journal of Botany* 9: 149, 1871.
7. Greville, 1824: 17.
8. Letchworth, Hertfordshire, May 2001.
9. Faversham, Kent, July 2015.

RESTHARROW
1. Pratt, 1857. 2: 37.
2. Marquand, 1906: 46.

RHUBARB
1. Horsted Keynes, West Sussex, February 1991.
2. Boyes, 1991: 44.
3. Llanuwchllyn, Gwynedd, April 1991.
4. Aultvaich, Inverness-shire, February 2012.

RIBWORT PLANTAIN
1. Scott & Palmer, 1987: 274.
2. Plymouth, Devon, May 1986.
3. Lenamore, Co. Longford, April 1991.
4. Limpsfield, Surrey, June 1993.
5. Wormwood Scrubs, London, June 2009.
6. Opie, 1969: 226–7.
7. Ringwood, Hampshire, November 1990.

8. Edinburgh, December 1991.
9. Hamworthy, Dorset, May 1991. It appears that this game is played wherever ribwort plantain grows, having been recorded from Croatia [East Ham, London, September 1914], New Zealand [Theberton, South Australia, March 1913] and Michigan, USA [Natural History Museum, London, July 2002].
10. Folklore Society, London, June 2010.
11. Woodbridge, Suffolk, July 2011.
12. Stewart, 1987: 96.
13. Wormwood Scrubs, London, June 2009.
14. IFCSS MSS 450: 55, Co. Kerry.
15. Daingean, Co. Offaly, January 1983.
16. New Cross Gate, London, October 2013; also recorded from North America [Lichfield, Staffordshire, September 2014].
17. Nixon, 1977: 181.

ROCK SAMPHIRE
1. Britten & Holland, 1886: 423.
2. Grigson, 1987: 216.
3. Parslow & Bennalick, 2017: 418.
4. McNeill, 1910: 128.
5. IFCSS MSS 1121: 425.

ROSE
1. *City Recorder*, 9 April 1987.
2. *Op. cit.*, 16 April 1987.
3. Letter from Tunley, Avon, *The Times*, 14 April 1989; 'many' was undoubtedly an exaggeration.
4. Letter from Bristol, Avon, *The Times*, 22 April 1989.
5. *The Times*, 1 August 1990.
6. Pers. obs.
7. Pers. obs.
8. Edition.cnn.com/2017/12/14/europe/grenfell-tower-memorial-intl/index.html, accessed December 2017. The *Daily Telegraph* of 20 December 2017 carried a photograph of the German Chancellor Angela Merkel stepping through white roses placed outside the Kaiser Wilhelm Memorial Church in Berlin, at the opening of a memorial to victims of a terrorist attack on a Christmas market the previous year.
9. *Daily Mail*, 14 February 1991.
10. *Daily Telegraph*, 12 February 1994.
11. Pers. obs., London, 2012–17.
12. Pers. obs., 20 June 1982 and 22 June 1986; J.R. Kell, Cathedral Warden, October 1977, and Eileen M. Mickleburgh, Parish Worker/Almoner, October 1990.
13. Pers. obs., 24 June 2017.
14. Hampstead, London, October 1993.
15. Roud, 2006: 221.
16. Prudhoe, Northumberland, March 2002.
17. *Daily Telegraph*, 25 March 1995.
18. *City of London Cemetery & Crematorium Newsletter*, 7: 15, 2002.

ROSEBAY WILLOWHERB
1. Curtis, 1798.2: [1].
2. Lees, 1867: 24.
3. Preston *et al.*, 2002: 418.
4. Corbridge, Northumberland, December 1995.
5. Email, December 2015.
6. Sisland, Norfolk, May 1999.
7. https://en.wikipedia.org/wiki/County_flowers_of_the_United_Kingdom; accessed March 2017.
8. Harrogate, North Yorkshire, October 1998.
9. East Bridgford, Nottinghamshire, March 1982.
10. Heatonmoor, Greater Manchester, October 1984.
11. Skipton, North Yorkshire, November 1991.
12. For example, Mabey, 2012: 204.
13. e.g. Magadan, Russia [Sutton, Surrey, September 2014].
14. Phillips, 1983: 77.
15. West Stow, Suffolk, November 1992.

ROSEMARY
1. Brown, 1971: 268.
2. Streatham, London, July 1992.
3. Streatham, London, April 1993.
4. Charmouth, Dorset, January 1994.
5. *N & Q* 5 ser. 11: 18, 1879.
6. Taylor MSS, Norwich, Norfolk.
7. L'Ancresse, Guernsey, April 1984.

8. Wicken, Cambridgeshire, April 1994.
9. Natural History Museum, London, December 1991.
10. Burton-on-Trent, Staffordshire, August 2003.
11. Stratford-upon-Avon, Warwickshire, February 1994.
12. Pers. obs., 23 April 1994.
13. Pers. obs., 23 April 2016.
14. Sutton, Surrey, August 1993.
15. Eardiston, Worcestershire, April 2012.

ROWAN
1. Aubrey, 1847: 56.
2. Lightfoot, 1777: 257.
3. Armstrong, 1976: 36.
4. Ashreigney, Devon, July 1983.
5. Dromsally, Co. Limerick, October 1984.
6. Aldbury, Hertfordshire, February 1998.
7. Sandiway, Cheshire, October 2004.
8. Ryan, 1993: 8.
9. Duncan, 1896: 182.
10. IFCSS MSS 50: 323.
11. ICSSS MSS 1000: 156.
12. Clark, 1882: 81.
13. Rhys, 1901: 85.
14. Stewart, 1823: 91.
15. Grigson, 1987: 172.
16. Worthing, West Sussex, March 1982.
17. Nelson, Lancashire, August 1983.
18. Wyse Jackson, 2014: 550.
19. Howkins, 1996: 8.
20. Coles, 1657: 305.
21. Britten & Holland, 1886: 112.
22. Moloney, 1919: 22.
23. Bromley, Kent, April 1991.
24. Barnstaple, Devon, October 1993.

ROYAL FERN
1. Britten, 1881: 178.
2. IFCSS MSS 50: 458.
3. IFCSS MSS 593: 43.

RUE
1. Fosbroke, 1821: 74.
2. Leather, 1912: 115.
3. Parsons MSS, 1952.
4. Steele MSS, 1978: 88.
5. Rolleston-on-Dove, Staffordshire, February 1998.
6. Marple Bridge, Cheshire, March 1998.

RUNNER BEAN
1. Laycock, 1940: 115.
2. Thorncombe, Dorset, March 1975.
3. Bexhill-on-Sea, East Sussex, February 1991.
4. October 1993.
5. Knaresborough, North Yorkshire, July 1997.
6. K. Palmer, 1976: 102.
7. SLF MSS, Havercroft, West Yorkshire, January 1969.
8. Wharton MSS, 1974: 35.

RUSH
1. O'Sullivan, 1977: 113.
2. IFCSS MSS 770: 143, Co. Longford.
3. O'Sullivan, 1973: 71.
4. Pers. obs., Dublin, May 1993.
5. Attwater, 1970: 75.
6. IFCSS MSS 975: 148, Co. Cavan.
7. Wyse Jackson, 2014: 365; see also O'Dowd, 2015, for an account of objects made from rushes in the National Museum of Ireland.
8. Evans, 1800: 55, 75 & 115.
9. IFCSS MSS 175: 250, Co. Sligo.
10. Chatfield, 2008.
11. Wyse Jackon, 2014: 369.
12. Simpson, 1931: 6.
13. *Ibid.*: 9.
14. Anon., n.d. (3).
15. Shuel, 1985: 86; pers. obs. 11 July 2015.
16. Hutton, 1996: 325.
17. Pers. obs. 3–4 September 2016, and 2016 Sowerby Bridge Rushbearing programme.
18. Sidmouth, Devon, October 1991.
19. Larne, Co. Antrim, November 1991.
20. Scalloway, Shetland, February 1994.
21. Clonlara, Co. Clare, September 1997.
22. Appleshaw, Hampshire, January 2014.
23. O'Dowd, 2015: 464.

RYE GRASS
1. Llantysilio, Denbyshire, February 1998; also recorded from Herefordshire [Leather, 1912: 63] and Sussex [Candlin, 1947: 131].
2. Cottram MSS, 1989: 27.

3. Robertsbridge, East Sussex, February 2003.

SAFFRON
1. Hirsh, 2009.
2. Townshed, 1908: 108
3. Wright, 1936: 74.

SAGE
1. Knight, 1945: 94
2. Chope, 1935: 132.
3. Friend, 1884: 8.
4. Plymouth, Devon, 1993.
5. Stockport, Greater Manchester, April 1991.
6. Barnstaple, Devon, May 1991.
7. North Litchfield, Hampshire, October 1993.
8. Leckhampton, Gloucestershire, September 1996.
9. Llanuwchllyn, Gwynedd, April 1991.
10. Fareham, Hampshire, November 1996.
11. Pocklington, York, December 1996.
12. Natural History Museum, London, July 2003.

ST FRANKIN'S DAYS
1. Amery, 1895: 120.
2. Amery, 1907: 108.

ST JOHN'S WORT
1. Robson, 1977: 293.
2. Douie & Farmer, 1962: 121.
3. Briggs, 1976: 346.
4. *The Times*, 12 March 1998; in rural Slovenia, people collect armfuls of St John's wort in summer, 'for drying to use in teas in winter to stave off the blues' [South Norwood, London, October 2011].
5. Lightfoot, 1777: 417.
6. Carmichael, 1900: 103.
7. *Ibid.*: 96.
8. Cinderford, Gloucestershire, November 1993.
9. Hole, 1977: 123.
10. Stowe, 1987: 193.
11. Deane & Shaw, 1975: 177.
12. Noall, 1977: 10.
13. Owen, 1978: 111.
14. St Clair, 1971: 41.
15. Wright, 1940: 16.
16. Trevelyan, 1909: 252.
17. Banks, 1941: 25.
18. Tongue, 1965: 35.

SCARLET PIMPERNEL
1. Barnstaple, Devon, May 1991.

SCOTS LOVAGE
1. Lightfoot, 1777: 160.

SCOTS PINE
1. Watts, 1989.
2. Briggs, 1974: 123.
3. Greville, 1824: 204.

SCURVY-GRASS
1. Brand, 1701: 80.

SEA BEANS
1. Nelson, 2000a.
2. Guppy, 1917: 33. According to 'local lore' in Madeira *Entada gigas* seeds known as *Castanha de Colombo*, chestnut of Colombus, 'led Colombus to understand the currents of the sea and discover America' [Porto Santo, Madeira, May 1998].
3. Nelson, 1978: 107.
4. St Mary's, Isles of Scilly, September 1992.
5. Carew, 1602: 27.
6. The Museum is currently (August 2017) closed.
7. Robinson, 1975: 58.
8. Lloyd, 1945: 307.
9. Pers. obs., Weymouth, Dorset, 12p each. July 1984; Scarborough, North Yorkshire, 35p, March 2002; St Ives, Cornwall, 50p, April 2007; Brixham, Devon, 79p, November 2018; Ambleside, Cumbria, 75p, July 2015.
10. Hemsley, 1892: 371.
11. Carmichael, 1928: 225.
12. *Proceedings of the Society of Antiquaries of Scotland* 27: 47, 1893.
13. Martin, 1703: 38.
14. *Ibid.*

SEA BEET
1. Bromfield, 1856: 421.
2. Woodnewton, Northamptonshire, June 1992.
3. Upminster, Essex, March 2011.

SEA BINDWEED
1. Fairweather, n.d.: 3.

SEA CAMPION
1. Edinburgh, December 1991.
2. St Saviour, Jersey, May 1993.
3. St Martin, Guernsey, April 2002.

SELFHEAL
1. Fowler, 1891: 193.
2. McNeill, 1910: 158.
3. IFCSS MSS 925: 7, Co. Wicklow.
4. IFCSS MSS 800: 53, Co. Offaly.

SERVICE-TREE
1. *Phytologist*, OS 4: 1102, 1853.
2. Maskew, 2014: 220.

SHALLOT
1. St Osyth, Essex, February 1989.
2. Thorncombe, Dorset, March 1975.
3. Bexhill-on-Sea, East Sussex, February 1991.

SHAMROCK
1. Colgan, 1896: 216.
2. Quoted in Colgan, *ibid.*: 217.
3. *Ibid.*: 218.
4. Gerard, 1597: 1017.
5. L'Obel, 1570: 380; translation from Colgan, 1896: 214.
6. Nelson, 1991: 34.
7. Colgan, 1896: 355.
8. Frazer, 1894: 135
9. See Seaby, 1970, for illustrations.
10. Colgan, 1896: 349.
11. Danaher, 1972: 62.
12. Colgan, 1896: 351.
13. Threlkeld, 1726: 160.
14. Colgan, 1892: 96.
15. Girton, Cambridge, March 1993.
16. Ashfield, New South Wales, Australia, April 1979.
17. Synott, 1979: 39.
18. *Ibid.*
19. Bicheno, 1831.
20. Britten & Holland, 1886: 597.
21. Colgan, 1896: 219.
22. Pers. obs., Dublin, February 1992.
23. William T Gillis, Michigan State University, March 1978.
24. Murphy, 1965: chap. 11.
25. Lightfoot, 1777: 405.
26. Colgan, 1893.
27. *Ibid.*
28. Britten & Holland, 1886: 425.
29. Pers. obs.
30. Pers. obs., 2012–17.
31. Pers. obs., April 2017.
32. Nelson, 1990.
33. Danaher, 1972: 64.
34. Synott, 1979: 39.
35. Daniel McCarthy, Instructor in Horticulture, Kanturk, Co. Cork, February 1980.
36. Pers. obs., 1995, and numerous press articles since. In 1995 lesser trefoil was used as shamrock, and examination of press photographs suggest that this is always the species used.

SHEEP'S BIT
1. Truro, Cornwall, December 1993.

SHEPHERD'S PURSE
1. Fowler, 1909: 302.
2. Johnston, 1853: 37.
3. Britten, 1878: 159; Britten was born in 1846.
4. Stevenage, Hertfordshire, May 1982.
5. Andreas, Isle of Man, May 1963 [Manx Folklife Survey].
6. Ipswich, Suffolk, August 2011.
7. Brockwell Park, London, February 2010.
8. Andover, Hampshire, December 2013; in 1995 shepherd's purse was being investigated in Nigeria for the treatment of high blood pressure [Geffyre Museum, London, August 1995].

SILVERWEED
1. Britten & Holland, 1886: 334.
2. Ray, 1670: 27.
3. Pratt, 1857, 1: 31.
4. McNeill, 1910: 119.
5. Carmichael, 1941: 119.
6. Pratt, 1857, 1: 32.

SKULLCAP
1. *Plants, People, Places* 2, June 1993.

SLENDER SPEEDWELL
1. Sinker *et al.*, 1985: 256.

SNOWDROP
1. *N & Q*, 160: 100, 1931.
2. Wiswell, Lancashire, April 1982; this story alerted the author to the wealth of plant-lore which could still be collected, thus stimulating an interest of which this book is a result.
3. Tooting, London, November 1999.
4. Helsington, Cumbria, January 2013.
5. Tregaer, Monmouthshire, October 2013.

SOAPWORT
1. Bailey, 1966: 19.
2. Moloney, 1919: 16.
3. Vesey-FitzGerald, 1944: 28.
4. St Saviour, Jersey, May 1993.

SOLOMON'S SEAL
1. Vesey-FitzGerald, 1944: 28.
2. Applethwaite, Cumbria, September 1996.

SORREL
1. Sidmouth, Devon, October 1991.
2. Email, September 2013.
3. Bromley, Kent, April 1991; London, SE1, February 2016.
4. Old Cleeve, Somerset, October 1993.

SOUTHERNWOOD
1. Porter, 1969: 2.
2. Histon, Cambridgeshire, January 1989.

SOWTHISTLE
1. IFCSS MSS 717: 217.
2. Natural History Museum, London, May 2003; cure also recorded from rural Victoria, Australia [Imperial College London, June 2015].

SPEAR THISTLE
1. Ballycastle, Co. Antrim, January 1991.
2. West Ealing, London, November 1991.

SPIGNEL
1. Edmondson, 1994: 46.
2. Halliday, 1997: 339.

SPINDLE
1. Grigson, 1987: 120.
2. Mabey, 1996: 244.
3. Archer Briggs, 1880: 78.
4. Grigson, 1987: 120.
5. I am indebted to J.B. Smith, of Bath, for this information; I have not been able to find the relevant passage in Evelyn's works.
6. Wyatt, [?1933]: 81.

SPRINGBEAUTY
1. Woodnewton, Northamptonshire, November 1976.
2. Pers. obs., 2012–17.

SPURGE LAUREL
1. Bromfield, 1856: 437.
2. Pratt, 1857, 1: 46.

STAG'S-HORN SUMACH
1. Great Barr, West Midlands, October 1982.

STAR OF BETHLEHEM
1. Felmersham, Bedfordshire, April 1993.

STINKING IRIS
1. Gerard, 1597: 54.

SUNDEW
1. Wilkinson, 1858: 33.
2. Bennett, 1991: 58.
3. Garrad, 1984: 79.
4. McNeil, 1910: 123.

SUN SPURGE
1. Douglas, Isle of Man, November 1988.
2. Spence, 1914: 103.
3. IFCSS MSS 750: 293.
4. Letchworth, Hertfordshire, May 2001.
5. Corbridge, Northumberland, February 1993.

SWEDE
1. Childwall, Liverpool, April 2013.
2. Bristol, November 2010.
3. Knight, 1947: 47.
4. SLF MSS, Aldbrough, Humberside, April 1972.
5. Llanuwchllyn, Gwynedd, April 1991.

SWEET CICELY
1. Halliday, 1997: 335.
2. Gullane, East Lothian, February 1997.
3. Norton Subcourse, Norfolk, July 2003.
4. Garrad, 1984: 75.
5. Natural History Museum, London, May 2003.
6. Pers.obs., Stoney Middleton,

Derbyshire, July 2015; Naylor & Porter, 2002: 96.

SWEET FLAG

1. *Plants, People, Places* 2, June 1993.

SWEET WILLIAM

1. Prior, 1863: 221.
2. Corbridge, Northumberland, January 1993.
3. Belfast, May 1994.

SYCAMORE

1. Mountrath, Co. Laois, September 1977.
2. Johns, 1847: 118.
3. Cooper, 1957: 170.
4. Whitlock, 1976: 163.
5. Postcard posted 13 January 1907 in author's collection.
6. *N & Q* 11 ser. 8: 425, 1913.
7. Letter from Santa Maria, California, USA, *This England,* Summer 2001; see *ibid.*, Winter 2000–1, for a photograph of the trees taken in 1948.
8. Pers. obs., October 2015.
9. *N & Q*, 11 ser. 8: 425, 1913.
10. *St Peter's Church, Tewin, A Visitor's Guide*, purchased October 2015.
11. Jones, L., 1980: 100.
12. Cunnack, 1971: 15.

TAMARISK

1. Marquand, 1906: 39.
2. De Garis, 1975: 41.

TANGERINE

1. Goody, 1993: 388.

TANSY

1. For example, Co. Cork [IFCSS MSS 313: 213]; Lerwick, Shetland, March 1994, and gypsies [Vesey-FitzGerald, 1944: 28].
2. Taylor MSS, East Harling, Norfolk.
3. Lerwick, Shetland, March 1994.
4. Cannock, Staffordshire, April 2012.
5. IFCSS MSS 350: 75, Co. Cork.
6. Porter, 1969: 10.

TEASEL

1. Shaw, P.J.A. & Shackleton, K. 2011. Carnivory in the Teasel *Dipsacus fullonum* – The effect of experimental feeding on growth and seed set. PloS ONE 6(3): e17935. https://doi.org/10.1371/journal.pone.0018935.
2. Ewen & Prime, 1975: 59.
3. Latham, 1878: 45.
4. Breage, Cornwall, October 1993.
5. Ryder, 1969: 117; Stace, 2010: 795.
6. Chard, Somerset, July 2013.

THISTLE

1. IFCSS MSS 750: 296, Co. Longford.
2. Moore, 1968: 369.
3. St Ervan, Cornwall, January 1994.
4. Dickson & Walker, 1981: 18.
5. Ballycastle, Co. Antrim, January 1991.
6. Dickson & Walker, 1981: 1.
7. *Ibid.*
8. *Ibid.*: 2.
9. For example, Martin, 1976: pl. 49; Webster, 1978: 365.
10. Stace, 2010: 695; Preston *et al.*, 2002: 614.
11. Denson, 1832: 356.

THORN-APPLE

1. Bonnard, 1993: 28.

THYME

1. Friend, 1884: 15.
2. Vesey-FitzGerald, 1944: 28.
3. Barnstaple, Devon, August 1992.
4. Sandiway, Cheshire, October 2004.
5. Colwyn Bay, Clwyd, June 1993.

TIMOTHY

1. Llantysilio, Denbighshire, February 1998.
2. Chiswick, London, July 1999.

TOMATO

1. See, for example, *The Times*, 26 August 1993; London *Evening Standard*, 28 August 1997; *Telegraph*, 17 August 2002; *Guardian*, 31 August 2006, and *Metro*, 27 August 2015.
2. *Europa* (Air Europa in-flight magazine), 143, August 2015.

TORMENTIL

1. Lovett, 1913: 121.
2. Johnston, 1853: 72.
3. Maloney, 1972: 74.
4. Barrington, 1984: 103.
5. Lerwick, Shetland, March 1994.

6. Marquand, 1906: 40.
7. Miller, 1858: 17.
8. McNeill, 1910: 118.
9. Nelson & McCracken, 1987: 11.
10. Evans & Laughlin, 1971: 85.

TRAVELLER'S JOY
1. Norman, 1969: 34.
2. Dorking, Surrey, November 2009.
3. Laver, 1990: 236.
4. Bloom, 1930: 245.

TREE MALLOW
1. Davey, 1909: 91.
2. Portesham, Dorset, January 2007.
3. St Mary's, Isles of Scilly, September 1992.
4. Le Sueur, 1984: 96.

TUFTED HAIR-GRASS
1. Teulon-Porter, 1956: 90.

TULIP
1. Akers, 2006: 7.

TURNIP
1. Daingean, Co. Offaly, January 1985.
2. Thorncombe, Dorset, September 1977.
3. IFCSS MSS 950: 248, Co. Monaghan.
4. Corbridge, Northumberland, January 1980.
5. Walthamstow, London, October 2015.
6. Lerwick, Shetland, March 1994.
7. West Stow, Suffolk, January 1991.
8. Email, December 2012.

TUTSAN
1. Boase, 1976: 114.
2. Grigson, 1987: 75.
3. Britten & Holland, 1886: 473.
4. Grigson, 1987: 75.
5. Allen & Hatfield, 2004: 103.
6. IFCSS MSS 200: 75, Co. Leitrim.

VALERIAN
1. Birdham, West Sussex, July 1993.
2. Greenway, Gloucestershire, September 1993.

VERVAIN
1. Pratt, 1857, 2: 184.
2. Latham, 1878: 38.

VIOLET
1. Towcester, Northamptonshire, August 1882.
2. Langtoft, Humberside, March 1885.
3. Plymouth, Devon, January 1993.
4. Farnham, Surrey, December 1985.
5. Tooting, London, December 2017.
6. IFCSS MSS 500: 74, Co. Limerick.
7. Llanellen, Monmouthshire, October 2013.
8. Grieve, 1931: 839.
9. Vesey-FitzGerald, 1944: 28.
10. UCL EFS MSS M9, Bromley, Kent, September 1963.

WALL BARLEY
1. Bristol, January 2013.
2. Totton, Hampshire, August 1993.
3. Lichfield, Staffordshire, September 2014.

WALNUT
1. Old Cleeve, Somerset, October 1993.
2. For example, Hilltop Garden Club, Eastcombe, Gloucestershire, November 2013.
3. Great Bedwyn, Wiltshire, May 2002; however, Joseph Wright in his *English Dialect Dictionary* (1898–1905) lists 'splash' as a widely used verb, meaning 'to beat or knock down walnuts, etc., with a pole', making no mention of May Day splashing.
4. Norton Subcourse, Norfolk, July 2002.
5. Correspondence August and September 2002.
6. Flowerdew, 1998: 133.
7. Eastwood, Essex. September 2004.
8. St Marychurch, Devon, August 2011.
9. Corbridge, Northumberland, March 1993.
10. Withall, 1978: 44.
11. Chelsea, London, May 2003.

WATERCRESS
1. See p. 688 in this volume.
2. Lenamore, Co. Longford, April 1991.
3. Thorncombe, Dorset, *c.*1962.
4. Shepherdswell, Kent, November 1979.
5. Barnstaple, Devon, May 1991.
6. St Albans, Hertfordshire, February 2004.

7. IFCSS MSS 98: 347, Co. Mayo.
8. ICFSS MSS 770: 63, Co. Longford.
9. IFCSS MSS 975: 27, Co. Cavan.
10. SSS MSS SA1963.32.A9.
11. Knockmult, Co. Londonderry, April 2002.

WATER-FIGWORT
1. Friend, 1882: 18.
2. Hart, 1898: 385.
3. Davey, 1909: 325.
4. Little Waltham, Essex, January 1978.

WELL-DRESSING
1. Hole, 1976: 212.
2. Porteous, 1973: 5.
3. *Ibid*, 1973: 1.
4. Vickery, 1975.
5. Womack, 1977: 26.
6. Vickery, 1975: 178.
7. Porteous, 1973: 10.
8. *Ibid.*: 16.
9. Meade-Waldo, 1902: 1; presumably the author meant alder fruits instead of 'elder catkins'.
10. Womack, 1977: 30.
11. Christian, 1991: 7.
12. Naylor & Porter, 2002: 105.
13. Smith, 2011: 7.
14. Buxton Well Dressing Festival, 87th Edition Souvenir Programme, 2013: 17.
15. New Southgate, London, May 1989.
16. Pers. obs., 24 May 2015; for other, some only one-off, well-dressings outside the Peak District, see Naylor & Porter, 2002: 76–85.
17. Hults, 1987.

WHEAT
1. Wormshill, Kent, May 2003.
2. *N & Q*, 7 ser. 4: 485, 1887.
3. Udal, 1922: 257.
4. Mary Vacca, Church Warden, St Margaret of Antioch Church, Sea Palling, Norfolk, April 2017.
5. Tull, 1976: 22.
6. Pers. obs., 7 January 2017.
7. Youngman, 1951: 423.
8. Hornby, Lancashire, November 1992.
9. Gwyndaf, 1989: 63.
10. Felmersham, Bedfordshire, April 1993.
11. O'Dowd, 2015: 98–133.
12. Estyn Evans, 1957: 286.
13. O'Dowd, 2015: 116.
14. National Botanic Garden, Glasnevin, Dublin, August 2014.
15. Frampton, 1989.
16. The 38th Whittlesea Straw Bear Festival Souvenir Programme, 2017.
17. Pers. obs., 14 January 2017.
18. Pers. obs., 2 September 2017; Hannant, n.d.: 96.

WHITEBEAM
1. See Rich *et al.*, 2010.
2. Letter from Drumshanbo, Co. Leitrim, *Ireland's Own*, 19 March 1993.
3. Wormshill, Kent, May 2003.

WHITE BRYONY
1. Threlkeld, 1726: 29.
2. *Folk-lore* 24: 240, 1913.
3. Porter, 1969: 46.
4. West Stow, Suffolk, March 1989.
5. Charlbury, Oxfordshire, January 1991.
6. Pratt, 1857, 2: 70.

WHITE CLOVER
1. Edinburgh, October 1991.
2. Brockwell Park, London, October 2016; practice also recorded from New Zealand [Tooting Common, London, August 2011].
3. Ballycastle, Co. Antrim, January 1991.

WHITE DEADNETTLE
1. Thomson, 2009: 13.
2. Lichfield, Staffordshire, February 2015.
3. Edinburgh, November 2015.
4. Johnston, 1853: 163.
5. Curtis, 1798, 2: text to plate 115.
6. Parsons MSS, 1952.
7. Tower Hamlets Cemetery Park, London, May 2016.
8. Brompton Cemetery, London, June 2010; practice also recorded from Belgium, Germany, Latvia and Poland [visitors to the Natural History Museum's Wildlife Garden, 2011–4].

WHITE FLOWERS
1. Wharton MSS, 1974: 34.
2. Stanton-on-the-Wolds, Nottinghamshire, January 1983.
3. Islington, London, May 1984.

WHITE HEATHER

1. Victoria, 1868: 154; see also Nelson, 2006.
2. Battiscombe, 1969: 36.
3. Victoria, 1884: 197.
4. McClintock, 1970: 159.
5. Waring, 1978: 118.
6. Nelson, 2006: 44.
7. McClintock, 1970: 159.
8. Towcester, Northamptonshire, August 1982.
9. Pers. obs., Balham, London, February 1982.
10. Putney, London, December 2016.
11. McClintock, 1970: 159.
12. Pers. obs., Edinburgh, November 2010, November 2015, and August 2017; and Glasgow, September 2017.

WHITE POPLAR

1. Johns, 1849: 357.

WHITE WATERLILY

1. McNeill, 1910: 97.
2. *Tocher* 36/37: 433, 1982.

WHITLOWGRASS

1. Mordiford, Herefordshire, December 1991.
2. Allen & Hatfield, 2004: 119.
3. Wyse Jackson, 2014: 303.

WILD CARROT

1. Wareham, Dorset, September 2011.
2. Martin, 1984: 124.
3. Goodrich-Freer, 1902: 45.

WILD CLARY

1. Briggs, 1974: 119.

WILD PANSY

1. Wharton MSS, 1974: 185.

WILD SERVICE-TREE

1. Bromfield, 1856: 166.
2. Roper MSS.
3. *Ibid.*

WILLOW

1. Hepper, 1980: 15.
2. Grigson, 1987: 256.
3. Lloyd, 1967: 184.
4. Trevelyan, 1909: 105.
5. Lloyd, 1967: 124.
6. Gunther, 1934: 75.
7. Two Locks, Gwent, March 1993.
8. Muchelney, Somerset, January 2007.
9. Stockwell, 1989: 67.
10. Morris, 1869: 79.

WINTER ACONITE

1. Hitchman, 1975: 22.

WOOD

1. Letter from Five Ashes, Sussex, *The Times*, 1 March 1929.
2. Opie & Tatem, 1989: 449; Roud, 2003: 484.
3. Northampton, January 1991.
4. Worcester Park, Surrey, February 1978.
5. SLF MSS, Sheffield, September 1972.

WOOD ANEMONE

1. Deacon, 1930: 26.

WOOD AVENS

1. Amphlett & Rea, 1909: 120.
2. Vesey-FitzGerald, 1944: 22.

WOOD SAGE

1. Lightfoot, 1777: 303.
2. Riddelsdell *et al.*, 1948: 398.
3. Gronant, Clwyd, April 1994.

WOOD SORREL

1. Bicheno, 1831.
2. Wilkinson, 1858: 54.
3. Pershore, Worcestershire, October 1991.
4. Merthyr Tydfil, Mid Glamorgan, October 2000.
5. Clonlara, Co. Clare, June 2002.

WORMWOOD

1. Ewen & Prime, 1975: 37.
2. Stoke Fleming, Devon, October 1996.
3. Bangor, Gwynedd, March 1993.
4. Gronant, Clwyd, April 1994.

YARROW

1. Driffield, Humberside, March 1985.
2. Email, July 2016.
3. Kinahan, 1884: 90.
4. Britten & Holland, 1886: 533.
5. Morris, 1925: 306.
6. Belfast, February 1991.
7. Britten, 1878: 156.

8. Rudkin, 1936: 26; the name nosebleed has been widely recorded.
9. Cinderford, Gloucestershire, November 1993.
10. Edinburgh, October 1991.
11. Lightfoot, 1777: 497.
12. IFCSS MSS 112: 48.
13. Taylor MSS.
14. Halesowen, West Midlands, October 1990.
15. Bettws, Gwent, February 1991.
16. Chester, Cheshire, July 1996.
17. Newcastle-on-Clun, Shropshire, November 2004.
18. Redcar, North Yorkshire, January 2014.
19. St Mary's, Isles of Scilly, September 1992.

YELLOW BARTSIA

1. Syston, Leicester, January 1991.

YELLOW HORNED-POPPY

1. Stoke, Plymouth, Devon, April 1993.
2. Newton, 1698: 263.

YELLOW IRIS

1. Gorleston, Norfolk, April 1991.
2. Scalloway, Shetland, February 1994.
3. Lerwick, Shetland, March 1994.
4. MacCulloch, 1903: 101.
5. Le Sueur, 1984: 184.
6. Lightfoot, 1777: 86.

YELLOW RATTLE

1. Britten & Holland, 1886: 312; see Matthew 3: 1–3.
2. Davey, 1909: 339.

YEW

1. Danaher, 1972: 68.
2. Daingean, Co. Offaly, January 1985.
3. Lenamore, Co. Longford, May 1991.
4. *The Times*, 3 October 1998.
5. Bevan-Jones, 2002: 32.
6. *Ibid.*: 30.
7. *Ibid.*: 44.
8. K. Palmer, 1976: 56.
9. Jeacock, 1982; Chandler, 1992: 5.
10. Chandler, 1992: 6.
11. Stoke, Devon, April 1993; *Sunday Mirror*, 30 March 1997.
12. Chandler, 1992: 6.
13. Pers. obs., August 2017.
14. NHM MSS; unfortunately and unusually the provenance of this specimen is unknown.
15. Smith, 1969: 12.
16. *Barnet First*, August–September 2003.
17. Bevan-Jones, 2002: 91.
18. http://bromyardhistorysociety.org.uk/newsletter_5.htm, accessed January 2009.
19. Piper, 1896: 141.
20. Tim Hills, Ancient Yews website, January 2009.
21. Wilks, 1972: 101.
22. http://www.rampantscotland.com/know.blknow_fortingall.htm, accessed January 2009.
23. Milner, 1992: 82.
24. *Mail on Sunday*, 1 November 2015; *The Times* and *Scotsman*, 2 November 2015; *Metro*, 3 November 2015; given the complexity of the tree, could it be that the fruit-producing branches are not part of the original, but belong to a separate tree growing amongst the trunks of the original?
25. Church guide book, January 1992.
26. Morton, 1998: 38.
27. Pers. obs., March 2011.
28. *The Times*, 3 January 1991.
29. Wilks, 1972: 131.
30. Honor Oak, London, April 2016.
31. Gerard, 1597: 1188.
32. Dorchester, Dorset, February 1992.
33. Maulden, Bedfordshire, April 1993.
34. Alton, Hampshire, June 1993.
35. *Southern Evening Echo*, 15 October 1993.
36. *The Times*, 12 August 2008; *Metro*, 6 August 2014.

YORKSHIRE FOG

1. Darlington, Co. Durham, May 2017.

YUCCA

1. Poole, Dorset, April 1992.

Bibliography

Published sources

Addison, J. 1985. *The Illustrated Plague* [illegible]

Aitken, J. 1944. *English Diaries of the* [illegible] *Century* [illegible] Penguin Books.

[illegible] (ed.) 2004. *Old Hames* [illegible] West Bretton: Yorkshire [illegible] England Folk Society.

Alexander, K. 2010. The Cotswold [illegible] in need of [illegible] conservation. *Gloucestershire Naturalist* [illegible]

Alexander, M. 2004. *A Surrey Garland*. [illegible] Surrey County [illegible]

Allen, D. E. [illegible] A possible scent [illegible] between [illegible]

Allen, D. E. & Hatfield, G. 2004. *Medicinal Plants in Folk Tradition*. Portland, Oregon: Timber Press.

Amery, P. F. S. 1898. Thirteenth report of the Committee on Devonshire folk-lore. *Report and Transactions of the Devonshire Association for the Advancement of Science* [illegible]

Amery, P. F. S. 1905. Twenty-second report of the Committee on Devonshire folk-lore. *Report and Transactions of the Devonshire Association for the Advancement of Science* [illegible]

Amery, P. F. S. 1907. Twenty-fourth report of the Committee on Devonshire folk-lore. *Report and Transactions of the Devonshire Association for the Advancement of Science* [illegible]

[illegible] & Rea, C. 1990. *The Book of* [illegible]

[illegible] *Here beginneth the* [illegible] & Peg [illegible]

Anon. [illegible] Empire Day, May 24th [illegible]

Anon. [illegible] Notes on Irish folk-lore [illegible]

Anon. [illegible] *Road Topography* [illegible] Britain and [illegible] London: Hartnell & Sons.

[illegible] Thor [illegible]

[illegible] Church, Glaston [illegible]

[illegible] Gabb [illegible]

[illegible] *The Cotswold of* [illegible] Publishing Ltd.

[illegible] *A Guide to Glastonbury* [illegible] Sons.

Anon. [illegible] Eastbourne [illegible] Published 2004)

Arber, A. 1938. *Herbals*, ed. 2, Oxford: Oxford University Press.

Bibliography

Published sources

Addison, J. 1985. *The Illustrated Plant Lore*, London: Sidgwick & Jackson.

Aitken, J. 1944. *English Diaries of the XIX Century, 1800–1850*, Harmondsworth: Penguin Books.

Akers, J. (ed.) 2006. *Old Flames: An Illustrated History of the Perfection of the English Florists' Tulip*, West Bretton: Yorkshire Sculpture Park & Wakefield and North of England Tulip Society.

Alexander, K. 2010. 'The Cotswold ash pollards: A unique heritage in need of active conservation', *Gloucestershire Naturalist* 21: 4–9.

Alexander, M. 2004. *A Surrey Garland*, Newbury: Countryside Books.

Allen, D.E. 1980. 'A possible scent difference between *Crataegus* species', *Watsonia* 13: 119–20.

Allen, D.E. & Hatfield, G. 2004. *Medicinal Plants in Folk Tradition*, Portland, Oregon: Timber Press.

Amery, P.F.S. 1895. Thirteenth report of the Committee on Devonshire folk-lore, *Report and Transactions of the Devonshire Association for the Advancement of Science* 27: 61–74.

Amery, P.F.S. 1905. Twenty-second report of the Committee on Devonshire folk-lore, *Report and Transactions of the Devonshire Association for the Advancement of Science* 37: 111–21.

Amery, P.F.S. 1907. Twenty-fourth report of the Committee on Devonshire folk-lore, *Report and Transactions of the Devonshire Association for the Advancement of Science* 39: 105–9.

Amphlett, J. & Rea, C. 1909. *The Botany of Worcestershire*, Birmingham: Cornish Bros Ltd.

Anon. 1520. *Here begynneth the lyfe of Joseph of Armathia*, London: R. Pynson.

Anon. [?1910]. 'Empire Day, May 24th', Empire Movement leaflet 14, London.

Anon. 1916. 'Notes on Irish folk-lore', *Folk-lore* 27: 419–26.

Anon. 1967. *Royal Pageantry: Customs and Festivals of Great Britain and Northern Ireland*, Paulton: Purnell & Sons.

Anon. 1977. *The Glastonbury Thorn* [leaflet produced in aid of the restoration of St John the Baptist church, Glastonbury].

Anon. 1985. *The Jersey Giant Cabbage*, St Ouen: L'Etacq Woodcrafts.

Anon. n.d. (1). *The Cathedral of Our Lady and St Philip Howard, Arundel*, St Ives: Photo Precision Ltd.

Anon. n.d. (2). *A Guide to Glastonbury and its Abbey*, Glastonbury: E.C. Helliker & Sons.

Anon. n.d. (3). *The Rushbearing in Grasmere*, Cross Hills: Kay Jay Print Ltd. [purchased 2014].

Arber, A. 1938. *Herbals*, ed. 2, Oxford: Oxford University Press.

Archer, F. 1990. *Country Sayings*, Stroud: Alan Sutton.
Archer Briggs, T.R. 1880. *Flora of Plymouth*, London: J. Van Voorst.
Armstrong, S. 1976. *A Croft in Clachan*, London: Hutchinson.
Asberg, M. & Stearn, W.T. 1973. 'Linnaeus's Öland and Gotland Journey 1741', *Biological Journal of the Linnean Society* 5: 1–107.
Atkins, E.A. 1986. *Tales from our Cornish Island*, London: Harrap.
Atkinson, T. 2003. *Napiers History of Herbal Healing, Ancient and Modern*, Edinburgh: Luath Press Ltd.
Attwater, D. 1970. *The Penguin Dictionary of Saints*, Harmondsworth: Penguin Books Ltd.
Aubrey, J. 1847. *The Natural History of Wiltshire*, ed. J. Britton, London: Wiltshire Topographical Society.
Aubrey, J. 1881. *Remaines of Gentilisme and Judaisme*, ed. J. Britten, London: The Folklore Society
Ayres, P. 2015. *Britain's Green Allies: Medicinal Plants in Wartime*, Knibworth Beauchamp: Matador.
Bacon, F. 1631. *Sylva Sylvarum*, 3rd ed, London: J.H. for William Lee.
Bailey, B. 1966. 'Bouncing Bet', *World of Interiors* February 1966: 19.
Baker, A.E. 1854. *Glossary of Northamptonshire Words and Phrases*, vol. 1, London: Russell Smith.
Baker, M. 1974. *Discovering the Folklore and Customs of Love and Marriage*, Princes Risborough: Shire Publications Ltd.
Baker, M. 1977. *Wedding Customs and Folklore*, Newton Abbot: David & Charles.
Baker, M. 1996. *Discovering the Folklore of Plants*, Princes Risborough: Shire Publications Ltd.
Balfour, M.C. 1904. *County Folk-lore*, vol. 4, *Examples of Printed Folk-lore concerning Northumberland*, London: David Nutt for the Folk-lore Society.
Banks, M.M. 1941. *British Calendar Customs: Scotland*, vol. 3, London: Wm Glaisher Ltd for the Folk-lore Society.
Barker, A. 2011. *Remembered Remedies: Scottish Traditional Plant Lore*, Edinburgh, Birlinn Ltd.
Barrett, H. 1967. *Early to Rise*, London: Faber.
Barrington, J. 1984. *Red Sky at Night*, London: Joseph.
Basford, K. 1978. *The Green Man*, Ipswich: Boydell & Brewer.
Batten, E.C. 1881. 'The Holy Thorn of Glastonbury', *Proceedings of the Somerset Archaeological and Natural History Society* 26 ,2: 118–25.
Battiscombe, G. 1969. *Queen Alexandra*, London: Constable.
Bean, W.J. 1914. *Trees and Shrubs Hardy in the British Isles*, London: John Murray.
Beavis, C. 2013. 'Lye – "this unlovely spot"', *The Blackcountryman* 47, 1: 24–5.
Beckett, G. & Bull, A. 1999. *A Flora of Norfolk*, privately published.
Beckwith, L. 1978. *Hebridean Cookbook*, London: Arrow Books.
Bennett, M. 1991. 'Plant lore in Gaelic Scotland', in R.J. Pankhurst & J.M. Mullin, *Flora of the Outer Hebrides.* London: Natural History Museum Publications: 56–60.
Bergamar, K. n.d. *Discovering Hill Figures*, Tring: Shire Publications.
Best, M.R. & Brightman, F.H. (eds) 1973. *The Book of Secrets of Albertus Magnus*, Oxford: Clarendon Press.
Bett, H. 1952. *English Myths and Traditions*, London: B.T. Batsford.
Bevan-Jones, R. 2002. *The Ancient Yew*, Bollington: Windgather Press.
Bicheno, J.E. 1831. 'On the plant intended by the shamrock of Ireland', *Journal of the Royal Institution of Great Britain* 1: 453–8.

Biden, W.D. 1852. *The History and Antiquities of the Ancient and Royal Town of Kingston-upon-Thames*, Kingston: William Lindsey.

Biggar, J. 2000. 'Badges of pride', *Scots Magazine*, April 2000: 380–3.

Bloom, J.H. 1930. *Folk Lore, Old Customs and Superstitions in Shakespeare Land*, London: Mitchell Hughes & Clarke.

Bloxam, C. & Picken, M. 1990. *Love and Marriage*, Exeter: Webb & Bower.

Boardman, J. & Scarisbrick, D. 1977. *The Ralph Harari Collection of Finger Rings*, London: Thames and Hudson.

Boase, W. 1976. *The Folklore of Hampshire and the Isle of Wight*, London: B.T. Batsford Ltd.

Bonnard, B. 1993. *Channel Island Plant Lore*, Guernsey: The Guernsey Press Co. Ltd.

Boulger, G.S. 1906. *Familiar Trees*, London: Cassell.

Bowman, M. 2006. 'The Holy Thorn Ceremony: Revival, rivalry and civil religion in Glastonbury', *Folklore* 117: 123–40.

Box, J. 2003. 'Dressing the Arbor Tree', *Folklore* 114: 13–28.

Boyes, G. 1991. 'Not quite blue: colour in the mock-obscene riddle', in J. Hutchings & J. Wood (eds), *Colour and Appearance in Folklore*, London: Folkore Society: 40–5.

Boys, W. 1792. *Collections for an History of Sandwich in Kent*, Canterbury: Simmons, Kirkby & Jones.

Brand, J. 1701. *A Brief Description of Orkney, Zetland, Pightland-Firth and Caithness*, Edinburgh: George Mosman.

Brand, J. 1813. *Observations on Popular Antiquities*, London: Rivington.

Brand, J. 1853. *Observations on the Antiquities of Great Britain*, revised by Sir H. Ellis, London: Henry G. Bohn.

Briggs, K.M. 1971. *A Dictionary of British Folk-tales*, part B, vol. 2, London: Routledge & Kegan Paul.

Briggs, K.M. 1974. *The Folklore of the Cotswolds*, London: B.T. Batsford Ltd.

Briggs, K.M. 1976. *A Dictionary of Fairies*, London: Allen Lane.

Briggs, K.M. & Tongue, R.L. 1965. *Folktales of England*, London: Routledge & Kegan Paul.

Brimble, L.J.F. 1944. *Flowers in Britain*, London: Macmillan & Co.

Britten, J. 1869. 'Spring flowers', *Hardwicke's Science Gossip*: 122.

Britten, J. 1878. 'Plant-lore notes to Mrs Latham's West Sussex Superstitions', *Folk-lore Record* 1: 155–9.

Britten, J. 1881. *European Ferns*, London: Cassell, Petter, Galpin & Co.

Britten, J. & Holland, R. [1878–]1886. *A Dictionary of English Plant-names*, London: Trubner, for the English Dialect Society.

Bromfield, W.A. 1850. 'Notes and occasional observations on some of the rarer British plants growing wild in Hampshire', *Phytologist* 3: 951–87.

Bromfield, W.A. 1856. *Flora Vectensis*, London: W. Pamplin.

Brown, P. 2016. *The Apple Orchard*, London: Particular Books.

Brown, T. 1952. Forty-ninth report on folk-lore, *Report and Transactions of the Devonshire Association for the Advancement of Science* 84: 296–301.

Brown, T. 1953. Fiftieth report on folk-lore, *Report and Transactions of the Devonshire Association for the Advancement of Science* 85: 217–25.

Brown, T. 1971. 68th report on folklore, *Report and Transactions of the Devonshire Association for the Advancement of Science* 103: 265–71.

Buczacki, S. 2002. *Fauna Britannica*, London: Hamlyn.

Burdy, S. 1792. *Life of the late Rev. Philip Skelton*, Dublin: William Jones for the author.

Burne, C.S. 1883. *Shropshire Folk-lore*, London: Trübner.

C. E., 1951. 'Fragments of Oxfordshire plant-lore', *Oxford and District Folklore Society Annual Record* 3: 11–13.
Calderbank, D.A. 1984. *Canny Leek Growing*, Wimborne: Right Angle Books.
Cameron, J. 1883. *Gaelic Names of Plants*, Edinburgh: William Blackwood and Sons.
Campbell-Culver, M. 2001. *The Origin of Plants*, London: Headline Book Publishing.
Campbell-Preston, F. 2006. *The Rich Spoils of Time*, Wimborne Minster: The Dovecote Press.
Candlin, L.N. 1947. 'Plant lore of Sussex', *Sussex County Magazine* 21: 130–1.
Cann, D. 2012. '*Sorbus devoniensis*: the "mast" of Heligan', *BSBI News* 121: 41–2.
Carew, R. 1602. *The Survey of Cornwall*, London: John Jaggard.
Carmichael, A. 1900. *Carmina Gadelica*, vol. 1, Edinburgh: Norman Macleod.
Carmichael, A. 1928. *Carmina Gadelica*, vol. 2, Edinburgh: Oliver & Boyd.
Carmichael, A. 1941. *Carmina Gadelica*, vol. 4, Edinburgh: Oliver & Boyd.
Carre, F. 1975. *Folklore of Lytchett Matravers Dorset*, St Peter Port: Toucan Press.
Challenger, F. 1955. 'Chemistry – the grand master key', *University of Leeds Review* 4(3): 264–72.
Chamberlain, E. 1990. *29 Inman Road*, London: Virago.
Chandler, J. 1992. 'Old men's fancies: the case of the churchyard yew', *FLS News* 15: 3-6.
Chandler, J. 1993. 'The days of May', *FLS News* 17: 11–12.
Chater, A.O. 2010. *Flora of Cardiganshire*, Aberystwyth: the author.
Chatfield, J, 2008. 'Rush lights', *The SLBI Gazette* ser. 2, 6: 10–11.
Child, F.J. (ed.). 1889. *The English and Scottish Popular Ballards*, vol. 3, Boston, Massachusetts: Little, Brown & Company.
Chope, P.R. 1929. Thirtieth report on Devonshire folk-lore, *Report and Transactions of the Devonshire Association for the Advancement of Science* 61: 125–31.
Chope, P.R. 1931. Thirty-first report on Devonshire folk-lore, *Report and Transactions of the Devonshire Association for the Advancement of Science* 63: 123–35.
Chope, P.R. 1932. Thirty-second report on Devonshire folk-lore, *Report and Transactions of the Devonshire Association for the Advancement of Science* 64: 153–68.
Chope, P.R. 1933. Thirty-third report on Devonshire folk-lore, *Report and Transactions of the Devonshire Association for the Advancement of Science* 65: 121–7.
Chope, P.R. 1935. Thirty-fifth report on Devonshire folk-lore, *Report and Transactions of the Devonshire Association for the Advancement of Science* 67: 131–44.
Chope, P.R. 1938. Devonshire calendar customs. Part II. Fixed festivals, *Report and Transactions of the Devonshire Association for the Advancement of Science* 70: 341–404.
Christensen, K.I. 1992. 'Revision of *Crataegus* sect. *Crataegus* and Nothosection Crataeguineae (Rosaceae – Maloideae) in the Old World', *Systematic Botany Monographs*, 35.
Christian, R. 1991. *Well-Dressing in Derbyshire*, Derby: Derbyshire Countryside Ltd.
Christy, M. 1928. 'On the variability and instability of coloration in the flowers of the primrose (*Primula vulgaris*) and cowslip (*P. veris*)', *Vasculum* 14: 89–94.
Clapham, A.R., Tutin, T.G. & Warburg, E.F. 1962. *Flora of the British Isles*, 2nd ed., Cambridge: Cambridge University Press.
Clark, R. 1882. 'Folk-lore collected in Co. Wexford', *Folk-lore Record* 5: 81–3.
Clifford, S. & King, A. 2006. *England in Particular*, London: Hodder & Stoughton.
Cloves, J. 1993. *The Official Conker Book*, London: Jonathan Cape.
Coles, W. 1657. *Adam in Eden*, London: Nathaniel Brooke.
Colgan, N. 1892. 'The shamrock: an attempt to fix its species', *Irish Naturalist* 1: 95–7.
Colgan, N. 1893. 'The shamrock: a further attempt to fix its species', *Irish Naturalist* 2: 207–11.

Colgan, N, 1896. 'The shamrock in literature: a critical chronology', *Journal of the Royal Society of Antiquaries of Ireland* 26: 211–26, 349–61.

Collinson, J. 1791. *The History and Antiquities of the County of Somerset*, Bath: R. Cruttwell.

Combermere, M. & Knollys, W.W. 1866. *Memories and Correspondence of Field-Marshall Viscount Combermere*, 2, London: Hurst & Blackett.

Conquer, L. 1970. 'Corn-dollies and "trees"', *Folklore* 81: 145–7.

Cooper, R.E. 1971. 'The sycamore tree', *Scottish Forestry* 11 (4): 169–76.

Copper, B. 1971. *A Song for Every Season*, London: William Heinemann.

Cornish, V. [1941]. *Historic Thorn Trees in the British Isles*, London: Country Life.

Court, T. 1967. '"Urt" picking on Exmoor', *Exmoor Review*, 1967: 42–3.

Crompton, G. 2003. *Catalogue of Cambridgeshire Plant Records since 1538*, part 2, Cambridge: the author.

Culpeper, N. 1652. *The English Physician*, London: Peter Cole.

Cunnack, E.M. 1971. *The Helston Furry Dance*, Helston: The Flora Day Association and the Stewards of the Helston Furry Dance.

Curtis, W. 1798. *Flora Londinensis*, London: the author/B. White & Son.

Curwen, J.F. 1899. *Historical Description of Levens Hall*, Kendal: T. Wilson.

Dacombe, M.R. (ed.). 1951. *Dorset Up Along and Down Along*, ed. 3, Dorchester: Dorset Federation of Women's Institutes.

Danaher, K. 1972. *The Year in Ireland*, Cork: Mercier Press.

Darlington, A. & Hirons, M.J.D. 1975. *The Pocket Encyclopaedia of Plant Galls in Colour*, Poole: Blandford Press.

Dartnell, G.E. & Goddard, E.H. 1894. *A Glossary of Words Used in the County of Wiltshire*, London: English Dialect Society.

Darwin, T. 1996. *The Scots Herbal: The Plant Lore of Scotland*, Edinburgh: Mercat Press.

Davey, F.H. 1909. *Flora of Cornwall*, Penryn: F. Chegwidden.

Davies, J. 2000. *Saying it with Flowers: The Story of the Flower Shop*, London: Headline Book Publishing.

Deacon, E. 1930. 'Some quaint customs and superstitions in north Staffordshire and elsewhere', *North Staffordshire Field Club Transactions and Annual Report* 64: 18–32.

Deane, T. & Shaw, T., 1975. *The Folklore of Cornwall*, London: B.T. Batsford Ltd.

De Garis, M. 1975. *Folklore of Guernsey*, St Pierre du Bois: the author.

Denson, J. 1832. 'The thistle of Scotland', *Gardener's Magazine* 8: 335–6.

de Sausmarez, R. 1970. *The Guernsey Lily*, St Peter Port: Guernsey Press Co. Ltd.

Dickson, J.H., 1991. *Wild Plants of Glasgow*, Aberdeen: Aberdeen University Press.

Dickson, J.H. & Walker, A. 1981. 'What is the Scottish Thistle?', *Glasgow Naturalist* 20 (2): 1–21.

Dinneen, P.S. 1927. *An Irish-English Dictionary*, Dublin: Irish Texts Society.

Dixon, D.D. 1890. 'Northumbrian plant names', *Nature Notes* 1: 110–11.

Donald, J. 1973. *Long Crendon: A Short History, part 2, 1800–1914*. Long Crendon: Long Crendon Preservation Society.

Dony, J.G. 1953. *Flora of Bedfordshire*, Luton: The Corporation of Luton Museum & Library.

Douglas, S. 1989. 'The hoodoo of the Hanging Tree', in G. Bennett & P. Smith (eds), *The Questing Beast* [Perspectives on Contemporary Legend, vol. 4], Sheffield: Sheffield Academic Press, 133–43.

Douie, D.L. & Farmer, D. H. 1962. *Magna Vita Sancti Hugonis*, London: Thomas Nelson and Sons Ltd.

Drury, S.M. 1984. 'The use of wild plants as famine foods in eigthteenth century Scotland and Ireland', in R. Vickery (ed.) *Plant-lore Studies*, London: Folklore Society: 43–60.

Duncan, L.L. 1896. 'Fairy beliefs and other folklore notes from County Leitrim', *Folk-lore* 7: 161–83.

Dunsford, M.E. 1978. 79th report on dialect, *Report and Transactions of the Devonshire Association for the Advancement of Science* 110: 208–9.

Dunsford, M.E. 1981. 23rd report of the Folklore Section, *Report and Transactions of the Devonshire Association for the Advancement of Science* 113: 173–6.

E. 1884. 'Oak and Nettle Day in Nottinghamshire', *Folk-lore Journal* 2: 381–2.

Eastwood, J., n.d. *The Mole Race*, Burton Bradstock: Winterbourne Publications.

Edlin, H.L. 1949. *Woodland Crafts in Britain*, London: B.T. Batsford.

Edmondson, J.R. 1994. 'Snuffed out for snuff: *Meum athamanticum* in the Roberts Leyland herbarium', *Naturalist* 119: 45–6.

Edom, G. 2010. *From Sting to Spin: A History of Nettle Fibre*, Bognor Regis: Urtica Books.

Elliott, B. 1984. 'The Victorian Language of Flowers', in R. Vickery (ed.), *Plant-lore Studies*: 61–5.

Elworthy, T.F. 1888. *The West Somerset Word Book*, London: Trubner.

[Emslie, J.P.J.] 1915. 'Scraps of folklore collected by John Philipps Emslie', *Folk-lore* 25: 153–70.

Estyn Evans, E. 1957. *Irish Folk Ways*, London: Routledge & Kegan Paul.

Evans, A.B. 1881. *Leicestershire Words, Phrases and Proverbs*, London: English Dialect Society.

Evans, A.J. 1895. 'The Rollright Stones and their folklore', *Folk-lore* 6: 5–50.

Evans, E.E. & Laughlin, S.J. 1971. 'A County Tyrone tan yard', *Ulster Folklife* 17: 85–7.

Evans, G.E. 1969. 'Aspects of oral tradition', *Folk Life* 7: 5–14.

Evans, G.E. 1971. *The Pattern under the Plough*, London: Faber.

Evans, J. 1800. *A Tour through Part of North Wales in the year 1798, and at other times*, London: J. White.

Evelyn, J., 1664. *Sylva, or, a Discourse on Forest-trees*, London: J. Martyn and J. Allestry.

Evelyn, J. 1706. *Silva, or a Discourse of Forest-trees*, London: for Robert Scott, Richard Chiswell *et al.*

Evershed, H. 1877. 'The Cedars of Lebanon', *Gardeners' Chronicle* ns 7: 39–40.

Ewen, A.H. & Prime, C.T., [1975]. *Ray's Flora of Cambridgeshire*, Hitchin: Wheldon & Wesley Ltd.

Fairweather, B. n.d. *Highland Plant Lore*, Glencoe: Glencoe and North Lorn Folk Museum.

Ffennell, M.C. 1898. 'The shrew ash in Richmond Park', *Folk-lore* 9: 330–6.

Flowerdew, B. 1998. *Bob Flowerdew's Organic Bible*, London: Kyle Cathie.

Foley, W. 1974. *Child in the Forest*, London: British Broadcasting Corporation.

Forby, R. 1830. *The Vocabulary of East Anglia*, London: Nichols.

Fosbroke, T.D. 1821. *Companion to the Wye Tour: Ariconesia, or Archaeological Sketches of Ross and Archenfield*, Ross: W. Farror.

Fowler, W.M.E. 1891. 'Superstitions regarding wild flowers in the Selborne country', *Nature Notes* 2: 193–4.

Fowler, W.M.E. 1909. 'Yorkshire folklore', in T.M. Fallow (ed.), *Memorials of Old Yorkshire*, London: Allen & Sons: 286–305.

Frampton, G. 1989. *Whittlesey Straw Bear*, Peterborough: Cambridgeshire Libraries Publications.

Fraser, A.S. 1973. *The Hills of Home*, London: Routledge & Kegan Paul.
Frazer, W. 1894. 'The shamrock: its history', *Journal of the Royal Society of Antiquaries of Ireland* 24: 132–5.
Freethy, R. 1985. *From Agar to Zenry*, Ramsbury: The Crowood Press.
Friend, H., 1882. *A Glossary of Devonshire Plant-names*, London: English Dialect Society.
Friend, H., 1884. *Flowers and Flower Lore*, London: Sonnerschein & Co.
Gamble, R. 1979. *Chelsea Child*, London: British Broadcasting Coporation.
Gardiner, W. 1843. 'Contributions towards a flora of the Breadalbane Mountains', *Phytologist* 1: 468–76.
Garrad, L.S. 1984. 'Some Manx plant-lore', in R. Vickery (ed.), *Plant-lore Studies*, London: Folklore Society: 75–83.
Gent, G. & Wilson, R. 2012. *The Flora of Northamptonshire and the Soke of Peterborough*, Rothwell: Robert Wilson Designs.
Gerard, J. 1597. *The Herball, or Generall Historie of Plants*, London: John Norton.
Gibson, J.W. 1853. 'Seven score superstitious sayings', *N & Q*, 1 ser. 7: 152–3.
Gilmore, L. & Oalcz, C. 1993. *The Blackthorn*, Belfast: the authors.
[Gloucester, Duchess of] 1983. *The Memoirs of Princess Alice, Duchess of Gloucester*, London: Collins.
Gmelch, G. & Kroup, B. 1978. *To Shorten the Road*, Dublin: O'Brien Press.
Gomme, A.B. 1894. *The Traditional Games of England, Scotland and Ireland*, 1, London: Nutt.
Gomme, G.L. (ed.) 1884. *The Gentleman's Magazine Library: Popular Superstitions*, London: Elliot Stock.
Goodrich-Freer, A. 1902. 'More folklore from the Hebrides', *Folk-lore* 13: 29–62.
Goody, J. 1993. *The Culture of Flowers*, Cambridge: Cambridge University Press.
Graves, R. 1948. *The White Goddess*, London: Faber & Faber.
Gregor, W. 1874. *An Echo of the Olden Time from the North of Scotland*, Edinburgh: John Menzies & Co.
Gregor, W. 1881. *Notes on the Folk-lore of the North-east of Scotland*, London: Folk-lore Society.
Gregor, W. 1884. 'Unspoken nettles', *Folk-lore Journal* 2: 377–8.
Gregor, W. 1889. 'Some folk-lore on trees, animals and river-fishing from the north-east of Scotland', *Folk-lore Journal* 7: 41–4.
Greville, R.K. 1824. *Flora Edinensis*, Edinburgh: William Blackwood.
Grieve, Mrs M. 1931. *A Modern Herbal*, London: Jonathan Cape Ltd.
Griffin-Kremer, C. 2011. 'French *Muguet* customs: Recording change', *Folk Life* 49: 75–87.
Griffin-Kremer, C. 2015. 'Tracking change: Lily-of-the-valley custom and festival in France', *Folk Life* 53: 1–18,
Grigson, G. 1987. *The Englishman's Flora*, London: J.M. Dent & Sons.
Gunther, R.W.T. 1934. *The Greek Herbal of Dioscorides*, Oxford: J. Johnson for the author.
Guppy, H.B. 1917. *Plants, Seeds and Currents in the West Indies and Azores*, London: Williams & Norgate.
Gupta, S.M. 1971. *Plant Myths and Traditions in India*, Leiden: E.J. Brill.
Gutch, E. 1901. *County Folk-lore, vol. 2, Printed Extracts concerning the North Riding of Yorkshire, York and the Ainsty*, London: David Nutt.
Gutch, E. 1912. *County Folk-lore, vol. 6, Printed Extracts concerning the East Riding of Yorkshire*, London: David Nutt.
Gwyndaf, R. 1989. Welsh Folk Tales, Cardiff, National Museum of Wales.

Hadfield, M. 1957. *British Trees: A Guide for Everyman*, London: J.M. Dent & Son.
Halliday, G. 1997. *A Flora of Cumbria*, Lancaster: Centre for North-West Regional Studies, University of Lancaster.
Halsband, R. (ed.). 1965. *The Complete Letters of Lady Mary Wortley Montagu*, vol. 1, Oxford: Clarendon Press.
Hammer, J. 2008. *Island of Flowers: A Seasonal Guide to Wild Flowers of the Isle of Wight*, Thorley: Ampersand.
Hannant, S. n.d. *Mummers, Maypoles and Milkmaids*, London: Merrell.
Hardy, J. (ed.) 1892–5. *The Denham Tracts*, 2 vols, London: Folklore Society.
Hardy, T. 1887. *The Woodlanders*, London: Macmillan & Co.
Harris, A. 1992. 'Gorse in the East Riding of Yorkshire', *Folk Life* 30: 17–29.
Hart, H.C. 1873. '*Euphorbia hyberna, Equisetum trachypodon* &c. in Co. Galway', *Journal of Botany* 11: 338–9.
Hart, H.C. 1898. *Flora of County Donegal*, Dublin: Sealy & Co.
Harte, J. 2004. *Explore Fairy Traditions*, Loughborough: Explore Books.
Hatfield, G. 1994. *Country Remedies: Traditional East Anglian Plant Remedies in the Twentieth Century*, Woodbridge: The Boydell Press.
Hatfield, G. 1998. *Warts: Summary of Wart-cure Survey of the Folklore Society*, London: The Folklore Society.
Hatfield, G. 1999. *Memory, Wisdom and Healing: The History of Domestic Plant Medicine*, Stroud: Sutton Publishing.
Heanley, R.M. 1901. 'The Vikings: traces of their folk-lore in Marshland', *Saga Book of the Viking Club* 3: 35–62.
Heather, P.J. 1940. Folk-lore Section, *Papers and Proceedings of the Hampshire Field Club* 14: 402–9.
Helm, A. 1981. *The English Mummers' Play*, Woodbridge: D.C. Brewer.
Hemsley, W.B. 1892. 'A drift-seed (*Ipomoea tuberosa* L.)', *Annals of Botany* 6: 369–72.
Henderson, G. 1856. *The Popular Rhymes, Sayings and Proverbs of the County of Berwick*, Newcastle-on-Tyne: David Wilson.
Hepper, F.N. 1980. *Bible Plants at Kew*, London: HMSO.
Hersom, K. 1973. 'Games with flora', *Countryman*, 78 (3): 79–85.
Hewins, A. 1981. *The Dillen: Memories of a Man of Stratford upon Avon*, Oxford: Oxford University Press.
Hewins, M.E. 1985. *Mary after the Queen: Memories of a Working Girl*, Oxford: Oxford University Press.
Hill, S. 2016. 'The white poppy', *The Friend*, 11 November 2016: 6–7.
Hillaby, J. 1983. *Journey through Britain*, London: Paladin.
Hirsh, J. (rev. M. Evans & M. Starte). 2009. *The Saffron Crocus: A Brief History*, Saffron Walden: Saffron Walden Museum.
Hitchman, J. 1975. *Such a Strange Lady: An Introduction to Dorothy L. Sayers, 1893–1957*, London: New English Library.
Hodson, R. 1917. 'Notes on Staffordshire folklore', *Folk-lore* 28: 452.
Hole, C. 1937. *Traditions and Customs of Cheshire*, London: Williams & Norgate.
Hole, C. 1950. *English Custom and Usage*, 3rd ed., London: B.T. Batsford Ltd.
Hole, C. 1965. *Saints in Folklore*, London: G. Bell & Sons.
Hole, C. 1976. *British Folk Customs*, London: Hutchinson & Co.
Hole, C. 1977. 'Protective symbols in the home', in H.R.E. Davidson (ed.), *Symbols of Power*, Cambridge: D.S. Brewer for the Folklore Society: 121–30.
Holt, J.C. 1983. *Robin Hood*, London: Thames and Hudson.
Howell, J. 1640. *Dodona's Grove*, London: T.B. for H. Mosley.

Howkins, C. 1996. *Rowan: Tree of Protection*, Addlestone: the author.
Howkins, C. 2006. *Poisonous Plants in Britain: A Celebration*, Addlestone: the author.
Howse, W.H. 1949. *Radnorshire*, Hereford: E.J. Thurston.
Hults, D.S. 1987. 'A Derbyshire custom in transition? Well dressing in Perth, Western Australia', *Australian Folklore* 1: 25–43.
Hunt, R. 1881. *Popular Romances of the West of England*, London: Chatto & Windus.
Hunt, R. 1930. *Popular Romances of the West of England*, London: Chatto & Windus.
Hutton, R. 1996. *The Stations of the Sun: A History of the Ritual Year in Britain*, Oxford: Oxford University Press.
Jackson, C. (ed.) 1870. *The Diary of Abraham de la Pryme, the Yorkshire Antiquary*, Durham: Andrews & Co.
Jackson, J.R. 1871. 'Applications of some British Plants', *Journal of Botany* 9: 173.
Jackson, T. 1873. *Recollections of My Own Life and Times*, London: Wesleyan Conference Office.
Jacobs, C.A. 2002. 'The games we played as children', *Norfolk Natterjack*, 38: 3.
James, E.O. 1961 *Seasonal Feasts and Festivals*, London: Thames & Hudson.
Jeacock, R. 1982. *Plants and Trees in Legend, Fact and Fiction*, Chester: Mothers' Union.
Jekyll, G. 1908. *Children and Gardens*, London: Offices of 'Country Life'.
Johns, C.A. 1847. *Forest Trees of Britain*, London: Society for Promoting Christian Knowledge.
Johnston, G. 1853. *The Botany of the Eastern Borders*, London: Van Voorst.
Jones, A.E. 1980. 'Folk medicine in living memory in Wales', *Folk Life* 18: 58–68.
Jones, B.H. 1908. 'Irish folklore from Cavan, Meath, Kerry and Limerick', *Folk-lore* 19: 315–23.
Jones, L. 1980. *Schoolin's Log*, London: Michael Joseph.
Judge, R. 1993. 'Fact and fancy in Tennyson's "May Queen" and in Flora Thompson's "May Day"', in T. Buckland & J. Wood (eds), *Aspects of British Calendar Customs*, Sheffield: Sheffield Academic Press: 167–83.
Judge, R. 2000. *The Jack-in-the-Green*, London: Folklore Society
Karpeles, M. 1987. *The Crystal Spring*, Oxford: Oxford University Press.
Kavanagh, P. 1975. *The Green Fool*, London: Penguin.
Keller, H.A. & Prance, G.T. 2015. 'The ethnobotany of ferns and lycophytes', *Fern Gazette* 20: 1–13.
K'Eogh, J. 1735. *Botanalogia Universalis Hibernica*, Cork: George Harrison.
Keyte, H. & Parrott, A. 1992. *The New Oxford Book of Carols*, Oxford: Oxford University Press.
Killip, M. 1975. *The Folklore of the Isle of Man*, London: B.T. Batsford Ltd.
Kinahan, G.H. 1881. 'Notes on Irish folklore', *Folk-lore Record* 4: 96–125.
Kinahan, G.H. 1884. 'May Eve', *Folk-lore Journal* 2: 90–1.
Kinahan, G.H. 1888. 'Irish plant-lore notes', *Folk-lore Journal* 6: 265–7.
King, R.J. 1877. Second report of the Committee on Devonshire folk-lore, *Report and Transactions of the Devonshire Association for the Advancement of Science* 9: 88–102.
Kitchen, P., 1990. *For Home and Country*, London: Ebury.
Knapp, J.L., 1829. *Journal of a Naturalist*, 2nd ed., London: John Murray.
Knight, W.F.G. 1945. Forty-fourth report on Devonshire folk-lore, *Report and Transactions of the Devonshire Association for the Advancement of Science* 79: 47–9.
Knight, W.F.G. 1947. Forty-second report on Devonshire folk-lore, *Report and Transactions of the Devonshire Association for the Advancement of Science* 77: 47–9.
Lafonte, A.M. 1984. *Herbal Folklore*, Bideford: Badger Books.

Laghetti, G. & Perrino, P. 1994. 'Utilization of *Silene vulgaris* (Moench) Garcke in Italy', *Economic Botany* 48: 337–9.

Lake, C. n.d. *The Battle of Flowers Story*, St John: Redberry Press Ltd.

Larkcom, J. 1982. 'The Kelsae Onion Festival', *The Garden* 107: 58–61.

Latham, C. 1878. 'Some west Sussex superstitions lingering in 1868', *Folk-lore Record* 1: 1–67.

Laver, F.J.M. 1990. 91st report on dialect, *Report and Transactions of the Devonshire Association for the Advancement of Science* 122: 233–8.

Laver, F.J.M. 1995. 96th report on dialect, *Report and Transactions of the Devonshire Association for the Advancement of Science* 127: 319–21.

Laycock, C.H. 1940. Thirty-ninth report on Devonshire folk-lore, *Report and Transactions of the Devonshire Association for the Advancement of Science* 72: 115–16.

Leather, E.M. 1912. *The Folk-lore of Herefordshire*, Hereford: Jakeman & Carver.

Lees, E, 1850. 'On the botanical features of the Great Orme's Head, with notices of some plants observed in other parts of North Wales during the summer of 1849', *Phytologist*, ser. 1, 3: 869–81.

Lees, E. 1867. *The Botany of Worcestershire*, Worcester: Worcestershire Naturalists' Club.

Legg, P. 1986. *So Merry Let Us Be – The Living Tradition of Somerset Cider*, Bridgwater: Somerset County Council Library Service.

Le Sueur, F. 1984. *Flora of Jersey*, Jersey: Société Jersiaise.

Lewis, D. 2001. *A-Z of Traditional Cures and Remedies*, Newbury: Countryside Books.

Lewis, G. 2009. *Balfour and Weizmann: The Zionist, the Zealot and the Emergence of Israel*, London: Continuum UK.

Lightfoot, J. 1777. *Flora Scotica*, London: B. White.

Lindegaard, P. 1978. 'The colliers' tale – a Bristol incident of 1753', *Journal of the Bath and Avon Family History Society* 1978: 8.

Lindsay,W.L. 1856. 'Notes on the flora of Holstein, relating to an expedition made in August 1850', *Phytologist* ser. 2, 1: 369–74.

Lloyd, A.L. 1967. *Folk Song in England*, London: Lawrence and Wishart.

Lloyd, B. 1945. 'Notes on Pembrokeshire folklore, superstitions, dialect words, etc.', *Folk-lore* 56: 307–20.

[Lloyd George, D.] 1938. *War Memoirs of David Lloyd George*, London: Oldhams Press.

L'Obel, M. de & Pena, P. 1570. *Stirpium adversaria nova*, London: Purfoot.

Lockton, A. & Whild, S. 2015. *The Flora and Vegetation of Shropshire*, Shrewsbury: Shropshire Botanical Society.

Loftus, B. 1994. *Mirrors Orange & Green*, Dundrum: Picture Press.

Logan, P. 1965. 'Folk medicine in the Cavan-Leitrim area, II', *Ulster Folklife* 11: 51–3.

Lomax, A. & Kennedy P. 1971. Notes to accompany the record *Songs of Ceremony* (Folk Songs of Britain, 9), London: Topic Records Ltd.

Lousley, J.E. 1971. *Flora of the Isles of Scilly*, Newton Abbot: David & Charles.

Lovett, E. 1913. 'Folk-medicine in London', *Folk-lore* 24: 120–1.

Lownes, A.E. 1940. 'The strange case of Coles vs Culpeper', *Journal of the New York Botanical Garden* 41: 158–66.

Lucas, A.T. 1960a. *Furze: A Survey and History of Its Uses in Ireland*, Dublin: National Museum of Ireland.

Lucas, A.T. 1960b. 'Irish food before the potato', *Gwerin* 3 (2): 8–43.

Lucas, A.T. 1979. 'Furze: a survey and history of its uses in Ireland', *Bealoideas* 45–47: 30–45.

Mabberley, D.J. 1997. *The Plant-book*, 2nd ed., Cambridge: Cambridge University Press.
Mabey, R. 1972/2012. *Food for Free,* London: Collins.
Mabey, R. 1996. *Flora Britannica*, London: Sinclair Stevenson.
Mabey, R. 2010. *Weeds*, London: Profile Books Ltd.
Mac Coitir, N., 2006. *Irish Wild Plants: Myths, Legends and Folklore*, Wilton, Cork: The Collins Press.
MacGregor, A. 1935. 'Scottish hill berries', *Deeside Field* 7: 13–17.
Mac Manus, D. 1973. *The Middle Kingdom: The Faerie World of Ireland*, Gerrards Cross: Colin Smythe.
Macmillan, A.S. 1922. *Popular Names of Flowers, Fruits, etc.*, Yeovil: Western Gazette.
MacNeil, M. 1962. *The Festival of Lughnasa*, Oxford: Oxford University Press.
MacNicholas, E. 1992. 'The four-leaved shamrock and the cock', *ARV* 47: 209–16.
Mahood, M.M. 2008. *The Poet as Botanist*, Cambridge: Cambridge University Press.
Malone, F.E., Kennedy, S., Reilly, G.A.C., & Woods, F.M. 1992. 'Bog asphodel (*Narthecium ossifragum*) poisoning in cattle', *Veterinary Record* 1 August 1992: 100–3.
Maloney, B. 1972. 'Traditional herbal cures in County Cavan: part 1', *Ulster Folklife* 18: 66–79.
Maloney, M.F. 1919. *Irish Ethnobotany*, Dublin: M.H. Gill & Son.
Maple, E. 1971. *Superstition and the Superstitious*, London: W.H. Allen.
Marquand, E.D. 1906. 'The Guernsey dialect and its plant names', *Transactions of the Guernsey Society of Natural Science and Local Studies* 5: 31–47.
Marren, P. 1996. 'Scottish vernacular plant names', in R. Mabey, *Flora Britannica*: 450–55.
Marson, C. 1904. Preface to A.A. Hilton, *In the Garden of God*, London: G.J. Palmer & Sons.
Martin, L.C. 1984. *Wildflower Folklore*, Charlotte, North Carolina: East Woods Press.
Martin, M. 1703. *Description of the Western Islands of Scotland*, London: A. Bell.
Martin, W.K. 1976. *The Concise British Flora in Colour*, 3rd ed., London: Ebury Press & Michael Joseph.
Maskew, R. 2014. *The Flora of Worcestershire*, Stoke Bliss: the author.
Mayhew, H. 1851. *London Labour and the London Poor*, vol. 1, London: George Woodfall & Son.
McBride, D. 1991. *What They Did with Plants*, Banbridge: Adare Press.
McClintock, D. 1970. 'Why is white heather lucky?', *Country Life*, 15 January: 159.
McClintock, D. 1975. *The Wild Flowers of Guernsey*, London: Collins.
McClintock, D. 1987. *Supplement to 'The Wild Flowers of Guernsey' (Collins 1975)*, St Peter Port: La Société Guernesiaise.
McKenna, N. 2004. *The Secret Life of Oscar Wilde*, London: Arrow.
McNeill, M. 1910. *Colonsay*, Edinburgh: David Douglas.
McNicholas, O Dulaing, D. & Ross, M., 1990. 'The legend of the four-leaved shamrock and the cock', *Sinsear* 6: 83–90.
Meade-Waldo, Mrs. 1902. 'Tissington well-dressing', *Journal of the Derbyshire Archaeological and Natural History Society* 24: 1–4.
Melton, J. 1620. *Astrologaster, or the Figure-Caster*, London: B. Alsop for E. Blackmore.
Miles, A. 2013. *The British Oak*, London: Constable.
Miles, C.A. 1912. *Christmas in Ritual and Tradition, Christian and Pagan*, London: Unwin.
Miller, H. 1858. *The Cruise of the Betsey*, London: Hamilton, Allen.

Milliken, W. & Bridgewater, S. 2004. *Flora Celtica*, Edinburgh: Birlinn Limited.
Milner, J.E. 1992. *The Tree Book*, London: Collins & Brown Limited.
Moloney, M.F. 1919. *Irish Ethno-botany*, Dublin: M.H. Gill & Son.
Moorcroft Wilson, J. 1998. *Siegfried Sassoon, The Making of a War Poet: A Biography (1886–1918)*, London: Gerald Duckworth & Co. Ltd.
Moore, A. 1989. *Where is Brown Hill?*, Victoria Park: Hesperian Press.
Moore, G.F. 1968. 71st report on dialect, *Report and Transactions of the Devonshire Association for the Advancement of Science* 100: 367–71.
Morris, J.P. 1869. *A Glossary of Words and Phrases of Furness (North Lancashire)*, London: J. Russell Smith.
Morris, R.E. 1925. 'Some old-time superstitions of Devon', *Report and Transactions of the Devonshire Association for the Advancement of Science*, 56: 305–8.
Morton, A. 1998. *Tree Heritage of Britain and Ireland*, Shrewsbury: Swan Hill Press.
Muggleton, W. 2016. 'The Black Worcester Pear', *Worcestershire Record* 40: 40–4.
Muggleton, W. 2017. 'Will the real Black Worcester Pear please step forward', *Worcestershire Record* 42: 40–2.
Murray, J. 1937. 'Cumbrian plant names', *North Western Naturalist* 12: 178–82.
Murphy, D. 1965. *Full Tilt – from Ireland to India with a Bicycle*, London: John Murray.
Naylor, P. & Porter, L. 2002. *Well Dressing*, Ashbourne: Landmark Publishing Ltd.
Nelson, E.C. 1978. 'Tropical drift fruits and seeds on the coasts of the British Isles and Western Europe, I. Irish beaches', *Watsonia*, 12: 103–12.
Nelson, E.C. 1990. Shamrock, 1988, *Ulster Folklife*, 36: 32–42.
Nelson, E.C. 1991. *Shamrock: Botany and History of an Irish Myth*, Aberystwyth: Boethius Press.
Nelson, E.C. 2000a. *Sea Beans and Nickar Nuts*, London: Botanical Society of the British Isles.
Nelson, E.C. 2000b. 'Viking ale and the quest for the impossible: some marginalia leading, perhaps, to "the most powerfullest drink ever known"', *Yearbook of the Heather Society* 2000: 25–33.
Nelson, E.C. 2006. 'Lucky white heather: a sesquicentennial review of a Scottish Victorian conceit', *Heathers* 3: 38–46.
Nelson, E.C. & McCracken, E.M. 1987. *The Brightest Jewel: A History of the National Botanic Gardens, Glasnevin*, Kilkenny: Boethius Press.
Newall, V. 1971. *An Egg at Easter*, London: Routledge & Kegan Paul.
Newton, J. 1698. 'An account of some effects of *Papaver corniculatum*, etc.', *Philosophical Transactions of the Royal Society* 20: 263–4.
Nicholson, C. 1861. *The Annals of Kendal*, ed. 2, London: Whitaker & Co.
Nicolson, J. 2006. *The Perfect Summer: Dancing into Shadow in 1911*, London: John Murray.
Nicolson, J.R. 1978. *Traditional Life in Shetland*, London: Robert Hale.
Nixon, B. 1977. *Walk Soft in the Fold*, London: Chatto & Windus.
Noall, C. 1977. *The Cornish Midsummer Eve Bonfire Celebrations*, Penzance: Federation of Old Cornwall Societies.
Norman, F. 1969. *Banana Boy*, London: Secker & Warburg.
Ö Danachair, C. 1970. 'The luck of the house', *Ulster Folklife* 15/16: 20–7.
O'Dowd, A. 2015. *Straw, Hay & Rushes in Irish Folk Tradition*, Newport: Irish Academic Press.
Ogden, J. 1978. 'Marbles and conkers', *Lore and Language*, 2(9): 71–2.
Ogilby, L. 1845. 'Notes of a botanical ramble in Connemara and Arran', *Phytologist* 2: 345–51.

Opie, I. & P. 1959. *The Lore and Language of Schoolchildren*, Oxford: Oxford University Press.
Opie, I. & P., 1969. *Children's Games in Street and Playground*, Oxford: Oxford University Press.
Opie, I. & Tatem, M. 1989. *A Dictionary of Superstitions*, Oxford: Oxford University Press.
Ó Suilleabhain, S. 1967. *Irish Folk Custom and Belief*, Dublin: Cultural Relations Committee of Ireland.
O'Sullivan, J.C. 1973. 'St Brigid's crosses', *Folk Life* 11: 60–81.
O'Sullivan, S. 1966. *Folktales of Ireland*, Chicago: University of Chicago Press.
O'Sullivan, S. 1977. *Legends from Ireland*, London: B.T. Batsford.
Owen, T.M. 1978. *Welsh Folk Customs*, Cardiff: National Museum of Wales.
Palmer, G. & Lloyd, N. 1972. *A Year of Festivals*, London: Frederick Warne & Co.
Palmer, K. 1973. *Oral Folk-Tales of Wessex*, Newton Abbot: David & Charles.
Palmer, K. 1976. *The Folklore of Somerset*, London: B.T. Batsford Ltd.
Palmer, R. 1976. *The Folklore of Warwickshire*, London: B.T. Batsford Ltd.
Palmer, R. 1979. *Everyman's Book of English Country Songs*, London: J.M. Dent & Sons.
Palmer, R. 1985. *The Folklore of Leicestershire and Rutland*, Wymondham: Tempus Publishing.
Parker, A. 1923. 'Oxfordshire village folklore, II', *Folk-lore* 34: 323–33.
Parker, S. & Stevens Cox, G. 1974. *The Giant Cabbage of the Channel Islands*, 2nd ed., St Peter Port: Toucan Press.
Parkinson, J. 1640. *Theatrum Botanicum*, London: Thomas Cotes.
Parslow, R. & Bennalick, I. 2017. *The New Flora of the Isles of Scilly*, privately published.
Partridge, J.B. 1917. 'Notes on English folklore', *Folk-lore* 28: 311–15.
Patten, R.W. 1974. *Exmoor Custom and Song*, Dulverton: Exmoor Press.
Pegg, B. 1981. *Rites and Riots: Folk Customs of Britain and Europe*, Poole: Blandford Press.
Peter, T. 1915. 'Cornish folklore notes', *Journal of the Royal Cornish Institution* 20: 117–33.
Phelps, H. 1977. *Just over Yonder*, London: Michael Joseph Ltd.
Philip, N. 1989. *The Cinderella Story*, London, Harmondsworth: Penguin Books.
Phillips, H. 1825. *Floral Emblems*, London: Saunders & Otley.
Phillips, R. 1983. *Wild Food*, London: Pan Books Ltd.
Phillips, S., 2012. *An Encyclopaedia of Plants in Myth, Legend, Magic and Lore*, London: Robert Hale Ltd.
Phipps, J.B. 2003. *Hawthorns and Medlars*, Portland, Oregon: Timber Press.
Pickering, I., 1995. *Some Goings On!: A Selection of Newspaper Articles about Fowey, Polruan and Lanteglos Districts from 1800–1899*, Fowey: the author.
Piper, G. 1896. 'The gospel yew', *Transactions of the Woolhope Naturalists' Field Club* 1896: 141.
Plomer, W. (ed.) 1977. *Kilvert's Diary: A Selection*, Harmondsworth: Penguin Books.
Plot, R. 1686. *The Natural History of Staffordshire*, Oxford: 'printed at the Theater'.
Pollock, A.J. 1960. 'Hallowe'en customs in Lecale, Co. Down', *Ulster Folklife* 6: 62–4.
Poole, C.H. 1877. *The Customs, Superstitions and Legends of the County of Somerset*, London: Sampson Low.
Porteous, C. 1971. *The Ancient Customs of Derbyshire*, Derby: Derbyshire Countryside Limited.
Porteous, C. 1973. *The Well-dressing Guide*, Derby: Derbyshire Countryside Limited.

Porter, E. 1958. 'Some folk beliefs of the Fens', *Folklore* 69: 112–22.

Porter, E. 1969. *Cambridgeshire Customs and Folklore*, London: Routledge & Kegan Paul.

Porter, E. 1974. *The Folklore of East Anglia*, London: B.T. Batsford Ltd.

Pratt, A. 1857. *Wild Flowers*, London: Society for Promoting Christian Knowledge.

Prendergast, H.D.V. 2002. 'Useful marine monocots – more collections than data', *Economic Botany* 56: 110–12.

Preston, C.D., Pearman, D.A. & Dines, T.D. 2002. *New Atlas of the British & Irish Flora*, Oxford: Oxford University Press.

Prime, C.T. 1960. *Lords and Ladies*, London: Collins.

Prior, R.C.A. 1863. *On the Popular Names of British Plants*, London: Williams & Norgate.

Purslow, F. 1972. *The Constant Lovers: More English Folk Songs from the Hammond and Gardiner Manuscripts*, London: E.F.D.S.S. Publications.

Quelch, M.T. 1941. *Herbs for Daily Use*, London: Faber.

Quennell., P. (ed.), 1984. *Mayhew's London*, London: Bracken Books.

Radford, E. & M.A. 1961. *Encyclopaedia of Superstitions*, ed. and rev. C. Hole, London: Hutchinson.

Radford, R. & U. 1998. *West Country Folklore*, Newton Abbot: Peninsula Press.

Raglan, Lady 1939. 'The Green Man in church architecture', *Folk-lore* 50: 45–57.

Randall, C. 2003. 'Historical and modern uses of *Urtica*', in G.M. Kavalali (ed.), *Urtica: Therapeutic and Nutritional Aspects of Stinging Nettles*, London & New York: Taylor & Francis: 12–24.

Ranson, F. 1949. *British Herbs*, Harmondsworth: Penguin Books.

Raphael, S. 1990. *An Oak Spring Pomona*, Upperville, Virginia: Oak Spring Garden Library.

Raven, J. 1978. *The Folklore of Staffordshire*, London: B.T. Batsford Ltd.

Rawlence, E.A. 1914. 'Folk-lore and superstitions still obtaining to Dorset', *Proceedings of the Dorset Natural History and Antiquarian Field Club*, 35: 81–7.

Rawlinson, R. 1722. *The History and Antiquities of Glastonbury*, Oxford: printed at the Theater.

Ray, J. 1670. *Catalogus Plantarum Angliae*, London: John Martyn.

Reynolds, S. 1993. 'A botanist's understanding of the names used for native Irish plants in the Schools' Collection,' Department of Irish Folklore, *Sinsear* 7: 22–8.

Rhys, J. 1901. *Celtic Folklore*, Oxford: Oxford University Press.

Rich, T.C.G. *et al.* 2010. *Whitebeams, Rowans and Service Trees of Britain and Ireland: A Monograph of British and Irish* "Sorbus" *L.*, London: Botanical Society of the British Isles.

Richards, D. 1979. 'Folklore and medicine', *WHEN* (World Health and Ecology News), 1(3): 13.

Riddelsdell, H.J. *et al.* 1948. *Flora of Gloucestershire*, Arbroath: Buncle.

Roberts, R. 1971. *The Classic Slum*, Manchester: Manchester University Press.

Robinson, J. 1975. *The Life and Times of Francie Nichol of South Shields*, London: Allen & Unwin.

Robson, N.K.B. 1977. 'Studies in the genus *Hypericum* L. I. Infrageneric classification', *Bulletin of the British Museum (Natural History), Botany* 5: 293–355.

Robson, P. 1993. 'Dorset Garland Days on the Chesil Coast', in T. Buckland & J. Wood (eds.), *Aspects of British Calendar Customs*, Sheffield: Sheffield Academic Press: 155–66.

Rolph, C.H. 1980. *London Particulars*, Oxford: Oxford University Press.

Ross, C.C.G. 1987. *The Story of Oak Apple Day in Wishford Magna*, Salisbury: Wishford Oak Apple Club.

Roud, S. 2003. *The Penguin Guide to the Superstitions of Britain and Ireland*, London: Penguin Books.

Roud, S. 2006. *The English Year*, London: Penguin Books.

Roud, S. 2010. *The Lore of the Playground*, London: Random House Books.

Rowbottom, S., 2008. *Edward Carpenter: A Life of Liberty and Love*, London: Verso.

Rowling, M. 1976. *The Folklore of the Lake District*, London: B.T. Batsford Ltd.

Rudkin, E.H. 1936. *Lincolnshire Folklore*, Gainsborough: Beltons.

Ryan, J.S. 1993. 'Halloween and other traditional customs in Scottish New Zealand', *FLS News* 18: 8–10.

Ryder, M.L. 1969. 'Teasel growing for cloth raising', *Folk Life* 7: 117–19.

Rymer, L., 1976. 'The history and ethnobotany of bracken', *Botanical Journal of the Linnean Society* 73: 151–76.

St Clair, S. 1971. *Folklore of the Ulster People*, Cork: Mercier Press.

Salaman, R.N. 1949. *The History and Social Influence of Potato*, Cambridge: Cambridge University Press.

Salmon, C.E., 1931. *Flora of Surrey*, London: G. Bell & Sons Ltd.

Salmon, W. 1710. *Botanologia: The English Herbal or History of Plants*, London: L. Dawks for H. Rhodes.

Scott, W. & Palmer, R. 1987. *The Flowering Plants and Ferns of the Shetland Islands*, Lerwick: The Shetland Times.

Seaby, P. 1970. *Coins and Tokens of Ireland*, London: B.A. Seaby.

Sharman, N. 1977. *Nothing to Steal*, London: Kaye & Ward.

Shirreffs, D.A. 2015. *The Wild Flowers of Jersey*, Taunton: Brambleby Books.

Short, E. 1983. *I Knew My Place*, London: Macdonald.

Shuel, B., 1985. *The National Trust Guide to Traditional Customs of Britain*, Exeter: Webb & Bower.

Sikes, W. 1880. *British Goblins*, London: Sampson Low.

Simpson, B. 1987. *Spalding in Flower*, Norwich: Spring Colour Publications.

Simpson, G.M. 1931. *The Rushbearing in Grasmere and Ambleside*, Manchester: John Heywood Ltd.

Simpson, Jacqueline 1973. *The Folklore of Sussex*, London: B.T. Batsford Ltd.

Simpson, Jacqueline 1976. *The Folklore of the Welsh Border*, London: B.T. Batsford Ltd.

Simpson, Jacqueline 2002. *Folklore of Sussex*, Stroud: Tempus Publishing Ltd.

Simpson, Jacqueline 2011. 'Glastonbury Thorn chopped', *FLS News* 63: 8–9.

Simpson, Jacqueline & Roud, S. 2000. *A Dictionary of English Folklore*, Oxford: Oxford University Press.

Simpson, John (ed.) 1982. *The Concise Oxford Dictionary of Proverbs*. Oxford: Oxford University Press.

Sinker, C.A. *et al.* 1985. *Ecological Flora of the Shropshire Region*, Shrewsbury: Shropshire Trust for Nature Conservation.

Smith, A. 1959. 'Some local lore collected in Essex', *Folklore* 70: 414–5.

Smith, A. 1969. *Discovering Folklore in Industry*, Tring: Shire Publications.

Smith, J. 1989. *Fairs Feasts and Frolics: Customs and Traditions in Yorkshire*, Otley: Smith Settle Ltd.

Smith, J.B. 1994. 'Comparative notes on traditional German and English names and uses of some plants representing the Palm-Sunday palm', *German Life and Letters* 47: 242–53.

Smith, J.B. 1996. 'Towards the demystification of Lawrence Lazy', *Folklore* 107: 101–5.

Smith, J.B. 2004. Ninety-fifth report on folklore, *Report and Transactions of the Devonshire Association for the Advancement of Science* 136: 300–14.

Smith, J.B. 2005. 'Bracken lore', *Pteridologist* 4, 4: 114–15.

Smith, R. 2011. *Well Dressing Guide & Souvenir*, Ashbourne: Ashbourne Editions.

Smith-Bendell, M. 2009. *Our Forgotten Years: A Gypsy Woman's Life on the Road*, Hatfield: University of Hertfordshire Press.

Spence, M. 1914. *Flora Orcadensis*, Kirkwall: D. Spence.

Spray, M., 2004. 'Two cheers for bracken', *Pteridologist* 4, 3: 70–4.

Stabursvik, A. 1959. 'A phytochemical study of *Narthecium ossifragum* (L.) Huds.', *Norges Tekniske Vitenskapsakademi*, ser. 2, 6.

Stace, C. 2010. *New Flora of the British Isles*, 3rd ed., Cambridge: Cambridge University Press.

Stace, C.A. & Crawley, M.J. 2015. *Alien Plants*, London: William Collins.

Stearn, W.T. 1976. 'From Theophrastus and Dioscorides to Sibthorp and Smith: the background and origin of the *Flora Graeca*', *Biological Journal of the Linnean Society* 8: 285–98.

Stearn, W.T. 2002. *Stearn's Dictionary of Plant Names for Gardeners*, London: Cassell.

Stevens Cox, J. 1970. *Mumming and the Mummers' Play of St. George*, St Peter Port: Toucan Press.

Stevens Cox, J. 1971. *Guernsey Folklore Recorded in the Summer of 1882*, St Peter Port: Toucan Press.

Stewart, S. 1987. *Lifting the Latch*, Oxford: Oxford University Press.

Stewart, W.G. 1823. *The Popular Superstitions and Festive Amusements of the Highlanders of Scotland*, Edinburgh: Archibald Constable & Co.

Stibbons, P. & Cleveland, D. 1990. *Poppyland*, North Walsham: Poppyland Publishing.

Stockwell, C. 1989. *Nature's Pharmacy*, London: Ebury Press.

Stout, A. 2007. *The Thorn and the Waters*, Glastonbury: The Library of Avalon.

Stow, J. 1987. *The Survey of London*, ed. E.B. Wheatley, London: Dent.

Studley, L. 1988, *My Story: Nigh 80 years in the Broadwindsor Area of the Hill Country of West Dorset*, Broadwindsor: the author.

Summers, J. 2015. *Jambusters: The Story of the Women's Institute in the Second World War*, London: Simon & Schuster UK Ltd.

Sykes, H. 1977. *Once a Year: Some Traditional British Customs*, London: The Gordon Fraser Gallery Ltd.

Sykes, M. 1987. 'Bracken: friend or foe?', *The Ecologist* 17, 6: 241–2.

Synott, D.M. 1979. 'Folk-lore, legend and Irish plants', in C. Nelson and A. Brady (eds), *Irish Gardening and Horticulture*, Dublin: Royal Horticultural Society of Ireland: 37–43.

Tait, R.W. 1947. 'Some Shetland plant names', *Shetland Folk Book* 1: 74–88.

Tangye, M. 2008. 'Traditional uses of Japanese knotweed', *Old Cornwall*, 13 (10): 46–7.

Taylor, J. 1722. *John Taylor's Wandering to see the Wonders of the West*, London: John Martyn.

Teulon-Porter, N. 1956. 'Bull-fonts as church hassocks, up to the mid-nineteenth century', *Gwerin* 1: 90–1.

Thiselton-Dyer, T.F. 1889. *The Folk-lore of Plants*, New York: D. Appleton & Co.

Thomas, K. 1971. *Religion and the Decline of Magic*, London: Weidenfeld and Nicolson.

Thompson, F. 1939. *Lark Rise*, London: Oxford University Press.

Thompson, T.W. 1925. 'English gypsy folk-medicine', *Journal of the Gypsy Lore Society*, ser. 3, 4: 159–72.

Thomson, S. 2009. 'Herefordshire healers', *The Flycatcher*, 75: 11–16.

Threlkeld, C. 1727. *Synopsis stirpium Hibernicarum*, Dublin: S. Powell for F. Davys [The pages are unnumbered in the original edition; the page numbers cited are those given in the Boethius Press, 1988 facsimile].
Thurston, E. 1930. *British and Foreign Trees and Shrubs in Cornwall*, Cambridge: Cambridge University Press for the Royal Institution of Cornwall.
Tighe, W. 1802. *Statistical Observations Relating to the County of Kilkenny*, Dublin: Graisberry & Campbell.
Tongue, R.L. 1965. *Somerset Folklore*, London: The Folklore Society.
Tongue, R.L. 1967. *The Chime Child*, London: Routledge & Kegan Paul.
Tongue, R.L. 1970. *Forgotten Folk-tales of the English Counties*, London: Routledge & Kegan Paul.
Townshend, D. 1908. 'Fishers' folklore', *Folk-lore* 19: 108.
Trevelyan, M. 1909. *Folk-lore and Folk-stories of Wales*, London: Elliot Stock.
Trueman, I., Morton, A. & Wainwright, M. 1995. *The Flora of Montgomeryshire*, Welshpool: Montgomeryshire Field Society and the Montgomeryshire Wildlife Trust.
Tull, G.F. 1976. *The Heritage of Centuries*, Ashford, Middlesex: The Manor Press.
Turner, E.S. 1961. *The Phoney War on the Home Front*, London: Michael Joseph.
Udal, J.S. 1922. *Dorsetshire Folk-lore*. Hertford: Stephen Austin & Sons.
Uí Chonchubhair, M. 1995. *Flóra Chorca Dhuibhne*, Trá Lí: Oidhreacht Chorca Dhuibhne.
Vaughan Williams, R. & Lloyd, A.L. 1968. *The Penguin Book of English Folk Songs*, Harmondsworth: Penguin Books Ltd.
Verrall, P. 1991. 'Queen Victoria's wedding bouquet – 1', *BSBI News* 58: 28.
Vesey-FitzGerald, B. 1944. 'Gypsy medicine', *Journal of the Gypsy Lore Society* 23: 21–33.
Vickery, A.R. 1975. 'The use of lichens in well-dressing', *Lichenologist* 7: 178–9.
Vickery, A.R. 1978. 'James Britten, a founder member of the Folklore Society', *Folklore* 89: 71–4.
Vickery, A.R. 1978. 'West Dorset folklore notes', *Folklore* 89: 154–9.
Vickery, A.R. 1983. '*Lemna minor* and Jenny Greenteeth', *Folklore* 94: 247–50.
Vickery, [A.] R. (ed.) 1984. *Plant-lore Studies*, London: The Folklore Society.
Vickery, [A.] R. 1985. *Unlucky Plants*, London: The Folklore Society.
Vickery, A.R. 1991. 'Early collections of the Holy Thorn (*Crataegus monogyna* cv. Biflora)', *Bulletin of the British Museum (Natural History) Botany* 21: 81–3.
Vickery, [A.] R., 1995. *A Dictionary of Plant-lore*, Oxford: Oxford University Press.
Vickery, A.R. 2004. 'Oaks in British and Irish folklore', *International Oaks* 15: 59–67.
Vickery, [A.] R. 2006. 'Remembered remedies', *Herbs* 31 (4): 18–19.
Vickery, [A.] R. 2008. *Naughty Man's Plaything – Folklore and Uses of Stinging Nettles in the British Isles*, London: the author.
Vickery, [A.] R. 2010. *Garlands, Conkers and Mother-die*, London: Continuum.
Vickery, [A.] R. 2013. 'Cowslips – Conservation icon and flower of folklore', *evolve* 15: 50–3.
Vickery, A.R. & Vickery, M.E. 1977. 'Memories of grottoes, 1905–1935', *Folklore* 88: 183–90.
Victoria, Queen 1868. *Leaves from the Journal of a Life in the Highlands from 1848 to 1861*, London: Smith, Elder & Co.
Victoria, Queen 1884. *More Leaves from the Journal of a Life in the Highlands*, London: Smith, Elder & Co.
Vince, J. 1979. *Discovering Saints in Britain*, Princes Risborough: Shire Publications Ltd.

Ward, H.G., 1891. 'Local plant and bird-names from North Marston, Bucks', *Hardwicke's Science-Gossip*, 27: 119.
Waring, E. 1977. *Ghosts and Legends of the Dorset Countryside*, Tisbury: Compton Press.
Waring, P. 1978. *A Dictionary of Omens and Superstitions*, London: Souvenir Press.
Waters, C. 1987. *Who was St Wite?*, Broadoak: C.J. Creed.
Watson, W.G.W. 1920. *Calendar of Customs, Superstitions, Weather-lore, Popular Sayings and Important Events Connected with the County of Somerset*, Taunton: Somerset County Herald.
Watts, K. 1989. 'Scots pine and droveways', *Wiltshire Folklife* 19: 3–6.
Weaver, O.J. 1987. *Boscobel House and White Ladies Priory*, London: English Heritage.
Webster, M.M. 1978. *Flora of Moray, Nairn and East Inverness*, Aberdeen: Aberdeen University Press.
Wentersdorf, K.P. 1978. '*Hamlet*: Ophelia's long purples', *Shakespeare Quarterly* 29: 413–17.
Wenis, E. and H. 1990. 'Multi-leaved clovers again', *BSBI News*, 56: 24.
Westwood, J. 1985. *Albion: A Guide to Legendary Britain*, London: Granada.
Westwood, J. & Simpson, J. 2005. *The Lore of the Land*, London: Penguin.
Wheeler, C. 2006. 'A herbalist's view – lesser celandine', *White Admiral* 63:22.
Whistler, C.W. 1908. 'Miscellaneous notes from west Somerset and Devon', *Folk-lore*, 19: 88–91.
White, G. 1822. *The Natural History of Selborne*, London : J. & A. Arch.
Whitlock, R. 1976. *The Folklore of Wiltshire*, London: B.T. Batsford Ltd.
Widdowson, J.D.A. 2010. 'Folklore studies in English Higher Education: Lost cause or new opportunity', *Folklore* 121: 125–42.
Widdowson, J.D.A. 2016. 'New beginnings: Towards a National Folklore Survey', *Folklore* 127: 257–69.
Wilde, O. & others 1999. *Teleny*, London: Prowler Books.
Wiliam, E. 1991. *The Welsh Folk Museum Visitor Guide*, Cardiff: National Museum of Wales.
Wilkinson, Lady. 1858. *Weeds and Wild Flowers: Their Uses, Legends, and Literature*, London: John Van Voorst.
Wilks, J.H. 1972. *Trees of the British Isles in History and Legend*, London: Frederick Muller Ltd.
Willey, G.R. 1983. 'Burning the ashen faggot: a surviving Somerset custom', *Folklore*, 94: 40–3.
Williams, A. 1922. *Round about the Upper Thames*, London: Duckworth.
Williams, D. 1987. *Festivals of Cornwall*, Bodmin: Bossiney Books.
Williams, F.R. 1944. 'Some Sussex customs and superstitions', *Sussex Notes and Queries*, 10: 58–62.
Williams-Davies, J. 1983. 'A time to sow and a time to reap: The Welsh farmer's calendar', *Folklore* 94: 229–34.
Wilson, E.M. 1940. 'A Westmorland initiation ceremony', *Folk-lore* 51: 74–6.
Wiltshire, K. 1975. *Wiltshire Folklore*, Salisbury: Compton Russell Ltd.
Withall, M.R. (ed.) 1978. *Somerset Remembers: Recollections of Country People Compiled by SFWI*, Taunton: Somerset Federation of Women's Institutes.
Withering, W. 1776. *A Botanical Arrangement of all the Vegetables growing naturally in Great Britain*, Birmingham: M. Swinney.
Withering, W. 1785. *An Account of the Foxglove and some of its Medical Uses*, Birmingham: M. Swinney.

Womack, J. 1977. *Well-Dressing in Derbyshire*, Clapham: Dalesman Publishing Company Ltd.

Wright, A.R. 1936. *British Calendar Customs, England*, vol. 1, *Movable Feasts*, London: Wm Glaisher Ltd for the Folk-lore Society.

Wright, A.R. 1938. *British Calendar Customs, England,* vol. 2, London: Wm Glaisher Ltd for the Folk-lore Society

Wright, A.R. 1940. *British Calendar Customs, England*, vol. 3, London: Wm Glaisher for the Folk-lore Society.

Wright, J., 1898–1905. *The English Dialect Dictionary*, Oxford: Oxford University Press.

Wyatt, I. [?1933]. *The Book of Huish: The Story of a Somerset Village from 200 BC to 1933 AD*, Yeovil: Western Gazette Co.

Wyllie-Echeverria & Cox, P.A. 1999. 'The seagrass (*Zostera marina* [Zosteraceae]) industry of Nova Scotia (1907–1960)', *Economic Botany* 53: 418–26.

Wyse Jackson, P. 2014. *Ireland's Generous Nature*, St Louis: Missouri Botanical Garden Press.

Yallop, H.J. 1984. 'An example of 17th century Honiton lace', *Devon Historian* 28: 27–31.

Youngman, B.J. 1951. 'Germination of old seeds', *Kew Bulletin* 6: 423–6.

Unpublished sources

Britten & Holland MSS – slips accumulated by James Britten (1846–1924) and Robert Holland (1829–93) intended for a supplement to their *Dictionary of English Plant-Names* (1878–86); now in the Library of the Natural History Museum, London.

Cottam MSS, 1989 – 'A survey of farming traditions and the role of animals in Retford and the surrounding district', student project by T. Cottam, the Centre for English Cultural Tradition and Language, University of Sheffield, 1989.

Dickinson MSS, 1974 – 'Ayton past and present', thesis for the examination of English Special Studies BA supervised by Mr Sanderson of the Folk Life Studies Department [University of Leeds], by M.E. Dickinson.

Hammond MSS, 1970 – 'The folklore of wild flowers in the parish of Leckhampton, Cheltenham, in Gloucestershire, collected in the summer of 1970', thesis by Penelope Ruth Hammond, University of Leeds, towards the degree of BA (Hons).

Hole MSS – notes accumulated by Christina Hole (1896–1986), now in the Archives of the Folklore Society, London.

IFC MSS – material collected by professional collectors working for the Irish Folklore Commission; now in the National Folklore Collection, University College Dublin, 1935–.

IFCSS MSS – material contributed to the Irish Folklore Commission's Schools' Scheme, 1937–9, in which children attending Irish primary schools collected and recorded local folklore; now in the National Folklore Collection, University College Dublin.

Macpherson MSS – card-index 'Collection of Folk Medicines' compiled by J. Harvey Macpherson; now in the archives of the Folklore Society, London.

Margaret Gale Collection – folklore relating to the Christian year, copy of plant-lore material in author's possession.

McKelvie MSS, 1963. – 'Some aspects of oral, social, and material tradition in an industrial urban area', a thesis presented for the degree of PhD, University of Leeds, by D. McKelvie.

Milner MSS, 1991–2 – transcripts of interviews made by J. Edward Milner in connection with his television series 'The Spirit of Trees', shown on Channel Four TV, October–December 1992; copy in author's collection.

Neale MSS – 'A new look at an old manuscript and its author', University of Leeds, 1975.

Newton MSS – 'Mr Newton's Mss Notes as set down in his Catalogus Pl. Angl.', copied by an unknown hand in an interleaved copy of John Ray's *Catalogus Plantarum Angliae*, London, 1677, now in the library of the Natural History Museum, London. James Newton (*c.*1639–1718) is believed to have compiled these notes in *c.*1683; the original notes do not seem to have been preserved.

NHM MSS, herb. . . . – information from notes on herbarium specimens, Department of Life Sciences, the Natural History Museum, London.

Parsons MSS, 1952 – 'Horseheath: some recollections of a Cambridgeshire village' by Catherine E. Parsons (1952), copy in the Cambridge Record Office.

Roper MSS – 'Report on the wild service tree, *Sorbus torminalis*: economics and sociology', by Patrick Roper, F.L.S.; copy in author's collection.

Rowe MSS – list of plant-names compiled by Mrs H.K. Rowe, of Minehead, Somerset, November 1993; in the author's possession.

SLF MSS – material contributed to the Survey of Language and Folklore; now in the University of Sheffield Library.

SSS MSS – transcriptions of summaries of tape-recordings in the School of Scottish Studies, Edinburgh.

Steele MSS, 1978 – 'The medicinal value and usage of plants', by Margaret Ann Steele, student project, the Centre for English Cultural Tradition and Language, University of Sheffield.

Taylor MSS – notes compiled in the 1920s by Dr Mark Taylor on East Anglian herbal remedies and folklore; in the Norfolk Record Office, Norwich.

UCL EFS MSS – material accumulated in the 1960s as a result of a Survey of English Folklore conducted by staff of the Department of English, University College London.

Wharton MSS, 1974 – 'The folklore of south Warwickshire: a field collection with comparative annotations and commentary', thesis presented for the degree of PhD in the Institute of Dialect and Folk Life Studies, School of English, University of Leeds, 1974, by C. Wharton.

Indexes

Geographical Index

Subject Index

LUCK, good or bad is not included in the index as it is mentioned throughout the book.